Lecture Notes in Computer Science 9618

Commenced Publication in 1973
Founding and Former Series Editors:
Gerhard Goos, Juris Hartmanis, and Jan van Leeuwen

More information about this series at http://www.springer.com/series/7407

Adrian-Horia Dediu · Jan Janoušek
Carlos Martín-Vide · Bianca Truthe (Eds.)

Language
and Automata Theory
and Applications

10th International Conference, LATA 2016
Prague, Czech Republic, March 14–18, 2016
Proceedings

 Springer

Editors
Adrian-Horia Dediu
Rovira i Virgili University
Tarragona
Spain

Jan Janoušek
Czech Technical University
Prague
Czech Republic

Carlos Martín-Vide
Rovira i Virgili University
Tarragona
Spain

Bianca Truthe
Justus Liebig University Giessen
Gießen
Germany

ISSN 0302-9743   ISSN 1611-3349   (electronic)
Lecture Notes in Computer Science
ISBN 978-3-319-29999-0   ISBN 978-3-319-30000-9   (eBook)
DOI 10.1007/978-3-319-30000-9

Library of Congress Control Number: 2016931197

LNCS Sublibrary: SL1 – Theoretical Computer Science and General Issues

Printed on acid-free paper

This Springer imprint is published by SpringerNature
The registered company is Springer International Publishing AG Switzerland

Preface

These proceedings contain the papers that were presented at the 10th International Conference on Language and Automata Theory and Applications (LATA 2016), held in Prague, Czech Republic, during March 14–18, 2016.

The scope of LATA is rather broad, including: algebraic language theory; algorithms for semi-structured data mining; algorithms on automata and words; automata and logic; automata for system analysis and program verification; automata networks; automata, concurrency, and Petri nets; automatic structures; cellular automata; codes; combinatorics on words; computational complexity; data and image compression; descriptional complexity; digital libraries and document engineering; foundations of finite-state technology; foundations of XML; fuzzy and rough languages; grammars (Chomsky hierarchy, contextual, unification, categorial, etc.); grammatical inference and algorithmic learning; graphs and graph transformation; language varieties and semigroups; language-based cryptography; mathematical and logical foundations of programming methodologies; parallel and regulated rewriting; parsing; patterns; power series; string and combinatorial issues in bioinformatics; string processing algorithms; symbolic dynamics; term rewriting; transducers; trees, tree languages, and tree automata; unconventional models of computation; weighted automata.

LATA 2016 received 119 submissions. Most of the papers were given at least three reviews by Program Committee members or by external referees. After a thorough and vivid discussion phase, the committee decided to accept 42 papers (which represents an acceptance rate of about 35 %). The conference program also included five invited talks. Part of the success in the management of such a large number of submissions was due to the excellent facilities provided by the EasyChair conference management system.

We would like to thank all invited speakers and authors for their contributions, the Program Committee and the reviewers for their cooperation, and Springer for its very professional publishing work.

December 2015

Adrian-Horia Dediu
Jan Janoušek
Carlos Martín-Vide
Bianca Truthe

Organization

LATA 2016 was organized by the Department of Theoretical Computer Science, Faculty of Information Technology, Czech Technical University in Prague, and the Research Group on Mathematical Linguistics – GRLMC, from Rovira i Virgili University, Tarragona.

Program Committee

Amihood Amir	Bar-Ilan University, Ramat Gan, Israel
Dana Angluin	Yale University, New Haven, USA
Franz Baader	Technical University of Dresden, Germany
Christel Baier	Technical University of Dresden, Germany
Hans L. Bodlaender	Utrecht University, The Netherlands
Jean-Marc Champarnaud	University of Rouen, France
Bruno Courcelle	LaBRI, Bordeaux, France
Rod Downey	Victoria University of Wellington, New Zealand
Frank Drewes	Umeå University, Sweden
Ding-Zhu Du	University of Texas, Dallas, USA
Javier Esparza	Technical University of Munich, Germany
Michael Fellows	Charles Darwin University, Darwin, Australia
Mohammad Taghi Hajiaghayi	University of Maryland, College Park, USA
Yo-Sub Han	Yonsei University, Seoul, South Korea
Markus Holzer	University of Gießen, Germany
Juraj Hromkovič	ETH Zurich, Switzerland
Oscar H. Ibarra	University of California, Santa Barbara, USA
Costas S. Iliopoulos	King's College London, UK
Jan Janoušek	Czech Technical University in Prague, Czech Republic
Galina Jirásková	Slovak Academy of Sciences, Košice, Slovakia
Ming-Yang Kao	Northwestern University, Evanston, USA
Juhani Karhumäki	University of Turku, Finland
Joost-Pieter Katoen	RWTH Aachen University, Germany
Martin Kutrib	University of Gießen, Germany
Zhiwu Li	Xidian University, Xi'an, China
Andreas Malcher	University of Gießen, Germany
Oded Maler	VERIMAG, Gières, France
Carlos Martín-Vide (Chair)	Rovira i Virgili University, Tarragona, Spain
Bořivoj Melichar	Czech Technical University in Prague, Czech Republic
Ugo Montanari	University of Pisa, Italy
František Mráz	Charles University in Prague, Czech Republic
Mitsunori Ogihara	University of Miami, Coral Gables, USA
Alexander Okhotin	University of Turku, Finland

Doron A. Peled	Bar-Ilan University, Ramat Gan, Israel
Martin Plátek	Charles University in Prague, Czech Republic
Alberto Policriti	University of Udine, Italy
Daniel Reidenbach	University of Loughborough, UK
Antonio Restivo	University of Palermo, Italy
Kai Salomaa	Queen's University, Kingston, Canada
Davide Sangiorgi	University of Bologna, Italy
Uli Sattler	University of Manchester, UK
Frits Vaandrager	Radboud University, Nijmegen, The Netherlands
Pierre Wolper	University of Liège, Belgium
Zhilin Wu	Chinese Academy of Sciences, Beijing, China
Mengchu Zhou	New Jersey Institute of Technology, Newark, USA

Organizing Committee

Adrian-Horia Dediu	Rovira i Virgili University, Tarragona, Spain
Jan Janoušek (Co-chair)	Czech Technical University in Prague, Czech Republic
Carlos Martín-Vide (Co-chair)	Rovira i Virgili University, Tarragona, Spain
Radomír Polách	Czech Technical University in Prague, Czech Republic
Eliška Šestáková	Czech Technical University in Prague, Czech Republic
Jan Trávníček	Czech Technical University in Prague, Czech Republic
Bianca Truthe	Justus Liebig University Giessen, Germany
Lilica Voicu	Rovira i Virgili University, Tarragona, Spain

Additional Reviewers

Eugene Asarin	Alberto Casagrande	Marcus Frean
Jean-Michel Autebert	Pierre Castéran	Leo Freitas
Holger Bock Axelsen	Giuseppa Castiglione	Dominik
Evangelos Bampas	Jeong-Won Cha	D. Freydenberger
Mikhail Barash	Souymodip Chakraborty	Fabio Gadducci
Nicolas Bedon	Taolue Chen	William Gasarch
Simon Beier	Salimur Choudhury	Thomas Genet
Eduard Bejček	Stefano Crespi Reghizzi	Nicola Gigante
Suna Bensch	Flavio D'Alessandro	Carlos Gómez-Rodríguez
Maria Paola Bianchi	Carsten Damm	Yaakov HaCohen-Kerner
Henrik Björklund	Jürgen Dassow	Vesa Halava
Johanna Björklund	Giuseppe De Giacomo	Tero Harju
Ondřej Bojar	Ziadi Djelloul	Rolf Hennicker
Henning Bordihn	Clemens Dubslaff	Mika Hirvensalo
Stefan Borgwardt	Rüdiger Ehlers	Jan Holub
Luca Bortolussi	Petter Ericson	David N. Jansen
Christopher Broadbent	Henning Fernau	Jozef Jirásek
Erik Cambria	Gabriele Fici	Timo Jolivet

Abstracts of Invited Talks

Reconstructing Preferences from Opaque Transactions

Avrim Blum[✉]

Department of Computer Science,
Carnegie Mellon University,
Pittsburgh, PA 15213, USA
http://www.cs.cmu.edu/~avrim

Abstract. There has been significant work on learning about utility functions of single agents from observing their behavior in isolation. In this talk, I will discuss the problem of learning about utilities of a collection of agents, when the only information available is some kind of overall outcome of a joint interaction.

For example, consider an auction of a single item where n agents each draw values from their own personal probability distributions D_i, and the only information that can be observed is the identity of the winner (highest bidder). From repeated observations, plus the ability to enter the auction (and win or lose) yourself, can you reconstruct the relevant parts of the individual distributions? Or consider a setting with multiple items where agents have combinatorial preferences, and where a seller is running a mechanism that you do not know. From observing the results of a sequence of these interactions, can you learn both the preferences of the buyers *and* the mechanism of the seller? In this talk I will discuss algorithms in the context of both of these problems. In the process we will see connections to decision-list learning in learning theory and Kaplan-Meier estimators in medical statistics.

This is joint work with Yishay Mansour and Jamie Morgenstern [1, 2].

References

1. Blum, A., Mansour, Y., Morgenstern, J.: Learning valuation distributions from partial observation. In: Proceedings of 29th AAAI Conference on Artificial Intelligence (AAAI), pp. 798–804 (2015)
2. Blum, A., Mansour, Y., Morgenstern, J.: Learning what's going on: reconstructing preferences and priorities from opaque transactions. In: Proceedings of 16th ACM Conference on Economics and Computation (EC), pp. 601–618 (2015)

Work supported in part by the National Science Foundation under grants CCF-1101215, CCF-1116892, and CCF-1331175.

Non-Zero Sum Games for Reactive Synthesis

Romain Brenguier[1], Lorenzo Clemente[2], Paul Hunter[3],
Guillermo A. Pérez[3], Mickael Randour[3], Jean-François Raskin[3,✉],
Ocan Sankur[4], and Mathieu Sassolas[5]

[1] University of Oxford, Oxford, UK
[2] University of Warsaw, Warsaw, Poland
[3] Université Libre de Bruxelles, Brussels, Belgium
[4] CNRS, IRISA, Rennes, France
[5] Université Paris-Est – Créteil, LACL, Créteil, France

Abstract. In this invited contribution, we summarize new solution concepts useful for the synthesis of reactive systems that we have introduced in several recent publications. These solution concepts are developed in the context of non-zero sum games played on graphs. They are part of the contributions obtained in the inVEST project funded by the European Research Council.

Work supported by the ERC starting grant inVEST (FP7-279499), G.A. Pérez is supported by F.R.S.-FNRS ASP fellowship, M. Randour is a F.R.S.-FNRS Postdoctoral Researcher.

Tangles and Connectivity in Graphs

Martin Grohe[✉]

RWTH Aachen University, Aachen, Germany
grohe@informatik.rwth-aachen.de

Abstract. This paper is a short introduction to the theory of tangles, both in graphs and general connectivity systems. An emphasis is put on the correspondence between tangles of order k and k-connected components. In particular, we prove that there is a one-to-one correspondence between the triconnected components of a graph and its tangles of order 3.

Restricted Turing Machines and Language Recognition

Giovanni Pighizzini[✉]

Dipartimento di Informatica,
Università degli Studi di Milano, Milano, Italy
pighizzini@di.unimi.it

Abstract. In 1965 Hennie proved that one-tape deterministic Turing machines working in linear time are equivalent to finite automata, namely they characterize regular languages. This result has been improved in different directions, by obtaining optimal lower bounds for the time that one-tape deterministic and nondeterministic Turing machines need to recognize nonregular languages. On the other hand, in 1964 Kuroda showed that one-tape Turing machines that are not allowed to use any extra space, besides the part of the tape which initially contains the input, namely linear bounded automata, recognize exactly context-sensitive languages. In 1967 Hibbard proved that for each integer $d \geq 2$, one-tape Turing machines that are allowed to rewrite each tape cell only in the first d visits are equivalent to pushdown automata. This gives a characterization of the class of context-free languages in terms of restricted Turing machines. We discuss these and other related models, by presenting an overview of some fundamental results related to them. Descriptional complexity aspects are also considered.

Automata for Ontologies

Frank Wolter[✉]

Department of Computer Science,
University of Liverpool, Liverpool, UK

Abstract. We present three reasoning problems for description logic ontologies and discuss how automata theory can be used to analyze them.

Contents

Grammars

Grammatical Inference, Algorithmic Learning, and Patterns

Graphs, Trees, and Weighted Automata

Invited Talks

Non-Zero Sum Games for Reactive Synthesis

Romain Brenguier[1], Lorenzo Clemente[2], Paul Hunter[3], Guillermo A. Pérez[3],
Mickael Randour[3], Jean-François Raskin[3(✉)], Ocan Sankur[4],
and Mathieu Sassolas[5]

[1] University of Oxford, Oxford, UK
[2] University of Warsaw, Warsaw, Poland
[3] Université Libre de Bruxelles, Brussels, Belgium
jraskin@ulb.ac.be
[4] CNRS, IRISA, Rennes, France
[5] Université Paris-Est – Créteil, LACL, Créteil, France

Abstract. In this invited contribution, we summarize new solution concepts useful for the synthesis of reactive systems that we have introduced in several recent publications. These solution concepts are developed in the context of non-zero sum games played on graphs. They are part of the contributions obtained in the INVEST project funded by the European Research Council.

1 Introduction

Reactive systems are computer systems that maintain a continuous interaction with the environment in which they operate. They usually exhibit characteristics, like real-time constraints, concurrency, parallelism, etc., that make them difficult to develop correctly. Therefore, formal techniques using mathematical models have been advocated to help to their systematic design.

One well-studied formal technique is *model checking* [2,20,39] which compares a model of a system with its specification. The main objective of this technique is to find design errors early in the development cycle. So model-checking can be considered as a sophisticated *debugging* method. A scientifically more challenging goal, called *synthesis*, is to design algorithms that, given a specification for a reactive system and a model of its environment, directly synthesize a correct system, i.e., a system that enforces the specification *no matter how* the environment behaves.

Synthesis can take different forms: from computing optimal values of parameters to the full-blown automatic synthesis of finite-state machine descriptions for components of the reactive system. The main mathematical models proposed for the synthesis problem are based on *two-player zero-sum games played on graphs* and the main solution concept for those games is the notion of *winning strategy*.

Work supported by the ERC starting grant INVEST (FP7-279499), G.A. Pérez is supported by F.R.S.-FNRS ASP fellowship, M. Randour is a F.R.S.-FNRS Postdoctoral Researcher.

© Springer International Publishing Switzerland 2016
A.-H. Dediu et al. (Eds.): LATA 2016, LNCS 9618, pp. 3–23, 2016.
DOI: 10.1007/978-3-319-30000-9_1

This model encompasses the situation where a *monolithic* controller has to be designed to interact with a *monolithic* environment that is supposed to be *fully antagonistic*. In the sequel, we call the two players Eve and Adam, Eve plays the role of the system and Adam plays the role of the environment.

A fully antagonistic environment is most often a *bold abstraction* of reality: the environment usually has its own goal which, in general, does not correspond to that of falsifying the specification of the reactive system. Nevertheless, this abstraction is popular because it is simple and sound: a winning strategy against an antagonistic environment is winning against any environment that pursues its own objective. However this approach may fail to find a winning strategy even if solutions exist when the objective of the environment are taken into account, or it may produce sub-optimal solutions because they are overcautious and do not exploit the fact the environment has its own objective. In several recent works, we have introduced new solution concepts for synthesis of reactive systems that take the objective of the environment into account or relax the fully adversarial assumption.

Assume Admissible Synthesis. In [7], we proposed a novel notion of synthesis where the objective of the environment can be captured using the concept of *admissible* strategies [3,5,8]. For a player with objective ϕ, a strategy σ is *dominated* by σ' if σ' does as well as σ w.r.t. ϕ against all strategies of the other players, and better for some of those strategies. A strategy σ is *admissible* if it is *not* dominated by another strategy. We use this notion to derive a meaningful notion to *synthesize* systems with several players, with the following idea. Only admissible strategies should be played by *rational* players as dominated strategies are clearly *sub-optimal options*. In *assume-admissible synthesis*, we make the assumption that both players play admissible strategies. Then, when synthesizing a controller, we search for an admissible strategy that is *winning* against all admissible strategies of the environment. Assume admissible synthesis is *sound*: if both players choose strategies that are winning against admissible strategies of the other player, the objectives of both players will be satisfied.

Regret Minimization: Best-Responses as Yardstick. In [32] we studied strategies for Eve which *minimize her regret*. The regret of a strategy σ of Eve corresponds to the difference between the value Eve achieves by playing σ against Adam and the value she could have ensured if she had known the strategy of Adam in advance. Regret is not a novel concept in game theory see, e.g., [30], but it was not explicitly used for games played on graphs before [28]. The complexity of deciding whether a regret-minimizing strategy for Eve exists, and the memory requirements for such strategies change depending on what type of behavior Adam can use. We have focused on three particular cases: arbitrary behaviors, positional behaviors, and time-dependent behaviors (otherwise known as *oblivious* environments). The latter class of regret games was shown in [32] to be related to the problem of determining whether an automaton has a certain form of determinism.

Games with an Expected Adversary. In [11,12,21], we combined the classical formalism of two-player zero-sum games (where the environment is considered to be completely antagonistic) with *Markov decision processes* (MDPs), a well-known model for decision-making inside a stochastic environment. The motivation is that one has often a good idea of the *expected behavior* (i.e., average-case) of the environment represented as a stochastic model based on statistical data such as the frequency of requests for a computer server, the average traffic in a town, etc. In this case, it makes sense to look for strategies that will maximize the *expected performance* of the system. This is the traditional approach for MDPs, but it gives no guarantee at all if the environment deviates from its expected behavior, which can happen, for example, if events with small probability happen, or if the statistical data upon which probabilities are estimated is noisy or unreliable. On the other hand, two-player zero-sum games lead to strategies guaranteeing a *worst-case performance* no matter how the environment behaves — however such strategies may be far from optimal against the expected behavior of the environment. With our new framework of *beyond worst-case* synthesis, we provide formal grounds to synthesize strategies that *both* guarantee some minimal performance against any adversary *and* provide an higher expected performance against a given expected behavior of the environment — thus essentially combining the two traditional standpoints from games and MDPs.

Structure of the Paper. Section 2 recalls preliminaries about games played on graphs while Sect. 3 recalls the classical setting of zero-sum two player games. Section 4 summarizes our recent works on the use of the notion of admissibility for synthesis of reactive systems. Section 5 summarizes our recent results on regret minimization for reactive synthesis. Section 6 summarizes our recent contributions on the synthesis of strategies that ensure good expected performance together with guarantees against their worst-case behaviors.

2 Preliminaries

We consider two-player turn-based games played on finite (weighted) graphs. Such games are played on so-called weighted game arenas.

Definition 1 (Weighted Game Arena). *A (turn-based) two-player weighted game arena is a tuple* $\mathcal{A} = \langle S_\exists, S_\forall, \mathsf{E}, s_{\mathsf{init}}, \mathsf{w} \rangle$ *where:*

- $S_\exists$ *is the finite set of states owned by* Eve, $S_\forall$ *is the finite set of states owned by* Adam, $S_\exists \cap S_\forall = \emptyset$ *and we denote* $S_\exists \cup S_\forall$ *by* S.
- $\mathsf{E} \subseteq S \times S$ *is a set of edges, we say that* E *is* total *whenever for all states* $s \in S$, *there exists* $s' \in S$ *such that* $(s, s') \in \mathsf{E}$ *(we often assume this w.l.o.g.).*
- $s_{\mathsf{init}} \in S$ *is the initial state.*
- $\mathsf{w} : \mathsf{E} \to \mathbb{Z}$ *is the weight function that assigns an integer weight to each edge.*

We do not always use the weight function defined on the edges of the weighted game arena and in these cases we simply omit it.

Unless otherwise stated, we consider for the rest of the paper a fixed weighted game arena $\mathcal{A} = \langle S_\exists, S_\forall, \mathsf{E}, s_{\mathsf{init}}, \mathsf{w} \rangle$.

A *play* in the arena $\mathcal{A}$ is an *infinite* sequence of states $\pi = s_0 s_1 \ldots s_n \ldots$ such that for all $i \geq 0$, $(s_i, s_{i+1}) \in \mathsf{E}$. A play $\pi = s_0 s_1 \ldots$ is *initial* when $s_0 = s_{\mathsf{init}}$. We denote by $\mathsf{Plays}(\mathcal{A})$ the set of plays in the arena $\mathcal{A}$, and by $\mathsf{InitPlays}(\mathcal{A})$ its subset of initial plays.

A *history* ρ is a finite sequence of states which is a *prefix* of a play in $\mathcal{A}$. We denote by $\mathsf{Pref}(\mathcal{A})$ the set of histories in $\mathcal{A}$, and the set of prefixes of initial plays is denoted by $\mathsf{InitPref}(\mathcal{A})$. Given an infinite sequence of states π, and two finite sequences of states ρ_1, ρ_2, we write $\rho_1 < \pi$ if ρ_1 is a prefix of π, and $\rho_2 \leq \rho_1$ if ρ_2 is a prefix of ρ_1. For a history $\rho = s_0 s_1 \ldots s_n$, we denote by $last(\rho)$ its last state s_n, and for all $i, j, 0 \leq i \leq j \leq n$, by $\rho(i..j)$ the infix of ρ between position i and position j, i.e., $\rho(i..j) = s_i s_{i+1} \ldots s_j$, and by $\rho(i)$ the position i of ρ, i.e., $\rho(i) = s_i$. The set of histories that belong to Eve, noted $\mathsf{Pref}_\exists(\mathcal{A})$ is the subset of histories $\rho \in \mathsf{Pref}(\mathcal{A})$ such that $last(\rho) \in S_\exists$, and the set of histories that belong to Adam, noted $\mathsf{Pref}_\forall(\mathcal{A})$ is the subset of histories $\rho \in \mathsf{Pref}(\mathcal{A})$ such that $last(\rho) \in S_\forall$.

Definition 2 (Strategy). *A strategy for* Eve *in the arena* $\mathcal{A}$ *is a function* $\sigma_\exists : \mathsf{Pref}_\exists(\mathcal{A}) \to S$ *such that for all* $\rho \in \mathsf{Pref}_\exists(\mathcal{A})$, $(last(\rho), \sigma_\exists(\rho)) \in \mathsf{E}$, *i.e., it assigns to each history of* $\mathcal{A}$ *that belongs to* Eve *a state which is a* E-*successor of the last state of the history. Symmetrically, a strategy for* Adam *in the arena* $\mathcal{A}$ *is a function* $\sigma_\forall : \mathsf{Pref}_\forall(\mathcal{A}) \to S$ *such that for all* $\rho \in \mathsf{Pref}_\forall(\mathcal{A})$, $(last(\rho), \sigma_\forall(\rho)) \in \mathsf{E}$. *The set of strategies for* Eve *is denoted by* $\Sigma_\exists$ *and the set of strategies of* Adam *by* $\Sigma_\forall$.

When we want to refer to a strategy of Eve or Adam, we write it σ. We denote by $\mathsf{Dom}(\sigma)$ the domain of definition of the strategy σ, i.e., for all strategies σ of Eve (resp. Adam), $\mathsf{Dom}(\sigma) = \mathsf{Pref}_\exists(\mathcal{A})$ (resp. $\mathsf{Dom}(\sigma) = \mathsf{Pref}_\forall(\mathcal{A})$).

A play $\pi = s_0 s_1 \ldots s_n \ldots$ is *compatible* with a strategy σ if for all $i \geq 0$ such that $\pi(0..i) \in \mathsf{Dom}(\sigma)$, we have that $s_{i+1} = \sigma(\rho(0..i))$. We denote by $\mathsf{Outcome}_s(\sigma)$ the set of plays that start in s and are compatible with the strategy σ. Given a strategy $\sigma_\exists$ for Eve and a strategy $\sigma_\forall$ for Adam, and a state s, we write $\mathsf{Outcome}_s(\sigma_\exists, \sigma_\forall)$ the unique play that starts in s and which is compatible both with $\sigma_\exists$ and $\sigma_\forall$.

A strategy σ is *memoryless* when for all histories $\rho_1, \rho_2 \in \mathsf{Dom}(\sigma)$, if we have that $last(\rho_1) = last(\rho_2)$ then $\sigma(\rho_1) = \sigma(\rho_2)$, i.e., memoryless strategies only depend on the last state of the history and so they can be seen as (partial) functions from S to S. $\Sigma_\exists^{\mathsf{ML}}$ and $\Sigma_\forall^{\mathsf{ML}}$ denotes memoryless strategies of Eve and of Adam, respectively. A strategy σ is *finite-memory* if there exists an equivalence relation $\sim \subseteq \mathsf{Dom}(\sigma) \times \mathsf{Dom}(\sigma)$ of *finite index* such that for all histories ρ_1, ρ_2 such that $\rho_1 \sim \rho_2$, we have that $\sigma(\rho_1) = \sigma(\rho_2)$. If the relation $\sim$ is *regular* (computable by a finite state machine) then the finite memory strategy can be modeled by a finite state transducer (a so-called *Moore* or *Mealy* machine). If a strategy is encoded by a machine with m states, we say that it has *memory size* m.

An *objective* Win $\subseteq$ Plays($\mathcal{A}$) is a subset of plays. A strategy σ is winning from state s if Outcome$_s(\sigma) \subseteq$ Win. We will consider both *qualitative objectives*, that do not depend on the weight function of the game arena, and *quantitative objectives* that depend on the weight function of the game arena.

Our qualitative objectives are defined with Muller conditions (which are a canonical way to represent all the regular sets of plays). Let $\pi \in S^\omega$, be a play, then inf$(\pi) = \{s \in S \mid \forall i \cdot \exists j \geq i \geq 0 : \pi(j) = s\}$ is the subset of elements of S that occur infinitely often along π. A Muller objective for a game arena $\mathcal{A}$ is a defined by a set of sets of states $\mathcal{F}$ and contains the plays $\{\pi \in S^\omega \mid \text{inf}(\pi) \in \mathcal{F}\}$. We sometimes take the liberty to define such regular sets using standard LTL syntax. For a formal definition of the syntax and semantics of LTL, we refer the interested reader to [2].

We associate, to each play π, an infinite sequence of weights, denoted w(π), and defined as follows:

$$\text{w}(\pi) = \text{w}(\pi(0), \pi(1))\text{w}(\pi(1), \pi(2)) \ldots \text{w}(\pi(i), \pi(i+1)) \cdots \in \mathbb{Z}^\omega.$$

To assign a value Val(π) to a play π, we classically use functions like sup (that returns the supremum of the values along the play), inf (that returns the infimum), limsup (that returns the limit superior), liminf (that returns the limit inferior), MP (that returns the limit of the average of the weights along the play), or dSum (that returns the discounted sum of the weights along the play). We only define the mean-payoff measure formally.

Let $\rho = s_0 s_1 \ldots s_n$ be s.t. $(s_i, s_{i+1}) \in$ E for all i, $0 \leq i < n$, the mean-payoff of this sequence of edges is

$$\text{MP}(\rho) = \frac{1}{n} \cdot \sum_{i=0}^{i=n-1} \text{w}(\rho(i), \rho(i+1)),$$

i.e., the mean-value of the weights of the edges traversed by the finite sequence ρ. The *mean-payoff* of an (infinite) play π, denoted MP(π), is a real number defined from the sequence of weights w(π) as follows:

$$\text{MP}(\pi) = \liminf_{n \to +\infty} \frac{1}{n} \cdot \sum_{i=0}^{i=n-1} \text{w}(\pi(i), \pi(i+1)),$$

i.e., MP(π) is the limit inferior of running averages of weights seen along the play π. Note that we need to use lim inf because the value of the running averages of weights may oscillate along π, and so the limit is not guaranteed to exist.

A game is defined by a (weighted) game arena, and objectives for Eve and Adam.

Definition 3 (Game). *A game* G $= (\mathcal{A}, \text{Win}_\exists, \text{Win}_\forall)$ *is defined by a game arena* $\mathcal{A}$, *an objective* Win$_\exists$ *for* Eve, *and an objective* Win$_\forall$ *for* Adam.

3 Classical Zero-Sum Setting

In zero sum games, players have antagonistic objectives.

Definition 4. *A game* $\mathsf{G} = (\mathcal{A}, \mathsf{Win}_\exists, \mathsf{Win}_\forall)$ *is* zero-sum *if* $\mathsf{Win}_\forall = \mathsf{Plays} \setminus \mathsf{Win}_\exists$.

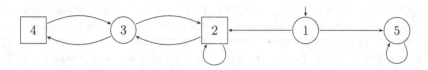

Fig. 1. An example of a two-player game arena. Rounded positions belong to Eve, and squared positions belong to Adam.

Example 1. Let us consider the example of Fig. 1. Assume that the objective of Eve is to visit 4 infinitely often, i.e., $\mathsf{Win}_\exists = \{\pi \in \mathsf{Plays} \mid \pi \models \Box\Diamond 4\}$, and that the objective of Adam is $\mathsf{Win}_\forall = \mathsf{Plays} \setminus \mathsf{Win}_\exists$. Then it should be clear that Eve does not have a strategy that enforces a play in $\mathsf{Win}_\exists$ no matter what Adam plays. Indeed, if Adam always chooses to stay at state 2, there is no way for Eve to visit 4 at all.

As we already said, zero-sum games are usually a bold abstraction of reality. This is because the system to synthesize usually interacts with an environment that has its own objective, and this objective is not necessarily the complement of the objective of the system. A classical way to handle this situation (see e.g., [4]) is to ask the system to win only when the environment meets its own objective.

Definition 5 (Win-Hyp). *Let* $\mathsf{G} = (\mathcal{A}, \mathsf{Win}_\exists, \mathsf{Win}_\forall)$ *be a game,* Eve *achieves* $\mathsf{Win}_\exists$ *from state* s *under hypothesis* $\mathsf{Win}_\forall$ *if there exists* $\sigma_\exists$ *such that*

$$\mathsf{Outcome}_s(\sigma_\exists) \subseteq \mathsf{Win}_\exists \cup \overline{\mathsf{Win}_\forall}.$$

The synthesis rule in the definition above is called *winning under hypothesis*, Win-Hyp for short.

Example 2. Let us consider the example of Fig. 1 again. But now assume that the objective of Adam is to visit 3 infinitely often, i.e., $\mathsf{Win}_\forall = \{\pi \in \mathsf{Plays} \mid \pi \models \Box\Diamond 3\}$. In this case, it should be clear then the strategy $1 \to 2$ and $3 \to 4$ for Eve is winning for the objective

$$\mathsf{Win\text{-}Hyp}_{\Box\Diamond 4 \vee \overline{\Box\Diamond 3}} = \{\pi \in \mathsf{Plays} \mid \pi \models \Box\Diamond 4\} \cup \overline{\{\pi \in \mathsf{Plays} \mid \pi \models \Box\Diamond 3\}}$$

i.e., under the hypothesis that the outcome satisfies the objective of Adam.

Unfortunately, there are strategies of Eve which are winning for the rule Win-Hyp but which are not desirable. As an example, consider the strategy that in 1 chooses to go to 5. In that case, the objective of Adam is unmet and so this strategy of Eve is winning for $\mathsf{Win\text{-}Hyp}_{\Box\Diamond 4 \vee \overline{\Box\Diamond 3}}$, but clearly such a strategy is not interesting as it excludes the possibility to meet the objective of Eve.

4   Assume Admissible Synthesis

To define the notion of admissible strategy, we first need to define when a strategy σ is dominated by a strategy σ'. We will define the notion for Eve, the definition for Adam is symmetric.

Let $\sigma_\exists$ and $\sigma'_\exists$ be two strategies of Eve in the game arena $\mathcal{A}$. We say that $\sigma'_\exists$ *dominates* $\sigma_\exists$ if the following two conditions hold:

1. $\forall \sigma_\forall \in \Sigma_\forall \cdot \text{Outcome}_{s_{\text{init}}}(\sigma_\exists, \sigma_\forall) \in \text{Win}_\exists \rightarrow \text{Outcome}_{s_{\text{init}}}(\sigma'_\exists, \sigma_\forall) \in \text{Win}_\exists$

2. $\exists \sigma_\forall \in \Sigma_\forall \cdot \text{Outcome}_{s_{\text{init}}}(\sigma_\exists, \sigma_\forall) \notin \text{Win}_\exists \wedge \text{Outcome}_{s_{\text{init}}}(\sigma'_\exists, \sigma_\forall) \in \text{Win}_\exists$

So a strategy $\sigma_\exists$ is dominated by $\sigma'_\exists$ if $\sigma'_\exists$ does as well as $\sigma_\exists$ against any strategy of Adam (condition 1), and there exists a strategy of Adam against which $\sigma'_\exists$ does better than $\sigma_\exists$ (condition 2).

Definition 6 (Admissible Strategy). *A strategy is admissible if there does not exist a strategy that dominates it.*

Let $G = (\mathcal{A}, \text{Win}_\exists, \text{Win}_\forall)$ be a game, the set of admissible strategies for Eve is noted $\text{Adm}_\exists$, and the set of admissible strategies for Adam is denoted $\text{Adm}_\forall$.

Clearly, a rational player should not play a dominated strategy as there always exists some strategy that behaves strictly better than the dominated strategy. So, a rational player only plays admissible strategies.

Example 3. Let us consider again the example of Fig. 1 with $\text{Win}_\exists = \{\pi \in \text{Plays} \mid \pi \models \Box\Diamond 4\}$ and $\text{Win}_\forall = \{\pi \in \text{Plays} \mid \pi \models \Box\Diamond 3\}$. We claim that the strategy $\sigma_\exists$ that plays $1 \rightarrow 5$ is not admissible in $\mathcal{A}$ from state 1. This is because the strategy $\sigma'_\exists$ that plays $1 \rightarrow 2$ and $4 \rightarrow 3$ dominates this strategy. Indeed, while $\sigma_\exists$ is always losing for the objective of Eve, the strategy $\sigma'_\exists$ wins for this objective whenever Adam eventually plays $2 \rightarrow 3$.

Definition 7 (AA). *Let* $G = (\mathcal{A}, \text{Win}_\exists, \text{Win}_\forall)$ *be a game,* Eve *achieves* $\text{Win}_\exists$ *from s under the hypothesis that* Adam *plays admissible strategies if*

$$\exists \sigma_\exists \in \text{Adm}_\exists \cdot \forall \sigma_\forall \in \text{Adm}_\forall \cdot \text{Outcome}_s(\sigma_\exists, \sigma_\forall) \in \text{Win}_\exists.$$

Example 4. Let us consider again the example of Fig. 1 with $\text{Win}_\exists = \{\pi \in \text{Plays} \mid \pi \models \Box\Diamond 4\}$ and $\text{Win}_\forall = \{\pi \in \text{Plays} \mid \pi \models \Box\Diamond 3\}$. We claim that the strategy $\sigma_\exists$ of Eve that plays $1 \rightarrow 2$ and $4 \rightarrow 3$ is admissible (see previous example) and winning against all the admissible strategies of Adam. This is a consequence of the fact that the strategy of Adam that always plays $2 \rightarrow 2$, and which is the only counter strategy of Adam against $\sigma_\exists$, is *not* admissible. Indeed, this strategy falsifies $\text{Win}_\forall$ while a strategy that always chooses $2 \rightarrow 3$ enforces the objective of Adam.

Theorem 1 ([3,7,8]). *For all games* $G = (\mathcal{A}, \text{Win}_\exists, \text{Win}_\forall)$, *if* $\text{Win}_\exists$ *and* $\text{Win}_\forall$ *are omega-regular sets of plays, then* $\text{Adm}_\exists$ *and* $\text{Adm}_\forall$ *are both non empty sets.*

The problem of deciding if a game $G = (\mathcal{A}, \text{Win}_\exists, \text{Win}_\forall)$, *where* $\text{Win}_\exists$ *and* $\text{Win}_\forall$ *are omega-regular sets of plays expressed as Muller objectives, satisfies*

$$\exists \sigma_\exists \in \text{Adm}_\exists \cdot \forall \sigma_\forall \in \text{Adm}_\forall \cdot \text{Outcome}_s(\sigma_\exists, \sigma_\forall) \in \text{Win}_\exists$$

is PSPACE-COMPLETE.

Additional Results. The assume-admissible setting we present here relies on procedures for *iterative elimination of dominated strategies* for multiple players which was studied in [3] on games played on graphs. In this context, dominated strategies are repeatedly eliminated for each player. Thus, with respect to the new set of strategies of its opponent, new strategies may become dominated, and will therefore be eliminated, and so on until the process stabilizes. In [8], we studied the algorithmic complexity of this problem and proved that for games with Muller objectives, deciding whether all outcomes compatible with iteratively admissible strategy profiles satisfy an omega-regular objective defined by a Muller condition is PSPACE-COMPLETE and in UP∩coUP for the special case of Büchi objectives.

The assume-admissible rule introduced in [7] is also defined for multiple players and corresponds, roughly, to the first iteration of the elimination procedure. We additionally prove that if players have Büchi objectives, then the rule can be decided in polynomial-time. One advantage of the assume-admissible rule is the *rectangularity* of the solution set: the set of strategy profiles that witness the rule can be written as a product of sets of strategies for each player. In particular, this means that a strategy witnessing the rule can be chosen separately for each player. Thus, the rule is *robust* in the sense that the players do not need to agree on a strategy profile, but only on the admissibility assumption on each other. In addition, we show in [7] that the rule is amenable to abstraction techniques: we show how state-space abstractions can be used to check a sufficient condition for assume-admissible, only doing computations on the abstract state space.

Related Works. The rule "winning under hypothesis" (Win-Hyp) and its weaknesses are discussed in [4]. We have illustrated the limitations of this rule in Example 2.

There are related works in the literature which propose concepts to model systems composed of several parts, each having their own objectives. The solutions that are proposed are based on *n*-players *non*-zero sum games. This is the case both for *assume-guarantee synthesis* [17] (AG), and for *rational synthesis* [29] (RS).

For the case of two player games, AG is based on the concept of *secure equilibria* [18] (SE), a refinement of Nash equilibria [37] (NE). In SE, objectives of the players are lexicographic: each player first tries to force his own objective, and then tries to falsify the objectives of the other players. It was shown in [18] that SE are the NE that form enforceable contracts between the two players. When the AG rule is extended to several players, as in [17], it no longer corresponds to secure equilibria. We gave a direct algorithm for multiple players in [7]. The difference between AG and SE is that AG strategies have to be resilient to deviations of all the other players, while SE profiles have to be resilient to deviations by only one player. A variant of the rule AG, called *Doomsday equilibria*, has been proposed in [14]. We have also studied quantitative extensions of the notion of secure equilibria in [13].

In the context of infinite games played on graphs, one well known limitation of NE is the existence of *non-credible threats*. Refinements of the notion of

NE, like *sub-game perfect equilibria* (SPE), have been proposed to overcome this limitation. SPE for games played on graphs have been studied in e.g., [9,42]. Admissibility does not suffer from this limitation.

In RS, the system is assumed to be monolithic and the environment is made of several components that are only *partially controllable*. In RS, we search for a profile of strategies in which the system forces its objective and the players that model the environment are given an "*acceptable*" strategy profile, from which it is assumed that they will not deviate. "Acceptable" can be formalized by any *solution concept*, e.g., by NE, *dominant* strategies, or *sub-game perfect equilibria*. This is the existential flavor of RS. More recently, Kupferman et al. have proposed in [34] a *universal* variant of this rule. In this variant, we search for a strategy of the system such that in all strategy profiles that extend this strategy for the system and that are NE, the outcome of the game satisfies the specification of the system.

In [25], Faella studies several alternatives to the notion of winning strategy including the notion of admissible strategy. His work is for two-players but only the objective of one player is taken into account, the objective of the other player is left unspecified. In that work, the notion of admissibility is used to define a notion of *best-effort* in synthesis.

The notion of admissible strategy is definable in strategy logics [19,36] and decision problems related to the assume-admissible rule can be reduced to satisfiability queries in such logics. This reduction does not lead to worst-case optimal algorithms; we presented worst-case optimal algorithms in [7] based on our previous work [8].

5   Regret Minimization

In the previous section, we have shown how the notion of admissible strategy can be used to relax the classical worst-case hypothesis made on the environment. In this section, we review another way to relax this worst-case hypothesis.

The idea is simple and intuitive. When looking for a strategy, instead of trying to find a strategy which is worst-case optimal, we search for a strategy that takes *best-responses* (against the behavior of the environment) as a yardstick. That is, we would like to find a strategy that behaves "not far" from an optimal response to the strategy of the environment — when the latter is fixed. The notion of regret minimization is naturally defined in a quantitative setting (although it also makes sense in a Boolean setting).

Let us now formally define the notion of regret associated to a strategy of Eve. This definition is parameterized by a set of strategies for Adam.

Definition 8 (Relative Regret). *Let* $\mathcal{A} = \langle S_\exists, S_\forall, E, s_{\text{init}}, \mathsf{w} \rangle$ *be a weighted game arena, let* $\sigma_\exists$ *be a strategy of* Eve, *the regret of this strategy relative to a set of strategies* $\mathsf{Str}_\forall \subseteq \Sigma_\forall$ *is defined as follows:*

$$\mathsf{Reg}(\sigma_\exists, \mathsf{Str}_\forall) = \sup_{\sigma_\forall \in \mathsf{Str}_\forall} \sup_{\sigma'_\exists \in \Sigma_\exists} \mathsf{Val}(\sigma_\forall, \sigma'_\exists) - \mathsf{Val}(\sigma_\forall, \sigma_\exists).$$

We interpret the sub-expression $\sup_{\sigma'_\exists \in \Sigma_\exists} \mathsf{Val}(\sigma_\forall, \sigma'_\exists)$ as the *best-response* of Eve against $\sigma_\forall$. Then, the relative regret of a strategy of Eve can be seen as the supremum of the differences between the value achieved by $\sigma_\exists$ against a strategy of Adam and the value achieved by the corresponding best-response.

We are now equipped to formally define the problem under study, which is parameterized by payoff function $\mathsf{Val}(\cdot)$ and a set $\mathsf{Str}_\forall$ of strategies of Adam.

Definition 9 (Regret Minimization). *Given a weighted game arena $\mathcal{A}$ and a rational threshold r, decide if there exists a strategy $\sigma_\exists$ for Eve such that*

$$\mathsf{Reg}(\sigma_\exists, \mathsf{Str}_\forall) \leq r$$

and synthesize such a strategy if one exists.

In [32], we have considered several types of strategies for Adam: the set $\Sigma_\forall$, i.e., any strategy, the set $\Sigma_\forall^{\mathsf{ML}}$, i.e., memoryless strategies for Adam, and the set $\Sigma_\forall^{\mathsf{W}}$, i.e., word strategies for Adam.[1] We will illustrate each of these cases on examples below.

Example 5. Let us consider the weighted game arena of Fig. 2, and let us assume that we want to synthesize a strategy for Eve that minimizes her mean-payoff regret against Adam playing a memoryless strategy. The memoryless restriction is useful when designing a system that needs to perform well in an environment which is only partially known. In practice, a controller may discover the environment with which it is interacting during run-time. Such a situation can be modeled by an arena in which choices in nodes of the environment model an entire family of environments and each memoryless strategy models a specific environment of the family. In such cases, if we want to design a controller that performs reasonably well against all the possible environments, we can consider each best-response of Eve for each environment and then try to choose one unique strategy for Eve that minimizes the difference in performance w.r.t. those best-responses: a regret-minimizing strategy.

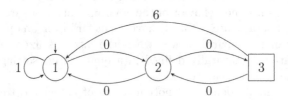

Fig. 2. An example of a two-player game arena with MP objective for Eve. Rounded positions belong to Eve, and squared positions belong to Adam.

---

[1] To define word strategies, it is convenient to consider game arenas where edges have labels called letters. In that case, when playing a word strategy, Adam commits to a sequence of letters (i.e., a word) and plays that word regardless of the exact state of the game. Word strategies are formally defined in [32] and below.

In our example, prior to a first visit to state 3, we do not know if the edge $3 \to 2$ or the edge $3 \to 1$ will be activated by Adam. But as Adam is bound to play a memoryless strategy, once he has chosen one of the two edges, we know that he will stick to this choice.

A regret-minimizing strategy in this example is as follows: play $1 \to 2$, then $2 \to 3$, if Adam plays $3 \to 2$, then play $2 \to 1$ and then $1 \to 1$ forever, otherwise Adam plays $3 \to 1$ and then Eve should continue to play $1 \to 2$ and $2 \to 3$ forever. This strategy has regret 0. Note that this strategy uses memory and that there is no memoryless strategy of Eve with regret 0 in this game.

Let us now illustrate the interest of the notion of regret minimization when Adam plays *word* strategies. When considering this restriction, it is convenient to consider letters that label the edges of the graph (Fig. 3). A word strategy for Adam is a function $w : \mathbb{N} \to \{a, b\}$. In this setting Adam plays a sequence of letters and this sequence is independent of the current state of the game. We have shown in [32] that the notion of regret minimization relative to word strategies is a generalization of the notion of *good-for-games automata* introduced by Henzinger and Piterman in [31].

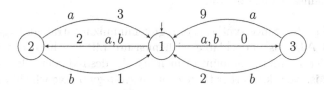

Fig. 3. An example of a two-player game arena with MP objective for Eve. Edges are annotated by letters: Adam chooses a word w and Eve resolves the non-determinism on edges.

Example 6. In this example, a strategy of Eve determines how to resolve non-determinism in state 1. The best strategy of Eve for mean-payoff regret minimization is to always take the edge $1 \to 3$. Indeed, let us consider all the sequences of two letters that Adam can choose and compute the regret of choosing $1 \to 2$ (left) and the regret of choosing $1 \to 3$ (right):

- $*a$ with $* \in \{a, b\}$, the regret of left is equal to 0, and the regret of right is $\frac{5-3}{2} = 1$.
- $*b$ with $* \in \{a, b\}$, the regret of left is equal to $\frac{9-3}{2} = 3$, and the regret of right is 0.

So the strategy that minimizes the regret of Eve is to always take the arrow $1 \to 3$ (right), the regret is then equal to 1.

In [32], we have studied the complexity of deciding the existence of strategies for Eve that have less than a given regret threshold. The results that we have obtained are summarized in the theorem below.

Theorem 2 ([32]). *Let* $\mathcal{A} = \langle S_\exists, S_\forall, \mathsf{E}, s_{\text{init}}, \mathsf{w} \rangle$ *be a weighted game arena, the complexity of deciding if* Eve *has a strategy with regret less than or equal to a threshold* $r \in \mathbb{Q}$ *against* Adam *playing:*

- *a strategy in* $\Sigma_\forall$, *is* PTIME-COMPLETE *for payoff functions* inf, sup, liminf, limsup, *and in* NP∩coNP *for* MP.
- *a strategy in* $\Sigma_\forall^{\text{ML}}$, *is in* PSPACE *for payoff functions* inf, sup, liminf, limsup, *and* MP, *and is* coNP-HARD *for* inf, sup, limsup, *and* PSPACE-HARD *for* liminf, *and* MP.
- *a strategy in* $\Sigma_\forall^{\text{W}}$, *is* EXPTIME-COMPLETE *for payoff functions* inf, sup, liminf, limsup, *and undecidable for* MP.

The above results are obtained by reducing the synthesis of regret-minimizing strategies to finding winning strategies in classical games. For instance, a strategy for Eve that minimizes regret against $\Sigma_\forall^{\text{ML}}$ for the mean-payoff measure corresponds to finding a winning strategy in a mean-payoff game played on a larger game arena which encodes the witnessed choices of Adam and forces him to play positionally. When minimizing regret against word strategies, for the decidable cases the reduction is done to parity games and is based on the *quantitative simulation* games defined in [15].

Additional Results. Since synthesis of regret-minimizing strategies against word strategies of Adam is undecidable with measure MP, we have considered the sub-case which limits the amount of memory the desired controller can use (as in [1]). That is, we ask whether there exists a strategy of Eve which uses at most memory m and ensures regret at most r. In [32] we showed that this problem is in NTIME($m^2|\mathcal{A}|^2$) for MP.

Theorem 3 ([32]). *Let* $\mathcal{A} = \langle S_\exists, S_\forall, \mathsf{E}, s_{\text{init}}, \mathsf{w} \rangle$ *be a weighted game arena, the complexity of deciding if* Eve *has a strategy using memory of at most* m *with regret less than or equal to a threshold* $\lambda \in \mathbb{Q}$ *against* Adam *playing a strategy in* $\Sigma_\forall^{\text{W}}$, *is in non-deterministic polynomial time w.r.t.* m *and* $|\mathcal{A}|$ *for* inf, sup, liminf, limsup, *and* MP.

Finally, we have established the equivalence of a quantitative extension of the notion of good-for-games automata [31] with determinization-by-pruning of the refinement of an automaton [1] and our regret games against word strategies of Adam. Before we can formally state these results, some definitions are needed.

Definition 10 (Weighted Automata). *A finite weighted automaton is a tuple* $\langle Q, q_{\text{init}}, A, \Delta, \mathsf{w} \rangle$ *where:* Q *is a finite set of states,* $q_{\text{init}} \in Q$ *is the initial state,* A *is a finite alphabet of actions or symbols,* $\Delta \subseteq Q \times A \times Q$ *is the transition relation, and* $\mathsf{w} : \Delta \to \mathbb{Z}$ *is the weight function.*

A *run* of an automaton on a word $a \in A^\omega$ is an infinite sequence of transitions $\rho = (q_0, a_0, q_1)(q_1, a_1, q_2) \cdots \in \Delta^\omega$ such that $q_0 = q_{\text{init}}$ and $a_i = a(i)$ for all $i \geq 0$. As with plays in a game, each run is assigned a *value* with a payoff function $\text{Val}(\cdot)$. A weighted automaton $\mathcal{M}$ defines a function $A^\omega \to \mathbb{R}$ by assigning to

$a \in A^\omega$ the supremum over all the values of its runs on a. The automaton is said to be *deterministic* if for all $q \in Q$ and $x \in A^\omega$ the set $\{q' \in Q \mid (q, x, q') \in \Delta\}$ is a singleton.

In [31], Henzinger and Piterman introduced the notion of *good-for-games automata*. A non-deterministic automaton is good for solving games if it fairly simulates the equivalent deterministic automaton.

Definition 11 (α-good-for-games). *A finite weighted automaton $\mathcal{M}$ is α-good-for-games if a player (Simulator), against any word $x \in A^\omega$ spelled by Spoiler, can resolve non-determinism in $\mathcal{M}$ so that the resulting run has value v and $\mathcal{M}(x) - v \leq \alpha$.*

The above definition is a quantitative generalization of the notion proposed in [31]. We link their class of automata with our regret games in the sequel.

Proposition 1 ([32]). *A weighted automaton $\mathcal{M} = \langle Q, q_{\text{init}}, A, \Delta, \mathsf{w} \rangle$ is α-good-for-games if and only if there exists a strategy $\sigma_\exists$ for Eve with relative regret of at most α against strategies $\Sigma_\forall^{\mathsf{W}}$ of Adam.*

Our definitions also suggest a natural notion of approximate determinization for weighted automata on infinite words. This is related to recent work by Aminof et al.: in [1], they introduce the notion of *approximate-determinization-by-pruning* for weighted sum automata over finite words. For $\alpha \in (0, 1]$, a weighted sum automaton is α-*determinizable-by-pruning* if there exists a finite state strategy to resolve non-determinism and that constructs a run whose value is at least α times the value of the maximal run of the given word. So, they consider a notion of approximation which is a *ratio*. Let us introduce some additional definitions required to formalize the notion of determinizable-by-pruning.

Consider two weighted automata $\mathcal{M} = \langle Q, q_{\text{init}}, A, \Delta, \mathsf{w} \rangle$ and $\mathcal{M}' = \langle Q', q'_{\text{init}}, A, \Delta', \mathsf{w}' \rangle$. We say that $\mathcal{M}'$ α-*approximates* $\mathcal{M}$ if $|\mathcal{M}(x) - \mathcal{M}'(x)| \leq \alpha$, for all $x \in A^\omega$. We say that $\mathcal{M}$ *embodies* $\mathcal{M}'$ if $Q' \subseteq Q$, $\Delta' \subseteq \Delta$, and w' agrees with w on Δ'. For an integer $k \geq 0$, the k-*refinement* of $\mathcal{M}$ is the automaton obtained by refining the state-space of $\mathcal{M}$ using k boolean variables.

Definition 12 ((α, k)-determinizable-by-pruning). *A finite weighted automaton $\mathcal{M}$ is (α, k)-determinizable-by-pruning if the k-refinement of $\mathcal{M}$ embodies a deterministic automaton which α-approximates $\mathcal{M}$.*

We show in [32] that when Adam plays word strategies only, our notion of regret defines a notion of approximation with respect to the *difference* metric for weighted automata (as defined above).

Proposition 2 ([32]). *A weighted automaton $\mathcal{M} = \langle Q, q_{\text{init}}, A, \Delta, \mathsf{w} \rangle$ is α-determinizable-by-pruning if and only if there exists a strategy $\sigma_\exists$ for Eve using memory at most 2^m with relative regret of at most α against strategies $\Sigma_\forall^{\mathsf{W}}$ of Adam.*

Related Works. The notion of regret minimization is important in game and decision theory, see e.g., [45] and additional bibliographical pointers there. The concept of *iterated* regret minimization has been recently proposed by Halpern et al. for *non-zero* sum games [30]. In [28], the concept is applied to games played on weighted graphs with shortest path objectives. Variants on the different sets of strategies considered for Adam were not considered there.

In [23], Damm and Finkbeiner introduce the notion of *remorse-free strategies*. The notion is introduced in order to define a notion of *best-effort* strategy when winning strategies do not exist. Remorse-free strategies are exactly the strategies which minimize regret in games with ω-regular objectives in which the environment (Adam) is playing word strategies only. The authors of [23] do not establish lower bounds on the complexity of the realizability and synthesis problems for remorse-free strategies.

A concept equivalent to good-for-games automata is that of history- determinism [22]. Proposition 1 thus allows us to generalize history-determinism to a quantitative setting via this relationship with good-for-games automata.

Finally, we would like to highlight some differences between our work and the study of Aminof et al. in [1] on determinization-by-pruning. First, we consider infinite words while they consider finite words. Second, we study a general notion of regret minimization problem in which Eve can use any strategy while they restrict their study to fixed memory strategies only and leave the problem open when the memory is not fixed a priori.

6   Game Arenas with Expected Adversary

In the two previous sections we have relaxed the worst-case hypothesis on the environment (modeled by the behavior of Adam) by either considering an explicit objective for the environment or by considering as yardsticks the best-responses to the strategies of Adam. Here, we introduce another model where the environment is modeled as a stochastic process (i.e., Adam is expected to play according to some known *randomized* strategy) and we are looking for strategies for Eve that ensure good expectation against this stochastic process while guaranteeing acceptable worst-case performance even if Adam deviates from his expected behavior.

To define formally this new framework, we need game arenas in which an expected behavior for Adam is given as a *memoryless randomized strategy*.[2] We first introduce some notation. Given a set A, let $\mathcal{D}(A)$ denote the set of rational probability distributions over A, and, for $d \in \mathcal{D}(A)$, we denote its support by $\mathsf{Supp}(d) = \{a \in A \mid d(a) > 0\} \subseteq A$.

---

[2] It should be noted that we can easily consider finite-memory randomized strategies for Adam, instead of memoryless randomized strategies. This is because we can always take the synchronized product of a finite-memory randomized strategy with the game arena to obtain a new game arena in which the finite-memory strategy on the original game arena is now equivalent to a memoryless strategy.

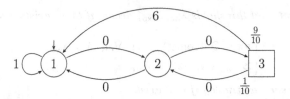

Fig. 4. A game arena associated with a memoryless randomized strategy for Adam can be seen as an MDP: the fractions represent the respective probability to take each outgoing edge when leaving state 3.

Definition 13. *Fix a weighted game arena* $\mathcal{A} = \langle S_\exists, S_\forall, \mathsf{E}, s_{\mathsf{init}}, \mathsf{w} \rangle$. *A memoryless randomized strategy for* Adam *is a function*

$$\sigma_\forall^{\mathsf{rnd}} : S_\forall \to \mathcal{D}(S)$$

such that for all $s \in S_\forall$, $\mathsf{Supp}(\sigma_\forall^{\mathsf{rnd}}(s)) \subseteq \{s' \in S \mid (s, s') \in \mathsf{E}\}$.

For the rest of this section, we model the expected behavior of Adam with a strategy $\sigma_\forall^{\mathsf{rnd}}$, given as part of the input for the problem we will consider. Given a weighted game arena $\mathcal{A}$ and a memoryless randomized strategy $\sigma_\forall^{\mathsf{rnd}}$ for Adam, we are left with a model with both non-deterministic choices (for Eve) and stochastic transitions (due to the randomized strategy of Adam). This is essentially what is known in the literature as a $1\frac{1}{2}$-player game or more commonly, a *Markov Decision Process* (MDP), see for example [26,38]. One can talk about plays, strategies and other notions in MDPs as introduced for games.

Consider the game in Fig. 4. We can see it as a classical two-player game if we forget about the fractions around state 3. Now assume that we fix the memoryless randomized strategy $\sigma_\forall^{\mathsf{rnd}}$ for Adam to be the one that, from 3, goes to 1 with probability $\frac{9}{10}$ and to 2 with the remaining probability, $\frac{1}{10}$. This is represented by the fractions on the corresponding outgoing edges. In the remaining model, only Eve still has to pick a strategy: it is an MDP. We denote this MDP by $\mathcal{A}[\sigma_\forall^{\mathsf{rnd}}]$.

Let us go one step further. Assume now that Eve also picks a strategy $\sigma_\exists$ in this MDP. Now we obtain a fully stochastic process called a *Markov Chain* (MC). We denote it by $\mathcal{A}[\sigma_\exists, \sigma_\forall^{\mathsf{rnd}}]$. In an MC, an *event* is a measurable set of plays. It is well-known from the literature [43] that every event has a uniquely defined probability (Carathéodory's extension theorem induces a unique probability measure on the Borel σ-algebra over plays in the MC). Given $\mathcal{E}$ a set of plays in $M = \mathcal{A}[\sigma_\exists, \sigma_\forall^{\mathsf{rnd}}]$, we denote by $\mathbb{P}_M(\mathcal{E})$ the probability that a play belongs to $\mathcal{E}$ when M is executed for an infinite number of steps. Given a measurable value function Val, we denote by $\mathbb{E}_M(\mathsf{Val})$ the *expected value* or *expectation* of Val over plays in M. In this paper, we focus on the mean-payoff function MP.

We are now finally equipped to formally define the problem under study.

Definition 14 (Beyond Worst-Case Synthesis). *Given a weighted game arena $\mathcal{A}$, a stochastic model of* Adam *given as a memoryless randomized strategy*

$\sigma_{\forall}^{\mathsf{rnd}}$, and two rational thresholds $\lambda_{\mathsf{wc}}, \lambda_{\mathsf{exp}}$, decide if there exists a strategy $\sigma_{\exists}$ for Eve such that

$$\begin{cases} \forall \pi \in \mathsf{Outcome}_{s_{init}}^{\mathcal{A}}(\sigma_{\exists}) \cdot \mathsf{Val}(\pi) > \lambda_{\mathsf{wc}} \\ \mathbb{E}_{\mathcal{A}[\sigma_{\exists}, \sigma_{\forall}^{\mathsf{rnd}}]}(\mathsf{Val}) > \lambda_{\mathsf{exp}} \end{cases}$$

and synthesize such a strategy if one exists.

Intuitively, we are looking for strategies that can *simultaneously* guarantee a worst-case performance higher than λ_{wc}, i.e., against any behavior of Adam in the game $\mathcal{A}$, *and* guarantee an expectation higher than λ_{exp} when faced to the expected behavior of Adam, i.e., when played in the MDP $\mathcal{A}[\sigma_{\forall}^{\mathsf{rnd}}]$. We can of course assume w.l.o.g. that $\lambda_{\mathsf{wc}} < \lambda_{\mathsf{exp}}$, otherwise the problem reduces trivially to just a worst-case requirement: any lower bound on the worst-case value is also a lower bound on the expected value.

Example 7. Consider the arena depicted in Fig. 4. As mentioned before, the probability distribution models the expected behavior of Adam. Assume that we want now to synthesize a strategy for Eve which ensures that (C_1) the mean-payoff will be at least $\frac{1}{3}$ no matter how Adam behaves (worst-case guarantee), and (C_2) at least $\frac{3}{2}$ if Adam plays according to his expected behavior (good expectation).

First, let us study whether this can be achieved through the two classical solution concepts used in games and MDPs respectively. We start by considering the arena as a traditional two-player zero-sum game: in this case, it is known that an optimal memoryless strategy exists [24]. Let $\sigma_{\exists}^{\mathsf{wc}}$ be the strategy of Eve that always plays $1 \rightarrow 1$ and $2 \rightarrow 1$. That strategy maximizes the worst-case mean-payoff, as it enforces a mean-payoff of 1 no matter how Adam behaves. Thus, (C_1) is satisfied. Observe that if we consider the arena as an MDP (i.e., taking the probabilities into account), this strategy yields an expected value of 1 as the unique possible play from state 1 is to take the self-loop forever. Hence this strategy does not satisfy (C_2).

Now, consider the arena as an MDP. Again, it is known that the expected value can be maximized by a memoryless strategy [26,38]. Let $\sigma_{\exists}^{\mathsf{exp}}$ be the strategy of Eve that always chooses the following edges: $1 \rightarrow 2$ and $2 \rightarrow 3$. Its expected mean-payoff can be calculated in two steps: first computing the probability vector that represents the limiting stationary distribution of the irreducible MC induced by this strategy, second multiplying it by the vector containing the expected weights over outgoing edges for each state. In this case, it can be shown that the expected value is equal to $\frac{54}{29}$, hence the strategy does satisfy (C_2). Unfortunately, it is clearly not acceptable for (C_1) as, if Adam does not behave according to the stochastic model and always chooses to play $3 \rightarrow 2$, the mean-payoff will be equal to zero.

Hence this shows that the classical solution concepts do not suffice if one wants to go beyond the worst-case and mix guarantees on the worst-case and the expected performance of strategies. In contrast, with the framework developed in [11,12], it is indeed possible for the considered arena (Fig. 4) to build a strategy for Eve that ensures the worst-case constraint (C_1) and at the same time, yields an expected value arbitrarily close to the optimal expectation achieved by strategy $\sigma_\exists^{\exp}$. In particular, one can build a finite-memory strategy that guarantees *both* (C_1) and (C_2). The general form of such strategies is a combination of $\sigma_\exists^{\exp}$ and $\sigma_\exists^{wc}$ in a well-chosen pattern. Let $\sigma_\exists^{cmb(K,L)}$ be a *combined strategy* parameterized by two integers $K, L \in \mathbb{N}$. The strategy is as follows.

1. Play according to $\sigma_\exists^{\exp}$ for K steps.
2. If the mean-payoff over the last K steps is larger than the worst-case threshold λ_{wc} (here $\frac{1}{3}$), then go to phase 1.
3. Otherwise, play according to $\sigma_\exists^{wc}$ for L steps, and then go to phase 1.

Intuitively, the strategy starts by mimicking $\sigma_\exists^{\exp}$ for a long time, and the witnessed mean-payoff over the K steps will be close to the optimal expectation with high probability. Thus, with high probability it will be higher than $\lambda_{\exp}$, and therefore higher than λ_{wc} — recall that we assumed $\lambda_{wc} < \lambda_{\exp}$. If this is not the case, then Eve has to switch to $\sigma_\exists^{wc}$ for sufficiently many steps L in order to make sure that the worst-case constraint (C_1) is satisfied before switching back to $\sigma_\exists^{\exp}$.

One of the key results of [12] is to show that for any $\lambda_{wc} < \mu$, where μ denotes the optimal worst-case value guaranteed by $\sigma_\exists^{wc}$, and for any expected value threshold $\lambda_{\exp} < \nu$, where ν denotes the optimal expected value guaranteed by $\sigma_\exists^{\exp}$, it is possible to compute values for K and L such that $\sigma_\exists^{cmb(K,L)}$ satisfies the beyond worst-case constraint for thresholds λ_{wc} and $\lambda_{\exp}$. For instance, in the example, where $\lambda_{wc} = \frac{1}{3} < 1$ and $\lambda_{\exp} = \frac{3}{2} < \frac{54}{29}$, one can compute appropriate values of the parameters following the technique presented in [12, Theorem 5]. The crux is proving that, for large enough values of K and L, the contribution to the expectation of the phases when $\sigma_\exists^{cmb(K,L)}$ mimics $\sigma_\exists^{wc}$ are negligible, and thus the expected value yield by $\sigma_\exists^{cmb(K,L)}$ tends to the optimal one given by $\sigma_\exists^{\exp}$, while at the same time the strategy ensures that the worst-case constraint is met.

In the next theorem, we sum up some of the main results that we have obtained for the beyond worst-case synthesis problem applied to the mean-payoff value function.

Theorem 4 ([11, 12, 21]). *The beyond worst-case synthesis problem for the mean-payoff is in NP $\cap$ coNP, and at least as hard as deciding the winner in two-player zero-sum mean-payoff games, both when looking for finite-memory or infinite-memory strategies of Eve. When restricted to finite-memory strategies, pseudo-polynomial memory is both sufficient and necessary.*

The NP $\cap$ coNP-membership is good news as it matches the long-standing complexity barrier for two-player zero-sum mean-payoff games [10, 16, 24, 46]: the beyond worst-case framework offers *additional modeling power for free* in terms of decision complexity. It is also interesting to note that in general, infinite-memory strategies are more powerful than finite-memory ones in the beyond worst-case setting, which is not the case for the classical problems in games and MDPs.

Looking carefully at the techniques from [11,12], it can be seen that the main bottleneck in complexity is solving mean-payoff games in order to check whether the worst-case constraint can be met. Therefore, a natural relaxation of the problem is to consider the *beyond almost-sure threshold problem* where the worst-case constraint is softened by only asking that a threshold is satisfied with probability one against the stochastic model given as the strategy $\sigma_\forall^{rnd}$ of Adam. In this case, the complexity is reduced.

Theorem 5 ([21]). *The beyond almost-sure threshold problem for the mean-payoff is in* PTIME *and finite-memory strategies are sufficient.*

Related Works We originally introduced the beyond worst-case framework in [12] where we studied both mean-payoff and shortest path objectives. This framework generalizes classical problems for two-player zero-sum games and MDPs. In mean-payoff games, optimal memoryless strategies exist and deciding the winner lies in NP ∩ coNP while no polynomial algorithm is known [10,16,24,46]. For shortest path games, where we consider game graphs with strictly positive weights and try to minimize the accumulated cost to target, it can be shown that memoryless strategies also suffice, and the problem is in PTIME [33]. In MDPs, optimal strategies for the expectation are studied in [26,38] for the mean-payoff and the shortest path: in both cases, memoryless strategies suffice and they can be computed in PTIME. While we saw that the beyond worst-case synthesis problem does not cost more than solving games for the mean-payoff, it is not the case anymore for the shortest path: we jump from PTIME to a pseudo-polynomial-time algorithm. We proved in [12, Theorem 11] that the problem is inherently harder as it is NP-hard.

The beyond worst-case framework was extended to the multi-dimensional setting — where edges are fitted with vectors of integer weights — in [21]. The general case is proved to be coNP-complete.

Our strategies can be considered as *strongly risk averse*: they avoid at all cost outcomes that are below a given threshold (no matter what is their probability), and inside the set of those *safe* strategies, we maximize the expectation. Other different notions of risk have been studied for MDPs: in [44], the authors want to find policies which minimize the probability (risk) that the total discounted rewards do not exceed a specified value (target); in [27] the authors want policies that achieve a specified value of the long-run limiting average reward at a specified probability level (percentile). The latter problem was recently extended significantly in the framework of *percentile queries*, which provide elaborate guarantees on the performance profile of strategies in multi-dimensional MDPs [40]. While all those strategies limit risk, they only ensure *low probability* for bad behaviors but they do not ensure their absence, furthermore, they do not ensure good expectation either.

Another body of work is the study of strategies in MDPs that achieve a trade-off between the expectation and the variance over the outcomes (e.g., [6] for the mean-payoff, [35] for the cumulative reward), giving a statistical measure of the stability of the performance. In our setting, we strengthen this requirement by asking for *strict guarantees on individual outcomes*, while maintaining an appropriate expected payoff.

A survey of rich behavioral models extending the classical approaches for MDPs—including the beyond worst-case framework presented here—was published in [41], with a focus on the shortest path problem.

References

1. Aminof, B., Kupferman, O., Lampert, R.: Reasoning about online algorithms with weighted automata. ACM Trans. Algorithms **6**(2), 28:1–28:36 (2010). doi:10.1145/1721837.1721844. http://doi.acm.org/10.1145/1721837.1721844
2. Baier, C., Katoen, J.-P.: Principles of Model Checking. MIT Press, Cambridge (2008)
3. Berwanger, D.: Admissibility in infinite games. In: Thomas, W., Weil, P. (eds.) STACS 2007. LNCS, vol. 4393, pp. 188–199. Springer, Heidelberg (2007)
4. Bloem, R., Ehlers, R., Jacobs, S., Könighofer, R.: How to handle assumptions in synthesis. In: Proceedings of SYNT. EPTCS, vol. 157, pp. 34–50 (2014)
5. Brandenburger, A., Friedenberg, A., Keisler, H.J.: Admissibility in games. Econometrica **76**(2), 307–352 (2008)
6. Brázdil, T., Chatterjee, K., Forejt, V., Kucera, A.: Trading performance for stability in Markov decision processes. In: Proceedings of LICS, pp. 331–340. IEEE (2013)
7. Brenguier, R., Raskin, J.-F., Sankur, O.: Assume-admissible synthesis. In: Proceedings of CONCUR. LIPIcs, vol. 42, pp. 100–113. Schloss Dagstuhl-LZI (2015)
8. Brenguier, R., Raskin, J.-F., Sassolas, M.: The complexity of admissibility in omega-regular games. In: Proceedings of CSL-LICS, pp. 23:1–23:10. ACM (2014)
9. Brihaye, T., Bruyère, V., Meunier, N., Raskin, J.-F.: Weak subgame perfect equilibria and their application to quantitative reachability. In: Proceedings of CSL. LIPIcs, vol. 41, pp. 504–518. Schloss Dagstuhl - LZI (2015)
10. Brim, L., Chaloupka, J., Doyen, L., Gentilini, R., Raskin, J.-F.: Faster algorithms for mean-payoff games. Formal Methods Syst. Des. **38**(2), 97–118 (2011)
11. Bruyère, V., Filiot, E., Randour, M., Raskin, J.-F.: Expectations or guarantees? I want it all! A crossroad between games and MDPs. In: Proceedings of SR. EPTCS, vol. 146, pp. 1–8 (2014)
12. Bruyère, V., Filiot, E., Randour, M., Raskin, J.-F.: Meet your expectations with guarantees: beyond worst-case synthesis in quantitative games. In: Proceedings of STACS. LIPIcs, vol. 25, pp. 199–213. Schloss Dagstuhl - LZI (2014)
13. Bruyère, V., Meunier, N., Raskin, J.-F.: Secure equilibria in weighted games. In: Proceedings of CSL-LICS, pp. 26:1–26:26. ACM (2014)
14. Chatterjee, K., Doyen, L., Filiot, E., Raskin, J.-F.: Doomsday equilibria for omega-regular games. In: McMillan, K.L., Rival, X. (eds.) VMCAI 2014. LNCS, vol. 8318, pp. 78–97. Springer, Heidelberg (2014)
15. Chatterjee, K., Doyen, L., Henzinger, T.A.: Quantitative languages. ACM Trans. Comput. Logic **11**(4), 23:1–23:38 (2010). doi:10.1145/1805950.1805953. http://doi.acm.org/10.1145/1805950.1805953
16. Chatterjee, K., Doyen, L., Randour, M., Raskin, J.-F.: Looking at mean-payoff and total-payoff through windows. Inf. Comput. **242**, 25–52 (2015)
17. Chatterjee, K., Henzinger, T.A.: Assume-guarantee synthesis. In: Grumberg, O., Huth, M. (eds.) TACAS 2007. LNCS, vol. 4424, pp. 261–275. Springer, Heidelberg (2007)

18. Chatterjee, K., Henzinger, T.A., Jurdziński, M.: Games with secure equilibria. Theoret. Comput. Sci. **365**(1), 67–82 (2006)
19. Chatterjee, K., Henzinger, T.A., Piterman, N.: Strategy logic. Inf. Comput. **208**(6), 677–693 (2010)
20. Clarke, E.M., Emerson, E.A.: Design and synthesis of synchronization skeletons using branching-time temporal logic. In: Kozen, D. (ed.) Logics of Programs. LNCS, vol. 131, pp. 52–71. Springer, Heidelberg (1981)
21. Clemente, L., Raskin, J.-F.: Multidimensional beyond worst-case and almost-sure problems for mean-payoff objectives. In: Proceedings of LICS, pp. 257–268. IEEE (2015)
22. Colcombet, T.: Forms of determinism for automata. In: Proceedings of STACS. LIPIcs, vol. 14, pp. 1–23. Schloss Dagstuhl - LZI (2012)
23. Damm, W., Finkbeiner, B.: Does it pay to extend the perimeter of a world model? In: Butler, M., Schulte, W. (eds.) FM 2011. LNCS, vol. 6664, pp. 12–26. Springer, Heidelberg (2011)
24. Ehrenfeucht, A., Mycielski, J.: Positional strategies for mean payoff games. Int. J. Game Theory **8**, 109–113 (1979)
25. Faella, M.: Admissible strategies in infinite games over graphs. In: Královič, R., Niwiński, D. (eds.) MFCS 2009. LNCS, vol. 5734, pp. 307–318. Springer, Heidelberg (2009)
26. Filar, J., Vrieze, K.: Competitive Markov decision processes. Springer, New York (1997)
27. Filar, J.A., Krass, D., Ross, K.W.: Percentile performance criteria for limiting average Markov decision processes. Trans. Autom. Control **40**, 2–10 (1995)
28. Filiot, E., Le Gall, T., Raskin, J.-F.: Iterated regret minimization in game graphs. In: Hliněný, P., Kučera, A. (eds.) MFCS 2010. LNCS, vol. 6281, pp. 342–354. Springer, Heidelberg (2010)
29. Fisman, D., Kupferman, O., Lustig, Y.: Rational synthesis. In: Esparza, J., Majumdar, R. (eds.) TACAS 2010. LNCS, vol. 6015, pp. 190–204. Springer, Heidelberg (2010)
30. Halpern, J.Y., Pass, R.: Iterated regret minimization: a new solution concept. Games Econ. Behav. **74**(1), 184–207 (2012)
31. Henzinger, T.A., Piterman, N.: Solving games without determinization. In: Ésik, Z. (ed.) CSL 2006. LNCS, vol. 4207, pp. 395–410. Springer, Heidelberg (2006)
32. Hunter, P., Pérez, G.A., Raskin, J.-F.: Reactive synthesis without regret. In: Proceedings of CONCUR. LIPIcs, vol. 42, pp. 114–127. Schloss Dagstuhl - LZI (2015)
33. Khachiyan, L., Boros, E., Borys, K., Elbassioni, K., Gurvich, V., Rudolf, G., Zhao, J.: On short paths interdiction problems: total and node-wise limited interdiction. Theory Comput. Syst. **43**, 204–233 (2008)
34. Kupferman, O., Perelli, G., Vardi, M.Y.: Synthesis with rational environments. In: Bulling, N. (ed.) EUMAS 2014. LNCS, vol. 8953, pp. 219–235. Springer, Heidelberg (2015)
35. Mannor, S., Tsitsiklis, J.: Mean-variance optimization in Markov decision processes. In: Proceedings of ICML, pp. 177–184. Omnipress (2011)
36. Mogavero, F., Murano, A., Vardi, M.Y.: Reasoning about strategies. In: Proceedings of FSTTCS, LIPIcs, vol. 8, pp. 133–144. Schloss Dagstuhl - LZI (2010)
37. Nash, J.: Equilibrium points in n-person games. PNAS **36**, 48–49 (1950)
38. Puterman, M.: Markov decision processes: discrete stochastic dynamic programming, 1st edn. Wiley, New York (1994)

39. Queille, J.-P., Sifakis, J.: Specification and verification of concurrent systems in CESAR. In: Dezani-Ciancaglini, M., Montanari, U. (eds.) International Symposium on Programming. LNCS, vol. 137, pp. 337–351. Springer, Heidelberg (1982)

40. Randour, M., Raskin, J.-F., Sankur, O.: Percentile queries in multi-dimensional Markov decision processes. In: Kroening, D., Păsăreanu, C.S. (eds.) CAV 2015. LNCS, vol. 9206, pp. 123–139. Springer, Heidelberg (2015)

41. Randour, M., Raskin, J.-F., Sankur, O.: Variations on the stochastic shortest path problem. In: D'Souza, D., Lal, A., Larsen, K.G. (eds.) VMCAI 2015. LNCS, vol. 8931, pp. 1–18. Springer, Heidelberg (2015)

42. Ummels, M.: Rational behaviour and strategy construction in infinite multiplayer games. In: Arun-Kumar, S., Garg, N. (eds.) FSTTCS 2006. LNCS, vol. 4337, pp. 212–223. Springer, Heidelberg (2006)

43. Vardi, M.Y.: Automatic verification of probabilistic concurrent finite state programs. In: Proceedings of FOCS, pp. 327–338. IEEE (1985)

44. Wu, C., Lin, Y.: Minimizing risk models in Markov decision processes with policies depending on target values. J. Math. Anal. Appl. **231**(1), 47–67 (1999)

45. Zinkevich, M., Johanson, M., Bowling, M., Piccione, C.: Regret minimization in games with incomplete information. In: Proceedings of NIPS, pp. 905–912 (2008)

46. Zwick, U., Paterson, M.: The complexity of mean payoff games on graphs. Theoret. Comput. Sci. **158**(1), 343–359 (1996)

Tangles and Connectivity in Graphs

Martin Grohe[(✉)]

RWTH Aachen University, Aachen, Germany
`grohe@informatik.rwth-aachen.de`

Abstract. This paper is a short introduction to the theory of tangles, both in graphs and general connectivity systems. An emphasis is put on the correspondence between tangles of order k and k-connected components. In particular, we prove that there is a one-to-one correspondence between the triconnected components of a graph and its tangles of order 3.

1 Introduction

Tangles, introduced by Robertson and Seymour in the tenth paper [21] of their graph minors series [20], have come to play an important part in structural graph theory. For example, Robertson and Seymour's structure theorem for graphs with excluded minors is phrased in terms of tangles in its general form [22]. Tangles have also played a role in algorithmic structural graph theory (for example in [3,7,8,11,14]).

Tangles describe highly connected regions in a graph. In a precise mathematical sense, they are "dual" to decompositions (see Theorem 23). Intuitively, a graph has a highly connected region described by a tangle if and only if it does not admit a decomposition along separators of low order. By decomposition I always mean a decomposition in a treelike fashion; formally, this is captured by the notions of tree decomposition or branch decomposition.

However, tangles describe regions of a graph in an indirect and elusive way. This is why we use the unusual term "region" instead of "subgraph" or "component". The idea is that a tangle describes a region by pointing to it. A bit more formally, a *tangle of order k* assigns a "big side" to every separation of order less than k. The big side is where the (imaginary) region described by the tangle is supposed to be. Of course this assignment of "big sides" to the separations is subject to certain consistency and nontriviality conditions, the "tangle axioms".

To understand why this way of describing a "region" is a good idea, let us review decompositions of graphs into their k-connected components. It is well known that every graph can be decomposed into its connected components and into its biconnected components. The former are the (inclusionwise) maximal connected subgraphs, and the latter the maximal 2-connected subgraphs. It is also well-known that a graph can be decomposed into its triconnected components, but the situation is more complicated here. Different from what one might guess, the triconnected components are not maximal 3-connected subgraphs; in fact they are not even subgraphs, but just topological subgraphs (see Sect. 2 for a definition of topological subgraphs). Then what about 4-connected components?

© Springer International Publishing Switzerland 2016
A.-H. Dediu et al. (Eds.): LATA 2016, LNCS 9618, pp. 24–41, 2016.
DOI: 10.1007/978-3-319-30000-9_2

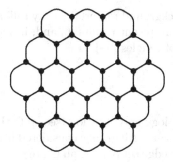

Fig. 1. A hexagonal grid

It turns out that in general a graph does not have a reasonable decomposition into 4-connected components (neither into k-connected components for any $k \geq 5$), at least if these components are supposed to be 4-connected and some kind of subgraph. To understand the difficulty, consider the hexagonal grid in Fig. 1. It is 3-connected, but not 4-connected. In fact, for any two nonadjacent vertices there is a separator of order 3 separating these two vertices. Thus it is not clear what the 4-connected components of a grid could possibly be (except, of course, just the single vertices, but this would not lead to a meaningful decomposition). But maybe we need to adjust our view on connectivity: a hexagonal grid is fairly highly connected in a "global sense". All its low-order separations are very unbalanced. In particular, all separations of order 3 have just a single vertex on one side and all other vertices on the other side. This type of global connectivity is what tangles are related to. For example, there is a unique tangle of order 4 in the hexagonal grid: the big side of a separation of order 3 is obviously the side that contains all but one vertex. The "region" this tangle describes is just the grid itself. This does not sound particularly interesting, but the grid could be a subgraph of a larger graph, and then the tangle would identify it as a highly connected region within that graph. A key theorem about tangles is that every graph admits a canonical tree decomposition into its tangles of order k [1, 21]. This can be seen as a generalisation of the decomposition of a graph into its 3-connected components. A different, but related generalisation has been given in [2].

The theory of tangles and decompositions generalises from graphs to an abstract setting of *connectivity systems*. This includes nonstandard notions of connectivity on graphs, such as the "cut-rank" function, which leads to the notion of "rank width" [16,17], and connectivity functions on other structures, for example matroids. Tangles give us an abstract notion of "k-connected components" for these connectivity systems. The canonical decomposition theorem can be generalised from graphs to this abstract setting [5,13].

This paper is a short introduction to the basic theory of tangles, both for graphs and for general connectivity systems. We put a particular emphasis on the correspondence between tangles of order k and k-connected components of a graph for $k \leq 3$, which gives some evidence to the claim that for all k, tangles of order k may be viewed as a formalisation of the intuitive notion of "k-connected component".

The paper provides background material for my talk at LATA. The talk itself will be concerned with more recent results [6] and, in particular, computational aspects and applications of tangles [9–11].

2   Preliminaries

We use a standard terminology and notation (see [4] for background); let me just review a few important notions. All graphs considered in this paper are finite and simple. The vertex set and edge set of a graph G are denoted by $V(G)$ and $E(G)$, respectively. The *order* of G is $|G| := |V(G)|$. For a set $W \subseteq V(G)$, we denote the *induced subgraph* of G with vertex set W by $G[W]$ and the induced subgraph with vertex set $V(G) \setminus W$ by $G \setminus W$. The *(open) neighbourhood* of a vertex v in G is denoted by $N^G(v)$, or just $N(v)$ if G is clear from the context. For a set $W \subseteq V(G)$ we let $N(W) := \left(\bigcup_{v \in W} N(v) \right) \setminus W$, and for a subgraph $H \subseteq G$ we let $N(H) := N(V(H))$. The *union* of two graphs A, B is the graph $A \cup B$ with vertex set $V(A) \cup V(B)$ and edge set $E(A) \cup E(B)$, and the *intersection* $A \cap B$ is defined similarly.

A *separation* of G is a pair (A, B) of subgraphs of G such that $A \cup B = G$ and $E(A) \cap E(B) = \emptyset$. The *order* of the separation (A, B) is $\mathrm{ord}(A, B) := |V(A) \cap V(B)|$. A separation (A, B) is *proper* if $V(A) \setminus V(B)$ and $V(B) \setminus V(A)$ are both nonempty. A graph G is *k-connected* if $|G| > k$ and G has no proper $(k - 1)$-separation.

A *subdivision* of G is a graph obtained from G by subdividing some (or all) of the edges, that is, replacing them by paths of length at least 2. A graph H is a *topological subgraph* of G if a subdivision of H is a subgraph of G.

3   Tangles in a Graph

In this section we introduce tangles of graphs, give a few examples, and review a few basic facts about tangles, all well-known and at least implicitly from Robertson and Seymour's fundamental paper on tangles [21] (except Theorem 7, which is due to Reed [19]).

Let G be a graph. A *G-tangle* of order k is a family $\mathcal{T}$ of separations of G satisfying the following conditions.

(GT.0) The order of all separations $(A, B) \in \mathcal{T}$ is less than k.
(GT.1) For all separations (A, B) of G of order less than k, either $(A, B) \in \mathcal{T}$ or $(B, A) \in \mathcal{T}$.
(GT.2) If $(A_1, B_1), (A_2, B_2), (A_3, B_3) \in \mathcal{T}$ then $A_1 \cup A_2 \cup A_3 \neq G$.
(GT.3) $V(A) \neq V(G)$ for all $(A, B) \in \mathcal{T}$.

Observe that (GT.1) and (GT.2) imply that for all separations (A, B) of G of order less than k, exactly one of the separations $(A, B), (B, A)$ is in $\mathcal{T}$.

We denote the order of a tangle $\mathcal{T}$ by $\mathrm{ord}(\mathcal{T})$.

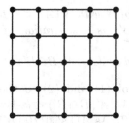

Fig. 2. A (5×5)-grid

Example 1. Let G be a graph and $C \subseteq G$ a cycle. Let $\mathcal{T}$ be the set of all separations (A, B) of G of order 1 such that $C \subseteq B$. Then $\mathcal{T}$ is a G-tangle of order 2.

To see this, note that $\mathcal{T}$ trivially satisfies (GT.0). It satisfies (GT.1), because for every separation (A, B) of G of order 1, either $C \subseteq A$ or $C \subseteq B$. To see that $\mathcal{T}$ satisfies (GT.3), let $(A_i, B_i) \in \mathcal{T}$ for $i = 1, 2, 3$. Note that it may happen that $V(A_1) \cup V(A_2) \cup V(A_3) = V(G)$ (if $|C| = 3$). However, no edge of C can be in $E(A_i)$ for any i, because $C \subseteq B_i$ and $|A_i \cap B_i| \leq 1$. Hence $E(A_1) \cup A(A_2) \cup E(A_3) \neq E(G)$, which implies (GT.2). Finally, $\mathcal{T}$ satisfies (GT.3), because $V(C) \setminus V(A) \neq \emptyset$ for all $(A, B) \in \mathcal{T}$.   ⌐

Example 2. Let G be a graph and $X \subseteq V(G)$ a clique in G. Note that for all separations (A, B) of G, either $X \subseteq V(A)$ or $X \subseteq V(B)$. For every $k \geq 1$, let $\mathcal{T}_k$ be the set of all separations (A, B) of G of order less than k such that $X \subseteq V(B)$.

Then if $k < \frac{2}{3}|X| + 1$, the set $\mathcal{T}_k$ is a G-tangle of order k. We omit the proof, which is similar to the proof in the previous example.

Instead, we prove that $\mathcal{T}_k$ is not necessarily a G-tangle if $k = \frac{2}{3}|X| + 1$. To see this, let G be a complete graph of order $3n$, $k := 2n + 1$, and $X := V(G)$. Suppose for contradiction that $\mathcal{T}_k$ is a G-tangle of order k. Partition X into three sets X_1, X_2, X_3 of size n. For $i \neq j$, let $A_{ij} := G[X_i \cup X_j]$ and $B_{ij} := G$. Then (A_{ij}, B_{ij}) is a separation of G of order $2n < k$. By (GT.1) and (GT.3), we have $(A_{ij}, B_{ij}) \in \mathcal{T}_k$. However, $A_{12} \cup A_{13} \cup A_{23} = G$, and this contradicts (GT.2).   ⌐

Example 3. Let G be a graph and $H \subseteq G$ a $(k \times k)$-grid (see Fig. 2). Let $\mathcal{T}$ be the set of all separations (A, B) of G of order at most $k - 1$ such that B contains some row of the grid. Then $\mathcal{T}$ is a G-tangle of order k. (See [21] for a proof.)   ⌐

The reader may wonder why in (GT.2) we take three separations, instead of two or four or seventeen. The following lemma gives (some kind of) an explanation: we want our tangles to be closed under intersection, in the weak form stated as assertion (3) of the lemma; this is why taking just two separations in (GT.2) would not be good enough. Three is just enough, and as we do not want to be unnecessarily restrictive, we do not take more than three separations.

Lemma 4. *Let $\mathcal{T}$ be a G-tangle of order k.*

(1) If (A, B) is a separation of G with $|V(A)| < k$ then $(A, B) \in \mathcal{T}$.

(2) If $(A, B) \in \mathcal{T}$ and (A', B') is a separation of G of order $< k$ such that $B' \supseteq B$, then $(A', B') \in \mathcal{T}$.

(3) If $(A, B), (A', B') \in \mathcal{T}$ and $\mathrm{ord}(A \cup A', B \cap B') < k$ then $(A \cup A', B \cap B') \in \mathcal{T}$.

Proof. We leave the proofs of (1) and (2) to the reader. To prove (3), let $(A, B), (A', B') \in \mathcal{T}$ and $\mathrm{ord}(A \cup A', B \cap B') < k$. By (GT.1), either $(A \cup A', B \cap B') \in \mathcal{T}$ or $(B \cup B', A \cap A') \in \mathcal{T}$. As $A \cup A' \cup (B \cup B') = G$, by (GT.2) we cannot have $(B \cup B', A \cap A') \in \mathcal{T}$. $\square$

Corollary 5. *Let $\mathcal{T}$ be a G-tangle of order k. Let $(A, B), (A', B') \in \mathcal{T}$. Then $|B \cap B'| \geq k$.*

The following lemma will allow us, among other things, to give an alternative characterisation of tangles in terms of so-called brambles.

Lemma 6. *Let $\mathcal{T}$ be a G-tangle of order k. Then for every set $S \subseteq V(G)$ of cardinality $|S| < k$ there is a unique connected component $C(\mathcal{T}, S)$ of $G \setminus S$ such that for all separations (A, B) of G with $V(A) \cap V(B) \subseteq S$ we have $(A, B) \in \mathcal{T} \iff C(\mathcal{T}, S) \subseteq B$.*

Proof. Let $C_1, \ldots, C_m$ be the set of all connected components of $G \setminus S$. For every $I \subseteq [m]$, let $C_I := \bigcup_{i \in I} C_i$. We define a separation (A_I, B_I) of G as follows. B_I is the graph with vertex set $S \cup V(C_I)$ and all edges that have at least one endvertex in $V(C_I)$, and A_I is the graph with vertex set $S \cup V(C_{[m] \setminus I})$ and edge set $E(G) \setminus E(B_I)$. Note that $V(A_I) \cap V(B_I) = S$ and thus $\mathrm{ord}(A_I, B_I) < k$. Thus for all I, either $(A_I, B_I) \in \mathcal{T}$ or $(B_I, A_I) \in \mathcal{T}$. It follows from Lemma 4(1) and (GT.2) that $(B_I, A_I) \in \mathcal{T}$ implies $(A_{[m] \setminus I}, B_{[m] \setminus I}) \in \mathcal{T}$, because $(G[S], G) \in \mathcal{T}$ and $B_I \cup B_{[m] \setminus I} \cup G[S] = G$. Furthermore, it follows from Lemma 4(3) that $(A_I, B_I), (A_J, B_J) \in \mathcal{T}$ implies $(A_{I \cap J}, B_{I \cap J}) \in \mathcal{T}$. By (GT.3) we have $(A_{[m]}, B_{[m]}) \in \mathcal{T}$ and $(A_\emptyset, B_\emptyset) \notin \mathcal{T}$.

Let $I \subseteq [m]$ be of minimum cardinality such that $(A_I, B_I) \in \mathcal{T}$. Since (A_I, B_I), $(A_J, B_J) \in \mathcal{T}$ implies $(A_{I \cap J}, B_{I \cap J}) \in \mathcal{T}$, the minimum set I is unique. If $|I| = 1$, then we let $C(\mathcal{T}, S) := C_i$ for the unique element $i \in I$. Suppose for contradiction that $|I| > 1$, and let $i \in I$. By the minimality of $|I|$ we have $(A_{\{i\}}, B_{\{i\}}) \notin \mathcal{T}$ and thus $(A_{[m] \setminus \{i\}}, B_{[m] \setminus \{i\}}) \in \mathcal{T}$. This implies $(A_{I \setminus \{i\}}, B_{I \setminus \{i\}}) \in \mathcal{T}$, contradicting the minimality of $|I|$. $\square$

Let G be a graph. We say that subgraphs $C_1, \ldots, C_m \subseteq G$ *touch* if there is a vertex $v \in \bigcap_{i=1}^m V(C_i)$ or an edge $e \in E(G)$ such that each C_i contains at least one endvertex of e. A family $\mathcal{C}$ of subgraphs of G *touches pairwise* if all $C_1, C_2 \in \mathcal{C}$ touch, and it *touches triplewise* if all $C_1, C_2, C_3 \in \mathcal{C}$ touch. A *vertex cover* (or *hitting set*) for $\mathcal{C}$ is a set $S \subseteq V(G)$ such that $S \cap V(C) \neq \emptyset$ for all $C \in \mathcal{C}$.

Theorem 7 (Reed [19]). *A graph G has a G-tangle of order k if and only if there is a family $\mathcal{C}$ of connected subgraphs of G that touches triplewise and has no vertex cover of cardinality less than k.*

In fact, Reed [19] defines a tangle of a graph G to be a family $\mathcal{C}$ of connected subgraphs of G that touches triplewise and its order to be the cardinality of a minimum vertex cover. A *bramble* is a family $\mathcal{C}$ of connected subgraphs of G that touches pairwise. In this sense, a tangle is a special bramble.

Proof (of Theorem 7). For the forward direction, let $\mathcal{T}$ be a G-tangle of order k. We let

$$\mathcal{C} := \{C(\mathcal{T}, S) \mid S \subseteq V(G) \text{ with } |S| < k\}.$$

$\mathcal{C}$ has no vertex cover of cardinality less than k, because if $S \subseteq V(G)$ with $|S| < k$ then $S \cap V(C(\mathcal{T}, S)) = \emptyset$. It remains to prove that $\mathcal{C}$ touches triplewise. For $i = 1, 2, 3$, let $C_i \in \mathcal{C}$ and $S_i \subseteq V(G)$ with $|S_i| < k$ such that $C_i = C(\mathcal{T}, S_i)$. Let B_i be the graph with vertex set $V(C_i) \cup S$ and all edges of G that have at least one vertex in $V(C_i)$, and let A_i be the graph with vertex set $V(G) \setminus V(C_i)$ and the remaining edges of G. Since $C(\mathcal{T}, S_i) = C_i \subseteq B_i$, we have $(A_i, B_i) \in \mathcal{T}$. Hence $A_1 \cup A_2 \cup A_3 \neq G$ by (GT.2), and this implies that C_1, C_2, C_3 touch.

For the backward direction, let $\mathcal{C}$ be a family of connected subgraphs of G that touches triplewise and has no vertex cover of cardinality less than k. We let $\mathcal{T}$ be the set of all separations (A, B) of G of order less than k such that $C \subseteq B \setminus V(A)$ for some $C \in \mathcal{C}$. It is easy to verify that $\mathcal{T}$ is a G-tangle of order k. □

Let $\mathcal{T}, \mathcal{T}'$ be κ-tangles. If $\mathcal{T}' \subseteq \mathcal{T}$, we say that $\mathcal{T}$ is an *extension* of $\mathcal{T}'$. The *truncation* of $\mathcal{T}$ to order $k \leq \mathrm{ord}(\mathcal{T})$ is the set $\{(A, B) \in \mathcal{T} \mid \mathrm{ord}(A, B) < k\}$, which is obviously a tangle of order k. Observe that if $\mathcal{T}$ is an extension of $\mathcal{T}'$, then $\mathrm{ord}(\mathcal{T}') \leq \mathrm{ord}(\mathcal{T})$, and $\mathcal{T}'$ is the truncation of $\mathcal{T}$ to order $\mathrm{ord}(\mathcal{T}')$.

4 Tangles and Components

In this section, we will show that there is a one-to-one correspondence between the tangles of order at most 3 and the connected, biconnected, and triconnected components of a graph. Robertson and Seymour [21] established a one-to-one correspondence between tangles of order 2 and biconnected component. Here, we extend the picture tangles of order 3.[1]

4.1 Biconnected and Triconnected Components

Let G be a graph. Following [2], we call a set $X \subseteq V(G)$ *k-inseparable* in G if $|X| > k$ and there is no separation (A, B) of G of order at most k such that $X \setminus V(B) \neq \emptyset$ and $X \setminus V(A) \neq \emptyset$. A *k-block* of G is an inclusionwise maximal k-inseparable subset of $V(G)$. We call a k-inseparable set of cardinality greater than $k + 1$ a *proper k-inseparable set* and, if it is a k-block, a *proper k-block*. (Recall that a $(k+1)$-connected graph has order greater than $k+1$ by definition.)

---

[1] My guess is that the result for tangles of order 3 is known to other researchers in the field, but I am not aware of it being published anywhere.

We observe that every vertex x in a proper k-inseparable set X has degree at least $(k+1)$, because it has $(k+1)$ internally disjoint paths to $X \setminus \{x\}$.

A *biconnected component* of G is a subgraph induced by a 1-block, which is usually just called a *block*.[2] It is easy to see that a biconnected component B either consists of a single edge that is a bridge of G, or it is 2-connected. In the latter case, we call B a *proper biconnected component*.

The definition of triconnected components is more complicated, because the subgraph induced by a 2-block is not necessarily 3-connected (even if it is a proper 2-block).

Example 8. Let G be a graph obtained from the complete graph K_4 by subdividing each edge once. Then the vertices of the original K_4, which are precisely the vertices of degree 3 in G, form a proper 2-block, but the subgraph they induce has no edges and thus is certainly not 3-connected.

It can be shown, however, that every proper 2-block of G is the vertex set of a 3-connected topological subgraph. For a subset $X \subseteq V(G)$, we define the *torso* of X in G to be the graph $G[\![X]\!]$ obtained from the induced subgraph $G[X]$ by adding an edge vw for all distinct $v, w \in X$ such that there is a connected component C of $G \setminus X$ with $v, w \in N(C)$. We call the edges in $E(G[\![X]\!]) \setminus E(G)$ the *virtual edges* of $G[\![X]\!]$. It is not hard to show that if X is a 2-block of G then for every connected component C of $G \setminus X$ it holds that $N(C) \leq 2$; otherwise X would not be an *inclusionwise maximal* 2-inseparable set. This implies that $G[\![X]\!]$ is a topological subgraph of G: if, for some connected component C of $G \setminus X$, $N(C) = \{v, w\}$ and hence vw is a virtual edge of the torso, then there is a path from v to w in C, which may be viewed as a subdivision of the edge vw of $G[\![X]\!]$. We call the torsos $G[\![X]\!]$ for the 2-blocks X the *triconnected components* of G. We call a triconnected component *proper* if its order is at least 4.

It is a well known fact, going back to MacLane [15] and Tutte [25], that all graphs admit tree decompositions into their biconnected and triconnected components. Hopcroft and Tarjan [12, 24] proved that the decompositions can be computed in linear time.

4.2   From Components to Tangles

Lemma 9. *Let G be a graph and $X \subseteq V(G)$ a $(k-1)$-inseparable set of order $|X| > \frac{3}{2} \cdot (k-1)$. Then*

$$\mathcal{T}^{(k)}(X) := \{(A, B) \mid (A, B)\text{separation of } G \text{ of order } < k \text{ with } X \subseteq V(B)\}$$

is a G-tangle of order k.

Proof. $\mathcal{T}^{(k)}(X)$ trivially satisfies (GT.0). It satisfies (GT.1), because the $(k-1)$-inseparability of X implies that for every separation (A, B) of G of order $< k$ either $X \subseteq V(A)$ or $X \subseteq V(B)$.

---

[2] There is a slight discrepancy to standard terminology here: a set consisting of a single isolated vertex is usually also called a block, but it is not a 1-block, because its size is not greater than 1.

To see that $\mathcal{T}^{(k)}(X)$ satisfies (GT.2), let $(A_i, B_i) \in \mathcal{T}^k(X)$ for $i = 1, 2, 3$. Then $|V(A_i) \cap X| \leq k-1$, because $V(A_i) \cap X \subseteq V(A_i) \cap V(B_i)$. As $|X| > \frac{3}{2} \cdot (k-1)$, there is a vertex $x \in X$ such that x is contained in at most one of the sets $V(A_i)$. Say, $x \notin V(A_2) \cup V(A_3)$. If $x \notin V(A_1)$, then $V(A_1) \cup V(A_2) \cup V(A_3) \neq V(G)$. So let us assume that $x \in V(A_1)$.

Let $y_1, \ldots, y_{k-1} \in X \setminus \{x\}$. As X is $(k-1)$-inseparable, for all i there is a path P_i from x to y_i such that $V(P_i) \cap V(P_j) = \{x\}$ for $i \neq j$. Let w_i be the last vertex of P_i (in the direction from x to y_i) that is in $V(A_1)$. We claim that $w_i \in V(B_1)$. This is the case if $w_i = y_i \in X \subseteq V(B_1)$. If $w_i \neq y_i$, let z_i be the successor of w_i on P_i. Then $z_i \in V(B_1) \setminus V(A_1)$, and as $w_i z_i \in E(G)$, it follows that $w_i \in V(B_1)$ as well.

Thus $\{x, w_1, \ldots, w_{k-1}\} \subseteq V(A_1) \cap V(B_1)$, and as $|V(A_1) \cap V(B_1)| \leq k - 1$, it follows that $w_i = x$ for some i. Consider the edge $e = xz_i$. We have $e \notin E(A_1)$ because $z_i \notin V(A_1)$ and $e \notin E(A_2) \cup E(A_3)$ because $x \notin V(A_2) \cup V(A_3)$. Hence $E(A_1) \cup E(A_2) \cup E(A_3) \neq E(G)$, and this completes the proof of (GT.2).

Finally, $\mathcal{T}^{(k)}(X)$ satisfies (GT.3), because for every $(A, B) \in \mathcal{T}$ we have $|V(A) \cap X| \leq k - 1 < |X|$. $\qquad\square$

Corollary 10. *Let G be a graph and $X \subseteq V(G)$.*

(1) If X is the vertex set of a connected component of G (that is, a 0-block), then $\mathcal{T}^1(X)$ is a G-tangle of order 1.

(2) If X is the vertex set of a biconnected component of G (that is, a 1-block), then $\mathcal{T}^2(X)$ is a G-tangle of order 2.

(3) If X is the vertex set of a proper triconnected component of G (that is, a 2-block of cardinality at least 4), then $\mathcal{T}^3(X)$ is a G-tangle of order 3.

Let us close this section by observing that the restriction to *proper* triconnected components in assertion (3) of the corollary is necessary.

Lemma 11. *Let G be a graph and $X \subseteq V(G)$ be a 2-block of cardinality 3. Then $\mathcal{T}^3(X)$ is not a tangle.*

Proof. Let $\mathcal{T} := \mathcal{T}^3(X)$. Suppose that $X = \{x_1, x_2, x_3\}$. For $i \neq j$, let $S_{ij} := \{x_i, x_j\}$, and let Y_{ij} be the union of the vertex sets of all connected components C of $G \setminus X$ with $N(C) \subseteq S_{ij}$, and let $Z_{ij} := V(G) \setminus (Y_{ij} \cup S_{ij})$. Let $A_{ij} := G[Y_{ij} \cup S_{ij}]$, and let B_{ij} be the graph with vertex set $S_{ij} \cup Z_{ij}$ and edge set $E(G) \setminus E(A_{ij})$. Then $(A_{ij}, B_{ij}) \in \mathcal{T}$, because $X \subseteq V(B_{ij})$. As X is a 2-block, for every connected component C of $G \setminus X$ it holds that $|N(C)| \leq 2$, and hence $C \subseteq A_{ij}$ for some i, j. It is not hard to see that this implies $A_{12} \cup A_{13} \cup A_{23} = G$. Thus $\mathcal{T}$ violates (GT.2). $\qquad\square$

4.3   From Tangles to Components

For a G-tangle $\mathcal{T}$, we let

$$X_{\mathcal{T}} := \bigcap_{(A,B) \in \mathcal{T}} V(B).$$

In general, X_T may be empty; an example is the tangle of order k associated with a $(k \times k)$-grid for $k \geq 5$ (see Example 3). However, it turns out that for tangles of order $k \leq 3$, the set X_T is a $(k-1)$-block. This will be the main result of this section.

Lemma 12. *Let T be a G-tangle of order k. If $|X_T| \geq k$, then X_T is a $(k-1)$-block of G and $T = T^k(X_T)$.*

Proof. Suppose that $|X_T| \geq k$. If (A, B) is a separation of G of order less than k then either $(A, B) \in T$ or $(B, A) \in T$, which implies $X_T \subseteq V(B)$ or $X_T \subseteq V(A)$. Thus X_T is $(k-1)$-inseparable. If $X \supset X_T$, say, with $x \in X \setminus X_T$, then there is some separation $(A, B) \in T$ with $x \in V(A) \setminus V(B)$ and $X_T \subseteq V(B)$, and this implies that X is not $(k-1)$-inseparable. Hence X_T is a k-block.

We have $T = T^k(X_T)$, because $X_T \subseteq V(B)$ for all $(A, B) \in T$, and for a separation (A, B) of order at most $k-1$ we cannot have $X_T \subseteq V(A) \cap V(B)$. □

Let T be a G-tangle. A separation $(A, B) \in T$ is *minimal* in T if there is no $(A', B') \in T$ such that $B' \subset B$. Clearly, X_T is the intersection of all sets $V(B)$ for minimal $(A, B) \in T$. Hence if we want to understand X_T, we can restrict our attention to the minimal separations in T. Let $(A, B) \in T$ be minimal and $S := V(A) \cap V(B)$. It follows from Lemma 6 that $B \setminus S = C := C(T, S)$, and it follows from the minimality that $S = N(C)$ and that $E(B)$ consists of all edges with one endvertex in $V(C)$. Hence B is connected.

Theorem 13 (Robertson and Seymour [21]). *Let G be a graph.*

(1) For every G-tangle T of order 1, the set X_T is a vertex set of a connected component of G, and we have $T = T^1(X_T)$.

(2) For every G-tangle T of order 2, the set X_T is the vertex set of a biconnected component of G, and we have $T = T^2(X_T)$.

Proof. To prove (1), let T be a G-tangle of order 1. Let $C = C(T, \emptyset)$. Then $(G \setminus V(C), C)$ is the unique minimal separation in T, and thus we have $X_T = V(C)$.

To prove (2), let T be a G-tangle of order 2. By Lemma 12, it suffices to prove that $|X_T| \geq 2$. Let T' be the truncation of T to order 1. Then $W := X_{T'}$ is the vertex set of a connected component C of G, and we have $X_T \subseteq W$. Moreover, for every minimal $(A, B) \in T$ we have $B \subseteq C$, because B is connected and $B \cap C \neq \emptyset$ by (GT.2).

Claim 1. Let $(A_1, B_1), (A_2, B_2) \in T$ be distinct and minimal in T. Then $A_1 \cap C \subseteq B_2$ and $A_2 \cap C \subseteq B_1$.

Proof. We have $\mathrm{ord}(A_1 \cup A_2, B_1 \cap B_2) \geq 2$, because otherwise $(A_1 \cup A_2, B_1 \cap B_2) \in T$ by Lemma 4(3), which contradicts the minimality of the separations (A_i, B_i). Suppose that $V(A_i) \cap V(B_i) = \{s_i\}$. As

$$V(A_1 \cup A_2) \cap (V(B_1 \cap B_2) \subseteq V(A_1 \cap B_1) \cup V(A_2 \cap B_2) = \{s_1, s_2\},$$

we must have $s_1 \neq s_2$ and $V(A_1 \cup A_2) \cap V(B_1 \cap B_2) = \{s_1, s_2\}$ (see Fig. 3). This implies $V(A_1 \cap A_2) \cap V(B_1 \cup B_2) = \emptyset$. Then $(A_1 \cap A_2, B_1 \cup B_2)$ is a separation of G

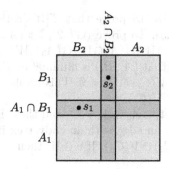

Fig. 3. Proof of Theorem 13

of order 0, and as C is connected and $(B_1 \cup B_2) \cap C \neq \emptyset$, we have $A_1 \cap A_2 \cap C = \emptyset$. The assertion of the claim follows. ⌐

Let $(A_1, B_1), \ldots, (A_m, B_m)$ be an enumeration of all minimal separations in T of order 1. Even if C is 1-inseparable, there is such a separation: $(G \setminus (V(C) \setminus \{v\}), C)$ for an arbitrary $v \in V(C)$. Thus $m \geq 1$. If $m = 1$, then $X_T = V(B_1)$ and thus $|X_T| \geq 2$ by Lemma 4(1).

If $m \geq 2$, let $A_i \cap B_i = \{s_i\}$. We can assume the s_i to be mutually distinct, because if $s_i = s_j$ then $B_i = B_j$. It follows from Claim 1 that $s_1, \ldots, s_m \in \bigcap_i V(B_i) = X_T$. This implies $|X_T| \geq 2$. □

To extend Theorem 13 to tangles of order 3, we first prove a lemma, which essentially says that we can restrict our attention to 2-connected graphs. Let G be graph and $X \subseteq V(G)$. For every $A \subseteq G$, let $A \cap X := A[V(A) \cap X]$. Note that if (A, B) is a separation of G, then $(A \cap X, B \cap X)$ is a separation of $G[X]$ with $\mathrm{ord}(A \cap X, B \cap X) \leq \mathrm{ord}(A, B)$.

Lemma 14. *Let T be a G-tangle of order 3. Let T' be the truncation of T to order 2, and let $W := X_{T'}$. Let $T[W]$ be the set of all separations $(A \cap W, B \cap W)$ of $G[W]$ where $(A, B) \in T$. Then $T[W]$ is a $G[W]$-tangle of order 3. Furthermore, $X_T = X_{T[W]}$.*

Proof. By Theorem 13, $G[W]$ is a biconnected component of G. This implies that $|W| \geq 2$ and $|N(C)| \leq 1$ for every connected component C of $G \setminus W$. For every $w \in W$, we let Y_w be union of the vertex sets of all connected components C of $G \setminus W$ with $N(C) \subseteq \{w\}$. Then $V(G) = W \cup \bigcup_{w \in W} Y_w$. Let $Z_w := V(G) \setminus (Y_w \cup \{w\})$. Let $A_w := G[Y_w \cup \{w\}]$ and $B_w := G[Z_w \cup \{w\}]$. Then $W \subseteq V(B_w)$ and thus $(A_w, B_w) \in T^2(W) = T' \subseteq T$.

Claim 1. Let $(A, B) \in T$. Then $W \setminus V(A) \neq \emptyset$.

Proof. Suppose for contradiction that $W \subseteq V(A)$. Let $S := V(A) \cap V(B)$ and suppose that $S = \{s_1, s_2\}$. Let $w_i \in W$ such that $s_i \in Y_{w_i} \cup \{w_i\}$. Then $A \cup A_{w_1} \cup A_{w_2} = G$, which contradicts (GT.2). This proves that $W \setminus V(A) \neq \emptyset$. ⌐

It is now straightforward to prove that $T[W]$ satisfies the tangle axioms (GT.0), (GT.1), and (GT.3). To prove (GT.2), let $(A_i, B_i) \in T$ for $i = 1, 2, 3$. We need to prove that $(A_1 \cap W) \cup (A_2 \cap W) \cup (A_3 \cap W) \neq G[W]$. Without loss of generality we may assume that (A_i, B_i) is minimal in T. Then $C_i := B_i \setminus V(A_i)$ is connected. By Claim 1, $V(C_i) \cap W \neq \emptyset$. This implies that if $V(C_i) \cap Y_w \neq \emptyset$ for some $w \in W$, then $w \in V(C_i)$.

As T satisfies (GT.2), $A_1 \cup A_2 \cup A_3 \neq G$, and thus there either is a vertex in $V(C_1) \cap V(C_2) \cap V(C_3)$ or an edge with an endvertex in every $V(C_i)$. Suppose first that $v \in V(C_1) \cap V(C_2) \cap V(C_3)$. If $v \in W$ then

$$V\big((A_1 \cap W) \cup (A_2 \cap W) \cup (A_3 \cap W)\big) \neq W = V\big(G[W]\big).$$

Otherwise, $v \in Y_w$ for some $w \in W$, and we have $w \in V(C_1) \cap V(C_2) \cap V(C_3)$. Similarly, if $e = vv'$ has an endvertex in every $V(C_i)$, then we distinguish between the case that $v, v' \in W$, which implies $E\big((A_1 \cap W) \cup (A_2 \cap W) \cup (A_3 \cap W)\big) \neq E(G[W])$, and the case that $e \in E(A_w)$ for some $w \in W$, which implies $w \in V(C_1) \cap V(C_2) \cap V(C_3)$ and thus $V\big((A_1 \cap W) \cup (A_2 \cap W) \cup (A_3 \cap W)\big) \neq W = V\big(G[W]\big)$. This proves (GT.2) and hence that $T[W]$ is a tangle.

The second assertion $X_T = X_{T[W]}$ follows from the fact that $X_T \subseteq X_{T'} = W$. □

Theorem 15. *Let G be a graph. For every G-tangle T of order 3, the set X_T is a vertex set of a proper triconnected component of G.*

Proof. Let T be a G-tangle of order 3. It suffices to prove that $|X_T| \geq 3$. Then by Lemma 12, X_T is a 3-block and $T = T^3(X_T)$, and by Lemma 11, X_T is proper 3-block, that is, the vertex set of a proper triconnected component.

By the previous lemma, we may assume without loss of generality that G is 2-connected. The rest of the proof follows the lines of the proof of Theorem 13. The core of the proof is again an "uncrossing argument" (this time a more complicated one) in Claim 1.

Claim 1. Let $(A_1, B_1), (A_2, B_2) \in T$ be distinct and minimal in T. Then $V(A_1) \subseteq V(B_2)$ and $V(A_2) \subseteq V(B_1)$.

Proof. Let $S_i := V(A_i) \cap V(B_i)$ and $Y_i := V(A_i) \setminus S_i$ and $Z_i := V(B_i) \setminus S_i$ (see Fig. 4(a)). By the minimality of (A_i, B_i), we have $Z_i = V\big(C(T, S_i)\big)$ and $S_i = N(Z_i)$. Thus $S_1 \neq S_2$ and $Z_1 \neq Z_2$, because the two separations are distinct.

It follows that $(A_1 \cup A_2, B_1 \cap B_2)$ is a separation with $B_1 \cap B_2 \subset B_i$, and by the minimality of (A_i, B_i) this separation is not in T. By Lemma 4(3), this means that its order is at least 3. Thus

$$|S_1 \cap Z_2| + |S_1 \cap S_2| + |Z_1 \cap S_2| = |V(A_1 \cup A_2) \cap V(B_1 \cap B_2)| \geq 3. \quad (\star)$$

As $|S_i| \leq 2$ and $S_1 \neq S_2$, it follows that

$$|S_1 \cap Y_2| + |S_1 \cap S_2| + |Y_1 \cap S_2| = |V(A_1 \cap A_2) \cap V(B_1 \cup B_2)| \leq 1.$$

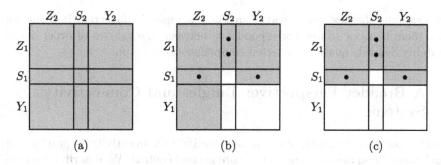

Fig. 4. Uncrossing minimal separations of order 2

Hence $(A_1 \cap A_2, B_1 \cup B_2)$ is a separation of order at most 1. As G is 2-connected, the separation is not proper, which means that either $V(A_1 \cap A_2) = V(G)$ or $V(B_1 \cup B_2) = V(G)$. By Lemma 4(2), we have $(A_1 \cap A_2, B_1 \cup B_2) \in \mathcal{T}$ and thus $V(A_1 \cap A_2) \neq V(G)$. Thus $V(B_1 \cup B_2) = V(G)$, and this implies $Y_1 \cap Y_2 = \emptyset$.

To prove that $V(A_i) = S_i \cup Y_i \subseteq V(B_{3-i}) = S_{3-i} \cup Z_{3-i}$, we still need to prove that $S_i \cap Y_{3-i} = \emptyset$. Suppose for contradiction that $S_1 \cap Y_2 \neq \emptyset$. Then $(\star)$ implies $|S_1 \cap Y_2| = 1$ and $|S_1 \cap Z_2| = 1$ and $|S_2 \cap Z_1| = 2$ and $S_1 \cap S_2 = Y_1 \cap S_2 = \emptyset$ (see Fig. 4(b)). Note that $(Y_1 \cup S_1) \cap Z_2 = V(A_1) \setminus V(A_2)$. It follows that $(A_1 \setminus V(A_2), B_1)$ is a separation of G of order 1, and we have $(A_1 \setminus V(A_2), B_1) \in \mathcal{T}$. Thus $Y_1 \cap Z_2 = \emptyset$, which implies $V(B_2) = Z_2 \cup S_2 \subset Z_1 \cup S_1 = V(B_1)$ (see Fig. 4(c)). This contradicts the minimality of (A_1, B_1). Hence $S_1 \cap Y_2 = \emptyset$, and similarly $Y_1 \cap S_2 = \emptyset$.

Let $(A_1, B_1), \ldots, (A_m, B_m)$ be an enumeration of all minimal separations in $\mathcal{T}$ of order 2. Note that there is at least one minimal separation of order 2 even if G has no proper separations of order 2. Thus $m \geq 1$.

Let $S_i := V(A_i) \cap V(B_i)$. Then the sets S_i are all distinct, because two minimal separations in $\mathcal{T}$ with the same separators are equal. It follows from Claim 1 that $S_i \subseteq V(B_j)$ for all $j \in [m]$ and thus

$$S_1 \cup \ldots \cup S_m \subseteq X_\mathcal{T}.$$

If $m \geq 2$ this implies $|X_\mathcal{T}| \geq 3$. If $m = 1$, then $X_\mathcal{T} = V(B_1)$ and thus $|X_\mathcal{T}| \geq 3$ by Lemma 4.   □

The results of this section clearly do not extend beyond tangles of order 3. For example, the hexagonal grid H in Fig. 1 has a (unique) tangle $\mathcal{T}$ of order 4. But the set $X_\mathcal{T}$ is empty, and the graph H has no 3-inseparable set of cardinality greater than 1.

Nevertheless, it is shown in [6] that there is an extension of the theorem to tangles of order 4 if we replace 4-connectivity by the slightly weaker "quasi-4-connectivity": a graph G is *quasi-4-connected* if it is 3-connected and for all separations (A, B) of order 3, either $|V(A) \setminus V(B)| \leq 1$ or $|V(B) \setminus V(A)| \leq 1$.

For example, the hexagonal grid H in Fig. 1 is quasi-4-connected. It turns out that there is a one-to-one correspondence between the tangles of order 4 and (suitably defined) quasi-4-connected components of a graph.

5   A Broader Perspective: Tangles and Connectivity Systems

Many aspects of "connectivity" are not specific to connectivity in graphs, but can be seen in an abstract and much more general context. We describe "connectivity" on some structure as a function that assigns an "order" (a nonnegative integer) to every "separation" of the structure. We study symmetric connectivity functions, where the separations (A, B) and (B, A) have the same order. The key property such connectivity functions need to satisfy is submodularity.

Separations can usually be described as partitions of a suitable set, the "universe". For example, the separations of graphs we considered in the previous sections are essentially partitions of the edge set. Technically, it will be convenient to identify a partition $(\overline{X}, X)$ with the set X, implicitly assuming that $\overline{X}$ is the complement of X. This leads to the following definition.

A *connectivity function* on a finite set U is a symmetric and submodular function $\kappa \colon 2^U \to \mathbb{N}$ with $\kappa(\emptyset) = 0$. *Symmetric* means that $\kappa(X) = \kappa(\overline{X})$ for all $X \subseteq U$; here and whenever the ground set U is clear from the context we write $\overline{X}$ to denote $U \setminus X$. *Submodular* means that $\kappa(X) + \kappa(Y) \geq \kappa(X \cap Y) + \kappa(X \cup Y)$ for all $X, Y \subseteq U$. The pair (U, κ) is sometimes called a *connectivity system*.

The following two examples capture what is known as *edge connectivity* and *vertex connectivity* in a graph.

Example 16 (Edge connectivity). Let G be a graph. We define the function $\nu_G : 2^{V(G)} \to \mathbb{N}$ by letting $\nu_G(X)$ be the number of edges between X and $\overline{X}$. Then ν_G is a connectivity function on $V(G)$.   ⌟

Example 17 (Vertex connectivity). Let G be a graph. We define the function $\kappa_G : 2^{E(G)} \to \mathbb{N}$ by letting $\kappa_G(X)$ be the number of vertices that are incident with an edge in X and an edge in $\overline{X}$. Then κ_G is a connectivity function on $E(G)$.

Note that for all separations (A, B) of G we have $\kappa_G(E(A)) = \kappa_G(E(B)) \leq \text{ord}(A, B)$, with equality if $V(A) \cap V(B)$ contains no isolated vertices of A or B. For $X \subseteq E(G)$, let us denote the set of endvertices of the edges in X by $V(X)$. Then for all $X \subseteq E(G)$ we have $\kappa_G(X) = \text{ord}(A_X, B_X)$, where $B_X = (V(X), X)$ and $A_X = (V(\overline{X}), \overline{X})$. The theory of tangles and decompositions of the connectivity function of κ_G is essentially the same as the theory of tangles and decompositions of G (partially developed in the previous sections).   ⌟

Example 18. Let G be a graph. For all subsets $X, Y \subseteq V(G)$, we let $M = M_G(X, Y)$ be the $X \times Y$-matrix over the 2-element field $\mathbb{F}_2$ with entries $M_{xy} = 1 \iff xy \in E(G)$. Now we define a connectivity function ρ_G on $V(G)$ by letting $\rho_G(X)$, known as the *cut rank* of X, be the row rank of the matrix $M_G(X, \overline{X})$. This connectivity function was introduced by Oum and Seymour [17] to define the *rank width* of graphs, which approximates the *clique width*, but has better algorithmic properties.   ⌐

Let us also give an example of a connectivity function not related to graphs.

Example 19. Let M be a matroid with ground set E and rank function r. (The rank of a set $X \subseteq E$ is defined to be the maximum size of an independent set contained in X.) The connectivity function of M is the set function $\kappa_M : E \to \mathbb{N}$ defined by $\kappa_M(X) = r(X) + r(\overline{X}) - r(E)$ (see, for example, [18]).   ⌐

5.1   Tangles

Let κ be a connectivity function on a set U. A κ-*tangle* of order $k \geq 0$ is a set $\mathcal{T} \subseteq 2^U$ satisfying the following conditions.

(T.0) $\kappa(X) < k$ for all $X \in \mathcal{T}$,
(T.1) For all $X \subseteq U$ with $\kappa(X) < k$, either $X \in \mathcal{T}$ or $\overline{X} \in \mathcal{T}$.
(T.2) $X_1 \cap X_2 \cap X_3 \neq \emptyset$ for all $X_1, X_2, X_3 \in \mathcal{T}$.
(T.3) $\mathcal{T}$ does not contain any singletons, that is, $\{a\} \notin \mathcal{T}$ for all $a \in U$.

We denote the order of a κ-tangle $\mathcal{T}$ by $\mathrm{ord}(\mathcal{T})$.

We mentioned in Example 17 that the theory of κ_G-tangles is essentially the same as the theory of tangles in a graph. Indeed, κ_G-tangles and G-tangles are "almost" the same. The following proposition makes this precise.

We call an edge of a graph *isolated* if both of its endvertices have degree 1. We call an edge *pendant* if it is not isolated and has one endvertex of degree 1.

Proposition 20. *Let G be a graph and $k \geq 0$.*

(1) If $\mathcal{T}$ is a κ_G-tangle of order k, then

$$\mathcal{S} := \big\{(A,B) \mid (A,B) \text{ separation of } G \text{ of order } < k \text{ with } E(B) \in \mathcal{T}\big\}$$

is a G-tangle of order k.
(2) If $\mathcal{S}$ is a G-tangle of order k, then

$$\mathcal{T} := \big\{E(B) \mid (A,B) \in \mathcal{S}\big\}$$

is a κ_G-tangle of order k, unless
(i) either $k = 1$ and there is an isolated vertex $v \in V(G)$ such that $\mathcal{S}$ is the set of all separations (A, B) of order 0 with $v \in V(B) \setminus V(A)$,
(ii) or $k = 1$ and there is an isolated edge $e \in E(G)$ such that $\mathcal{S}$ is the set of all separations (A, B) of order 0 with $e \in E(B)$,
(iii) or $k = 2$ and there is an isolated or pendant edge $e = vw \in E(G)$ and $\mathcal{S}$ is the set of all separations (A, B) of order at most 1 with $e \in E(B)$.

We omit the straightforward (albeit tedious) proof.

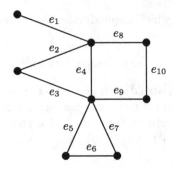

Fig. 5. A graph G with three G tangles of order 2 and two κ_G-tangles of order 2

Example 21. Let G be the graph shown in Fig. 5. G has one tangle of order 1 (since it is connected) and three tangles of order 2 corresponding to the three biconnected components. The G-tangle corresponding to the "improper" biconnected component consisting of the edge e_1 and its endvertices does not correspond to a κ_G-tangle (by Proposition 20(2-iii)).   ⌐

A *star* is a connected graph in which at most 1 vertex has degree greater than 1. Note that we admit degenerate stars consisting of a single vertex or a single edge.

Corollary 22. *Let G be a graph that has a G-tangle of order k. Then G has a κ_G-tangle of order k, unless $k = 1$ and G only has isolated edges or $k = 2$ and all connected components of G are stars.*

6   Decompositions and Duality

A *cubic tree* is a tree where every node that is not a leaf has degree 3. An *oriented edge* of a tree T is a pair (s, t), where $st \in E(T)$. We denote the set of all oriented edges of T by $\overrightarrow{E}(T)$ and the set of leaves of T by $L(T)$. A *branch decomposition* of a connectivity function κ over U is a pair (T, ξ), where T is a cubic tree and ξ a bijective mapping from $L(T)$ to U. For every oriented edge $(s, t) \in \overrightarrow{E}(T)$ we define $\widetilde{\xi}(s, t)$ to be the set of all $\xi(u)$ for leaves $u \in L(T)$ contained in the same connected component of $T - \{st\}$ as t. Note that $\widetilde{\xi}(s, t) = \overline{\widetilde{\xi}(t, s)}$. We define *width* of the decomposition (T, ξ) be the maximum of the values $\kappa(\widetilde{\xi}(t, u))$ for $(t, u) \in \overrightarrow{E}(T)$. The *branch width* of κ, denoted by bw(κ), is the minimum of the widths of all its branch decompositions.

The following fundamental result relates tangles and branch decompositions; it is one of the reasons why tangles are such interesting objects.

Theorem 23 (Duality Theorem; Robertson and Seymour [21]). *The branch width of a connectivity function κ equals the maximum order of a κ-tangle.*

We omit the proof.

Let G be a graph. A *branch decomposition* of G is defined to be a branch decomposition of κ_G, and the *branch width* of G, denoted by $\mathrm{bw}(G)$, is the branch width of κ_G.

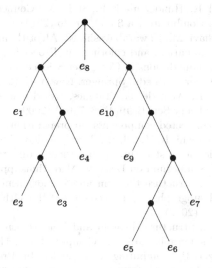

Fig. 6. A branch decomposition of width 2 of the graph shown in Fig. 5

Example 24. Let G be the graph shown in Fig. 5. Figure 6 shows a branch decomposition of G of width 2. Thus $\mathrm{bw}(G) \leq 2$. As G has a tangle of order 2 (see Example 21), by the Duality Theorem we have $\mathrm{bw}(G) = 2$.

The branch width of a graph is closely related to the better-known *tree width* $\mathrm{tw}(G)$: it is not difficult to prove that

$$\mathrm{bw}(G) \leq \mathrm{tw}(G) + 1 \leq \max\left\{\frac{3}{2}\,\mathrm{bw}(\kappa_G),\, 2\right\}$$

(Robertson and Seymour [21]). Both inequalities are tight. For example, a complete graph K_{3n} has branch width $2n$ and tree width $3n - 1$, and a path of length 3 has branch width 2 and tree width 1. There is also a related duality theorem for tree width, due to Seymour and Thomas [23]: $\mathrm{tw}(G) + 1$ equals the maximum order of bramble of G. (Recall the characterisation of tangles that we gave in Theorem 7 and the definition of brambles right after the theorem.)

Acknowledgements. I thank Pascal Schweitzer and Konstantinos Stavropoulos for helpful comments on a earlier version of the paper.

References

1. Carmesin, J., Diestel, R., Hamann, M., Hundertmark, F.: Canonical tree-decompositions of finite graphs I. Existence and algorithms (2013). arxiv:1305. 4668v3
2. Carmesin, J., Diestel, R., Hundertmark, F., Stein, M.: Connectivity and tree structure in finite graphs. Combinatorica **34**(1), 11–46 (2014)
3. Demaine, E., Hajiaghayi, M., Kawarabayashi, K.: Algorithmic graph minor theory: decomposition, approximation, and coloring. In: Proceedings of the 45th Annual IEEE Symposium on Foundations of Computer Science, pp. 637–646 (2005)
4. Diestel, R.: Graph Theory, 4th edn. Springer, New York (2010)
5. Geelen, J., Gerards, B., Whittle, G.: Tangles, tree-decompositions and grids in matroids. J. Comb. Theory Ser. B **99**(4), 657–667 (2009)
6. Grohe, M.: Quasi-4-connected components (in preparation)
7. Grohe, M., Kawarabayashi, K., Reed, B.: A simple algorithm for the graph minor decomposition - logic meets structural graph theory. In: Proceedings of the 24th Annual ACM-SIAM Symposium on Discrete Algorithms, pp. 414–431 (2013)
8. Grohe, M., Marx, D.: Structure theorem and isomorphism test for graphs with excluded topological subgraphs. In: Proceedings of the 44th ACM Symposium on Theory of Computing (2012)
9. Grohe, M., Marx, D.: Structure theorem and isomorphism test for graphs with excluded topological subgraphs. SIAM J. Comput. **44**(1), 114–159 (2015)
10. Grohe, M., Schweitzer, P.: Computing with tangles. In: Proceedings of the 47th ACM Symposium on Theory of Computing, pp. 683–692 (2015)
11. Grohe, M., Schweitzer, P.: Isomorphism testing for graphs of bounded rank width. In: Proceedings of the 55th Annual IEEE Symposium on Foundations of Computer Science (2015)
12. Hopcroft, J.E., Tarjan, R.: Dividing a graph into triconnected components. SIAM J. Comput. **2**(2), 135–158 (1973)
13. Hundertmark, F.: Profiles. An algebraic approach to combinatorial connectivity (2011). arxiv:1110.6207v1
14. Kawarabayashi, K.I., Wollan, P.: A simpler algorithm and shorter proof for the graph minor decomposition. In: Proceedings of the 43rd ACM Symposium on Theory of Computing, pp. 451–458 (2011)
15. MacLane, S.: A structural characterization of planar combinatorial graphs. Duke Math. J. **3**(3), 460–472 (1937)
16. Oum, S.I.: Rank-width and vertex-minors. J. Comb. Theory Ser. B **95**, 79–100 (2005)
17. Oum, S.I., Seymour, P.: Approximating clique-width and branch-width. J. Comb. Theory Ser. B **96**, 514–528 (2006)
18. Oxley, J.: Matroid Theory, 2nd edn. Cambridge University Press, Cambridge (2011)
19. Reed, B.: Tree width and tangles: a new connectivity measure and some applications. In: Bailey, R. (ed.) Surveys in Combinatorics. LMS, vol. 241, pp. 87–162. Cambridge University Press, Cambridge (1997)
20. Robertson, N., Seymour, P.: Graph minors I-XXIII. J. Comb. Theory Ser. B (1982–2012)
21. Robertson, N., Seymour, P.: Graph minors X. Obstructions to tree-decomposition. J. Comb. Theory Ser. B **52**, 153–190 (1991)
22. Robertson, N., Seymour, P.: Graph minors XVI. Excluding a non-planar graph. J. Comb. Theory Ser. B **77**, 1–27 (1999)

23. Seymour, P., Thomas, R.: Graph searching and a min-max theorem for tree-width. J. Comb. Theory Ser. B **58**, 22–33 (1993)
24. Tarjan, R.: Depth-first search and linear graph algorithms. SIAM J. Comput. **1**(2), 146–160 (1972)
25. Tutte, W.: Graph Theory. Addison-Wesley, Reading (1984)

Restricted Turing Machines and Language Recognition

Giovanni Pighizzini[⊠]

Dipartimento di Informatica, Università degli Studi di Milano, Milano, Italy
pighizzini@di.unimi.it

Abstract. In 1965 Hennie proved that one-tape deterministic Turing machines working in linear time are equivalent to finite automata, namely they characterize regular languages. This result has been improved in different directions, by obtaining optimal lower bounds for the time that one-tape deterministic and nondeterministic Turing machines need to recognize nonregular languages. On the other hand, in 1964 Kuroda showed that one-tape Turing machines that are not allowed to use any extra space, besides the part of the tape which initially contains the input, namely linear bounded automata, recognize exactly context-sensitive languages. In 1967 Hibbard proved that for each integer $d \geq 2$, one-tape Turing machines that are allowed to rewrite each tape cell only in the first d visits are equivalent to pushdown automata. This gives a characterization of the class of context-free languages in terms of restricted Turing machines. We discuss these and other related models, by presenting an overview of some fundamental results related to them. Descriptional complexity aspects are also considered.

Keywords: Models of computation · Turing machines · Descriptional complexity · Chomsky hierarchy · Context-free languages

1 Introduction

It is well-known that each class of the *Chomsky hierarchy* can be characterized by a family of acceptors. Usually the following families are considered: unrestricted *Turing machines* for recursive enumerable languages (type 0), *linear bounded automata* for context-sensitive languages (type 1), *pushdown automata* for context-free languages (type 2), and *finite automata* for regular languages (type 3). It can be observed that these families of devices do not define a hierarchy of acceptors. In fact, according to the original definition by Myhill [20] (see also [13]), linear bounded automata are two-way nondeterministic finite automata extended with the capability of rewriting the tape content, while pushdown automata are defined by adding to one-way nondeterministic finite automata an extra storage, with a restricted access. Hence, pushdown automata cannot seen as a special case of linear bounded automata.

However, using different acceptor characterizations for the classes of languages in the Chomsky hierarchy, we can obtain machine hierarchies. Let us mention two possibilities.

© Springer International Publishing Switzerland 2016
A.-H. Dediu et al. (Eds.): LATA 2016, LNCS 9618, pp. 42–56, 2016.
DOI: 10.1007/978-3-319-30000-9_3

First, we can consider machines with a read-only input tape and a separate work-tape. Clearly, these devices have the computational power of unrestricted Turing machines, namely they characterize type 0 languages. When restricted to use a linear amount of space, they have the same power as linear bounded automata, so characterizing context-sensitive languages. On the other hand, each context-free language can be accepted by a pushdown automaton such that the height of the pushdown store is bounded by the length of the input. Hence, type 2 languages are accepted by machines working in linear space with the additional restrictions that the work-tape should be used as a stack, and that on the input tape the head cannot be moved to the left, i.e., it is one-way. (Notice that restricting the input tape of Turing machines and of Turing machines working in linear space to be one-way, the classes of accepted languages remain type 0 and type 1, respectively.) Since finite automata with a one-way input tape are clearly a restriction of pushdown automata, this gives a hierarchy of Turing machines corresponding to the Chomsky hierarchy.

Another characterization can be obtained in terms of restrictions of *one-tape Turing machines*. It is well-known that, without any bound on the resources, these devices still accept all recursive enumerable languages. If the *space* is restricted to be linear then it is possible to accept all context-sensitive languages. If these machines can use only the tape cells which initially contain the input, namely the space is limited by the input length, then we obtain linear bounded automata. Hence, under this further restriction, the corresponding class of languages is still that of context-sensitive. On the other hand, if the *time* is restricted to be linear and the transitions are deterministic, then these machines recognize only regular languages, as proved in 1965 by Hennie [8], namely they are equivalent to finite automata. A characterization of the class of context-free languages in terms of one-tape Turing machines was obtained by Hibbard in 1967, by introducing *limited automata* [9]. These devices are defined by restricting the "active visits" to tape cells, namely the visits that can rewrite cell contents. Fixed an integer $d \geq 0$, a d-limited automaton is a linear bounded automaton that can rewrite each tape cell only in the first d visits. Hibbard proved that for each $d \geq 2$, d-limited automata are equivalent to pushdown automata. Hence, considering one-tape Turing machines, linear bounded automata, limited automata and finite automata, we have another hierarchy of devices corresponding to the Chomsky hierarchy.

In this work we mainly focus on one-tape Turing machines operating with restricted resources. In the first part we consider time bounds. As we already mentioned, in his seminal paper Hennie proved that deterministic machines required to be "fast", namely to work in linear time, are no more powerful than finite automata [8]. Several extensions of this result has been obtained, by increasing the time bound and by considering nondeterministic computations. We discuss some of them. The second part of the work is devoted to the presentation of limited automata and of other related models characterizing context-free languages.

As usual, we denote by Σ^* the set of all strings over a finite alphabet Σ and by $|x|$ the length of a string $x \in \Sigma^*$. We assume that, besides a finite state control, one-tape Turing machines are equipped with a semi-infinite tape. At the beginning of the computation the input string is written on the tape starting from the leftmost cell, while the remaining cells contain the blank symbol. At each step of the computation, the machine writes a symbol on the currently scanned cell of the tape (possibly changing its content), and moves its head to the left, to the right, or keeps it stationary, according to the transition function. Special states are designed as accepting and rejecting states. We assume that in these states the computation stops.

2   Fast One-Tape Turing Machines

As mentioned in the Introduction, one-tape Turing machines which are fast recognize only regular languages. In this section we will explain the meaning of "fast" by presenting an overview of time requirements for nonregular language recognition. For more details, and a more extended bibliography, we point the reader to [22].

The notion of *crossing sequence*, introduced by Rabin and Scott to study the behavior of two-way finite automata [28], turns out to be useful in the investigation of one-tape Turing machines. Given a computation $\mathcal{C}$ of a machine M, the crossing sequence defined by $\mathcal{C}$ at a boundary b between two consecutive tape cells is the sequence of the states of the finite control of M when in the computation $\mathcal{C}$ the head crosses b.

For each computation $\mathcal{C}$ of a deterministic or nondeterministic one-tape Turing machine M, we consider the following resources:

- The *time*, denoted as $t(\mathcal{C})$, is the number of moves in the computation $\mathcal{C}$.
- The *length of the crossing sequences*, denoted as $c(\mathcal{C})$, is the number of the states in the longest crossing sequence used by $\mathcal{C}$.

In the case of deterministic machines, on each input string there is only one computation. So the measures t and c are trivially defined by taking into account such computation. On the other hand, in the case of nondeterministic machines, many different computations could be possible on a same string. This leads to several measures. We now present and briefly discuss the ones considered in the paper.

We say that machine M uses $r(x)$ of a resource $r \in \{t, c\}$ (time, length of crossing sequences, resp.), on an input x if and only if

- **strong** *measure:*

$$r(x) = \max\{r(\mathcal{C}) \mid \mathcal{C} \text{ is a computation on } x\}$$

- **accept** *measure:*

$$r(x) = \begin{cases} \max\{r(\mathcal{C}) \mid \mathcal{C} \text{ is an accepting computation on } x\} & \text{if } x \in L \\ 0 & \text{otherwise} \end{cases}$$

– weak *measure:*

$$r(x) = \begin{cases} \min\{r(\mathcal{C}) \mid \mathcal{C} \text{ is an accepting computation on } x\} & \text{if } x \in L \\ 0 & \text{otherwise} \end{cases}$$

The weak measure corresponds to an optimistic view related to the idea of nondeterminism: a nondeterministic machine, besides choosing an accepting computation, if any, is able to choose that of minimal cost. On the opposite side, the strong measure keeps into account the costs of all possible computations. Between these two measures, the accept measure keeps into account the costs of *all* accepting computations. (For technical reasons, for inputs which are not in the language it is suitable to set the accept and the weak measure to 0). These notions have been proved to be different, for example in the context of space bounded computations [17].

As usual, we will mainly measure complexities with respect to input lengths. This is done by considering the worst case among all possible inputs of the same length. Hence, under the strong, accept, and weak measures, for $r \in \{t, c\}$, we define

$$r(n) = \max\{r(x) \mid x \in \Sigma^*, |x| = n\}.$$

When an input string w is accepted by a Turing machine M without reading all its content, namely without reaching its right end, then any string having w as a prefix should be also accepted by the same machine. Refining this simple observation, it is not difficult to prove that each Turing machine working in sublinear time accepts a regular language. Actually, a Turing machine working in sublinear time works in constant time, namely it inspects only input prefixes of length bounded by a constant. This simple result holds also for nondeterministic machines under the weak time measure (for a proof see [22]). What happens when the time is linear?

As mentioned in the introduction, for *deterministic* one-tape machines the language is still regular, as proved by Hennie [8]. This result was independently improved by Hartmanis and Trakhtenbrot, by increasing the time. In fact, they showed that in order to recognize a nonregular language, the running time of a deterministic one-tape machine should grow at least as $n \log n$. Furthermore, the bound is optimal, namely there are nonregular languages accepted in time $O(n \log n)$ [7, 32].

The situation for *nondeterministic* one-tape machines is more complicated. First of all, the previous lower bound has been extended by Tadaki, Yamakami and Lin, in the case the time is measured by considering all possible computations, namely under the strong measure [31]. However, if the time is measured in an "optimistic" way, that is by taking into account, for each accepted input, only the shortest accepting computation (weak measure) then the same lower bound does not hold. Indeed, under the weak measure even in linear time it is possible to recognize nonregular languages. This is a corollary of a result proved by Michel, stating the existence of an NP-complete language accepted by a nondeterministic machine in linear time under the weak measure [18]. The bound

Table 1. Lower bounds on $t(n)$ and $c(n)$ for nonregular language recognition. Each cell of the table contains in the first line the lower bound on $t(n)$ and in the second line that on $c(n)$ for machines indicated at the left of the row, under the measure indicated at the top of the column. The bounds for *deterministic machines* under the strong measure have been proved by Trakhtenbrot [32] and Hartmanis [7]. (A different proof of this result was obtained by Kobayashi [12].) Hennie [8] proved the same lower bound for $c(n)$ and a smaller lower bound for the time. The bounds for *nondeterministic machines* under the strong measure have been proved by Tadaki, Yamakami, and Lin [31] and extended to the accept measure in [22]. This also gives, as special cases, the bounds for deterministic machines under the accept and weak measures (since on each input a deterministic machine can have at most one accepting computation, for deterministic machines these two cases are the same). Finally, for nondeterministic machines in the weak case the lower bound for $t(n)$ is "folklore", while that for $c(n)$ is from [22]

	strong	accept	weak
One-tape	$n \log n$	$n \log n$	$n \log n$
deterministic	$\log n$	$\log n$	$\log n$
One-tape	$n \log n$	$n \log n$	n
nondeterministic	$\log n$	$\log n$	$\log \log n$

for the strong measure has been extended to the intermediate case of the accept measure in [22].

Together with the above mentioned lower bounds for the time necessary to recognize nonregular languages, lower bounds for the maximal length of crossing sequences (measure $c(n)$) have been also discovered. These bounds are summarized in Table 1.

We briefly discuss the optimality of these lower bounds. As observed in [7], the set of powers 2 written in unary notation can be accepted by a one-tape deterministic machine M which works as follows. Suppose that at the beginning of the computation the tape contains the input a^n. The head completely scans the tape from left to right by rewriting each second a with a fixed symbol X from the working alphabet. If n is odd then M rejects. Otherwise, M starts a sweep of the tape from right to left again rewriting with X each second a left on the tape. If the number of as, namely $n/2$, was odd, then M rejects, otherwise it iterates the same process, with another sweep of the tape and so on. M accepts the input when only one a is left on the tape. Since after the kth sweep the number of as left on the tape is $a/2^k$, if n is a power of 2 this happens after $\log n$ sweeps of the input. Hence, the length of the crossing sequences of M is logarithmic with respect to the length of the input and the time is $O(n \log n)$.

This proves the optimality of all the bounds in the table, with the only exception of those for nondeterministic machines under the weak measure, for which the optimality for $t(n)$ follows from the above mentioned result by Michel [18], while the optimality for $c(n)$ has been proved in [22]. The optimality has been proved using unary witnesses, i.e., defined over a one-letter alphabet, with the only exception of the lower bound for $t(n)$ in the nondeterministic case under the weak measure, for which the witness given in [18] has been obtained by a padding technique which relies on the use of an input alphabet with more than one symbol. The existence of a unary witness for this case is an open problem. We know the existence of a nonregular unary language accepted in time $O(n \log \log n)$ by a one-tape nondeterministic machine under the weak measure [22]. We conjecture that each unary language accepted within a smaller amount of time should be regular. This would imply that the time lower bound in the unary case is higher than in the general case.

We conclude this section by shortly mentioning some recent results in this area. First of all, Průša investigated some aspects of *Hennie machines*. These devices are linear bound automata visiting each tape cell a number of times bounded by a constant [27]. Průša proved that it is undecidable whether or not a Turing machine is a Hennie machine. For a constructive variant of Hennie machines, where each symbol written in a tape cell keeps track of the number of visits already spent in the cell, he studied the cost, in terms of description sizes, of the simulation by several variants of finite automata. Gajser proved that for all "reasonable" functions $T(n) = o(n \log n)$ it is possible to decide whether of not a given one-tape Turing machine works in time at most $T(n)$ [4].

3   One-Tape Turing Machines with Rewriting Restrictions

In this part of the paper we consider one-tape Turing machines that can use only the portion of the tape initially containing the input, namely linear bounded automata, and are subject to some further restrictions. The restrictions we are going to consider are mainly related to the use of transitions rewriting the content of tape cells. Notice that, without any restrictions on rewritings, these machines characterize context-sensitive languages. On the other hand, if rewritings are completely forbidden, we obtain two-way finite automata that, as well-known, recognize only regular languages. A more extended overview can be found in [23].

Limited Automata

Let us start by considering *limited automata*, introduced by Hibbard in 1967 [9]. Given an integer $d \geq 0$, a *d-limited automaton* is allowed to rewrite the content of each tape cell only during the first d visits, after that the content of a cell is "frozen".[1] A few technical details are useful for the next discussion. At the beginning of the computation the input is stored onto the tape surrounded by the two special symbols ▷ and ◁, called the left and the right end-marker, respectively, the head of the automaton is on the cell containing the first input symbol, and the finite control contains, as usual, the initial state. At each step the head can be moved to the left or to the right. However it cannot violate the

---

[1] For technical reasons actually we count the scans from left to right and from right to left on each cell. Hence, each transition reversing the head direction is counted as a double visit.

end-markers, except at the end of computation, where the machine can accept the input by moving the head to the right from the cell containing the right end-marker while entering a final state.

Hibbard proved that for each integer $d \geq 2$, d-limited automata recognize exactly the class of context-free languages. 0-limited automata are two-way finite automata, hence they characterize regular languages. Furthermore, also 1-limited automata characterize regular languages. The situation is summarized in the following theorem, where the deterministic versions of these devices are also mentioned:

Theorem 1. (i) *For each $d \geq 2$, the class of languages accepted by d-limited automata coincides with the class of context-free languages [9].*

(ii) *The class of languages accepted by deterministic 2-limited automata coincides with the class of deterministic context-free languages [26].*

(iii) *For each $d \geq 2$, there exists a language which is accepted by a deterministic d-limited automaton, which cannot be accepted by any deterministic $(d-1)$-limited automaton [9].*

(iv) *The class of languages accepted by 1-limited automata coincides with the class of regular languages [33].*

We now present two examples.

Example 2. For each integer $k \geq 1$, we denote by Ω_k the alphabet of k types of brackets, which will be represented as $\{(_1,)_1, (_2,)_2, \ldots, (_k,)_k\}$. The *Dyck language* D_k over the alphabet Ω_k is the set of strings representing well balanced sequences of brackets. We will refer to the $(_i$ symbols as "open brackets" and the $)_i$ symbols as "closed brackets", i.e. opening and closing brackets.

The *Dyck language* D_k can be recognized by a 2-limited automaton A_D which starts having the input on its tape, surrounded by two end-markers $\triangleright$ and $\triangleleft$, with the head on the first input symbol. From this configuration, A_D moves to the right to find a closed bracket $)_i$, $1 \leq i \leq k$. Then A_D replaces $)_i$ with a symbol $X \notin \Omega_k$ and changes the head direction, moving to the left. In a similar way, it stops when during this scan it meets for the first time a left bracket $(_j$. If $i \neq j$, i.e., the two brackets are not of the same type, then A_D rejects. Otherwise, A_D writes X on the cell and changes again the head direction moving to the right. This procedure is repeated until the head of A_D reaches one of the end-markers.

- If the left end-marker is reached, then it means that at least one of the right brackets in the input does not have a matching left bracket. Hence, A_D rejects.
- If instead the right end-marker is reached, then A_D has to make sure that every left bracket has a matching right one. In order to do this, it scans the entire tape from the right to the left and if it finds a left bracket not marked with X then A_D rejects. On the other hand, if A_D reaches the left end-marker reading only Xs, then it can accept the input.   □

Example 3. For each integer n, let us denote by K_n the set of all strings over the alphabet $\{0, 1\}$ consisting of the concatenation of blocks of length n, such that at least n blocks are equal to the last one. Formally:

$$K_n = \{x_1 x_2 \cdots x_k x \mid k \geq 0, \ x_1, x_2, \ldots, x_k, x \in \{0, 1\}^n,$$
$$\exists i_1 < i_2 < \cdots < i_n \in \{1, \ldots, k\}, \ x_{i_1} = x_{i_2} = \ldots = x_{i_n} = x\}.$$

We now describe a 2-limited automaton M accepting K_n. Notice that the parameter n is fixed. Suppose M receives an input string w of length N.

1. First, M scans the input tape from left to right, to reach the right end-marker.
2. M moves its head $n + 1$ positions to the left, namely to the cell $i = N - n$, the one immediately to the left of the input suffix x of length n.
3. Starting from this position i, M counts how many blocks of length n coincide with x. This is done as follows.
   When M, arriving from the right, visits a position $i \leq N - n$ for the first time, it replaces the content a by a special symbol X, after copying a in the finite control. Hence, M starts to move to the right, in order to compare the symbol removed from the cell with the corresponding symbol in the block x. While moving to the right, M counts modulo n and stops when the counter is 0 and a cell containing a symbol other than X is reached. The symbol of x in this cell has to be compared with a. Then, M moves to the left until it reaches cell $i - 1$, namely the first cell which does not contain X, immediately to the left of cells containing X.
   We observe that the end of a block is reached each time a symbol a copied from the tape is compared with the leftmost symbol of x, which lies immediately to the right of a cell containing X. If in the block just inspected no mismatches have been discovered then the counter of blocks matching with x is incremented (unless its value was already n).
4. When the left end-marker is reached, M accepts if and only if the input length is a multiple of n and the value of the counter of blocks matching with x is n.

We can easily observe that the above strategy can modify tape cells only in the first two visits. Hence, it can be implemented by a *deterministic* 2-limited automaton. Such an automaton uses $O(n^2)$ states and a constant size alphabet.

Actually, using nondeterminism, it is possible to recognize the language K_n using $O(n)$ states and modifying tape cells only in the first visit, namely K_n is accepted by a nondeterministic 1-limited automaton M (see [23] and, for a slightly different example, [25]). □

The argument used by Hibbard to prove Theorem 1(i) is very difficult. He provided some constructions to transform a kind of rewriting system, equivalent to pushdown automata, to 2-limited automata and vice versa, together with reductions from $(d + 1)$-limited automata to d-limited automata, for $d \geq 2$ [9].

A different construction of 2-limited automata from context-free languages, based on the Chomsky-Schützenberger representation theorem for context-free languages [3], has been obtained in [25]. We remind the reader that this theorem

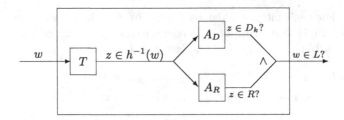

Fig. 1. A machine accepting $L = h(D_k \cap R)$

states that each context-free language can be obtained by selecting in a Dyck language D_k, with k kinds of brackets, only the strings belonging to a regular language R, and then renaming the symbols in the remaining strings according to a homomorphism L. More precisely, every context-free language $L \subseteq \Sigma^*$ can be expressed as $L = h(D_k \cap R)$, where $D_k \subseteq \Omega_k^*$, $k \geq 1$, is a Dyck language, $R \subseteq \Omega_k^*$ is a regular language, and $h : \Omega_k \to \Sigma^*$ is a homomorphism.

Given $L \subseteq \Sigma^*$ context-free, we can consider the following machines:

– A nondeterministic transducer T computing h^{-1}.
– The 2-limited automaton A_D described in Example 2 recognizing the Dyck language D_k.
– A finite automaton A_R accepting the regular language R.

To decide if a string $w \in \Sigma^*$, we can suitably combine these machines as in Fig. 1, in such a way that the resulting machine is a 2-limited automaton.

The results in Theorem 1 have been revisited by also taking into account the size of the descriptions. Concerning the case $d \geq 2$, we have the following results:

Theorem 4. *(i) Each n-state d-limited automaton can be simulated by a pushdown automaton of size exponential in a polynomial in n.*

(ii) The previous upper bound becomes a double exponential when a deterministic 2-limited automaton is simulated by a deterministic pushdown automaton, however it remains a single exponential if the input of the deterministic pushdown automaton is given with a symbol to mark the right end.

(iii) Each pushdown automaton M can be simulated by a 2-limited automaton whose size is polynomial with respect to the size of M.

(iv) The previous upper bound remains polynomial when a deterministic pushdown automaton is simulated by a deterministic 2-limited automaton.

Statements (ii), (iii), (iv) have been proved in [26], with statement (i) for the case $d = 2$. The proof has been recently extended to $d > 2$ in [14].

The exponential gap for the conversion of 2-limited automata into equivalent pushdown automata cannot be reduced. In fact, the language K_n presented in Example 3 is accepted by a (deterministic) 2-limited automaton with $O(n^2)$ states and a constant size alphabet, while the size of each pushdown automaton accepting it must be at least exponential in n [26]. This also implies that the simulation of deterministic 2-limited automata by deterministic pushdown automata is exponential in size. Actually, we conjecture that this simulation costs a double exponential, namely it matches the upper bound in Theorem 4(ii).

For 1-limited automata we have the following costs:

Theorem 5 ([25]). *Each n-state 1-limited automaton M can be simulated by a nondeterministic automaton with $n \cdot 2^{n^2}$ states and by a deterministic automaton with $2^{n \cdot 2^{n^2}}$ states. Furthermore, if M is deterministic then an equivalent one-way deterministic finite automaton with no more than $n \cdot (n+1)^n$ states can be obtained.*

The doubly exponential upper bound for the conversion of nondeterministic 1-limited automata into deterministic automata is related to a double role of nondeterminism in 1-limited automata. When a 1-limited automaton visits one cell after the first rewriting, the possible nondeterministic transitions depend on the symbol that has been written in the cell in the first visit, which, in turns, depends on the nondeterministic choice taken in the first visit. This double exponential cannot be avoided. In fact, as already observed, the language K_n of Example 3 is accepted by a nondeterministic 1-limited automaton with $O(n)$ states while, using standard distinguishability arguments, it can be shown that each deterministic automaton accepting it requires a number of states doubly exponential in n. As noticed in [25], even the simulation of deterministic 1-limited automata by *two-way nondeterministic automata* is exponential in size.

We conclude this section by briefly mentioning the case of unary languages. It is well-known that in this case each context-free language is regular [6]. Hence, for each $d \geq 0$, d-limited automata with a one letter input alphabet recognize only regular languages. In [25], a result comparing the size of unary 1-limited automata with the size of equivalent two-way nondeterministic finite automata has been obtained. Recently, Kutrib and Wendlandt proved state lower bounds for the simulation of unary d-limited automata by different variants of finite automata [15].

Strongly Limited Automata

We observed that, using the Chomsky-Schützenberger representation theorem for context-free languages, given a context-free language it is possible to construct a 2-limited automaton accepting it. The main component of such automaton is the 2-limited automaton A_D accepting the Dyck language D_k (Example 2). Actually, A_D does not use all the capabilities of 2-limited automata. For instance, it does not need to rewrite each tape cell two times, but only while moving from right to left during the second visit. So, we can ask if it is possible to further restrict the moves of 2-limited automata, without reducing the computational power.

In [24] we gave a positive answer to this question, by introducing *strongly limited automata*, a restriction of limited automata which closely imitates the moves that are used by the 2-limited automaton A_D in Example 2. In particular, these machines satisfy the following restrictions:

- While moving to the right, a strongly limited automaton always uses the same state q_0 until the content of a cell (which has not been yet rewritten) is modified. Then it changes its internal state and starts to move to the left.
- While moving to the left, the automaton, without changing its state, rewrites each cell it meets that is not yet rewritten up to some position where it re-enters the state q_0 and starts again to move to the right.
- In the final phase of the computation, the automaton inspects all tape cells, to check whether or not the final content belongs to a given 2-strictly locally testable language. Roughly, this means that all the factors of two letters of the string which is finally written on the tape (including the end-markers) should belong to a given set.

We already mentioned that strongly limited automata have the same computational power as limited automata, namely they characterize context-free languages. This equivalence has been studied also considering descriptional complexity aspects:

Theorem 6 ([24]).

(i) Each context-free language L is accepted by a strongly limited automaton whose description has a size which is polynomial with respect to the size of a given context-free grammar generating L or of a given pushdown automaton accepting L.

(ii) Each strongly limited automaton $\mathcal{M}$ can be simulated by a pushdown automaton of size polynomial with respect the size of $\mathcal{M}$.

Example 7 The deterministic context-free language $\{a^n b^{2n} \mid n \geq 0\}$ is accepted by a strongly limited automaton which guesses each second b. While moving from left to right and reading b, the automaton makes a nondeterministic choice between further moving to the right or rewriting the cell by X and turning to the left. Furthermore, while moving to the left, the content of each cell containing b which is visited is rewritten by Y, still moving to the left, and when a cell containing a is visited, its content is replaced by Z, turning to the right. In the final scan the machine accepts if and only if the string on the tape belongs to $\triangleright Z^*(YX)^* \triangleleft$.

We can modify the above algorithm to recognize the language $\{a^n b^n \mid n \geq 0\} \cup \{a^n b^{2n} \mid n \geq 0\}$. While moving from left to right, when the head reaches a cell containing b three actions are possible: either the automaton continues to move to the right, without any rewriting, or it rewrites the cell by X, turning to the left, or it rewrites the cell by W, also turning to the left. While moving from right to left, the automaton behaves as the one above described for $\{a^n b^{2n} \mid n \geq 0\}$. The input is accepted if and only if the string which is finally on the tape belongs to $\triangleright Z^* W^* \triangleleft + \triangleright Z^*(YX)^* \triangleleft$. □

Concerning deterministic computations, it is not difficult to observe that deterministic strongly limited automata cannot recognize all deterministic context-free languages. Consider, for instance, the deterministic language $L = \{ca^n b^n \mid n \geq 0\} \cup \{da^{2n} b^n \mid n \geq 0\}$. While moving from left to right, a strongly limited automaton can use only the state q_0. Hence, it cannot remember if the first symbol of the input is a c or a d and, then, if it has to check whether the number of as is equal to the number of bs or whether the number of as is two times the number of bs. A formal proof that the language L, and also the language $\{a^n b^{2n} \mid n \geq 0\}$ (Example 7), are not accepted by any deterministic strongly limited automaton is presented in [24]. In that paper it was also proposed to slightly relax the definition of strongly limited automata, by allowing state changes while moving to the left and to the right, but still forbidding them on rewritten cells and by keeping all the other restrictions. This model, called *almost strongly limited automata*, still characterizes the context-free languages. Furthermore, the two above mentioned deterministic context-free languages can be easily recognized by almost strongly limited automata having only deterministic transitions.

It would be interesting to know if almost strongly limited automata are able to accept all deterministic context-free languages without taking nondeterministic decisions.

Forgetting Automata Deleting Automata

In 1996 Jancar, Mráz, and Plátek introduced *forgetting automata* [11]. These devices can erase tape cells by rewriting their contents with a special symbol. However, rewritten cells are kept on the tape and are still considered during the computation. For instance, the state can be changed while visiting an erased cell. In a variant of forgetting automata that characterizes context-free languages, when a cell which contains an input symbol is visited while moving to the left, its content is rewritten, while no changes can be done while moving to the right. This way of operating is very close to that of strongly limited automata. However, in strongly limited automata the rewriting alphabet can contain more than one symbol. Furthermore, rewritten cells are completely ignored (namely, the head direction and the state cannot be changed while visiting them) except in the final scan of the tape from the right to the left end-marker. So the two models are different. For example, to recognize the set of palindromes, a strongly limited automaton needs a working alphabet of at least 3 symbols while, by definition, to rewrite tape cells forgetting automata use only one symbol [24].

If erased cells are removed from the tape of a forgetting automaton, we obtain another computational model called *deleting automata*. This model is less powerful. In fact it is not able to recognize all context-free languages [11].

Wechsung's Model, Return Complexity, Dual Return Complexity

In 1975 Wechsung considered another variant of one-tape Turing machines which is defined, as limited automata, by restricting active visits: for any fixed d, each tape cell can be rewritten only in the *last* d visits [34]. It should be clear that for $d = 1$ only regular languages can be accepted (each cell, after the first rewriting, will be never visited again, hence the rewriting is useless). However, for each fixed $d \geq 2$, these models still characterize context-free languages. Even in this case, there exists a hierarchy of deterministic languages (cf. Theorem 1(iii) in the case of limited automata). However, this hierarchy is not comparable with the class of deterministic context-free languages. For instance, it can be easily seen that the set of palindromes, which is not a deterministic context-free language, can be recognized with $d = 2$, by using only deterministic transitions. However, there are deterministic context-free languages that cannot be recognized by any deterministic machine of this kind, for any integer d [21].

The maximum number of visits to a tape cell, counted starting from the first visit which modifies the cell content, is also called *return complexity* [34,35]. Notice that this measure is dual with respect to the one considered to define limited automata. The maximum number of visits to a cell up to the last rewriting, namely the measure used to define limited automata, is sometimes called *dual return complexity* [33].

4   Final Remarks

We discussed restricted versions of one-tape Turing machines. It is suitable to remind the reader that many interesting results have been also obtained for Turing machines having a read-only input tape and one or several work-tapes, mainly considering the amount of work space used in the computations.

It is well-known that if the space is restricted to be linear then these models characterize context-sensitive languages, while if the space is constant then they are equivalent to finite automata. In their pioneering papers, Hartmanis, Stearns, and Lewis investigated the minimal amount of work space that deterministic Turing machines need to recognize nonregular languages [16,29]. They proved that if the input tape is *one-way*, then a logarithmic amount of space is necessary.

In the case of *two-way* machines, the lower bound reduces to a function growing as $\log \log n$. These results have been generalized to nondeterministic machines by Hopcroft and Ullman under the **strong** space measure, namely by taking into account all computations [10]. The optimal space lower bound for nonregular acceptance on one-way nondeterministic machines reduces to $\log \log n$, if on each accepted input the computation using minimum space is considered (**weak** space), as proved by Alberts [1]. In [2], the number of reversals of the head on the input tape has been studied, obtaining a lower bound for the product of the space by the number of reversals for nonregular language recognition. For a survey on these lower bounds we point the reader to [17].

Many interesting results concerning machine working in "low" space have been proved in the literature. For surveys, see, e.g., the monograph by Szepietowski [30] and the papers by Michel [19] and Geffert [5].

References

1. Alberts, M.: Space complexity of alternating Turing machines. In: Budach, L. (ed.) FCT. LNCS, vol. 199, pp. 1–7. Springer, Heidelberg (1985)
2. Bertoni, A., Mereghetti, C., Pighizzini, G.: An optimal lower bound for nonregular languages. Inf. Process. Lett. **50**(6), 289–292 (1994). Corrigendum. ibid. 52(6), 339
3. Chomsky, N., Schützenberger, M.: The algebraic theory of context-free languages. In: Braffort, P., Hirschberg, D. (eds.) Computer Programming and Formal Systems, Studies in Logic and the Foundations of Mathematics, vol. 35, pp. 118–161. Elsevier, Amsterdam (1963)
4. Gajser, D.: Verifying time complexity of Turing machines. Theor. Comput. Sci. **600**, 86–97 (2015)
5. Geffert, V.: Bridging across the log(n) space frontier. Inf. Comput. **142**(2), 127–158 (1998)
6. Ginsburg, S., Rice, H.G.: Two families of languages related to ALGOL. J. ACM **9**(3), 350–371 (1962)
7. Hartmanis, J.: Computational complexity of one-tape Turing machine computations. J. ACM **15**(2), 325–339 (1968)
8. Hennie, F.C.: One-tape, off-line Turing machine computations. Inf. Control **8**(6), 553–578 (1965)
9. Hibbard, T.N.: A generalization of context-free determinism. Inf. Control **11**(1/2), 196–238 (1967)
10. Hopcroft, J.E., Ullman, J.D.: Some results on tape-bounded Turing machines. J. ACM **16**(1), 168–177 (1969)
11. Jancar, P., Mráz, F., Plátek, M.: Forgetting automata and context-free languages. Acta Inf. **33**(5), 409–420 (1996)
12. Kobayashi, K.: On the structure of one-tape nondeterministic Turing machine time hierarchy. Theor. Comput. Sci. **40**, 175–193 (1985)
13. Kuroda, S.: Classes of languages and linear-bounded automata. Inf. Control **7**(2), 207–223 (1964)
14. Kutrib, M., Pighizzini, G., Wendlandt, M.: Descriptional complexity of limited automata (submitted)
15. Kutrib, M., Wendlandt, M.: On simulation cost of unary limited automata. In: Shallit, J., Okhotin, A. (eds.) DCFS 2015. LNCS, vol. 9118, pp. 153–164. Springer, Heidelberg (2015)
16. Lewis II., P.M., Stearns, R.E., Hartmanis, J.: Memory bounds for recognition of context-free and context-sensitive languages. In: FOCS, pp. 191–202. IEEE (1965)
17. Mereghetti, C.: Testing the descriptional power of small Turing machines on non-regular language acceptance. Int. J. Found. Comput. Sci. **19**(4), 827–843 (2008)
18. Michel, P.: An NP-complete language accepted in linear time by a one-tape Turing machine. Theor. Comput. Sci. **85**(1), 205–212 (1991)
19. Michel, P.: A survey of space complexity. Theor. Comput. Sci. **101**(1), 99–132 (1992)
20. Mill, J.R.: Linear bounded automata. WADD Tech. Note, pp. 60–165, Wright Patterson Air Force Base, Ohio (1960)

21. Peckel, J.: On a deterministic subclass of context-free languages. In: Gruska, J. (ed.) Mathematical Foundations of Computer Science 1977. LNCS, vol. 53, pp. 430–434. Springer, Heidelberg (1977)

22. Pighizzini, G.: Nondeterministic one-tape off-line Turing machines. J. Automata, Lang. Combinatorics **14**(1), 107–124 (2009). arxiv:0905.1271

23. Pighizzini, G.: Guest column: one-tape Turing machine variants and language recognition. SIGACT News **46**(3), 37–55 (2015). arxiv:1507.08582

24. Pighizzini, G.: Strongly limited automata. Fundam. Inform. (to appear). A preliminaryversion appeared In: Bensch, S., Freund, R., Otto, F. (eds.) Proceedings of the Sixth Workshop on Non-Classical Models for Automata and Applications - NCMA 2014, 28–29 July 2014, Kassel, Germany, vol. 304, pp. 191–206. Österreichische Computer Gesellschaft (2014). books@ocg.at

25. Pighizzini, G., Pisoni, A.: Limited automata and regular languages. Int. J. Found. Comput. Sci. **25**(7), 897–916 (2014)

26. Pighizzini, G., Pisoni, A.: Limited automata and context-free languages. Fundam. Inform. **136**(1–2), 157–176 (2015)

27. Průša, D.: Weight-Reducing hennie machines and their descriptional complexity. In: Dediu, A.-H., Martín-Vide, C., Sierra-Rodríguez, J.-L., Truthe, B. (eds.) LATA 2014. LNCS, vol. 8370, pp. 553–564. Springer, Heidelberg (2014)

28. Rabin, M.O., Scott, D.: Finite automata and their decision problems. IBM J. Res. Dev. **3**(2), 114–125 (1959)

29. Stearns, R.E., Hartmanis, J., Lewis II., P.M.: Hierarchies of memory limited computations. In: FOCS, pp. 179–190. IEEE (1965)

30. Szepietowski, A.: Turing Machines with Sublogarithmic Space. LNCS, vol. 843. Springer, Heidelberg (1994)

31. Tadaki, K., Yamakami, T., Lin, J.C.H.: Theory of one-tape linear-time Turing machines. Theor. Comput. Sci. **411**(1), 22–43 (2010)

32. Trakhtenbrot, B.A.: Turing machine computations with logarithmic delay (in russian). Algebra I Logica **3**, 33–48 (1964)

33. Wagner, K.W., Wechsung, G.: Computational Complexity. D. Reidel Publishing Company, Dordrecht (1986)

34. Wechsung, G.: Characterization of some classes of context-free languages in terms of complexity classes. In: Becvár, J. (ed.) Mathematical Foundations of Computer Science 1975. LNCS, vol. 32, pp. 457–461. Springer, Heidelberg (1975)

35. Wechsung, G., Brandstädt, A.: A relation between space, return and dual return complexities. Theor. Comput. Sci. **9**, 127–140 (1979)

Automata for Ontologies

Frank Wolter[✉]

Department of Computer Science, University of Liverpool, Liverpool, UK
wolter@liverpool.ac.uk

Abstract. We present three reasoning problems for description logic ontologies and discuss how automata theory can be used to analyze them.

In computer science, ontologies are used to provide a vocabulary for a domain of interest and define the logical relationships that hold between the vocabulary items. Ontologies are employed in many areas. For example, they provide vocabularies for structured data markup on web pages (an example is the Schema.org ontology initiated by Google, Microsoft, Yahoo, and Yandex), they provide large scale terminologies in the life sciences and in healthcare (an example is SNOMED CT, the Systematized Nomenclature of Medicine, which is a comprehensive clinical terminology used in many international and national healthcare systems), and they provide conceptual models that can be used to access incomplete and distributed data in large-scale data integration projects.

The majority of ontologies in computing are given in description logics, a family of fragments of first-order predicate logic developed in Artificial Intelligence in which typical reasoning tasks are, in contrast to full first-order predicate logic, decidable and efficiently implementable.

In this talk I present three important reasoning tasks in description logic which can partly be analyzed using tools from automata theory. The tasks are:

- deciding the existence of uniform interpolants of description logic ontologies for a given signature of vocabulary items;
- deciding query emptiness and containment of ontology-mediated queries for databases over a given signature of vocabulary items;
- deciding the rewritability of ontology-mediated queries into equivalent first-order predicate logic queries (equivalently, SQL queries) over a given signature of vocabulary items.

In what follows we discuss these three reasoning tasks in more detail. Note that there are many more examples of applications of automata theory in description logic research which we cannot discuss in detail [1,4,5,9,11,12].

An ontology $\mathcal{O}$ in a description logic $\mathcal{L}$ is a finite set of axioms given in the fragment of first-order logic corresponding to $\mathcal{L}$. Ontologies can be very large. For example, the ontology SNOMED CT mentioned above has more than 300 000 axioms. In many applications of such a large ontology $\mathcal{O}$ only a small set S of predicate symbols from $\mathcal{O}$ is relevant. In this case, instead of working with $\mathcal{O}$ itself, one would like to work with a *uniform interpolant of $\mathcal{O}$ w.r.t. S*; i.e., a set of axioms $\mathcal{O}_S$ in $\mathcal{L}$ such that

© Springer International Publishing Switzerland 2016
A.-H. Dediu et al. (Eds.): LATA 2016, LNCS 9618, pp. 57–60, 2016.
DOI: 10.1007/978-3-319-30000-9_4

- $\mathcal{O} \models \mathcal{O}_S$;
- the set of all predicate symbols in $\mathcal{O}_S$ is included in S;
- for any sentence φ in $\mathcal{L}$ not using any predicate symbols from $\mathcal{O}$ that do not occur in S: if $\mathcal{O} \models \varphi$, then $\mathcal{O}_S \models \varphi$.

Intuitively, it is often more appropriate to use a uniform interpolant $\mathcal{O}_S$ rather than the original ontology $\mathcal{O}$ since $\mathcal{O}_S$ only uses predicate symbols that are relevant for the application but still has the same relevant consequences as $\mathcal{O}$. Unfortunately, for many description logics $\mathcal{L}$ (and for first-order predicate logic itself), uniform interpolants do not always exist. It is, therefore, an interesting problem to decide for which input ontology $\mathcal{O}$ and signature S a uniform interpolant of $\mathcal{O}$ w.r.t. S exists in a description logic $\mathcal{L}$. For the fundamental description logic $\mathcal{ALC}$, for example, one can provide a model-theoretic characterization of the existence of uniform interpolants based on bisimulations and then use an exponential encoding in alternating parity tree automata, results on the closure properties of languages recognized by such automata, and the fact that emptiness can be decided in exponential time [15], to obtain an algorithm deciding the existence of uniform interpolants in 2ExpTime (which is optimal) [14]. The encoding in tree automata can also be used to represent a uniform interpolant even if it does not exist in $\mathcal{ALC}$. For the weaker description logic $\mathcal{EL}$ (which underpins large-scale ontologies such as SNOMED CT), one can characterize the existence of uniform interpolants using a weaker notion of simulation and then provide an appropriate encoding in a modified version of tree automata to obtain an algorithm deciding the existence of uniform interpolants in ExpTime (which is again optimal) [13].

In many applications, ontologies are used to query incomplete data. In what follows we assume that a database $\mathcal{D}$ is a finite set of ground sentences of the form $P(c_1, \ldots, c_n)$, where P is a predicate symbol of arity n and $c_1, \ldots, c_n$ are individual constants. We also assume that our database queries are conjunctive queries; i.e., first-order formulas φ constructed from atomic formulas $P(x_1, \ldots, x_n)$ using conjunction and existential quantification. Given an ontology $\mathcal{O}$ and a conjunctive query $q(\boldsymbol{x})$, the (certain) answer to $\mathcal{O}$ and $q(\boldsymbol{x})$ in $\mathcal{D}$ is the set of all tuples $\boldsymbol{c}$ in $\mathcal{D}$ of the same length as $\boldsymbol{x}$ such that

$$\mathcal{O} \cup \mathcal{D} \models q(\boldsymbol{c}).$$

The pair $(\mathcal{O}, q(\boldsymbol{x}))$ is often called an ontology-mediated query and can be regarded as a database query for which a tuple $\boldsymbol{c}$ is an answer over data $\mathcal{D}$ if $\mathcal{O} \cup \mathcal{D} \models q(\boldsymbol{c})$. As answering ontology-mediated queries is the main algorithmic problem when ontologies are used to access data, it is of interest to investigate the following two fundamental properties of ontology-mediated queries (well known from database research). An ontology-mediated query $(\mathcal{O}, q(\boldsymbol{x}))$ is *empty w.r.t. signature S* if $\mathcal{O} \cup \mathcal{D} \not\models q(\boldsymbol{c})$ for any database $\mathcal{D}$ over S and any tuple $\boldsymbol{c}$ of individual constants in $\mathcal{D}$. For ontology-mediated queries $(\mathcal{O}_1, q_1(\boldsymbol{x}))$ and $(\mathcal{O}_2, q_2(\boldsymbol{x}))$, we say that $(\mathcal{O}_1, q_1(\boldsymbol{x}))$ *is contained in* $(\mathcal{O}_2, q_2(\boldsymbol{x}))$ *w.r.t. signature S*, if

$$\mathcal{O}_1 \cup \mathcal{D} \models q(\boldsymbol{c}) \quad \Rightarrow \quad \mathcal{O}_2 \cup \mathcal{D} \models q(\boldsymbol{c})$$

holds for all databases $\mathcal{D}$ over S and all tuples c. Both the emptiness problem and the containment problem can be characterized model-theoretically. For the important class of horn description logics one can prove that it is sufficient to consider tree-shaped databases when deciding emptiness and containment. In this case the computational complexity of both problems can therefore be analyzed using non-deterministic bottom-up automata on finite ranked trees [2,3,7].

In general, answering ontology-mediated queries is computationally much harder than standard relational query answering [6,10]. In fact, even for rather weak description logics this problem can be non-tractable in data complexity. On the other hand, the most popular technique for answering ontology-mediated queries is by reduction to query answering for relational databases. Thus, one is interested in the following notion of rewritability which ensures that queries can be answered using relational database management systems: an ontology-mediated query $(\mathcal{O}, q(x))$ is *first-order rewritable w.r.t. signature S* if there exists a first-order formula $\varphi(x)$ such that for any database $\mathcal{D}$ over S:

$$\mathcal{O} \cup \mathcal{D} \models q(c) \quad \Leftrightarrow \quad \mathcal{D} \models \varphi(c),$$

where on the right hand of this equivalence we identify $\mathcal{D}$ with the relational structure given by the ground atoms in $\mathcal{D}$. As query answering can be harder (in data complexity) for ontology-mediated queries than for standard relational queries, sometimes no first-order rewriting exists. It is thus of fundamental importance for ontology-based data access to decide whether a given ontology-mediated query is first-order rewritable. Again, for horn description logics the existence of first-order rewritings can be characterized using tree-shaped databases only which enables the use of tools from automata theory to prove ExpTime upper bounds for the computational complexity of this problem [8].

References

1. Artale, A., Kontchakov, R., Kovtunova, A., Ryzhikov, V., Wolter, F., Zakharyaschev, M.: First-order rewritability of temporal ontology-mediated queries. In: Proceedings of the Twenty-Fourth International Joint Conference on Artificial Intelligence, IJCAI 2015, 25–31 July 2015, Buenos Aires, Argentina, pp. 2706–2712 (2015)
2. Baader, F., Bienvenu, M., Lutz, C., Wolter, F.: Query and predicate emptiness in description logics. In: Proceedings of the Twelfth International Conference on Principles of Knowledge Representation and Reasoning, KR 2010, 9–13 May 2010, Toronto, Ontario, Canada (2010)
3. Baader, F., Bienvenu, M., Lutz, C., Wolter, F.: Query and predicate emptiness in description logics. J. Artif. Intell. Res. (2016, to appear)
4. Baader, F., Hladik, J., Lutz, C., Wolter, F.: From tableaux to automata for description logics. Fundam. Inform. **57**(2–4), 247–279 (2003)
5. Baader, F., Peñaloza, R.: Automata-based axiom pinpointing. J. Autom. Reasoning **45**(2), 91–129 (2010)

6. Bienvenu, M., ten Cate, B., Lutz, C., Wolter, F.: Ontology-based data access: a study through disjunctive datalog, CSP, and MMSNP. ACM Trans. Database Syst. **39**(4), 33:1–33:44 (2014)

7. Bienvenu, M., Lutz, C., Wolter, F.: Query containment in description logics reconsidered. In: Proceedings of the Thirteenth International Conference Principles of Knowledge Representation and Reasoning, KR 2012, 10–14 June 2012, Rome, Italy (2012)

8. Bienvenu, M., Lutz, C., Wolter, F.: First-order rewritability of atomic queries in horn description logics. In: Proceedings of the 23rd International Joint Conference on Artificial Intelligence, IJCAI 2013, 3–9 August 2013, Beijing, China (2013)

9. Calvanese, D., Carbotta, D., Ortiz, M.: A practical automata-based technique for reasoning in expressive description logics. In: Proceedings of the 22nd International Joint Conference on Artificial Intelligence, IJCAI 2011, 16–22 July 2011, Barcelona, Catalonia, Spain, pp. 798–804 (2011)

10. Calvanese, D., De Giacomo, G., Lembo, D., Lenzerini, M., Rosati, R.: Data complexity of query answering in description logics. Artif. Intell. **195**, 335–360 (2013)

11. Calvanese, D., De Giacomo, G., Lenzerini, M.: Reasoning in expressive description logics with fixpoints based on automata on infinite trees. In: Proceedings of the Sixteenth International Joint Conference on Artificial Intelligence, IJCAI 1999, July 31 - August 6, 1999, Stockholm, Sweden, vol 2, pp. 84–89 (1999). 1450 pages

12. Hladik, J., Sattler, U.: A translation of looping alternating automata into description logics. In: Baader, F. (ed.) CADE 2003. LNCS (LNAI), vol. 2741, pp. 90–105. Springer, Heidelberg (2003)

13. Lutz, C., Seylan, I., Wolter, F.: An automata-theoretic approach to uniform interpolation and approximation in the description logic EL. In: Proceedings of the Thirteenth International Conference on Principles of Knowledge Representation and Reasoning, KR 2012, 10–14 June 2012, Rome, Italy (2012)

14. Lutz, C., Wolter, F.: Foundations for uniform interpolation and forgetting in expressive description logics. In: Proceedings of the 22nd International Joint Conference on Artificial Intelligence, IJCAI 2011, 16–22 July 2011, Barcelona, Catalonia, Spain, pp. 989–995 (2011)

15. Wilke, T.: Alternating tree automata, parity games, and modal-calculus. Bull. Belg. Math. Soc. **8**, 359–391 (2001)

Automata and Logic

Reasoning with Prototypes in the Description Logic $\mathcal{ALC}$ Using Weighted Tree Automata

Franz Baader and Andreas Ecke[(⊠)]

Theoretical Computer Science, TU Dresden, Dresden, Germany
{baader,ecke}@tcs.inf.tu-dresden.de

Abstract. We introduce an extension to Description Logics that allows us to use prototypes to define concepts. To accomplish this, we introduce the notion of prototype distance functions (pdfs), which assign to each element of an interpretation a distance value. Based on this, we define a new concept constructor of the form $P_{\sim n}(d)$ for $\sim\ \in \{<, \leq, >, \geq\}$, which is interpreted as the set of all elements with a distance $\sim n$ according to the pdf d. We show how weighted alternating parity tree automata (wapta) over the non-negative integers can be used to define pdfs, and how this allows us to use both concepts and pointed interpretations as prototypes. Finally, we investigate the complexity of reasoning in $\mathcal{ALCP}(\text{wapta})$, which extends the Description Logic $\mathcal{ALC}$ with the constructors $P_{\sim n}(d)$ for pdfs defined using wapta.

1 Introduction

Description Logics (DLs) [3] can be used to formalize the important notions of an application domain as *concepts*, by formulating necessary and sufficient conditions for an individual to belong to the concept. Basically, such conditions can be (Boolean combinations of) atomic properties required for the individual (expressed by concept names) and properties that refer to relationships with other individuals and their properties (expressed as role restrictions). The expressivity of a particular DL depends on what kind of properties can be required and how they can be combined. Given an interpretation of the atomic entities (concept and role names), the semantics of a DL determines, for each concept expressed in this DL, its extension, i.e., the set of individuals satisfying all the conditions stated in the definition of the concept. Knowledge about the application domain is then represented by stating subconcept-superconcept relationships between concepts within a terminology (TBox). Given such a TBox, reasoning procedures can be used to derive implicit knowledge from the explicitly represented knowledge. For example, the satisfiability tests checks whether a given concept is non-contradictory w.r.t. the knowledge represented in the TBox.

In many applications, it is quite hard to give exact definitions of certain concepts. In fact, cognitive psychologists [9] argue that humans recognize categories by prototypes rather than concepts. For example, assume that we want to

---

A. Ecke—Supported by DFG in the Research Training Group QuantLA (GRK 1763).

A.-H. Dediu et al. (Eds.): LATA 2016, LNCS 9618, pp. 63–75, 2016.
DOI: 10.1007/978-3-319-30000-9_5

define the concept of a human from an anatomical point of view. One would be tempted to require two arms and two legs, five fingers on each hand, a heart on the left side, etc. However, none of these conditions are necessarily satisfied by an individual human being, though most of them should probably be satisfied to be categorized as human being. Thus, an anatomical description talking about arms and legs etc. describes a prototypical human being rather than necessary and sufficient conditions for being human. As an other example, taken from [8], consider the notion of a cup: we can say that cups are small, cylindrical, concave containers with handles, whose top side is open; they can hold liquids and are used for drinking; and they are made of plastic or porcelain. But again, this describes a prototypical cup rather than stating necessary and sufficient conditions for being a cup: square metal cups are easily imaginable, measuring cups are not used for drinking and may hold non-liquids such as flour, while sippy cups for toddlers are not open on the top. One could, of course, try to capture all such exceptional cups by using a big disjunction of (exactly defined) concepts, but this would obviously be rather clumsy and with high likelihood one would overlook some exceptions.

In order to be used within a formal knowledge representation language with automated reasoning capabilities, prototypes need to be equipped with a formal semantics. To obtain such a semantics, we use the ideas underlying Gärdenfors' conceptual spaces [6], where categories are explained in terms of convex regions, which are defined using the distance from a focal point. To obtain a concrete representation language, we need to define what are focal points and how to define the distance of an individual to such a focal point. Instead of employing prototypical individuals or concepts as focal points, we take a more abstract approach based on automata, which is inspired by the automata-approach for reasoning in DLs (see Sect. 3.2 in [1] for a gentle introduction). Basically, in this approach, a given concept C and a TBox $\mathcal{T}$ are translated into a tree automaton $\mathcal{A}_{C,\mathcal{T}}$ that accepts all the tree-shaped models of $\mathcal{T}$ whose root belongs to C. Testing satisfiability of C w.r.t. $\mathcal{T}$ then boils down to the emptiness test for $\mathcal{A}_{C,\mathcal{T}}$, i.e., checking whether there is a tree accepted by $\mathcal{A}_{C,\mathcal{T}}$. Instead of using a classical automaton that returns 1 (accepted) or 0 (not accepted) for an input tree, we propose to use a weighted automaton [5]. Intuitively, this automaton receives as input a tree-shaped interpretation and returns as output a non-negative integer, which we interpret as the distance of the individual at the root of the tree to the prototype (focal point) described by the automaton. This approach can be applied to non-tree-shaped models by the usual unraveling operation. In order to integrate such prototypes into a Description Logic, we propose to use thresholds to derive concepts from prototypes. More precisely, the threshold concept $P_{\sim n}(\mathcal{A})$ for $\sim \in \{<, \leq, >, \geq\}$ is interpreted as the set of all elements with a distance $\sim n$ according to the weighted automaton $\mathcal{A}$. The concepts obtained this way can then be used like atomic concepts within a DL.

It might appear to be more intuitive to use concepts or individuals rather than automata to describe prototypes. However, in these alternative settings, one then needs to give formal definitions of the distance between two individuals or between an individual and a concept, whereas in our approach this comes for free by the definition of the semantics of weighted automata. We show that these alternative settings can actually be seen as instances of our weighted automata approach.

In this paper, we investigate the extension $\mathcal{ALCP}$(wapta) of the DL $\mathcal{ALC}$ by threshold concepts defined using weighted alternating parity tree automata. In order to obtain inference procedures for the extended DL, the weighted automata are turned into automata that accept the cut-point language consisting of the trees whose distance (computed by the weighted automaton) is below a given threshold. In fact, this cut-point construction yields languages accepted by unweighted alternating parity tree automata, for which the emptiness problem is decidable. This allows us to extend the automata-approach for reasoning in $\mathcal{ALC}$ to $\mathcal{ALCP}$(wapta).

Regarding related work, non-monotonic logics are sometimes also used to formalize prototypes. However, there one usually tries to maximize typicality, i.e. one assumes that an individual stated to belong to a prototype concept has all the properties of the prototype, unless one is forced by other knowledge to retract this assumption. In contrast, our new logic is monotonic and we only conclude that an individual belongs to a threshold concept $P_{\sim n}(d)$ if this follows from the available knowledge. The work that comes closest to this paper is [2], where concepts of the lightweight DL $\mathcal{EL}$ are used to describe prototypes. To be more precise, the paper introduces a graded membership function, which for a given $\mathcal{EL}$-concept C and an individual d of an interpretation returns a membership degree in the interval $[0, 1]$. This is then used as "distance" to define threshold concepts and an extension of $\mathcal{EL}$ by such concepts basically in the same way as sketched above. The difference to the present work is, on the one hand, that prototypes are given by concepts rather than weighted automata and that the interval $[0, 1]$ is used in place of the non-negative integers. On the other hand, we consider a more expressive DL ($\mathcal{ALC}$ rather than $\mathcal{EL}$), and we can reason w.r.t. general TBoxes in the extended language, whereas the results in [2] are restricted to reasoning without a TBox.

2 Preliminaries

The Description Logic $\mathcal{ALC}$. $\mathcal{ALC}$-concepts are built from two disjoint sets N_C of concept names and N_R of role names using concept constructors. Every concept name is a basic $\mathcal{ALC}$-concept. Furthermore, one can construct complex $\mathcal{ALC}$-concepts using conjunction, disjunction, negation, and existential and universal restrictions as shown in Table 1. As usual, we use $\top$ as abbreviation for the concept $A \sqcup \neg A$, where A is an arbitrary concept name. The semantics of $\mathcal{ALC}$-concepts is defined using *interpretations* $\mathcal{I}$ consisting of a non-empty *interpretation domain* $\Delta^{\mathcal{I}}$ and an *interpretation function* $\cdot^{\mathcal{I}}$, which assigns to all concept names $A \in N_C$ a subset $A^{\mathcal{I}} \subseteq \Delta^{\mathcal{I}}$, and to all role names $r \in N_R$ a binary relation $r^{\mathcal{I}} \subseteq \Delta^{\mathcal{I}} \times \Delta^{\mathcal{I}}$ on the domain. The interpretation function is extended to complex concepts as shown in the last column of Table 1. Note that $\top$ is interpreted as $\Delta^{\mathcal{I}}$.

Table 1. Concept constructors for $\mathcal{ALC}$.

Constructor	Syntax	Semantics
conjunction	$C \sqcap D$	$C^{\mathcal{I}} \cap D^{\mathcal{I}}$
disjunction	$C \sqcup D$	$C^{\mathcal{I}} \cup D^{\mathcal{I}}$
negation	$\neg C$	$\Delta^{\mathcal{I}} \setminus C^{\mathcal{I}}$
existential restriction	$\exists r.C$	$\{d \in \Delta^{\mathcal{I}} \mid \exists e.(d,e) \in r^{\mathcal{I}} \wedge e \in C^{\mathcal{I}}\}$
universal restriction	$\forall r.C$	$\{d \in \Delta^{\mathcal{I}} \mid \forall e.(d,e) \in r^{\mathcal{I}} \Rightarrow e \in C^{\mathcal{I}}\}$

Terminological knowledge can be expressed using *general concept inclusions* (GCIs) of the form $C \sqsubseteq D$, where C and D are $\mathcal{ALC}$-concepts. A GCI $C \sqsubseteq D$ is satisfied by an interpretation $\mathcal{I}$ if $C^{\mathcal{I}} \subseteq D^{\mathcal{I}}$. A *TBox* $\mathcal{T}$ is a set of GCIs, and we call an interpretation $\mathcal{I}$ a *model* of $\mathcal{T}$ if it satisfies all GCIs in $\mathcal{T}$.

For example, to express that every container that has a handle and is only used to hold liquids is either a cup or a jug, one could use the GCI

$$\mathsf{Container} \sqcap \exists \mathsf{hasPart.Handle} \sqcap \forall \mathsf{holds.Liquid} \sqsubseteq \mathsf{Cup} \sqcup \mathsf{Jug}.$$

DL systems usually come equipped with a range of reasoning services. Standard inferences provided by most DL systems include concept satisfiability and subsumption. We say that a concept C is satisfiable w.r.t. a TBox $\mathcal{T}$ if there exists a model $\mathcal{I}$ of $\mathcal{T}$ with $C^{\mathcal{I}} \neq \emptyset$; and a concept C is subsumed by a concept D w.r.t. $\mathcal{T}$ ($C \sqsubseteq_{\mathcal{T}} D$) if for all models $\mathcal{I}$ of $\mathcal{T}$ we have $C^{\mathcal{I}} \subseteq D^{\mathcal{I}}$. Subsumption can be reduced to concept satisfiability. Indeed, we have $C \sqsubseteq_{\mathcal{T}} D$ iff $C \sqcap \neg D$ is unsatisfiable in $\mathcal{T}$. Therefore, an algorithm that decides concept satisfiability can also be used to decide subsumption. It is well-known [1] that $\mathcal{ALC}$ has the tree model property, i.e., every satisfiable $\mathcal{ALC}$-concept C has a tree-shaped model in which the root of the tree is an instance of C. Thus, to decide concept satisfiability, it is enough to consider tree-shaped interpretations. In fact, we will show that the tree model property still holds for the extended logic with prototypes, which is important for our approach to work.

Deciding Concept Satisfiability Using Alternating Parity Tree Automata. We show how tree automata can be used to decide satisfiability of $\mathcal{ALC}$-concepts w.r.t. $\mathcal{ALC}$-TBoxes. This result is a simple adaptation of the approach in [10] to $\mathcal{ALC}$. This approach requires the concept and the TBox to be in negation normal form. Recall that an $\mathcal{ALC}$-concept C is in negation normal form if negation occurs only directly in front of concept names. Any concept can be transformed in linear time into an equivalent concept in negation normal form [1].

We can transform a TBox $\mathcal{T}$ into a single concept $C_{\mathcal{T}} = \bigsqcap_{C \sqsubseteq D \in \mathcal{T}} \neg C \sqcup D$; then an interpretation satisfies $\mathcal{T}$ iff it satisfies the GCI $\top \sqsubseteq C_{\mathcal{T}}$. In order to decide $\mathcal{ALC}$-concept satisfiability using tree automata, we first need to introduce the relevant notions from automata theory.

A *tree domain* is a prefix-closed, non-empty set $D \subseteq \mathbb{N}^*$, i.e., for every $ui \in D$ with $u \in \mathbb{N}^*$ and $i \in \mathbb{N}$ we also have $u \in D$. The elements of D are called nodes, the node ε is the root of D, and for every $u \in D$, the nodes $ui \in D$ are called children of u. A node is called a leaf, if it has no children. A path π in D is a subset $\pi \subseteq D$ such that $\varepsilon \in \pi$ and for every $u \in \pi$, u is either a leaf or there is a unique $i \in \mathbb{N}$ with $ui \in \pi$. Given an alphabet Σ, a Σ-labeled tree is a pair (dom_T, T) consisting of a tree domain dom_T and a labeling function $T : \text{dom}_T \to \Sigma$. Instead of the pair (dom_T, T) we often use only T to denote a labeled tree. With $\text{Tree}(\Sigma)$ we denote the set of all Σ-labeled trees.

The automata type we introduce now is based mainly on the alternating tree automata defined by Wilke [11], which are working on $\mathcal{P}(\Sigma)$-Trees, which are labeled with the power set of some finite alphabet Σ. Given such a Σ and a set of states Q, a transition condition $\text{TC}(\Sigma, Q)$ is one of the following: true; false; σ or $\neg\sigma$ for $\sigma \in \Sigma$; $q_1 \wedge q_2$ or $q_1 \vee q_2$ for $q_1, q_2 \in Q$; or $\square q$ or $\lozenge q$ for $q \in Q$.

Definition 1 (alternating parity tree automaton). *An alternating parity tree automaton (apta) $\mathcal{A}$ working on $\mathcal{P}(\Sigma)$-trees is a tuple $\mathcal{A} = (\Sigma, Q, q_0, \delta, \Omega)$, where 1. Σ is a finite alphabet; 2. Q is a finite set of states and $q_0 \in Q$ is the initial state; 3. $\delta : Q \to TC(\Sigma, Q)$ is the transition function; and 4. $\Omega : Q \to \mathbb{N}$ is the priority function that specifies the parity acceptance condition.*

Given a $\mathcal{P}(\Sigma)$-labeled tree T, a *run* is a $(\text{dom}_T \times Q)$-labeled tree R such that $\varepsilon \in \text{dom}_R$, $R(\varepsilon) = (\varepsilon, q_0)$, and for all $u \in \text{dom}_R$ with $R(u) = (v, q)$ we have:

- $\delta(q) \neq \mathsf{false}$
- if $\delta(q) = \sigma$, then $\sigma \in T(v)$; and if $\delta(q) = \neg\sigma$, then $\sigma \notin T(v)$;
- if $\delta(q) = q_1 \wedge q_2$, then there exists $i_1, i_2 \in \mathbb{N}$ such that $R(ui_1) = (v, q_1)$ and $R(ui_2) = (v, q_2)$;
- if $\delta(q) = q_1 \vee q_2$, then there exists $i \in \mathbb{N}$ such that $R(ui) = (v, q_1)$ or $R(ui) = (v, q_2)$;
- if $\delta(q) = \lozenge q'$, then there exists $i, j \in \mathbb{N}$ with $R(ui) = (vj, q')$; and
- if $\delta(q) = \square q'$, then for every $j \in \mathbb{N}$ with $vj \in \text{dom}_T$ there exists $i \in \mathbb{N}$ with $R(ui) = (vj, q')$.

A run is *accepting*, if every infinite path in R satisfies the *parity acceptance condition* specified by Ω, i.e., the largest priority occurring infinitely often along the branch is even. The language accepted by an apta $\mathcal{A}$, $L(\mathcal{A})$, is the set of all $\mathcal{P}(\Sigma)$-trees T for which there exists an accepting run R of $\mathcal{A}$ on T.

The emptiness problem for apta, i.e., deciding whether $L(\mathcal{A}) = \emptyset$, is in EXPTIME; the complement automaton which accepts the complement language $\text{Tree}(\Sigma) \setminus L(\mathcal{A})$ can be constructed in linear time [11]. Note that, instead of only the transition conditions mentioned above, one could allow for complex transition conditions like $\square(q_1 \wedge \neg B) \vee q_2$. Automata with complex transition conditions can be transformed into equivalent automata using only simple transition conditions by introducing new states for each subformula of the transition condition [11].

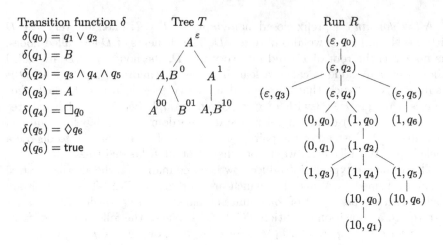

Fig. 1. Transition function δ, $\mathcal{P}(\Sigma)$-tree T, and accepting run R of $\mathcal{A}_{\mathrm{ex}}$ on T.

Example 2. Let $\mathcal{A}_{\mathrm{ex}} = (\Sigma, Q, q_0, \delta, \Omega)$ be an apta with alphabet $\Sigma = \{A, B\}$, with states $Q = \{q_0, \dots, q_6\}$, initial state q_0, transition function δ as given in Fig. 1, and priority function Ω with $\Omega(q) = 1$ for all $q \in Q$.

This automaton accepts only trees where the root label contains B (state q_1), or it is labeled with A and all of its successors (at least one) are again of this form. Since the parity function prohibits infinite paths in the run (though not in the input tree), $\mathcal{A}_{\mathrm{ex}}$ accepts exactly those trees where all paths start with nodes labeled with A until eventually a node with a label containing B is encountered. Figure 1 shows such a tree T and an accepting run R of $\mathcal{A}_{\mathrm{ex}}$ on T.

We now show how to construct an automaton that decides concept satisfiability in $\mathcal{ALC}$. Given a TBox $\mathcal{T}$ and a concept C, the idea underlying this approach is that the constructed automaton will accept exactly the tree models of $\mathcal{T}$ for which the root is an instance of C. Note that the trees introduced above do not have labeled edges, while interpretations do. To overcome this, we push role names into the labels of the children. Thus, the alphabet Σ consists of all concept and role names of C and $C_{\mathcal{T}}$ (called the signature, sig). The automaton contains a state for each subconcept of C and $\mathcal{T}$, denoted $\mathrm{sub}(C)$ and $\mathrm{sub}(C_{\mathcal{T}})$, which are used to simulate the semantics of $\mathcal{ALC}$. Cycles in $\mathcal{T}$ can enforce infinite tree models; infinite paths are always accepting if they satisfy the axioms in $\mathcal{T}$.

Definition 3. *Let $\mathcal{T}$ be an $\mathcal{ALC}$-TBox of the form $\{\top \sqsubseteq C_{\mathcal{T}}\}$ and C an $\mathcal{ALC}$-concept with both C and $C_{\mathcal{T}}$ in negation normal form. We define the automaton $\mathcal{A}_{C,\mathcal{T}} = (\Sigma, Q, q_0, \delta, \Omega)$ as follows:*

- *$\Sigma = \mathrm{sig}(C) \cup \mathrm{sig}(C_{\mathcal{T}})$ and $\Omega(q) = 0$ for all $q \in Q$,*
- *$Q = \{q_D \mid D \in \mathrm{sub}(C) \cup \mathrm{sub}(C_{\mathcal{T}})\} \cup \{q_r, q_{\neg r} \mid r \in \mathrm{sig}(C) \cup \mathrm{sig}(C_{\mathcal{T}})\} \cup \{q_0, q_{\mathcal{T}}\}$,*
- *the transition function δ is defined as follows (where $\sigma \in N_C \cup N_R$):*

$$\delta(q_0) = q_C \wedge q_T \qquad \delta(q_T) = q_{C_T} \wedge \Box q_T$$
$$\delta(q_\sigma) = \sigma \qquad \delta(q_{\neg\sigma}) = \neg\sigma$$
$$\delta(q_{C_1 \sqcap C_2}) = q_{C_1} \wedge q_{C_2} \qquad \delta(q_{C_1 \sqcup C_2}) = q_{C_1} \vee q_{C_2}$$
$$\delta(q_{\exists r.C}) = \Diamond(q_r \wedge q_C) \qquad \delta(q_{\forall r.C}) = \Box(q_{\neg r} \vee q_C)$$

The proof of the following proposition is similar to the one in [7, pp. 59–62]. It relies on the fact that any tree with accepting run can be interpreted as a model of the TBox with the root being an instance of C, and any model of the TBox can be unraveled into a tree for which an accepting run can be inductively constructed.

Proposition 4. *Given an $\mathcal{ALC}$-TBox $\mathcal{T}$ and an $\mathcal{ALC}$-concept C, the concept C is satisfiable w.r.t. $\mathcal{T}$ iff $L(\mathcal{A}_{C,\mathcal{T}}) \neq \emptyset$.*

Since the automaton $\mathcal{A}_{C,\mathcal{T}}$ is polynomial in the size of the TBox $\mathcal{T}$ and the concept C, this approach yields an ExpTime algorithm for concept satisfiability, which is worst-case optimal [3].

3   Prototypes and Weighted Tree Automata

In general, a prototype is some kind of structure that can be compared to elements of an interpretation, distinguishing elements that are closer (more similar or related) to the prototype from elements that are further away (dissimilar or different). More specifically, one may view a prototype as a function that assigns to each element a distance value from the focal point, where small distances correspond to similar elements, and large distances to dissimilar elements.

Definition 5. *A* prototype distance function *(pdf) d is a function that assigns to each element e of an interpretation $\mathcal{I}$ a distance value $d_{\mathcal{I}}(e) \in \mathbb{N}$. The constructor $P_{\sim n}(d)$ for a threshold $n \in \mathbb{N}$ is interpreted in an interpretation $\mathcal{I}$ as the set of all elements $e \in \Delta^{\mathcal{I}}$ such that $d_{\mathcal{I}}(e) \sim n$, for $\sim \in \{<, \leq, >, \geq\}$. If D is a set of pdfs, we use $\mathcal{ALCP}(D)$ to denote the Description Logic $\mathcal{ALC}$ extended by the prototype constructor for pdfs from D.*

As explained before, we will use weighted alternating tree automata to define pdfs. These automata can express distance functions from trees (in our case, tree-shaped pointed interpretations[1]) to the non-negative integers $\mathbb{N}$. By unraveling pointed interpretations we can extend this to a function from arbitrary pointed interpretations to $\mathbb{N}$, i.e., a prototype distance function.

The main idea behind the use of weighted automata to describe pdfs is that the automaton can punish a pointed interpretation by increasing the distance value whenever a feature described by the automaton is not as expected. For example, the automaton can require the current node to be labeled with the

---

[1] Recall that a pointed interpretation is an interpretation together with an element of the interpretation domain.

concept name Cup, and increase the distance by some number if this is not the case. Using this idea, the most natural interpretation of the transition conditions in the weighted setting is as follows: $q_1 \wedge q_2$ will compute the sum of the distances for q_1 and q_2 (both features should be present), $\vee$ will be interpreted as the minimum (one of the feature should be present), $\exists$ will also be interpreted as the minimum (one of the successors should have the feature, i.e., we choose the best one); and $\forall$ will be interpreted as the maximum (all successors should have the feature; if not, we take the distance of the worst).

A weighted alternating parity tree automaton is nearly the same as in the unweighted case, with the exception that the transition function may also contain non-negative integers. Given an alphabet Σ and a set of states Q, a weighted transition condition $\mathrm{wTC}(\Sigma, Q)$ is one of the following: $n \in \mathbb{N}$; σ or $\neg\sigma$ for $\sigma \in \Sigma$; $q_1 \wedge q_2$ or $q_1 \vee q_2$ for $q_1, q_2 \in Q$; or $\Box q$ or $\Diamond q$ for $q \in Q$.

Definition 6 (weighted alternating parity tree automaton). *A weighted alternating parity tree automaton (wapta) $\mathcal{A}$ working on $\mathcal{P}(\Sigma)$-trees is a tuple $\mathcal{A} = (\Sigma, Q, q_0, \delta, \Omega)$, where 1. Σ is a finite alphabet; 2. Q is a finite set of states and $q_0 \in Q$ is the initial state; 3. $\delta : Q \to \mathrm{wTC}(\Sigma, Q)$ is the transition function; and 4. $\Omega : Q \to \mathbb{N}$ is the priority function.*

Runs are defined as in the unweighted case, where nodes labeled with a state for which the transition function yields a number do not need to satisfy any additional conditions, they can be leafs in the run. In order to define the behavior of such a weighted automaton on a tree, we need to define the $\Box$-fixation of a run, which basically chooses for a $\Box$-operator a single successor node (instead of all of them). Given a run R, a $\Box$-fixation is a tree R' with $\mathrm{dom}_{R'} \subseteq \mathrm{dom}_R$, which can be obtained from R as follows: starting with the root, we keep all the successors for nodes where the transition function does not yield a box; for nodes u labeled with a state q for which the transition function is of the form $\delta(q) = \Box q'$, the $\Box$-fixation R' keeps at most one successor $ui \in \mathrm{dom}_R$. All nodes $u \in \mathrm{dom}_{R'}$ have the same label $R'(u) = R(u)$ as in R.

Then, we can define the behavior of the automaton as a function $\|\mathcal{A}\| :$ $\mathrm{Tree}(\mathcal{P}(\Sigma)) \to \mathbb{N}$. The weight of a $\Box$-fixation R' of a run R is defined as

$$\mathrm{weight}_{\mathcal{A}}(R') = \sum_{u \in \mathrm{dom}_{R'}, R'(u) = (d,q), \delta(q, T(u)) = n \in \mathbb{N}} n.$$

Note that this (possibly infinite) sum is well-defined: If infinitely many values $n > 0$ occur in R', the weight of R' is ∞; otherwise it is the finite sum of all weights in R'. The weight of a run R on T is $\mathrm{weight}_{\mathcal{A}}(R) = \sup_{R' \Box \text{-fixation of } R} \mathrm{weight}_{\mathcal{A}}(R')$, and the behavior of $\mathcal{A}$ is $\|\mathcal{A}\|(T) = \min_{R \text{ accepting run on } T} \mathrm{weight}_{\mathcal{A}}(R)$.

Constructions of Prototype Automata. In the following we will give a concrete example of how a weighted automaton can be constructed from an $\mathcal{ALC}$-concept. Recall from the introduction that a prototypical cup is a small container with handles, which can hold liquids and is made of plastic or porcelain. We can express this as an $\mathcal{ALC}$-concept:

Container ⊓ Small ⊓ ∃hasPart.Handle ⊓ ∀holds.Liquid ⊓ ∀material.(Glass ⊔ Porcelain)

This concept can directly be translated into a complex transition condition for an alternating tree automaton:

Container ∧ Small ∧ ◊(hasPart ∧ Handle)

   ∧ □(¬holds ∨ Liquid) ∧ □(¬material ∨ (Glass ∨ Porcelain))

Now, we can add weights in order to punish missing features:

(Container ∨ 3) ∧ (Small ∨ 1) ∧ (◊(hasPart ∧ Handle) ∨ 1)

   ∧ □(¬holds ∨ (Liquid ∨ 2)) ∧ □(¬material ∨ ((Glass ∨ 1) ∨ (Porcelain ∨ 1)))

The meaning of this weighted transition condition is as follows: If an element is not a container, it will be punished with a weight of 3 since there cannot be a run that uses the option Container at the root. Otherwise, there is such a run, which does not contribute a weight. Accordingly, the absence of the feature small is punished with weight 1. If the cup does not have a successor that is labeled with both hasPart and Handle, then a weight of 1 is added. Finally, if there is a material-successor that is not labeled with Glass or Porcelain, then this is punished with weight 1. If the cup does not have any material-successors, or all of them are glass or porcelain, no weight is added. Similarly for holding only liquids. In general, choosing the weights appropriately allows us to punish the absence of different features by different values.

For universal restrictions, the weights of several offending successors are not added up, but rather the supremum is taken. As a consequence, equivalent concepts may not yield equivalent wapta using this approach. For example, $\forall r.(A \sqcap B) \equiv \forall r.A \sqcap \forall r.B$, but the corresponding transition conditions after adding weights may lead to different results. However, one can argue that, when viewed as prototype descriptions, these two concept descriptions do actually encode different intentions. While in the first case we want to make sure that all r-successors are instance of A and B simultaneously (and pick the weight of the worst offender if there is one), in the second case we want to enforce both features separately, and punish for the worst offenders separately.

From the above example, it should be clear how a translation from $\mathcal{ALC}$-concepts to wapta works in general. On the other hand, one can also create prototypes from finite pointed interpretations. For this, one introduces a state for each element of the interpretation, and as transition condition for each state one simply conjoins all the concept names the element is instance of, negations of all concept names it is not instance of, and a ◊-transition for each successor in the interpretation, labeled with both the role name and the state of the successor-element. If one also introduces a □-transition with a disjunction of all possible successor-states and adds positive weights as in the above example, this weighted automaton will only give distance 0 to pointed interpretations that are bisimilar to the prototypical interpretation, and otherwise punish each difference by increasing the distance accordingly.

4   Reasoning with Prototype Automata in $\mathcal{ALC}$

To reason in $\mathcal{ALC}$ with prototypes, we have to achieve two things: First, for each prototype constructor $P_{\leq n}(\mathcal{A})$, we have to transform the wapta $\mathcal{A}$ into an unweighted automaton that accept exactly those trees T for which $\|\mathcal{A}\|(T) \leq n$. Then we need to combine the alternating tree automaton $\mathcal{A}_{C,\mathcal{T}}$ from Sect. 2 with the unweighted automata for the prototypes such that the resulting automaton accepts exactly the tree models of C w.r.t. $\mathcal{T}$. An emptiness test can then be used to decide (un-)satisfiability.

Cut-Point Automata. Given a weighted alternating parity tree automaton $\mathcal{A}$ and a threshold value $n \in \mathbb{N}$, we want to construct an unweighted automaton $\mathcal{A}_{\leq n}$ that accepts exactly the cut-point language, i.e. $L(\mathcal{A}_{\leq n}) = \{T \in \text{Tree}(\mathcal{P}(\Sigma)) \mid \|\mathcal{A}\|(T) \leq n\}$. In this cut-point automaton, each state needs to keep track of both the weight and the current state of the corresponding weighted automaton. However, instead of tracking the weight that has already been accumulated, it needs to track the weight that the automaton is still allowed to spend. The reason for this is that for trees each state can have multiple successors, and thus we have to budget the allowed weight for each of the successors so that the sum is not greater than the threshold.

Definition 7. *Given a wapta $\mathcal{A} = (\Sigma, Q, q_0, \delta, \Omega)$, the cut-point automaton $\mathcal{A}_{\leq n} = (\Sigma, Q', q_0', \delta', \Omega')$ for the threshold $n \in \mathbb{N}$ is an apta defined as follows:*

$$Q' = \{(q, i) \in Q \times \mathbb{N} \mid i \leq n\} \cup \{q_0'\} \quad \text{and} \quad \Omega'((q, i)) = \Omega(q)$$

$$\delta'(q_0') = \bigvee_{0 \leq i \leq n} (q_0, i) \qquad\qquad \delta'((q, i)) = \delta(q) \ \text{if} \ \delta(q) = \sigma, \neg\sigma$$

$$\delta'((q, i)) = \textsf{true} \ \text{if} \ \delta(q) = j \leq i \qquad \delta'((q, i)) = \textsf{false} \ \text{if} \ \delta(q) = j > i$$

$$\delta'((q, i)) = \Diamond(q', i) \ \text{if} \ \delta(q) = \Diamond q' \qquad \delta'((q, i)) = \Box(q', i) \ \text{if} \ \delta(q) = \Box q'$$

$$\delta'((q, i)) = (q_1, i) \vee (q_2, i) \ \text{if} \ \delta(q) = q_1 \vee q_2$$

$$\delta'((q, i)) = \bigvee_{0 \leq j \leq i} (q_1, j) \wedge (q_2, i - j) \ \text{if} \ \delta(q) = q_1 \wedge q_2$$

Proposition 8. *Let $\mathcal{A}$ be a wapta and $\mathcal{A}_{\leq n}$ the cut-point automaton derived from $\mathcal{A}$ using the threshold $n \in \mathbb{N}$. Then $\mathcal{A}_{\leq n}$ accepts the cut-point language, i.e., $L(\mathcal{A}_{\leq n}) = \{T \in \text{Tree}(\mathcal{P}(\Sigma)) \mid \|\mathcal{A}\|(T) \leq n\}$.*

Proof (sketch). We have to prove both directions. Given a tree $T \in L(\mathcal{A}_{\leq n})$, and an accepting run R of $\mathcal{A}_{\leq n}$ on T, we can construct a run R' of $\mathcal{A}$ on T by removing all weights from the labels of R. By induction on the weight i, we can then show that whenever we have $R(u) = (v, (q, i))$ for some node $u \in \text{dom}_R$, all

$\square$-fixations of R' starting from u will have a weight at most i. This follows from the claim that the sum of the weights of the children of a node v is never larger than the weight of v itself for all $\square$-fixations. Since the first successor of the root of R is labeled with $R(0) = (\varepsilon, (q_0, n))$, this means that $\text{weight}_{\mathcal{A}}(R') \leq n$.

Similarly, if we have a tree $T \in \text{Tree}(\mathcal{P}(\Sigma))$ with $\|\mathcal{A}\|(T) \leq n$, and a run R of $\mathcal{A}$ on T with $\text{weight}_{\mathcal{A}}(R) \leq n$, we can construct a run R' of $\mathcal{A}_{\leq n}$ on T by setting $R'(u) = (v, (q, i))$ where $R(u) = (v, q)$ and i is the weight assigned by $\mathcal{A}$ to the subtree of R rooted at u, starting in state q. It can then be shown that the run R' obtained this way is an accepting run of $\mathcal{A}_{\leq n}$ on T. $\qquad\square$

The cut-point automaton $\mathcal{A}_{\leq n}$ has $O(n \cdot q)$ states, where q is the number of states of the weighted automaton $\mathcal{A}$. Thus, if n is encoded in unary, this construction is polynomial, otherwise it is exponential.

Combined Reasoning Using Alternating Automata. We want to combine the cut-point automata constructed from prototype concepts with the automaton from Definition 3 in order to decide the concept satisfiability problem in $\mathcal{ALCP}$(wapta); more specifically, we want to construct an automaton $\mathcal{A}$ that accepts all those (tree-shaped) pointed interpretations that are instances of an $\mathcal{ALCP}$(wapta)-concept w.r.t. an $\mathcal{ALCP}$(wapta)-TBox.

For $\mathcal{ALCP}$(wapta)-concepts, one can again define a normal form. This extends the negation normal form used in Sect. 2 by requiring that prototype constructors occur only in the form $P_{\leq n}(\mathcal{A})$, possibly negated. For example, one can transform $P_{\geq n}(\mathcal{A})$ for $n \geq 1$ into negation normal form by replacing it with $\neg P_{\leq n-1}(\mathcal{A})$; $P_{\geq 0}(\mathcal{A})$ can be replaced by $\top$. The set of subconcepts now contains such prototype concepts as well.

In case a prototype constructor occurs negated, the complement automaton $\bar{\mathcal{A}}$ for a cut-point automaton $\mathcal{A}$ can be constructed in linear time, by exchanging true and false, $\vee$ and $\wedge$, $\square$ and $\lozenge$, and σ and $\neg\sigma$ for all $\sigma \in \Sigma$ in all transition conditions, as well as adding one to the priority of all states [11].

Definition 9. *Let T be an $\mathcal{ALCP}$(wapta)-TBox of the form $\{\top \sqsubseteq C_T\}$ and C an $\mathcal{ALCP}$(wapta)-concept, with both C and C_T in negation normal form, and let $\mathcal{A}_{i,\leq n}$ be the cut-point automaton of the wapta $\mathcal{A}_i$ for each prototype constructor $P_{\leq n}(\mathcal{A}_i)$ occurring in C or T.*

The apta $\mathcal{A}_{P,C,T}$ is the disjoint union of $\mathcal{A}_{C,T}$ from Definition 3, all automata $\mathcal{A}_{i,\leq n}$ for prototypes $P_{\leq n}(\mathcal{A}_i)$ occurring in C or C_T, and all automata $\bar{\mathcal{A}}_{i,\leq n}$ for negated prototypes $\neg P_{\leq n}(\mathcal{A}_i)$ occurring in C or C_T, such that the transition function additionally is defined for subconcepts of the form $P_{\leq n}(\mathcal{A}_i)$ and $\neg P_{\leq n}(\mathcal{A}_i)$ as follows:

$$\delta(q_{P_{\leq n}(\mathcal{A}_i)}) = q_i \text{ where } q_i \text{ is the initial state of } \mathcal{A}_{i,\leq n}$$
$$\delta(q_{\neg P_{\leq n}(\mathcal{A}_i)}) = q_i \text{ where } q_i \text{ is the initial state of } \bar{\mathcal{A}}_{i,\leq n}$$

The following theorem is an easy consequence of Propositions 4 and 8.

Theorem 10. *Given an $\mathcal{ALCP}(wapta)$-TBox $\mathcal{T}$ and an $\mathcal{ALCP}(wapta)$-concept C, the concept C is satisfiable w.r.t. $\mathcal{T}$ iff $L(\mathcal{A}_{P,C,\mathcal{T}}) \neq \emptyset$.*

Because of the size of the cut-point automata and the EXPTIME-emptiness test for alternating tree automata, concept satisfiability can thus be deciding in EXPTIME if the numbers are given in unary. This is worst-case optimal. If the numbers are given in binary, the complexity of the algorithm increases to 2EXPTIME. It is an open problem whether this second exponential blowup can be avoided.

5   Conclusions

We have introduced an extension to Description Logics that allows to define prototypes and reason over them. In particular, we have introduced the proto-type constructors $P_{\sim n}(d)$ that are interpreted as the set of all elements of the interpretation with distance $\sim n$ according to the prototype distance functions d. We have shown that pdfs can be defined using waptas, and that reasoning in $\mathcal{ALCP}$(wapta) has he same complexity as reasoning in $\mathcal{ALC}$ (if the threshold numbers n are coded in unary).

Of course, this approach has some limitations. As mentioned in Sect. 3, the pdfs obtained through a straightforward translation of $\mathcal{ALC}$-concepts into waptas are not equivalence invariant. This is due to the fact that we use the supremum rather than the sum to combine the weights obtained from different $\square$-fixations. However, replacing supremum by sum has the disadvantage that the cut-point language need no longer be recognizable by an apta. We conjecture that in that case, the cut-point language can actually be accepted by a graded apta [4], but the construction to be developed would definitely be considerably more complex than the one used in this paper. More generally, one could of course also look at weighted automata using other domains for weights and other operations combining them.

Finally, we are interested in adding prototypes to other DLs. Since prototypes can be used to express negation, considering less expressive DLs does not make sense. But adding nominals and quantified number restrictions would be inter-esting, as would be considering the instance problem and answering conjunctive queries.

References

1. Baader, F.: Description logics. In: Tessaris, S., Franconi, E., Eiter, T., Gutierrez, C., Handschuh, S., Rousset, M.-C., Schmidt, R.A. (eds.) Reasoning Web. LNCS, vol. 5689, pp. 1–39. Springer, Heidelberg (2009)
2. Baader, F., Brewka, G., Gil, O.F.: Adding threshold concepts to the description logic $\mathcal{EL}$. In: Lutz, C., Ranise, S. (eds.) FroCoS 2015, vol. 9322, pp. 33–48. Springer, Heidelberg (2015)

3. Baader, F., Calvanese, D., McGuinness, D.L., Nardi, D., Patel-Schneider, P.F. (eds.): The Description Logic Handbook: Theory, Implementation, and Applications. Cambridge University Press, Cambridge (2003)
4. Bonatti, P.A., Lutz, C., Murano, A., Vardi, M.Y.: The complexity of enriched μ-calculi. In: Bugliesi, M., Preneel, B., Sassone, V., Wegener, I. (eds.) ICALP 2006. LNCS, vol. 4052, pp. 540–551. Springer, Heidelberg (2006)
5. Droste, M., Kuich, W., Vogler, H.: Handbook of Weighted Automata. Springer, New York (2009)
6. Gärdenfors, P.: Conceptual Spaces - The Geometry of Thought. MIT Press, Cambridge (2000)
7. Hladik, J.: To and Fro Between Tableaus and Automata for Description Logics. Dissertation, TU Dresden, Germany (2007). http://www.qucosa.de/fileadmin/data/qucosa/documents/846/1201792812059-1908.pdf
8. Labov, W.: The boundaries of words and their meanings. In: Bailey, C.J.N., Shuy, R.W. (eds.) New Ways of Analyzing Variation in English. Georgetown University Press, Washington, DC (1973)
9. Rosch, E., Lloyd, B.B.: Cognition and Categorization. Lawrence Erlbaum Associates, Hillsdale (1978)
10. Sattler, U., Vardi, M.Y.: The hybrid μ-calculus. In: Goré, R.P., Leitsch, A., Nipkow, T. (eds.) IJCAR 2001. LNCS (LNAI), vol. 2083, p. 76. Springer, Heidelberg (2001)
11. Wilke, T.: Alternating tree automata, parity games, and modal μ-calculus. Bull. Belg. Math. Soc. **8**, 359–391 (2001)

+ω-Picture Languages Recognizable by Büchi-Tiling Systems

Parvaneh Babari[1] and Nicole Schweikardt[2]([✉])

[1] Institut für Informatik, Universität Leipzig, Leipzig, Germany
babari@informatik.uni-leipzig.de
[2] Institut für Informatik, Humboldt-Universität zu Berlin, Berlin, Germany
schweikn@informatik.hu-berlin.de

Abstract. We consider +ω-pictures, i.e., 2-dimensional pictures with a finite number of rows and a countably infinite number of columns. We extend conventional tiling systems with a Büchi acceptance condition and define the class of Büchi-tiling recognizable +ω-picture languages. We show that this class has the same closure properties as the class of tiling recognizable languages of finite pictures. We characterize the class of Büchi-tiling recognizable +ω-picture languages by generalized Büchi-tiling systems and by the logic EMSO$^\infty$, an extension of existential monadic second-order logic with quantification of infinite sets. The Büchi characterization theorem (stating that the ω-regular languages are finite unions of languages of the form $L_1 \cdot L_2^\omega$, for regular languages L_1 and L_2), however, does not carry over from regular ω-languages to Büchi-tiling recognizable languages of +ω-pictures.

Keywords: Automata and logic · Picture languages · Tiling systems · Existential monadic second-order logic

1 Introduction

The theory of two-dimensional languages as a generalization of formal languages of words was motivated by problems arising from image processing and pattern recognition [6], and also plays a role in the theory of cellular automata and other devices of parallel computing [16,17]. In the 1990s, Giammarresi and Restivo introduced the family of recognizable languages of finite pictures [7,9]. This family is very robust and has been characterized by many different devices, including automata, tiling systems, rational operations, and existential monadic second-order logic [7–10,13–15]. Notions of recognizability have also been studied for languages of $\omega\omega$-pictures, i.e., analogues of ω-words that are infinite in two dimensions [1,3,5,11,12,18].

In this paper, we study +ω-pictures, i.e., pictures that have a finite number of rows and an infinite number of columns. As a motivation for studying these pictures, consider for example the potentially infinite streams of taped videos

P. Babari—Supported by DFG Graduiertenkolleg 1763 (QuantLA)

A.-H. Dediu et al. (Eds.): LATA 2016, LNCS 9618, pp. 76–88, 2016.
DOI: 10.1007/978-3-319-30000-9_6

captured by digital security cameras. To the best of our knowledge, languages of $+\omega$-pictures have not been studied before.

We summarize our main contributions as follows: To obtain a notion of recognizability of languages of $+\omega$-pictures, we introduce the Büchi-tiling systems. These are an extension of the classical tiling systems [9] by a Büchi acceptance condition. This way, the Büchi-tiling recognizable $+\omega$-picture languages can be viewed as a natural generalization of the ω-regular languages. We show that the class of Büchi-tiling recognizable $+\omega$-picture languages has the same closure properties as the class of tiling recognizable languages of finite pictures [10]: it is closed under projection, union, and intersection, but not under complementation. We show that the class of Büchi-tiling recognizable $+\omega$-picture languages is robust, as it (1) can be characterized by generalized Büchi-tiling systems, and (2) has a logical characterization via an extension of existential monadic second-order logic by existential quantification of infinite sets.

The results mentioned so far witness that many results and techniques that have been developed for the tiling recognizable languages of finite pictures or for the ω-regular languages recognized by Büchi-automata, can be transferred to the Büchi-tiling recognizable $+\omega$-picture languages. However, using combinatorial arguments, we show that the well-known Büchi characterization theorem (stating that the ω-regular languages are unions of finitely many languages of the form $L_1 \cdot L_2^\omega$, for regular languages L_1 and L_2) does not carry over to the Büchi-tiling recognizable languages of $+\omega$-pictures.

The remainder of the paper is structured as follows. Section 2 fixes the basic notation. Section 3 introduces the Büchi-tiling systems, presents the mentioned closure properties, and establishes the characterization by generalized Büchi-tiling systems. Section 4 is devoted to the logical characterization. Section 5 shows that there is no Büchi characterization theorem for the Büchi-tiling recognizable $+\omega$-picture languages. Section 6 concludes the paper. Due to space limitations, most proof details had to be omitted. Detailed proofs will be provided in the paper's full version.

2 Preliminaries

We write $\mathbb{N}$ for the set of non-negative integers, and we let $\mathbb{N}_{\geqslant 1} := \mathbb{N} \setminus \{0\}$ and $[n] := \{1, \ldots, n\}$, for any $n \in \mathbb{N}_{\geqslant 1}$. Somewhat abusing notation, $[\omega]$ will denote the set $\mathbb{N}_{\geqslant 1}$. Throughout this paper, *alphabets* are finite non-empty sets.

A *finite picture* over Σ is a finite rectangular array of elements of Σ. Formally, for $m, n \in \mathbb{N}_{\geqslant 1}$, a picture of *size* (m, n) over Σ is a mapping $p : [m] \times [n] \to \Sigma$. The number $\ell_1(p) := m$ of rows is called the *height* of p, and the number $\ell_2(p) := n$ of columns is called the *width* of p. For $i \in [\ell_1(p)]$ and $j \in [\ell_2(p)]$ we write $p_{ij} := p(i, j)$ to denote the letter of p in row i and column j. For $m, n \in \mathbb{N}_{\geqslant 1}$, we write $\Sigma^{m,n}$ for the set of all pictures over Σ of size (m, n), and we let $\Sigma^{++} := \bigcup_{m,n \in \mathbb{N}_{\geqslant 1}} \Sigma^{m,n}$ be the set of all finite pictures over Σ. Often, we will refer to finite pictures as $++$-*pictures*. A $++$-*picture language* over Σ is a subset of Σ^{++}.

We generalize these notions to pictures that have an infinite number of columns: For $m \in \mathbb{N}_{\geqslant 1}$, an $m\omega$-*picture* over Σ is a mapping $p : [m] \times \mathbb{N}_{\geqslant 1} \to \Sigma$. Thus, p has exactly m rows and an infinite number of columns. We write $\Sigma^{m\omega}$ to denote the set of all $m\omega$-pictures over Σ. The set of all $+\omega$-*pictures* over Σ is the set $\Sigma^{+\omega} := \bigcup_{m\in\mathbb{N}_{\geqslant 1}} \Sigma^{m\omega}$. A $+\omega$-*picture language* over Σ is a subset of $\Sigma^{+\omega}$. Similarly as for finite pictures, for an $m\omega$-picture p we write $\ell_1(p)$ to denote the number m of rows of p. We write $\ell_2(p) = \omega$ to indicate that there is an infinite number of columns. The *size* of p is (m, ω).

We identify finite pictures of height 1 over Σ with finite non-empty words over Σ (i.e., with elements in Σ^+), and we identify $+\omega$-pictures of height 1 over Σ with ω-words over Σ (i.e., with elements in Σ^ω).

When given a (finite or $+\omega$-)picture p and numbers j_1, j_2 with $1 \leqslant j_1 \leqslant j_2 \leqslant \ell_2(p)$ we write $p[j_1, j_2]$ for the picture obtained from p by deleting all columns j with $j < j_1$ or $j > j_2$.

The (column) *concatenation* of a finite picture p and a (finite or $+\omega$-)picture q of the same height $m = \ell_1(p) = \ell_1(q)$ yields the picture $p \oslash q$ of height m whose first $\ell_2(p)$ columns are identical with p and whose remaining columns $\ell_2(p) + 1$, $\ell_2(p) + 2, \ldots$ are identical with the columns $1, 2, \ldots$ of q. I.e., $(p \oslash q)[1, \ell_2(p)] = p$ and $(p \oslash q)[\ell_2(p)+1, \ell_2(p)+\ell_2(q)] = q$ (using the convention "$n+\omega = \omega$"). The concatenation of a language $L \subseteq \Sigma^{++}$ of finite pictures and a language $L' \subseteq \Sigma^{+\omega}$ of $+\omega$-pictures is defined as $L \oslash L' := \{ p \oslash p' : p \in L, p' \in L', \ell_1(p) = \ell_1(p') \}$. I.e., a picture q belongs to $L \oslash L'$ if, and only if, some initial segment of q belongs to L and the rest of q belongs to L'. Let $L, L' \subseteq \Sigma^{++}$ be two languages of finite pictures. The concatenation $L \oslash L'$ is the $++$-picture language $L \oslash L' := \{p \oslash p' : p \in L, p' \in L', \ell_1(p) = \ell_1(p')\}$. We define the iterated concatenation $L^{\oslash n}$ via $L^{\oslash 1} := L$ and $L^{\oslash n} := L \oslash L^{\oslash(n-1)}$ for every $n \in \mathbb{N}$ with $n \geqslant 2$. Clearly, for every $n \in \mathbb{N}_{\geqslant 1}$, $L^{\oslash n}$ is a language of finite pictures. We let $L^{\oslash \omega}$ be the $+\omega$-picture language that consists of all $+\omega$-pictures p for which there is an infinite sequence $1 = j_1 < j_2 < \cdots$ of integers such that for every $i \in \mathbb{N}_{\geqslant 1}$, the picture $p[j_i, j_{i+1}-1]$ belongs to L. I.e., $L^{\oslash \omega}$ consists of all $+\omega$ pictures of the form $p_1 \oslash p_2 \oslash p_3 \oslash \cdots$ where $p_i \in L$ for every $i \in \mathbb{N}_{\geqslant 1}$, and all p_i have the same height. For a finite picture $p \in \Sigma^{++}$ we write $p^{\oslash \omega}$ for the unique $+\omega$-picture in $\{p\}^{\oslash \omega}$. Accordingly, for $n \in \mathbb{N}_{\geqslant 1}$ we write $p^{\oslash n}$ for the unique $++$-picture in the set $\{p\}^{\oslash n}$.

3 Büchi-Tiling Recognizable $+\omega$-Picture Languages

Local sets of words play an important role in the theory of regular string languages. The notion has been generalized to languages of finite pictures [7] and to $\omega\omega$-picture languages [3]. In this section, we extend this notion to $+\omega$-picture languages and introduce tiling systems with a Büchi acceptance condition. We exhibit closure properties of the class of $+\omega$-picture languages that are recognizable with these Büchi-tiling systems.

For a (finite or $+\omega$-)picture p over Σ and for numbers i_1, i_2, j_1, j_2 with $1 \leqslant i_1 \leqslant i_2 \leqslant \ell_1(p)$ and $1 \leqslant j_1 \leqslant j_2 \leqslant \ell_2(p)$ we write $p_{i_2}^{i_1}[j_1, j_2]$ for the subpicture of

p at rows $i_1, \ldots, i_2$ and columns $j_1, \ldots, j_2$. I.e., $p_{i_2}^{i_1}[j_1, j_2]$ is the picture obtained from p by deleting all rows i with $i < i_1$ or $i > i_2$ and deleting all columns j with $j < j_1$ or $j > j_2$. For numbers $m, n \in \mathbb{N}_{\geqslant 1}$ we write $T_{m,n}(p)$ for the set of all subpictures of p of size (m, n), i.e.,

$$T_{m,n}(p) = \left\{ p_{i+m-1}^{i}[j, j+n-1] : 1 \leqslant i \leqslant \ell_1(p)-m+1, \ 1 \leqslant j \leqslant \ell_2(p)-n+1 \right\}$$

(with the convention "$\omega - n = \omega$").

For an alphabet Γ, we write $\hat{\Gamma}$ for the alphabet $\Gamma \cup \{\#\}$, where $\#$ is a special boundary symbol that does not belong to Γ. For a finite picture q of size (m, n) over Γ, we write $\hat{q}$ for the picture of size $(m+2, n+2)$ over $\hat{\Gamma}$, obtained by surrounding q with the boundary symbol $\#$. Accordingly, for a +ω-picture q over Γ we write $\hat{q}$ for the $(m+2)\omega$-picture over $\hat{\Gamma}$, obtained by surrounding q with the boundary symbol $\#$ from the left, top and bottom. A *tile* is a picture of size $(2, 2)$ over the alphabet $\hat{\Gamma}$.

Definition 1. Let Γ be an alphabet and let $\bullet$ be one of the symbols $+$ or ω. The +$\bullet$-picture language *recognized* by a set $\Theta \subseteq \hat{\Gamma}^{2,2}$ of tiles is $L^{+\bullet}(\Theta) := \{q \in \Gamma^{+\bullet} : T_{2,2}(\hat{q}) \subseteq \Theta\}$. A +$\bullet$-picture language L over Γ is called *local* if there exists a set $\Theta \subseteq \hat{\Gamma}^{2,2}$ of tiles such that $L = L^{+\bullet}(\Theta)$.

In the literature, mappings $\pi : \Gamma \to \Sigma$, for alphabets Γ and Σ, are called *projections*. Such mappings are lifted to pictures and picture languages in the canonical way: for a picture q over Γ, $\pi(q)$ is the picture p over Σ of the same height and width as q, where for each row i and each column j, the letter p_{ij} in row i and column j is $\pi(q_{ij})$. For a picture language L over Γ, we let $\pi(L) = \{\pi(q) : q \in L\}$.

A language $L \subseteq \Sigma^{++}$ of finite pictures is called *tiling recognizable* [7] if there exists an alphabet Γ, a local ++-picture language L' over Γ, and a projection $\pi : \Gamma \to \Sigma$, such that $L = \pi(L')$. When dealing with recognizability, it is often convenient to assume that the alphabet Γ has the special form $\Gamma = \Sigma \times Q$, and the projection $\pi : \Gamma \to \Sigma$ just cancels the Q-component. It is straightforward to see that this assumption can be made without loss of generality (cf., e.g., [10]). Thus, a language $L \subseteq \Sigma^{++}$ is tiling recognizable if, and only if, there exists a *tiling system* $\mathcal{T}$ with $L = L^{++}(\mathcal{T})$ in the following sense.

Definition 2 (Tiling system).

(a) A *tiling system* is a 3-tuple $\mathcal{T} = (\Sigma, Q, \Theta)$ where Σ and Q are alphabets, and $\Theta \subseteq \hat{\Gamma}^{2,2}$ for $\Gamma := \Sigma \times Q$. The elements of Q are called *states* of $\mathcal{T}$.

(b) Let Σ and Q be alphabets. For a (finite or +ω-)picture p over Σ and a picture r over Q of the same size as p, we write $(p \times r)$ for the picture q over $\Gamma := \Sigma \times Q$ that has the same size as p, and where for every row i and every column j, the entry q_{ij} in row i and column j is (p_{ij}, r_{ij}).

(c) Let $\mathcal{T} = (\Sigma, Q, \Theta)$ be a tiling system, let $\bullet$ be one of the symbols $+$ or ω, and let p be a +$\bullet$-picture over Σ.

A *run* of $\mathcal{T}$ on p is a picture r over Q of the same size as p, such that the picture $q := (p \times r)$ has the following property: every subpicture of size $(2,2)$ of $\hat{q}$ belongs to Θ, i.e., $T_{2,2}(\hat{q}) \subseteq \Theta$.

The picture p is *accepted* by $\mathcal{T}$ if there exists a run of $\mathcal{T}$ on p.

The $+\bullet$-picture language *recognized* by $\mathcal{T}$ is the set $L^{+\bullet}(\mathcal{T})$ of all $+\bullet$-pictures p over Σ that are accepted by $\mathcal{T}$.

A picture language $L \subseteq \Sigma^{+\bullet}$ is *tiling recognizable* if there is a tiling system $\mathcal{T}$ with $L = L^{+\bullet}(\mathcal{T})$.

We now extend tiling systems with a *Büchi acceptance condition*. This will lead to a notion of *Büchi-tiling recognizable* $+\omega$-picture languages that can be viewed as a 2-dimensional generalization of the ω-regular languages (i.e., the languages of ω-words recognized by Büchi-automata).

Definition 3 (Büchi-tiling system).

(a) A *Büchi-tiling system* is a 4-tuple $\mathcal{S} = (\Sigma, Q, \Theta, F)$, where (Σ, Q, Θ) is a tiling system and $F \subseteq Q$. The elements of F are called *accepting states*; the set F is called the *acceptance condition*.

(b) Let $\mathcal{S} = (\Sigma, Q, \Theta, F)$ be a Büchi-tiling system and let $p \in \Sigma^{+\omega}$.
A *run* of $\mathcal{S}$ on p is a run of the tiling system $\mathcal{T} := (\Sigma, Q, \Theta)$ on p.
For a run r of $\mathcal{S}$ on p we write $\inf_1(r)$ for the set of states that occur infinitely often in the first row of r.
A run r of $\mathcal{S}$ on p is *accepting*, if $\inf_1(r) \cap F \neq \emptyset$ (i.e., there is an accepting state that occurs infinitely often in the first row of the run).
The $+\omega$-picture p is *accepted* by $\mathcal{S}$ if there exists an accepting run of $\mathcal{S}$ on p.
The $+\omega$-picture language *recognized* by $\mathcal{S}$ is the set $L^{+\omega}(\mathcal{S})$ of all $+\omega$-pictures p over Σ that are accepted by $\mathcal{S}$.

(c) A $+\omega$-picture language L over Σ is *Büchi-tiling recognizable* if there is a Büchi-tiling system $\mathcal{S} = (\Sigma, Q, \Theta, F)$ with $L = L^{+\omega}(\mathcal{S})$.

The interested reader may want to consider a variant of Büchi-tiling systems where the acceptance condition does not only refer to the first row, but to the entire run — i.e., a variant where a run is called accepting iff there is an accepting state that occurs infinitely often in the run. It is not difficult to transform this variant into our version of Büchi-tiling systems, and vice versa (in fact, this an easy consequence of our logical characterization provided in Theorem 11).

A simple counting argument shows that Büchi-tiling systems are strictly stronger than tiling systems:

Proposition 4. *Let $\Sigma := \{a, b\}$ consist of two distinct letters.*
The $+\omega$-picture language L over Σ which consists of all $p \in \Sigma^{1\omega}$ that contain an infinite number of $a's$, is Büchi-tiling recognizable, but not tiling recognizable.

Proof. The Büchi-tiling recognizability of L is straightforward.

Assume for contradiction that L is also tiling recognizable. I.e., assume that $\mathcal{T} = (\Sigma, Q, \Theta)$ is a tiling system with $L = L^{+\omega}(\mathcal{T})$. Let $n := |Q| + 1$ and let p be the $+\omega$-picture of height 1 that corresponds to the ω-word $(b^n a)^\omega$.

Since $p \in L$ and $L = L^{+\omega}(\mathcal{T})$, there exists a run r of $\mathcal{T}$ on p. Consider the 1ω-picture $q := (p \times r)$. By our choice of n, there must be two columns $j, j' \in [n]$ with $j < j'$ such that $q_{1j} = q_{1j'}$. Now consider the 1ω-pictures $\tilde{p} := p[1, j] \oplus p[j+1, j']^{\oplus\omega}$ and $\tilde{r} := r[1, j] \oplus r[j+1, j']^{\oplus\omega}$. It is straightforward to check that $\tilde{r}$ is a run of $\mathcal{T}$ on $\tilde{p}$. Hence, $\tilde{p} \in L^{+\omega}(\mathcal{T})$. However, $\tilde{p}$ does not contain any a and therefore $\tilde{p} \notin L$. A contradiction! $\qquad\qquad\square$

Büchi-tiling systems and the Büchi-tiling recognizable +ω-picture languages can be viewed as generalizations of Büchi-automata and the ω-regular languages. Recall that a Büchi-automaton $\mathcal{B} = (\Sigma, Q, \Delta, q_0, F)$ consists of the same components as a conventional non-deterministic finite automaton with transition relation $\Delta \subseteq Q \times \Sigma \times Q$. A run r of $\mathcal{B}$ on an ω-word $w \in \Sigma^\omega$ is *accepting* if it visits at least one of the states in F infinitely often. The ω-language *recognized* by $\mathcal{B}$ is the set $L^\omega(\mathcal{B})$ of all ω-words $w \in \Sigma^\omega$ on which $\mathcal{B}$ has an accepting run.

We identify ω-words $w \in \Sigma^\omega$ with 1ω-pictures over Σ, and we identify languages $L \subseteq \Sigma^\omega$ of ω-words with +ω-picture languages that contain pictures of height 1 only. It is straightforward to see that for languages of +ω-pictures of height 1, recognizability by Büchi-tiling systems is equivalent to recognizability by Büchi-automata.

It is well-known that Büchi-automata are equivalent to *generalized Büchi-automata*, i.e., Büchi-automata where the acceptance condition F is replaced by an acceptance condition of the form $\{F_1, \ldots, F_k\}$ with $k \in \mathbb{N}_{\geq 1}$ and $F_i \subseteq Q$ for every $i \in [k]$. A run r of such a generalized Büchi-automaton on an ω-word w is called *accepting* if for each $i \in [k]$ at least one of the states of F_i occurs infinitely often in r. We use the same generalization for Büchi-tiling systems.

Definition 5 (Generalized Büchi-tiling system). A *generalized Büchi-tiling system* is a 4-tuple $\mathcal{S} = (\Sigma, Q, \Theta, \tilde{F})$, where (Σ, Q, Θ) is a tiling system, and $\tilde{F} = \{F_1, \ldots, F_k\}$ for a $k \in \mathbb{N}_{\geq 1}$ and sets $F_1, \ldots, F_k \subseteq Q$. The set $\tilde{F}$ is called the *acceptance condition*.

A *run* of $\mathcal{S}$ on a +ω-picture p over Σ is a run of the tiling system $\mathcal{T} = (\Sigma, Q, \Theta)$ on p. A run r is *accepting* if $\inf_1(r) \cap F_i \neq \emptyset$ for *every* $i \in [k]$.

A +ω-picture w over Σ is *accepted* by $\mathcal{S}$ if there exists an accepting run of $\mathcal{S}$ on p. The +ω-picture language *recognized* by $\mathcal{S}$ is the set $L^{+\omega}(\mathcal{S})$ of all +ω-pictures over Σ that are accepted by $\mathcal{S}$.

For translating a generalized Büchi-automaton $\mathcal{B} = (\Sigma, Q, \Delta, q_0, \{F_1, \ldots, F_k\})$ into equivalent Büchi-automaton $\mathcal{B} = (\Sigma, Q', \Delta', q_0', F')$, one uses the well-known *counting construction*: it suffices to choose $Q' := [k] \times Q$, $q_0' := (1, q_0)$, and $F' := [1] \times F_1$, and to let Δ' be the set consisting of all transitions of the form $((i, q), a, (j, q'))$, where the following is true: $(q, a, q') \in \Delta$, and $j \equiv i+1 \mod k$ if $q \in F_i$, and $j = i$ otherwise. This construction can easily be adapted to obtain:

Proposition 6. *Let Σ be an alphabet and let $L \subseteq \Sigma^{+\omega}$.*
L is Büchi-tiling recognizable if, and only if, there is a generalized Büchi-tiling system $\mathcal{S}$ with $L = L^{+\omega}(\mathcal{S})$.

It is an easy (but somewhat tedious) exercise to show:

Lemma 7. *Let Σ be an alphabet and let $L \subseteq \Sigma^{++}$ be a tiling recognizable language of finite pictures over Σ. Then, the following is true:*

(a) The $+\omega$-picture language $L^{\oplus\omega}$ is Büchi-tiling recognizable.
(b) For every Büchi-tiling recognizable $+\omega$-picture language $L' \subseteq \Sigma^{+\omega}$, the $+\omega$-picture language $L \oplus L'$ is Büchi-tiling recognizable.

Techniques known for tiling systems over finite pictures (see [9,10]) can easily be adapted to show that the class of Büchi-tiling recognizable $+\omega$-picture languages has the same closure properties as the class of tiling recognizable languages of finite pictures:

Proposition 8. *Let Σ be an alphabet of size $|\Sigma| \geqslant 2$.*
The family of Büchi-tiling recognizable $+\omega$-picture languages over Σ is closed under projection, union, and intersection, but not under complementation.

An example for a $+\omega$-picture language that witnesses the non-closure under complementation is the following language. Let $\Sigma = \{a, b\}$ consist of two distinct letters. Let $L_1 \subseteq \Sigma^{++}$ be the language of all *finite* pictures of the form $s \oplus s$ for all $s \in \bigcup_{m \in \mathbb{N}_{\geqslant 1}} \Sigma^{m,m}$, and let $L_{all} := \Sigma^{\omega+}$ be the language of *all* $+\omega$-pictures over Σ. From [10] we know that the $++$-picture language L_1 is *not* tiling recognizable, while its complement $\Sigma^{++} \setminus L_1$ is tiling recognizable. An easy adaptation of their proof shows that the $+\omega$-picture language $L := L_1 \oplus L_{all}$ is *not* Büchi-tiling recognizable, while its complement $\Sigma^{+\omega} \setminus L$ *is* Büchi-tiling recognizable.

4   A Logical Characterization of the Büchi-Tiling Recognizable $+\omega$-Picture Languages

The well-known Büchi-Elgot-Trakhtenbrot Theorem establishes a bridge between logic and automata by showing that a language L of finite words is regular if, and only if, it is definable in monadic second-order logic MSO, and that MSO-definability of L coincides with definability of L in existential monadic second-order logic EMSO. This result has been extended to various structures including ω-words and finite and infinite trees (for an overview, see [19]).

As the class of tiling recognizable picture languages is not closed under complementation, but MSO is closed under negation, a characterization of the tiling recognizable picture languages by MSO is not conceivable. A characterization by EMSO, however, has been obtained in [10]: the tiling recognizable $++$-picture languages are exactly the $++$-picture languages that are definable in EMSO. For this, the authors of [10] use the signature $\tau_\Sigma = \{S_1, S_2\} \cup \{P_a : a \in \Sigma\}$ which consists of two binary relation symbols S_1 and S_2 and a unary relation symbol P_a for every letter $a \in \Sigma$. A finite picture p of size (m, n) over Σ is represented by a finite relational structure $\underline{p}$ of signature τ_Σ as follows:
$$\underline{p} := \big(\mathrm{dom}(p), S_1^p, S_2^p, (P_a^p)_{a \in \Sigma} \big), \text{ where}$$

- the domain dom(p) := $[m] \times [n]$ consists of all "positions" of p,
- for every $a \in \Sigma$, the relation P_a^p consists of all positions $(i,j) \in \text{dom}(p)$ with $p_{ij} = a$ (i.e., a is the letter in row i and column j of p),
- S_1^p is the vertical successor relation on the positions of p, i.e., it consists of all tuples of positions in dom(p) of the form $\big((i,j), (i{+}1,j)\big)$,
- S_2^p is the horizontal successor relation on the positions of p, i.e., it consists of all tuples of positions in dom(p) the form $\big((i,j), (i,j{+}1)\big)$.

We use the same representation for +ω-pictures p, where the domain dom(p) of a +ω-picture of height m is defined as dom(p) := $[m] \times \mathbb{N}_{\geqslant 1}$.

For describing picture languages by logical formulas, we use a countably infinite set Var_i of so-called *individual variables* and a countably infinite set Var_s of so-called *set variables*. Individual variables will always be interpreted with positions of a picture (i.e., with elements in dom(p)), while set variables will be interpreted with sets of positions of a picture (i.e., with subsets of dom(p)). We will use letters like $x, y, z, x_1, x_2, \ldots$ to denote individual variables, and we will use letters like $X, Y, Z, X_1, X_2, \ldots$ to denote set variables. The set FO[τ_Σ] of all *first-order formulas* of signature τ_Σ is inductively defined as follows:

FO[τ_Σ] contains all *atomic formulas* of the form $x{=}y$, $P_a(x)$, $X(x)$, $S_1(x,y)$, and $S_2(x,y)$, for all individual variables $x, y \in \text{Var}_i$, all letters $a \in \Sigma$, and all set variables $X \in \text{Var}_s$. The intended meaning of these formulas is "x and y are interpreted by the same position", "the letter at position x is a", "position x belongs to the set X", "position y is the vertical successor of position x" (same column, next row), and "position y is the horizontal successor of position x" (same row, next column). FO[τ_Σ] is closed under Boolean combinations, i.e., whenever φ and ψ belong to FO[τ_Σ], then FO[τ_Σ] also contains the formulas $\neg\varphi$, $(\varphi \wedge \psi)$, $(\varphi \vee \psi)$, $(\varphi \rightarrow \psi)$, and $(\varphi \leftrightarrow \psi)$. FO[$\tau_\Sigma$] is closed under existential and universal quantification of individual variables, i.e., whenever φ belongs to FO[τ_Σ] and $x \in \text{Var}_i$, then FO[τ_Σ] also contains the formulas $\exists x \, \varphi$ and $\forall x \, \varphi$. The intended meaning of these formulas is "there exists a position x such that the statement made by φ is true" and "for all positions x, the statement made by φ is true".

The set *free*(φ) of all *free variables* of φ consists of all set variables occurring in φ and all individual variables x that have at least one free occurrence in φ, i.e., an occurrence that is not within the range of a quantifier of the form $\exists x$ or $\forall x$. If *free*(φ) $\subseteq \{X_1, \ldots, X_k, x_1, \ldots, x_\ell\}$, p is a (finite or +ω-)picture over Σ, $A_1, \ldots, A_k$ are subsets of dom(p), and $a_1, \ldots, a_\ell$ are elements in dom(p), we write $(\underline{p}, A_1, \ldots, A_k, a_1, \ldots, a_\ell) \models \varphi$ to indicate that the statement made by φ is true in $\underline{p}$ when interpreting the set variable X_i with the set A_i and interpreting the free occurrences of x_j with the position a_j, for all $i \in [k]$, $j \in [\ell]$. We often abbreviate sequences $A_1, \ldots, A_k$ and $a_1, \ldots, a_\ell$ by $\overline{A}$ and $\overline{a}$.

The set EMSO[τ_Σ] of *existential monadic second-order formulas* of signature τ_Σ consists of all formulas Φ of the form $\exists X_1 \cdots \exists X_k \, \varphi$, where $k \geqslant 0$ and $\varphi \in \text{FO}[\tau_\Sigma]$. The set of free variables of Φ is *free*(Φ) := *free*(φ) $\setminus \{X_1, \ldots, X_k\}$. If *free*($\Phi$) $\subseteq \{X_{k+1}, \ldots, X_{k+k'}, x_1, \ldots, x_\ell\}$, p is a (finite or +ω-)picture over Σ, $\overline{A} = A_{k+1}, \ldots, A_{k+k'} \subseteq \text{dom}(p)$, and $\overline{a} = a_1, \ldots, a_\ell \in \text{dom}(p)$, then $(\underline{p}, \overline{A}, \overline{a})$

satisfies Φ (in symbols: $(\underline{p}, \overline{A}, \overline{a}) \models \Phi$) if there exist sets $A_1, \ldots, A_k \subseteq \text{dom}(p)$ such that $(\underline{p}, A_1, \ldots, A_k, \overline{A}, \overline{a}) \models \varphi$. *Sentences* are formulas Φ with *free*$(\Phi) = \emptyset$. For a *sentence* Φ we write $\underline{p} \models \Phi$ instead of $(\underline{p}) \models \Phi$.

Definition 9 (EMSO-definable picture language).
Let Σ be an alphabet and let • be one of the symbols $+$ or ω.
The $+$•-picture language *defined* by an EMSO$[\tau_\Sigma]$-sentence Φ is $L^{+\bullet}(\Phi) := \{p \in \Sigma^{+\bullet} : \underline{p} \models \Phi\}$. Let $\mathcal{L} \subseteq \text{EMSO}[\tau_\Sigma]$. A $+$•-picture language $L \subseteq \Sigma^{+\bullet}$ is $\mathcal{L}$*-definable* if there exists a sentence $\Phi \in \mathcal{L}$ such that $L = L^{+\bullet}(\Phi)$.

Giammarresi et al. [10] have shown that a language of *finite* pictures is tiling recognizable if, and only if, it is EMSO$[\tau_\Sigma]$-definable. Their characterization does *not* carry over to languages of $+\omega$-pictures:

Proposition 10. *Let* $\Sigma := \{a, b\}$ *consist of two distinct letters.*
Let $L := \{p \in \Sigma^{1\omega} : p$ *contains at least one occurrence of the letter* $a\}$, *and*
let $L' := \{p \in \Sigma^{1\omega} : p$ *contains infinitely many* $a's\}$.
L is FO$[\tau_\Sigma]$-definable, but not tiling recognizable.
L' is Büchi-tiling recognizable, but not EMSO$[\tau_\Sigma]$-definable.

Proof (idea). L is defined by the FO$[\tau_\Sigma]$-sentence $\exists x \, P_a(x)$. The proof that L is not tiling recognizable can be taken verbatim from the proof of Proposition 4. The Büchi-tiling recognizability of L' was already observed in Proposition 4. For proving that L' is not EMSO$[\tau_\Sigma]$-definable, one can use a standard tool from mathematical logic: a Hanf-locality argument (cf., e.g., [4]). $\square$

To obtain a logical characterization of the Büchi-tiling recognizable $\Sigma^{+\omega}$-picture languages, we extend EMSO by quantifiers of the form $\exists^\infty X$, for set variables $X \in \text{Var}_s$, with the intended meaning "there exists an *infinite* set X". We write EMSO$^\infty[\tau_\Sigma]$ for the set of all formulas Ψ of the form $\exists^\infty X_1 \cdots \exists^\infty X_k \, \Phi$ where $k \geq 0$, $X_1, \ldots, X_k \in \text{Var}_s$, and $\Phi \in \text{EMSO}[\tau_\Sigma]$. The set of free variables of Ψ is *free*$(\Psi) := $ *free*$(\Phi) \backslash \{X_1, \ldots, X_k\}$. If *free*$(\Psi) \subseteq \{X_{k+1}, \ldots, X_{k+k'}, x_1, \ldots, x_\ell\}$, p is a $+\omega$-picture over Σ, $\overline{A} = A_{k+1}, \ldots, A_{k+k'} \subseteq \text{dom}(p)$, and $\overline{a} = a_1, \ldots, a_\ell \in \text{dom}(p)$, then $(\underline{p}, \overline{A}, \overline{a})$ *satisfies* Ψ (in symbols: $(\underline{p}, \overline{A}, \overline{a}) \models \Psi$) if there exist *infinite* sets $A_1, \ldots, A_k \subseteq \text{dom}(p)$ such that $(\underline{p}, A_1, \ldots, A_k, \overline{A}, \overline{a}) \models \Phi$.

It is not difficult to see that EMSO$^\infty[\tau_\Sigma]$ is expressive enough to describe all Büchi-tiling recognizable $+\omega$-picture languages. For the opposite direction, we follow the overall approach of Giammarresi et al. [10]. The main step is to translate a given FO$[\tau_\Sigma]$-formula $\varphi(X_1, \ldots, X_k)$ (with free set variables $X_1, \ldots, X_k$) into a generalized Büchi-tiling system over the extended alphabet $\Sigma \times \{0, 1\}^k$ (a position that carries a letter $(a, (\alpha_1, \ldots, \alpha_k))$ of this extended alphabet corresponds to a position that carries the letter $a \in \Sigma$ and, for each $i \in [k]$, belongs to the set X_i iff $\alpha_i = 1$). Afterwards, we lift the translation so that it applies also to EMSO$^\infty[\tau_\Sigma]$-sentences. Due to the equivalence of generalized Büchi-tiling systems and Büchi-tiling systems, we then obtain the following:

Theorem 11. *Let* Σ *be an alphabet and let* $L \subseteq \Sigma^{+\omega}$.
L is Büchi-tiling recognizable if, and only if, L is EMSO$^\infty[\tau_\Sigma]$-definable.

5 No Büchi Characterization Theorem For +ω-Picture Languages

It is well-known (see e.g., [19]) that the ω-regular word-languages are exactly the languages of ω-words that are unions of finitely many ω-languages of the form $L_1 \cdot L_2^\omega$, where L_1, L_2 are regular languages of finite words. It is tempting to conjecture that the same holds true for Büchi-tiling recognizable languages of +ω-pictures. Indeed, by Lemma 7 and Proposition 8 we obtain the "easy direction" of the characterization theorem: If L is a +ω-picture language that is the union of a finite number of sets of the form $L_1 \oplus L_2^{\oplus\omega}$, where $L_1, L_2 \subseteq \Sigma^{++}$ are tiling recognizable sets of finite pictures, then L is Büchi-tiling recognizable. The opposite direction, however, is not true — even if we drop the requirement that L_1 and L_2 are tiling recognizable:

Theorem 12. *Let Σ be an alphabet with $|\Sigma| \geqslant 2$.*
For every $m \in \mathbb{N}_{\geqslant 1}$ let L_m be the language consisting of all +ω-pictures over Σ that are of the form $s_1 \oplus s_2 \oplus s_3 \oplus \cdots$ where $s_\nu \in \Sigma^{m,m}$ for every $\nu \in \mathbb{N}_{\geqslant 1}$ and $s_\nu \neq s_1$ for infinitely many $\nu \in \mathbb{N}_{\geqslant 1}$. Then, the +ω-picture language $L := \bigcup_{m \in \mathbb{N}_{\geqslant 1}} L_m$ is Büchi-tiling recognizable, but not equal to any union of a finite number of sets of the form $L_1 \oplus L_2^{\oplus\omega}$ with $L_1, L_2 \subseteq \Sigma^{++}$.

Proof (sketch). For proving the first statement, we show that L is $\mathrm{EMSO}^\infty[\tau_\Sigma]$-definable and then use Theorem 11. The essential idea for constructing the $\mathrm{EMSO}^\infty[\tau_\Sigma]$-formula is to "guess" a position $z = (i,j)$ of s_1 such that for infinitely many ν the letter of s_ν at position (i,j) is different from the letter of s_1 at position (i,j). To do this, we use a quantifier $\exists^\infty Z$ for the set of positions (i,j) in s_ν for the suitable ν. To make sure that z indeed belongs to s_1, we use further existential quantifiers X_1, X_2, X_3 with the intended meaning that X_1 consists of all positions on the diagonal of s_1, X_2 consists of all positions in the rightmost column of s_1, and X_3 consists of all positions of s_1. To make sure that the set Z only contains positions in row i and in columns of the form $m \cdot k + j$, for $k \geqslant 1$, we use additional existential quantifiers Y_r, Y_c, Z_c, Z_d with the intended meaning that Y_r consists of all positions in the same row as z (i.e., row i), Y_c consists of all positions in the same column as z (i.e., column j), Z_c consists of all positions in columns $j + km$ for all $k \geqslant 0$ (where m is the height of the considered picture), and Z_d consists of all positions in the particular diagonals that start at positions directly to the right of top-row positions in Z_c and always proceed from one position to the one in the next row and next column. It is not difficult (but somewhat tedious) to construct an $\mathrm{EMSO}^\infty[\tau_\Sigma]$-formula with the intended meaning.

For proving the second statement, assume for contradiction that $L = \bigcup_{\kappa=1}^k L_{\kappa 1} \oplus L_{\kappa 2}^{\oplus\omega}$, where $k \in \mathbb{N}_{\geqslant 1}$, and $L_{\kappa j} \in \Sigma^{++}$ for every $\kappa \in [k]$ and $j \in \{1, 2\}$. Using a combinatorial argument, we can show the following:

Claim 13. *Let $m \in \mathbb{N}_{\geqslant 1}$ and let $s, t \in \Sigma^{m,m}$ with $s \neq t$.*
There exist numbers $\kappa \in [k]$ and $r \in [m]$ such that $\left(t[r, m] \oplus t^{\oplus\omega} \right) \in L_{\kappa 2}^{\oplus\omega}$.

Furthermore, there exist numbers $\ell, n \in \mathbb{N}_{\geqslant 1}$ *and a finite picture* u *of height* m *such that the picture* $(s \oplus u)$ *belongs to* $L_{\kappa 1} \oplus L_{\kappa 2}^{\oplus \ell}$ *and has width* $m \cdot n + (r - 1)$.

For each picture $t \in \Sigma^{m,m}$ let $K(t)$ be the set of all $(\kappa, r) \in [k] \times [m]$ such that $(t[r, m] \oplus t^{\oplus \omega}) \in L_{\kappa 2}^{\oplus \omega}$. Claim 13 implies for all $t \in \Sigma^{m,m}$ that $K(t) \neq \emptyset$ (note that here we use that $|\Sigma| \geqslant 2$).

Let us now choose a number $m \in \mathbb{N}_{\geqslant 1}$ such that $|\Sigma|^{(m^2)} > 2^{k \cdot m}$. The pigeon hole principle shows that there are pictures $t_1, t_2 \in \Sigma^{m,m}$ with $t_1 \neq t_2$ and $K(t_1) = K(t_2)$.

Now apply Claim 13 for $s := t_1$ and $t := t_2$. This yields numbers $(\kappa, r) \in K(t_2)$, a finite picture u of height m, and numbers $\ell, n \in \mathbb{N}_{\geqslant 1}$ such that the picture $(t_1 \oplus u)$ belongs to $L_{\kappa 1} \oplus L_{\kappa 2}^{\oplus \ell}$ and has width $m \cdot n + (r - 1)$.
Since $(\kappa, r) \in K(t_2) = K(t_1)$, we know that $(t_1[r, m] \oplus t_1^{\oplus \omega}) \in L_{\kappa 2}^{\oplus \omega}$.
Hence, also the $+\omega$-picture $p := (t_1 \oplus u) \oplus t_1[r, m] \oplus t_1^{\oplus \omega}$ belongs to $L_{\kappa 1} \oplus L_{\kappa 2}^{\oplus \omega}$. However, $p \notin L$. A contradiction! □

6   Conclusion

We introduced Büchi-tiling systems and Büchi-tiling recognizable $+\omega$-languages. We showed that the class of all Büchi-tiling recognizable $+\omega$-picture languages has the same closure properties as the class of tiling recognizable languages of finite pictures: it is closed under projection, union, and intersection, but not under complementation (see Proposition 8).

While for languages of *finite* pictures, tiling recognizability coincides with EMSO$[\tau_\Sigma]$-definability [10], the situation is quite different for languages of $+\omega$-pictures: In this setting, the notion of tiling recognizability does not even cover the language of all $+\omega$-pictures over $\Sigma = \{a, b\}$ in which the letter a occurs at least once — a picture-language that can easily be defined in first-order logic FO$[\tau_\Sigma]$. As a consequence, EMSO$[\tau_\Sigma]$ is too strong for capturing the class of tiling recognizable $+\omega$-picture languages. On the other hand, EMSO$[\tau_\Sigma]$ is too weak for capturing the class of all *Büchi*-tiling recognizable $+\omega$-picture languages (see Proposition 10). To obtain a logical characterization of this class, we introduced the logic EMSO$^\infty$, which extends EMSO with existential quantification of *infinite* sets. Our main characterization results are summarized in the following theorem.

Theorem 14. *Let* Σ *be an alphabet and let* $L \subseteq \Sigma^{+\omega}$.
The following are equivalent:

(a) L is Büchi-tiling recognizable,
(b) $L = L^{+\omega}(\mathcal{S})$ for a generalized Büchi-tiling system $\mathcal{S}$,
(c) L is EMSO$^\infty[\tau_\Sigma]$-definable.

The equivalence of (a) and (b) is provided by Proposition 6, equivalence of (a) and (c) is provided by Theorem 11. Using combinatorial arguments, we showed that the Büchi characterization theorem for ω-regular languages does not carry over to the Büchi-tiling recognizable $+\omega$-picture languages (see Theorem 12).

Concerning future work, a generalization of our results to the quantitative setting would be interesting. Recently, in [2], the equivalence of a quantitative automaton model over finite pictures and a fragment of quantitative monadic second-order logic has been studied. Can these results be extended to the setting of $+\omega$-picture languages?

References

1. Altenbernd, J.-H., Thomas, W., Wöhrle, S.: Tiling systems over infinite pictures and their acceptance conditions. In: Ito, M., Toyama, M. (eds.) DLT 2002. LNCS, vol. 2450, pp. 297–306. Springer, Heidelberg (2003)
2. Babari, P., Droste, M.: A Nivat theorem for weighted picture automata and weighted MSO logic. In: Dediu, A.-H., Formenti, E., Martín-Vide, C., Truthe, B. (eds.) LATA 2015. LNCS, vol. 8977, pp. 703–715. Springer, Heidelberg (2015)
3. Dare, V.R., Subramanian, K.G., Thomas, D.G., Siromoney, R., Saec, B.L.: Infinite arrays and recognizability. Int. J. Pattern Recognit. Artif. Intell. **14**(4), 525–536 (2000)
4. Ebbinghaus, H.-D., Flum, J.: Finite Model Theory. Springer Monographs in Mathematics. Springer, Heidelberg (1995)
5. Finkel, O.: On recognizable languages of infinite pictures. Int. J. Found. Comput. Sci. **15**(6), 823–840 (2004)
6. Fu, K.S.: Syntactic Methods in Pattern Recognition. Academic Press, New York (1974)
7. Giammarresi, D., Restivo, A.: Recognizable picture languages. Int. J. Pattern Recognit. Artif. Intell. **6**(2&3), 241–256 (1992)
8. Giammarresi, D., Restivo, A.: Two-dimensional finite state recognizability. Fundam. Inform. **25**(3), 399–422 (1996)
9. Giammarresi, D., Restivo, A.: Two-dimensional languages. In: Rozenberg, G., Salomaa, A. (eds.) Handbook of Formal Languages, vol. 3, pp. 215–267. Springer, Heidelberg (1997)
10. Giammarresi, D., Restivo, A., Seibert, S., Thomas, W.: Monadic second-order logic over rectangular pictures and recognizability by tiling systems. Inf. Comput. **125**(1), 32–45 (1996)
11. Gnanasekaran, S., Dare, V.R.: Infinite arrays and domino systems. Electron. Notes Discrete Math. **12**, 349–359 (2003)
12. Gnanasekaran, S., Dare, V.R.: On recognizable infinite array languages. In: Klette, R., Žunić, J. (eds.) IWCIA 2004. LNCS, vol. 3322, pp. 209–218. Springer, Heidelberg (2004)
13. Inoue, K., Takanami, I.: A survey of two-dimensional automata theory. Inf. Sci. **55**(1–3), 99–121 (1991)
14. Inoue, K., Takanami, I.: A characterization of recognizable picture languages. In: 2nd International Conference Parallel Image Analysis, ICPIA 1992, pp. 133–143 (1992)
15. Latteux, M., Simplot, D.: Recognizable picture languages and domino tiling. Theor. Comput. Sci. **178**(1–2), 275–283 (1997)
16. Lindgren, K., Moore, C., Nordahl, M.: Complexity of two-dimensional patterns. J. Stat. Phys. **91**(5–6), 909–951 (1998)
17. Smith, A.R.: Two-dimensional formal languages and pattern recognition by cellular automata. In: 12th Annual Symposium on Switching and Automata Theory, FOCS 1971, East Lansing, Michigan, USA, October 13–15, pp. 144–152 (1971)

18. Thomas, W.: On logics, tilings, and automata. In: Albert, J.L., Monien, B., Artalejo, M.R. (eds.) Automata, Languages and Programming. LNCS, vol. 510, pp. 441–454. Springer, Heidelberg (1991)
19. Thomas, W.: Languages, automata, and logic. In: Rozenberg, G., Salomaa, A. (eds.) Handbook of Formal Languages, vol. 3, pp. 389–455. Springer, Heidelberg (1997)

A Logical Characterization for Dense-Time Visibly Pushdown Automata

Devendra Bhave[1]([✉]), Vrunda Dave[1], Shankara Narayanan Krishna[1], Ramchandra Phawade[1], and Ashutosh Trivedi[1,2]

[1] Indian Institute of Technology Bombay, Mumbai, India
devendra@cse.iitb.ac.in
[2] University of Colorado Boulder, Boulder, USA

Abstract. Two of the most celebrated results that effectively exploit visual representation to give logical characterization and decidable model-checking include visibly pushdown automata (VPA) by Alur and Madhusudan and event-clock automata (ECA) by Alur, Fix and Henzinger. VPA and ECA—by making the call-return edges visible and by making the clock-reset operation visible, respectively—recover decidability for the verification problem for pushdown automata implementation against visibly pushdown automata specification and timed automata implementation against event-clock timed automata specification, respectively. In this work we combine and extend these two works to introduce dense-time visibly pushdown automata that make both the call-return as well as resets visible. We present MSO logic characterization of these automata and prove the decidability of the emptiness problem for these automata paving way for verification problem for dense-timed pushdown automata against dense-time visibly pushdown automata specification.

Keywords: Visibly pushdown · Event-clock · Logical characterization

1 Introduction

Timed automata [2] are simple yet powerful generalization of finite automata where a finite set of continuous variables with uniform rates, aptly named clocks, are used to measure critical timing constraints among various events by permitting reset of these clocks to remember occurrence of an event. Due to the carefully crafted dynamics, the emptiness of timed automata is a decidable problem using a technique known as region-construction that computes their time-abstract finitary bisimulation. Timed automata are closed under union and intersection, but not under complementation and determinization, which makes it impossible to verify timed automata implementation against timed automata specifications.

Event-clock automata [3] are a determinizable subclass of timed automata that enjoy a nice set of closure properties: they are closed under union, intersection, complementation, and determinization. Event-clock automata achieve

A.-H. Dediu et al. (Eds.): LATA 2016, LNCS 9618, pp. 89–101, 2016.
DOI: 10.1007/978-3-319-30000-9_7

the closure under determinization by making clock resets visible—the reset of each clock variable is determined by a fixed class of event and hence visible just by looking at the input word. Partially thanks to these closure properties and closure under projection of certain "quasi" subclass, they are known to be precisely capture timed languages defined by an appropriate class of monadic second-order logic [7].

Recursive timed automata (RTA) [9] and dense-time pushdown automata (dtPDA) [1] are generalization of timed automata that accept certain real-time extensions of context-free languages. In general, the emptiness problem for the RTA in undecidable, however [9] characterizes classes of RTA with decidable emptiness problem. The emptiness problem for the dtPDA is known to be decidable. RTA and dtPDA naturally model the flow of control in time-critical software systems with potentially recursive procedure calls. Alur and Madhusudan [4] argued the need for context-free (representable using pushdown automata) specification while verifying systems modeled as pushdown systems. The goal of this paper is to develop decidable verification framework for RTA and dtPDA by introducing an appropriate class of specification formalism for context-free and time-critical properties that permit decidable verification.

Already for untimed pushdown automata it is not feasible to verify against general context-free specification, since pushdown automata are not closed under determinization and complementation. Alur and Madhusudan [4] introduced *visibly pushdown automata* as a specification formalism where the call and return edges are made visible in a structure of the word. This visibility enabled closure of these automata under determinization and hence complementation, and allowed them to be used in a decidable verification framework. Also, again owing to these closure properties, visibly pushdown automata are known to precisely capture the context-free languages definable by an appropriate class of monadic second order (MSO) logic [4].

In this paper we present dense-time visibly pushdown automata (dtVPA) that form a subclass of dense-time pushdown automata of Abdulla, Atig, and Stenman [1] and generalize both visibly pushdown automata and event-clock automata. We show that dtVPA are determinizable, closed under Boolean operations (union, intersection, and complementation), and a subclass is closed under projection. We build on these closure properties to give a logical characterization of the timed languages captured by dtVPA.

Related Work. Tang and Ogawa in [10] proposed a model called *event-clock visibly pushdown automata* (ECVPA) that generalized both ECA and VPA. For the proposed model they showed determinizability as well as closure under boolean operations, and proved the decidability of the verification problem for timed visibly pushdown automata against such event-clock visibly pushdown automata specifications. However, unlike dtVPAs, ECVPAs do not permit pushing the clocks on the stack and hence dtVPA capture a larger specification class than ECVPA. Moreover [10] did not explore any logical characterization of ECVPA. Our paper builds upon the ideas presented in D'Souza [7] for event-clock automata and Alur and Madhusudan [4] to present a visualized specification framework for dense-time pushdown automata. For the decidability of the emptiness problem, we exploit the recent untiming construction proposed by Clemente and Lasota [6]. For a survey of models related to recursive timed automata and dense-time pushdown automata we refer the reader to [1,9].

2   Preliminaries

We assume that the reader is comfortable with standard concepts from automata theory (such as context-free languages, pushdown automata, MSO logic), concepts from timed automata (such as clocks, event clocks, clock constraints, and valuations), and visibly pushdown automata. Due to space limitation, we only give a very brief introduction of required concepts in this section, and for a detailed background on these concepts we refer the reader to [2–4,7].

A finite timed word over Σ is a sequence $(a_1, t_1), (a_2, t_2), \ldots, (a_n, t_n) \in (\Sigma \times \mathbb{R}_{\geq 0})^*$ such that $t_i \leq t_{i+1}$ for all $1 \leq i \leq n - 1$. Alternatively, we can represent timed words as tuple $(\langle a_1, \ldots, a_n \rangle, \langle t_1, \ldots, t_n \rangle)$. We use both of these formats depending on technical convenience. We represent the set of finite timed words over Σ by $T\Sigma^*$. Before we introduce dtVPA in the next section, let us recall the basic notions of event-clock automata and visibly pushdown automata.

Event-Clock Automata. Event-clock automata (ECA) [3] are a determinizable subclass of timed automata [2] that for every action $a \in \Sigma$ implicitly associate two clocks x_a and y_a, where the "recorder" clock x_a records the time of the last occurrence of action a, and the "predictor" clock y_a predicts the time of the next occurrence of action a. Hence, event-clock automata do not permit explicit reset of clocks and it is implicitly governed by the input timed word. This property makes ECA determinizable and closed under all Boolean operations. However, ECAs are not closed under projection.

In order to develop a logical characterization of ECA D'Souza [7] required a class of ECA that is closed under projections. For this purpose, he introduced an equi-expressive generalization of event-clock automata – called quasi-event clock automata (qECA) – where event recorders and predictors are associated with a set of actions rather than a single action. Here, the finite alphabet Σ is partitioned into finitely many classes via a ranking function $\rho : \Sigma \to \mathbb{N}$ giving rise to finitely many partitions $P_1, \ldots, P_k$ of Σ where $P_i = \{a \in \Sigma \mid \rho(a) = i\}$. The event recorder x_{P_i} records the time elapsed since the last occurrence of some action in P_i, while the event predictor y_{P_i} predicts the time required for any action of P_i to occur.

Notice that since clock resets are "visible" in input timed word, the clock valuations after reading a prefix of the word is also determined by the timed word. For example, for a timed word $w = (a_1, t_1), (a_2, t_2), \ldots, (a_n, t_n)$, the value of the event clock $x_{\rho(a)}$ at position j is $t_j - t_i$ where i is the largest position preceding j where an action of $P_{\rho(a)}$ has occurred. If no symbols from $P_{\rho(a)}$ have occurred before the jth position, then the value of $x_{\rho(a)}$ is undefined denoted by a special symbol $\vdash$. Similarly, he value of $y_{\rho(a)}$ at position j of w is undefined if no symbols of $P_{\rho(a)}$ occur in w after the jth position. Otherwise, it is defined as $t_k - t_j$, where k is the first position after j where a symbol of $P_{\rho(a)}$ occurs. We write C_ρ for the set of all event clocks for a ranking function ρ and we use $\mathbb{R}_{>0}^\vdash$ for the set $\mathbb{R}_{>0} \cup \{\vdash\}$. Formally, the clock valuation after reading j-th prefix of the input timed word w, $\nu_j^w : C_\rho \mapsto \mathbb{R}_{>0}^\vdash$, is defined in the following fashion: $\nu_j^w(x_q) = t_j - t_i$ if there exists an $0 \leq i < j$ such that $\rho(a_i) = q$ and $a_k \notin P_q$ for all $i < k < j$, otherwise $\nu_j^w(x_q) = \vdash$ (undefined). Similarly, $\nu_j^w(y_q) = t_m - t_j$ if there is $j < m$ such that $\rho(a_m) = q$ and $a_l \notin P_q$ for all $j < l < m$, otherwise $\nu_j^w(y_q) = \vdash$.

A quasi-event clock automaton [7] is a tuple $A = (L, \Sigma, \rho, L^0, F, E)$ where L is a set of finite locations, Σ is a finite alphabet, ρ is the alphabet ranking function, $L^0 \in L$ is the set of initial locations, $F \in L$ is the set of final locations, and E is a finite set of edges of the form $(\ell, \ell', a, \varphi)$ where ℓ, ℓ' are locations, $a \in \Sigma$, and φ is a clock constraint over the clocks C_ρ. A clock constraint over C_ρ is a boolean combination of constraints of the form $z \sim c$ where $z \in C_\rho$, $c \in \mathbb{N}$ and $\sim \in \{\leq, \geq\}$. Event clock automata are a special kind of quasi-event clock automata when the ranking function ρ is a one-to-one function.

Quasi event-clock automata and event-clock automata are known to be equi-expressive [3,7]. Quasi event-clock automata are determinizable and closed under Boolean operations, concatenation, Kleene closure, and projection. The language accepted by (quasi) event-clock automata can be characterized by MSO logic over timed words augmented with timed modalities.

Visibly Pushdown Automata. Visibly pushdown automata [4] are a determinizable subclass of pushdown automata that operate over words that dictate the stack operations. This notion is formalized by giving an explicit partition of the alphabet into three disjoint sets of *call*, *return*, and *local* symbols and the visibly pushdown automata must push one symbol to stack while reading a call symbol, and must pop one symbol (given stack is non-empty) while reading a return symbol, and must not touch the stack while reading the local symbol.

A visibly pushdown alphabet is a tuple $\Sigma = \langle \Sigma_c, \Sigma_r, \Sigma_l \rangle$ where Σ is partitioned into a *call* alphabet Σ_c, a *return* alphabet Σ_r, and a *local* alphabet Σ_l. A visibly pushdown automata over $\Sigma = \langle \Sigma_c, \Sigma_r, \Sigma_l \rangle$ is a tuple $(L, \Sigma, \Gamma, L^0, \delta, F)$ where L is a finite set of locations including a set $L^0 \subseteq L$ of initial locations, a finite stack alphabet Γ with special end-of-stack symbol $\perp$, and $\Delta \subseteq (L \times \Sigma_c \times L \times (\Gamma \setminus \perp)) \cup (L \times \Sigma_r \times \Gamma \times L) \cup (L \times \Sigma_l \times L)$ and $F \subseteq L$ is final locations.

Alur and Madhusudan [4] showed that visibly pushdown automata are determinizable and closed under Boolean operations, concatenation, Kleene closure, and projection. They also showed that the language accepted by visibly pushdown automata can be characterized by MSO over words augmented with a binary matching predicate first studied in [8].

3   Dense-Time Visibly Pushdown Automata (dtVPA)

We introduce the dense-time visibly pushdown automata as an event-clock automaton equipped with a timed stack along with visibly pushdown alphabet $\Sigma = \langle \Sigma_c, \Sigma_r, \Sigma_l \rangle$. Due to space limitation and notational convenience, we assume that the partitioning function is one-to-one, i.e. each symbol $a \in \Sigma$ has unique recorder x_a and predictor y_a clocks assigned to it. This permits us to drop the ranking function ρ for the further discussion. However, note that for closure under projection, we require the definition with such ranking function. We refer the reader to [5] for more details.

Syntax and Semantics. Let C_Σ (or C when Σ is clear) be a finite set of event clocks. Let $\Phi(C)$ be the set of *clock constraints* over C and $\mathcal{I}$ be the set of intervals of the form $\langle a, b \rangle$ with $a \in \mathbb{N}$, $a \leq b$ and $b \in \mathbb{N} \cup \{\infty\}$.

Definition 1. *A dense-time visibly pushdown automata over $\Sigma = \{\Sigma_c, \Sigma_r, \Sigma_l\}$ is a tuple $M = (L, \Sigma, \Gamma, L^0, F, \Delta = \Delta_c \cup \Delta_r \cup \Delta_l)$, where L is a finite set of locations including a set $L^0 \subseteq L$ of initial locations, Γ is a finite stack alphabet with special end-of-stack symbol $\bot$, $\Delta_c = (L \times \Sigma_c \times \Phi(C) \times L \times (\Gamma \backslash \bot))$ is the set of call transitions, $\Delta_r = (L \times \Sigma_r \times \mathcal{I} \times \Gamma \times \Phi(C) \times L)$ is set of return transitions, $\Delta_l = (L \times \Sigma_l \times \Phi(C) \times L)$ is set of local transitions, and $F \subseteq L$ is final locations set.*

Let $w = (a_0, t_0), \ldots, (a_n, t_n)$ be a timed word. A configuration of the dtVPA is a tuple $(\ell, \nu_i^w, (\gamma\sigma, age(\gamma\sigma)))$ where ℓ is the current location of the dtVPA, ν_i^w gives the valuation of all the event clocks at position $i \leq |w|$, $\gamma\sigma \in \Gamma\Gamma^*$ is the content of the stack with γ being the topmost symbol and σ is the string representing the stack content below γ, while $age(\gamma\sigma)$ is a sequence of real numbers encoding the ages of all the stack symbols (the time elapsed since each of them was pushed on to the stack). We follow that assumption that $age(\bot) = \langle \vdash \rangle$ (undefined). If for some string $\sigma \in \Gamma^*$ we have that $age(\sigma) = \langle t_1, t_2, \ldots, t_n \rangle$ and for $\tau \in \mathbb{R}_{\geq 0}$ we write $age(\sigma) + \tau$ for the sequence $\langle t_1 + \tau, t_2 + \tau, \ldots, t_n + \tau \rangle$. For a sequence $\sigma = \langle \gamma_1, \ldots, \gamma_n \rangle$ and a member γ we write $\gamma :: \sigma$ for $\langle \gamma, \gamma_1, \ldots, \gamma_n \rangle$.

The run of a dtVPA on $w = (a_0, t_0), \ldots, (a_n, t_n)$ is a sequence of configurations $(\ell_0, \nu_0^w, ((\langle \bot \rangle, \langle \vdash \rangle))), (\ell_1, \nu_1^w, (\sigma_1, age(\sigma_1))), \ldots, (\ell_{n+1}, \nu_{n+1}^w, (\sigma_{n+1}, age(\sigma_{n+1})))$ where $\ell_i \in L, \sigma_i \in \Gamma \cup \{\bot\}, \ell_0 \in L^0$, and for each i, $0 \leq i \leq n$, we have:

- If $a_i \in \Sigma_c$, then there is a transition $(\ell_i, a_i, \varphi, \ell_{i+1}, \gamma) \in \Delta$ s.t. $\nu_i^w \models \varphi$. The symbol $\gamma \in \Gamma \backslash \{\bot\}$ is then pushed onto the stack, and its age is initialized to zero, obtaining $(\sigma_{i+1}, age(\sigma_{i+1})) = (\gamma :: \sigma_i, 0 :: (age(\sigma_i) + (t_i - t_{i-1})))$. Note that all symbols in the stack excluding the topmost age by $t_i - t_{i-1}$.
- If $a_i \in \Sigma_r$, then there is a transition $(\ell_i, a_i, I, \gamma, \varphi_i, \ell_{i+1}) \in \Delta$. The configuration $(\ell_i, \nu_i, (\sigma_i, age(\sigma_i)))$ evolves to $(\ell_{i+1}, \nu_{i+1}, (\sigma_{i+1}, age(\sigma_{i+1})))$ iff $\nu_i^w \models \varphi_i$, $\sigma_i = \gamma :: \kappa \in \Gamma\Gamma^*$ and $age(\gamma) + (t_i - t_{i-1}) \in I$. Then we obtain $\sigma_{i+1} = \kappa$, with $age(\sigma_{i+1}) = age(\kappa) + (t_i - t_{i-1})$. However, if $\gamma = \langle \bot \rangle$, the symbol is not popped, and the attached interval I is irrelevant.
- If $a_i \in \Sigma_l$, then there is a transition $(\ell_i, a_i, \varphi_i, \ell_{i+1}) \in \Delta$ such that $\nu_i^w \models \varphi_i$. In this case stack remains unchanged i.e. $\sigma_i = \sigma_{i+1}$, and $age(\sigma_{i+1}) = age(\sigma_i) + (t_i - t_{i-1})$. All symbols in the stack age by $t_i - t_{i-1}$.

A run ρ of a dtVPA M is accepting if it terminates in a final location. A timed word w is an accepting word if there is an accepting run of M on w. The language $L(M)$ of a dtVPA M, is the set of all timed words w accepted by M.

Deterministic dtVPA. A dtVPA $M = (L, \Sigma, L^0, F, \Delta)$ is said to be *deterministic* if it has exactly one start location, and for every configuration and input action exactly one transition is enabled. Formally, we have the following conditions: for every $(\ell, a, \phi_1, \ell', \gamma_1), (\ell, a, \phi_2, \ell'', \gamma_2) \in \Delta_c$, $\phi_1 \wedge \phi_2$ is unsatisfiable; for every $(\ell, a, I_1, \gamma, \phi_1, \ell'), (\ell, a, I_2, \gamma, \phi_2, \ell'') \in \Delta_r$, either $\phi_1 \wedge \phi_2$ is unsatisfiable or $I_1 \cap I_2 = \emptyset$; and for every $(\ell, a, \phi_1, \ell'), (\ell, a, \phi_2, \ell') \in \Delta_l$, $\phi_1 \wedge \phi_2$ is unsatisfiable.

Example 2. Consider the timed languages of the form $a^n b c^n d$ where the first c comes precisely 1 time-unit after last a and the first a and the last c are 2 time-units apart, and every other matching a and c are within $(1, 2)$ time-unit apart, i.e. $\big\{ (a^n b c^n d, \langle t_1, \ldots, t_n, t, t'_n, \ldots, t'_1, t' \rangle) \mid t'_n - t_n = 1, t'_1 - t_1 = 2, t'_i - t_i \in (1, 2)$ for all $i \leq n \big\}$. Given a partition $\Sigma_c = \{a\}$, $\Sigma_l = \{b, d\}$, $\Sigma_r = \{c\}$ and $\Gamma = \{\alpha\}$ this language can be accepted by the dtVPA shown below.

Here l_0 is the initial location and l_4 is only accepting location. The transitions relation contains the following transitions: the call transition $(l_0, a, \text{true}, l_0, \alpha) \in \Delta_c$, the local transition $(l_0, b, \text{true}, l_1) \in \Delta_l$ and the following set of return transitions $(l_1, c, [1, 1], \alpha, x_b \leq 1, l_2)$, $(l_2, c, (1, 2), \alpha, \text{true}, l_2)$, $(l_2, c, [2, 2], \alpha, \text{true}, l_3)$, $(l_3, d, \text{true}, \bot, \text{true}, l_4) \in \Delta_r$. In the figure we have omitted clock constraints that are logically true and depicted testing the age of the top symbol as $pop(\cdot) \in I$.

The following is one of the central result of the paper.

Theorem 3 (Determinizability, Emptiness and Closure). *Dense-time visibly pushdown automata are determinizable and closed under Boolean operations, concatenation, Kleene closure and an appropriate extension is closed under projection. Their emptiness is also decidable.*

The proofs for the union, intersection, concatenation, and Kleene closure are straightforward extensions of the closure of visibly pushdown automata and event-clock automata under these operations. The proof for the closure under projection, like [7], uses the extension of the model with the ranking function and the proof details can be found in [5]. The proof for the determinizability (and hence the complementation) is slightly more involved. In the next section we present a proof for the determinizability as well as decidability of the emptiness problem for dtVPA. Section 5 presents a logical characterization of dtVPA.

4   Emptiness and Determinizability

Event-clock visibly-pushdown automata (ECVPA) [10] can be considered as a subclass of dtVPA where the ages are not pushed on the stack. Hence a dtVPA $M = (L, \Sigma, L^0, F, \Delta)$ is an ECVPA if for every $(\ell, a, I, \gamma, \phi, \ell') \in \Delta_r$ we have that $I = [-\infty, +\infty]$. Tang and Ogawa, in [10], proved the following for ECVPA.

Theorem 4. *ECVPAs are determinizable and closed under Boolean operations.*

We now describe the *untiming-the-stack* construction to obtain from a dtVPA M over Σ, an ECVPA M' over an extended alphabet Σ' such that $L(M) = h(L(M'))$ where h is a homomorphism $h : \Sigma' \times \mathbb{R}^{\geq 0} \to \Sigma \times \mathbb{R}^{\geq 0}$ defined as $h(a, t) = (a, t)$ for $a \in \Sigma$ and $h(a, t) = \varepsilon$ for $a \notin \Sigma$. Our construction builds upon that of [6]. However, [6] cannot directly be used here since [6] introduces extra clocks that require resets which is not available under event-clock restriction.

Untiming Construction. We will first sketch the construction informally. In the following we write k for the maximum constant used in the dtVPA M within any interval I to check the age of a popped symbol. Let us first consider a call transition $(l, a, \varphi, l', \gamma)$ encountered in M. To construct an ECVPA M' from M, we guess the interval used in a constraint in return transition when γ will be popped from the stack. Assume the guess is an interval of the form $[0, \kappa)$. This amounts to checking that the age of γ at the time of popping is $< \kappa$. In M', the control switches from l to a special location $(l'_{a, <\kappa}, \{<\kappa\})$, and the symbol $(\gamma, <\kappa, \mathtt{first})$ is pushed onto the stack. Let $Z_{\tilde{k}} = \{\sim c \mid c \in \mathbb{N}, c \leq k, \sim \in \{<, \leq, >, \geq, =\}\}$ and let $\Sigma' = \Sigma \cup Z_{\tilde{k}}$ be the extended alphabet. All symbols of $Z_{\tilde{k}}$ are local symbols in M' i.e. $\Sigma' = \{\Sigma_c, \Sigma_l \cup Z_{\tilde{k}}, \Sigma_r\}$. At location $(l'_{a, <\kappa}, \{<\kappa\})$, the new symbol $<\kappa$ is read and we have the following transition: $((l'_{a, <\kappa}, \{<\kappa\}), <\kappa, x_a = 0, (l', \{<\kappa\}))$, which results in resetting the event recorder $x_{<\kappa}$ corresponding to the new symbol $<\kappa$. The constraint $x_a = 0$ ensures that no time is elapsed by the new transition. The information $<\kappa$ is retained in the control state until $(\gamma, <\kappa, \mathtt{first})$ is popped. At $(l', \{<\kappa\}))$, we continue the simulation of M from l'. Assume that we have another push operation at l' of the form (l', b, ψ, q, β). In M', from $(l', \{<\kappa\})$, we first guess the constraint that will be checked when β will be popped from the stack. If the guessed constraint is again $<\kappa$, then control switches from $(l', \{<\kappa\})$ to $(q, \{<\kappa\})$, and $(\beta, <\kappa, -)$ is pushed onto the stack and simulation continues from $(q, \{<\kappa\})$. However, if the guessed pop constraint is $<\zeta$ for $\zeta \neq \kappa$, then control switches from $(l', \{<\kappa\})$ to $(q_{b, <\zeta}, \{<\kappa, <\zeta\})$. The new obligation $<\zeta$ is also remembered in the control state. From $(q_{b, <\zeta}, \{<\kappa, <\zeta\})$, we read the new symbol $<\zeta$ which resets the event predictor $x_{<\zeta}$ and control switches to $(q, \{<\kappa, <\zeta\})$, pushing $(\beta, <\zeta, \mathtt{first})$ on to the stack. The idea thus is to keep the obligation $<\kappa$ alive in the control state until γ is popped; the value of $x_{<\kappa}$ at the time of the pop determines whether the pop is successful or not. If a further $<\kappa$ constraint is encountered while the obligation $<\kappa$ is already alive, then we do not reset the event clock $x_{<\kappa}$. The $x_{<\kappa}$ is reset only at the next call transition after $(\gamma, <\kappa, \mathtt{first})$ is popped, when $<\kappa$ is again guessed. The case when the guessed popped constraint is of the form $>\kappa$ is similar. In this case, each time the guess is made, we reset the event recorder $x_{>\kappa}$ at the time of the push. If the age of a symbol pushed later is $>\kappa$, so will be the age of a symbol pushed earlier. In this case, the obligation $>\kappa$ is remembered only in the stack. Handling guesses of the form $\geq \zeta \wedge \leq \kappa$ is similar, and we combine the ideas discussed above.

Now consider a return transition $(l, a, I, \gamma, \varphi, l')$ in M. In M', we are at some control state (l, P). On reading a, we check the top of stack symbol in M'.

It is of the form $(\gamma, S, \texttt{first})$ or $(\gamma, S, -)$, where S is either a singleton set of the form $\{<\kappa\}$ or $\{>\zeta\}$, or a set of the form $\{<\kappa, >\zeta\}$. Consider the case when the top of stack symbol is $(\gamma, \{<\kappa, >\zeta\}, \texttt{first})$. In M', on reading a, the control switches from (l, P) to (l', P') for $P' = P\backslash\{<\kappa\}$ iff the guard φ evaluates to true, the interval I is (ζ, κ) (this validates our guess made at the time of push) and the value of clock $x_{<\kappa}$ is $<\kappa$, and the value of clock $x_{>\zeta}$ is $>\zeta$. Note that the third component $\texttt{first}$ says that there are no symbols in the stack below $(\gamma, \{<\kappa, >\zeta\}, \texttt{first})$ whose pop constraint is $<\kappa$. Hence, we can remove the obligation $<\kappa$ from P in the control state. If the top of stack symbol was $(\gamma, \{<\kappa, >\zeta\}, -)$, then we know that the pop constraint $<\kappa$ is still alive. That is, there is some stack symbol below $(\gamma, \{<\kappa, >\zeta\}, -)$ of the form $(\beta, S, \texttt{first})$ such that $<\kappa \in S$. In this case, we keep P unchanged and control switches to (l', P).

We now give the formal construction. Given dtVPA $M = (L, \Sigma, \Gamma, L^0, F, \Delta)$ with max constant k used in return transitions, we construct ECVPA $M' = (L', \Sigma', \Gamma', L'^0, F', \Delta')$ where $L' = (L \times 2^{Z_{\widetilde{k}}}) \cup (L_{\Sigma} \times Z_{\widetilde{k}} \times 2^{Z_{\widetilde{k}}}) \cup (L_{\Sigma} \times Z_{\widetilde{k}} \times Z_{\widetilde{k}} \times 2^{Z_{\widetilde{k}}})$ $\Sigma' = (\Sigma_c, \Sigma_l \cup Z_{\widetilde{k}}, \Sigma_r)$ and $\Gamma' = \Gamma \times 2^{Z_{\widetilde{k}}} \times \{\texttt{first}, -\}$, $L^0 = \{(l^0, \emptyset) \mid l^0 \in L^0\}$, and $F = \{(l^f, \emptyset) \mid l^f \in F\}$. The transitions Δ' are defined as follows. For every $(l, a, \varphi, l', \gamma) \in \Delta_c$, we have the following classes of transitions in M'.

1. The first class of transitions correspond to the guessed pop constraint being $<\kappa$. In the first case, $<\kappa$ is alive, and hence there is no need to reset the clock $x_{<\kappa}$. In the second case, the obligation $<\kappa$ is fresh and hence it is remembered as $\texttt{first}$ in the stack, and the clock $x_{<\kappa}$ is reset.

$$((l, P), a, \varphi, (l', P), (\gamma, \{<\kappa\}, -)) \in \Delta'_c \ \text{ if } <\kappa \in P$$
$$((l, P), a, \varphi, (l'_{a,<\kappa}, P'), (\gamma, \{<\kappa\}, \texttt{first})) \in \Delta'_c \ \text{ if } <\kappa \notin P \text{ and } P' = P \cup \{<\kappa\}$$
$$((l'_{a,<\kappa}, P'), <\kappa, x_a = 0, (l', P')) \in \Delta'_l$$

2. The second class of transitions correspond to the case when the guessed pop constraint is $>\kappa$. The clock $x_{>\kappa}$ is reset, and obligation is stored in stack.

$$((l, P), a, \varphi, (l'_{a,>\kappa}, P), (\gamma, \{>\kappa\}, -)) \in \Delta'_c \text{ and } ((l'_{a,>\kappa}, P), >\kappa, x_a = 0, (l', P)) \in \Delta'_l$$

3. Finally the following transitions consider the case when the guessed pop constraint is $>\zeta$ and $<\kappa$. Depending on whether $<\kappa$ is alive or not, we have two cases. If alive, then we simply reset the clock $x_{>\zeta}$ and remember both the obligations in the stack. If $<\kappa$ is fresh, then we reset both clocks $x_{>\zeta}$ and $x_{<\kappa}$ and remember both obligations in the stack, and $<\kappa$ in the state.

$$((l, P), a, \varphi, (l'_{a,<\kappa,>\zeta}, P'), (\gamma, \{<\kappa, >\zeta\}, \texttt{first})) \in \Delta'_c \ \text{ if } <\kappa \notin P, P' = P \cup \{<\kappa, >\zeta\}$$
$$((l'_{a,<\kappa,>\zeta}, P'), >\zeta, x_a = 0, (l'_{a,<\kappa}, P')) \in \Delta'_l$$
$$((l, P), a, \varphi, (l'_{a,>\zeta}, P), (\gamma, \{<\kappa, >\zeta\}, -)) \in \Delta'_c \ \text{ if } <\kappa \in P$$

For every $(l, a, \varphi, l') \in \Delta_l$ we have the set of transition $((l, P), a, \varphi, (l', P)) \in \Delta'_l$, and for every $(l, a, I, \gamma, \varphi, l') \in \Delta_r$, we have following transitions in Δ'_r.

1. $((l, P), a, (\gamma, \{<\kappa, >\zeta\}, -), \varphi \wedge x_{<\kappa} < \kappa \wedge x_{>\zeta} > \zeta, (l', P))$ if $I = (\zeta, \kappa)$.
2. $((l, P), a, (\gamma, \{<\kappa, >\zeta\}, \texttt{first}), \varphi \wedge x_{<\kappa} < \kappa \wedge x_{>\zeta} > \zeta, (l', P'))$
   where $P' = P \backslash \{<\kappa\}$, if $I = (\zeta, \kappa)$.
3. $((l, P), a, (\gamma, \{<\kappa\}, -), \varphi \wedge x_{<\kappa} < \kappa, (l', P))$ if $I = [0, \kappa)$.
4. $((l, P), a, (\gamma, \{<\kappa\}, \texttt{first}), \varphi \wedge x_{<\kappa} < \kappa, (l', P'))$ with $P' = P \backslash \{<\kappa\}$ if $I = [0, \kappa)$.
5. $((l, P), a, (\gamma, \{>\zeta\}, -), \varphi \wedge x_{>\zeta} > \zeta, (l', P))$ if $I = (\zeta, \infty)$.

For the pop to be successful in M', the guess made at the time of the push must be correct, and indeed at the time of the pop, the age must match the constraint. The control state (l^f, P) is reached in M' on reading a word w' iff M accepts a string w and reaches l^f. Accepting locations of M' are of the form (l^f, P) for $P \subseteq Z_k^{\sim}$. For any $w = (a_1, t_1) \dots (a_n, t_n) \in L(M)$, we have $w' = (a_1, t_1) T_1 (a_2, t_2) T_2 \dots (a_n t_n) T_n$ accepted by $L(M')$, where for $1 \leq l \leq n$, $|T_l| \leq 2k$, and T_l is a timed word $(b_1, t_l) \dots (b_j, t_l)$ where $j \leq 2k$ and $b_i \in Z_k^{\sim}$ for $1 \leq i \leq j$ and the only time stamp used in T_i is t_i, since no time elapses in M' while remembering obligations and resetting the appropriate clocks.

Emptiness and Determinizability. In the construction above, it can shown by inducting on the length of words accepted that $h(L(M')) = L(M)$. Thus, $L(M') \neq \emptyset$ iff $L(M) \neq \emptyset$. Since M' is ECVPA, we can apply the standard region construction of event clock automata [3] to obtain a PDA preserving emptiness.

Next, we focus on the determinizability of dtVPA. Consider a dtVPA $M = (L, \Sigma, \Gamma, L^0, F, \Delta)$ and the corresponding ECVPA $M' = (L', \Sigma', \Gamma', L'^0, F', \Delta')$ as constructed in Sect. 4. From Theorem 4 we know that M' is determinizable. Let $Det(M')$ be the determinized automaton such that $L(Det(M')) = L(M')$. That is, $L(M) = h(L(Det(M')))$. By construction of M', we know that the new symbols introduced in Σ' are $Z_k^{\sim}$ ($\Sigma' = \Sigma \cup Z_k^{\sim}$) and (i) no time elapse happens on reading these symbols, and (ii) no stack operations happen on reading these symbols. Consider any transition in $Det(M')$ involving the new symbols. Since $Det(M')$ is deterministic, let $(s_1, \alpha, \varphi, s_2)$ be the unique transition on $\alpha \in Z_k^{\sim}$. In the following, we eliminate these transitions on $Z_k^{\sim}$ preserving the language accepted by M and the determinism of $det(M')$. In doing so, we will construct a dtVPA M'' which is deterministic, and which preserves the language of M. We now analyze various types for $\alpha \in Z_k^{\sim}$.

1. Assume that α is of the form $>\zeta$. Let $(s_1, \alpha, \varphi, s_2)$ be the unique transition on $\alpha \in Z_k^{\sim}$. By construction of M' (and hence $det(M')$), we know that φ has the form $x_a = 0$ for some $a \in \Sigma$. We also know that in $Det(M')$, there is a unique transition $(s_0, a, \psi, s_1, (\gamma, \alpha, -))$ preceding $(s_1, \alpha, \varphi, s_2)$. Since $(s_1, \alpha, \varphi, s_2)$ is a no time elapse transition, and does not touch the stack, we can combine the two transitions from s_0 to s_1 and s_1 to s_2 to obtain the call transition $(s_0, a, \psi, s_2, (\gamma, \alpha, -))$. This eliminates transition on $>\zeta$.
2. Assume that α is of the form $<\kappa$. Let $(s_1, \alpha, \varphi, s_2)$ be the unique transition on $\alpha \in Z_k^{\sim}$. We know that φ has the form $x_a = 0$ for some $a \in \Sigma$. From M', we also know that in $Det(M')$, there is a unique transition of one of the following forms preceding $(s_1, \alpha, \varphi, s_2)$:
   (a) $(s_0, a, \psi, s_1, (\gamma, \alpha, -))$, (b) $(s_0, a, \psi, s_1, (\gamma, \alpha, \texttt{first}))$, or

(c) $(s_0, >\zeta, \varphi, s_1)$ where it is preceded by $(s_0', a, \psi, s_0, (\gamma, \{\alpha, >\zeta\}, X))$ for $X \in \{\texttt{first}, -\}$.

Since $(s_1, \alpha, \varphi, s_2)$ is a no time elapse transition, and does not touch the stack, we can combine two transitions from s_0 to s_1 (cases (a), (b)) and s_1 to s_2 to obtain the call transition $(s_0, a, \psi, s_2, (\gamma, \alpha, -))$ or $(s_0, a, \psi, s_2, (\gamma, \alpha, \texttt{first}))$. This eliminates the transition on $<\kappa$. In case of transition (c), we first eliminate the local transition on $>\zeta$ obtaining $(s_0', a, \psi, s_1, (\gamma, \{\alpha, >\zeta\}, X))$ and then obtain the call transitions $(s_0', a, \psi, s_2, (\gamma, \{\alpha, >\zeta\}, X))$. We have thus eliminated local transitions on $<\kappa$.

Merging transitions as done here does not affect transitions on Σ as they simply eliminate the newly added transitions on $\Sigma' - \Sigma$. Recall that checking constraints on these clocks were required during return transitions. We now modify the pop operations in $Det(M')$ as follows: Return transitions have the following forms, and in all of these, φ is a constraint checked on the clocks of C_Σ in M during return:

- transitions $(s, a, (\gamma, \{<\kappa\}, X), \varphi \wedge x_{<\kappa} < \kappa, s')$ for $X \in \{-, \texttt{first}\}$ are modified to $(s, a, [0, \kappa), (\gamma, \{<\kappa\}, X), \varphi, s')$;
- transitions $(s, a, (\gamma, \{<\kappa, >\zeta\}, X), \varphi \wedge x_{>\zeta} > \zeta \wedge x_{<\kappa} < \kappa, s')$ for $X \in \{-, \texttt{first}\}$ are modified to $(s, a, (\zeta, \kappa), (\gamma, \{<\kappa, >\zeta\}, X), \varphi, s')$; and
- transition $(s, a, (\gamma, \{>\zeta\}, -), \varphi \wedge x_{>\zeta} > \zeta, s')$ are modified to the transitions $(s, a, (\zeta, \infty), (\gamma, \{>\zeta\}, -), \varphi, s')$.

Now it is straight forward to verify that the deterministic dtVPA M'' obtained from $det(M')$ is such that $L(M'') = L(M)$ and $h(L(M'')) = L(det(M'))$. This completes the proof of determinizability of dtVPA.

5   Logical Characterization of dtVPA

Monadic Second-Order Logic on Timed Words. We consider a timed word $w = (a_0, t_0), (a_1, t_1), \ldots, (a_m, t_m)$ over Σ as a *word structure* over the universe $U = \{1, 2, \ldots, |w|\}$ of positions in the timed word. The predicates in the word structure are $Q_a(i)$ which evaluates to true at position i iff $w[i] = a$, where $w[i]$ denotes the ith position of w. Following [4], we use the matching binary relation $\mu(i, j)$ which evaluates to true iff the ith position is a call and the jth position is its matching return. We also introduce three predicates $\lhd_a$, $\rhd_a$, and θ capturing the following relations. For an interval I, the predicate $\lhd_a(i) \in I$ evaluates to true on the word structure iff $\nu_i^w(x_a) \in I$ for recorder clock x_a. For an interval I, the predicate $\rhd_a(i) \in I$ evaluates to true on the word structure iff $\nu_i^w(y_a) \in I$ for predictor clock y_a. For an interval I, the predicate $\theta(i) \in I$ evaluates to true on the word structure iff $w[i] \in \Sigma_r$, and there is some $k < i$ such that $\mu(k, i)$ evaluates to true and $t_i - t_k \in I$. The predicate $\theta(i)$ measures the time elapse between position k where a call was made, and position i, its matching return. This time elapse is the age of the symbol pushed on to the stack during the call at position k. Since position i is the matching return, this symbol is popped at position i; if the age lies in the interval I, the predicate evaluates to true. We define MSO(Σ), the MSO logic over Σ, as:

$$\varphi := Q_a(x) \mid x{\in}X \mid \mu(x,y) \mid \lhd_a(x){\in}I \mid \rhd_a(x){\in}I \mid \theta(x){\in}I \mid \neg\varphi \mid \varphi{\vee}\varphi \mid \exists x.\varphi \mid \exists X.\varphi$$

where $a{\in}\Sigma$, $x_a{\in}C_\Sigma$, x is a first order variable and X is a second order variable. The models of a formula $\phi \in$ MSO(Σ) are timed words w over Σ. The semantics of these logic is standard where first order variables are interpreted over positions of w and second order variables over subsets of positions. As an example consider the formula $\varphi = \forall x(Q_a(x) \rightarrow \exists y[Q_c(y)\wedge\theta(y) \in (1,2)])$ over $\Sigma = (\{a\}, \{b\}, \{c\})$. It expresses that for every $a \in \Sigma_c$, there exists a $c \in \Sigma_r$ as the matching return such that the time elapse between the call and return is in the interval $(1,2)$. The word $w = (a,0)(a,0.2)(b,0.5)(c,1.3)(b,1.7)(b,1.9)(c,1.99)$ satisfies φ. We define the language $L(\varphi)$ of an MSO sentence φ as the set of all words satisfying φ.

Logic to automata. We first show that for any MSO formula φ over $\Sigma = (\Sigma_c, \Sigma_l, \Sigma_r)$, $L(\varphi)$ is accepted by a dtVPA. Let $Z = (x_1,\ldots,x_m,X_1,\ldots,X_n)$ be the free variables in φ. We work on the extended alphabet $\Sigma' = (\Sigma'_c, \Sigma'_l, \Sigma'_r)$ where $\Sigma'_s{=}\Sigma_s{\times}(Val : Z \rightarrow \{0,1\}^{m+n})$, for $s \in \{c,l,r\}$. A word w' over Σ' encodes a word over Σ along with the valuation of all first order and second order variables. Thus Σ' consists of all symbols (a,v) where $a \in \Sigma$ is such that $v(x) = 1$ means that x is assigned the position i of a in the word w, while $v(x) = 0$ means that x is not assigned the position of a in w. Similarly, $v(X) = 1$ means that the position i of a in w belongs to the set X. Next we use quasi-event clocks for Σ' by assigning suitable ranking function. We partition Σ' such that for a fixed $a \in \Sigma$, all symbols of the form $(a,d_1,\ldots,d_{m+n})$ and $d_i \in \{0,1\}$ lie in the same partition (a determines their partition). Let $\rho' : \Sigma' \rightarrow \mathbb{N}$ be the ranking function of Σ' wrt above partitioning scheme.

Let $L(\psi)$ be the set of all words w' over Σ' such that the underlying word w over Σ satisfies formula ψ along with the valuation Val. Structurally inducting over ψ, we show that $L(\psi)$ is accepted by a dtVPA. The cases $Q_a(x), \mu(x,y)$ are exactly as in [4]. We only discuss the new predicates here.

Consider the atomic formula $\lhd_a(x) \in I$. We construct a dtVPA that on reading a symbol $(b, v) \in \Sigma'$ with $v(x) = 1$ checks the constraint $x_a \in I$ for acceptance. The case of $\rhd_a(x) \in I$ is similar, and the check is done on clock y_a. Consider the atomic formula $\theta(x) \in I$. To handle this, we build a dtVPA that keeps pushing symbols (a, v) onto the stack whenever $a \in \Sigma_c$, initializing the age to 0 on push. It keeps popping the stack on reading return symbols (a', v'), and checks whether $v'(x) = 1$ and $age((a', v')) \in I$. It accepts on finding such a pop. The check $v'(x) = 1$ ensures that this is the matching return of the call made at position x. The check $age((a', v')) \in I$ confirms that the age of this symbol pushed at position x is indeed in the interval I. Negations, conjunctions and disjunctions follow from the closure properties of dtVPA. Existential quantifications correspond to projection by excluding the chosen variable from the valuation and renaming the alphabet Σ'. Let M be an dtVPA constructed for $\varphi(x_1,\ldots,x_n,X_1,\ldots,X_m)$ over Σ'. Consider $\exists x_i.\varphi(x_1,\ldots,x_n,X_1,\ldots,X_m)$ for some first order variable x_i. Let $Z_i = (x_1,\ldots,x_{i-1},x_{i+1},\ldots,x_n,X_1,\ldots,X_m)$ by removing x_i from Z.

We simply work on the alphabet $\Sigma_i' = \Sigma \times (Val : Z_i \to \{0,1\}^{m+n-1})$. Note that Σ_i' is partitioned exactly in the same way as Σ'. For a fixed $a \in \Sigma$, all symbols $(a, d_1, \ldots, d_{m+n-1})$ for $d_i \in \{0,1\}$ lie in the same partition. Thus, Σ' and Σ_i' have exactly the same number of partitions, namely $|\Sigma|$. Thus, an event clock $x_a = x_{(a,d_1,\ldots,d_{m+n})}$ used in M can be used the same way while constructing the automaton for $\exists x_i.\varphi(x_1, \ldots, x_n, X_1, \ldots, X_m)$. The case of $\exists X_i.\varphi(x_1, \ldots, x_n, X_1, \ldots, X_m)$ is similar. Hence we obtain in all cases, a dtVPA that accepts $L(\psi)$ when ψ is an MSO sentence.

Automata to logic. Consider a dtVPA $M = (L, \Sigma, \Gamma, L^0, F, \Delta)$. Let $L = \{l_1, \ldots l_n\}$ and $\Gamma = \{\gamma_1, \ldots, \gamma_m\}$. The MSO formula encoding accepting runs of dtVPA is:
$\exists X_{l_1} \ldots X_{l_n} C_{\gamma_1} \ldots C_{\gamma_m} R_{\gamma_1} \ldots R_{\gamma_m} \ \varphi(X_{l_1}, \ldots, X_{l_n}, C_{\gamma_1}, \ldots, C_{\gamma_m}, R_{\gamma_1}, \ldots, R_{\gamma_m})$,
where X_q denotes the set of positions in the word where the run is in location q, C_γ, R_γ stand for the set of positions in the run where γ is pushed and popped from the stack respectively. We assert that the starting position must belong to X_l for some $l \in L^0$. Successive positions must be connected by an appropriate transition. To complete the reduction we list these constraints. For call transitions $(\ell_i, a, \psi, \ell_j, \gamma) \in \Delta_c$, for positions x, y, we assert that $\quad X_{\ell_i}(x) \wedge X_{\ell_j}(y) \wedge Q_a(x) \wedge C_\gamma(x) \wedge \bigwedge_{b \in \Sigma} \left(\left(\bigwedge_{(x_b \in I) \in \psi} \lhd_b(x) \in I \right) \wedge \left(\bigwedge_{(y_b \in I) \in \psi} \rhd_b(x) \in I \right) \right)$.
For return transitions $(\ell_i, a, I, \gamma, \psi, \ell_j) \in \Delta_r$ for positions x and y we assert that
$X_{\ell_i}(x) \wedge X_{\ell_j}(y) \wedge Q_a(x) \wedge R_\gamma(x) \wedge \theta(x) \in I \wedge \bigwedge_{b \in \Sigma} \left(\left(\bigwedge_{(x_b \in I) \in \psi} \lhd_b(x) \in I \right) \wedge \left(\bigwedge_{(y_b \in I) \in \psi} \rhd_b(x) \in I \right) \right)$.
Finally, for local transitions $(\ell_i, a, \psi, \ell_j) \in \Delta_l$ for positions x and y we assert $X_{\ell_i}(x) \wedge X_{\ell_j}(y) \wedge Q_a(x) \wedge \bigwedge_{b \in \Sigma} \left(\left(\bigwedge_{(x_b \in I) \in \psi} \lhd_b(x) \in I \right) \wedge \left(\bigwedge_{(y_b \in I) \in \psi} \rhd_b(x) \in I \right) \right)$. We also assert that the last position of the word belongs to some X_l such that there is a transition (call, return, local) from l to an accepting location. The encoding of all 3 kinds of transitions is as above. Additionally, we assert that corresponding call and return positions should match, i.e. $\forall x \forall y \, \mu(x, y) \Rightarrow \bigvee_{\gamma \in \Gamma \setminus \bot} C_\gamma(x) \wedge R_\gamma(y)$.

These two parts together finish the proof of the main result of the paper.

Theorem 5. *A language L over Σ is accepted by an dtVPA iff there is a MSO sentence φ over Σ such that $L(\varphi) = L$.*

References

1. Abdulla, P., Atig, M., Stenman, J.: Dense-timed pushdown automata. In: LICS, pp. 35–44 (2012)
2. Alur, R., Dill, D.: A theory of timed automata. Theor. Comput. Sci. **126**, 183–235 (1994)
3. Alur, R., Fix, L., Henzinger, T.A.: Event-clock automata: a determinizable class of timed automata. TCS **211**(1–2), 253–273 (1999)
4. Alur, R., Madhusudan, P.: Visibly pushdown languages. In: Symposium on Theory of Computing, pp. 202–211 (2004)
5. Bhave, D., Dave, V., Krishna, S.N., Phawade, R., Trivedi, A.: A logical characterization for dense-time visibly pushdown automata. Technical report, IIT Bombay (2015). http://www.cse.iitb.ac.in/internal/techreports/reports/TR-CSE-2015-77.pdf

6. Clemente, L., Lasota, S.: Timed pushdown automata revisited. In: LICS, pp. 738–749 (2015)
7. D'Souza, D.: A logical characterisation of event clock automata. Int. J. Found. Comput. Sci. **14**(4), 625–640 (2003). http://dx.doi.org/10.1142/S0129054103001923
8. Lautemann, C., Schwentick, T., Thérien, D.: Logics for context-free languages. In: Pacholski, L., Tiuryn, J. (eds.) Computer Science Logic. LNCS, vol. 933, pp. 205–216. Springer, Heidelberg (1995)
9. Trivedi, A., Wojtczak, D.: Recursive timed automata. In: Bouajjani, A., Chin, W.-N. (eds.) ATVA 2010. LNCS, vol. 6252, pp. 306–324. Springer, Heidelberg (2010)
10. Van Tang, N., Ogawa, M.: Event-clock visibly pushdown automata. In: Nielsen, M., Kučera, A., Miltersen, P.B., Palamidessi, C., Tůma, P., Valencia, F. (eds.) SOFSEM 2009. LNCS, vol. 5404, pp. 558–569. Springer, Heidelberg (2009)

A Complexity Measure on Büchi Automata

Dana Fisman[(✉)]

University of Pennsylvania, Philadelphia, PA, USA
fisman@seas.upenn.edu

Abstract. We define a complexity measure on non-deterministic Büchi automata, based on the notion of the width of the skeleton tree introduced by Kähler and Wilke. We show that the induced hierarchy tightly correlates to the Wagner Hierarchy, a corner stone in the theory of regular ω-languages that is derived from a complexity measure on deterministic Muller automata. The relation between the hierarchies entails, for instance, that a nondeterministic Büchi automaton of width k can be translated to a deterministic parity automaton of degree at most $2k + 1$.

Keywords: Automata and logic · Automata for system analysis and program verification · Classification of regular ω-languages

1 Introduction

There are various way to define acceptance on infinite words, deriving different types of ω-automata. Büchi automata have the simplest acceptance criterion: a run is declared accepting if the set of states visited infinitely often intersect a designated set of *accepting* states F. Their dual, co-Büchi automata declare a run accepting if the set of states visited infinitely often does not intersect a designated set of *rejecting* states F. Muller automata have the most general acceptance criterion: a run is accepting if the set of states visited infinitely often is exactly one of a set of designated subsets of states $F_1, F_2, \ldots, F_k$. Other types of ω-automata include Rabin, Streett and parity.

The different ω-automata have varying levels of complexity and expressivity. Using the convention that $\mathbb{D}\mathbb{T}$ ($\mathbb{N}\mathbb{T}$) for $\mathbb{T} \in \{\mathbb{B}, \mathbb{C}, \mathbb{R}, \mathbb{S}, \mathbb{M}, \mathbb{P}\}$ denotes the class of languages accepted by deterministic (nondeterministic) automata of type Büchi, co-Büchi, Rabin, Street, Muller or parity, resp., the following relations are known. The class $\mathbb{D}\mathbb{B}$ is less expressive than $\mathbb{N}\mathbb{B}$ which is as expressive as $\mathbb{D}\mathbb{M}$, $\mathbb{D}\mathbb{P}$, $\mathbb{D}\mathbb{R}$, $\mathbb{D}\mathbb{S}$, which recognize all regular ω-languages. The classes $\mathbb{D}\mathbb{B}$ and $\mathbb{D}\mathbb{C}$ have incomparable expressive power, in the sense that there exists languages in $\mathbb{D}\mathbb{B} \setminus \mathbb{D}\mathbb{C}$ and in $\mathbb{D}\mathbb{C} \setminus \mathbb{D}\mathbb{B}$.

Wagner [13] has suggested a complexity measure on Muller automata, and showed that this complexity measure is language-specific and is invariant over all automata accepting the same language. This result, referred to as the Wagner Hierarchy, is a corner stone in the theory of regular ω-languages. The classes $\mathbb{D}\mathbb{B}$

This research was supported by US NSF grant CCF-1138996.

A.-H. Dediu et al. (Eds.): LATA 2016, LNCS 9618, pp. 102–113, 2016.
DOI: 10.1007/978-3-319-30000-9_8

and $\mathbb{DC}$ coincide with classes of the lower levels of this hierarchy. For a language L the minimal number of colors required by a recognizing deterministic parity automaton, and the minimal number of pairs required by a deterministic Rabin or Streett automaton tightly correlates to the minimal class in the hierarchy in which the recognized language L resides.

In this paper we define a complexity measure on nondeterministic Büchi automata (NBA). The measure is based on the notion of the width of the *skeleton-tree*, a structure for summarizing runs of NBAs, introduced by Kähler and Wilke [7]. A language L is said to be of degree k, or belong to $\mathbb{NB}_k$, if there exists an NBA $\mathcal{N}$ recognizing L such that the width of the skeleton-tree of all words with respect to $\mathcal{N}$ is at most k. We show that this measure induces a strict hierarchy of classes of languages, and that this hierarchy tightly correlates to the Wagner hierarchy. The relation between the hierarchies entails, for instance, that an NBA of width k can be translated to a parity automaton using at most $2k+1$ colors, and a language $L \in \mathbb{NB}_{k+1} \setminus \mathbb{NB}_k$ cannot be translated to a parity automaton using less than $2k$ colors.

We provide definitions for ω-automata, a summary of the Wagner hierarchy, and a definition of the skeleton-tree in Sect. 2. The contribution of the paper starts in Sect. 3 where we define the complexity measure on Büchi automata, and state the main theorem of the paper, that the induced hierarchy tightly correlates to the Wagner hierarchy. Section 4 proves the direction from the proposed hierarchy to the Wagner hierarchy and Sect. 5 the other direction. In a sense, the direction from the Wagner hierarchy to the proposed hierarchy provides an insight on the need for non-determinism in the Büchi model, and a quantification of the amount of non-determinism needed relative to the complexity of the Muller automaton.

2   Preliminaries

Automata on Infinite Words. An *automaton* is a tuple $\mathcal{A} = \langle \Sigma, Q, q_0, \delta, \alpha \rangle$ consisting of a finite alphabet Σ of symbols, a finite set Q of states, an initial state q_0, a transition function $\delta : Q \times \Sigma \to 2^Q$, and an acceptance condition α. A run of an automaton on an infinite word $v = a_1 a_2 \ldots$ is an infinite sequence of states $p_0, p_1, p_2 \ldots$, such that $p_0 = q_0$ and $p_{i+1} \in \delta(p_i, a_i)$ for every $i \in \mathbb{N}$. The transition function can be extended to a function from $Q \times \Sigma^*$ (to Q) by defining $\delta(q, \epsilon) = q$ and $\delta(q, av) = \delta(\delta(q, a), v)$ for $q \in Q$, $a \in \Sigma$ and $v \in \Sigma^*$, where ϵ denotes the empty word. We say that $\mathcal{A}$ is *deterministic* if $|\delta(q, a)| \leq 1$ for every $q \in Q$ and $a \in \Sigma$.

An automaton accepts a word if at least one of the runs on that word is accepting. We use $[\![\mathcal{A}]\!]$ to denote the set of words accepted by $\mathcal{A}$. We use $\overline{L}$ to denote the language $\Sigma^\omega \setminus L$. For finite words the acceptance condition is a set $F \subseteq Q$ and a run on v is accepting if it ends in an accepting state, i.e., if $\delta(q_0, v) \in F$. For infinite words, there are many acceptance conditions in the literature; here we mention four: Büchi, co-Büchi, parity and Muller. In *Büchi* and *co-Büchi* automata the acceptance condition refers to a subset F of the states set. In *parity automata* the acceptance condition refers to a mapping $\kappa : Q \mapsto [1..k]$ from the set of states to the set of *colors* $[1..k]$.[1] In *Muller* automata the acceptance condition is a function $\tau : 2^Q \rightarrow \{+, -\}$ assigning positive/negative polarity to subsets of states.

Let ρ be an infinite sequence of states $s_0, s_1, s_2 \ldots$. We use $\kappa(\rho)$ to denote the respective sequence of colors $\kappa(s_0), \kappa(s_1), \kappa(s_2) \ldots$. We use $\mathsf{Inf}(\rho)$ to denote the set of states that appear infinitely often in ρ. An infinite path ρ satisfies the

- Büchi condition w.r.t F iff $\mathsf{Inf}(\rho) \cap F \neq \emptyset$.
- co-Büchi condition w.r.t F iff $\mathsf{Inf}(\rho) \cap F = \emptyset$.
- parity condition w.r.t κ iff $\min(\mathsf{Inf}(\kappa(\rho)))$ is odd.
- Muller condition w.r.t τ iff $\tau(\mathsf{Inf}(\rho)) = +$.

We use DBA, DCA, DPA, and DMA, to denote deterministic Büchi, co-Büchi, parity and Muller automata and NBA, NCA, NPA, and NMA to denote the respective non-deterministic automata. Similarly, we use $\mathbb{DB}$, $\mathbb{DC}$, $\mathbb{DP}$ and $\mathbb{DM}$ to denote the class of languages recognized by DBA, DCA, DPA and DMA, respectively, and $\mathbb{NB}$, $\mathbb{NC}$, $\mathbb{NP}$, $\mathbb{NM}$ to denote the class of languages accepted by NBA, NCA, NPA, and NMA, respectively. For parity automata we use $\mathbb{DP}_k$ to denote the set of languages recognized by a DPA with k colors, and refer to it as the $\mathbb{DP}_k$ hierarchy.

The Chain Measure. Let $\mathcal{D} = \langle \Sigma, Q, q_0, \delta, \tau \rangle$ be a complete DMA (i.e. a DMA where $|\delta(q, a)| = 1$ for every $q \in Q$ and $a \in \Sigma$). A set of states $S \subseteq Q$ is said to be *admissible* if S is a reachable strongly connected component (SCC) of $\mathcal{D}$. An admissible set S is said to be *positive* or *accepting* (resp. *negative* or *rejecting*) iff $\tau(S) = +$ (resp. $\tau(S) = -$). A chain of admissible sets $S_0 \subset S_1 \subset \cdots \subset S_{m-1}$ is a $\mathcal{D}$-*chain* iff the sets are alternately positive and negative. The *length* of such a chain is m. The polarity of a $\mathcal{D}$-chain is determined according to the polarity of its bottom set. That is, the above $\mathcal{D}$-chain is said to be positive (resp. negative) iff S_0 is positive (resp. negative). We use $k(\mathcal{D})$, $k^+(\mathcal{D})$ and $k^-(\mathcal{D})$ to denote the maximal length of a $\mathcal{D}$-chain, a positive $\mathcal{D}$-chain and a negative $\mathcal{D}$-chain, resp. In the sequel we will focus on $k^+(\mathcal{D})$ to which we refer as the *positive-chain measure*. Note that $0 \leq k(\mathcal{D}) \leq |Q|$. Also $|k^+(\mathcal{D}) - k^-(\mathcal{D})| \leq 1$ simply by omitting the bottom set.

Definition 1 (The Classes $\mathbb{DM}_k^+$ and $\mathbb{DM}_k^-$ [13]). *Let $k \in \mathbb{N}$. The class of languages $\mathbb{DM}_k^+$ and $\mathbb{DM}_k^-$ are defined as follows.*

- $\mathbb{DM}_k^+ = \{ L \mid \exists \text{ DMA } \mathcal{D} : \quad L = [\![\mathcal{D}]\!], \quad k^+(\mathcal{D}) \leq k \}$
- $\mathbb{DM}_k^- = \{ L \mid \exists \text{ DMA } \mathcal{D} : \quad L = [\![\mathcal{D}]\!], \quad k^-(\mathcal{D}) \leq k \}$

The Wagner hierarchy consists of an additional measure, the length of the longest sequence of chains that are reachable from each other and have alternating polarities, and takes into account also the polarity of the chains. We omit the details for lack of space and since this measure is of less importance to this paper. Wagner [13] has shown that the hierarchy is strict and that these measures are invariant overall DMAs accepting the same language. He further showed

---

[1] For $j, k \in \mathbb{N}$ s.t. $j \leq k$, we use $[j..k]$ to denote the set $\{j, j+1, j+2, \ldots, k\}$.

that the chain hierarchy tightly correlates to the minimal numbers of chains in a Rabin automaton. A similar result regarding parity automata (see f.g. [2,3]) states that $L \in \mathbb{DM}_k^+$ iff $L \in \mathbb{DP}_k$. The relation between DBAs and the class $\mathbb{DM}_1^+$ was already shown by Landweber [10]. The Wagner hierarchy has been rediscovered several times [1,8].

Theorem 1 [2,3,10,13].

1. $L \in \mathbb{DM}_k^+ \iff \overline{L} \in \mathbb{DM}_k^-$
2. $\mathbb{DM}_k^- \subsetneq \mathbb{DM}_{k+1}^-$ and $\mathbb{DM}_k^+ \subsetneq \mathbb{DM}_{k+1}^+$
3. $L \in \mathbb{DM}_1^+ \iff L \in \mathbb{DB}$ and $L \in \mathbb{DM}_1^- \iff L \in \mathbb{DC}$.
4. $L \in \mathbb{DM}_k^+ \iff L \in \mathbb{DP}_k$

The Skeleton Tree. The complexity measure we propose is based on the notion of the *width* of the *skeleton tree* introduced by Kähler and Wilke [7]. Kähler and Wilke, aiming to provide constructions unifying Büchi determinization, complementation and disambiguation introduced the notions of the *split tree*, the *reduced tree* and the *skeleton tree*, where the latter has been said to be identified from the work of Muller and Schupp [11]. All three are mechanisms to summarize runs of NBAs. For lack of space we suffice here with an informal description.

The *split-*, *reduced-* and *skeleton-trees* are defined per a given word w and w.r.t. a given NBA $\mathcal{N}$. A key invariant that is maintained is that if there exists an accepting run of $\mathcal{N}$ on w then there is an accepting infinite path in all of these trees. Roughly speaking, the split tree refines the subset construction by separating accepting and non-accepting states. From each node of the tree the left son holds its accepting successors and the right son its non-accepting successor. Thus, an accepting path has infinitely many left turns, and is also referred to as *left recurring*. The width of a layer of the split-tree is generally unbounded. The reduced tree bounds the number of nodes on a layer of the tree to n, the number of states of the given Büchi automaton $\mathcal{N}$, by eliminating from a node of the tree all states that appeared in a node to its left. The skeleton-tree is the smallest sub-tree of the reduced-tree that contains all its infinite paths. The *width* of the skeleton tree thus equals the number of infinite paths in the reduced tree. We use $width(\mathcal{N}, w)$ to denote the width of the skeleton tree for w w.r.t to $\mathcal{N}$.

3   A Complexity Measure on NBAs

Given an NBA $\mathcal{N}$ we say that the *width* of $\mathcal{N}$ is the maximal width of a skeleton tree on any given word. Formally

$$width(\mathcal{N}) = \max \{width(\mathcal{N}, w) \mid w \in \Sigma^\omega\}$$

It is not hard to see that two NBAs $\mathcal{N}_1$ and $\mathcal{N}_2$ may have different widths even if $[\![\mathcal{N}_1]\!] = [\![\mathcal{N}_2]\!]$ since we can add nondeterministic transitions to accepting and non-accepting states without changing the language. For instance, consider the NBAs $\mathcal{N}_1$ and $\mathcal{N}_2$ over alphabet $\Sigma = \{a\}$ depicted in Fig. 1. We have $[\![\mathcal{N}_1]\!] = [\![\mathcal{N}_2]\!] = \{a^\omega\}$, yet $width(\mathcal{N}_1, a^w) = 1$ and $width(\mathcal{N}_2, a^w) = 2$. Moreover, since a^ω is the only word in Σ^ω we have $width(\mathcal{N}_1) = 1$ and $width(\mathcal{N}_2) = 2$.

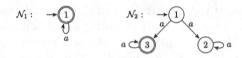

Fig. 1. Two NBAs $\mathcal{N}_1$ and $\mathcal{N}_2$ such that $[\![\mathcal{N}_1]\!] = [\![\mathcal{N}_2]\!]$ yet $width(\mathcal{N}_1) \neq width(\mathcal{N}_2)$.

Definition 2 (The Classes $\mathbb{NB}_k$). *Let $k \geq 1$ be a natural number. The class of languages $\mathbb{NB}_k$ is defined as follows.*

$$\mathbb{NB}_k = \{L \mid \exists \text{ NBA } \mathcal{N} : \ L = [\![\mathcal{N}]\!], \ width(\mathcal{N}) \leq k\}$$

The main contribution of this paper can be summarized by the following two theorems. Theorem 2 states that the hierarchy is strict. For every level k of the hierarchy there exists a language L such that L cannot be recognized by any NBA of width k or smaller. Theorem 3 states that this hierarchy is tightly correlated to the Wagner hierarchy.

Theorem 2. $\mathbb{NB}_k \subsetneq \mathbb{NB}_{k+1}$

Proof. This is a corollary of Theorem 1 and the forthcoming Theorem 3.   □

Theorem 3. *Let L be an ω-language, and let $k \geq 0$ be a natural number. Then*

- $L \in \mathbb{NB}_k \Longrightarrow L \in \mathbb{DM}^+_{2k+1}$
- $L \in \mathbb{DM}^+_k \Longrightarrow L \in \mathbb{NB}_{\lceil \frac{k}{2} \rceil + 1}$

Proof. The first and second items are respectively given by the forthcoming Corollary 7 and Proposition 12.   □

Deciding the Width. In view of these relations, an important question is to find the width of a given NBA $\mathcal{N}$.

Proposition 4. *The problem of finding the width of an NBA is solvable in time $n^{O(n)}$ where n is the number of states in the given NBA.*

Proof. Fisman and Lustig [4, Proposition 2] provide a construction for a DBA $\mathcal{B}_k$ that accepts a word w iff $width(\mathcal{N}, w) < k$ when $\mathcal{N}$ is a given NBA. A DBA can be seen as a DPA with 2 colors (accepting states are colored 1 and non-accepting states are colored 2). Given a DPA $\mathcal{P}_k$ one can construct a DPA $\overline{\mathcal{P}}_k$ for its complement by assigning $\overline{\kappa}(q) = \kappa(q) + 1$ (where κ is the coloring function of $\mathcal{P}_k$ and $\overline{\kappa}$ is the coloring function of $\overline{\mathcal{P}}_k$). The constructed DPA $\overline{\mathcal{P}}_k$ accepts a word w iff $width(\mathcal{N}, w) \geq k$. Thus, $width(\mathcal{N}) < k$ iff $\overline{\mathcal{P}}_k$ is empty. Emptiness of a DPA can be solved in time polynomial in the number of states and colors. Applying a binary search would add a factor of $\log n$ where n is the number of states in $\mathcal{N}$. When n is the number of states in $\mathcal{N}$, the DBA $\mathcal{B}_k$ may have upto $n^{O(n)}$ states. Since the DPA $\overline{\mathcal{P}}_k$ has the same number of states as $\mathcal{B}_k$ (and the time spent to build $\mathcal{B}_k$ is polynomial in the size of $\mathcal{B}_k$), we can decide the width in time $n^{O(n)}$.   □

Krishnan, Puri and Brayton [9] show that determining the Rabin index (Streett index) of a given a DRA (DSA) is NP-*complete*, yet it is decidable in polynomial time whether a DRA (DSA) with f pairs has Rabin index (Streett index) f or any constant c.[2] Wilke and Yoo [14] show that the chain measure of a given DMA over an alphabet of size ℓ with n states, m strongly connected components, and f accepting sets can be decided in time $O(f^2 n\ell + m \log m)$. Note that m and f may be exponential in n. Since the width is not an invariant among all NBAs accepting a given language, the question of finding the minimal k for which $[\![\mathcal{N}]\!] \in \mathbb{NB}_k$ where the input is $\mathcal{N}$ is not answered by Proposition 4. It can be answered e.g. by translating it to a DRA or DMA and inferring the result using the above mentioned results and the relations of the hierarchies, but a lower bound remains open.

4   From $\mathbb{NB}_k$ to $\mathbb{DM}_{k'}$

The proof of the first part of Theorem 3 goes via DPAs, using the relations between their rank and the chain measure as stated in Theorem 1.

Proposition 5. $L \in \mathbb{NB}_k \implies L \in \mathbb{DP}_{2k+1}$.

Proof. The known constructions from NBA to DPA, given an NBA $\mathcal{N}$ with n states produce a DPA of rank $2n$ [4,12]. The construction of Fisman and Lustig [4] is based on the notion of the width of the skeleton tree. Given an NBA $\mathcal{N}$, they first show how to construct a DPA $\mathcal{P}_k$ that provides a correct answer only for words of width exactly k w.r.t. $\mathcal{N}$, as stated in Lemma 6.

Lemma 6 [4, Proposition 4]. Let $\mathcal{N}$ be an NBA with n states, and let $k \in [1..n]$. There exists a DPA $\mathcal{P}_k$ using colors $\{0,1,2\}$ such that for any word w

– if $width(\mathcal{N}, w) = k$ then $\mathcal{P}_k$ accepts w iff $\mathcal{N}$ does,
– if $width(\mathcal{N}, w) < k$ then $\mathcal{P}_k$ rejects w,
– if $\mathcal{P}_k$ accepts w then w is accepted by $\mathcal{N}$, and
– $\mathcal{P}_k$ visits 0 infinitely often iff $width(\mathcal{N}, w) < k$.

They then show that a DPA recognizing the same language as $\mathcal{N}$ can be constructed by running the DPAs for width 1 to n in parallel, where n is the size of $\mathcal{N}$. The colors are distributed so that $\mathcal{P}_k$ uses colors $2k$, $2k + 1$ and $2n + 2$, and the color of the compound state $(s_1, s_2, \ldots, s_n)$ where s_i is the state of $\mathcal{P}_i$ is $\min\{c_1, c_2, \ldots, c_n\}$ where c_i is the color of s_i. The obtained DPA has rank $2n + 2$. Clearly the same reasoning shows that given the width of $\mathcal{N}$ is at most k, the DPA obtained by running in parallel the automata $\mathcal{P}_1, \mathcal{P}_2, \ldots, \mathcal{P}_k$ provides a correct result, and uses $2k + 1$ colors. $\square$

The following is a direct corollary of Proposition 5 and Theorem 1.

Corollary 7. $L \in \mathbb{NB}_k \implies L \in \mathbb{DM}_{2k+1}^+$.

---

[2] The Rabin index (Streett index) is the least possible number of accepting pairs used in a DRA (DSA) recognizing the language.

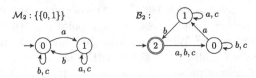

Fig. 2. A DMA $\mathcal{M}_2$ in $\mathbb{DM}_1^+$ and a minimal equivalent DBA $\mathcal{B}_2$.

5   From $\mathbb{DM}_k^+$ to $\mathbb{NB}_{k'}$

We show here the second item of Theorem 3. We start with the simple cases $\mathbb{DM}_1^+$ and $\mathbb{DM}_1^-$, and then generalize the ideas of these constructions to obtain the construction for $\mathbb{DM}_k^+$ for arbitrary k.

By Theorem 1, $L \in \mathbb{DM}_1^+ \iff L \in \mathbb{DB}$. Any DBA has width 1 when regarded as an NBA. It follows that $L \in \mathbb{DM}_1^+$ implies $L \in \mathbb{NB}_1$. We note, however, that given a DMA $\mathcal{M}$ such that $\mathcal{M} \in \mathbb{DM}_1^+$ it is not always the case that we can define a DBA on the same structure as $\mathcal{M}$. Consider, for instance, the DMA $\mathcal{M}_2$ of Fig. 2 defined over alphabet $\Sigma = \{a, b, c\}$. It accepts the language $(\Sigma^* a \Sigma^* b)^\omega$. No rejecting set subsumes an accepting set, thus the maximal length of a positive chain is 1 and $[\![\mathcal{M}_2]\!] \in \mathbb{DM}_1^+$. While $\mathcal{M}_2$ has two states there is no DBA with two states that accepts the same language. A minimal DBA for $[\![\mathcal{M}_2]\!]$ requires at least 3 states. The DBA $\mathcal{B}_2$ of the same figure is a minimal DBA for $[\![\mathcal{M}_2]\!]$.

Proposition 8 below follows from the fact that $L \in \mathbb{DM}_1^+ \iff L \in \mathbb{DB}$. We provide a direct construction based on the LAR (*latest appearance record*) data structure due to Gurevich and Harrington [6], for completeness and to introduce LAR which will be used in the proof of the relations of higher levels of the hierarchy as well.

Proposition 8. $L \in \mathbb{DM}_1^+$ *implies* $L \in \mathbb{NB}_1$.

Proof. Let $\mathcal{M}$ be a DMA in $\mathbb{DM}_1^+$. From the definition of the hierarchy class $\mathbb{DM}_1^+$ it follows that no superset of an accepting SCC in $\mathcal{M}$ can be rejecting. On the other hand, a subset of an accepting set may be rejecting. To be able to use a Büchi condition we need to define a set of states F such that a visit to F guarantees that *all* states of some accepting set were visited. The solution uses the LAR (*latest appearance record*) data structure due to Gurevich and Harrington [6]. The idea is to construct a deterministic automaton whose states are permutations of the states of $\mathcal{M}$ augmented by a hit position, denoted $\sharp$. If the current state is $p_1 p_2 \ldots p_i \sharp p_{i+1} \ldots p_n$ and $\delta(p_n, a) = q$ and $q = p_j$, then

$$\delta_\mathcal{B}(\langle p_1 p_2 \ldots p_i \sharp p_{i+1} \ldots p_n \rangle, a) = \langle p_1 p_2 \ldots p_{j-1} \sharp p_{j+1} \ldots p_n p_j \rangle$$

That is, the transition relation moves the state p_j currently visited by $\mathcal{M}$ to the rightmost position in the list, and moves the $\sharp$ symbol to the position on the list where p_j resided previously.

We formalize this using the following definitions, that will also be used in later proofs. Formally, let $Q = \{q_0, q_1, \ldots, q_{n-1}\}$, $\delta : Q \times \Sigma \to Q$, $\tau : 2^Q \to \{+, -\}$, $\sharp$ a symbol not in Q, and $A \subseteq Q$ a subset of cardinality ℓ with states $p_1, p_2, \ldots, p_\ell$ so that if $p_i = q_j$ and $p_{i+1} = q_k$ then $j < k$. Let

- $\text{LARSET}(A) = \{w \in (A \cup \{\sharp\})^* \mid \forall q \in A \cup \{\sharp\}, |w|_q = 1\}^3$
- $\text{LARINIT}(A, p_i) = \langle p_1 p_2 \ldots p_{i-1} \sharp p_{i+1} p_{i+2} \ldots p_\ell p_i \rangle$
- $\text{LARTRANS}(A) = \{(\langle p_1 p_2 \ldots p_i \sharp p_{i+1} \ldots p_\ell \rangle, a, \langle p_1 p_2 \ldots p_{j-1} \sharp p_{j+1} \ldots p_\ell p_j \rangle) \mid \delta(p_\ell, a) = p_j\}$
- $\text{LARACC}(A, \tau) = \{u \sharp v \mid set(uv) = A, \ \tau(set(v)) = +\}$ where $set(v)$ denotes the set of states in the word v.

Let $\mathcal{M} = \langle \Sigma, Q, q_0, \delta, \tau \rangle$. Let $\mathcal{B}$ be the DBA $\langle \Sigma, \text{LARSET}(Q), \text{LARINIT}(Q, q_0), \text{LARTRANS}(Q), \text{LARACC}(Q, \tau) \rangle$.

Lemma 9 [5, Lemma 1.21]. *Let ρ be a run of $\mathcal{M}$ on a given word w and let $\rho_\mathcal{B} = s_0 s_1 s_2 \ldots$ be the run of $\mathcal{B}$ on w where $s_i = u_i \sharp v_i$. Then $\mathsf{Inf}(\rho) = S$ iff the following conditions hold*

- *for some $i_0 \in \mathbb{N}$ for all $i > i_0$ we have $set(v_i) \subseteq S$ and*
- *for infinitely many i's we have $set(v_i) = S$*

Since $\mathcal{M} \in \mathbb{DM}_1^+$ guarantees that no superset of an accepting set may be rejecting, it is enough to require that we infinitely often visit a state $u \sharp v$ where all states of v form an accepting set of $\mathcal{M}$, this is exactly the acceptance condition of $\mathcal{B}$. Therefore, $[\![\mathcal{B}]\!] = [\![\mathcal{M}]\!]$. Since $\mathcal{B}$ is deterministic, the width of $\mathcal{B}$ is one. Hence, $L \in \mathbb{NB}_1$. □

For $L \in \mathbb{DM}_1^-$ there is no guarantee that an equivalent DBA exists. We are guaranteed, though, that an equivalent DCA exists. In Proposition 11 we show that this entails that $L \in \mathbb{NB}_2$.

First we need some terms and a lemma.

Let $\rho = q_0 q_1 q_2 \ldots$ be a run of a given automaton $\mathcal{A}$ on a given word w. We say that the run ρ *gets trapped* in an SCC S if starting from some $z \in \mathbb{N}$ for every $z' > z$ the state $q_{z'}$ of this run belongs to S. In an automaton with finitely many states, every run on an infinite word should eventually get trapped in some SCC. We let $trap(\rho)$ denote the minimal SCC that ρ gets trapped in. If the automaton is non-deterministic, there may be several runs on a given word and each run may get trapped in a different SCC. For a skeleton path $\varrho = Q_0 Q_1 Q_2 \ldots$ we say that it gets trapped in $\{S_1, \ldots, S_k\}$ if starting from some $z \in \mathbb{N}$ for every $z' > z$ we have $Q_{z'} \subseteq \cup_{i \in [1..k]} S_i$. We let $trap(\varrho) = \{S_1, \ldots, S_k\}$ denote the minimal set of minimal SCCs that ϱ gets trapped in.

Lemma 10. *Let $\mathcal{N} = \langle \Sigma, Q, q_0, \delta, F \rangle$ be an NBA with an SCC S satisfying the following two conditions:*

- *For every letter $a \in \Sigma$ and every state $q \in S$, $|\delta(q, a)| = 1$.*
- *For every pair of states $q, p \in S$ we have $[\![\mathcal{N}_{q_0, q}]\!] \cap [\![\mathcal{N}_{q_0, p}]\!] = \emptyset$.*[4]

---

[3] We use $|w|_a$ to denote the number of occurrences of the letter a in w.

[4] The notation $\mathcal{N}_{q, q'}$ is used to denote the NFA $\langle \Sigma, Q, q, \delta, \{q'\} \rangle$ obtained from $\mathcal{N}$ by making q the initial state, q' the final state, and regarding it as a nondeterminsitc automaton on finite words.

Let $\mathbf{SK}_w^A$ be the skeleton-tree of w w.r.t $\mathcal{N}$. Let ϱ_1 and ϱ_2 be two skeleton paths in $\mathbf{SK}_w^A$. Then $S \in \mathrm{trap}(\varrho_1)$ implies $S \notin \mathrm{trap}(\varrho_2)$.

Proof. Consider a word $w = a_1 a_2 a_3 \ldots$, and let ρ_1 and ρ_2 be two different runs of $\mathcal{N}$ on w. Assume now both ρ_1 and ρ_2 get trapped in S. We claim that there exists a point in which both runs reach the same state. That is, if $\rho_1 = q_0 q_1 q_2 \ldots$ and $\rho_2 = p_0 p_1 p_2 \ldots$ then there exists $z \in \mathbb{N}$ such that for every $z' > z$ we have $q_{z'} = p_{z'}$. Assume ρ_1, ρ_2 enter the SCC S at time points z_1, z_2, resp. and w.l.o.g. $z_1 \leq z_2$. Then ρ_1 (resp. ρ_2) entered S after reading the prefix $a_1 a_2 \ldots a_{z_1}$ of w (resp. $a_1 a_2 \ldots a_{z_2}$). We claim that $p_{z_2} = q_{z_2}$. If not, then exists two states $p, q \in S$ such that $p = p_{z_2} \neq q_{z_2} = q$ and $a_1 a_2 \ldots a_{z_2} \in [\![\mathcal{N}_{q_0, p}]\!] \cap [\![\mathcal{N}_{q_0, q}]\!]$ contradicting the second premise of the lemma. Now that we established $p_{z_2} = q_{z_2}$, since by the first premise all transitions within S are deterministic, it follows that $p_{z'} = q_{z'}$ for every $z' > z_2$. Since all runs that get trapped in S eventually reach the same state, it follows that they must conjoin to the same path ϱ of $\mathbf{SK}_w^A$.   □

Proposition 11. $L \in \mathbb{DM}_1^-$ *implies* $L \in \mathbb{NB}_2$.

Proof. Let $\mathcal{M}$ be a DMA in $\mathbb{DM}_1^-$. From the definition of the hierarchy class $\mathbb{DM}_1^-$ it follows that no subset of an accepting SCC in $\mathcal{M}$ can be rejecting. We show that we can build an equivalent NBA $\mathcal{B}$ of width 2 using the following idea. The NBA $\mathcal{B}$ will consists of all of $\mathcal{M}$'s states and transitions. In addition, for each maximal accepting SCC A of M, $\mathcal{B}$ will have the ability to non-deterministically transit to a copy A' of A. The primed copy of A will have all the inner transitions of A, but no transitions out of A'. This way $\mathcal{B}$ can at any point during the run choose to move to a primed copy A' of one of the accepting SCCs A. Once this choice was made, $\mathcal{B}$ can only remain in A' or fall off the automaton. The set of accepting states of $\mathcal{B}$ consists of all states in the primed copies. If $\mathcal{B}$ visits such a state in say the accepting SCC A_i' infinitely often then the run of $\mathcal{M}$ gets trapped in A_i, and if the run of $\mathcal{M}$ gets trapped in A_i, then there exists a run of $\mathcal{B}$ that will get trapped in A_i'. Therefore, both recognize the same language.

Formally, let $\mathcal{M} = \langle \Sigma, Q, \lambda, \delta, \tau \rangle$. Let $A_1, A_2, \ldots, A_k$ be the maximal accepting SCCs of $\mathcal{M}$.[5] The NBA $\mathcal{B} = \langle \Sigma, Q_\mathcal{B}, q_{0\mathcal{B}}, \delta_\mathcal{B}, F_\mathcal{B} \rangle$ is defined as follows. The states of $Q_\mathcal{B}$ are $Q \cup A_1' \cup A_2' \ldots \cup A_k'$ where $A_i' = \{q' \mid q \in A_i\}$. The initial state $q_{0\mathcal{B}}$ is q_0. The accepting states $F_\mathcal{B}$ are $\cup_{i \in [1..k]} A_i'$. The transitions $\delta_\mathcal{B}$ are defined as follows. If $(q, a, p) \in \delta$ then $(q, a, p) \in \delta_\mathcal{B}$. In addition, if $q \in Q$ and $p \in A_i$ then $(q, a, p') \in \delta_\mathcal{B}$. If $q \in A_i$ and $p \in A_i$ then $(q', a, p') \in \delta_\mathcal{B}$.

Suppose a run $q_0 q_1 q_2 \ldots$ of $\mathcal{M}$ on a given word w gets trapped in an accepting SCC A_i. Since no subset of A_i is rejecting, this run is accepting. Let z_0 be such that $q_z \in A_i$ for every $z > z_0$. Then $\mathcal{B}$ can mimic $\mathcal{M}$ up to point $z_0 - 1$, at z_0 it can make a nondeterministic transition to A_i' and then stay there forever long. That is, the run $q_0 q_1 q_2 \ldots q_{z_0 - 1} q_{z_0}' q_{z_0 + 1}' q_{z_0 + 2}' \ldots$ where $q_{z_0 + i}'$ is the primed version of $q_{z_0 + i}$ is a run of $\mathcal{B}$, and this run is accepting. For the other direction, if

---

[5] An SCC A of $\mathcal{M}$ is said to be *maximal accepting* if no accepting SCC A' subsumes it. Note that A is not required to be an MSCC. E.g. if $R = A \cup \{q\}$ is an MSCC, and R is rejecting and A is accepting, then A is a maximal accepting SCC, but not an MSCC.

there exists an accepting run of $\mathcal{B}$ on a given word w, it means that starting from some point the run moved to one of the A_i''s and stayed there forever long. This entails that when $\mathcal{M}$ reads w, it will get trapped in the SCC A_i. (If this was not the case, that run of $\mathcal{B}$ on w would encounter a letter for which no transitions is available.) Thus $[\![\mathcal{M}]\!] = [\![\mathcal{B}]\!]$.

Next we show that the width of $\mathcal{B}$ is 2. Let ρ_w be the run of $\mathcal{M}$ on a given word w and let $trap(\rho_w) = T$. If T is rejecting then $\mathcal{B}$ can only get trapped in T, since on any choice to move to some A_i' it will end up falling of A_i' since $\mathcal{M} \in \mathrm{DM}_1^-$ implies $T \not\subseteq A_i$. (If this was not the case then, since for any transition in A_i' there is a transition from A_i to A_i, this would entail that $\mathcal{M}$ would have got trapped in A_i which is not the case.) If T is accepting then it equals some A_j for $j \in [1..k]$. Then a run of $\mathcal{B}$ on w can get trapped either in A_j or in A_j', but it cannot get trapped in any other A_i or A_i'. Since, from the same arguments as in the rejecting case, if at some point $\mathcal{B}$ chooses to transit to A_i' it will eventually fall off of it. We have shown that each run of $\mathcal{B}$ may get trapped in at most 2 MSCCs. Since all MSCCs of $\mathcal{B}$ (the originals of $\mathcal{M}$ and the new MSCCs A_i') satisfy the premises of Lemma 10, and since for each skeleton path ϱ we have $trap(\varrho) \neq \emptyset$, the maximum width of any run of $\mathcal{B}$ is 2. Thus $[\![\mathcal{M}]\!] \in \mathrm{NB}_2$.   □

We can now generalize these two ideas to obtain a construction for DM_k^+ for arbitrary k.

Proposition 12. $L \in \mathrm{DM}_k^+ \implies L \in \mathrm{NB}_{\lceil \frac{k}{2} \rceil + 1}$.

Proof. Let $\mathcal{M}$ be a DMA in DM_k^+. From the definition of the hierarchy classes DM_k^+ it follows that the maximum length of a positive chain is k. We build an equivalent NBA $\mathcal{B}$ of width at most $\lceil \frac{k}{2} \rceil + 1$ using the following idea. As in the proof of Proposition 11, $\mathcal{B}$ will consists of all states of the given DMA $\mathcal{M}$, and will have nondeterministic transitions to copies of accepting SCCs. Unlike in that proof, we will need to consider not just the maximal accepting SCCs but all accepting SCCs. As in the proof of Proposition 8, since an accepting SCC A may contain rejecting SCCs, we will use a LAR construction to make sure that all states of an accepting SCC are visited infinitely often.

Formally, let $\mathcal{M} = \langle \Sigma, Q, \lambda, \delta, \tau \rangle$. Assume $A_1, A_2, \ldots, A_m$ are the accepting SCCs of $\mathcal{M}$. We define the NBA $\mathcal{B} = \langle \Sigma, Q_\mathcal{B}, q_{0\mathcal{B}}, \delta_\mathcal{B}, F_\mathcal{B} \rangle$ as follows.

- $q_{0\mathcal{B}} = q_0$
- $Q_\mathcal{B} = Q \uplus_{i \in [1..m]} \mathrm{LARSET}(A_i)$
- $\delta_\mathcal{B} = \delta \uplus_{i \in [1..m]} \mathrm{LARTRANS}(A_i) \cup$
    $\{(q, a, \mathrm{LARINIT}(A_i, p)) \mid (q, a, p) \in \delta, \ q \in Q \setminus A_i, \ p \in A_i\}$
- $F_\mathcal{B} = \uplus_{i \in [1..m]} \mathrm{LARACC}(A_i, \tau_i)$ where $\tau_i(S) = +$ iff $S = A_i$.

We claim that $\mathcal{B}$ recognizes the same language as $\mathcal{M}$. Let w be a word and let ρ be the run of $\mathcal{M}$ on w. Let $\mathsf{Inf}(\rho) = S$. If $\tau(S) = +$ then $S = A_i$ for some $i \in [1..m]$. Assume ρ gets trapped in S after time point z. Thus $\mathcal{B}$, at some time point after z, can choose a transition of the form $(q, a, \mathrm{LARINIT}(A_i, p))$ and move to $\mathrm{LARSET}(A_i)$. Since $\mathcal{M}$ gets trapped in A_i, $\mathcal{B}$ will not fall off $\mathrm{LARSET}(A_i)$, and by Lemma 9, infinitely often $\mathcal{B}$ will visit a state of the form $\natural_i v$ such that $set(v) = A_i$. Thus $\mathcal{B}$ will accept. Suppose now $\tau(S) = -$. If $\mathcal{B}$ did not move to any of the LARSET(A_i), clearly it will not accept. Suppose it did move to LARSET(A_i) for some i. If A_i does not subsume S then $\mathcal{B}$ will fall off LARSET(A_i). Assume thus $A_i \supseteq S$. Then by Lemma 9, eventually only states of the form $u\natural_i v$ where $set(v) \subseteq S$ will be visited. Since $A_i \supsetneq S$, by Lemma 9, it will not be the case that infinitely often states of the form $\natural_i v$ with $set(v) = A_i$ will be visited, thus $\mathcal{B}$ will reject. Hence, $[\![\mathcal{B}]\!] = [\![\mathcal{M}]\!]$.

We turn to reason about $\mathcal{B}$'s width. Let ρ_w be the run of $\mathcal{M}$ on w and assume $trap(\rho_w) = T$. Let ρ be a run of $\mathcal{B}$, then from the same arguments as in the proof of Proposition 11 we have that either $trap(\rho) = T$ or $trap(\rho) =$ LARSET(A_i) for some $A_i \supseteq T$. Note that by definition of $\mathcal{B}$, for every $i \in [1..m]$ the set LARSET(A_i) is an MSCC, and all the MSCCs of $\mathcal{B}$ satisfy the premises of Lemma 10. Thus if ϱ_1 and ϱ_2 are two skeleton paths of $\mathbf{SK}_w^{\mathcal{B}}$ then $trap(\varrho_1) \cap trap(\varrho_2) = \emptyset$. We claim further that if LARSET(A_1) $\in trap(\varrho_1)$ and LARSET(A_2) $\in trap(\varrho_2)$ then either $A_1 \subseteq A_2$ or $A_2 \subseteq A_1$. Assume this is not the case. Note that $\exists \varrho \in \mathbf{SK}_w^{\mathcal{B}}$ such that $T \in trap(\varrho)$. Since there is a single run leading to T, and all of its states are non-accepting, ϱ must be the rightmost skeleton-path in $\mathbf{SK}_w^{\mathcal{B}}$. Since Q is the only non sink MSCC of $\mathcal{B}$, all skeleton paths split from ϱ at some point. Assume ϱ_1 and ϱ_2 split from ϱ at time points z_1 and z_2, resp. The nodes at the splits must be accepting (since their sibling in ϱ is non-accepting). Thus at z_1 it must be that ϱ_1 recently visited all states of A_1 and at z_2 it must be that ϱ_2 recently visited all states of A_2. Assume w.l.o.g. $z_1 < z_2$. If at some time point between z_1 and z_2, the original path of $\mathcal{M}$, ρ_w visited a node in $A_2 \setminus A_1$ then ϱ_1 will not have any descendants (since LARSET(A_1) has no corresponding transitions). Assume thus all nodes of $A_2 \setminus A_1$ have recently been visited before z_1. Then at z_1 not only all A_1 are visited but also all of A_2, thus there exists a split corresponding to LARSET(A_2) before or at z_2, which contradicts the split at z_2 since LARSET(A_2) can only be trapped in one skeleton-path. Therefore if the set of skeleton paths of $\mathbf{SK}_w^{\mathcal{B}}$ is $\{\varrho\} \cup \{\varrho_1, \ldots, \varrho_\ell\}$ where ϱ_i split from ϱ before ϱ_{i+1} then $trap(\varrho_1), \ldots, trap(\varrho_\ell)$ will correspond to accepting sets $A_1, \ldots, A_\ell$ along one inclusion chain of $\mathcal{M}$. Since the length of a positive chain is bounded by k, the number of accepting sets along such a chain is bounded by $\lceil \frac{k}{2} \rceil$. Adding ϱ, the rightmost branch, we obtain that the width of $\mathcal{B}$ is $\lceil \frac{k}{2} \rceil + 1$.   □

This completes the proof of Theorem 3.

Acknowledgments. I would like to thank Javier Esparza for asking me if I can characterize words/languages of certain skeleton-width, when I presented [4] at CONCUR'15. This question initiated this study. I would like to thank Yoad Lustig for many interesting discussions on Büchi determinization, Dana Angluin for many interesting discussions on the Wagner Hierarchy, and Orna Kupferman for important comments on an early draft.

References

1. Barua, R.: The Hausdorff-Kuratowski hierarchy of omega-regular languages and a hierarchy of Muller automata. Theor. Comput. Sci. **96**(2), 345–360 (1992)
2. Carton, O., Perrin, D.: The Wadge-Wagner hierarchy of omega-rational sets. In: ICALP, pp. 17–35 (1997)
3. Perrin, D., Pin, J.E.: Infinite Words: Automata, Semigroups, Logic and Games. Elsevier, Amsterdam (2004)
4. Fisman, D., Lustig, Y.: A modular approach for Büchi determinization. In: 26th International Conference on Concurrency Theory, pp. 368–382 (2015)
5. Grädel, E., Thomas, W., Wilke, T. (eds.): A Guide to Current Research. LNCS, vol. 2500. Springer, Heidelberg (2002)
6. Gurevich, Y., Harrington, L.: Trees, automata, and games. In: Proceedings of the 14th ACM Symposium on Theory of Computing, pp. 60–65. ACM Press (1982)
7. Kähler, D., Wilke, T.: Complementation, disambiguation, and determinization of Büchi automata unified. In: Aceto, L., Damgård, I., Goldberg, L.A., Halldórsson, M.M., Ingólfsdóttir, A., Walukiewicz, I. (eds.) ICALP 2008, Part I. LNCS, vol. 5125, pp. 724–735. Springer, Heidelberg (2008)
8. Kaminski, M.: A classification of ω-regular languages. Theor. Comput. Sci. **36**, 217–229 (1985)
9. Krishnan, S.C., Puri, A., Brayton, R.K.: Structural complexity of ω-automata. In: Mayr, E.W., Puech, C. (eds.) STACS 1995. LNCS, vol. 900, pp. 143–156. Springer, Heidelberg (1995)
10. Landweber, L.: Decision problems for ω-automata. Math. Syst. Theor. **3**, 376–384 (1969)
11. Muller, D., Schupp, P.: Simulating alternating tree automata by nondeterministic automata: new results and new proofs of theorems of Rabin, McNaughton and Safra. Theor. Comput. Sci. **141**, 69–107 (1995)
12. Piterman, N.: From nondeterministic Büchi and Streett automata to deterministic parity automata. In: Proceedings of the 21st IEEE Symposium on Logic in Computer Science, IEEE Computer Society Press (2006)
13. Wagner, K.: A hierarchy of regular sequence sets. In: Bečvář, J. (ed.) Mathematical Foundations of Computer Science 1975 4th Symposium, Mariánské Lázně, September 1–5 1975. Lecture Notes in Computer Science, vol. 32, pp. 445–449. Springer, Heidelberg (1975)
14. Wilke, T., Yoo, H.: Computing the Wadge degree, the Lifschitz degree, and the Rabin index of a regular language of infinite words in polynomial time. In: Proceedings of the 6th Joint Conference on Theory and Practice of Software Development, pp. 288–302 (1995)

Compositional Bisimulation Minimization for Interval Markov Decision Processes

Vahid Hashemi[1,2]([✉]), Holger Hermanns[2], Lei Song[3,4], K. Subramani[5], Andrea Turrini[4], and Piotr Wojciechowski[5]

[1] Max Planck Institute for Informatics, Saarbrücken, Germany
hashemi@mpi-inf.mpg.de
[2] Department of Computer Science, Saarland University, Saarbrücken, Germany
[3] University of Technology Sydney, Sydney, Australia
[4] State Key Laboratory of Computer Science, Institute of Software, CAS, Beijing, China
[5] LDCSEE, West Virginia University, Morgantown, WV, USA

Abstract. Formal verification of PCTL properties of MDPs with convex uncertainties has been recently investigated by Puggelli et al. However, model checking algorithms typically suffer from state space explosion. In this paper, we address probabilistic bisimulation to reduce the size of such an MDP while preserving PCTL properties it satisfies. We give a compositional reasoning over interval models to understand better the ways how large models with interval uncertainties can be composed. Afterwards, we discuss computational complexity of the bisimulation minimization and show that the problem is coNP-complete. Finally, we show that, under a mild condition, bisimulation can be computed in polynomial time.

Keywords: Markov decision process · Interval MDP · Compositionality · Bisimulation · Complexity

1 Introduction

Probability, nondeterminism, and uncertainty are three core aspects of real systems. *Probability* arises when a system, performing an action, is able to reach more than one state and we can estimate the proportion between reaching each of such states: probability can model both specific system choices (such as flipping a coin, commonly used in randomized distributed algorithms) and general system properties (such as message loss probabilities when sending a message over a wireless medium). *Nondeterminism* represents behaviors that we can not or we do not want to attach a precise (possibly probabilistic) outcome to. This might reflect the concurrent execution of several components at unknown (relative) speeds or behaviors we keep undetermined for simplifying the system or allowing for different implementations. *Uncertainty* relates to the fact that not all system parameters may be known exactly, including exact probability values.

Probabilistic automata (*PAs*) [32] extend classical concurrency models in a simple yet conservative fashion. In probabilistic automata, concurrent processes

© Springer International Publishing Switzerland 2016
A.-H. Dediu et al. (Eds.): LATA 2016, LNCS 9618, pp. 114–126, 2016.
DOI: 10.1007/978-3-319-30000-9_9

may perform probabilistic experiments inside a transition. Labeled transition systems are instances of this model family, obtained by restricting to Dirac distributions (assigning full probability to single states). Thus, foundational concepts and results of standard concurrency theory are retained in full and extend smoothly to the PA model. PAs are akin to *Markov decision processes* (*MDPs*), their fundamental beauty can be paired with powerful model checking techniques, as implemented for instance in the PRISM tool [27].

In PAs and MDPs, probability values need to be specified precisely. This is often an impediment to their applicability to real systems. Instead it appears more viable to specify ranges of probabilities, so as to reflect the uncertainty in these values. This leads to a model where intervals of probability values replace probabilities. This is the model studied in this paper, we call it *interval Markov decision processes*, IMDPs.

In standard concurrency theory, *bisimulation* plays a central role as the undisputed reference for distinguishing the behaviour of systems. Besides for distinguishing systems, bisimulation relations conceptually allow us to reduce the size of a behaviour representation without changing its properties (i.e., with respect to logic formulae the representation satisfies). This is particularly useful to alleviate the state explosion problem notoriously encountered in model checking. If the bisimulation is a congruence with respect to a parallel composition operator used to build up the model out of smaller ones, this can give rise to a compositional strategy to associate a small model to a large system without intermediate state space explosion. In several related settings, this strategy has been proven very effective [9,18]. In order to be of practical use also for IMDPs, efficient bisimulation decision procedures are required.

This paper discusses the key ingredients for this to work for bisimulation on IMDPs. On the one hand, we discuss congruence properties, on the other hand we discuss the complexity of deciding bisimulation on IMDPs. More precisely, on the one hand we show that the probabilistic bisimulation for IMDPs proposed in [16] for the cooperative resolution of the nondeterminism in a dynamic setting is preserved by parallel composition and that it is transitive, so it is indeed a congruence. On the other hand, we show that deciding probabilistic bisimulation for IMDPs is in general coNP-complete; under a mild restriction on the amount of nondeterminism in the IMDP, the bisimulation becomes polynomial.

Related Work. Various probabilistic formalisms with uncertain transitions are studied in the literature. Uncertain MDPs [28,30,37] allow more general sets of distributions to be associated with each transition, not only those described by intervals. Usually, they are restricted to *rectangular uncertainty sets* requiring that the uncertainty is linear and independent for any two transitions of any two states. Our general algorithm working with polytopes can be easily adapted to this setting. Parametric MDPs [14] instead allow such dependencies as every probability is described as a rational function of a finite set of global parameters.

From the compositional specification point of view, Interval MCs [21,25] and Abstract PAs [10] serve as specification theories for MC and PAs featuring satisfaction relation, and various refinement relations. In order to be closed under parallel composition, Abstract *PA*s allow general polynomial constraints on probabilities instead of interval bounds. Since for Interval *MC*s it is not possible to explicitly construct parallel composition, the problem of whether there is a common implementation of a set of Interval *MC*s is addressed instead [11]. To the contrary, interval bounds on *rates* of outgoing transitions work well with parallel composition in the continuous-time setting of Abstract Interactive *MC*s [24]. The reason is that unlike probabilities, rates do not need to sum up to 1. Authors of [39] successfully define parallel composition for interval models by separating synchronizing transitions from the transitions with uncertain probabilities.

Probabilistic bisimulation for uncertain probabilistic models has been studied quite recently in [16]. To the best of our knowledge, we are not aware of any other existing results on probabilistic bisimulations for uncertain or parametric models. Among similar concepts studied in the literature are simulation [39] and refinement [10,21] relations for the previously mentioned models.

Many new verification algorithms for interval models appeared in last few years. Reachability and expected total reward is addressed for Interval *MC*s [8] as well as *IMDP*s [38]. PCTL model checking and PLTL model checking are studied for Interval *MC*s [6,8] and also for *IMDP*s [30,37]. Among other technical tools, all these approaches make use of (robust) dynamic programming relying on the fact that transition probability distributions are resolved dynamically. For the static resolution of distributions, adaptive discretization technique for PCTL parameter synthesis is given in [14]. Uncertain models are also widely studied in the control community (see, e.g. [28,38]), mainly interested in maximal expected finite-horizon/discounted reward.

Organization of the paper. We start with necessary preliminaries in Sect. 2. In Sect. 3, we give the definition of probabilistic bisimulation for *IMDP*s and discuss the main results of [16]. Furthermore, we show that the probabilistic bisimulation over *IMDP*s is compositional and transitive. In Sect. 4, we discuss the hardness of deciding probabilistic bisimulation and also show that polynomiality is achievable under a mild condition. Finally, in Sect. 5 we conclude the paper.

2   Preliminaries

Given $n \in \mathbb{N}$, we denote by $\mathbf{1} \in \mathbb{R}^n$ the unit vector and by $\mathbf{1}^T$ its transpose. In the sequel, the comparison between vectors is element-wise and all vectors are column ones unless otherwise stated. For a given set $P \subseteq \mathbb{R}^n$, we denote by $\mathrm{CH}(P)$ the convex hull of P. We denote by $\mathbb{I}$ is a set of closed subintervals of $[0,1]$ and, for a given $[a,b] \in \mathbb{I}$, we let $\inf[a,b] = a$ and $\sup[a,b] = b$.

For a given set X, we denote by $\Delta(X)$ the set of discrete probability distributions over X. For an equivalence relation $\mathcal{R}$ on X and $\rho_1, \rho_2 \in \Delta(X)$, we write $\rho_1 \, \mathcal{L}(\mathcal{R}) \, \rho_2$ if for each $\mathcal{C} \in X/\mathcal{R}$, it holds that $\rho_1(\mathcal{C}) = \rho_2(\mathcal{C})$. By abuse of notation, we extend $\mathcal{L}(\mathcal{R})$ to distributions over $X/\mathcal{R}$, i.e., for $\rho_1, \rho_2 \in \Delta(X/\mathcal{R})$, we write $\rho_1 \, \mathcal{L}(\mathcal{R}) \, \rho_2$ if for each $\mathcal{C} \in X/\mathcal{R}$, it holds that $\rho_1(\mathcal{C}) = \rho_2(\mathcal{C})$.

2.1   Interval Markov Decision Processes

Let us formally define Interval Markov Decision Processes.

Definition 1. *An* Interval Markov Decision Process *(IMDP)* $\mathcal{M}$ *is a tuple* $\mathcal{M} = (S, \bar{s}, \mathcal{A}, \text{AP}, L, I)$*, where S is a finite set of* states*, $\bar{s} \in S$ is the* initial state*, $\mathcal{A}$ is a finite set of* actions*, AP is a finite set of atomic propositions, $L\colon S \to 2^{\text{AP}}$ is a* labelling function*, and $I\colon S \times \mathcal{A} \times S \to \mathbb{I}$ is an* interval transition probability function *such that for each s, there exist a and s' such that $I(s, a, s') \neq [0, 0]$.*

We denote by $\mathcal{A}(s)$ the set of actions that are enabled from state s, i.e., $\mathcal{A}(s) = \{ a \in \mathcal{A} \mid \exists s' \in S.I(s, a, s') \neq [0, 0] \}$. Furthermore, for each state s and action $a \in \mathcal{A}(s)$, we denote by $s \xrightarrow{a} \mu_s$ that $\mu_s \in \Delta(S)$ is a *feasible distribution*, i.e., for each state s' we have $\mu_s(s') \in I(s, a, s')$. We require that the set $\{ \mu_s \mid s \xrightarrow{a} \mu_s \}$, also denoted by $\mathcal{P}^{s,a}$, is non-empty for each state s and action $a \in \mathcal{A}(s)$. We denote by $b_{\mathcal{M}}$ the *branching* of $\mathcal{M}$, where $b_{\mathcal{M}} = \max_{s \in S} \{|\mathcal{A}(s)|\}$.

An *IMDP* is initiated in some state s_1 and then moves in discrete steps from state to state forming an infinite path $s_1 \, s_2 \, s_3 \ldots$. One step, say from state s_i, is performed as follows. First, an action $a \in \mathcal{A}(s)$ is chosen nondeterministically by *scheduler*. Then, *nature* resolves the uncertainty and chooses nondeterministically one corresponding feasible distribution $\mu_{s_i} \in \mathcal{P}^{s_i, a}$. Finally, the next state s_{i+1} is chosen randomly according to the distribution μ_{s_i}.

Let us define the semantics of an *IMDP* formally. A *path* is a finite or infinite sequence of states $\omega = s_1 \, s_2 \cdots$. For a finite path ω, we denote by $last(\omega)$ the last state of ω. The set of all finite paths and the set of all infinite paths are denoted by $Paths_{fin}$ and $Paths_{inf}$, respectively. Furthermore, let $Paths_\omega = \{\omega\omega' \mid \omega' \in Paths_{inf}\}$ denote the set of paths that have the finite prefix $\omega \in Paths_{fin}$.

Definition 2. *A* scheduler *is a function* $\sigma\colon Paths_{fin} \to \Delta(\mathcal{A})$ *that to each finite path ω assigns a distribution over the set of actions. A* nature *is a function* $\pi\colon Paths_{fin} \times \mathcal{A} \to \Delta(S)$ *that to each finite path ω and action $a \in \mathcal{A}(last(\omega))$ assigns a feasible distribution, i.e., an element of $\mathcal{P}^{s,a}$ where $s = last(\omega)$. We denote by Σ the set of all schedulers and by Π the set of all natures.*

For a state s, a scheduler σ, and a nature π, let $\Pr_s^{\sigma, \pi}$ denote the unique probability measure over $(Paths_{inf}, \mathcal{B})^1$ such that the probability $\Pr_s^{\sigma, \pi}[Paths_{s'}]$ of starting in s' equals 1 if $s' = s$ and 0, otherwise; and the probability $\Pr_s^{\sigma, \pi}[Paths_{\omega s'}]$ of traversing a finite path $\omega s'$ equals $\Pr_s^{\sigma, \pi}[Paths_\omega] \cdot \sum_{a \in \mathcal{A}} \sigma(\omega)(a) \cdot \pi(\omega, a)(s')$.

Observe that the scheduler does not choose an action but a *distribution* over actions. It is well-known [32] that such randomization brings more power in the context of bisimulations. Note that for nature this is not the case, since $\mathcal{P}^{s,a}$ is closed under convex combinations, thus nature can choose all distributions.

---

[1] Here, $\mathcal{B}$ is the standard σ-algebra over $Paths_{inf}$ generated from the set of all cylinder sets $\{Paths_\omega \mid \omega \in Paths_{fin}\}$. The unique probability measure is obtained by the application of the extension theorem (see, e.g. [3]).

2.2 Probabilistic Computation Tree Logic (PCTL)

There are various ways how to describe properties of interval *MDP*s. Here we focus on *probabilistic CTL* (PCTL) [15]. The syntax of PCTL state formulas φ and PCTL path formulas ψ is given by:

$$\varphi := true \mid x \mid \neg\varphi \mid \varphi_1 \wedge \varphi_2 \mid \mathsf{P}_{\bowtie p}(\psi)$$
$$\psi := \mathsf{X}\varphi \mid \varphi_1\mathsf{U}\varphi_2 \mid \varphi_1\mathsf{U}^{\leq k}\varphi_2$$

where $x \in AP$, $p \in [0,1]$ is a rational constant, $\bowtie \in \{\leq, <, \geq, >\}$, and $k \in \mathbb{N}$.

The satisfaction relation for PCTL formulae depends on the way how nondeterminism is resolved for the probabilistic operator $\mathsf{P}_{\bowtie p}(\psi)$. When quantifying both the nondeterminisms universally, we define the satisfaction relation $s \models_c \varphi$ as follows: $s \models_c x$ if $x \in L(s)$; $s \models_c \neg\varphi$ if not $s \models_c \varphi$; $s \models_c \varphi_1 \wedge \varphi_2$ if both $s \models_c \varphi_1$ and $s \models_c \varphi_2$; and

$$s \models_c \mathsf{P}_{\bowtie p}(\psi) \quad \text{if } \forall \sigma \in \Sigma \;\; \forall \pi \in \Pi : \mathrm{Pr}_s^{\sigma,\pi}\left[\{\,\omega \in Paths_{inf} \mid \omega \models_c \psi\,\}\right] \bowtie p$$

where the satisfaction relation $\omega \models_c \psi$ for an infinite path $\omega = s_1 s_2 \cdots$ and a path formula ψ is given by:

$\omega \models_c \mathsf{X}\varphi \qquad$ if $s_2 \models \varphi$;
$\omega \models_c \varphi_1\mathsf{U}^{\leq k}\varphi_2$ if there exists $i \leq k$ such that $s_i \models_c \varphi_2$ and $s_j \models_c \varphi_1$
$\qquad\qquad\qquad$ for each $1 \leq j < i$;
$\omega \models_c \varphi_1\mathsf{U}\varphi_2 \quad$ if there exists $k \in \mathbb{N}$ such that $\omega \models_c \varphi_1\mathsf{U}^{\leq k}\varphi_2$.

It is easy to show that the set $\models_c \psi$ is measurable for any path formula ψ, hence the definition is mathematically well defined.

3 Probabilistic Bisimulation for Interval MDPs

We now recall the main results on probabilistic bisimulation for *IMDP*s, as developed in [16]. In this work, we consider the notion of probabilistic bisimulation for the cooperative resolution of nondeterminism. This semantics is very natural in the context of verification of parallel systems with uncertain transition probabilities in which we assume that scheduler and nature are resolved *cooperatively* in the most *adversarial* way. Moreover, resolution of a feasible probability distribution respecting the interval constraints can be either done *statically* [21], i.e., at the beginning once for all, or *dynamically* [20,33], i.e., independently for each computation step. In this paper, we focus on dynamic approach in resolving the stochastic nondeterminism that is easier to work with algorithmically and can be seen as a relaxation of the static approach that is often intractable [2,6].

Let $s \longrightarrow \mu_s$ denote that a transition from s to μ_s can be taken cooperatively, i.e., that there is a scheduler $\sigma \in \Sigma$ and a nature $\pi \in \Pi$ such that $\mu_s = \sum_{a \in \mathcal{A}(s)} \sigma(s)(a) \cdot \pi(s,a)$. In other words, $s \longrightarrow \mu_s$ if $\mu_s \in \mathrm{CH}(\bigcup_{a \in \mathcal{A}(s)} \mathcal{P}^{s,a})$.

Definition 3 (cf. [16]). *Given an IMDP $\mathcal{M}$, let $\mathcal{R} \subseteq S \times S$ be an equivalence relation. We say that $\mathcal{R}$ is a* probabilistic bisimulation *if for each $(s,t) \in \mathcal{R}$ we have that $L(s) = L(t)$ and for each $s \longrightarrow \mu_s$ there exists $t \longrightarrow \mu_t$ such that $\mu_s \mathcal{L}(\mathcal{R}) \mu_t$. Furthermore, we write $s \sim_c t$ if there is a probabilistic bisimulation $\mathcal{R}$ such that $(s,t) \in \mathcal{R}$.*

Intuitively, each (cooperative) step of scheduler and nature from state s needs to be matched by a (cooperative) step of scheduler and nature from state t; symmetrically, s also needs to match t. It is shown in [16] that $\sim_c$ preserves the (cooperative) universally quantified PCTL satisfaction $\models_c$. More precisely,

Theorem 4 (cf. [16]). *For states $s \sim_c t$ and any PCTL formula φ, we have $s \models_c \varphi$ if and only if $t \models_c \varphi$.*

Computation of probabilistic bisimulation for *IMDP*s follows the standard partition refinement approach [22,29]. However, the core part of the algorithm is to find out whether two states "violate the definition of bisimulation". Verification of this violation, however, amounts to check the inclusion of polytopes defined as follows. For $s \in S$ and an action $a \in \mathcal{A}$, recall that $\mathcal{P}^{s,a}$ denotes the polytope of feasible successor distributions over *states* with respect to taking the action a in the state s. By $\mathcal{P}_{\mathcal{R}}^{s,a}$, we denote the polytope of feasible successor distributions over *equivalence classes* of $\mathcal{R}$ with respect to taking the action a in the state s. Formally, for $\mu \in \Delta(S/\mathcal{R})$ we set $\mu \in \mathcal{P}_{\mathcal{R}}^{s,a}$ if

$$\mu(\mathcal{C}) \in \left[\sum_{s' \in \mathcal{C}} \inf I(s,a,s'), \sum_{s' \in \mathcal{C}} \sup I(s,a,s') \right] \qquad \text{for each } \mathcal{C} \in S/\mathcal{R}.$$

Furthermore, we define $\mathcal{P}_{\mathcal{R}}^s = \mathrm{CH}(\bigcup_{a \in \mathcal{A}(s)} \mathcal{P}_{\mathcal{R}}^{s,a})$, the set of feasible successor distributions over $S/\mathcal{R}$ with respect to taking an *arbitrary* distribution over enabled actions in state s. As specified in [16], checking violation of a given pair of states amounts to check equality of the corresponding constructed polytopes for the states. As regards the computational complexity of the proposed algorithm, the following theorem indicates that it is fixed parameter tractable. Formally,

Theorem 5 (cf. [16]). *Computing $\sim_c$ on an IMDP $\mathcal{M}$ is in $|\mathcal{M}|^{\mathcal{O}(1)} \cdot 2^{\mathcal{O}(f)}$ where $|\mathcal{M}| = |S|^2 \cdot |\mathcal{A}|$ and $f = \max_{s \in S, a \in \mathcal{A}(s)} |\{ s' \in S \mid I(s,a,s') \neq [0,0] \}|$.*

3.1   Compositional Reasoning

The compositional reasoning is a widely used technique (see, e.g., [7,18,23]) that permits to deal with large systems. In particular, a large system is decomposed into multiple components running in parallel; such components are then minimized by replacing each of them by a bisimilar but smaller one so that the overall behaviour remains unchanged. In order to apply this technique, bisimulation has first to be extended to pairs of components and then to be shown to be transitive and preserved by the parallel composition operator. The extension to a pair of components is trivial and commonly done (see, e.g., [4,32]):

Definition 6. *Given two IMDPs $\mathcal{M}_1$ and $\mathcal{M}_2$, we say that they are probabilistic bisimilar, denoted by $\mathcal{M}_1 \sim_c \mathcal{M}_2$, if there exists a probabilistic bisimulation on the disjoint union of $\mathcal{M}_1$ and $\mathcal{M}_2$ such that $\bar{s}_1 \sim_c \bar{s}_2$.*

The next step is to define the parallel composition for *IMDPs*:

Definition 7. *Given two IMDPs $\mathcal{M}_1$ and $\mathcal{M}_2$ with $\mathcal{M}_i = (S_i, \bar{s}_i, \mathcal{A}_i, \mathsf{AP}_i, L_i, I_i)$, $i = 1, 2$, we define the parallel composition of $\mathcal{M}_1$ and $\mathcal{M}_2$, denoted by $\mathcal{M}_1 \parallel \mathcal{M}_2$ as the IMDP $(S, \bar{s}, \mathcal{A}, \mathsf{AP}, L, I)$ where $S = S_1 \times S_2$, $\bar{s} = (\bar{s}_1, \bar{s}_2)$, $\mathcal{A} = \mathcal{A}_1 \times \mathcal{A}_2$, $\mathsf{AP} = \mathsf{AP}_1 \cup \mathsf{AP}_2$, for each $(s_1, s_2) \in S$, $L(s_1, s_2) = L_1(s_1) \cup L_2(s_2)$, and for each $(s_1, s_2) \in S$, $(a_1, a_2) \in \mathcal{A}$, and $(t_1, t_2) \in S$, $I((s_1, s_2), (a_1, a_2), (t_1, t_2)) = I_1(s_1, a_1, t_1) \times I_2(s_2, a_2, t_2) = [l_1 \cdot l_2, u_1 \cdot u_2]$ where $[l_1, u_1] = I_1(s_1, a_1, t_1)$ and $[l_2, u_2] = I_2(s_2, a_2, t_2)$.*

Proposition 8. *For each IMDP $\mathcal{M}_1$, $\mathcal{M}_2$, and $\mathcal{M}_3$, if $\mathcal{M}_1 \sim_c \mathcal{M}_2$, then $\mathcal{M}_1 \parallel \mathcal{M}_3 \sim_c \mathcal{M}_2 \parallel \mathcal{M}_3$.*

Another property that is needed for $\sim_c$ in order to support the compositional reasoning is that it has to be transitive. This is indeed a property of $\sim_c$, as stated by the following proposition:

Proposition 9. *For each IMDP $\mathcal{M}_1$, $\mathcal{M}_2$, and $\mathcal{M}_3$, if $\mathcal{M}_1 \sim_c \mathcal{M}_2$ and $\mathcal{M}_2 \sim_c \mathcal{M}_3$, then $\mathcal{M}_1 \sim_c \mathcal{M}_3$.*

By being transitive and preserved by parallel composition, $\sim_c$ fully supports the compositional verification of complex systems.

4   Hardness and Tractability

Definition 3 is the central definition around which the paper revolves. Given an *IMDP* $\mathcal{M}$, the complexity of computing $\sim_c$ strictly depends on checking bisimilarity of a pair of states as a core part: in this section we will show that this verification routine is coNP-complete and therefore, the computation of $\sim_c$ as a whole is coNP-complete. Later on, we show how an equivalence relation can be computed in polynomial time under a mild condition. The definition of bisimulation can be reformulated equivalently as follows:

Definition 10. *Let $\mathcal{R} \subseteq S \times S$ be an equivalence relation. We say that $\mathcal{R}$ is a probabilistic bisimulation if $(s, t) \in \mathcal{R}$ implies that $L(s) = L(t)$ and $\mathcal{P}_{\mathcal{R}}^s = \mathcal{P}_{\mathcal{R}}^t$.*

As it is clear from Definition 10, the complexity of verifying bisimilarity of a pair of states s and t strictly depends on the complexity of the *Convex Hull Equivalence (CHE)* problem stated as follows:

Definition 11. *Given an IMDP $\mathcal{M}$, a pair of states s and t, $n, n_s, n_t \in \mathbb{N}$, two sets $\{ P^{s,i} \mid i \in \{1, \ldots, n_s\} \}$ and $\{ P^{t,i} \mid i \in \{1, \ldots, n_t\} \}$ where for each $r \in \{s, t\}$ and $i \in \{1, \ldots, n_r\}$, for given $l^{r,i}, u^{r,i} \in \mathbb{R}^n$, $P^{r,i}$ is the convex polyhedron*

$$P^{r,i} = \left\{ x^{r,i} \in \mathbb{R}^n \ \middle| \ \begin{array}{c} l^{r,i} \leq x^{r,i} \leq u^{r,i} \\ \mathbf{1}^T x^{r,i} = 1 \end{array} \right\},$$

the CHE problem asks to determine whether $\mathrm{CH}(\bigcup_{i=1}^{n_s} P^{s,i}) = \mathrm{CH}(\bigcup_{i=1}^{n_t} P^{t,i})$.

Theorem 12. *The CHE problem is coNP-complete.*

Proof (Sketch). We show that CHE problem is in coNP by reducing it to the coNP quantified linear implication problem [12] and that it is coNP-hard by reducing the coNP-hard tautology problem for 3DNF [13] to CHE. □

With this result at hand, together with Definition 10, we know that checking the bisimilarity of two states of an *IMDP* $\mathcal{M}$ is coNP-complete. Since the standard partition refinement algorithm performs this check a polynomial number of times (see, e.g., [4,5,22,29,35]), it follows that also computing $\sim_c$ is coNP-complete.

Theorem 13. *Given an IMDP $\mathcal{M}$, computing $\sim_c$ is coNP-complete.*

4.1 Decision Algorithm

In the previous section we have shown that the problem of deciding whether two states of an *IMDP* are bisimilar is coNP-complete. In this section we show that, under a mild restriction on *IMDP*s, the problem is solvable in polynomial time.

In the following we present a polynomial algorithm for deciding $\sim_c$ on *IMDP*s, provided its branching is considered as a constant. Given an *IMDP* $\mathcal{M}$ as in Definition 1 and an equivalence relation $\mathcal{R} \subseteq S \times S$, according to Definition 10, $\mathcal{R}$ is a probabilistic bisimulation iff for any $(s, t) \in \mathcal{R}$, $\mathcal{P}_{\mathcal{R}}^s = \mathcal{P}_{\mathcal{R}}^t$. Our algorithm relies on an algorithm in [34] to check efficiently whether $\mathcal{P}_{\mathcal{R}}^s = \mathcal{P}_{\mathcal{R}}^t$. However, the algorithm in [34] requires that both $\mathcal{P}_{\mathcal{R}}^s$ and $\mathcal{P}_{\mathcal{R}}^t$ are represented in form of $\mathcal{H}$-polytopes, i.e., defined as intersection of a finite number of closed half-spaces. Note $\mathcal{P}_{\mathcal{R}}^s = \mathrm{CH}(\bigcup_{a \in \mathcal{A}(s)} \mathcal{P}_{\mathcal{R}}^{s,a})$. Therefore, we focus on the procedure of obtaining an equivalent $\mathcal{H}$-polytope of $\mathcal{P}_{\mathcal{R}}^s$ given all $\mathcal{P}_{\mathcal{R}}^{s,a}$ being represented as $\mathcal{H}$-polytopes.

Now we show how to represent each $\mathcal{P}_{\mathcal{R}}^{s,a}$ as an $\mathcal{H}$-polytope. To simplify our presentation, we shall fix an order over all equivalence classes in $S/\mathcal{R}$. By doing so, any distribution $\rho \in \Delta(S/\mathcal{R})$ can be seen as a vector $\boldsymbol{v}$ such that $\boldsymbol{v}_i = \rho(\mathcal{C}_i)$ for each $1 \leq i \leq n$, where $n = |S/\mathcal{R}|$, $\mathcal{C}_i$ is the i-th equivalence class, and $\boldsymbol{v}_i$ the i-th element in $\boldsymbol{v}$. For the above discussion, $\rho \in \mathcal{P}_{\mathcal{R}}^{s,a}$ iff $\rho(\mathcal{C}_i) \in [l_i^a, u_i^a]$ for any $1 \leq i \leq n$ and $\rho \in \Delta(S/\mathcal{R})$, where $\boldsymbol{l}^a$ and $\boldsymbol{u}^a$ are vectors whose components are $\boldsymbol{l}_i^a = \sum_{s' \in \mathcal{C}_i} \inf I(s, a, s')$ and $\boldsymbol{u}_i^a = \sum_{s' \in \mathcal{C}_i} \sup I(s, a, s')$ for each $1 \leq i \leq n$, respectively. Therefore, $\mathcal{P}_{\mathcal{R}}^{s,a}$ corresponds to an $\mathcal{H}$-polytope defined by:

$$\left\{ \boldsymbol{x}^a \in \mathbb{R}^n \;\middle|\; \begin{array}{c} \boldsymbol{l}^a \leq \boldsymbol{x}^a \leq \boldsymbol{u}^a \\ \mathbf{1}^T \boldsymbol{x}^a = 1 \end{array} \right\}. \tag{1}$$

We have represented each $\mathcal{P}_{\mathcal{R}}^{s,a}$ for $a \in \mathcal{A}(s)$ as an $\mathcal{H}$-polytope. It is remained to show that $\mathcal{P}_{\mathcal{R}}^s = \mathrm{CH}(\bigcup_{a \in \mathcal{A}(s)} \mathcal{P}_{\mathcal{R}}^{s,a})$ can be represented by an $\mathcal{H}$-polytope as well. For this, we use the Balas Extension Theorem [1] in order to compute the $\mathcal{H}$-representation of the convex hull in polynomial time. The theorem allows for additional variables and thus maps the original space of variables to a higher dimensional space. By definition, $\boldsymbol{x} \in \mathcal{P}_{\mathcal{R}}^s$ iff there exist $\lambda^a \in \mathbb{R}$ and $\boldsymbol{x}^a \in \mathcal{P}_{\mathcal{R}}^{s,a}$

for each $a \in \mathcal{A}(s)$ such that $\boldsymbol{x} = \sum_{a \in \mathcal{A}(s)} \lambda^a \boldsymbol{x}^a$, $\lambda^a \geq 0$, and $\sum_{a \in \mathcal{A}(s)} \lambda^a = 1$. Therefore, together with Eq. (1), we have

$$
\mathcal{P}_{\mathcal{R}}^s = \left\{ \boldsymbol{x} \in \mathbb{R}^n \middle| \begin{array}{ll} \exists \lambda^a \geq 0. \exists \boldsymbol{x}^a. & \forall a \in \mathcal{A}(s) \\ \boldsymbol{x} = \sum_{a \in \mathcal{A}(s)} \lambda^a \boldsymbol{x}^a & \\ \boldsymbol{1}^T \boldsymbol{x}^a = 1 & \forall a \in \mathcal{A}(s) \\ \boldsymbol{l}^a \leq \boldsymbol{x}^a \leq \boldsymbol{u}^a & \forall a \in \mathcal{A}(s) \\ \sum_{a \in \mathcal{A}(s)} \lambda^a = 1 & \end{array} \right\}. \tag{2}
$$

Unfortunately, the constraint $\boldsymbol{x} = \sum_{a \in \mathcal{A}(s)} \lambda^a \boldsymbol{x}^a$ of Eq. (2) is not linear, hence is not a desirable $\mathcal{H}$-representation of $\mathcal{P}_{\mathcal{R}}^s$. In order to avoid non-linearity in the representation, we let $\boldsymbol{t}^a = \lambda^a \boldsymbol{x}^a$ for each $a \in \mathcal{A}(s)$. Constraints in Eq. (2) can be rewritten to an equivalent form as follows:

$$
\mathcal{P}_{\mathcal{R}}^s = \left\{ \boldsymbol{x} \in \mathbb{R}^n \middle| \begin{array}{ll} \exists \lambda^a \geq 0. \exists \boldsymbol{t}^a. & \forall a \in \mathcal{A}(s) \\ \boldsymbol{x} = \sum_{a \in \mathcal{A}(s)} \boldsymbol{t}^a & \\ \boldsymbol{1}^T \boldsymbol{t}^a = \lambda^a & \forall a \in \mathcal{A}(s) \\ \lambda^a \boldsymbol{l}^a \leq \boldsymbol{t}^a \leq \lambda^a \boldsymbol{u}^a & \forall a \in \mathcal{A}(s) \\ \sum_{a \in \mathcal{A}(s)} \lambda^a = 1 & \end{array} \right\}. \tag{3}
$$

Alternatively, Eq. (3) can be seen as an $\mathcal{H}$-polytope over variables $\boldsymbol{x}$, $\boldsymbol{t}^a$, and λ^a (where $a \in \mathcal{A}(s)$), whose projection on the space of variables $\boldsymbol{x}$ gives exactly the $\mathcal{H}$-polytope corresponding to $\mathcal{P}_{\mathcal{R}}^s$. We first try to project out variables $\boldsymbol{t}^a$ for all $a \in \mathcal{A}(s)$. According to Eq. (3), $\boldsymbol{x} = \sum_{a \in \mathcal{A}(s)} \boldsymbol{t}^a$, so all constraints related to $\boldsymbol{t}^a$ can be combined. As a result, we obtain:

$$
\mathcal{P}_{\mathcal{R}}^s = \left\{ \boldsymbol{x} \in \mathbb{R}^n \middle| \begin{array}{ll} \exists \lambda^a \geq 0. & \forall a \in \mathcal{A}(s) \\ \boldsymbol{1}^T \boldsymbol{x} = 1 & \\ \sum_{a \in \mathcal{A}(s)} \lambda^a \boldsymbol{l}^a \leq \boldsymbol{x} \leq \sum_{a \in \mathcal{A}(s)} \lambda^a \boldsymbol{u}^a & \\ \sum_{a \in \mathcal{A}(s)} \lambda^a = 1 & \end{array} \right\}. \tag{4}
$$

In Eq. (4), we still have extra variables λ^a, which should be eliminated in order to obtain an $\mathcal{H}$-polytope representing $\mathcal{P}_{\mathcal{R}}^s$. For this, we apply the well-known Fourier-Motzkin (FM) elimination method. The main idea of the FM elimination is to partition all inequalities relevant to y into two sets: $\{\sum_{1 \leq j \leq n} e_{ij} x_j \leq y\}_{1 \leq i \leq m_1}$ and $\{\sum_{1 \leq j \leq n} e_{ij} x_j \geq y\}_{m_1 < i \leq m}$, where e_{ij} ($1 \leq i \leq m, 1 \leq j \leq n$) are coefficients and y is the variable to be eliminated. The resultant set of inequalities will contain those in form of $\sum_{1 \leq j \leq n} e_{ij} x_j \leq \sum_{1 \leq j \leq n} e_{i'j} x_j$ for each $1 \leq i \leq m_1$ and $m_1 < i' \leq m$, which defines a projection of the original $\mathcal{H}$-polytope on variables $\{x_j\}_{1 \leq j \leq n} \cup \{y\}$ to an $\mathcal{H}$-polytope on $\{x_j\}_{1 \leq j \leq n}$. More details can be found in e.g. [26].

However, the FM elimination causes an exponential blow-up and results in $4(\frac{m}{4})^{2^d}$ inequalities in the worst case, where m is the number of inequalities in the original representation and d the number of variables having been eliminated [31]. Given an *IMDP* $\mathcal{M}$, the number of inequalities in Eq. (4) is in $\mathcal{O}(S)$ and the number of λ^a is upper-bounded by the branching of $\mathcal{M}$. Therefore, if we assume

$b_{\mathcal{M}}$ to be a constant, FM elimination will not cause exponentially blow-up after removing all λ^a in Eq. (4) and we obtain an $\mathcal{H}$-polytope representing $\mathcal{P}_{\mathcal{R}}^s$, the number of inequalities in which is at most $4(\frac{|S|}{4})^{2^d}$ with $d = b_{\mathcal{M}}$.

Given we have obtained $\mathcal{H}$-polytopes of both $\mathcal{P}_{\mathcal{R}}^s$ and $\mathcal{P}_{\mathcal{R}}^t$, to check whether $\mathcal{P}_{\mathcal{R}}^s = \mathcal{P}_{\mathcal{R}}^t$, the convex set inclusion checking algorithm in [34, Algorithm 1] can be applied to decide whether $\mathcal{P}_{\mathcal{R}}^s \subseteq \mathcal{P}_{\mathcal{R}}^t$ and $\mathcal{P}_{\mathcal{R}}^t \subseteq \mathcal{P}_{\mathcal{R}}^s$ hold, which can be done in polynomial time. Our algorithm for deciding $\sim_c$ mainly follows the partition-refinement approach [4,17,22,35], the key ingredient of which is a refinement procedure that keeps refining a relation $\mathcal{R}$ until for every pair of states s and t in a same equivalence class, $\mathcal{P}_{\mathcal{R}}^s = \mathcal{P}_{\mathcal{R}}^t$. Let $L(n, m)$ denote the polynomial running time of an LP algorithm on n inequalities and m variables [36]. The complexity of the algorithm for deciding $\sim_c$ is shown in the following theorem.

Theorem 14. *Given an IMDP $\mathcal{M}$, $\sim_c$ on $\mathcal{M}$ can be decided in time $\mathcal{O}(|S|^3 \times \eta \times L(\eta, |S|))$, where $\eta = 4(\frac{|S|}{4})^{2^d}$ with $d = b_{\mathcal{M}}$.*

In practice it often holds that the branching of an *IMDP* is constant. As an evidence, all *MDPs* in PRISM [27] benchmarks have constant branching. Thus, our algorithm for computing $\sim_c$ will often terminate in polynomial time for practical models. It is worthwhile to mention that even in the case where $b_{\mathcal{M}}$ is a constant for an *IMDP* $\mathcal{M}$, its equivalent *MDP* can be exponentially larger than $\mathcal{M}$. Consequently, for such models our algorithm can avoid an exponential blow-up comparing to the bisimulation decision algorithm for *MDPs* in [4].

5   Concluding Remarks

In this paper, we have studied the probabilistic bisimulation problem for interval *MDPs* in order to speed up the run time of model checking algorithms that often suffer from the state space explosion. Interval MDPs include two sources of nondeterminism for which we have considered the cooperative resolution in a dynamic setting. We have extended the results in [16] by further investigating two core aspects of the defined bisimulation relation: compositional reasoning and complexity analysis of the decision algorithm. As regards the former, we have established a framework for compositional verification of complex systems with interval uncertainty. As regards the latter, we have shown that deciding probabilistic bisimulation for *IMDPs* is coNP-complete and also shown that tractability is guaranteed under a mild restriction on the class of *IMDP* models. There are various promising directions for future work. From modelling viewpoint, it is worthwhile to address more expressive formalisms to encode uncertainties (such as polynomial constraints or even parameters appearing in multiple states/actions). Moreover, from semantics viewpoint, it would be interesting to extend the current results for the competitive semantics of resolving nondeterminism. Finally, from algorithmic viewpoint, we conjecture that probabilistic bisimulation problem can be decided in polynomial time for *IMDPs* if a proper model of uncertainty is considered. More precisely, the probabilistic bisimulation problem can be modeled as an instance of uncertain LPs based on the techniques in [17,35]. This class of uncertain LPs is computationally tractable [19], under some technical assumption regarding a proper model of uncertainty.

Acknowledgments. This work is supported by the EU 7th Framework Programme under grant agreements 295261 (MEALS) and 318490 (SENSATION), by the DFG as part of SFB/TR 14 AVACS, by the CAS/SAFEA International Partnership Program for Creative Research Teams, by the National Natural Science Foundation of China (Grants 61472473 and 61550110249), by the Chinese Academy of Sciences Fellowship for International Young Scientists (Grant 2015VTC029), and by the CDZ project CAP (GZ 1023). This research is supported in part by the National Science Foundation through Award CCF-1305054.

References

1. Balas, E.: Disjunctive programming: properties of the convex hull of feasible points. Discrete Appl. Math. **89**(1), 3–44 (1998)
2. Benedikt, M., Lenhardt, R., Worrell, J.: LTL model checking of interval Markov chains. In: Piterman, N., Smolka, S.A. (eds.) TACAS 2013 (ETAPS 2013). LNCS, vol. 7795, pp. 32–46. Springer, Heidelberg (2013)
3. Billingsley, P.: Probability and Measure. John Wiley and Sons, New York (1979)
4. Cattani, S., Segala, R.: Decision algorithms for probabilistic bisimulation. In: Brim, L., Jančar, P., Křetínský, M., Kučera, A. (eds.) CONCUR 2002. LNCS, vol. 2421, pp. 371–385. Springer, Heidelberg (2002)
5. Cattani, S., Segala, R., Kwiatkowska, M., Norman, G.: Stochastic transition systems for continuous state spaces and non-determinism. In: Sassone, V. (ed.) FOSSACS 2005. LNCS, vol. 3441, pp. 125–139. Springer, Heidelberg (2005)
6. Chatterjee, K., Sen, K., Henzinger, T.A.: Model-checking ω-regular properties of interval Markov chains. In: Amadio, R.M. (ed.) FOSSACS 2008. LNCS, vol. 4962, pp. 302–317. Springer, Heidelberg (2008)
7. Chehaibar, G., Garavel, H., Mounier, L., Tawbi, N., Zulian, F.: Specification and verification of the PowerScale® busarbitration protocol: an industrial experiment with LOTOS. In: FORTE, pp. 435–450 (1996)
8. Chen, T., Han, T., Kwiatkowska, M.: On the complexity of model checking interval-valued discrete time Markov chains. Inf. Process. Lett. **113**(7), 210–216 (2013)
9. Coste, N., Hermanns, H., Lantreibecq, E., Serwe, W.: Towards performance prediction of compositional models in industrial GALS designs. In: Bouajjani, A., Maler, O. (eds.) CAV 2009. LNCS, vol. 5643, pp. 204–218. Springer, Heidelberg (2009)
10. Delahaye, B., Katoen, J.-P., Larsen, K.G., Legay, A., Pedersen, M.L., Sher, F., Wąsowski, A.: Abstract probabilistic automata. In: Jhala, R., Schmidt, D. (eds.) VMCAI 2011. LNCS, vol. 6538, pp. 324–339. Springer, Heidelberg (2011)
11. Delahaye, B., Larsen, K.G., Legay, A., Pedersen, M.L., Wąsowski, A.: Decision problems for interval Markov chains. In: Dediu, A.-H., Inenaga, S., Martín-Vide, C. (eds.) LATA 2011. LNCS, vol. 6638, pp. 274–285. Springer, Heidelberg (2011)
12. Eirinakis, P., Ruggieri, S., Subramani, K., Wojciechowski, P.: On quantified linear implications. AMAI **71**(4), 301–325 (2014)
13. Garey, M.R., Johnson, D.S.: Computers and Intractability; A Guide to the Theory of NP-Completeness. W. H. Freeman & Co., New York (1990)

14. Hahn, E.M., Han, T., Zhang, L.: Synthesis for PCTL in parametric Markov decision processes. In: Bobaru, M., Havelund, K., Holzmann, G.J., Joshi, R. (eds.) NFM 2011. LNCS, vol. 6617, pp. 146–161. Springer, Heidelberg (2011)
15. Hansson, H., Jonsson, B.: A logic for reasoning about time and reliability. Formal Aspects Comput. 6(5), 512–535 (1994)
16. Hashemi, V., Hatefi, H., Krčál, J.: Probabilistic bisimulations for PCTL model checking of interval MDPs. SynCoP EPTCS 145, 19–33 (2014)
17. Hashemi, V., Hermanns, H., Turrini, A.: On the efficiency of deciding probabilistic automata weak bisimulation. ECEASST, 66 (2013)
18. Hermanns, H., Katoen, J.-P.: Automated compositional Markov chain generation for a plain-old telephone system. Sci. Comp. Progr. 36(1), 97–127 (2000)
19. Iancu, D.A., Sharma, M., Sviridenko, M.: Supermodularity and affine policies in dynamic robust optimization. Oper. Res. 61(4), 941–956 (2013)
20. Iyengar, G.N.: Robust dynamic programming. Math. Oper. Res. 30(2), 257–280 (2005)
21. Jonsson, B., Larsen, K.G.: Specification and refinement of probabilistic processes. In: LICS, pp. 266–277 (1991)
22. Kanellakis, P.C., Smolka, S.A.: CCS expressions, finite state processes, and three problems of equivalence. Inf. Comput. 86(1), 43–68 (1990)
23. Katoen, J.-P., Kemna, T., Zapreev, I., Jansen, D.N.: Bisimulation minimisation mostly speeds up probabilistic model checking. In: Grumberg, O., Huth, M. (eds.) TACAS 2007. LNCS, vol. 4424, pp. 87–101. Springer, Heidelberg (2007)
24. Katoen, J.-P., Klink, D., Neuhäußer, M.R.: Compositional abstraction for stochastic systems. In: Ouaknine, J., Vaandrager, F.W. (eds.) FORMATS 2009. LNCS, vol. 5813, pp. 195–211. Springer, Heidelberg (2009)
25. Kozine, I., Utkin, L.V.: Interval-valued finite Markov chains. Reliable Comput. 8(2), 97–113 (2002)
26. Kroening, D., Strichman, O.: Decision Procedures: An Algorithmic Point of View, 1st edn. Springer, Heidelberg (2008)
27. Kwiatkowska, M., Norman, G., Parker, D.: PRISM 4.0: verification of probabilistic real-time systems. In: Gopalakrishnan, G., Qadeer, S. (eds.) CAV 2011. LNCS, vol. 6806, pp. 585–591. Springer, Heidelberg (2011)
28. Nilim, A., Ghaoui, L.E.: Robust control of Markov decision processes with uncertain transition matrices. Oper. Res. 53(5), 780–798 (2005)
29. Paige, R., Tarjan, R.E.: Three partition refinement algorithms. SIAM J. Comput. 16(6), 973–989 (1987)
30. Puggelli, A., Li, W., Sangiovanni-Vincentelli, A.L., Seshia, S.A.: Polynomial-time verification of PCTL properties of MDPs with convex uncertainties. In: Sharygina, N., Veith, H. (eds.) CAV 2013. LNCS, vol. 8044, pp. 527–542. Springer, Heidelberg (2013)
31. Schrijver, A.: Theory of Linear and Integer Programming. John Wiley & Sons, Amsterdam (1998)
32. Segala, R.: Modeling and verification of randomized distributed real-time systems. Ph.D. thesis, MIT (1995)
33. Sen, K., Viswanathan, M., Agha, G.: Model-checking Markov chains in the presence of uncertainties. In: Hermanns, H., Palsberg, J. (eds.) TACAS 2006. LNCS, vol. 3920, pp. 394–410. Springer, Heidelberg (2006)
34. Subramani, K.: On the complexities of selected satisfiability and equivalence queries over boolean formulas and inclusion queries over hulls. JAMDS, 2009 (2009)
35. Turrini, A., Hermanns, H.: Polynomial time decision algorithms for probabilistic automata. Inf. Comput. 244, 134–171 (2015)

36. Vaidya, P.M.: An algorithm for linear programming which requires $O(((m+n)n^2)+ (m+n)^{1.5})n)L)$ arithmetic operations. Math. Program. **47**, 175–201 (1990)
37. Wolff, E.M., Topcu, U., Murray, R.M.: Robust control of uncertain Markov decision processes with temporal logic specifications. In: CDC, pp. 3372–3379 (2012)
38. Wu, D., Koutsoukos, X.D.: Reachability analysis of uncertain systems using bounded-parameter Markov decision processes. AI **172**(8–9), 945–954 (2008)
39. Yi, W.: Algebraic reasoning for real-time probabilistic processes with uncertain information. In: Langmaack, H., de Roever, W.-P., Vytopil, J. (eds.) FTRTFT 1994 and ProCoS 1994. LNCS, vol. 863, pp. 680–693. Springer, Heidelberg (1994)

A Weighted MSO Logic with Storage Behaviour and Its Büchi-Elgot-Trakhtenbrot Theorem

Heiko Vogler[1], Manfred Droste[2], and Luisa Herrmann[1(✉)]

[1] Faculty of Computer Science, Technische Universität Dresden, Dresden, Germany
{Heiko.Vogler,Luisa.Herrmann}@tu-dresden.de
[2] Institut Für Informatik, Universität Leipzig, Leipzig, Germany
Droste@informatik.uni-leipzig.de

Abstract. We introduce a weighted MSO-logic in which one outermost existential quantification over behaviours of a storage type is allowed. As weight structures we take unital valuation monoids which include all semirings, bounded lattices, and computations of average or discounted costs. Each formula is interpreted over finite words yielding elements in the weight structure. We prove that this logic is expressively equivalent to weighted automata with storage. In particular, this implies a Büchi-Elgot-Trakhtenbrot Theorem for weighted iterated pushdown languages. For this choice of storage type, the satisfiability problem of the logic is decidable for each bounded lattice provided that its infimum is computable.

1 Introduction

The Büchi-Elgot-Trakhtenbrot Theorem [1,2,10,24] (for short: BET-theorem) states that the languages definable by monadic second-order logic (MSO-logic) coincide with the recognizable languages. This theorem has been extended to the weighted setting where the weight structures are semirings [4,5], valuation monoids [6], bounded lattices [8], and multioperator monoids (for trees) [15,16].

Another direction of extension was taken in [20] where context-free languages (CFL) were characterized by an extension of MSO-logic in which formulas of the form $\exists M.\varphi$ are allowed and M is a matching and φ is a formula of first-order logic. Roughly speaking, a matching is a set of pairs of positions of the given word which can be regarded as a word from the Dyck-language or, in automata terms, as an executable sequence of instructions of a pushdown automaton, e.g.,

$$\text{push}(\alpha); \text{push}(\beta); \text{pop}(\beta); \text{push}(\gamma); \text{pop}(\gamma); \text{pop}(\alpha).$$

The quantification over matchings allows to express non-local properties which go beyond regularity and capture precisely context-freeness. A semiring-weighted version of the BET-theorem for CFL was proved in [22]. In the unweighted case, this direction was further investigated in [14] where a BET-theorem for realtime indexed languages was proved.

L. Herrmann—Supported by DFG Graduiertenkolleg 1763 (QuantLA).

A.-H. Dediu et al. (Eds.): LATA 2016, LNCS 9618, pp. 127–139, 2016.
DOI: 10.1007/978-3-319-30000-9_10

A sequence of instructions as above can be considered as a behaviour of the pushdown storage. Scott [23] introduced the general concept of (unweighted) automata with storage type. As in Engelfriet [12], a storage type S consists of a set C of configurations, a set P of predicates on C, a set F of instructions transforming the configurations, and an initial configuration c_0. A behaviour is an executable sequence

$$(p_1, f_1)(p_2, f_2) \ldots (p_n, f_n)$$

of predicates p_i and instructions f_i starting from c_0.

Inspired by the logic in [20], we introduce a new weighted MSO-logic which is based on [15,16]. Additionally, we allow one outermost second-order existential quantification over behaviours of an arbitrary storage type. As new atomic formulas we introduce equalities $B(x) = (p, f)$ describing that the x-th element of a behaviour chosen for the variable B is the pair (p, f). As weight structure we employ unital valuation monoids [6,9]. These include all semirings and bounded lattices but also computations of average or of discounted costs.

In this paper, we prove the following BET-theorem (Theorem 8): our new weighted MSO-logic is expressively equivalent to weighted automata with storage type S [18]. The idea for this result derives from the setting of [20] as follows. The class CFL of languages accepted by pushdown automata is a full principle abstract family of languages (fp-AFL) generated by the Dyck language. Unweighted automata with arbitrary storage type can be considered as a finitely-encoded abstract family of acceptors (fe-AFA). By [17, Theorem 5.2.1], each class of languages accepted by an fe-AFA is a fp-AFL and consequently generated by one language: the set of behaviours of the storage [17, Lemma 4.2.4]. Our new MSO-logic results from the one of [20] by replacing the existential quantification over the generator for CFL by the existential quantification over the generator for an arbitrary fe-AFA.

In our proof of the equivalence, first we describe the computations of weighted automata with storage by usual logic formulas, but employing the new atomic formulas $B(x) = (p, f)$. For the converse, we first transform formulas without the quantification over behaviours inductively into usual weighted automata, i.e. without storage, taking care of the behaviours in a symbolic way. Then, for the outermost existential quantification, we apply a projection to obtain weighted automata with storage (cf. Lemma 15).

By choosing the n-iterated pushdown storage type [3,11,12,21], our equivalence result yields a BET-theorem for weighted n-iterated pushdown languages. For this choice, our result implies the decidability of the satisfiability problem of our logic for a large class of zero-sum-free commutative strong bimonoids, in particular, the class of all bounded lattices with computable infimum.

2   Preliminaries

Notations and Notions. We denote the set of all non-negative integers (including 0) by $\mathbb{N}$, and the set $\mathbb{N} \setminus \{0\}$ by $\mathbb{N}_+$. For $n \in \mathbb{N}$ we let $[n] = \{i \in \mathbb{N} \mid 1 \leq i \leq n\}$.

Let A be a set. We denote the set of all subsets of A by $\mathcal{P}(A)$. We denote the set of all words over A by A^*; ε is the *empty word*; the length of a word $u \in A^*$ is denoted by $|u|$. Let B be a set and $\nu : A \to B$. Then we denote the unique extension of ν to a monoid morphism from $(A^*, \cdot, \varepsilon)$ to $(B^*, \cdot, \varepsilon)$ where $\cdot$ denotes concatenation of words, also by ν.

Let $A_1, \ldots, A_n$ be sets and $a = (a_1, \ldots, a_n) \in A_1 \times \ldots \times A_n$. Then, for each $i \in [n]$, the *i-th projection of a*, denoted by $\mathrm{pr}_i(a)$, is defined by $\mathrm{pr}_i(a) = a_i$.

An *alphabet Σ* is a non-empty, finite set. Let $u \in \Sigma^*$. We denote the i-th symbol of u by u_i; thus, $u = u_1 \ldots u_n$ for some $n \geq 0$ and $u_i \in \Sigma$. We denote the set $[n]$ of *positions* of u by $\mathrm{pos}(u)$. For $\alpha \in \Sigma$ and $u \in \Sigma^*$ we denote the number of occurrences of α in u by $|u|_\alpha$.

Unital Valuation Monoids. A *unital valuation monoid* [6,9,18] is an algebra $(K, +, \mathrm{val}, 0, 1)$ such that $(K, +, 0)$ is a commutative monoid and $\mathrm{val} \colon K^* \to K$ is a mapping such that (i) $\mathrm{val}(k) = k$ for each $k \in K$, (ii) $\mathrm{val}(k) = 0$ for each $k \in K^*$ such that $k_i = 0$ for some $i \in [|k|]$, and (iii) $\mathrm{val}(k \, 1 \, k') = \mathrm{val}(k \, k')$ for every $k, k' \in K^*$. It follows that $\mathrm{val}(\varepsilon) = 1$.

Strong bimonoids [7] $(K, +, \cdot, 0, 1)$ are algebras where $(K, +, 0)$ is a commutative monoid, $(K, \cdot, 1)$ is a monoid, and $0 \cdot k = k \cdot 0 = 0$ for each $k \in K$. Clearly, all semirings, in particular, the Boolean semiring $\mathbb{B} = (\{0, 1\}, \vee, \wedge, 0, 1)$, and all bounded lattices are strong bimonoids. Each strong bimonoid naturally becomes a unital valuation monoid by letting val be the product of finite sequences of elements. Other important examples of unital valuation monoids arise by taking the average or discounting (see below) operations on real numbers, see [9, Example 1] for further examples.

Example 1. Let $\widetilde{\mathbb{R}} = \mathbb{R}_{\geq 0} \cup \{-\infty\}$. We define the unital valuation monoid $(\widetilde{\mathbb{R}}, \max, \mathrm{val}_{\mathrm{disc}}, -\infty, 0)$ such that for each $n \in \mathbb{N}$ and $k_1, \ldots, k_n \in \widetilde{\mathbb{R}} \setminus \{0\}$ we have $\mathrm{val}_{\mathrm{disc}}(k_1 \ldots k_n) = 0.5^0 \cdot k_1 + \ldots + 0.5^{n-1} \cdot k_n$. For instance $\mathrm{val}_{\mathrm{disc}}(2\,1\,2) = 0.5^0 \cdot 2 + 0.5^1 \cdot 1 + 0.5^2 \cdot 2 = 3$. □

A *(K-)weighted language (over Σ)* is a mapping $s : \Sigma^* \to K$. The *support of* s is the set $\mathrm{supp}(s) = \{u \in \Sigma^* \mid s(u) \neq 0\}$.

3 Weighted Automata with Storage

Storage Types. A *storage type S* [12,23] is a tuple (C, P, F, c_0) where C is a set (*configurations*), P is a set of total functions each having the type $p \colon C \to \{\mathrm{true}, \mathrm{false}\}$ (*predicates*), F is a set of partial functions each having the type $f \colon C \to C$ (*instructions*), and $c_0 \in C$ (*initial configuration*).

Example 2. Let c be an arbitrary, fixed symbol. The *trivial storage type* is the storage type $\mathrm{TRIV} = (\{c\}, \{p_{\mathrm{true}}\}, \{f_{\mathrm{id}}\}, c)$ where $p_{\mathrm{true}}(c) = \mathrm{true}$ and $f_{\mathrm{id}}(c) = c$.

Next we recall the pushdown operator P from [12, Definition 5.1] and [13, Definition 3.28]: $P(S)$ is the storage type of which the configurations have the form of a pushdown; each cell contains a pushdown symbol and a configuration of S. Formally, the *pushdown of* S is the storage type $P(S) = (C', P', F', c'_0)$ where

- $C' = (\Gamma \times C)^+$ and Γ is the pushdown alphabet and $c'_0 = (\gamma_0, c_0)$,
- $P' = \{\text{bottom}\} \cup \{(\text{top} = \gamma) \mid \gamma \in \Gamma\} \cup \{\text{test}(p) \mid p \in P\}$, and
- $F' = \{\text{pop}\} \cup \{\text{stay}(\gamma) \mid \gamma \in \Gamma\} \cup \{\text{push}(\gamma, f) \mid \gamma \in \Gamma, f \in F\}$

with the usual interpretation. For each $n \geq 0$ we define $P^n(S)$ inductively as follows: $P^0(S) = S$ and $P^n(S) = P(P^{n-1}(S))$ for each $n \geq 1$.

Example 3. Intuitively, $P(\text{TRIV})$ corresponds to the usual pushdown storage except that there is no empty pushdown. For $n \geq 0$, we abbreviate $P^n(\text{TRIV})$ by P^n and call it the *n-iterated pushdown storage*.   □

Behaviour. Let Ω be a finite subset of $P \times F$. Moreover, let $n \geq 0$ and $b = (p_1, f_1) \ldots (p_n, f_n) \in \Omega^*$. We call b an *Ω-behaviour of length n* if for every $i \in [n]$ we have (i) $p_i(c') = \text{true}$ and (ii) $f_i(c')$ is defined where $c' = f_{i-1}(\ldots f_1(c_0) \ldots)$ (note that $c' = c_0$ for $i = 1$). We denote the set of all Ω-behaviours of length n by $B(\Omega, n)$, and we let $B(\Omega) = \bigcup_{n \in \mathbb{N}} B(\Omega, n)$. We note that each behaviour of c is a path in the approximation of c_0 according to [13, Definition 3.23].

Automata: An (S, Σ, K)-*automaton* [18] is a tuple $\mathcal{A} = (Q, Q_i, Q_f, T, \text{wt})$ where Q is a finite set (*states*), $Q_i \subseteq Q$ (*initial states*), $Q_f \subseteq Q$ (*final states*), $T \subseteq Q \times \Sigma \times P \times Q \times F$ is a finite set (*transitions*), and $\text{wt}: T \to K$.

The set $Q \times \Sigma^* \times C$ is the set of $\mathcal{A}$-*configurations*. For every transition $\tau = (q, \alpha, p, q', f)$ in T we define the binary relation $\vdash^\tau$ on the set of $\mathcal{A}$-configurations: for every $u \in \Sigma^*$ and $c \in C$, we let $(q, \alpha u, c) \vdash^\tau (q', u, f(c))$ if $p(c)$ is true and $f(c)$ is defined. The *computation relation of* $\mathcal{A}$ is the binary relation $\vdash = \bigcup_{\tau \in T} \vdash^\tau$.

A *computation* is a sequence $\theta = (d_0 \vdash^{\tau_1} d_1 \cdots \vdash^{\tau_n} d_n)$ where $n \in \mathbb{N}$ and d_i are $\mathcal{A}$-configurations with $d_{i-1} \vdash^{\tau_i} d_i$. We abbreviate this computation by $\theta = (d_0 \vdash^* d_n)$. For each $u \in \Sigma^*$, we denote the (finite) set of computations $\{\theta \mid \theta = ((q, u, c_0) \vdash^* (q', \varepsilon, c')), q \in Q_i, q' \in Q_f, c' \in C\}$ by $\Theta_{\mathcal{A}}(u)$. Thus, $\Theta_{\mathcal{A}}(\varepsilon) = \{(q, \varepsilon, c_0) \mid q \in Q_i \cap Q_f\}$. Obviously, each computation of $\mathcal{A}$ induces an $\Omega_{\mathcal{A}}$-behaviour where $\Omega_{\mathcal{A}} = \{(p, f) \mid (q, \alpha, p, q', f) \in T\}$.

Let $\theta = (d_0 \vdash^{\tau_1} d_1 \cdots \vdash^{\tau_n} d_n)$ be a computation of $\mathcal{A}$. The *weight of* θ is the element in K defined by $\text{wt}(\theta) = \text{val}(\text{wt}(\tau_1) \ldots \text{wt}(\tau_n))$. Note that $\text{wt}(\theta) = \text{val}(\varepsilon) = 1$ for each $\theta \in \Theta_{\mathcal{A}}(\varepsilon)$. The *weighted language recognized by* $\mathcal{A}$ is the weighted language $[\![\mathcal{A}]\!]: \Sigma^* \to K$ defined for every $u \in \Sigma^*$ by $[\![\mathcal{A}]\!](u) = \sum_{\theta \in \Theta_{\mathcal{A}}(u)} \text{wt}(\theta)$. Thus, $[\![\mathcal{A}]\!](\varepsilon) = \sum_{q \in Q_i \cap Q_f} 1$. A weighted language $r: \Sigma^* \to K$ is (S, Σ, K)-*recognizable* if there is an (S, Σ, K)-automaton $\mathcal{A}$ such that $r = [\![\mathcal{A}]\!]$. We denote the class of (S, Σ, K)-recognizable weighted languages by $\text{Rec}(S, \Sigma, K)$.

For each $n \geq 0$, we call a (P^n, Σ, K)-recognizable weighted language a *weighted n-iterated pushdown language over Σ and K*.

Example 4. We consider the alphabet $\Sigma = \{\alpha, \beta, \#\}$ and the unital valuation monoid K of Example 1. Moreover, we consider the weighted language

$s : \Sigma^* \to K$ such that for each $u \in \Sigma^*$: if $u = w\#$ for some $w \in \{\alpha, \beta\}^*$, and $|w|_\alpha = |w|_\beta$, then $s(u) = 0.5^0 \cdot \widetilde{wt}(w_1) + \ldots + 0.5^{n-1} \cdot \widetilde{wt}(w_n)$ where n is the length of w and $\widetilde{wt}(\alpha) = 2$ and $\widetilde{wt}(\beta) = 1$, and otherwise we have $s(u) = -\infty$.

Indeed, s is recognizable by a (P^1, Σ, K)-automaton. For this, we can choose a deterministic pushdown automaton accepting $\{w\# \in \Sigma^* \mid w \in \{\alpha, \beta\}^*, |w|_\alpha = |w|_\beta\}$ and then weight its transitions according to the function $\widetilde{wt}$ and such that the transition which reads $\#$ is of weight 0.

Special Cases. If $S = \mathrm{TRIV}$, then we drop all references to S from the concepts introduced for (S, Σ, K)-automata. Thus $T \subseteq Q \times \Sigma \times Q$, and we speak about (Σ, K)-*automata* and (Σ, K)-*recognizable*. We denote the class of (Σ, K)-recognizable weighted languages by $\mathrm{Rec}(\Sigma, K)$. If K is a semiring (or unital valuation monoid), then (Σ, K)-automata coincide with the weighted automata of [4] apart from initial and final weights (respectively, of [9]).

If $K = \mathbb{B}$, then we drop wt from the automaton specification, and we speak about (S, Σ)-automata. They coincide with $\mathrm{REG}(S)$ r-acceptors [12, p. 11]. The *language recognized by* an (S, Σ)-automaton $\mathcal{A}$ is the set $\mathcal{L}(\mathcal{A}) = \{u \in \Sigma^* \mid \Theta_\mathcal{A}(u) \neq \emptyset\}$. A language $L \subseteq \Sigma^*$ is (S, Σ)-*recognizable* if there is an (S, Σ)-automaton $\mathcal{A}$ such that $\mathcal{L}(\mathcal{A}) = L$. We denote the class of all (S, Σ)-recognizable languages by $\mathrm{Rec}(S, \Sigma)$. For instance, (P^1, Σ)-automata are essentially pushdown automata (apart from ε-moves), and for each $n \geq 1$, (P^n, Σ)-automata correspond to n-iterated pushdown automata of [3,11,21]; for $n = 2$, they accept the OI-macro languages, cf. [3] (equivalently, the indexed languages, cf. [12, p. 18], and nested stack languages, cf. [18, Example 3]).

If $S = \mathrm{TRIV}$ and $K = \mathbb{B}$, then we obtain the classical finite state automata.

4   Weighted Logic with Storage Behaviour

Here we will introduce our new weighted MSO-logic with storage behaviour. Technically the logic is based on the weighted MSO-logic of [15,16].

As usual, we use first-order variables, like $x, x_1, x_2, \ldots, y, z$ to denote single positions of a given word, and second-order variables, like $X, X_1, X_2, \ldots, Y, Z$ to denote sets of positions of a given word.

We recall the (unweighted) MSO-logic where we parameterize the set of atomic formulas. For a set $\mathcal{C}$, we define the *set of formulas of MSO-logic over* Σ *with atomic formulas in* $\mathcal{C}$, denoted by $\mathrm{MSO}(\Sigma, \mathcal{C})$, as the set generated by the EBNF: $\varphi ::= \psi \mid \neg\varphi \mid (\varphi \vee \varphi) \mid (\varphi \wedge \varphi) \mid \exists x.\varphi \mid \forall x.\varphi \mid \exists X.\varphi \mid \forall X.\varphi$ where $\psi \in \mathcal{C}$.

The *set of MSO-logic formulas over* Σ, denoted by $\mathrm{MSO}(\Sigma)$, is the set $\mathrm{MSO}(\Sigma, \mathcal{C}_{\mathrm{MSO}})$ where $\mathcal{C}_{\mathrm{MSO}}$ contains all formulas of either of the following three forms: (i) $P_\alpha(x)$ where $\alpha \in \Sigma$, (ii) $\mathrm{next}(x, y)$, or (iii) $(x \in X)$. In the usual way we define the set $\mathrm{Free}(\varphi)$ of free variables of a formula φ.

Let $\varphi \in \mathrm{MSO}(\Sigma)$ and $\mathcal{V}$ be a set of variables such that $\mathrm{Free}(\varphi) \subseteq \mathcal{V}$. In the usual way, we define the set $\mathcal{L}_\mathcal{V}(\varphi) = \{(u, \sigma) \mid u \in \Sigma^*, \sigma \text{ is a } \mathcal{V}\text{-variable}$ assignment of u, $(u, \sigma) \models \varphi\}$ of models of φ. We will use macros like $\mathrm{first}(x)$, $\mathrm{last}(x)$, and $\varphi \to \psi$ with their obvious meaning.

Our new weighted logic extends M-expressions [15, 16] by allowing one outermost existential quantification over a *second-order behaviour variable B*. Intuitively, a variable assignment σ interprets B as an Ω-behaviour for some finite subset Ω of $P \times F$. Atomic expressions of our new logic have the form Val_κ (cf. [16, Sect. 4.2]) where κ maps symbols from an *extended alphabet* $\Sigma_\mathcal{U}$ to K. For each finite set $\mathcal{U}$ of variables with $B \in \mathcal{U}$ we let

$$\Sigma_\mathcal{U} = \Sigma \times \mathcal{P}(\mathrm{fo}(\mathcal{U}) \cup \mathrm{so}(\mathcal{U})) \times \Omega$$

where $\mathrm{fo}(\mathcal{U})$ and $\mathrm{so}(\mathcal{U})$ are the subsets of first-order variables and second-order variables occurring in $\mathcal{U}$, respectively. Moreover, in expressions of the form $\varphi \triangleright e$, in φ additionally atomic formulas of the form $B(x) = (p, f)$ are allowed; they can be used to retrieve the predicate and the instruction assigned by the behaviour $\sigma(B)$ to the position $\sigma(x)$.

Definition 5. Let Ω be a finite subset of $P \times F$. We define the set $\mathrm{BExp}(\Omega, \Sigma, K)$ of *B-expressions over* (Ω, Σ, K) to be the set generated by the EBNF:

$$e ::= \mathrm{Val}_\kappa \mid (e + e) \mid (\varphi \triangleright e) \mid \sum_x e \mid \sum_X e$$

where $\kappa : \Sigma_\mathcal{U} \to K$ for some finite set $\mathcal{U}$ of variables with $B \in \mathcal{U}$, and $\varphi \in \mathrm{MSO}(\Sigma, \mathcal{C}_{\mathrm{MSO}} \cup \mathcal{C}_\mathcal{B})$; the set $\mathcal{C}_\mathcal{B}$ contains all formulas of the form $B(x) = (p, f)$ where $(p, f) \in \Omega$.

Let $e \in \mathrm{BExp}(\Omega, \Sigma, K)$. The set of *free variables of* e, denoted by $\mathrm{Free}(e)$, is defined as usual where we set $\mathrm{Free}(B(x) = (p, f)) = \{x, B\}$ and $\mathrm{Free}(\mathrm{Val}_\kappa) = \mathcal{U}$. (Note that $B \in \mathrm{Free}(e)$ for each $e \in \mathrm{BExp}(\Omega, \Sigma, K)$.)

We define the set $\mathrm{Exp}(\Omega, \Sigma, K)$ of *expressions over* (Ω, Σ, K) to be the set of all expressions of the form

$$\sum_B^{\mathrm{beh}} e$$

where $e \in \mathrm{BExp}(\Omega, \Sigma, K)$ with $\mathrm{Free}(e) = \{B\}$. An *expression over* (S, Σ, K) is an expression over (Ω, Σ, K) for some finite set $\Omega \subseteq P \times F$.

Let $\mathcal{V}$ be a finite set of variables with $B \in \mathcal{V}$. Let $u \in \Sigma^*$. A $\mathcal{V}$-*assignment for* u is a function with domain $\mathcal{V}$ which maps each element in $\mathrm{fo}(\mathcal{V})$ to an element of $\mathrm{pos}(u)$, each element in $\mathrm{so}(\mathcal{V})$ to a subset of $\mathrm{pos}(u)$, and B to an Ω-behaviour of length $|u|$. We let $\Phi_{\mathcal{V},u}$ denote the set of all $\mathcal{V}$-assignments for u.

In the usual way we define updates of $\mathcal{V}$-assignments. Let $\sigma \in \Phi_{\mathcal{V},u}$ and $i \in \mathrm{pos}(u)$. By $\sigma[x \mapsto i]$ we denote the $(\mathcal{V} \cup \{x\})$-assignment for u that agrees with σ on $\mathcal{V} \setminus \{x\}$ and that satisfies $\sigma[x \mapsto i](x) = i$. Similarly, we define the update $\sigma[X \mapsto I]$ for a set $I \subseteq \mathrm{pos}(u)$.

Extending the usual technique (cf., e.g., [4]), we can encode a pair (u, σ), where $u \in \Sigma^*$ and σ is a $\mathcal{V}$-assignment for u, as a word over the extended alphabet $\Sigma_\mathcal{V}$. A word $\zeta \in \Sigma_\mathcal{V}^*$ is called *valid* if (i) for each $x \in \mathrm{fo}(\mathcal{V})$ there is a unique $i \in \mathrm{pos}(\zeta)$ such that x occurs in the second component of ζ_i and (ii) the word $\mathrm{pr}_3(\zeta_1) \ldots \mathrm{pr}_3(\zeta_n)$ is an Ω-behaviour of length n where $n = |\zeta|$. We denote the set of all valid words in $\Sigma_\mathcal{V}^*$ by $\Sigma_\mathcal{V}^{*\mathrm{v}}$.

It is clear that, for each finite set $\mathcal{V}$ of variables with $B \in \mathcal{V}$, there is a one-to-one correspondence between the set $\{(u, \sigma) \mid u \in \Sigma^*, \sigma \in \Phi_{\mathcal{V},u}\}$ and the set $\Sigma_{\mathcal{V}}^{*\mathcal{V}}$. Thus, as usual, we will not distinguish between the pair (u, σ) and the corresponding word $\zeta \in \Sigma_{\mathcal{V}}^{*\mathcal{V}}$.

Definition 6. Let $e \in \mathrm{BExp}(\Omega, \Sigma, K)$ and $\mathcal{V}$ be a finite set of variables containing $\mathrm{Free}(e)$. The *semantics of e with respect to* $\mathcal{V}$ is the weighted language $[\![e]\!]_{\mathcal{V}} : \Sigma_{\mathcal{V}}^* \to K$ such that $\mathrm{supp}([\![e]\!]_{\mathcal{V}}) \subseteq \Sigma_{\mathcal{V}}^{*\mathcal{V}}$ and for each $\zeta = (u, \sigma) \in \Sigma_{\mathcal{V}}^{*\mathcal{V}}$ we define $[\![e]\!]_{\mathcal{V}}(\zeta)$ inductively as follows:

- for every $\mathcal{U} \subseteq \mathcal{V}$ with $B \in \mathcal{U}$ and every $\kappa : \Sigma_{\mathcal{U}} \to K :$ $[\![\mathrm{Val}_\kappa]\!]_{\mathcal{V}}(\zeta) = \mathrm{val}(\kappa(\zeta_{\mathcal{U}}))$ where $\zeta_{\mathcal{U}}$ is obtained from ζ by replacing each symbol (a, V, ω) by $(a, V \cap (\mathrm{fo}(\mathcal{U}) \cup \mathrm{so}(\mathcal{U})), \omega)$,
- for every $e_1, e_2 \in \mathrm{BExp}(\Omega, \Sigma, K) :$ $[\![e_1 + e_2]\!]_{\mathcal{V}}(\zeta) = [\![e_1]\!]_{\mathcal{V}}(\zeta) + [\![e_2]\!]_{\mathcal{V}}(\zeta)$,
- for every $\varphi \in \mathrm{MSO}(\Sigma, \mathcal{C}_{\mathrm{MSO}} \cup \mathcal{C}_{\mathcal{B}})$ and $e \in \mathrm{BExp}(\Omega, \Sigma, K)$:
  $[\![\varphi \rhd e]\!]_{\mathcal{V}}(\zeta) = [\![e]\!]_{\mathcal{V}}(\zeta)$, if $\zeta \in \mathcal{L}_{\mathcal{V}}(\varphi)$, and 0 otherwise
  where $\mathcal{L}_{\mathcal{V}}(\varphi) = \{\xi \in \Sigma_{\mathcal{V}}^{*\mathcal{V}} \mid \xi \models \varphi\}$ and $\models$ extends the usual models operator of (classical) MSO-formulas in $\mathrm{MSO}(\Sigma, \mathcal{C}_{\mathrm{MSO}})$ by defining $(u, \sigma) \models (B(x) = (p, f))$ is true if $\sigma(B)_{\sigma(x)} = (p, f)$, otherwise it is false,
- for every first-order variable x and $e \in \mathrm{BExp}(\Omega, \Sigma, K)$:
  $[\![\textstyle\sum_x e]\!]_{\mathcal{V}}(\zeta) = \sum_{i \in \mathrm{pos}(\zeta)} [\![e]\!]_{\mathcal{V} \cup \{x\}}(u, \sigma[x \mapsto i])$,
- for every second-order variable X and $e \in \mathrm{BExp}(\Omega, \Sigma, K)$:
  $[\![\textstyle\sum_X e]\!]_{\mathcal{V}}(\zeta) = \sum_{I \subseteq \mathrm{pos}(\zeta)} [\![e]\!]_{\mathcal{V} \cup \{X\}}(u, \sigma[X \mapsto I])$.

Now let $e = \sum_B^{\mathrm{beh}} e'$ be an expression over (Ω, Σ, K). Then we define the weighted language $[\![e]\!] : \Sigma^* \to K$ for each $u \in \Sigma^*$ by:
$$[\![\textstyle\sum_B^{\mathrm{beh}} e']\!](u) = \sum_{b \in \mathrm{B}(\Omega, |u|)} [\![e']\!]_{\{B\}}(u, [B \mapsto b]).$$
We say that $s : \Sigma^* \to K$ is *definable by an expression over* (S, Σ, K) if there is an expression e over (S, Σ, K) such that $s = [\![e]\!]$.

Example 7. We consider the weighted language s of Example 4 and construct an expression $e = \sum_B^{\mathrm{beh}} (\varphi \rhd \mathrm{Val}_\kappa)$ over $(\mathrm{P}^1, \Sigma, K)$ such that $[\![e]\!] = s$. For this, we let Ω consist of the following seven elements.

$$\begin{aligned}
\omega_1 &= (\mathrm{top} = \gamma_0, \mathrm{push}(\beta, \mathrm{f_{id}})), & \omega_4 &= (\mathrm{top} = \gamma_0, \mathrm{push}(\alpha, \mathrm{f_{id}})), \\
\omega_2 &= (\mathrm{top} = \beta, \mathrm{push}(\beta, \mathrm{f_{id}})), & \omega_5 &= (\mathrm{top} = \alpha, \mathrm{push}(\alpha, \mathrm{f_{id}})), \\
\omega_3 &= (\mathrm{top} = \beta, \mathrm{pop}), & \omega_6 &= (\mathrm{top} = \alpha, \mathrm{pop}), \\
& & \omega_7 &= (\mathrm{bottom}, \mathrm{stay}(\$)).
\end{aligned}$$

Recall that γ_0 is the initial pushdown symbol of P^1; we let $\$$ be an arbitrary pushdown symbol different from α, β, and γ_0.

Then, using a rational expression over singletons and context-free sets, it is easy to see that $\mathrm{B}(\Omega) = \mathrm{Pref}((\omega_1 \cdot C_{2,3} + \omega_4 \cdot C_{5,6})^* \cdot \omega_7)$ where $\mathrm{Pref}(D)$ is the set of all prefixes of any set $D \subseteq \Omega^*$ and

$$C_{2,3} = \{b \in \{\omega_2, \omega_3\}^* \mid |b|_{\omega_2} + 1 = |b|_{\omega_3}, \forall b' \text{ prefix of } b : |b'|_{\omega_2} + 1 \geq |b'|_{\omega_3}\},$$
$$C_{5,6} = \{b \in \{\omega_5, \omega_6\}^* \mid |b|_{\omega_5} + 1 = |b|_{\omega_6}, \forall b' \text{ prefix of } b : |b'|_{\omega_5} + 1 \geq |b'|_{\omega_6}\}.$$

Next we define the MSO-formula $\varphi \in \text{MSO}(\Sigma_{\{B\}}, \mathcal{C}_{\text{MSO}} \cup \mathcal{C}_\mathcal{B})$ which checks for some given $b \in B(\Omega)$ whether it ends with ω_7. Moreover, φ forces a labeling with symbols from Σ. We let $\varphi = \varphi_{\text{last}} \wedge \varphi_{\text{symb}}$ where

- $\varphi_{\text{last}} = \forall x.(\text{last}(x) \to B(x) = \omega_7)$
- $\varphi_{\text{symb}} = \forall x. \bigwedge_{\omega \in \Omega}(B(x) = \omega \to \text{P}_{\rho(\omega)}(x))$
  where $\rho(\omega) = \alpha$ if $\omega \in \{\omega_1, \omega_2, \omega_6\}$, $\rho(\omega) = \beta$ if $\omega \in \{\omega_4, \omega_5, \omega_3\}$, $\rho(\omega_7) = \#$.

We note that for each $u \in \Sigma^*$ there is at most one $b \in B(\Omega)$ such that $(u, [B \mapsto b]) \models \varphi$; let us call this b_u.

Moreover, we define $\kappa \colon \Sigma_{\{B\}} \to K$ for each $\omega \in \Omega$ by $\kappa((\alpha, \emptyset, \omega)) = 2$, $\kappa((\beta, \emptyset, \omega)) = 1$, and $\kappa((\#, \emptyset, \omega)) = 0$. Then we obtain

$$
\begin{aligned}
[\![e]\!](u) &= \sum_{b \in B(\Omega, |u|)} [\![\varphi \triangleright \text{Val}_\kappa]\!]_{\{B\}}(u, [B \mapsto b]) \\
&= [\![\text{Val}_\kappa]\!]_{\{B\}}(u, [B \mapsto b_u]) = \text{val}_{\text{disc}}(\kappa((u, [B \mapsto b_u]))) = s(u).
\end{aligned}
$$

The above example employed for each word only a single behaviour. By starting with pushdown automata for inherently ambiguous context-free languages and introducing weights for the transitions, we could obtain expressions whose semantics use several behaviours.

5   Results

The next theorem is our main result and it follows from Lemmas 9 and 16.

Theorem 8. *Let K be a unital valuation monoid, Σ an alphabet, and S a storage type. Moreover, let $s : \Sigma^* \to K$ be a weighted language. Then the following are equivalent:*

1. *s is (S, Σ, K)-recognizable.*
2. *s is definable by an expression over (S, Σ, K).*

The constructions in both directions are effective.

By taking $S = \text{P}^n$, Theorem 8 shows a BET-theorem for weighted n-iterated pushdown languages over Σ and K. For instance, by taking $S = \text{P}^2$ and $K = \mathbb{B}$, Theorem 8 shows a BET-theorem for OI-macro languages; these were investigated e.g. in [3]. Since OI-macro languages coincide with indexed languages, Theorem 8 might be considered as an alternative to the BET-theorem in [14].

From Automata to Expressions. For an arbitrary (S, Σ, K)-automaton $\mathcal{A}$ we will construct an equivalent expression e over (S, Σ, K). The construction follows the usual idea, but additionally the existential quantifier over the second-order behaviour variable B is employed to ensure that e simulates behaviours of the storage.

Lemma 9. For each (S, Σ, K)-automaton $\mathcal{A}$, the weighted language $[\![\mathcal{A}]\!]$ is definable by an expression over (S, Σ, K).

Proof. Let $\mathcal{A} = (Q, Q_i, Q_f, T, \mathrm{wt})$ and $T = \{\tau_1, \ldots, \tau_m\}$. We define the set $\mathcal{V} = \{X_\tau \mid \tau \in T\}$ and consider each element of $\mathcal{V}$ to be a second-order variable. Recall $\Omega_\mathcal{A}$ from Sect. 3. We define the expression e over $(\Omega_\mathcal{A}, \Sigma, K)$ as follows:

$$e = \sum_B^{\mathrm{beh}} \sum_{X_{\tau_1}} \cdots \sum_{X_{\tau_m}} \left((\varphi \rhd \mathrm{Val}_\kappa) + (\varphi_\varepsilon \rhd \underbrace{\mathrm{Val}_\kappa + \ldots + \mathrm{Val}_\kappa}_{k}) \right),$$

where $k = |Q_i \cap Q_f|$ and $\varphi = \varphi_{\mathrm{part}} \wedge \varphi_{\mathrm{comp}} \wedge \neg \varphi_\varepsilon$ and

- $\varphi_{\mathrm{part}} = \forall x. \bigvee_{\tau \in T} \left((x \in X_\tau) \wedge \bigwedge_{\substack{\tau' \in T: \\ \tau \neq \tau'}} \neg(x \in X_{\tau'}) \right),$
- $\varphi_{\mathrm{comp}} = \forall x. \psi \wedge \psi' \wedge \bigwedge_{\tau = (q,a,p,q',f) \in T} \left((x \in X_\tau) \to (\psi_1 \wedge \psi_2 \wedge \psi_3) \right)$ and
  - $\psi = \mathrm{first}(x) \to \bigvee_{\tau' \in T: \mathrm{pr}_1(\tau') \in Q_i} (x \in X_{\tau'}),$
  - $\psi' = \mathrm{last}(x) \to \bigvee_{\tau' \in T: \mathrm{pr}_4(\tau') \in Q_f} (x \in X_{\tau'})$
  - $\psi_1 = P_a(x)$ and $\psi_2 = (B(x) = (p, f)),$
  - $\psi_3 = \forall y. \left(\mathrm{next}(x, y) \to \bigvee_{\tau' \in T: \mathrm{pr}_1(\tau') = q'} (y \in X_{\tau'}) \right),$
- $\varphi_\varepsilon = \forall x. \mathrm{next}(x, x)$. Note: $\mathcal{L}_{\mathcal{V} \cup \{B\}}(\varphi_\varepsilon) = \{(\varepsilon, \sigma_\varepsilon)\}$ for $\sigma_\varepsilon(X_{\tau_i}) = \emptyset$, $\sigma_\varepsilon(B) = \varepsilon$.

Intuitively, φ models the computations of $\mathcal{A}$ as it is usual for weighted automata without storage; additionally, ψ_2 assures that the $\Omega_\mathcal{A}$-behaviour guessed by the semantics of $\sum_B^{\mathrm{beh}}$ fits to the behaviour guessed by the semantics of the sequence $\sum_{X_{\tau_1}} \cdots \sum_{X_{\tau_m}}$. We use φ_ε to handle the empty input word appropriately.

We define the mapping $\kappa : \Sigma_{\mathcal{V} \cup \{B\}} \to K$ for each $(a, V, \omega) \in \Sigma_{\mathcal{V} \cup \{B\}}$ by: $\kappa((a, V, \omega)) = \mathrm{wt}(\tau)$ if $V = \{X_\tau\}$, and 0 otherwise.

Finally, we can compute for each $u \in \Sigma^* \setminus \{\varepsilon\}$ of some length $n > 0$ as follows:

$$[\![e]\!](u) = \sum_{b \in B(\Omega_\mathcal{A}, n)} \sum_{I_1, \ldots, I_m \subseteq \mathrm{pos}(u)} [\![\varphi \rhd \mathrm{Val}_\kappa]\!]_{\mathcal{V} \cup \{B\}}(u, \sigma) = \sum_{\theta \in \Theta_\mathcal{A}(u)} \mathrm{wt}(\theta) = [\![\mathcal{A}]\!](u)$$

where $\sigma(B) = b$ and $\sigma(X_{\tau_i}) = I_i$ for each $1 \leq i \leq n$. Moreover,

$$[\![e]\!](\varepsilon) = [\![\underbrace{\mathrm{Val}_\kappa + \ldots + \mathrm{Val}_\kappa}_{k = |Q_i \cap Q_f|}]\!]_{\mathcal{V} \cup \{B\}}(\varepsilon, \sigma_\varepsilon) = \sum_{q \in Q_i \cap Q_f} 1 = [\![\mathcal{A}]\!](\varepsilon),$$

where the last but one equation holds, because $[\![\mathrm{Val}_\kappa]\!]_{\mathcal{V} \cup \{B\}}(\varepsilon, \sigma_\varepsilon) = 1$. Thus, $[\![\mathcal{A}]\!]$ is definable by an expression over (S, Σ, K). □

From Expressions to Automata. The proof consists of a first part, which is an induction on B-expressions (similar to [15, Sect. 4.2]) and a second part which transforms B-expressions into weighted automata with storage (see Lemma 15). In the sequel we let Ω be a finite subset of $P \times F$.

In Lemmas 10–14 we will deal with the class $\mathrm{Rec}(\Sigma_\mathcal{V}, K)$. Since $(\Sigma_\mathcal{V}, K)$-automata cannot check whether a word $b \in \Omega^*$ is an Ω-behaviour, we always have to intersect with $\Sigma_\mathcal{V}^{*\mathcal{V}}$ in the following sense. Let $s : \Sigma^* \to K$ and $L \subseteq$

Σ^*. We define the weighted language $(s \cap L) : \Sigma^* \to K$ for every $u \in \Sigma^*$ by $(s \cap L)(u) = s(u)$ if $u \in L$, and 0 otherwise. We extend this intersection to a set Ψ of weighted languages of type $\Sigma^* \to K$ by letting $\Psi \cap L = \{(s \cap L) \mid s \in \Psi\}$.

Lemma 10 (compare [4, Proposition 3.3] and [15, Lemma 3.8]). Let e be a B-expression over (Ω, Σ, K) and let $\mathcal{V}$ and $\mathcal{W}$ be finite sets of variables with Free$(e) \subseteq \mathcal{W} \subseteq \mathcal{V}$. Then, $[\![e]\!]_{\mathcal{W}} \in \text{Rec}(\Sigma_{\mathcal{W}}, K) \cap \Sigma_{\mathcal{W}}^{*\text{v}}$ if and only if $[\![e]\!]_{\mathcal{V}} \in \text{Rec}(\Sigma_{\mathcal{V}}, K) \cap \Sigma_{\mathcal{V}}^{*\text{v}}$.

Lemma 11 (cf. [15, Lemma 4.5]). Let $\mathcal{U}$ and $\mathcal{V}$ be finite sets of variables such that $B \in \mathcal{U}$ and $\mathcal{U} \subseteq \mathcal{V}$. Moreover, let $\kappa : \Sigma_{\mathcal{U}} \to K$. Then $[\![\text{Val}_\kappa]\!]_{\mathcal{V}} \in \text{Rec}(\Sigma_{\mathcal{V}}, K) \cap \Sigma_{\mathcal{V}}^{*\text{v}}$.

Lemma 12 (cf. [15, Lemma 4.6]). Let $\mathcal{V}$ be a finite set of variables such that $B \in \mathcal{V}$. Moreover, let e_1, e_2 be two B-expressions over $(\Omega, \Sigma_{\mathcal{V}}, K)$ such that Free$(e_1) \cup$ Free$(e_2) \subseteq \mathcal{V}$. If $[\![e_i]\!]_{\mathcal{V}} \in \text{Rec}(\Sigma_{\mathcal{V}}, K) \cap \Sigma_{\mathcal{V}}^{*\text{v}}$ for $i \in \{1, 2\}$, then $[\![e_1 + e_2]\!]_{\mathcal{V}} \in \text{Rec}(\Sigma_{\mathcal{V}}, K) \cap \Sigma_{\mathcal{V}}^{*\text{v}}$.

Lemma 13 (cf. [15, Lemma 4.10]). Let $\mathcal{V}$ be a finite set of variables such that $B \in \mathcal{V}$. Moreover, let e be a B-expression over $(\Omega, \Sigma_{\mathcal{V}}, K)$ such that Free$(e) \subseteq \mathcal{V}$. Also let $\varphi \in \text{MSO}(\Sigma, \mathcal{C}_{\text{MSO}} \cup \mathcal{C}_B)$ with Free$(\varphi) \subseteq \mathcal{V}$. If $[\![e]\!]_{\mathcal{V}} \in \text{Rec}(\Sigma_{\mathcal{V}}, K) \cap \Sigma_{\mathcal{V}}^{*\text{v}}$, then $[\![\varphi \triangleright e]\!]_{\mathcal{V}} \in \text{Rec}(\Sigma_{\mathcal{V}}, K) \cap \Sigma_{\mathcal{V}}^{*\text{v}}$.

Lemma 14 (cf. [4, Lemma 4.3] and [15, Lemma 4.9]). Let $\mathcal{V}$ be a finite set of variables such that $B \in \mathcal{V}$.

1. If e is a B-expression over $(\Omega, \Sigma_{\mathcal{V} \cup \{x\}}, K)$, Free$(e) \subseteq \mathcal{V} \cup \{x\}$, and $[\![e]\!]_{\mathcal{V} \cup \{x\}} \in \text{Rec}(\Sigma_{\mathcal{V} \cup \{x\}}, K) \cap \Sigma_{\mathcal{V} \cup \{x\}}^{*\text{v}}$, then $[\![\sum_x e]\!]_{\mathcal{V}} \in \text{Rec}(\Sigma_{\mathcal{V}}, K) \cap \Sigma_{\mathcal{V}}^{*\text{v}}$.
2. If e is a B-expression over $(\Omega, \Sigma_{\mathcal{V} \cup \{X\}}, K)$, Free$(e) \subseteq \mathcal{V} \cup \{X\}$, and $[\![e]\!]_{\mathcal{V} \cup \{X\}} \in \text{Rec}(\Sigma_{\mathcal{V} \cup \{X\}}, K) \cap \Sigma_{\mathcal{V} \cup \{X\}}^{*\text{v}}$, then $[\![\sum_X e]\!]_{\mathcal{V}} \in \text{Rec}(\Sigma_{\mathcal{V}}, K) \cap \Sigma_{\mathcal{V}}^{*\text{v}}$.

In the next lemma we obtain the required (S, Σ, K)-automaton $\mathcal{A}'$ as a suitable projection of the given $(\Sigma_{\{B\}}, K)$-automaton $\mathcal{A}$ by incorporating the Ω-information contained in its extended alphabet into the transitions of $\mathcal{A}'$.

Lemma 15. Let e be a B-expression over $(\Omega, \Sigma_{\{B\}}, K)$ with Free$(e) = \{B\}$. If $[\![e]\!]_{\{B\}} \in \text{Rec}(\Sigma_{\{B\}}, K) \cap \Sigma_{\{B\}}^{*\text{v}}$, then $[\![\sum_B^{\text{beh}} e]\!] \in \text{Rec}(S, \Sigma, K)$.

Proof. Let $\mathcal{A} = (Q, Q_i, Q_f, T, \text{wt})$ be a $(\Sigma_{\{B\}}, K)$-automaton such that $[\![e]\!]_{\{B\}} = [\![\mathcal{A}]\!] \cap \Sigma_{\{B\}}^{*\text{v}}$. We construct the (S, Σ, K)-automaton $\mathcal{A}' = (Q, Q_i, Q_f, T', \text{wt}')$ as follows. If $\tau = (q, (a, \emptyset, (p, f)), q')$ is in T, then $\tau' = (q, a, p, q', f)$ is in T' and we let $\text{wt}'(\tau') = \text{wt}(\tau)$.

Let $u \in \Sigma^*$ of some length $n \geq 0$. We define the mapping $\nu : \Theta_{\mathcal{A}'}(u) \to B(\Omega, n)$ for each $\theta = (d_0 \vdash^{\tau_1'} d_1 \cdots \vdash^{\tau_n'} d_n)$ by $\nu(\theta) = b_1 \ldots b_n$ where $b_i = (p_i, f_i)$ if $p_i = \text{pr}_3(\tau_i')$ and $f_i = \text{pr}_5(\tau_i')$.

Then, for each $b \in B(\Omega, n)$, the mapping $\delta : \Theta_{\mathcal{A}}((u, [B \mapsto b])) \to \Theta_{\mathcal{A}'}(u) \cap \nu^{-1}(b)$ induced by $\delta(\tau) = \tau'$ is a bijection, and $\mathrm{wt}(\theta) = \mathrm{wt}'(\delta(\theta))$ for each $\theta \in \Theta_{\mathcal{A}}((u, [B \mapsto b]))$. Then we can calculate as follows:

$$
\begin{aligned}
[\![\textstyle\sum_B^{\mathrm{beh}} e]\!](u) &= \sum_{b \in B(\Omega, n)} [\![e]\!]_{\{B\}}(u, [B \mapsto b]) = \sum_{b \in B(\Omega, n)} \left([\![\mathcal{A}]\!] \cap \Sigma_{\{B\}}^{*v} \right)(u, [B \mapsto b]) \\
&= \sum_{b \in B(\Omega, n)} [\![\mathcal{A}]\!](u, [B \mapsto b]) = \sum_{b \in B(\Omega, n)} \sum_{\theta \in \Theta_{\mathcal{A}}(u, [B \mapsto b])} \mathrm{wt}(\theta) \\
&= \sum_{b \in B(\Omega, n)} \sum_{\theta' \in \Theta_{\mathcal{A}'}(u) \cap \nu^{-1}(b)} \mathrm{wt}'(\theta') = \sum_{\theta' \in \Theta_{\mathcal{A}'}(u)} \mathrm{wt}'(\theta') = [\![\mathcal{A}']\!](u).
\end{aligned}
$$

Lemma 16. *Let $s : \Sigma^* \to K$. If s is definable by some expression over (S, Σ, K), then s is (S, Σ, K)-recognizable.*

Proof. Using Lemmas 10–14 we can prove by induction that for each finite set $\mathcal{V}$ of variables such that $B \in \mathcal{V}$ and for each B-expression e over $(\Omega, \Sigma_{\mathcal{V}}, K)$ with $\mathrm{Free}(e) \subseteq \mathcal{V}$, we have $[\![e]\!]_{\mathcal{V}} \in \mathrm{Rec}(\Sigma_{\mathcal{V}}, K) \cap \Sigma_{\mathcal{V}}^{*v}$.

Now let $s : \Sigma^* \to K$ be definable by some expression over (S, Σ, K). Then there is a finite set $\Omega \subseteq P \times F$ and a B-expression e over $(\Omega, \Sigma_{\{B\}}, K)$ such that $\mathrm{Free}(e) = \{B\}$ and $s = [\![\textstyle\sum_B^{\mathrm{beh}} e]\!]$. Since $[\![e]\!]_{\{B\}} \in \mathrm{Rec}(\Sigma_{\{B\}}, K) \cap \Sigma_{\{B\}}^{*v}$ by the above, we obtain from Lemma 15 that $s \in \mathrm{Rec}(S, \Sigma, K)$. $\square$

Support and Decidability. Let $(K, \cdot, 1)$ be a monoid. For $k_1, \ldots, k_n \in K$, we let $\langle k_1, \ldots, k_n \rangle$ denote the smallest submonoid of K containing $k_1, \ldots, k_n$. If $k \in K$ and $A \subseteq K$, let $k \cdot A = \{k \cdot a \mid a \in A\}$. As defined by Kirsten [19], the *zero generation problem* (ZGP) for a monoid $(K, \cdot, 1)$ with zero 0 consists of two integers $m, n \in \mathbb{N}$, elements $k_1, \ldots, k_m, k_1', \ldots, k_n' \in K$ and the question whether $0 \in k_1 \cdot \ldots \cdot k_m \cdot \langle k_1', \ldots, k_n' \rangle$. For instance, each idempotent monoid has a decidable ZGP problem. A strong bimonoid $(K, +, \cdot, 0, 1)$ [8] is called *zero-sum-free*, if $k + k' = 0$ implies $k = k' = 0$, and *commutative*, if $k \cdot k' = k' \cdot k$ (for $k, k' \in K$).

Theorem 17. *Let K be a zero-sum-free commutative strong bimonoid.*

1. *For every (S, Σ, K)-recognizable series $s : \Sigma^* \to K$, $\mathrm{supp}(s)$ is (S, Σ)-recognizable.*
2. *Let $|\Sigma| \geq 2$. There is an effective construction of an (S, Σ)-automaton recognizing $\mathrm{supp}([\![\mathcal{A}]\!])$ from any given (S, Σ, K)-automaton $\mathcal{A}$ iff $(K, \cdot, 1)$ has a decidable ZGP.*

From Theorems 8 and 17 and the fact that the emptiness problem of iterated pushdown automata is decidable [11, Corollary 1], we obtain the following result:

Corollary 18. *Let $|\Sigma| \geq 2$, K be a zero-sum-free commutative strong bimonoid with a decidable ZGP, and $s : \Sigma^* \to K$ be definable by an expression over $(\mathrm{P}^n, \Sigma, K)$ for some $n \geq 0$. Then it is decidable whether $\mathrm{supp}(s) = \emptyset$.*

This shows the decidability of the satisfiability problem of the logic given by (P^n, Σ, K)-expressions for Σ and K as indicated. In particular, each bounded lattice with computable infimum is a zero-sum-free commutative strong bimonoid with a decidable ZGP.

References

1. Büchi, J.: Weak second-order arithmetic and finite automata. Zeitschr. für math. Logik und Grundl. der Mathem. **6**, 66–92 (1960)
2. Büchi, J.: On a decision method in restricted second-order arithmetic. In: Proceedings of 1960 International Congress for Logic, Methodology and Philosophy of Science, pp. 1–11. Stanford University Press, Stanford (1962)
3. Damm, W., Goerdt, A.: An automata-theoretical characterization of the OI-hierarchy. Inf. Control **71**, 1–32 (1986)
4. Droste, M., Gastin, P.: Weighted automata and weighted logics. Theoret. Comput. Sci. **380**(1–2), 69–86 (2007)
5. Droste, M., Gastin, P.: Weighted automata and weighted logics. In: Droste, M., Kuich, W., Vogler, H. (eds.) Handbook of Weighted Automata, Chap. 5. Springer, Heidelberg (2009)
6. Droste, M., Meinecke, I.: Weighted automata and weighted MSO logics for average and long-time behaviors. Inf. Comput. **220–221**, 44–59 (2012)
7. Droste, M., Stüber, T., Vogler, H.: Weighted finite automata over strong bimonoids. Inf. Sci. **180**, 156–166 (2010)
8. Droste, M., Vogler, H.: Weighted automata and multi-valued logics over arbitrary bounded lattices. Theoret. Comput. Sci. **418**, 14–36 (2012)
9. Droste, M., Vogler, H.: The Chomsky-Schützenberger theorem for quantitative context-free languages. Int. J. Found. Comput. Sci. **25**(8), 955–969 (2014)
10. Elgot, C.: Decision problems of finite automata design and related arithmetics. Trans. Am. Math. Soc. **98**, 21–52 (1961)
11. Engelfriet, J.: Iterated pushdown automata and complexity classes. In: Proceedings of 15th Annual ACM Symposium on Theory of Computing (STOCS), pp. 365–373 (1983)
12. Engelfriet, J.: Context-free grammars with storage. Technical Report 86–11, University of Leiden (1986). arxiv:1408.0683 [cs.FL] (2014)
13. Engelfriet, J., Vogler, H.: Pushdown machines for the macro tree transducer. Theoret. Comput. Sci. **42**(3), 251–368 (1986)
14. Fratani, S., Voundy, E.: Context-free characterization of indexed languages. CoRR - Computing Research Repository arXiv:1409.6112 (2014)
15. Fülöp, Z., Stüber, T., Vogler, H.: A Büchi-like theorem for weighted tree automata over multioperator monoids. Theory Comput. Syst. **50**(2), 241–278 (2012). doi:10.1007/s00224-010-9296-1. Published online 28.10.2010
16. Fülöp, Z., Vogler, H.: Characterizations of recognizable weighted tree languages by logic and bimorphisms. Soft Comput. (2015). doi:10.1007/s00500-015-1717-2
17. Ginsburg, S.: Algebraic and Automata-theoretic Properties of Formal Languages. North-Holland Publishing, Amsterdam (1975)
18. Herrmann, L., Vogler, H.: A Chomsky-Schützenberger theorem for weighted automata with storage. In: Maletti, A. (ed.) CAI 2015. LNCS, vol. 9270, pp. 115–127. Springer, Heidelberg (2015)

19. Kirsten, D.: The support of a recognizable series over a zero-sum free, commutative semiring is recognizable. Acta Cybern. **20**(2), 211–221 (2011)
20. Lautemann, C., Schwentick, T., Therien, D.: Logics for context-free languages. In: Pacholski, L., Tiuryn, J. (eds.) CSL 1994. LNCS, vol. 933, pp. 205–216. Springer, Heidelberg (1995)
21. Maslov, A.N.: Multilevel stack automata. Probl. Inf. Transm. **12**, 38–43 (1976)
22. Mathissen, C.: Weighted logics for nested words and algebraic formal power series. Log. Meth. Comput. Sci. **5**, 1–34 (2010)
23. Scott, D.: Some definitional suggestions for automata theory. J. Comput. Syst. Sci. **1**, 187–212 (1967)
24. Trakhtenbrot, B.: Finite automata and logic of monadic predicates. Doklady Akademii Nauk SSSR **149**, 326–329 (1961). in Russian

Automata and Words

Colored Nested Words

Rajeev Alur and Dana Fisman[⊠]

Department of Computer and Information Sciences, University of Pennsylvania,
Philadelphia, PA, USA
{alur,fisman}@cis.upenn.edu

Abstract. Nested words allow modeling of linear and hierarchical structure in data, and nested word automata are special kinds of pushdown automata whose push/pop actions are directed by the hierarchical structure in the input nested word. The resulting class of regular languages of nested words has many appealing theoretical properties, and has found many applications, including model checking of procedural programs. In the nested word model, the hierarchical matching of open- and close- tags must be properly nested, and this is not the case, for instance, in program executions in presence of exceptions. This limitation of nested words narrows its model checking applications to programs with no exceptions.

We introduce the model of *colored nested words* which allows such hierarchical structures with mismatches. We say that a language of colored nested words is *regular* if the language obtained by inserting the missing closing tags is a well-colored regular language of nested words. We define an automata model that accepts regular languages of colored nested words. These automata can execute restricted forms of ε-pop transitions. We provide an equivalent grammar characterization and show that the class of regular languages of colored nested words has the same appealing closure and decidability properties as nested words, thus removing the restriction of programs to be exception-free in order to be amenable for model checking, via the nested words paradigm.

1 Introduction

Nested words, introduced in [4], are a data model capturing both a linear ordering and a hierarchically nested matching of items. Examples for data with both of these characteristics include executions of structured programs, annotated linguistic data, and documents in marked-up languages such as XML. While *regular languages of nested words* allow capturing of more expressive structure than traditional words, they retain all the good properties of regular languages. In particular, deterministic nested word automata are as expressive as their non-deterministic counterparts; the class is closed under the following operations: union, intersection, complementation, concatenation, Kleene-*, prefixes and language homomorphism; and the following problems are decidable: emptiness, membership, language inclusion and language equivalence.

This research was supported by US NSF grant CCF-1138996.

```
p0: P(n) {              q0: Q(n) {              r0: R(n) {
p1:     try {           q1:     y = n / 2      r1:     z = n-1
p2:         x = 2*n     q2:     y = R(y)       r2:     if (z<0)
p3:         x = Q(x)    q3:     y = y+1        r3:         throw 0
p4:     }               q4:     return y }     r4:     return z }
p5:     catch (int) {}
p6:     return x }
```

(a)

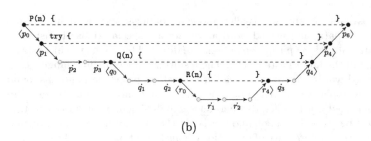

(b)

Fig. 1. (a) A procedural program. (b) An illustration of an execution of the code of the program where function calls are captured hierarchically.

Many algorithms for problems concerning such data can be formalized and solved using constructions for basic operations and algorithms for decision problems. This fact led the way to many interesting applications and tools. Two prominent areas are XML processing (see e.g. [8,14,15]) and model checking of procedural programs (see e.g. [1–3,6,9,20]). By modeling executions of structured programs as nested words, one can algorithmically verify/refute various aspects of program correctness. Consider for instance, the program in Fig. 1a. An example execution is illustrated in Fig. 1b. Each step of the execution is mapped to the program counter line, and in addition, function calls create hierarchical connections to their respective returns. In the illustrations calls are depicted with down arrows, returns with up arrows, and internal code with horizontal arrows.

Nested words can be represented by graphs as in Fig. 1b or via an implicit representation using words over an alphabet $\Sigma \times \{\langle, \cdot, \rangle\}$. We use $\langle a, \dot{a}$ and $a \rangle$ as abbreviations for $(a, \langle), (a, \cdot)$ and $(a, \rangle)$, respectively.[1] For the program in Fig. 1a, we can define the first component of the alphabet to be the set of possible program counter lines $\{p_0, p_1, \ldots, r_4\}$. Then the call to Q(), for instance, will be modeled by the letter $\langle q_0$. The implicit representation for the nested word in Fig. 1b is the word obtained by concatenating the letters on the path consisting of all solid edges. The fact that the hierarchical matching between calls and returns is explicitly captured (in comparison to treating them as a linear sequence of instructions) can be

---

[1] Our notation for internal letters, marking a letter with a dot as in $\dot{a}$, differs slightly from nested words literature which uses simply a. When there is no risk of confusion we may use un-dotted versions too.

```
        1   2   3   4   5
           def isPrime(n):
              if n<2: return False
              elif n==2: return True
              else:
                 i = 2
                 while i < sqrt(n)+1:
                    if n%i==0:
                       return False
                    else:
                       i = i+1
              return True

        Closes scopes 5,4,3
```

```
        <div id="s">
           <ol>
              <li> ML  </li>
              <li> Haskell </li>
              <li> Scheme
           </ol>
        </div>
        <div id="w">
           <ol>
              <li> C
              <li> C++ </li>
              <li> Perl </li>
           </ol>
        </div>
```

(a)   (b)

Fig. 2. (a) A Python program. (b) An excerpt from an HTML document.

exploited for writing more expressive specifications of procedural programs such as pre/post conditions that can be algorithmically verified — see [17] for details.

But what happens if an exception is thrown? Then a call (or several calls) will not have a matching return. Viewing the run as a nested word might match the thrown exception with the most recent call, but this is not what we want.

A similar situation happens in parsing programs written in programming languages like *Python* or *Haskell*, that use whitespace to delimit program blocks and deduce variables' scope. In such programming languages, a new block begins by a line starting at a column greater than that of the previous line. If the current line starts at column n (i.e. after n spaces from a new line) and the following line starts at the same column n it is considered on the same block. If the next line starts at a column $n' > n$ it is considered a new block. Last, if the next line starts in a column $n' < n$ then it is considered in the block that started at n' and this implicitly closes all blocks that were opened in between. (If no block started at column n' this would be a syntax error.) If we were to model this with nested words, we can only close the last block, but here we need to close as many blocks as needed.

For strongly matched languages, such as XML, one might want to use this principle to help recover un-closed tags, in cases where this will not result in a confusion, but rather help processing the rest of the document. Consider for instance, the example of Fig. 2b. In this example we have a list with a couple of well-matched list items, and one list item that has no closing tag. We would like to be able to process it and recover from the unmatched list item. If we consider </ol> in a way similar to a thrown exception, we can achieve this task.

If we can't model exceptions correctly, we cannot use model checking to formally prove/refute properties about them, and a fundamental property such as "if a certain condition occurs in a program, an exception is thrown and properly caught" is left beyond the scope of verification.

In this work we suggest to augment the nested words model with *colors*. Each call and return, or opening and closing tags, are associated with some color. The

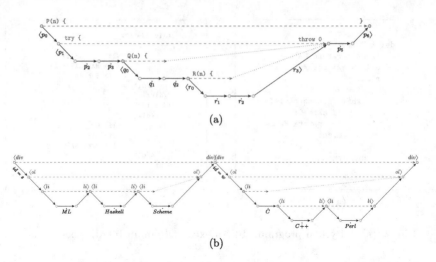

(a)

(b)

Fig. 3. (a) a colored nested word corresponding to an execution of the code in Fig. 1a where an exception is thrown. (b) a colored nested word corresponding to the HTML excerpt in Fig. 2b (Color figure online).

hierarchical structure matches only nodes of the same color. This allows relaxing the requirements on the hierarchical edges. A hierarchical edge of a certain color may be unmatched if it is encapsulated by a matched hierarchical edge of a different color. This models catching thrown exceptions, closing as many blocks as needed, or recovering from unmatched tags. Figure 3a depicts *colored nested words* for the two elaborated examples.

Following a formalization of *colored nested words*, we ask ourselves whether we can use existing machinery of nested word automata and/or nested words tranducers to process colored nested words. Realizing that this is unfeasible, we present *colored nested word automata* (CNA). These automata augment automata for nested words with restricted forms of ε-pop transitions. These ε-transitions enable the automaton to read all the information on the stack that was pushed on un-matched calls. We study also a *blind* version (BCNA), that can see just the information recorded on the matched call. We show that although there could be unboundedly many stack symbols that it cannot observe in comparison to the first automata model, their expressive power is the same.

We show that CNA recognize exactly the class of *regular languages of colored nested words*. We then show that this class of languages is as robust as regular languages: Deterministic CNA are as expressive as their non-deterministic counterparts. It is closed under the following operations: union, intersection, complementation, concatenation, Kleene-*, prefixes, suffixes, reversal, homomorphism and inverse homomorphism. The following problems are decidable: emptiness, membership, language inclusion and language equivalence. We conclude with a grammar characterization. Due to lack of space proofs are omitted; they can be found in the full version on the authors' homepages.

Related Work. The key idea in the model of nested words as well as colored nested words is to expose the hierarchical matching between the open and close tags, and the corresponding automata models are really processing the input DAG. To understand the relationship of these automata with classical formalisms such as context-free languages, we can view the input as a sequence of symbols, with the hierarchical structure only implicit, and measure expressiveness by the class of languages of words they define. With this interpretation, the class of regular languages of nested words is a strict superset of regular word languages and a strict subset of DCFLs. This relationship has led to renewed interest in finding classes between regular languages and DCFLs such as realtime height deterministic PDAs [16], synchronized grammars [5], and Floyd grammars/automata [7] (and some interested in languages accepted by higher order pushdown automata, e.g. [13], which are not CFL.) The class of regular languages of colored nested words is a strict superset of regular nested-word-languages and a strict subset of Floyd grammars. While a CNA can be encoded as a Floyd automaton by defining a suitable dependency matrix between input symbols to dictate the stack operations, the view that CNAs are finite-state machines operating over the DAG structure of the input colored nested word leads to a clean theory of regular languages of colored nested words.

2   Colored Nested Words

As is the case in nested words, colored nested words can be represented explicitly using graphs as in Fig. 3a or implicitly using words over an augmented alphabet. We start with the implicit representation. Formally, we define *colored nested words* to be words over alphabets of the form $A \cup A \times C \times \{+, -\}$. Given a triple $\langle a, c, h \rangle$, the first component $a \in A$ provides some content, the second component $c \in C$ provides a *color* and the third component h indicates whether a hierarchical connection starts $(+)$ or ends $(-)$. Letters in A do not influence the hierarchical structure. Letters of the form $\langle a, red, + \rangle$ and $\langle a, red, - \rangle$, can be abbreviated using $(\!(a$ and $a)\!)$, respectively, and similarly for other colors. One can use instead, different parenthesis types for the different colors and form abbreviations such as $\{a, [a, (\!(a \text{ and } a)\!), a], a\}$. More generally, we can use $(\!(_c a, a_c)\!)$ to abbreviate $\langle a, c, + \rangle$ and $\langle a, c, - \rangle$, respectively. When A and C are clear from the context we use $(\!(\Sigma, \dot{\Sigma} \text{ and } \Sigma)\!)$ for $A \times C \times \{+\}$, A and $A \times C \times \{-\}$, respectively. For a given color $c \in C$ we use $(\!(_c \Sigma \text{ and } \Sigma_c)\!)$ for the sets $A \times \{c\} \times \{+\}$ and $A \times \{c\} \times \{-\}$, respectively. Finally, we use $\hat{\Sigma}$ for $(\!(\Sigma \cup \dot{\Sigma} \cup \Sigma)\!)$, and a, b, c, and w, u, v for letters and words in $\hat{\Sigma}$, respectively.

Explicit Representation. A colored nested word of length n can be represented by explicitly a tuple $(w, \kappa, \leftrightharpoons, \rightleftharpoons, \leftharpoonup, \rightharpoonup)$ where w is a word of length n over a finite set of symbols A; κ maps nodes in $[0..n]$ to colors in C; $\leftrightharpoons, \rightleftharpoons$ are binary relations in $[0..n-1] \times [1..n]$ and $\rightharpoonup$ and $\leftharpoonup$ are unary relations over $[0..n-1]$ and $[1..n]$, resp. The relations $\leftrightharpoons, \rightleftharpoons, \leftharpoonup, \rightharpoonup$ describe the *hierarchical edges*. The *linear edges* are implicit; there is a linear edge from every $i \in [0..n-1]$ to $i + 1$, and w maps the linear edges to the A-symbols, as shown in Fig. 3a.

We refer to $\leftrightharpoons$ and $\Rightarrow$ as *matched* edges, and *recovered* edges, and to $\leftarrow$ and $\rightarrow$ as *pending calls* and *pending returns*. The following conditions must be satisfied, where for uniformity we view the relations $\leftarrow$ and $\rightarrow$ as binary by interpreting $i \leftarrow$ and $\rightarrow j$ as $i \leftarrow \infty$ and $-\infty \rightarrow j$.

1. Edges point forward: if $i \rightsquigarrow j$ for some $\rightsquigarrow \in \{\leftrightharpoons, \Rightarrow, \leftarrow, \rightarrow\}$ then $i < j$.
2. Edges do not cross: if $i \rightsquigarrow j$ and $i' \rightsquigarrow' j'$ for some $\rightsquigarrow, \rightsquigarrow' \in \{\leftrightharpoons, \Rightarrow, \leftarrow, \rightarrow\}$ then it is not the case that $i < i' < j < j'$.
3. Source positions may not be shared: if $i \rightsquigarrow j$ and $i' \rightsquigarrow' j'$ for some $\rightsquigarrow, \rightsquigarrow' \in \{\leftrightharpoons, \Rightarrow, \leftarrow\}$ then $i \neq i'$.[2]
4. Target positions join at a match: for every $j \in [1..n]$ if the set $\{i \rightsquigarrow j \mid \rightsquigarrow \in \{\leftrightharpoons, \Rightarrow\}\}$ is non-empty then it contains exactly one $\leftrightharpoons$ edge.

A colored nested word is said to be *well-colored* if (a) matched edges are monochromatic, i.e. if $i \leftrightharpoons j$ then $\kappa(i) = \kappa(j)$, and (b) recovered edges are bi-chromatic, i.e. if $i \Rightarrow j$ then $\kappa(i) \neq \kappa(j)$ and there exists $i' < i$ such that $i' \leftrightharpoons j$. A well-colored colored nested word is said to be *well-matched* if it has no pending calls, no pending returns, and no recovered edges. It is said to be *weakly-matched* if it has no pending returns and no pending calls (but it may have recovered edges). A weakly-matched colored nested word is said to be *rooted* if the first letter is in $(\Sigma$ and the last letter is in $\Sigma)$. It is said to be *c-rooted* if it is rooted and the first and last letters are colored c. The *outer level* of a colored nested word is the word obtained by omitting all weakly-matched proper infixes. For instance, if w $= (a[bb]c(d[ef]g)$ its outer level is (acg).

For executions of programs with exceptions (Figs. 1b and 3a), a word is well-matched if no exceptions are thrown. For the HTML example, being well-matched means that all open tags are closed in the correct order. If not, as is the case Fig. 2b, the explicit representation of the nested word will contain bi-chromatic edges. In the case of Python programs, being well matched means that after a block ends, there are always some lines of code before the outermost block ends, which is very unlikely.

3   Regularity

Next, we define a notion of *regularity* for colored nested words. We would like to say that a language of colored nested words is *regular* if the language obtained by inserting the missing closing tags is a well-colored regular language of nested words. First we need to define this mapping from a colored nested word to the uncolored nested word obtained by adding the missing closing tags.

Assume our colored alphabet is $\hat{\Sigma} = A \cup A \times C \times \{+, -\}$. We can define the uncolored alphabet $\tilde{\Sigma} = A \cup (A \times C \times \{+, -\}) \cup (\_ \times C \times \{-\})$. A letter a $\in \hat{\Sigma}$ can be mapped to a letter $\tilde{\text{a}}$ in $\tilde{\Sigma}$ as follows. An opening letter $(a, c, +)$ can be mapped to $\langle (a, c)$, an internal letter a to itself, and a closing letter $(a, c, -)$ can be mapped to $(a, c)\rangle$. We will use letters of the form $(\_, c)\rangle$ to fill the gap of

---

[2] Pending returns ($\rightarrow$) by definition share a source.

"missing" closing letters. We say that a language over $\tilde{\Sigma}$ is *well-colored* if it can be accepted by a product of two nested word automata (NWA) $\mathcal{A}$ and $\mathcal{A}_c$ where $\mathcal{A}_c$ is a fixed two-state NWA that upon reading $(\!(_c a$ letters pushes the color c to the stack and upon reading $b_d)\!)$ checks that the color on the stack is d, and if it is not goes to its rejecting state.

Adding the missing closing letters is the tricky part. The opening letter corresponding to a missing closing letter is some index i such that in the explicit representation $i \Rrightarrow j$ for some j. For each $j \leqslant |w|$ let I_j be the set of indices i such that $i \Rrightarrow j$. That is, I_j is the set of indices i that are recovered by j. Assume $I_j = \{i_1, i_2, \ldots, i_{\ell_j}\}$ where $i_1 < i_2 < \ldots i_{\ell_j}$. Define $\mathrm{u}_j = \mathrm{c}_{\ell_j} \cdots \mathrm{c}_2 \mathrm{c}_1$ where $\mathrm{c}_k = (\_, \kappa(i_k)))\rangle$ for $k \in [1..\ell_j]$. Note that if I_j is empty then $\mathrm{u}_j = \epsilon$. Then adding u_j just before the j-th letter will close the missing parenthesis recovered by j (if such exist). Formally, for a word $\mathrm{w} \in \tilde{\Sigma}$ we define the mapping

$$f(\mathrm{a}_1 \mathrm{a}_2 \cdots \mathrm{a}_n) = \mathrm{u}_1 \tilde{\mathrm{a}}_1 \mathrm{u}_2 \tilde{\mathrm{a}}_2 \ldots \mathrm{u}_n \tilde{\mathrm{a}}_n$$

Definition 1. *A language L of colored nested words is* regular *if the language $f(L) = \{f(\mathrm{w}) \mid \mathrm{w} \in L\}$ is a well-colored regular language of nested words.*

Now that we have a definition of regularity in place, we can ask what machinery can we use to process regular languages of colored nested words. If we can define a transducer machine $\mathcal{M}$ that implements f then we can feed its output $f(\mathrm{w})$ to a nested word automaton and process it instead of w. But such a transducer machine $\mathcal{M}$ won't be a finite state transducer, nor it will be a nested words transducer (NWT) [10–12,18,19]. Intuitively, since it needs to map a return letter to several return letters, in fact to an unbounded number of return letters, dependent on the number of unmatched call letters, and while the stack can be used to store this information, an NWT can only inspect the top symbol of the stack.

Therefore we need new machinery to process colored nested words. We can either define a new transducer model that will allow implementing the desired transformation or we can simply define a new automata model that directly process colored nested words. We pursue the second option, which generalizes nested words, and can serve as a base line for a respective transducer model.

4   Colored Nested Word Automata

A *colored nested word automaton* (CNA), is a pushdown automaton that operates in a certain manner, capturing the colored nested structure of the read word. A CNA over $\tilde{\Sigma}$ uses some set of stack symbols P to record information on the hierarchical structure. As in the case of nested word automata, opening letters always cause a push, closing letters always cause a pop, and internal letters do not affect the stack. For CNAs, when a symbol is pushed to the stack, it is automatically colored by the color of the opening letter. Formally on reading $(\!(_c a$ a letter in $P \times \{c\}$ is pushed. When reading a closing letter $b_c)\!)$ the CNA will pop from the stack symbol after symbol until reaching the most recent stack symbol which is c-colored, and make its final move on this letter. We can see this move as composed of several ε-transitions. Note, though, that these are the only possible ε-transitions; the CNA can and must apply an ε-transition only when reading a closing letter of color c, and until a symbol colored c is visible on the stack, but it may not apply an ε-transition at any other time.

We assume a default color $\bot \in C$ for coloring the bottom of the stack. We use Γ to denote stack pairs, i.e. symbols in $P \times C$. For $c \in C$ we use Γ_c and Γ_{-c} for $P \times \{c\}$ an $P \times (C \backslash \{c\})$, respectively. A *configuration* of the automaton is a string γq where q is a state and $\gamma \in \Gamma_\bot \Gamma^*$. The *frontier* of a configuration $s = \gamma q$, denoted $frnt(\gamma q)$ is the pair (q, p) where (p, c) is the top pair of γ for some $c \in C$. We use the term frontiers also for arbitrary pairs in $Q \times P$.

Definition 2 (Colored Nested Word Automaton (CNA)). *A* CNA *over alphabet* $A \cup A \times C \times \{+, -\}$ *is a tuple* $\mathcal{A} = (Q, P, I, F, \delta^\langle, \dot{\delta}, \delta^\rangle, \delta^\varepsilon)$ *where Q is a finite set of states, P is a finite set of stack symbols, $I \subseteq Q \times P$ is a set of initial frontiers, $F \subseteq Q \times P$ is a set of final frontiers. The transition relation is split into four components $\delta^\langle, \dot{\delta}, \delta^\rangle, \delta^\varepsilon$. Letters in $\langle\!\langle \Sigma$ and $\dot{\Sigma}$ are processed by $\delta^\langle$ and $\dot{\delta}$, respectively. Letters in $\Sigma\rangle\!\rangle$ are processed by both $\delta^\rangle$ and δ^ε. The types of the different δ's are as follows: $\delta^\langle : Q \times \langle\!\langle \Sigma \to 2^{Q \times P}$, $\dot{\delta} : Q \times \dot{\Sigma} \to 2^Q$, $\delta^\rangle : Q \times \Sigma\rangle\!\rangle \times P \to 2^Q$ and $\delta^\varepsilon : Q \times P \to 2^Q$.*

From δ we can infer the evolution of the configuration of the automaton, η as follows.

– Case $a \in \dot{\Sigma}$: $\eta(\gamma q, \dot{a}) = \{\gamma q' \mid q' \in \dot{\delta}(q, \dot{a})\}$

– Case $a \in \langle\!\langle \Sigma$: $\eta(\gamma q, \langle\!\langle_c a) = \{\gamma(p', c)q' \mid (q', p') \in \delta^\langle(q, \langle\!\langle_c a)\}$

– Case $a \in \Sigma\rangle\!\rangle$:

$$\eta(\gamma q, a_c\rangle\!\rangle) = \left\{ \gamma' q' \left| \begin{array}{l} \exists k \geq 0, \quad q_0, \ldots, q_k, \quad p_0, \ldots, p_k, \quad c_0, \ldots, c_k \\ \text{s.t. } q_k = q, \quad c_0 = c, \quad c_k, \ldots, c_1 \neq c, \\ \gamma = \gamma'(p_0, c_0)(p_1, c_1) \ldots (p_k, c_k), \\ \forall 0 \leq i < k : q_i \in \delta^\varepsilon(q_{i+1}, p_{i+1}) \text{ and } q' \in \delta^\rangle(q_0, p_0, a_c\rangle\!\rangle) \end{array} \right. \right\}$$
$$\cup \left\{ (p_0, \bot)q_0 \left| \begin{array}{l} \exists k \geq 0, \quad q_0, \ldots, q_k, \quad p_0, \ldots, p_k, \quad c_1, \ldots, c_k \\ \text{s.t. } q_k = q, \quad c_k, \ldots, c_1 \neq c, \\ \gamma = (p_0, \bot)(p_1, c_1) \ldots (p_k, c_k) \in \Gamma^*_{-c}, \\ \forall 0 \leq i < k : q_i \in \delta^\varepsilon(q_{i+1}, p_{i+1}) \end{array} \right. \right\}$$

A *run* of the automaton on a $\dot{\Sigma}$-word $\mathbf{w} = a_1 \ldots a_n$ is a sequence of configurations $s_0 s_1 \ldots s_n$ such that s_0 is an initial frontier and $s_{i+1} \in \eta(s_i, a_{i+1})$ for every $0 \leq i < n$. A run is *accepting* if $frnt(s_n) \in F$. The automaton accepts a word $\mathbf{w}$ if there exists an accepting run on $\mathbf{w}$. We also use $(q, p) \overset{\mathbf{w}}{\Longrightarrow}_{\mathcal{A}} (q', p')$ if $\mathcal{A}$ starting from configuration (q, p) and reading $\mathbf{w}$ may reach a configuration whose frontier is (q', p'). Thus $\mathbf{w}$ is accepted by $\mathcal{A}$ if $(q, p) \overset{\mathbf{w}}{\Longrightarrow}_{\mathcal{A}} (q', p')$ for some $(q, p) \in I$ and $(q', p') \in F$. We use $\mathcal{L}(\mathcal{A})$ to denote the set of words accepted by $\mathcal{A}$. An automaton is *deterministic* if I is a singleton and the right hand side of all the δ's are singletons. We use DCNA and NCNA for deterministic and nondeterministic CNAs, respectively.

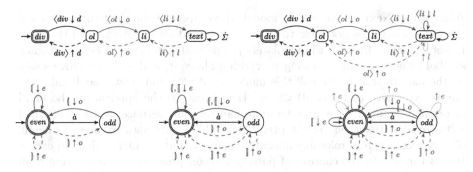

Fig. 4. Some examples of CNAs.

Figure 4 provides some examples of CNAs. Push and pop transitions are colored by the respective color, whereas internal transitions are colored black. Push edges have labels of the form $⟨_c b ↓ p$ signifying that p is pushed to the stack, pop edges have labels of the form $b_c⟩ ↑ p$ signifying that the top symbol of the respective color is p. To ease distinction between push and pop transitions, pop edges are dashed. Finally, ε-transitions use grey dotted edges. The initial frontiers in first and second line are all $(div, ⊥)$ and $(even, ⊥)$, respectively, and the final frontier of each is the same as its initial frontier.

In the first line we have a CNA recognizing a subset of HTML with div, ol and li tags requiring the document to be well matched (left), and a CNA allowing li to be unmatched if recovered by ol (right).

In the second line the left and middle automata are actually nested word automata — no use of the color is made. The left recognizes all words over $\{a, (,)\}$ where the number of a's within any $()$ and within the outer level, is even. The middle recognized words over $\{[,], a, (,)\}$ where in addition the number of a's between any $[]$ is odd. When we say here "the number of a's in the word" we mean in the outer level of the word as defined in Sect. 2. The language recognized by the right automaton allows also unmatched $($ if it is recovered by an encapsulating $[]$ in which case the number of a letters in between should be odd. For instance $[a(a[aa(a]$ should be accepted whereas $[a(a[aa(aa]$ should not.

Theorem 1. *A language of colored nested words is regular iff it is accepted by a DCNA.*

5   Equivalent Models

We show that as is the case in finite automata and nested word automata, non-determinism does not add expressive power. The proof goes via a generalization of the subset construction. The states of the DCNA are sets of pairs of states. A run of the DCNA on word w will reach state $\{(q_1, q_1'), (q_2, q_2'), \ldots, (q_k, q_k')\}$ iff any run of the NCNA on w reaches one of the states q_i' for $i ∈ [1..k]$ and for every $i ∈ [1..k]$ the respective run entered the current hierarchical level at state q_i.

Theorem 2. DCNAs *have the same expressive power as* NCNAs.

Blind CNA. Next we consider a model where upon reading a return letter a_c the automaton does not have the privilege to read all the stack until the most recent c-symbol. Instead it immediately jumps to the most recent c-colored stack symbol p, popping and ignoring everything above it, and makes a move solely on the base of that p. We call this model *blind* CNA and show that blind CNAs are as expressive as (sighted) CNAs.[3] Dependent on the application the blind or original (sighted) CNA may be more natural. For instance, in the context of software executions, one might prefer the sighted automata to allow modeling of operations such as releasing allocated memory that are taken when an exception is thrown. In the context of parsing Python programs, or recovering from unmatched HTML tags, the blind model may be more natural.

Definition 3 (Blind Colored Nested Word Automaton (BCNA)). *A BCNA is a tuple* $\mathcal{B} = (Q, P, I, F, \delta^{(}, \dot{\delta}, \delta^{)})$ *where all the components are as in the definition of a CNA. The evolution of the configuration of the automaton for* $\dot{\delta}$ *and* $\delta^{(}$ *is the same as in CNAs. For* $\delta^{)}$ *we have that* $\eta(\gamma q, a_c) = \{\gamma' q' \mid \gamma = \gamma'(p, c)\gamma'' \text{ where } \gamma'' \in \Gamma^*_{-c} \text{ and } q' \in \delta^{)}(q, p, a_c)\} \cup \{(p_0, \bot)q \mid \gamma = (p_0, \bot)\gamma' \in \Gamma^*_{-c}\}$. *As in CNAs a run of a BCNA is a sequence of configurations which adheres to* η *and whose first element is an initial frontier.*

Clearly every BCNA can be simulated by a CNA whose epsilon transitions do not change the state of the automaton. Some CNAs are naturally blind. For instance, the CNA at the top right of Fig. 4 can be made blind by omitting the ε-transition. Simulating the CNA at the bottom right of Fig. 4 by a blind CNA requires adding more states to account for the computations done by the ε-transitions. The proof of the following theorem provides a constructive way to perform such a simulation. The idea is that the states and stack symbols carry an additional component recording a function $\varphi : Q \times C \to Q$ such that $\varphi(q, c)$ tells to which state the CNA will get after popping all non c-stack symbols if the current state is q.

Theorem 3. *Given a deterministic CNA* $\mathcal{A}$ *with* n *states,* k *colors and* m *stack symbols, one can effectively construct a deterministic BCNA* $\mathcal{B}$ *such that* $\mathcal{L}(\mathcal{B}) = \mathcal{L}(\mathcal{A})$ *with* kn^{n+1} *states and* kmn^n *stack symbols.*

6   Closure Properties and Decision Problems

Theorem 4. *Regular languages of colored nested words are closed under complementation, intersection and union.*

Complementation is done by complementing the set of final frontiers, and intersection and union are done via a product construction.

Theorem 5. *Regular languages of colored nested words are closed under concatenation and Kleene-**.

---

[3] Note that a blind CNA is still different than a traditional nested word automaton, as it has the means to skip all the unmatched calls and arrive to the matching call.

Concatenation is proved by guessing a split point and simulating the automata for the operands on each part of the split word. The second automaton treats stack symbols of the first automaton as the bottom of the stack. For Kleene-* the idea is similar, but requires two copies of the simulated automaton to distinguish different sub-words of the split.

Theorem 6. *Regular languages of colored nested words are closed under the operations of prefix, suffix and reversal.*

Closure under reversal is done by dualizing the transitions and switching initial and final frontiers. Closure under prefixes requires guessing an accepting frontier for an extension of the word and tracking it via the transitions. It relies on decidability of emptiness that is provided in Theorem 8. Closure for suffix follows from these two: $suff(L) = (pref(L^R))^R$.

Let $\hat{\Sigma}$ and $\hat{\Sigma}'$ be two colored alphabets. For every $a \in \hat{\Sigma}$ let $H(a)$ be a language of colored nested words over $\hat{\Sigma}'$. We call H a *substitution*. We say that substitution H is *color-respecting* if the following three conditions hold: (1) for every $\dot{a} \in \hat{\Sigma}$ any word $w' \in H(\dot{a})$ is weakly-matched, (2) for every $[_c a \in [\Sigma$ any word $w' \in H([_c a)$ is of the form $[_{c'} b\, v$ where v is weakly-matched (3) for every $a_c] \in \Sigma]$ any word $w' \in H(a_c])$ is of the form $v\, b_{c'}]$ where v is weakly-matched. When H maps every a to a singleton set, we refer to H as an *homomorphism*, and usually denote it with small h. A homomorphism thus maps letters to strings. If h is a homomorphism from $\hat{\Sigma}$ to $\hat{\Sigma}'$, given a language L' over $\hat{\Sigma}'$ we can define its inverse-homomorphic image as $h^{-1}(L') = \{w \mid \exists w' \in L'.\ h(w) = w'\}$.

Theorem 7. *Regular languages of colored nested words are closed under color-respecting substitution, homomorphism and inverse homomorphism.*

The idea in these proofs is to use an NCNA that guesses a substituted letter a and then runs in parallel the DCNA for L on the guessed letter a and the DCNA $\mathcal{M}_a$ for the substitution $H(a)$.

We note that in general, it may be that $h(h^{-1}(L)) \neq L$ and $h^{-1}(h(L)) \neq L$, or moreover that, $h^{-1}(h(L)) \nsubseteq L$. Examples are given in the full version.

The following theorem follows from the result on emptiness of pushdown automata and from the closure under complementation and intersection.

Theorem 8. *Emptiness of NCNAs can be solved in polynomial time. Inclusion, universality and equivalence of NCNAs are EXPTIME-complete.*

The *membership* problem for non-deterministic pushdown automata too is solvable in polynomial time. It thus follows that we can decide on NCNA's membership in polynomial time.

In some contexts, it makes sense to ask about the complexity of membership when the given CNA is fixed. This is the case, for instance, in parsing programs in a given programming language. A CNA for this will stay valid as long as the programming language syntax has not changed. If $\mathcal{A}$ is fixed, we can construct the equivalent deterministic automaton $\mathcal{D}$ using the method in Theorem 2 and then simulate it on the given membership query w. This yields a *streaming algorithm*—an algorithm which reads the input in one pass from left to right (and cannot traverse it again). Thus, this would take $O(|w|)$ time and $O(d(w))$ space where $d(w)$ is the hierarchical nesting depth of the colored nested word w.

Theorem 9. *Membership of* NCNA*s can be solved in polynomial time. For a fixed* CNA $\mathcal{A}$ *and colored nested word* w *of length* ℓ *and depth* d, *the membership problem can be solved in time* $O(\ell)$ *and space* $O(d)$.

7 Grammar Characterization

In the following we provide a grammar characterization for regular languages of colored nested words. We first recall some basic definitions. We assume familiarity of basic definition of context-free grammar (and provide it explicitly in the full version.)

Definition 4. *A grammar* $(\mathcal{V}, S, Prod)$ *is said to be a* CNW *grammar w.r.t a set* $C = \{c_1, c_2, \ldots, c_k\}$ *of colors if its variables can be partitioned into sets* $\mathcal{V}^{\langle}, \mathcal{V}^{\rangle}, \mathcal{V}^{c_1}, \mathcal{V}^{c_2}, \ldots, \mathcal{V}^{c_k}$ *such that the production rules of the grammar are in one of the following forms, where* $X^{\langle}, Y^{\langle}, Z^{\langle} \in \mathcal{V}^{\langle}$, $X^{\rangle} \in \mathcal{V}^{\rangle}$, $Y, Z \in \mathcal{V}^{\rangle} \cup \mathcal{V}^{\langle}$, $X^c, Y^c, Z^c \in \mathcal{V}^c$ *and* $X \in \mathcal{V}$:

- $X^{\langle} \longrightarrow a Y^{\langle}$ *for* $a \in \dot{\Sigma} \cup (\!\langle \Sigma$
- $X^{\langle} \longrightarrow (\!\langle_c a \ Y^c \ b_c]\!\rangle Z^{\langle}$
- $X^{\rangle} \longrightarrow a Y$ *for* $a \in \Sigma]\!\rangle \cup \dot{\Sigma}$
- $X^{\rangle} \longrightarrow (\!\langle_c a \ Y^c \ b_c]\!\rangle Z$

- $X \longrightarrow \varepsilon$
- $X^c \longrightarrow \dot{a} Y^c$
- $X^c \longrightarrow (\!\langle_{c'} a \ Y^c \ b_{c'}]\!\rangle Z^c$
- $X^c \longrightarrow (\!\langle_{c'} a \ Y^c$ *for* $c' \neq c$

Intuitively, variables in $\mathcal{V}^{\langle}$ and $\mathcal{V}^{\rangle}$ derive words with no pending returns and no pending calls, respectively, and variables in $\mathcal{V}^{c_i}$ derive weakly matched c_i-rooted words. Note that while the grammar characterization of nested words partitions the grammar variables into two categories, one that disallows pending calls and one that disallows pending returns. For colored nested words, we have additional categories, one per each color. The variables in the category of color c derive weakly matched c-rooted words, thus allowing pending calls of any color other than c.

Theorem 10. *A language* L *is derived by a* CNW*-grammar iff* L *is recognized by a* CNA.

References

1. Alur, R., Chaudhuri, S.: Temporal reasoning for procedural programs. In: Barthe, G., Hermenegildo, M. (eds.) VMCAI 2010. LNCS, vol. 5944, pp. 45–60. Springer, Heidelberg (2010)
2. Alur, R., Chaudhuri, S., Madhusudan, P.: A fixpoint calculus for local and global program flows. In: POPL, pp. 153–165 (2006)

3. Alur, R., Chaudhuri, S., Madhusudan, P.: Software model checking using languages of nested trees. ACM Trans. Program. Lang. Syst. **33**(5), 15 (2011)

4. Alur, R., Madhusudan, P.: Visibly pushdown languages. In: STOC, pp. 202–211 (2004)

5. Caucal, D., Hassen, S.: Synchronization of grammars. In: Hirsch, E.A., Razborov, A.A., Semenov, A., Slissenko, A. (eds.) Computer Science – Theory and Applications. LNCS, vol. 5010, pp. 110–121. Springer, Heidelberg (2008)

6. Chaudhuri, S., Alur, R.: Instrumenting C programs with nested word monitors. In: Bošnački, D., Edelkamp, S. (eds.) SPIN 2007. LNCS, vol. 4595, pp. 279–283. Springer, Heidelberg (2007)

7. Crespi-Reghizzi, S., Mandrioli, D.: Operator precedence and the visibly pushdown property. J. Comput. Syst. Sci. **78**(6), 1837–1867 (2012)

8. Debarbieux, D., Gauwin, O., Niehren, J., Sebastian, T., Zergaoui, M.: Early nested word automata for xpath query answering on XML streams. In: CIAA 2013

9. Driscoll, E., Burton, A., Reps, T.W.: Checking conformance of a producer and a consumer. In: SIGSOFT/FSE, pp. 113–123 (2011)

10. Filiot, E., Gauwin, O., Reynier, P-A., Servais, F.: Streamability of nested word transductions. In: Annual Conference on Foundations of Software Technology and Theoretical Computer Science, FSTTCS, pp. 312–324 (2011)

11. Filiot, E., Raskin, J.-F., Reynier, P.-A., Servais, F., Talbot, J.-M.: Properties of visibly pushdown transducers. In: Hliněný, P., Kučera, A. (eds.) MFCS 2010. LNCS, vol. 6281, pp. 355–367. Springer, Heidelberg (2010)

12. Filiot, E., Servais, F.: Visibly pushdown transducers with look-ahead. In: Bieliková, M., Friedrich, G., Gottlob, G., Katzenbeisser, S., Turán, G. (eds.) SOFSEM 2012. LNCS, vol. 7147, pp. 251–263. Springer, Heidelberg (2012)

13. Hague, M., Murawski, A.S., Ong, C.-H.L., Serre, O.: Collapsible pushdown automata and recursion schemes. In: LICS, pp. 452–461 (2008)

14. Madhusudan, P., Viswanathan, M.: Query automata for nested words. In: Královič, R., Niwiński, D. (eds.) MFCS 2009. LNCS, vol. 5734, pp. 561–573. Springer, Heidelberg (2009)

15. Mozafari, B., Zeng, K., Zaniolo, C.: High-performance complex event processing over xml streams. In: SIGMOD Conference, pp. 253–264 (2012)

16. Nowotka, D., Srba, J.: Height-deterministic pushdown automata. In: Kučera, L., Kučera, A. (eds.) MFCS 2007. LNCS, vol. 4708, pp. 125–134. Springer, Heidelberg (2007)

17. Esparza, J., Alur, R., Bouajjani, A.: Model checking of procedural programs. In: Clarke, E.M., Henzinger, T.A., Veith, H. (eds.) Handbook of Model Checking. Springer, Heidelberg (2015)

18. Raskin, J.-F., Servais, F.: Visibly pushdown transducers. In: Aceto, L., Damgård, I., Goldberg, L.A., Halldórsson, M.M., Ingólfsdóttir, A., Walukiewicz, I. (eds.) ICALP 2008, Part II. LNCS, vol. 5126, pp. 386–397. Springer, Heidelberg (2008)

19. Staworko, S., Laurence, G., Lemay, A., Niehren, J.: Equivalence of deterministic nested word to word transducers. In: Kutyłowski, M., Charatonik, W., Gębala, M. (eds.) FCT 2009. LNCS, vol. 5699, pp. 310–322. Springer, Heidelberg (2009)

20. Thomo, A., Venkatesh, S.: Rewriting of visibly pushdown languages for XML data integration. Theoret. Comput. Sci. **412**(39), 5285–5297 (2011)

Input-Driven Queue Automata with Internal Transductions

Martin Kutrib, Andreas Malcher$^{(\boxtimes)}$, and Matthias Wendlandt

Institut für Informatik, Universität Giessen, Arndtstr. 2, 35392 Giessen, Germany
{kutrib,malcher,matthias.wendlandt}@informatik.uni-giessen.de

Abstract. For input-driven queue automata (IDQA) the input alpha-bet is divided into three distinct classes and the actions on the queue (enter, remove, nothing) are solely governed by the input symbols. Here, this model is extended in such a way that the input of an IDQA is preprocessed by an internal deterministic sequential transducer. These automata are called tinput-driven queue automata (TDQA). It turns out that even TDQAs with weak, that is, deterministic injective and length-preserving, internal transducers are more powerful than IDQAs. We study closure properties of the family of languages accepted by TDQAs. For example, for compatible signatures the closure under the Boolean operations union, intersection, and complementation is shown. For incompatible signatures and the operations reversal, concatenation, iteration, and length-preserving homomorphism non-closure results are obtained. Depending on the working mode of the transducer and the IDQA, there are three nondeterministic working modes for tinput-driven queue automata. It is shown that for devices with nondeterministic transducers the nondeterministic IDQA can be determinized. The other classes form a strict hierarchy. Finally, several decidability problems are addressed.

Keywords: Automata and formal languages · Input-driven automata · Queue automata · Closure properties · Decidability

1 Introduction

Finite automata possess many nice properties such as equivalence of nondeter-ministic and deterministic models, existence of minimization algorithms, closure under many operations, and decidable questions such as emptiness, inclusion, or equivalence. On the other hand, their computational power is quite low since only regular languages are accepted. It is therefore natural to consider extensions of the model featuring additional storage media such as pushdown stores, stacks, or queues. In general, such extensions lead to a broader family of accepted lan-guages, but also to a weaker manageability of the models since certain closure properties do not longer hold, minimization algorithms do not exist, and formerly decidable questions become undecidable. Thus, there is an obvious interest in extensions which enlarge the language family, but keep as many of the 'good' properties as possible.

© Springer International Publishing Switzerland 2016
A.-H. Dediu et al. (Eds.): LATA 2016, LNCS 9618, pp. 156–167, 2016.
DOI: 10.1007/978-3-319-30000-9_12

One such extension is represented by input-driven automata. Basically, for such devices the operations on the storage medium are dictated by the input symbols. The first references date back to [6,12], where input-driven pushdown automata are introduced in which the input symbols define whether a push operation, a pop operation, or no operation on the pushdown store has to be performed. A recent survey with many valuable references on complexity aspects of input-driven pushdown automata may be found in [13]. Extensions of the model with respect to multiple pushdown stores or more general auxiliary storages are introduced in [9,11]. Recently, the computational power of input-driven automata using the storage medium of a stack and a queue, respectively, have been investigated in [3,8]. Here we are particularly interested in *queue automata*. Some contributions relate these devices to other well-known concepts in formal language theory and theoretical computer science. For instance, in [5] queue automata (there called Post machines) with certain features are shown to characterize the class of languages accepted by multi-reset machines [4], as well as some classes of languages defined by equality sets. In [7], a restricted version of context-free grammars, called breadth-first grammars, are provided as a generating system for certain classes of languages accepted by queue automata. The computational capacity and complexity of queues are compared with the storage types stack and tapes in [10].

The edge between languages that are accepted by input-driven queue automata (IDQA) or not is very small. For example, language $\{\, a^n \$ b^n \mid n \geq 1 \,\}$ is accepted by an IDQA where an a means an enter operation, b means a remove operation, and a $\$$ leaves the queue unchanged. On the other hand, the very similar language $\{\, a^n \$ a^n \mid n \geq 1 \,\}$ is not accepted by any IDQA. Similarly, the language $\{\, w\$w \mid w \in \{a,b\}^+ \,\}$ is not accepted by any IDQA, but if the second w is written down with some marked alphabet $\{\hat{a},\hat{b}\}$, then language $\{\, w\$\hat{w} \mid w \in \{a,b\}^+ \,\}$ is accepted by an IDQA. To overcome these obstacles we provide the input-driven queue automaton with an internal sequential transducer that preprocesses the input. In the first example above such a transducer translates every a before reading $\$$ to a and after reading $\$$ to b. An IDQA with internal transducer is said to be *tinput-driven* (TDQA). While an internal transducer does not affect the computational capacity of general queue automata, it clearly does for *input-driven* versions. To implement the idea without giving the transducers too much power for the overall computation, essentially, we will consider only deterministic injective and length-preserving transducers.

2 Preliminaries

Let Σ^* denote the set of all words over the finite alphabet Σ. The *empty word* is denoted by λ, and $\Sigma^+ = \Sigma^* \setminus \{\lambda\}$. The *reversal* of a word w is denoted by w^R. For the *length* of w we write $|w|$. We use $\subseteq$ for *inclusions* and $\subset$ for *strict inclusions*.

A classical deterministic queue automaton is called input-driven if the next input symbol defines the next action on the queue, that is, entering a symbol at the end of the queue, removing a symbol from the front of the queue, or changing the internal state without modifying the queue content. To this end, we assume that the input alphabet Σ is partitioned into the sets Σ_D, Σ_R, and Σ_N, that control the actions enter (D), remove (R), and state change only (N). Such a partition is called a *signature*. A formal definition is:

Definition 1. *A deterministic input-driven queue automaton, abbreviated as IDPDA, is a system $M = \langle Q, \Sigma, \Gamma, q_0, F, \perp, \delta_D, \delta_R, \delta_N \rangle$, where*

1. *Q is the finite set of* internal states,
2. *Σ is the finite set of* input symbols *consisting of the disjoint union of sets Σ_D, Σ_R, and Σ_N,*
3. *Γ is the finite set of* queue symbols,
4. *$q_0 \in Q$ is the* initial state,
5. *$F \subseteq Q$ is the set of* accepting states,
6. *$\perp \notin \Gamma$ is the* empty-queue symbol,
7. *δ_D is the partial transition function mapping $Q \times \Sigma_D \times (\Gamma \cup \{\perp\})$ to $Q \times \Gamma$,*
8. *δ_R is the partial transition function mapping $Q \times \Sigma_R \times (\Gamma \cup \{\perp\})$ to Q,*
9. *δ_N is the partial transition function mapping $Q \times \Sigma_N \times (\Gamma \cup \{\perp\})$ to Q.*

A *configuration* of an IDQA $M = \langle Q, \Sigma, \Gamma, q_0, F, \perp, \delta_D, \delta_R, \delta_N \rangle$ is a triple (q, w, s), where $q \in Q$ is the current state, $w \in \Sigma^*$ is the unread part of the input, and $s \in \Gamma^*$ denotes the current queue content, where the leftmost symbol is at the front. Thus, the *initial configuration* for an input string w is set to (q_0, w, λ). During the course of its computation, M runs through a sequence of configurations. One step from a configuration to its successor configuration is denoted by $\vdash$. Let $a \in \Sigma$, $w \in \Sigma^*$, $z, z' \in \Gamma$, and $s \in \Gamma^*$. We set

1. $(q, aw, zs) \vdash (q', w, zsz')$, if $a \in \Sigma_D$ and $(q', z') = \delta_D(q, a, z)$,
2. $(q, aw, \lambda) \vdash (q', w, z')$, if $a \in \Sigma_D$ and $(q', z') = \delta_D(q, a, \perp)$,
3. $(q, aw, zs) \vdash (q', w, s)$, if $a \in \Sigma_R$ and $q' = \delta_R(q, a, z)$,
4. $(q, aw, \lambda) \vdash (q', w, \lambda)$, if $a \in \Sigma_R$ and $q' = \delta_R(q, a, \perp)$,
5. $(q, aw, zs) \vdash (q', w, zs)$, if $a \in \Sigma_N$ and $q' = \delta_N(q, a, z)$,
6. $(q, aw, \lambda) \vdash (q', w, \lambda)$, if $a \in \Sigma_N$ and $q' = \delta_N(q, a, \perp)$.

So, whenever the queue is empty, the successor configuration is computed according to the definition of the transition functions on the special empty-queue symbol $\perp$. We denote the reflexive and transitive (resp. transitive) closure of $\vdash$ by $\vdash^*$ (resp. $\vdash^+$). The language accepted by the IDQA M is the set $L(M)$ of words for which there exists a computation beginning in the initial configuration and ending in a configuration in which the whole input is read and an accepting state is entered. Formally:

$$L(M) = \{ w \in \Sigma^* \mid (q_0, w, \lambda) \vdash^* (q, \lambda, s) \text{ with } q \in F, s \in \Gamma^* \}.$$

For the definition of tinput-driven queue automata we need the notion of *deterministic one-way sequential transducers* (DST) which are basically deterministic finite automata equipped with an initially empty output tape. In every transition a DST appends a string over the output alphabet to the output tape. The transduction defined by a DST is the set of all pairs (w, v), where w is the input and v is the output produced after having read w completely. Formally, a DST is a system $T = \langle Q, \Sigma, \Delta, q_0, \delta \rangle$, where Q is the finite set of internal states, Σ is the finite set of input symbols, Δ is the finite set of output symbols, $q_0 \in Q$ is the initial state, and δ is the total transition function mapping $Q \times \Sigma$ to $Q \times \Delta^*$. By $T(w) \in \Delta^*$ we denote the output produced by T on input $w \in \Sigma^*$. Here we will consider only injective and length-preserving DSTs which are also known as injective Mealy machines.

Let M be an IDQA and T be an injective and length-preserving DST so that the output alphabet of T is the input alphabet of M. The pair (M, T) is called a *tinput-driven queue automaton* (TDQA) and the language accepted by (M, T) is $L(M, T) = \{ w \in \Sigma^* \mid T(w) \in L(M) \}$.

For a computation of a queue automaton, a turn is a phase in which the length of the queue first increases and then decreases. Formally, a sequence of at least three configurations $(q_1, w_1, s_1) \vdash (q_2, w_2, s_2) \vdash \cdots \vdash (q_m, w_m, s_m)$ is a *turn* if $|s_1| < |s_2| = \cdots = |s_{m-1}| > |s_m|$. For any given $k \geq 0$, a *k-turn* computation is any computation containing exactly k turns.

A TDQA performing *at most* k turns in *any* computation is called k-turn TDQA and will be denoted by TDQA_k. Analogously, k-turn IDQA are defined, and will be denoted by IDQA_k.

In order to clarify this notion we continue with an example.

Example 2. Language $L_1 = \{ a^n \$ a^n \mid n \geq 1 \}$ is accepted by a TDQA. Before reading symbol $\$$ the transducer maps an a to an a, and after reading $\$$ it maps an a to a b. Thus, L_1 is translated to $\{ a^n \$ b^n \mid n \geq 1 \}$ which is accepted by some IDQA.

Similarly, $L_2 = \{ w \$ w \mid w \in \{a, b\}^* \}$ can be accepted by some TDQA. Here, the transducer maps any a, b to a, b before reading $\$$ and to $\hat{a}, \hat{b}$ after reading $\$$. This gives the language $\{ w \$ \hat{w} \mid w \in \{a, b\}^* \}$ which clearly is accepted by some IDQA.

Finally, consider $L_3 = \{ a^n b^{2n} \mid n \geq 1 \}$. Here, the transducer maps an a to a and every b alternately to b and c. This gives language $\{ a^n (bc)^n \mid n \geq 1 \}$ which is accepted by some IDQA: every a implies an enter-operation, every b implies a remove, and every c leaves the queue unchanged. ∎

3   Closure Properties

The property of working input-driven forces the device to perform the operation on its storage medium that is associated with the current input symbol. Considering input-driven automata and closure properties under binary operations like for union, intersection, and concatenation, it may happen that certain letters have to be associated with different operations to accept the both languages involved. In the worst case this implies that the input-driven language families are not closed under these operations [8]. On the other hand, for input-driven queue (resp. pushdown) automata, strong closure properties have been derived in [8] (resp. [1]) *provided that* all automata involved share the same partition of the input alphabet. Here we distinguish this important special case from the general one. We call the partition of an input alphabet a *signature*, and say that two signatures $\Sigma = \Sigma_D \cup \Sigma_R \cup \Sigma_N$ and $\Sigma' = \Sigma'_D \cup \Sigma'_R \cup \Sigma'_N$ are *compatible* if and only if $\bigcup_{j \in \{D,R,N\}} (\Sigma_j \setminus \Sigma'_j) \cap \Sigma' = \emptyset$ and $\bigcup_{j \in \{D,R,N\}} (\Sigma'_j \setminus \Sigma_j) \cap \Sigma = \emptyset$.

The following technical lemma allows to earn several non-closure properties.

Lemma 3. *The concatenation of the input-driven queue automata languages* $L = \{a^n b^n \mid n \geq 1\}$ *and* $L' = \{b^m a^m \mid m \geq 1\}$ *is not accepted by any TDQA.*

Proof. In contrast to the assertion, assume that the concatenation LL' is accepted by some TDQA (M, T) with $M = \langle Q, \Sigma, \Gamma, q_0, F, \bot, \delta_D, \delta_R, \delta_N \rangle$ and $T = \langle Q', \Sigma', \Sigma, q'_0, \delta' \rangle$. The transducer T translates an input word $r = uvw$, where $u = a^i$, $v = b^i b^j$, $w = a^j$, to some word $r_T = u_T v_T w_T$ with $|u| = |u_T|$, $|v| = |v_T|$, $|w| = |w_T|$. Since the subwords u, v, w are unary, the transducer runs into a loop while each of them is processed. So, each of u_T, v_T, and w_T has the form

$$x_1 x_2 \cdots x_k (y_1 y_2 \cdots y_\ell)^c y_1 y_2 \cdots y_m$$

where $k \leq |Q'|$ is the length of the initial part until T enters the loop, $\ell \leq |Q'|$ is the length of the loop, $c \geq 0$ is some constant, $m < \ell$ is the length of a possible incomplete loop at the end, and $x_i, y_j \in \Sigma$, $1 \leq i \leq k$, $1 \leq j \leq \ell$. Since M is input driven, for a given u_T (resp. v_T, w_T) it can be determined whether the computation of M on u_T (resp. v_T, w_T) for growing c increases the length of the queue, decreases the length of the queue, or leaves the length unchanged up to a constant. In the following, these three cases are distinguished for growing c.

First we consider u_T and assume that the length of the queue does not increase for growing c. Then there is a constant s_u so that the length of the queue content is at most s_u after (M, T) has processed any input prefix of the form a^i, $i \geq 1$. So, there are two different prefixes a^i and $a^{i'}$ with $i \neq i'$ that drive (M, T) into the same states with the same queue contents. Since $a^i b^i b^j a^j$ and $a^{i'} b^{i'} b^j a^j$ are accepted, $a^{i'} b^i b^j a^j$ is accepted as well, a contradiction. Thus, for growing c, M has to increase the length of the queue while processing u_T.

Next we consider v_T and assume that the length of the queue does not increase for growing c. Let $i \geq 1$ be fixed. Then there is a constant s_{uv} so that the length of the queue content is at most s_{uv} after (M, T) has processed any input prefix of the form $a^i b^{i+j}$, $j \geq 1$. So, there are two different prefixes $a^i b^{i+j}$ and $a^i b^{i+j'}$ with $j \neq j'$ that drive (M, T) into the same states with the same queue contents. Since $a^i b^{i+j} a^j$ and $a^i b^{i+j'} a^{j'}$ are accepted, $a^i b^{i+j'} a^j$ is accepted as well, a contradiction. Thus, for growing c, M has to increase the length of the queue while processing v_T.

Finally, the subword z_T remains to be considered. Let $i \geq 1$ be large compared with $j \geq 1$. We choose the word $a^i b^i b^j a^j$. From above it is known that the length of the queue increases while processing $a^i b^i b^*$. So, there is another $j' \neq j$ where the two different prefixes $a^i b^{i+j}$ and $a^i b^{i+j'}$ drive (M, T) into the same states with the same content at the front of the queue. Moreover, the length of the matching queue content exceeds j. Since $a^i b^{i+j} a^j$ and $a^i b^{i+j'} a^{j'}$ are accepted, $a^i b^{i+j'} a^j$ is accepted as well. The reason is that M cannot access the possibly different part of the queue content during the last j time steps. Now the assertion follows from the contradiction.   □

As an immediate consequence we obtain the non-closure under concatenation.

Theorem 4. *The family of languages accepted by tinput-driven queue automata is not closed under concatenation.*

Proof. By Example 2 the two languages of Lemma 3 are accepted by TDQAs. So, the theorem follows by Lemma 3.   □

Theorem 5. *The family of languages accepted by tinput-driven queue automata is not closed under length-preserving homomorphism.*

Proof. It is not hard to construct a tinput-driven queue automaton that accepts language $L = \{ a^n b^n c b^{m-1} a^m \mid m, n \geq 1 \}$.

Let $h : \{a, b, c\}^* \to \{a, b\}^*$ be the length-preserving homomorphism defined by $h(a) = a$, $h(b) = b$, $h(c) = b$. Then $h(L) = \{ a^n b^n b^m a^m \mid m, n \geq 1 \}$ which is not accepted by any TDQA by Lemma 3.   □

Theorem 6. *The family of languages accepted by tinput-driven queue automata is not closed under iteration.*

Proof. We consider the language $L = \{ a^n b^n \mid n \geq 1 \} \cup \{ b^n a^n \mid n \geq 1 \}$, which is clearly accepted by some tinput-driven queue automaton.

Words of the form $a^+ b^+ a^+$ that belong to L^* must be from the concatenation $\{ a^n b^n \mid n \geq 1 \} \cdot \{ b^n a^n \mid n \geq 1 \}$, that is, from the two-fold iteration of L. However, the proof of Lemma 3 shows that if these words are accepted by some TDQA, then also some words not belonging to L^* are accepted.   □

Now we turn to the Boolean operations and obtain a first positive closure under complementation.

Theorem 7. *Let (M, T) be a TDQA. Then a TDQA accepting the complement of $L(M, T)$ can effectively be constructed.*

For the remaining Boolean operations union and intersection one has to distinguish whether or not the given TDQAs have identical or at least compatible signatures. Clearly, two TDQAs M and M' have identical signatures if they are defined over the same input alphabet and the behavior of the queue of M and M' is identical for all input symbols. In case of *compatible* signatures, the input alphabets of M and M' may differ, but the behavior of the queue of M and M' is identical for all input symbols belonging to the intersection of both input alphabets. We consider first TDQAs having compatible signatures and identical translations. Later, we will see that TDQAs lose some positive closure properties if the signatures are no longer compatible.

Theorem 8. *Let (M,T) and (M',T) be two TDQAs with compatible signatures. Then a TDQA accepting the intersection $L(M,T) \cap L(M',T)$, can effectively be constructed.*

The effective closure under union follows from the effective closure under intersection and complementation.

Corollary 9. *Let (M,T) and (M',T) be two TDQAs with compatible signatures. Then a TDQA accepting the union $L(M,T) \cup L(M',T)$, can effectively be constructed.*

In contrast, TDQAs are *not* closed under union and intersection in case of incompatible signatures.

Theorem 10. *The family of languages accepted by TDQAs is not closed under union.*

Proof. As witnesses for the non-closure we consider $L_1 = \{ ba^n ca^n b \mid n \geq 0 \}$ and $L_2 = \{ ba^n ba^m ca^m b \mid m,n \geq 0 \}$. Both languages are accepted by some TDQA, but it is shown in [7] that the union $L_1 \cup L_2$ is not even accepted by any realtime deterministic queue automaton. This shows the non-closure under union.   □

Since the family of languages accepted by TDQAs is closed under complementation by Theorem 7, we obtain non-closure under intersection as well.

Corollary 11. *The family of languages accepted by TDQAs is not closed under intersection.*

Similarly, the non-closure under reversal can be derived.

Theorem 12. *The family of languages accepted by TDQAs is not closed under reversal.*

Proof. The reversal of $L = \{ ba^n ca^n b \mid n \geq 0 \} \cup \{ ba^n ba^m ca^m b \mid m,n \geq 0 \}$ is accepted by some TDQA, but L itself is not even accepted by any realtime deterministic queue automaton [7].   □

Finally it should be noted that TDQAs are closed under intersection with regular languages.

Theorem 13. *Let (M,T) be a TDQA and M' be a deterministic finite automaton. Then a TDQA accepting the intersection $L(M,T) \cap L(M')$ can effectively be constructed.*

The examples given in Example 2 showed that the computational capacity of TDQAs is larger than that of IDQAs. In Lemma 3 it was shown that language $\{ a^n b^n b^m a^m \mid n,m \geq 1 \}$ is not accepted by any TDQA. However, this language can easily be accepted by a realtime deterministic queue automaton. Let us denote the latter type of queue automata with $rt{-}QA$. Then, we obtain the following hierarchy:

$$\mathscr{L}(\text{IDQA}) \subset \mathscr{L}(\text{TDQA}) \subset \mathscr{L}(rt\text{-QA}).$$

4   Determinization

Each language accepted by a nondeterministic input-driven pushdown automaton is also accepted by a deterministic input-driven pushdown automaton. The simulation costs for an n-state nondeterministic input-driven pushdown automaton are $2^{\Theta(n^2)}$ states [6]. It is shown in [2] that this size is necessary in the worst case.

The basic idea of the proof is that the automaton stores applicable transition rules onto the pushdown store, when a push operation should be done, instead of pushing the appropriate symbol. The actual push operation is simulated at the time at which the symbol to be pushed is popped. In this way, it may happen that some transition rule pushed does not belong to a valid computation. However, the construction allows to distinguish these cases and, thus, only valid transitions will be evaluated. This technique cannot be assigned to input-driven queue automata. The reason is that on input-driven pushdown automata the last symbol pushed is used first. Thus it can be determined whether it belongs to a valid computation or not. In contrast, input-driven queue automata work according to the FIFO principle. Therefore, there could be many symbols in between the queue symbol currently to be removed and the symbol that has been entered last. So, the technique does not allow to verify that the remove operation simulates only enter operations belonging to a valid computation.

This brings us to the question whether the nondeterministic version of a tinput-driven queue automaton can be determinized as well. There are four different working modes for a tinput-driven queue automaton. The sequential transducer can be deterministic or nondeterministic and also the input-driven queue automaton may be deterministic or nondeterministic. We use the notation $\mathrm{TDQA}_{x,y}$ with $x, y \in \{d, n\}$ where x stands for the working mode of the transducer and y for the mode of the input-driven queue automaton. For example, $\mathrm{TDQA}_{n,d}$ is a tinput-driven queue automaton with a nondeterministic sequential transducer and a deterministic input-driven queue automaton. We require here that nondeterministic sequential transducers are injective and length-preserving as their deterministic variant. The first result reveals that, for nondeterministic transducers, the nondeterministic IDQA can be determinized.

Theorem 14. *The family of languages accepted by $TDQA_{n,n}$ and $TDQA_{n,d}$ coincide.*

Proof. In the following we use the abbreviation δ for the union of the transition functions δ_D, δ_R, and δ_N of an IDQA. Given a $TDQA_{n,n}$ with nondeterministic transducer $T = \langle Q, \Sigma, \Delta, q_0, \sigma \rangle$ and nondeterministic IDQA $M = \langle P, \Delta, \Gamma, p_0, F, \bot, \delta_D, \delta_R, \delta_N \rangle$, the basic idea of the construction of a nondeterministic transducer T' and a deterministic IDQA M' so that $L(M', T') = L(M, T)$ is to let T' simulate T and additionally guess which transition M' would perform dependent on the current symbol at the front of the queue. The guess is the output symbol of T'.

At first we construct the nondeterministic transducer T' as $\langle Q, \Sigma, \Delta', q_0, \sigma' \rangle$, where Δ' is the set of deterministic transition functions of an IDQA with state set P, a fixed input symbol from Δ, and set of queue symbols Γ. Formally, for $d \in \Delta$, let $\hat{\delta}_d$ be a mapping from $P \times \{d\} \times (\Gamma \cup \{\bot\})$ to $P \times \Gamma \cup P$, and Δ' be the union of all such mappings $\hat{\delta}_d$ for all $d \in \Delta$. Now, it is sufficient to define σ': for all $p \in Q$, $a \in \Sigma$,

$$\sigma'(q, a) = \{(q', \hat{\delta}_b) \mid (q', b) \in \sigma(q, a) \text{ and}$$

$$\hat{\delta}_b \text{ so that } \hat{\delta}_b(s, b, \gamma) \in \delta(s, b, \gamma), \text{ for all } s \in P \text{ and } \gamma \in \Gamma \cup \{\bot\}\}.$$

The deterministic IDQA $M' = \langle P, \Delta', \Gamma, p_0, F, \bot, \delta'_D, \delta'_R, \delta'_N \rangle$ is now defined through $\delta'(p, \hat{\delta}_b, \gamma) = \hat{\delta}_b(p, b, \gamma)$, for all $p \in P$ and $\gamma \in \Gamma \cup \{\bot\}$. Clearly, M' is length-preserving and M' is injective since M is.

In order to show that (M', T') and (M, T) are equivalent, assume that both TDQAs M and M' are in the same state $p \in P$, have the same queue content $s \in \Gamma^*$, and that both transducer T and T' are in the same state $q \in Q$. By definition and construction this is true for the initial configurations.

Let (M, T) perform a transition on some input symbol $a \in \Sigma$ that drives T nondeterministically into state q' whereby symbol $b \in \Delta$ is output. Furthermore, let M apply $\delta(p, b, \gamma)$ so that its successor state is nondeterministically chosen to be $p' \in P$ and the queue is manipulated by some operation op. By construction, then and only then T' may perform a nondeterministic transition on a that yields state q' as well, and outputs the deterministic transition function $\hat{\delta}$ so that $\hat{\delta}(p, b, \gamma)$ gives the successor state p' and manipulates the queue by the operation op. So, after processing the next input symbol, M and M' are again in the same state, have again the same queue content, and T and T' are again in the same state. Therefore, (M', T') and (M, T) accept the same language. $\square$

So, the nondeterminism of the transducer is a powerful resource. Once it is available, it does not matter whether the IDQA is nondeterministic or not. In both cases the same language family is accepted. However, the next result shows that the absence of nondeterminism of the transducer strictly weakens the TDQA, regardless whether or not nondeterminism is provided for the IDQA.

Theorem 15. *The family of languages accepted by $TDQA_{d,n}$ is properly included in the family of languages accepted by $TDQA_{n,d}$.*

The previous theorem showed once more that the nondeterminism of the transducer is a powerful resource. Even if the associated IDQA is deterministic, the nondeterminism of the transducer cannot be compensated by a nondeterministic IDQA. On the other hand, the next result reveals that nondeterminism for the IDQA is better than determinism if the transducers are deterministic.

Theorem 16. *The family of languages accepted by $TDQA_{d,d}$ is properly included in the family of languages accepted by $TDQA_{d,n}$.*

Proof. The inclusion $\mathscr{L}(\text{TDQA}_{d,d}) \subset \mathscr{L}(\text{TDQA}_{d,n})$ follows immediately for structural reasons. The properness of the inclusion is witnessed by language

$$L = \{ a^n \$h(w_1)\$h(w_2)\$ \cdots \$h(w_m) \mid m, n \geq 1, w_k \in \{a, b\}^n, 1 \leq k \leq m,$$
$$\text{and there exist } 1 \leq i < j \leq m \text{ so that } w_i = w_j \},$$

where h is the homomorphism that maps a to #a and b to #b. It is accepted by a $\text{TDQA}_{d,n}$ with $\Sigma_D = \{a, b\}$, $\Sigma_R = \{\#\}$ and $\Sigma_N = \{\$\}$. The transducer can be chosen to compute the identity mapping. Then, the accepting nondeterministic IDQA M first stores the prefix a^n into its queue, whereby the first symbol is marked. Subsequently, whenever a $ appears in the input, M guesses whether the following subword is $h(w_i)$. If not, for every input symbol # one queue symbol is removed from the queue and stored in the state, and for every symbol from $\{a, b\}$ the symbol stored in the state is entered in the queue again. If yes, instead of the symbols removed from the queue, the current input symbols are entered in the queue, whereby in both cases the first symbol is marked, again. In this way, M verifies that the lengths of the subwords read are correct. After storing w_i into the queue, the IDQA M continues to read subwords and to verify their lengths whereby the queue content is shifted circularly through the queue. Moreover, whenever a $ appears in the input, M guesses whether the following subword is $h(w_j)$. If yes, it compares the subwords symbol by symbol with the current queue content and accepts if and only if both match.

In order to show that L is not accepted by any $\text{TDQA}_{d,d}$ we assume in contrast to the assertion that it is accepted by some $\text{TDQA}_{d,d}$ (M', T). Let $n \geq 1$ be a fixed and long enough integer. There are 2^n different words over the alphabet $\{a, b\}$ whose length is n. We consider a subset $W = \{w_1, w_2, \ldots, w_\ell\}$ of such words and associate an input $w = a^n\$h(w_1)\$h(w_2)\$\cdots\$h(w_\ell)\$h(x)$ with it, where $x \in \{a, b\}^n$. When the computation on w has reached the last $ symbol, the remaining computation depends only on the current state of T, the current state of M', and the queue content that is still accessible in the last $|h(x)| = 2n$ time steps. For the latter we have the first $2n$ symbols in the queue. So, there are at most $|Q| \cdot |P| \cdot |\Gamma|^{2n} \in o(2^{2^n})$ different possibilities for such situations. On the other hand, there are 2^{2^n} different subsets of binary words of length n. This implies that there are two different subsets, say $W = \{w_1, w_2, \ldots, w_\ell\}$ and $W' = \{w'_1, w'_2, \ldots, w'_{\ell'}\}$ whose associated words drive the $\text{TDQA}_{d,d}$ (M', T) in the same situation. We conclude that, for all $x \in \{a, b\}^n$, the input $a^n\$h(w_1)\$h(w_2)\$\cdots\$h(w_\ell)\$h(x)$ is accepted if and only if the input $a^n\$h(w'_1)\$h(w'_2)\$\cdots\$h(w'_{\ell'})\$h(x)$ is accepted. However, without loss of generality we may assume that there is a word in W, say w_i, that does not belong to W'. Since $a^n\$h(w_1)\$h(w_2)\$\cdots\$h(w_\ell)\$h(w_i)$ is accepted, we conclude that the input $a^n\$h(w'_1)\$h(w'_2)\$\cdots\$h(w'_{\ell'})\$h(w_i)$ is accepted as well. But the latter does not belong to L since all subwords appearing are different. The contradiction shows that L is not accepted by any $\text{TDQA}_{d,d}$. $\qquad\square$

Summarizing the results regarding determinization, we end up with the following hierarchy:

$$\mathscr{L}(\text{TDQA}_{d,d}) \subset \mathscr{L}(\text{TDQA}_{d,n}) \subset \mathscr{L}(\text{TDQA}_{n,d}) = \mathscr{L}(\text{TDQA}_{n,n}).$$

5   Decidability Questions

It has been shown in [8] that IDQAs with an unbounded number of turns are a powerful model, since they can encode in a suitable way the computations of linear bounded automata. Owing to this encoding it is possible to reduce decidability questions for linear bounded automata to decidability questions for IDQAs. Since it is known that for the former class of automata many decidability questions are undecidable and not even semidecidable, we obtain by our reduction that the commonly studied questions of emptiness, finiteness, inclusion, equivalence, regularity, and context-freeness are all undecidable and not semidecidable for IDQAs.

Theorem 17. *[8] The questions of emptiness, finiteness, infiniteness, universality, inclusion, equivalence, regularity, and context-freeness are not semidecidable for IDQA0s.*

Since every IDQA can be considered as a TDQA such that its corresponding injective and length-preserving deterministic sequential transducer simply realizes the identity map, we immediately obtain that all the above questions are not semidecidable for TDQAs as well.

Corollary 18. *The questions of emptiness, finiteness, infiniteness, universality, inclusion, equivalence, regularity, and context-freeness are not semidecidable for TDQAs.*

In the turn-bounded case, it has been shown in [8] that the questions of emptiness, finiteness, and equivalence with regular languages are decidable for k-turn DQAs which are not necessarily input-driven. These results help to show the following decidability results for TDQAs.

Theorem 19. *Let $k \geq 0$ be a constant and (M, T) be a k-turn TDQA. Then emptiness and finiteness of $L(M, T)$ is decidable. Furthermore, equivalence with regular sets and, in particular, universality is decidable for (M, T) as well.*

Proof. Let (M, T) be a k-turn TDQA. Then the DST T is in particular length-preserving and maps every letter of the input alphabet Σ to another letter of M's input alphabet Δ. Now, we construct a DQA M' which translates in its state set every input symbol $a \in \Sigma$ to $T(a) \in \Delta$ and simulates M on input $T(a)$. Thus, $w \in L(M')$ if and only if $w \in L(M, T)$ and we obtain that $L(M') = L(M, T)$. Moreover, M' is k-turn since M is k-turn. Thus, we obtain our statements by applying the decidability results for k-turn DQAs [8].   □

Now, we turn to show more decidable questions for k-turn TDQAs. We obtain that inclusion and equivalence is decidable as long as the signatures are compatible.

Theorem 20. *Let $k \geq 0$ be a constant and (M, T) as well as (M', T) be k-turn TDQA with compatible signatures. Then the inclusion and the equivalence of $L(M, T)$ and $L(M', T)$ is decidable.*

The positive decidability of inclusion gets lost if the signatures are no longer compatible. In [8] an example is given which shows that the inclusion problem for two 1-turn IDQAs with incompatible signatures is not semidecidable. Clearly, this result holds for TDQAs as well. Thus, we obtain the following corollary.

Corollary 21. *Let $k \geq 1$ be a constant and (M,T) as well as (M',T) be two k-turn TDQA. Then the inclusion $L(M,T) \subseteq L(M',T)$ is not semidecidable.*

It should be noted that it is currently an open question whether the equivalence of k-turn IDQAs or k-turn TDQAs becomes undecidable in case of incompatible signatures. Now, we turn back to TDQAs with an unbounded number of turns. Here, we get the interesting result that inclusion and equivalence remain not semidecidable even if compatible signatures are provided.

Theorem 22. *Let (M,T) and (M',T) be two TDQAs with compatible signatures. Then the inclusion $L(M,T) \subseteq L(M',T)$ and the equivalence $L(M,T) = L(M',T)$ is not semidecidable.*

References

1. Alur, R., Madhusudan, P.: Visibly pushdown languages. In: Babai, L. (ed.) Symposium on Theory of Computing (STOC 2004), pp. 202–211. ACM (2004)
2. Alur, R., Madhusudan, P.: Adding nesting structure to words. J. ACM **56**, 1–43 (2009)
3. Bensch, S., Holzer, M., Kutrib, M., Malcher, A.: Input-driven stack automata. In: Baeten, J.C.M., Ball, T., de Boer, F.S. (eds.) TCS 2012. LNCS, vol. 7604, pp. 28–42. Springer, Heidelberg (2012)
4. Book, R.V., Greibach, S.A., Wrathall, C.: Reset machines. J. Comput. Syst. Sci. **19**, 256–276 (1979)
5. Brandenburg, F.: Multiple equality sets and Post machines. J. Comput. Syst. Sci. **21**, 292–316 (1980)
6. von Braunmühl, B., Verbeek, R.: Input-driven languages are recognized in $\log n$ space. In: Karpinski, M., van Leeuwen, J. (eds.) Topics in the Theory of Computation, Mathematics Studies, vol. 102, pp. 1–19. North-Holland, Amsterdam (1985)
7. Cherubini, A., Citrini, C., Crespi-Reghizzi, S., Mandrioli, D.: QRT FIFO automata, breadth-first grammars and their relations. Theoret. Comput. Sci. **85**, 171–203 (1991)
8. Kutrib, M., Malcher, A., Mereghetti, C., Palano, B., Wendlandt, M.: Input-driven queue automata: Finite turns, decidability, and closure properties. Theoret. Comput. Sci. **578**, 58–71 (2015)
9. La Torre, S., Madhusudan, P., Parlato, G.: A robust class of context-sensitive languages. In: Logic in Computer Science (LICS 2007), pp. 161–170. IEEE Computer Society (2007)
10. Li, M., Longpré, L., Vitányi, P.M.B.: The power of the queue. SIAM J. Comput. **21**, 697–712 (1992)
11. Madhusudan, P., Parlato, G.: The tree width of auxiliary storage. In: Ball, T., Sagiv, M. (eds.) Principles of Programming Languages, (POpPL 2011), pp. 283–294. ACM (2011)
12. Mehlhorn, K.: Pebbling moutain ranges and its application of DCFL-recognition. In: de Bakker, J.W., van Leeuwen, J. (eds.) ICALP 1980, vol. 85, pp. 422–435. Springer, Heidelberg (1980)
13. Okhotin, A., Salomaa, K.: Complexity of input-driven pushdown automata. SIGACT News **45**, 47–67 (2014)

Periodic Generalized Automata over the Reals

Klaus Meer[1]([✉]) and Ameen Naif[1]

Computer Science Institute, Brandenburg University of Technology
Cottbus-Senftenberg, Platz der Deutschen Einheit 1, 03046 Cottbus, Germany
{meer,naif}@b-tu.de

Abstract. In [4] Gandhi, Khoussainov, and Liu introduced and studied a generalized model of finite automata able to work over arbitrary
structures. The model mimics the finite automata over finite structures
but has an additional ability to perform in a restricted way operations
attached to the structure under consideration. As one relevant area of
investigations for this model the authors of [4] identified studying the
new automata over uncountable structures such as the real numbers.
This research was started in [7]. However, there it turned out that many
elementary properties known from classical finite automata are lost. This
refers both to structural properties of accepted languages and to decidability and computability questions. The intrinsic reason for this is that
the computational abilities of the new model turn out to be too strong.

We therefore propose a restricted version of the model which we call
periodic GKL automata. The new model still has certain computational
abilities which, however, are restricted in that computed information is
deleted again after a fixed period in time. We show that this limitation
regains a lot of classical properties including the pumping lemma and
many decidability results. Thus the new model seems to reflect more
adequately what might be considered as a finite automata over the reals
and similar structures. Though our results resemble classical properties,
for proving them other techniques are necessary. One fundamental proof
ingredient will be quantifier elimination over real closed fields.

Keywords: Unconventional models of computation · Computational
complexity

1 Introduction

In recent work Gandhi, Khoussainov, and Liu [4] introduced a generalized model
of finite automata called $(\mathcal{S}, k)$-automata. It is able to work over an arbitrary
structure $\mathcal{S}$, and here in particular over infinite alphabets like the real numbers.
A structure is characterized by an alphabet (also called universe) together with
a finite number of binary functions and relations over that alphabet. Intuitively
the model processes words over the underlying alphabet componentwise. Each

---

A. Naif was supported by a 'Promotionsstipendium nach Graduiertenförderungsver-
ordnung des Landes Brandenburg GradV'. The support is gratefully acknowledged.

A.-H. Dediu et al. (Eds.): LATA 2016, LNCS 9618, pp. 168–180, 2016.
DOI: 10.1007/978-3-319-30000-9_13

single step is made of finitely many test operations relying on the fixed relations as well as finitely many computational operations relying on the fixed functions. For performing the latter an $(\mathcal{S}, k)$-automaton can use a finite number k of registers. It moves between finitely many states and finally accepts or rejects an input.

The motivation to study such generalizations is manifold. In [4] the authors discuss different previous approaches to design finite automata over infinite alphabets and their role in program verification and database theory. One goal is to look for a generalized framework that is able to homogenize at least some of these approaches. As the authors remark, many classical automata models like pushdown automata, Petri nets, visible pushdown automata can be simulated by the new model. Another major motivation results from work on algebraic models of computation over structures like the real and complex numbers. Here, the authors suggested their model as a finite automata variant of the Blum-Shub-Smale BSS model [1, 2]. They then ask to analyze such automata over structures like real or algebraically closed fields.

This line of research has been started recently by the present authors in [7]. Given the tremendous impact finite automata have in classical computability theory it looks promising to introduce and study a similar concept as restriction of the real computational model introduced by Blum, Shub, and Smale. However, the main lesson from [7] is that the general automata model by Gandhi et al. turns out to be too strong when applied to computations over the real or complex numbers. Almost all basic automata problems turn out to be undecidable in the BSS framework for these two uncountable structures. As another consequence, only weak structural properties can be derived for real languages accepted by such an automaton. The intrinsic reason for this is the automata's ability to store intermediate results during an entire computation, something obviously not possible in the finite automata world. We therefore suggest a restricted version of the Ghandi-Khoussainov-Liu (for short: GKL) model. The basic idea is to force the automaton to periodically forget after a constant number of steps its intermediate results. This still enables the automaton to perform operations present in the given structure, but in a limited way.

For the resulting restricted periodic version of real GKL automata we shall prove both structural as well as several decidability results. Since the automata can perform basic arithmetic operations over $\mathbb{R}$ semi-algebraic sets naturally show up; computability and decidability results then naturally rely on quantifier elimination algorithms for the first order theory of the reals.

Our results hopefully indicate that the restricted model is a meaningful alternative giving back many of the features finite automata have, but non-trivially related to the uncountable structure under consideration. Some of the further questions arising are discussed at the end.

In the next section we recall the automata model by Gandhi et al., equipped with the additional restriction of periodicity. The main results are proved in Sects. 3 and 4, which collect both decidability results for several questions about our periodic automata and structural properties of the languages accepted by them.

The paper intends to be a further step towards the development of a generalized model of finite automata. It might be promising to analyse our variant as well for the many further scenarios treated in [4].

2   Periodic GKL Automata

We suppose the reader to be familiar with the basics of the Blum-Shub-Smale model of computation and complexity over $\mathbb{R}$. Very roughly, algorithms in this model work over finite strings of real numbers. The operations either can be computational, in which case addition, subtraction, and multiplication are allowed; without much loss of generality we do not consider divisions in this paper to avoid technical inconveniences. Or an algorithm can branch depending on the result of a binary test operation. The latter will be inequality tests of form 'is $x \geq 0$?' The size of a string is the number of components it has, the cost of an algorithm is the number of operations it performs until it halts. For more details see [1].

The generalized finite automata introduced in [4] work over structures. Here, a structure $\mathcal{S}$ consists of a universe D together with finite sets of (binary) functions and relations over the universe. An automaton informally works as follows. It reads a word from the universe, i.e., a finite string of components from D and processes each component once. Reading a component the automaton can set up some tests using the relations in the structure. The tests might involve a fixed set of constants from the universe D that the automaton can use. It can as well perform in a limited way computations on the current component. Towards this aim, there is a fixed number k of registers that can store elements from D. Those registers can be changed using their current value, the current input and the functions related to $\mathcal{S}$. The new aspect we include here is that such computations cannot be performed unlimited; after a fixed number of steps performed all register values are reset to the intial assignment 0, thus forgetting intermediate results. After having read the entire input word the automaton accepts or rejects it depending on the state in which the computation stops. These automata can both be deterministic and non-deterministic.

The approach in particular easily can be adapted to define generalized finite automata over structures like $\mathbb{R}$ and $\mathbb{C}$. We shall in the rest of the paper focus on the real numbers, though all our results hold as well in their corresponding variant for the complex numbers. Statements about computability and decidability refer to the real BSS model of computation.

We consider exclusively the structure $\mathcal{S}_{\mathbb{R}} := (\mathbb{R}, +, -, \bullet, pr_1, pr_2, \geq, =)$ of reals as ring with order. As in the original work [4] we include the projection operators pr_1, pr_2 which give back the first and the second component of a tuple, respectively. In order to avoid technicalities for the subtraction operation we allow both orders of the involved arguments, i.e., applying $-$ to two values x, v can mean $x - v$ or $v - x$. Similarly, the order test can be performed both as $x \leq v$? and $v \leq x$? We do not include division as an operation. This will not significantly change our results.

The following definition makes these ideas precise. We alternatively call the resulting automata real periodic $(S_\mathbb{R}, k, T)$-automata or (a bit less clumsy) real periodic GKL automata, where GKL refers to the initials of the authors of [4]. The definition below adapts the general automata definition in [4] and its real number version from [7].

Definition 1. *(Periodic GKL automata over $\mathbb{R}$) Let $k, T \in \mathbb{N}$ be fixed.*

(a) *A deterministic periodic $(S_\mathbb{R}, k, T)$-automaton $\mathcal{A}$, also called real periodic GKL automaton, consists of the following objects:*
  - *a finite state space Q and an initial state $q_0 \in Q$,*
  - *a set $F \subseteq Q$ of final (accepting) states,*
  - *a set of ℓ registers which contain fixed given constants $c_1, \ldots, c_\ell \in \mathbb{R}$,*
  - *a set of k registers which can store real numbers denoted by $v_1, \ldots, v_k$,*
  - *a counter containing a number $t \in \{0, 1, \ldots, T-1\}$; we call T the periodicity of the automaton,*
  - *a transition function $\delta : Q \times \mathbb{R} \times \mathbb{R}^k \times \{0,1\}^{k+\ell} \times \{0,1,\ldots,T-1\} \mapsto Q \times \mathbb{R}^k \times \{0,1,\ldots,T-1\}$.*

  The automaton processes elements of $\mathbb{R}^* := \bigsqcup_{n \geq 1} \mathbb{R}^n$, i.e., words of finite length with real components. For such an $(x_1, \ldots, x_n) \in \mathbb{R}^n$ it works as follows. The computation starts in q_0 with initial assignment $0 \in \mathbb{R}$ for the values $v_1, \ldots, v_k \in \mathbb{R}$. The automaton has a counter which stores an integer $t \in \{0, 1, \ldots, T-1\}$. At the beginning of a computation its value is 0. $\mathcal{A}$ reads the input components step by step. Suppose a value x is read in state $q \in Q$ with counter value t. The next state together with an update of the values v_i and t is computed as follows:
    - *$\mathcal{A}$ performs the $k + \ell$ comparisons $x\sigma_1 v_1?, x\sigma_2 v_2?, \ldots, x\sigma_k v_k?, x\sigma_{k+1} c_1?,$ $\ldots, x\sigma_{k+\ell} c_\ell?$, where $\sigma_i \in \{\geq, \leq, =\}$. This gives a vector $b \in \{0,1\}^{k+\ell}$, where a component 0 indicates that the comparison that was tested is violated whereas 1 codes that it is valid;*
    - *depending on state q and b the automaton moves to a state $q' \in Q$ (which could again be q);*
    - *if the value of the counter is $t = T-1$, then the counter as well as all register entries v_i are reset to 0. Otherwise, the counter value is increased by 1 and the values of all v_i are updated applying one of the operations in the structure: $v_i \leftarrow x \circ_i v_i$. Here, $\circ_i \in \{+, -, \bullet, pr_1, pr_2\}, 1 \leq i \leq k$ depends on q and b only.*

  When the final component of an input is read $\mathcal{A}$ performs the tests for this component and moves to its last state without any further computation. It accepts the input if this final state belongs to F, otherwise $\mathcal{A}$ rejects.

(b) *Non-deterministic $(S_\mathbb{R}, k, T)$-automata are defined similarly with the only difference that δ becomes a relation in the following sense: If in state q the tests result in $b \in \{0,1\}^{k+\ell}$ the automaton can non-deterministically choose for the next state and the update operations one among finitely many tuples $(q', \circ_1, \ldots, \circ_k) \in Q \times \{+, -, \bullet, pr_1, pr_2\}^k$. The counter, however, is changed as in the deterministic case and if $t = T-1$ the register values have to be reset to 0 as well in the non-deterministic case.*

As usual, a non-deterministic automaton accepts an input if there is at least
one accepting computation.

(c) The language of finite strings accepted by $\mathcal{A}$ is denoted by $L(\mathcal{A}) \subseteq \mathbb{R}^*$.

(d) A configuration of $\mathcal{A}$ is a tuple from $Q \times \mathbb{R}^k \times \{0, \ldots, T-1\}$ specifying the
current data during a computation.

Example 2. (a) Every classically regular language $L \subseteq \{0,1\}^*$ when considered
as language in $\mathbb{R}^*$ can be accepted by a real periodic GKL automaton. This can
be seen easily by interpreting a corresponding finite automaton for L as GKL
automaton which does not use its registers.

(b) For every semi-algebraic set $S \in\subseteq \mathbb{R}^n, n \in \mathbb{N}$ there is an $m \in \mathbb{N}$ and a real
periodic GKL automaton $\mathcal{A}$ such that S is the projection of the set $L(\mathcal{A}) \cap \mathbb{R}^m$
onto its first n components. This was shown in [7] for the real GKL automata
model, but the proof applies as well in the periodic case; this is true because the
main point is to give a bound on the number of steps needed to evaluate the
polynomial conditions defining S. The peridocity of the new automaton then
should be larger than this bound.

Below in Sect. 4 we outline a more systematic description of acceptable
languages which resembles the structure of regular sets over finite alphabets.

3   Decidability

Let us start with the study of some typical decision problems for periodic GKL
automata. To each periodic $(S_\mathbb{R}, k, T)$-automaton $\mathcal{A}$ we attach a directed graph
$G_\mathcal{A}$. Additionally, the edges are labeled by certain semi-algebraic sets. Both the
graph and those sets turn out to be crucial for solving many of the fundamental
decision problems about periodic GKL automata. Before being more precise let
us describe the intuition behind those definitions. Since $\mathcal{A}$ has periodicity T we
are naturally led to consider the following two questions: Starting in a state q
with register values 0, which other states are reachable within precisely a number
of t steps from q, where $1 \leq t \leq T$? And this happens when processing which
inputs from $\mathbb{R}^t$? We therefore split any computation on inputs $x \in \mathbb{R}^n$ in s blocks
of length T and t remaining steps, where $n = s \cdot T + t, s \in \mathbb{N}_0, 0 \leq t < T$.

Definition 3. Let $\mathcal{A}$ be a deterministic $(S_\mathbb{R}, k, T)$-automaton with state set Q,
initial state q_0 and final states $F \subseteq Q$.

(a) The directed graph $G_\mathcal{A} = (V, E)$ attached to $\mathcal{A}$ is defined as follows: For each
$q \in Q$ the set V contains a vertex q and vertices $q(t)$ for $1 \leq t < T$. $G_\mathcal{A}$
has an edge (p, q) iff there is a computation of $\mathcal{A}$ that when starting in p
with register values 0 reaches q after T steps. $G_\mathcal{A}$ has an edge $(p, q(t))$ for
an $1 \leq t < T$ iff there is a computation of $\mathcal{A}$ that when starting in p with
register values 0 reaches q after t steps. No other edges are present in $G_\mathcal{A}$.

(b) *For edges (p, q) and $(p, q(t))$ of $G_\mathcal{A}$, respectively, define $S(p, q) \subseteq \mathbb{R}^T$ and $S(p, q(t)) \subseteq \mathbb{R}^t$ as set of those $x \in \mathbb{R}^T$ or $x \in \mathbb{R}^t$, respectively, for which $\mathcal{A}$ moves from p to q when reading x according to the conditions under (a).*

Intuitively, vertices named by $p \in Q$ are used for dealing with sequences of T computational steps of $\mathcal{A}$, whereas the copies $q(t), 1 \leq t < T$ are used to reflect the final t steps of a computation. Therefore, there are no directed edges of form $(q(t), p)$.

Theorem 4. *Let $\mathcal{A}$ be an $(S_\mathbb{R}, k, T)$-automaton with attached directed graph $G_\mathcal{A} = (V, E), S(u, v)$ be defined as above for vertices $u, v \in V$. Then the following holds:*

(a) *All $S(p, q)$ are semi-algebraic in $\mathbb{R}^T$, all $S(p, q(t))$ are semi-algebraic in $\mathbb{R}^t$.[1]*

(b) *The edge relation of $G_\mathcal{A}$ is BSS-computable, i.e., for given $u, v \in V$ one can decide by a BSS algorithm whether $(u, v) \in E$.*

(c) *For each $q \in Q, 0 \leq t \leq T - 1$ the set $V(q, t)$ of register values $v \in \mathbb{R}^k$ that occur as valid entries during a computation of $\mathcal{A}$ which reaches q such that the counter contains t is semi-algebraic (and thus BSS decidable).*

Proof. We are in this paper not interested in efficiency results and thus only argue how the questions under consideration lead to certain quantifier elimination tasks in the first order theory of the reals. Since this theory is BSS decidable, see [8] for more on the history of respective algorithms, the claimed results then follow.

Ad (a) Let $p, q \in Q$ be fixed. The arguments below will be the same for a pair $(p, q(t))$ of vertices. Suppose $\mathcal{A}$ uses ℓ constants and is in state p with all register values equal to 0. We enumerate all sequences $\mathcal{P} := (p_0, p_1, p_2, \ldots, p_T)$ of states $p_i \in Q$, where $p_0 = p$ and $p_T = q$, together with sequences $\beta := (b_1, \ldots, b_T), b_i \in \{0, 1\}^{k+\ell}$ of decision vectors such that in $\mathcal{A}$ it is possible to move from state p_i to p_{i+1} if the actual input component x_{i+1} yields the test results coded by the components of b_i.

We now construct for each such pair $(\mathcal{P}, \beta)$ a first order formula $\Phi_{(\mathcal{P}, \beta)}(x)$ expressing for which inputs $x \in \mathbb{R}^T$ the path $\mathcal{P}$ is representing $\mathcal{A}$'s computation on x. Towards this aim we must record the changes of the register values as well as check the related test results. For each $1 \leq i \leq T$ let $h_1^{(i)}, \ldots, h_k^{(i)} : \mathbb{R}^i \mapsto \mathbb{R}$ be polynomials such that $h_j^{(i)}(x_1, \ldots, x_i)$ is the entry in register v_j if the computation follows $(\mathcal{P}, \beta)$ on input $x_1, \ldots, x_i$ starting from register values 0. The $h_j^{(i)}$ clearly are polynomials of degree at most i given the way $\mathcal{A}$ can compute.

Next, first order formulas $\varphi_i(x_1, \ldots, x_i)$ express that if $\mathcal{A}$ is in state p_{i-1} with register values $v_1 = h_1^{(i-1)}(x_1, \ldots, x_{i-1}), \ldots, v_k = h_k^{(i-1)}(x_1, \ldots, x_{i-1})$ and reads x_i, then all performed tests give a result according to b_i, $\mathcal{A}$ moves to state p_i, and the new register values are $v_1 = h_1^{(i)}(x_1, \ldots, x_i), \ldots, v_k = h_k^{(i)}(x_1, \ldots, x_i)$. Once again, all above conditions can be expressed in first order logic over $\mathbb{R}$

---

[1] A semi-algebraic set in $\mathbb{R}^n$ is a set that can be defined as a finite union of solution sets of polynomial equalities and inequalities.

due to the form of the tests that can be performed. $\Phi_{(\mathcal{P},\beta)}(x)$ now is the conjunction of all these formulas for all $1 \leq i \leq T$. It follows that $S(p,q) = \{x \in \mathbb{R}^T \mid \bigcup_{(\mathcal{P},\beta) \in \Gamma} \Phi_{(\mathcal{P},\beta)}(x)\}$, where Γ contains all suitable sequences leading in T steps from p to q respecting the constraints described above. Since there are only finitely many suitable pairs and for each of it the set of x satisfying $\Phi_{(\mathcal{P},\beta)}(x)$ is semi-algebraic, the claim follows. The argument for vertices $(p, q(t))$ is the same.

Ad (b)   Given the result in (a) for each pair (u, v) of vertices of the graph G_A it is decidable whether $S(u,v) \neq \emptyset$ by (one of) the well known algorithms for quantifier elimination over real closed fields [8]. Therefore, the edge relation of G_A is BSS computable.

Ad (c)   The argument is similar to those in (a). First, any computation of n steps that reaches state q such that the counter has value t can be decomposed into s blocks of T steps followed by t final steps, where $n = sT + t$. In order to determine the realizable register assignments we have to figure out for which states $p \in Q$ automaton A can reach q in t steps when starting with register values 0. Among those states we are only interested in the ones reachable in sT steps, i.e., those p for which there is a path in G_A from q_0 to p. The latter can be checked by a usual search algorithm on directed graphs once G_A has been computed according to (b). Let $H(q,t)$ be the set of states p reachable in the above sense and such that $S(p, q(t)) \neq \emptyset$. Then a $v \in \mathbb{R}^k$ is in $V(q,t)$ iff there is a $p \in H(q,t)$ and an $x \in S(p, q(t))$ with $v = (h_1^{(t)}(x), \ldots, h_k^{(t)}(x))$, where $h_j^{(t)}$ are as defined in the proof of part (a). Clearly, this set is again semi-algebraic.   □

The above theorem implies several decidability results of fundamental questions about real periodic GKL automata. Decidability here refers to the real number BSS model. Whereas for most of the problems treated below decidability follows relatively straightforwardly from the proof of Theorem 4, the equivalence problem is a bit harder to handle. Note that we do not currently know about a state minimization algorithm, thus the idea behind the classical algorithm for deciding equivalence of deterministic finite automata using minimal ones is not applicable. Note also that given the significantly extended computability features of the original definition of real GKL automata in [4] none of the problems listed below is decidable in this more general model, as was shown in [7].

Theorem 5. *The problems below are decidable in the real BSS model:*

(a) *Emptiness: Given an $(S_\mathbb{R}, k, T)$-automaton A, is $L(A) = \emptyset$?*

(b) *Reachability I: Given A as in (a) with state set Q together with a state $q \in Q$, is there a computation starting in A's initial state with register entries 0 that reaches q?*

(c) *Reachability II: Given an automaton A, a state q, a counter value $t \in \{0, \ldots, T-1\}$, and a $v \in \mathbb{R}^k$, is there a computation starting in A's initial state with register entries 0 that reaches p such that the counter's value is t and the register values equal v?*

(d) Reachability III: Similar to Reachability II, but without t being specified?
(e) Equivalence: Given two real periodic GKL automata A_1, A_2, is $L(A_1) = L(A_2)$?

Proof. Decidability of the questions under consideration is a relative immediate consequence of (the proof of) Theorem 4.

Ad (a)   For emptiness we compute a set H of states reachable by A from its starting configuration in some $s \cdot T$ steps, where $s \in \mathbb{N}_0$. The computation of H can be done using the arguments from the proof of parts (a) and (b) in Theorem 4. If $H \cap F \neq \emptyset$ it follows $L(A) \neq \emptyset$. Otherwise, for each $p \in H, 1 \leq t < T$, and $q \in Q$ we compute a description of the semi-algebraic set $S(p, q(t))$ and check whether it is empty or not using quantifier elimination. It follows that now $L(A) = \emptyset$ iff all those $S(p, q(t))$ are empty.

Ad (b)   Using G_A and the corresponding sets $S(u, v)$ it is easy to check whether a state q or one of its copies $q(t), 1 \leq t < T$ are reachable in G_A from the starting state. This can be done, for example, by a breadth-first search.

Ad (c)   Check whether q (if $t = 0$) or $q(t)$ (if $t > 0$) are reachable in G_A. In case it is we analyze the set of attainable register values $V(q, t)$ as in the proof of Theorem 4.

Ad (d)   As in (c), but instead of checking one fixed t the algorithm has to be performed for all $0 \leq t \leq T - 1$.

Ad (e)   Let A_1, A_2 be given with state sets Q_1, Q_2 and parameters $(k_i, T_i), i \in \{1, 2\}$, respectively. The key idea is to give a bound N for the dimension of an $x \in \mathbb{R}^N$ which is accepted by exactly one of the two automata in case they are not equivalent. Knowing such a bound we can search for x by using once more the previous arguments together with quantifier elimination algorithms.

For both automata we consider computational blocks of length $T := T_1 \cdot T_2$. It then follows that for each integer $r \in \mathbb{N}$ after $r \cdot T$ steps both automata have 0 as its register assignments. Suppose $L(A_1) \neq L(A_2)$, then there exist an $N \in \mathbb{N}$ and an $x \in \mathbb{R}^N$ such that without loss of generality $x \in L(A_1) \setminus L(A_2)$. Decompose $N = s \cdot T + t, s \in \mathbb{N}_0, 0 \leq t < T$. We want to find an absolute upper bound s_0 for s such that a witness x can be proved to exist up to dimension $s_0 \cdot T + t < (s_0 + 1)T$ if one exists at all.

In order to determine s_0 define once again directed graphs G_{A_1}, G_{A_2} attached to the given automata as in Definition 3. However, this time both graphs are defined with respect to computational blocks of length T instead of taking the respective periodicities of the two automata. In the first reasoning below we want to decide existence of a witness $x \in \mathbb{R}^N$ for some $N = s \cdot T$, i.e., if $t = 0$.

For $0 \leq i \leq s$ let $(p_i^{(1)}, p_i^{(2)}) \in Q_1 \times Q_2$ denote the pairs of states that A_1, A_2 attain after $i \cdot T$ steps of processing input x. Define $s_0 := |Q_1| \cdot |Q_2| - 1$. Then no matter how x looks like after at most $s_0 \cdot T$ steps a pair of states occurs for the second time. Since the configurations of both automata at these steps are the same, we can neglect the corresponding part of the computation. By removing other loops along the computation path in a similar way it follows that if a witness x exists at all (for choice $t = 0$) there is one of dimension at most $N := s_0 \cdot T$. This reduces the question to a finite set of quantifier elimination problems in the spaces $\mathbb{R}^T, \mathbb{R}^{2T}, \ldots, \mathbb{R}^{s_0 T}$. For each $\mathbb{R}^{iT}, 1 \leq i \leq s_0$ express the question whether there is an $x \in \mathbb{R}^{iT}$ such that $x \in L(\mathcal{A}_1) \setminus L(\mathcal{A}_2)$ as closed existential first order formula. This can be done using the arguments in Theorem 4. We then decide truth of the formula by quantifier elimination.

If this decision procedure shows that there is no x of a dimension $N \in \{T, 2T, 3T, \ldots\}$ witnessing the difference of the two automata we next decide the respective question for dimensions of the form $N = sT + t$ with $1 \leq t < T$. As before, we first compute the set H of all pairs $(p, q) \in Q_1 \times Q_2$ that can be reached by $\mathcal{A}_1, \mathcal{A}_2$ in a sequence of some sT steps. For each fixed $1 \leq t < T$ and all pairs $(p', q') \in F_1 \times F_2$ of final states we decide, whether there are edges in $G_{\mathcal{A}_1}$ and $G_{\mathcal{A}_2}$, respectively, from p to $p'(t)$ and from q to $q'(t)$. In order to guarantee that both corresponding computations by $\mathcal{A}_1, \mathcal{A}_2$ are followed for the same input $x \in \mathbb{R}^t$ we consider the conjunction of the first order formulas expressing reachability of p' from p and of q' from q and only then apply quantifier elimination. If pairs $(p, q) \in H, (p', q') \in F_1 \times F_2, t \in \{1, \ldots, T-1\}$ are found it follows $L(\mathcal{A}_1) \neq L(\mathcal{A}_2)$, otherwise the two automata are equivalent.   □

4   Structural Results

In this section the focus is on elaborating elementary structural results for languages acceptable by real periodic GKL automata. They extend the corresponding ones for discrete finite automata and include a pumping lemma, the equivalence of non-deterministic and deterministic periodic GKL automata, and some initial ideas about a generalization of regular expressions to our setting. Note that in its usual form the pumping lemma does not hold for general GKL automata over $\mathbb{R}$ and no variant is known to be true; similarly, non-deterministic real GKL automata are strictly more powerful than deterministic ones, see [7] for both issues and [4] for similar results over other structures. Thus, periodicity significantly reduces the computational power of GKL automata and brings them probably closer to what one would expect from a real version of finite automata.

Lemma 6 (Pumping Lemma for Real Periodic GKL Automata). *Let $L \subseteq \mathbb{R}^*$ be accepted by a real periodic GKL automaton $\mathcal{A}$ with periodicity T, k registers and s states. Then for all $w \in L$ of algebraic size $|w| \geq sT$ there exist $x, y, z \in \mathbb{R}^*$ such that $w = xyz, |y| \geq T, |xy| \leq sT$, and $\forall i \in \mathbb{N}_0 \; xy^i z \in L$.*

Proof. Periodicity of $\mathcal{A}$ implies that for all $r \in \mathbb{N}$ after rT steps of any computation the register values are all 0. The actual configuration thus after rT steps only depends on the current state. For inputs w of length at least sT there occurs at least one state twice among the current states after one of the time steps $\{0, T, 2T, \ldots, sT\}$. The rest of the proof is as usual: If p is a state occuring twice, x the prefix of w processed by $\mathcal{A}$ until p is reached for first time, y the part processed until p occurs for the second time, and z the rest, then $w = xyz$ by definition and $xy^i z \in L$ for all $i \in \mathbb{N}_0$ because $\mathcal{A}$'s configuration is $(p, 0, 0)$ both when it starts to process y and when it has finished. The remaining conditions obviously are satisfied.   □

The lemma shows that the language $\{0^n1^n|n \in \mathbb{N}\}$, as in the finite automata world, is neither acceptable by a periodic real GKL automaton. It might be interesting to sharpen the statement by not only considering loops having a length being an integer multiple of the periodicity; this, however, would require to control the evolvement of the register entries, something that seemed to hinder a sharper structural result in [7].

Without requiring periodicity non-deterministic real GKL automata are more powerful than deterministic ones [7]. This difference vanishes if we include the periodicity requirement into the model.

Theorem 7. *The classes of languages over* $\mathbb{R}^*$ *acceptable by non-deterministic and by deterministic periodic real GKL automata are the same.*

Proof. Let $\mathcal{A}$ be a periodic non-deterministic $(S_\mathbb{R}, k, T)$-automaton with state set Q and final states $F \subseteq Q$. The construction of an equivalent deterministic automaton $\mathcal{A}'$ in principle uses the classical powerset idea; however, it is more complicated because of the automata's ability to compute. The new automaton has not only to record states that can be reached non-deterministically, but also register entries. In particular, it might be possible that $\mathcal{A}$ in a state has several choices how to continue to the same successor state but with different computations performed on the registers. Therefore, instead of (unordered) subsets of Q we are lead to consider (ordered) tuples of elements in Q as states of the new automaton $\mathcal{A}'$.

Let $M \in \mathbb{N}$ upper bound the maximal number of non-deterministic transitions $\mathcal{A}$ can choose from for any of its states. $\mathcal{A}$ begins its computations in a start state q_0 with register assignment 0. For $t < T$ steps $\mathcal{A}$ can achieve at most M^{t-1} different configurations. After T steps, i.e., one sweep of length the periodicity, there are at most $|Q|$ different configurations since the register values are reset. For the following sweeps of length T the same reasoning shows that at most $m := |Q| \cdot M^{T-1}$ different configurations can occur at any stage of a computation of $\mathcal{A}$.

This motivates the definition of the state set Q' of $\mathcal{A}'$: Q' will be used to code the multi-set of states of Q that can occur with potentially different register values at a certain point of a computation. Since not necessarily m different configurations are realizable let $q^* \notin Q$ denote a new dummy symbol and define Q' as a subset of $(Q \cup \{q^*\})^m$; a state in Q' lists with the respective cardinalities the set of states in Q reachable non-deterministically within a given number of steps. If less than m different configurations are reachable the lacking components are filled with q^*.[2] The starting state of $\mathcal{A}'$ is $(q_0, q^*, q^*, \ldots, q^*) \in (Q \cup \{q^*\})^m$, final states in Q' are those which contain at least one state from F as a component. $\mathcal{A}'$ uses km registers which are divided into m blocks of length k; each block is used to code the evolvement of $\mathcal{A}$'s k registers during one possible computation.

---

[2] In order to make the coding unique we could order the components of any tuple in Q' according to an order of $Q \cup \{q^*\}$, but we refrain from elaborating on this because it will likely not increase understandability.

The transition function can now be defined straightforwardly; in order not to overload the presentation notationally we just describe its functioning informally. If $\mathcal{A}$ after t steps on input x can reach a configuration $(p, v_1, \ldots, v_k, t)$, then $\mathcal{A}'$ after t steps is in a state p' which has p as one of its components; attached to each component is one block of k registers - and the one attached to the component containing the above p has $(v_1, \ldots, v_k)$ as its register values. Now each of the at most M many non-deterministic transitions $\mathcal{A}$ can perform next is coded via changing the corresponding components of p' and their attached register blocks accordingly. If $\mathcal{A}'$ has less than M many choices, in p the lacking components are set to q^* and the corresponding register values to 0. Of course, it has to be specified which components of a state p' and which register blocks of $\mathcal{A}'$ code which potential computation path of $\mathcal{A}$. However, it should be obvious that this easily can be done. Finally, $\mathcal{A}$ has an accepting computation on x ending in a state $q_f \in F$ if and only if $\mathcal{A}'$ reaches at the end a state that has q_f as one of its components, i.e., $\mathcal{A}'$ accepts x. □

The results especially of Theorem 4 resemble a strong similarity between the structures of real languages acceptable by periodic GKL automata and of regular languages. Due to space limitations we just outline this similarity and postpone a more complete treatment to a full version.

It has been shown in Sect. 3 that computation cycles of length the periodicity T play a crucial role in the analysis of periodic automata. If such an automaton can move in T steps from state p to q assuming all registers have been initialized to 0, then the (non-empty) semi-algebraic sets $S(p, q)$ introduced in Definition 3 constitute building blocks of words being processed by the automaton; similarly at the end of a computation with sets of form $S(p, q(t))$. Thus, for the development of a theory of 'regular expressions' in our context such sets could serve as elementary objects. Of course, this requires as well taking more care about what kind of semi-algbraic sets can be allowed here.

Next, there is as well a natural way to define a Kleene-$*$ operation on those sets. Starting from the automata side, each cycle in the directed graph $G_{\mathcal{A}}$ indicates that a corresponding sequence of operations can be performed by $\mathcal{A}$ arbitrarily many times, each time processing a word from a corresponding semi-algebraic set; the latter has the concatenated structure $S(p_1, p_2)S(p_2, p_3) \ldots S(p_r, p_1)$, where $p_1 \to p_2 \to \ldots \to p_r \to p_1$ denotes the underlying cycle. That way, we obtain a recursive construction of expressions being built on base of certain semi-algebraic sets. It is then not hard to work out a decomposition result for all real languages L accepted by a real periodic GKL automaton. Each such L can be decomposed using the operations of concatenation of fundamental semi-algebraic sets together with the $*$-operation related to the cycle structure in $G_{\mathcal{A}}$. It is probably more demanding to work out the other direction: Which kind of semi-algebraic ground-pieces can be allowed to obtain in the above way exactly those languages that are accepted by real periodic GKL automata? Nevertheless, the structural similarity to the concept of regular languages seems striking.

5   Further Questions

There are several ways to go. First, one could continue the line of research in this paper and analyze further properties of real languages acceptable by periodic GKL automata. For example, is there a way to minimize the number of states and of registers, and what impact has the choice of the periodicity? Next, above we completely disregarded complexity issues and only applied elimination results in their general form. In many of our problems requiring quantifier elimination the formulas obtained have a very specific structure because the automata process an input component only once and it is only influencing T steps of a computation. Therefore, one might ask which of the problems treated above are efficiently solvable, which are $NP_{\mathbb{R}}$-complete in the BSS framework? Especially finding new $NP_{\mathbb{R}}$-complete problems is interesting since the list of known such problems still is relatively limited.

A third area of further research is considering other underlying structures than the reals. One major motivation of [4] was to have a generalized automata model for many different structures which also homogenizes existing ones. So the question is in how far our restricted version of GKL automata also is meaningful for further structures like those considered in [4]? Finally, it seems promising to analyze the new automata model as well for infinite computations over countably infinite sequences of reals. The classical theory initiated by Büchi, see [9] for a longer introduction into the topic, established a close relation between such infinite automata and logic. For working with infinite structures meta-finite model theory was developed in [5] and applied to BSS computability theory in [3,6]. We believe it to be interesting to investigate potential links between the latter and a kind of periodic real Büchi automata.

References

1. Blum, L., Cucker, F., Shub, M., Smale, S.: Complexity and Real Computation. Springer, Berlin (1998)
2. Blum, L., Shub, M., Smale, S.: On a theory of computation and complexity over the real numbers: NP-completeness, recursive functions and universal machines. Bull. Am. Math. Soc. New Ser. **21**(1), 1–46 (1989)
3. Cucker, F., Meer, K.: Logics which capture complexity classes over the reals. J. Symb. Log. **64**(1), 363–390 (1999)
4. Gandhi, A., Khoussainov, B., Liu, J.: Finite automata over structures. In: Agrawal, M., Cooper, S.B., Li, A. (eds.) TAMC 2012. LNCS, vol. 7287, pp. 373–384. Springer, Heidelberg (2012)
5. Grädel, E., Gurevich, Y.: Metafinite model theory. Inf. Comput. **140**(1), 26–81 (1998)
6. Grädel, E., Meer, K.: Descriptive complexity theory over the real numbers. In: Leighton, F.T., Borodin, A. (eds.) Proceedings of the Twenty-Seventh Annual ACM Symposium on Theory of Computing, Las Vegas, Nevada, USA, pp. 315–324. ACM, 29 May–1 June 1995
7. Meer, K., Naif, A.: Generalized finite automata over real and complex numbers. Theor. Comput. Sci. **591**, 85–98 (2015)

8. Renegar, J.: On the computational complexity and geometry of the first-order theory of the reals, parts 1–3. J. Symb. Comput. **13**(3), 255–352 (1992)
9. Thomas, W.: Automata on infinite objects. In: van Leeuven, J. (ed.) Handbook of Theoretical Computer Science, Volume B: Formal Models and Sematics (B), pp. 133–192. Elsevier, Amsterdam (1990)

Minimal Separating Sequences for All Pairs of States

Rick Smetsers[(✉)], Joshua Moerman, and David N. Jansen

Institute for Computing and Information Sciences, Radboud University,
Toernooiveld 212, 6525 EC Nijmegen, The Netherlands
{r.smetsers,joshua.moerman,dnjansen}@cs.ru.nl

Abstract. Finding minimal separating sequences for all pairs of inequivalent states in a finite state machine is a classic problem in automata theory. Sets of minimal separating sequences, for instance, play a central role in many conformance testing methods. Moore has already outlined a partition refinement algorithm that constructs such a set of sequences in $\mathcal{O}(mn)$ time, where m is the number of transitions and n is the number of states. In this paper, we present an improved algorithm based on the minimization algorithm of Hopcroft that runs in $\mathcal{O}(m \log n)$ time. The efficiency of our algorithm is empirically verified and compared to the traditional algorithm.

Keywords: Algorithms on automata and words · Partition refinement

1 Introduction

In diverse areas of computer science and engineering, systems can be modelled by *finite state machines* (FSMs). One of the cornerstones of automata theory is minimization of such machines (and many variation thereof). In this process one obtains an equivalent minimal FSM, where states are different if and only if they have different behaviour. The first to develop an algorithm for minimization was Moore [9]. His algorithm has a time complexity of $\mathcal{O}(mn)$, where m is the number of transitions, and n is the number of states of the FSM. Later, Hopcroft improved this bound to $\mathcal{O}(m \log n)$ [6].

Minimization algorithms can be used as a framework for deriving a set of *separating sequences* that show *why* states are inequivalent. The separating sequences in Moore's framework are of minimal length [3]. Obtaining minimal separating sequences in Hopcroft's framework, however, is a non-trivial task. In this paper, we present an algorithm for finding such minimal separating sequences for all pairs of inequivalent states of a FSM in $\mathcal{O}(m \log n)$ time.

Coincidentally, Bonchi and Pous recently introduced a new algorithm for the equally fundamental problem of proving equivalence of states in non-deterministic automata [1]. As both their and our work demonstrate, even classical problems in automata theory can still offer surprising research opportunities.

Supported by NWO project 628.001.009 on Learning Extended State Machines for Malware Analysis (LEMMA).

A.-H. Dediu et al. (Eds.): LATA 2016, LNCS 9618, pp. 181–193, 2016.
DOI: 10.1007/978-3-319-30000-9_14

Moreover, new ideas for well-studied problems may lead to algorithmic improvements that are of practical importance in a variety of applications.

One such application for our work is in *conformance testing*. Here, the goal is to test if a black box implementation of a system is functioning as described by a given FSM. It consists of applying sequences of inputs to the implementation, and comparing the output of the system to the output prescribed by the FSM. Minimal separating sequences are used in many test generation methods [2]. Therefore, our algorithm can be used to improve these methods.

2   Preliminaries

We define a *FSM* as a Mealy machine $M = (I, O, S, \delta, \lambda)$, where I, O and S are finite sets of *inputs*, *outputs* and *states* respectively, $\delta : S \times I \rightarrow S$ is a *transition function* and $\lambda : S \times I \rightarrow O$ is an *output function*. The functions δ and λ are naturally extended to $\delta : S \times I^* \rightarrow S$ and $\lambda : S \times I^* \rightarrow O^*$. Moreover, given a set of states $S' \subseteq S$ and a sequence $x \in I^*$, we define $\delta(S', x) = \{\delta(s, x) | s \in S'\}$ and $\lambda(S', x) = \{\lambda(s, x) | s \in S'\}$. The *inverse transition function* $\delta^{-1} : S \times I \rightarrow \mathcal{P}(S)$ is defined as $\delta^{-1}(s, a) = \{t \in S | \delta(t, a) = s\}$. Observe that Mealy machines are deterministic and input-enabled (i.e. complete) by definition. The initial state is not specified because it is of no importance in what follows. For the remainder of this paper we fix a machine $M = (I, O, S, \delta, \lambda)$. We use n to denote its number of states, i.e. $n = |S|$, and m to denote its number of transitions, i.e. $m = |S| \times |I|$.

Definition 1. *States s and t are* equivalent *if $\lambda(s, x) = \lambda(t, x)$ for all x in I^*.*

We are interested in the case where s and t are not equivalent, i.e. *inequivalent*. If all pairs of distinct states of a machine M are inequivalent, then M is *minimal*. An example of a minimal FSM is given in Fig. 1.

Definition 2. *a* separating sequence *for states s and t in s is a sequence $x \in i^*$ such that $\lambda(s, x) \neq \lambda(t, x)$. We say x is* minimal *if $|y| \geq |x|$ for all separating sequences y for s and t.*

A separating sequence always exists if two states are inequivalent, and there might be multiple minimal separating sequences. Our goal is to obtain minimal separating sequences for all pairs of inequivalent states of M.

2.1   Partition Refinement

In this section we will discuss the basics of minimization. Both Moore's algorithm and Hopcroft's algorithm work by means of partition refinement. A similar treatment (for DFAs) is given in [4].

A *partition* P of S is a set of pairwise disjoint non-empty subsets of S whose union is exactly S. Elements in P are called *blocks*. If P and P' are partitions of S, then P' is a *refinement* of P if every block of P' is contained in a block of P. A partition refinement algorithm constructs the finest partition under some constraint. In our context the constraint is that equivalent states belong to the same block.

Definition 3. *A partition is* valid *if equivalent states are in the same block.*

Partition refinement algorithms for FSMs start with the trivial partition $P = \{S\}$, and iteratively refine P until it is the finest valid partition (where all states in a block are equivalent). The blocks of such a *complete* partition form the states of the minimized FSM, whose transition and output functions are well-defined because states in the same block are equivalent.

Let B be a block and a be an input. There are two possible reasons to split B (and hence refine the partition). First, we can *split B with respect to output after a* if the set $\lambda(B, a)$ contains more than one output. Second, we can *split B with respect to the state after a* if there is no single block B' containing the set $\delta(B, a)$. In both cases it is obvious what the new blocks are: in the first case each output in $\lambda(B, a)$ defines a new block, in the second case each block containing a state in $\delta(B, a)$ defines a new block. Both types of refinement preserve validity.

Partition refinement algorithms for FSMs first perform splits w.r.t. output, until there are no such splits to be performed. This is precisely the case when the partition is *acceptable*.

Definition 4. *A partition is* acceptable *if for all pairs s, t of states contained in the same block and for all inputs a in I, $\lambda(s, a) = \lambda(t, a)$.*

Any refinement of an acceptable partition is again acceptable. The algorithm continues performing splits w.r.t. state, until no such splits can be performed. This is exactly the case when the partition is stable.

Definition 5. *A partition is* stable *if it is acceptable and for any input a in I and states s and t that are in the same block, states $\delta(s, a)$ and $\delta(t, a)$ are also in the same block.*

Since an FSM has only finitely many states, partition refinement will terminate. The output is the finest valid partition which is acceptable and stable. For a more formal treatment on partition refinement we refer to [4].

2.2 Splitting Trees and Refinable Partitions

Both types of splits described above can be used to construct a separating sequence for the states that are split. In a split w.r.t. the output after a, this sequence is simply a. In a split w.r.t. the state after a, the sequence starts with an a and continues with the separating sequence for states in $\delta(B, a)$. In order to systematically keep track of this information, we maintain a *splitting tree*. The splitting tree was introduced by Lee and Yannakakis [8] as a data structure for maintaining the operational history of a partition refinement algorithm.

Definition 6. *A splitting tree for M is a rooted tree T with a finite set of nodes with the following properties:*

– *Each node u in T is labelled by a subset of S, denoted $l(u)$.*
– *The root is labelled by S.*

- *For each inner node u, $l(u)$ is partitioned by the labels of its children.*
- *Each inner node u is associated with a sequence $\sigma(u)$ that separates states contained in different children of u.*

We use $C(u)$ to denote the set of children of a node u. The *lowest common ancestor* (lca) for a set $S' \subseteq S$ is the node u such that $S' \subseteq l(u)$ and $S' \not\subseteq l(v)$ for all $v \in C(u)$ and is denoted by $\mathrm{lca}(S')$. For a pair of states s and t we use the shorthand $\mathrm{lca}(s, t)$ for $\mathrm{lca}(\{s, t\})$.

The labels $l(u)$ can be stored as a *refinable partition* data structure [11]. This is an array containing a permutation of the states, ordered so that states in the same block are adjacent. The label $l(u)$ of a node then can be indicated by a slice of this array. If node u is split, some states in the *slice* $l(u)$ may be moved to create the labels of its children, but this will not change the *set* $l(u)$.

A splitting tree T can be used to record the history of a partition refinement algorithm because at any time the leaves of T define a partition on S, denoted $P(T)$. We say a splitting tree T is valid (resp. acceptable, stable, complete) if $P(T)$ is as such. A leaf can be expanded in one of two ways, corresponding to the two ways a block can be split. Given a leaf u and its block $B = l(u)$ we define the following two splits:

split-output. Suppose there is an input a such that B can be split w.r.t output after a. Then we set $\sigma(u) = a$, and we create a node for each subset of B that produces the same output x on a. These nodes are set to be children of u.

split-state. Suppose there is an input a such that B can be split w.r.t. the state after a. Then instead of splitting B as described before, we proceed as follows. First, we locate the node $v = \mathrm{lca}(\delta(B, a))$. Since v cannot be a leaf, it has at least two children whose labels contain elements of $\delta(B, a)$. We can use this information to expand the tree as follows. For each node w in $C(v)$ we create a child of u labelled $\{s \in B | \delta(s, a) \in l(w)\}$ if the label contains at least one state. Finally, we set $\sigma(u) = a\sigma(v)$.

A straight-forward adaptation of partition refinement for constructing a stable splitting tree for M is shown in Algorithm 1. The termination and the correctness of the algorithm outlined in Sect. 2.1 are preserved. It follows directly that states are equivalent if and only if they are in the same label of a leaf node.

Example 7. Figure 1 shows a FSM and a complete splitting tree for it. This tree is constructed by Algorithm 1 as follows. First, the root node is labelled by $\{s_0, \ldots, s_5\}$. The even and uneven states produce different outputs after a, hence the root node is split. Then we note that s_4 produces a different output after b than s_0 and s_2, so $\{s_0, s_2, s_4\}$ is split as well. At this point T is acceptable: no more leaves can be split w.r.t. output. Now, the states $\delta(\{s_1, s_3, s_5\}, a)$ are contained in different leaves of T. Therefore, $\{s_1, s_3, s_5\}$ is split into $\{s_1, s_5\}$ and $\{s_3\}$ and associated with sequence ab. At this point, $\delta(\{s_0, s_2\}, a)$ contains states that are in both children of $\{s_1, s_3, s_5\}$, so $\{s_0, s_2\}$ is split and the associated sequence is aab. We continue until T is complete.

Input: A FSM M
Result: A valid and stable splitting tree T
initialize T to be a tree with a single node labeled S
repeat
> find $a \in I, B \in P(T)$ such that we can split B w.r.t. output $\lambda(\cdot, a)$
> expand the $u \in T$ with $l(u) = B$ as described in (split-output)

until $P(T)$ *is acceptable*
repeat
> find $a \in I, B \in P(T)$ such that we can split B w.r.t. state $\delta(\cdot, a)$
> expand the $u \in T$ with $l(u) = B$ as described in (split-state)

until $P(T)$ *is stable*

Algorithm 1. Constructing a stable splitting tree

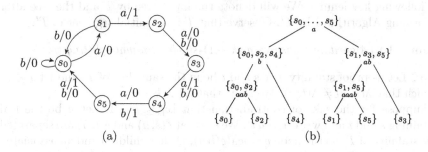

(a)   (b)

Fig. 1. A FSM (a) and a complete splitting tree for it (b)

3   Minimal Separating Sequences

In Sect. 2.2 we have described an algorithm for constructing a complete splitting tree. This algorithm is non-deterministic, as there is no prescribed order on the splits. In this section we order them to obtain minimal separating sequences.

Let u be a non-root inner node in a splitting tree, then the sequence $\sigma(u)$ can also be used to split the parent of u. This allows us to construct splitting trees where children will never have shorter sequences than their parents, as we can always split with those sequences first. Trees obtained in this way are guaranteed to be *layered*, which means that for all nodes u and all $u' \in C(u)$, $|\sigma(u)| \le |\sigma(u')|$. Each layer consists of nodes for which the associated separating sequences have the same length.

Our approach for constructing minimal sequences is to ensure that each layer is as large as possible before continuing to the next one. This idea is expressed formally by the following definitions.

Definition 8. *A splitting tree T is k-stable if for all states s and t in the same leaf we have $\lambda(s, x) = \lambda(t, x)$ for all $x \in I^{\le k}$.*

Definition 9. *A splitting tree T is minimal if for all states s and t in different leaves $\lambda(s, x) \ne \lambda(t, x)$ implies $|x| \ge |\sigma(\mathrm{lca}(s, t))|$ for all $x \in I^*$.*

Minimality of a splitting tree can be used to obtain minimal separating sequences for pairs of states. If the tree is in addition stable, we obtain minimal separating sequences for all inequivalent pairs of states. Note that if a minimal splitting tree is $(n-1)$-stable (n is the number of states of M), then it is stable (Definition 5). This follows from the well-known fact that $n-1$ is an upper bound for the length of a minimal separating sequence [9].

Algorithm 2 ensures a stable and minimal splitting tree. The first repeat-loop is the same as before (in Algorithm 1). Clearly, we obtain a 1-stable and minimal splitting tree here. It remains to show that we can extend this to a stable and minimal splitting tree. Algorithm 3 will perform precisely one such step towards stability, while maintaining minimality. Termination follows from the same reason as for Algorithm 1. Correctness for this algorithm is shown by the following key lemma. We will denote the input tree by T and the tree after performing Algorithm 3 by T'. Observe that T is an initial segment of T'.

Lemma 10. *Algorithm 3 ensures a $(k+1)$-stable minimal splitting tree.*

Proof. Let us proof stability. Let s and t be in the same leaf of T' and let $x \in I^*$ be such that $\lambda(s, x) \neq \lambda(t, x)$. We show that $|x| > k + 1$.

Suppose for the sake of contradiction that $|x| \leq k + 1$. Let u be the leaf containing s and t and write $x = ax'$. We see that $\delta(s, a)$ and $\delta(t, a)$ are separated by k-stability of T. So the node $v = \mathrm{lca}(\delta(l(u), a))$ has children and an associated sequence $\sigma(v)$. There are two cases:

- $|\sigma(v)| < k$, then $a\sigma(v)$ separates s and t and is of length $\leq k$. This case contradicts the k-stability of T.
- $|\sigma(v)| = k$, then the loop in Algorithm 3 will consider this case and split. Note that this may not split s and t (it may occur that $a\sigma(v)$ splits different elements in $l(u)$). We can repeat the above argument inductively for the newly created leaf containing s and t. By finiteness of $l(u)$, the induction will stop and, in the end, s and t are split.

Both cases end in contradiction, so we conclude that $|x| > k + 1$.

Let us now prove minimality. It suffices to consider only newly split states in T'. Let s and t be two states with $|\sigma(\mathrm{lca}(s, t))| = k + 1$. Let $x \in I^*$ be a sequence such that $\lambda(s, x) \neq \lambda(t, x)$. We need to show that $|x| \geq k + 1$. Since $x \neq \epsilon$ we can write $x = ax'$ and consider the states $s' = \delta(s, a)$ and $t' = \delta(t, a)$ which are separated by x'. Two things can happen:

- The states s' and t' are in the same leaf in T. Then by k-stability of T we get $\lambda(s', y) = \lambda(t', y)$ for all $y \in I^{\leq k}$. So $|x'| > k$.
- The states s' and t' are in different leaves in T and let $u = \mathrm{lca}(s', t')$. Then $a\sigma(u)$ separates s and t. Since s and t are in the same leaf in T we get $|a\sigma(u)| \geq k + 1$ by k-stability. This means that $|\sigma(u)| \geq k$ and by minimality of T we get $|x'| \geq k$.

In both cases we have shown that $|x| \geq k + 1$ as required.   □

Input: A FSM M with n states
Result: A stable, minimal splitting tree T
initialize T to be a tree with a single node labeled S
repeat
| find $a \in I, B \in P(T)$ such that we can split B w.r.t. output $\lambda(\cdot, a)$
| expand the $u \in T$ with $l(u) = B$ as described in (split-output)
until $P(T)$ *is acceptable*
for $k = 1$ *to* $n - 1$ **do**
| perform Algorithm 3 or Algorithm 4 on T for k

Algorithm 2. Constructing a stable and minimal splitting tree

Input: a k-stable and minimal splitting tree T
Result: T is a $(k + 1)$-stable, minimal splitting tree
forall the *leaves $u \in T$ and all inputs a* **do**
| locate $v = \mathrm{lca}(\delta(l(u), a))$
| **if** v *is an inner node and* $|\sigma(v)| = k$ **then**
| | expand u as described in (split-state) (which generates new leaves)

Algorithm 3. A step towards the stability of a splitting tree

Example 11. Figure 2a shows a stable and minimal splitting tree T for the machine in Fig. 1a. This tree is constructed by Algorithm 2 as follows. It executes the same as Algorithm 1 until we consider the node labeled $\{s_0, s_2\}$. At this point $k = 1$. We observe that the sequence of $\mathrm{lca}(\delta(\{s_0, s_2\}, a))$ has length 2, which is too long, so we continue with the next input. We find that we can indeed split w.r.t. the state after b, so the associated sequence is ba. Continuing, we obtain the same partition as before, but with smaller witnesses.

The internal data structure (a refinable partition) is shown in Fig. 2b: the array with the permutation of the states is at the bottom, and every block includes an indication of the slice containing its label and a pointer to its parent (as our final algorithm needs to find the parent block, but never the child blocks).

4   Optimizing the Algorithm

In this section, we present an improvement on Algorithm 3 that uses two ideas described by Hopcroft in his seminal paper on minimizing finite automata [6]: *using the inverse transition set*, and *processing the smaller half*. The algorithm that we present is a drop-in replacement, so that Algorithm 2 stays the same except for some bookkeeping. This way, we can establish correctness of the new algorithms more easily. The variant presented in this section reduces the amount of redundant computations that were made in Algorithm 3.

Using Hopcroft's first idea, we turn our algorithm upside down: instead of searching for the lca for each leaf, we search for the leaves u for which $l(u) \subseteq \delta^{-1}(l(v), a)$, for each potential lca v and input a. To keep the order of splits as before, we define *k-candidates*.

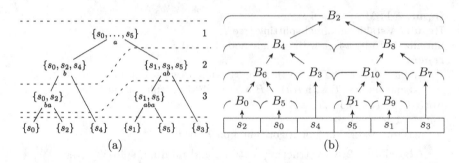

Fig. 2. A complete and minimal splitting tree for the FSM in Fig. 1a (a) and its internal refinable partition data structure (b)

Definition 12. *A k-candidate is a node v with $|\sigma(v)| = k$.*

A k-candidate v and an input a can be used to split a leaf u if $v = \mathrm{lca}(\delta(l(u), a))$, because in this case there are at least two states s, t in $l(u)$ such that $\delta(s, a)$ and $\delta(t, a)$ are in labels of different nodes in $C(v)$. Refining u this way is called *splitting u with respect to* (v, a). The set $C(u)$ is constructed according to (split-state), where each child $w \in C(v)$ defines a child u_w of u with states

$$l(u_w) = \{s \in l(u) \mid \delta(s, a) \in l(w)\} \qquad (1)$$
$$= l(u) \cap \delta^{-1}(l(w), a)$$

In order to perform the same splits in each layer as before, we maintain a list L_k of k-candidates. We keep the list in order of the construction of nodes, because when we split w.r.t. a child of a node u before we split w.r.t. u, the result is not well-defined. Indeed, the order on L_k is the same as the order used by Algorithm 2. So far, the improved algorithm still would have time complexity $\mathcal{O}(mn)$.

   To reduce the complexity we have to use Hopcroft's second idea of *processing the smaller half*. The key idea is that, when we fix a k-candidate v, all leaves are split with respect to (v, a) simultaneously. Instead of iterating over of all leaves to refine them, we iterate over $s \in \delta^{-1}(l(w), a)$ for all w in $C(v)$ and look up in which leaf it is contained to move s out of it. From Lemma 8 in [7] it follows that we can skip one of the children of v. This lowers the time complexity to $\mathcal{O}(m \log n)$. In order to move s out of its leaf, each leaf u is associated with a set of temporary children $C'(u)$ that is initially empty, and will be finalized after iterating over all s and w.

   In Algorithm 4 we use the ideas described above. For each k-candidate v and input a, we consider all children w of v, except for the largest one (in case of multiple largest children, we skip one of these arbitrarily). For each state $s \in \delta^{-1}(l(w), a)$ we consider the leaf u containing it. If this leaf does not have an associated temporary child for w we create such a child (line 9), if this child exists we move s into that child (line 10).

Input: a k-stable and minimal splitting tree T, and a list L_k
Result: T is a $(k+1)$-stable and minimal splitting tree, and a list L_{k+1}

1   $L_{k+1} \leftarrow \emptyset$
2   **forall the** k-candidates v in L_k in order **do**
3   let w' be a node in $C(v)$ such that $|l(w')| \geq |l(w)|$ for all nodes w in $C(v)$
4   **forall the** inputs a in I **do**
5   **forall the** nodes w in $C(v) \setminus w'$ **do**
6   **forall the** states s in $\delta^{-1}(l(w), a)$ **do**
7   locate leaf u such that $s \in l(u)$
8   **if** $C'(u)$ does not contain node u_w **then**
9   add a new node u_w to $C'(u)$
10   move s from $l(u)$ to $l(u_w)$

11   **foreach** leaf u with $C'(u) \neq \emptyset$ **do**
12   **if** $|l(u)| = 0$ **then**
13   **if** $|C'(u)| = 1$ **then**
14   recover u by moving its elements back and clear $C'(u)$
15   **continue** with the next leaf
16   set $p = u$ and $C(u) = C'(u)$
17   **else**
18   construct a new node p and set $C(p) = C'(u) \cup \{u\}$
19   insert p in the tree in the place where u was
20   set $\sigma(p) = a\sigma(v)$
21   append p to L_{k+1} and clear $C'(u)$

Algorithm 4. A better step towards the stability of a splitting tree

Once we have done the simultaneous splitting for the candidate v and input a, we finalize the temporary children. This is done at lines 11–21. If there is only one temporary child with all the states, no split has been made and we recover this node (line 14). In the other case we make the temporary children permanent.

The states remaining in u are those for which $\delta(s, a)$ is in the child of v that we have skipped; therefore we will call it the *implicit child*. We should not touch these states to keep the theoretical time bound. Therefore, we construct a new parent node p that will "adopt" the children in $C'(u)$ together with u (line 16).

We will now explain why considering all but the largest children of a node lowers the algorithm's time complexity. Let T be a splitting tree in which we color all children of each node blue, except for the largest one. Then:

Lemma 13. *A state s is in at most $(\log_2 n) - 1$ labels of blue nodes.*

Proof. Observe that every blue node u has a sibling u' such that $|l(u')| \geq |l(u)|$. So the parent $p(u)$ has at least $2|l(u)|$ states in its label, and the largest blue node has at most $n/2$ states.

Suppose a state s is contained in m blue nodes. When we walk up the tree starting at the leaf containing s, we will visit these m blue nodes. With each visit we can double the lower bound of the number of states. Hence $n/2 \geq 2^m$ and $m \leq (\log_2 n) - 1$. □

Corollary 14. *A state s is in at most $\log_2 n$ sets $\delta^{-1}(l(u), a)$, where u is a blue node and a is an input in I.*

If we now quantify over all transitions, we immediately get the following result. We note that the number of blue nodes is at most $n - 1$, but since this fact is not used, we leave this to the reader.

Corollary 15. *Let $\mathcal{B}$ denote the set of blue nodes and define*

$$\mathcal{X} = \{(b, a, s) \mid b \in \mathcal{B}, a \in I, s \in \delta^{-1}(l(b), a)\}.$$

Then $\mathcal{X}$ has at most $m \log_2 n$ elements.

The important observation is that when using Algorithm 4 we iterate in total over every element in $\mathcal{X}$ at most once.

Theorem 16. *Algorithm 2 using Algorithm 4 runs in $\mathcal{O}(m \log n)$ time.*

Proof. We prove that bookkeeping does not increase time complexity by discussing the implementation.

Inverse transition. δ^{-1} can be constructed as a preprocessing step in $\mathcal{O}(m)$.
State sorting. As described in Sect. 2.2, we maintain a refinable partition data structure. Each time new pair of a k-candidate v and input a is considered, leaves are split by performing a bucket sort.
First, buckets are created for each node in $w \in C(v) \setminus w'$ and each leaf u that contains one or more elements from $\delta^{-1}(l(w), a)$, where w' is a largest child of v. The buckets are filled by iterating over the states in $\delta^{-1}(l(w), a)$ for all w. Then, a pivot is set for each leaf u such that exactly the states that have been placed in a bucket can be moved right of the pivot (and untouched states in $\delta^{-1}(l(w'), a)$ end up left of the pivot). For each leaf u, we iterate over the states in its buckets and the corresponding indices right of its pivot, and we swap the current state with the one that is at the current index. For each bucket a new leaf node is created. The refinable partition is updated such that the current state points to the most recently created leaf. This way, we assure constant time lookup of the leaf for a state, and we can update the array in constant time when we move elements out of a leaf.
Largest child. For finding the largest child, we maintain counts for the temporary children and a current biggest one. On finalizing the temporary children we store (a reference to) the biggest child in the node, so that we can skip this node later in the algorithm.
Storing sequences. The operation on line 20 is done in constant time by using a linked list. □

5 Application in Conformance Testing

A splitting tree can be used to extract relevant information for two classical test generation methods: a *characterization set* for the W-method and a *separating family* for the HSI-method. For an introduction and comparison of FSM-based test generation methods we refer to [2].

Definition 17. *A set $W \subset I^*$ is called a* characterization set *if for every pair of inequivalent states s, t there is a sequence $w \in W$ such that $\lambda(s, w) \neq \lambda(t, w)$.*

Lemma 18. *Let T be a complete splitting tree, then $\{\sigma(u)|u \in T\}$ is a characterization set.*

Proof. Let $W = \{\sigma(u)|u \in T\}$. Let $s, t \in S$ be inequivalent states, then by completeness s and t are contained in different leaves of T. Hence $u = lca(s, t)$ exists and $\sigma(u)$ separates s and t. Furthermore $\sigma(u) \in W$. This shows that W is a characterisation set. □

Lemma 19. *A characterization set with minimal length sequences can be constructed in time $\mathcal{O}(m \log n)$.*

Proof. By Lemma 18 the sequences associated with the inner nodes of a splitting tree form a characterization set. By Theorem 16, such a tree can be constructed in time $\mathcal{O}(m \log n)$. Traversing the tree to obtain the characterization set is linear in the number of nodes (and hence linear in the number of states). □

Definition 20. *A collection of sets $\{H_s\}_{s \in S}$ is called a* separating family *if for every pair of inequivalent states s, t there is a sequence h such that $\lambda(s, h) \neq \lambda(t, h)$ and h is a prefix of some $h_s \in H_s$ and some $h_t \in H_t$.*

Lemma 21. *Let T be a complete splitting tree, the sets $\{\sigma(u)|s \in l(u), u \in T\}_{s \in S}$ form a separating family.*

Proof. Let $H_s = \{\sigma(u)|s \in l(u)\}$. Let $s, t \in S$ be inequivalent states, then by completeness s and t are contained in different leaves of T. Hence $u = lca(s, t)$ exists. Since both s and t are contained in $l(u)$, the separating sequence $\sigma(u)$ is contained in both sets H_s and H_t. Therefore, it is a (trivial) prefix of some word $h_s \in H_s$ and some $h_t \in H_t$. Hence $\{H_s\}_{s \in S}$ is a separating family. □

Lemma 22. *A separating family with minimal length sequences can be constructed in time $\mathcal{O}(m \log n + n^2)$.*

Proof. The separating family can be constructed from the splitting tree by collecting all sequences of all parents of a state (by Lemma 21). Since we have to do this for every state, this takes $\mathcal{O}(n^2)$ time. □

For test generation one moreover needs a transition cover. This can be constructed in linear time with a breadth first search. We conclude that we can construct all necessary information for the W-method in time $\mathcal{O}(m \log n)$ as opposed the $\mathcal{O}(mn)$ algorithm used in [2]. Furthermore, we conclude that we can construct all the necessary information for the HSI-method in time $\mathcal{O}(m \log n + n^2)$, improving on the reported bound $\mathcal{O}(mn^3)$ in [5]. The original HSI-method was formulated differently and might generate smaller sets. We conjecture that our separating family has the same size if we furthermore remove redundant prefixes. This can be done in $\mathcal{O}(n^2)$ time using a trie data structure.

6   Experimental Results

We have implemented Algorithms 3 and 4 in Go, and we have compared their running time on two sets of FSMs.[1] The first set is from [10], where FSMs for embedded control software were automatically constructed. These FSMs are of increasing size, varying from 546 to 3 410 states, with 78 inputs and up to 151 outputs. The second set is inferred from [6], where two classes of finite automata, A and B, are described that serve as a worst case for Algorithms 3 and 4 respectively. The FSMs that we have constructed for these automata have 1 input, 2 outputs, and $2^2 - 2^{15}$ states. The running times in seconds on an Intel Core i5-2500 are plotted in Fig. 3. We note that different slopes imply different complexity classes, since both axes have a logarithmic scale.

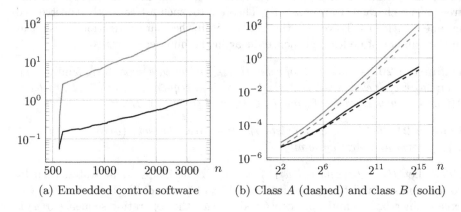

(a) Embedded control software   (b) Class A (dashed) and class B (solid)

Fig. 3. Running time in seconds of Algorithms 3 (gray) and 4 (black)

7   Conclusion

In this paper we have described an efficient algorithm for constructing a set of minimal-length sequences that pairwise distinguish all states of a finite state machine. By extending Hopcroft's minimization algorithm, we are able to construct such sequences in $\mathcal{O}(m \log n)$ for a machine with m transitions and n states. This improves on the traditional $\mathcal{O}(mn)$ method that is based on the

---

[1] Available at https://gitlab.science.ru.nl/rick/partition/.

classic algorithm by Moore. As an upshot, the sequences obtained form a characterization set and a separating family, which play a crucial in conformance testing.

Two key observations were required for a correct adaptation of Hopcroft's algorithm. First, it is required to perform splits in order of the length of their associated sequences. This guarantees minimality of the obtained separating sequences. Second, it is required to consider nodes as a candidate before any one of its children are considered as a candidate. This order follows naturally from the construction of a splitting tree.

Experimental results show that our algorithm outperforms the classic approach for both worst-case finite state machines and models of embedded control software. Applications of minimal separating sequences such as the ones occurring in [2,10] therefore show that our algorithm is useful in practice.

References

1. Bonchi, F., Pous, D.: Checking NFA equivalence with bisimulations up to congruence. In: POPL, pp. 457–468 (2013)
2. Dorofeeva, R., El-Fakih, K., Maag, S., Cavalli, A., Yevtushenko, N.: FSM-based conformance testing methods: a survey annotated with experimental evaluation. Inf. Softw. Technol. **52**(12), 1286–1297 (2010)
3. Gill, A.: Introduction to the Theory of Finite-state Machines. McGraw-Hill, New York (1962)
4. Gries, D.: Describing an algorithm by Hopcroft. Acta Informatica **2**(2), 97–109 (1973)
5. Hierons, R.M., Türker, U.C.: Incomplete distinguishing sequences for finite state machines. Comput. J. **58**, 1–25 (2015)
6. Hopcroft, J.E.: An n log n algorithm for minimizing states in a finite automaton. In: Theory of Machines and Computations, pp. 189–196 (1971)
7. Knuutila, T.: Re-describing an algorithm by Hopcroft. Theoret. Comput. Sci. **250**(1–2), 333–363 (2001)
8. Lee, D., Yannakakis, M.: Testing finite-state machines: state identification and verification. Computers **43**(3), 306–320 (1994)
9. Moore, E.F.: Gedanken-experiments on sequential machines. Automata Stud. **34**, 129–153 (1956)
10. Smeenk, W., Moerman, J., Vaandrager, F., Jansen, D.N.: Applying automata learning to embedded control software. In: Butler, M., et al. (eds.) ICFEM 2015. LNCS, vol. 9407, pp. 67–83. Springer, Heidelberg (2015). doi:10.1007/978-3-319-25423-4_5
11. Valmari, A., Lehtinen, P.: Efficient minimization of DFAs with partial transition functions. In: STACS, pp. 645–656 (2008)

Forkable Regular Expressions

Martin Sulzmann[1]([⊠]) and Peter Thiemann[2]

[1] Faculty of Computer Science and Business Information Systems, Karlsruhe University of Applied Sciences, Moltkestrasse 30, 76133 Karlsruhe, Germany
martin.sulzmann@hs-karlsruhe.de
[2] Faculty of Engineering, University of Freiburg, Georges-Köhler-Allee 079, 79110 Freiburg, Germany
thiemann@acm.org

Abstract. We consider *forkable regular expressions*, which enrich regular expressions with a fork operator, to establish a formal basis for static and dynamic analysis of the communication behavior of concurrent programs. We define a novel compositional semantics for forkable expressions, establish their fundamental properties, and define derivatives for them as a basis for the generation of automata, for matching, and for language containment tests.

Forkable expressions may give rise to non-regular languages, in general, but we identify sufficient conditions on expressions that guarantee finiteness of the automata construction via derivatives.

Keywords: Automata and logic · Forkable expressions · Derivatives

1 Introduction

Languages like Concurrent ML and Go come with built-in support for fine-grained concurrency, dynamic thread creation, and channel-based communication. Analyzing the communication behavior of programs in these languages may be done by an *effect system*. Such a system computes an abstraction of the sequences of events (i.e., the communication traces with events like communication actions or synchronizations) that a program may exhibit.

Effect systems for concurrent programs have been explored by the Nielsons [12], who proposed to model event traces with "behaviors" which are regular expressions extended with a fork operator that encapsulates the behavior of a newly created thread. While their work enables the analysis of finiteness properties of the communication topology, it stops short of providing a semantics of behaviors in terms of effect traces. Subsequent work by the same authors [1,2,11] concentrates on subtyping and automatic inference of effects.

We take up the Nielsons' notion of behavior and tackle the problem of defining a compositional semantics for behaviors in terms of effect traces. Our novel definition yields a semantic basis for the static and dynamic analysis of concurrent languages with dynamic thread creation and other communication effects.

© Springer International Publishing Switzerland 2016
A.-H. Dediu et al. (Eds.): LATA 2016, LNCS 9618, pp. 194–206, 2016.
DOI: 10.1007/978-3-319-30000-9_15

We show that, in general, a behavior may give rise to a *non-regular* trace language. This observation is in line with previous work on concurrent regular expressions [4], flow expressions [13], and shuffle expressions [6] all of which augment regular expressions with (at least) shuffle and shuffle closure operators. The shuffle closure is also referred to as the iterated shuffle [7].

We explore two application areas for forkable expressions. In a run-time verification setting (aka dynamic analysis), we are interested in matching traces against behaviors, either at run time or post-mortem. To this end, we extend Brzozowski derivatives to forkable expressions. Although Brzozowski's construction no longer gives rise to a finite automaton, in general, derivatives can still be used to solve instances of the word problem (which is hence decidable).

For static analysis, we are interested in approximation and testing language containment in a specification. For this use case, we give a decidable criterium that guarantees finiteness of (our extension of) Brzozowski's automaton construction. This criterium essentially requires a finite communication topology, that is, it forbids that new communicating threads are created in loops. We conjecture that this property can be established with the Nielsons' original analysis [12].

In summary our contributions are:

- In Sect. 3, we define a novel trace semantics of behaviors (i.e., regular expressions with fork), and establish their fundamental properties.
- Section 4 extends Brzozowski's derivative operation to behaviors.
- In Sect. 5, we characterize a class of behaviors with regular trace languages. For these behaviors, Brzozowski's construction yields finite automata.

Related work is discussed in Sect. 6.

The online version of this paper contains an appendix with all proofs.[1]

2   Preliminaries

For a set X, we write $\sharp X$ for the cardinality of X and $\wp(X)$ for its powerset. If $F : \wp(X) \to \wp(X)$ is a monotone function (i.e., $X \subseteq Y$ implies $F(X) \subseteq F(Y)$), then we write μF for the least fixpoint of this function, which is uniquely defined due to Tarski's theorem. We write $\mu X.e$ for the fixpoint $\mu(\lambda X.e)$ where e is a set-valued expression composed of monotone functions assuming that X is a set with the same type of elements. (The scope of the μ operator extends as far to the right as possible.) We will employ this operator to define the meaning of forkable Kleene star behaviors [10].

Let Σ be a finite set, the *alphabet of primitive events*. We write Σ^* for the set of finite words over Σ and denote with $v \cdot w$ the concatenation of words $v, w \in \Sigma^*$. For languages $L, M \subseteq \wp(\Sigma^*)$, we write $L \cdot M = \{v \cdot w \mid v \in L, w \in M\}$ for the set of all pairwise concatenations. We write $v \cdot M$ as a shorthand for $\{v\} \cdot M$. The (asynchronous) shuffle operation $v \| w \subseteq \Sigma^*$ on words is the set of all interleavings of words v and w. It is defined inductively by

$$\varepsilon \| w = \{w\} \qquad v \| \varepsilon = \{v\} \qquad xv \| yw = \{x\} \cdot (v \| yw) \cup \{y\} \cdot (xv \| w)$$

---

[1] http://arxiv.org/abs/1510.07293.

$$L(r) = L(r, \{\varepsilon\}) \qquad \begin{aligned} L(\phi, K) &= \emptyset \\ L(\varepsilon, K) &= K \\ L(x, K) &= \{x\} \cdot K \end{aligned} \qquad \begin{aligned} L(r + s, K) &= L(r, K) \cup L(s, K) \\ L(r \cdot s, K) &= L(r, L(s, K)) \\ L(r^*, K) &= \mu X.L(r, X) \cup K \\ L(Fork(r), K) &= L(r) \| K \end{aligned}$$

Fig. 1. Trace language of a behavior

The shuffle operation is lifted to languages by $L \| M = \bigcup \{ v \| w \mid v \in L, w \in M \}$.

We write $L_1 \backslash L_2$ to denote the left quotient of L_2 with L_1 where $L_1 \backslash L_2 = \{ w \mid \exists v \in L_1.v \cdot w \in L_2 \}$. We write $x \backslash L$ as a shorthand for $\{x\} \backslash L$.

3 Behaviors

Recall that Σ is the alphabet of primitive events. Intuitively, a primitive event $x \in \Sigma$ is a globally visible side effect like sending or receiving a message. A *behavior* is a regular expression over Σ extended with a new fork operator.

$$r, s, t ::= \phi \mid \varepsilon \mid x \mid r + s \mid r \cdot s \mid r^* \mid Fork(r) \mid (r)$$

As usual, we assume that $\cdot$ binds tighter than $+$.

The semantics of a behavior r is going to be a *trace language* $L(r) \subseteq \Sigma^*$. However, due to the presence of the fork operator, its definition is not a simple extension of the standard semantics $\llbracket \cdot \rrbracket$ of a regular expression.

Definition 1. *Figure 1 defines, for a behavior r and a continuation language $K \subseteq \Sigma^*$, the trace languages $L(r) \subseteq \Sigma^*$ and $L(r, K) \subseteq \Sigma^*$.*

By induction on r, we can show that the mapping $K \mapsto L(r, K)$ in $\wp(\Sigma^*) \rightarrow \wp(\Sigma^*)$ is monotone, so that L is well-defined. For fork-free behaviors that do not make use of the $Fork(r)$ operator, the trace language is regular and coincides with the standard semantics $\llbracket r \rrbracket$ of a regular expression.

Theorem 2. *If r is fork-free, then $L(r)$ is regular and $L(r) = \llbracket r \rrbracket$.*

It is known that the regular languages are closed under the shuffle operation [5]. However, for forkable expressions the semantics of $Fork(r)$ is defined by *shuffling with the continuation language* so that the language defined by a behavior need not be regular as the following example shows.

Example 3. Consider the behavior $Fork(s)^*$ for a fork-free regular expression s. Its semantics is the shuffle closure of $\llbracket s \rrbracket$ as demonstrated by the following calculation

$$L(Fork(s)^*, \{\varepsilon\}) = \mu X.L(Fork(s), X) \cup \{\varepsilon\} = \mu X.L(s) \| X \cup \{\varepsilon\} \qquad (1)$$
$$= \{\varepsilon\} \cup L(s) \cup L(s) \| L(s) \cup \ldots$$

where we assume that $\|$ binds tighter than $\cup$.

$$
\begin{array}{llll}
\mathcal{C}(\phi) & = \phi & \mathcal{S}(\phi) & = \phi \\
\mathcal{C}(\varepsilon) & = \varepsilon & \mathcal{S}(\varepsilon) & = \phi \\
\mathcal{C}(x) & = \phi & \mathcal{S}(x) & = x \\
\mathcal{C}(r+s) & = \mathcal{C}(r) + \mathcal{C}(s) & \mathcal{S}(r+s) & = \mathcal{S}(r) + \mathcal{S}(s) \\
\mathcal{C}(r \cdot s) & = \mathcal{C}(r) \cdot \mathcal{C}(s) & \mathcal{S}(r \cdot s) & = \mathcal{S}(r) \cdot s + \mathcal{C}(r) \cdot \mathcal{S}(s) \\
\mathcal{C}(r^*) & = \mathcal{C}(r)^* & \mathcal{S}(r^*) & = \mathcal{C}(r)^* \cdot \mathcal{S}(r) \cdot r^* \\
\mathcal{C}(Fork(r)) & = Fork(r) & \mathcal{S}(Fork(r)) & = \phi
\end{array}
$$

Fig. 2. Concurrent and sequential part of a behavior

In general, the shuffle closure is not regular [7] as the following concrete instance shows. Consider the behavior $r = Fork(x \cdot y + y \cdot x)^*$. By the calculation in (1), $L(r)$ is the shuffle closure of $\{x \cdot y, y \cdot x\}$ which happens to be the context-free language $\{w \in \{x, y\}^* \mid \sharp(x, w) = \sharp(y, w)\}$ of words that contain the same number of xs and ys. This language is not regular.

Some of our proofs rely on semantic equivalence and employ identities from Kleene algebra [9] that hold for standard regular expressions. Hence, we need to establish that forkable expressions also form a Kleene algebra.

Definition 4 (Semantic equality and containment)

1. *Behaviors r and s are equal, $r \equiv s$, if $L(r, K) = L(s, K)$, for all K.*
2. *Behaviors r and s are contained, $r \leq s$, if $L(r, K) \subseteq L(s, K)$, for all K.*

Theorem 5. *The set of forkable expressions with semantic equality and containment is a Kleene algebra.*

Each behavior r can be decomposed into a sequential part $\mathcal{S}(r)$ and a concurrent part $\mathcal{C}(r)$, which are defined by induction on r in Fig. 2. The intuition is that the sequential part of a behavior describes what must happen next, inevitably, whereas the concurrent part describes behavior that happens eventually and concurrent to the sequential behavior. For example, in case of concatenation $r \cdot s$, the sequential part must either start with $\mathcal{S}(r)$, or must end with $\mathcal{S}(s)$. For Kleene star r^* it is similar, we simply consider the possible unrolling of the underlying expression r.

Our decomposition theorem proves that every behavior is semantically equivalent to the union of its concurrent part and its sequential part. Its proof requires the Kleene identity $r^* \equiv \varepsilon + r \cdot r^*$.[2]

Theorem 6. *For all r, $r \equiv \mathcal{C}(r) + \mathcal{S}(r)$.*

The next lemma establishes some algebraic properties of the functions $\mathcal{C}()$ and $\mathcal{S}()$ that we need in subsequent proofs.

---

[2] We generally write $r = s$ for *syntactic equality* of expressions and use other symbols like $r \equiv s$ for equivalences where some additional reasoning may be involved.

Lemma 7. *For all r: 1. $\mathcal{C}(\mathcal{C}(r)) = \mathcal{C}(r)$ (syntactic equality); 2. $\mathcal{C}(\mathcal{S}(r)) \equiv \phi$; 3. $\mathcal{S}(\mathcal{C}(r)) \equiv \phi$; 4. $\mathcal{S}(\mathcal{S}(r)) \equiv \mathcal{S}(r)$.*

Proof. The proof for part 1 is by trivial induction on r. See the online version for the remaining parts; they are not needed in the rest of this paper. □

Lemma 8. *For all r, $\varepsilon \in L(r)$ iff $\varepsilon \leq \mathcal{C}(r)$.*

4   Derivatives

We want to use Brzozowski's derivative operation [3] to translate behaviors to automata and to create algorithms for checking language containment and matching. To this end, we extend derivatives to forkable expressions. The derivative of r w.r.t. some symbol x, written $d_x(r)$, yields the new behavior after consumption of the leading symbol x. The derivative operation for behaviors is defined by structural induction. In addition to the regular operators, the derivative needs to deal with $Fork(r)$ expressions and the case of concatenated expressions $r \cdot s$ requires special attention.

Definition 9 (Derivatives). *The derivative of behavior r w.r.t. some symbol x is defined inductively as follows:*

$$d_x(\phi) = \phi \qquad\qquad d_x(r+s) \;= d_x(r) + d_x(s)$$
$$d_x(\varepsilon) = \phi \qquad\qquad d_x(r \cdot s) \;= d_x(r) \cdot s + \mathcal{C}(r) \cdot d_x(s)$$
$$d_x(y) = \begin{cases} \varepsilon & if\ x = y \\ \phi & otherwise \end{cases} \qquad \begin{array}{l} d_x(r^*) \;= d_x(r) \cdot r^* \\ d_x(Fork(r)) = Fork(d_x(r)) \end{array}$$

We just explain the cases that differ from Brzozowski's definition. The derivative of a fork, $Fork(r)$, is simply pushed down to the underlying expression. The derivative of $r \cdot s$ consists of two components. The first one, $d_x(r) \cdot s$, is identical to the standard definition: it computes the derivative of r and continues with s. The second one covers symbols that may reach s. In a fork-free regular expression, a symbol in s can only be consumed if r is nullable, i.e. $\varepsilon \in L(r)$. For forkable behaviors, a symbol in s can also be consumed if r exhibits concurrent behavior. Hence, we extract the concurrent behavior $\mathcal{C}(r)$ and concatenate it with the derivative of s. The concurrent behavior generalizes nullability in the sense that $\varepsilon \in \mathcal{C}(r)$ iff $\varepsilon \in L(r)$. See Lemma 8.

Next, we verify that the derivative operation is correct in the sense that the resulting expression $d_x(r)$ denotes the left quotient of r by x.

Theorem 10 (Left Quotients). *Let r be a behavior and x be a symbol. Then, we have that $L(d_x(r)) = x \backslash L(r)$.*

Proof. We need to expand the definition of $L(r) = L(r, \{\varepsilon\})$ and generalize the statement to an arbitrary continuation language $K \subseteq \Sigma^*$:

$$\forall r. \forall K. \; L(d_x(r), K) \cup L(\mathcal{C}(r)) \| (x \backslash K) = x \backslash L(r, K) \qquad (2)$$

Recall that $L\|\emptyset = \emptyset$ so that the original statement follows by setting $K = \{\varepsilon\}$:

$$L(d_x(r)) = L(d_x(r)) \cup L(\mathcal{C}(r))\|\emptyset$$
$$= L(d_x(r), \{\varepsilon\}) \cup L(\mathcal{C}(r))\|(x\backslash\{\varepsilon\})$$
$$\overset{(2)}{=} x\backslash L(r, \{\varepsilon\}) = x\backslash L(r)$$

The proof of (2) proceeds by induction on r.   □

Like in the standard regular expression case, we can conclude (based on the above result) that each behavior can be represented as a sum of its derivatives.

Theorem 11 (Representation). *For any behavior r, we have $L(r) = (\varepsilon \in L(r) \implies \{\varepsilon\}) \cup \bigcup_{x \in \Sigma} x \cdot L(d_x(r))$.*

Expression $(\varepsilon \in L(r) \implies \{\varepsilon\})$ denotes $\{\}$ if $\varepsilon \in L(r)$, otherwise, $\{\}$.

The representation theorem is the basis for solving the word problem with derivatives. Here, we extend the derivative operation to words as usual by $d_\varepsilon(r) = r$ and $d_{aw}(r) = d_w(d_a(r))$.

Corollary 12. *For a behavior r and $w \in \Sigma^*$, $w \in L(r)$ iff $\varepsilon \in L(d_w(r))$.*

This corollary implies decidability of the word problem for forkable expressions: the derivative is computable and the nullability test $\varepsilon \in L(d_w(r))$ is a syntactic test as for standard regular expressions. Full details how to compute all dissimilar derivatives can be found in the online version.

To construct an automaton from an expression r, Brzozowski repeatedly takes the derivative with respect to all symbols $x \in \Sigma$. We call these derivatives *descendants*.

Definition 13 (Descendants). *A descendant s of a behavior r is either r itself, a derivative of r, or the derivative of a descendant. We write $s \sqsubset r$, if s is a direct descendant of r, that is, if $s = d_x(r)$, for some x. The "is descendant of" relation is the reflexive, transitive closure of the direct descendant relation: $s \preceq r = s \sqsubset^* r$. The "is a true descendant of" relation is the transitive closure of the direct descendant relation: $s \prec r = s \sqsubset^+ r$. We define $d(r) = \{s \mid s \preceq r\}$ as the set of descendants of r.*

For standard regular expressions, Brzozowski showed that the set of descendants of an expression is finite up to *similarity*. Two expressions are similar if they are equal modulo associativity, commutativity, and idempotence. This result no longer holds in our setting.

In the following, we write $r \overset{x}{\longrightarrow} s$ if $s = d_x(r)$. Subterms on which the derivation operation is applied are underlined.

Example 14. Let $r = (Fork(x \cdot y))^*$ and take the derivative by x repeatedly.

$$\underline{(Fork(x \cdot y))^*}$$
$$\overset{x}{\longrightarrow} Fork(\underline{y}) \cdot r$$
$$\overset{x}{\longrightarrow} Fork(\phi) \cdot r + Fork(y) \cdot Fork(y) \cdot r$$
$$\overset{x}{\longrightarrow} \cdots + Fork(\phi) \cdot Fork(y) \cdot r + Fork(y) \cdot (Fork(\phi) \cdot r + Fork(y) \cdot Fork(y) \cdot r)$$
$$\overset{x}{\longrightarrow} \cdots$$

$$(\text{Refl, Trans, Sym, Comp}) \quad r \approx r \qquad \frac{r \approx s \quad s \approx t}{r \approx t} \qquad \frac{s \approx t}{t \approx s} \qquad \frac{s \approx t}{E[s] \approx E[t]}$$

$$(\text{Assoc, Comm}) \qquad r + (s + t) \approx (r + s) + t \qquad r + s \approx s + r$$

$$(\text{Idem, Unit}) \qquad r + r \approx r \qquad r + \phi \approx r \qquad \phi + r \approx r$$

$$(\text{Empty Word}) \qquad \varepsilon \cdot r \approx r \qquad r \cdot \varepsilon \approx r \qquad \varepsilon^* \approx \varepsilon \qquad Fork(\varepsilon) \approx \varepsilon$$

$$(\text{Empty Language}) \qquad \phi \cdot r \approx \phi \qquad r \cdot \phi \approx \phi \qquad \phi^* \approx \varepsilon \qquad Fork(\phi) \approx \phi$$

$$(\text{Regular Contexts}) \qquad E ::= [] \mid E^* \mid E \cdot s \mid r \cdot E \mid E + s \mid r + E \mid Fork(E)$$

Fig. 3. Rules and axioms for similarity

Here we omit parentheses (assuming associativity) and apply equivalences such as $\mathcal{C}(\mathcal{C}(r)) = \mathcal{C}(r)$ (Lemma 7). Clearly, we obtain an increasing sequence of behaviors of the form $Fork(y) \cdot \ldots \cdot Fork(y) \cdot r$. Hence, the set of descendants of r is infinite even if we consider behaviors equal modulo associativity, commutativity, and idempotence of alternatives.

This observation is no surprise, given that behaviors may give rise to non-regular languages (cf. Example 3). In general, there is no hope to retain Brzozowski's result, but it turns out that we can find a well-behavedness condition for behaviors that is sufficient to retain finiteness of descendants.

5   Well-Behaved Behaviors

In this section, we develop a criterion to guarantee that a forkable expression only gives rise to a finite set of dissimilar descendants. To start with, we adapt Brzozowski's notion of similarity to our setting. In addition to associativity, commutativity, and idempotence we introduce simplification rules that implement further Kleene identities and that deal with forks.

Definition 15 (Similarity). *Behaviors r and s are similar, if $r \approx s$ is derivable using the rules and axioms in Fig. 3.*

The compatibility rule (Comp) uses regular contexts E, which are regular expressions with a single hole $[]$. In the rule, we write $E[t]$ to denote the expression with the hole replaced by t.

We establish some basic results for similar behaviors, all with straightforward inductive proofs: Similarity implies semantic equivalence, it is complete for recognizing ε and ϕ, and it is compatible with derivatives and extraction of concurrent parts.

Lemma 16. *If $r \approx s$, then $r \equiv s$.*

Lemma 17. *1. If $L(r) = \{\varepsilon\}$ then $r \approx \varepsilon$. 2. If $L(r) = \{\}$ then $r \approx \phi$.*

Lemma 18. *1. If $r \approx r'$, then $d_x(r) \approx d_x(r')$, for all $x \in \Sigma$.
2. If $r \approx r'$, then $C(r) \approx C(r')$.*

Similarity is an equivalence relation. We write $[s] = \{t \mid t \approx s\}$ to denote the equivalence class of all expressions similar to s. If R is a set of behaviors, we write $R/\approx = \{[r] \mid r \in R\}$ for the set of equivalence classes of elements of R.

To identify the set of well-behaved behaviors, we need to characterize the set of dissimilar descendants. First, we establish that each composition of derivatives and applications of $C()$ that finishes in some $C(r)$ may be compressed to the composition of the derivatives applied to the remaining $C(r)$.

Lemma 19. *For a behavior r and symbol x, $C(d_x(C(r))) = d_x(C(r))$, syntactically.*

The above result makes it easier to classify the forms of dissimilar descendants.

The Kleene star case is clearly highly relevant. The following statement confirms the observation in Example 14.

Lemma 20. *For $w \in \Sigma^+$, $d_w(r^*) \approx d_w(r) \cdot r^* + t$ where t is a possibly empty sum of terms of the form $\quad s_1 \cdot \ldots \cdot s_n \cdot r' \cdot r^* \quad$ where $r' \prec r$, $n \geq 1$, and for each s_i, $s_i \preceq C(s)$ for some descendant $s \prec r$.*

Proof. Induction on w.
    Case x: $d_x(r^*) = d_x(r) \cdot r^* \approx d_x(r) \cdot r^* + \phi$.
    Case wx: $d_{wx}(r^*) = d_x(d_w(r^*))$. By induction for w, $d_w(r^*) \approx d_w(r) \cdot r^* + t$ where each summand of t has the form $s_1 \cdot \ldots \cdot s_n \cdot r' \cdot r^*$. First, observe that $d_x(d_w(r) \cdot r^* + t) = d_{wx}(r) \cdot r^* + C(d_w(r)) \cdot d_x(r) \cdot r^* + d_x(t)$. We show by auxiliary induction on n that the derivative of t is a sum of terms of the desired form.
    Case 0:
$$d_x(r' \cdot r^*) = d_x(r') \cdot r^* + C(r') \cdot d_x(r^*)$$
$$= d_x(r') \cdot r^* + C(r') \cdot d_x(r) \cdot r^*$$

    which has the desired format.
    Case $n > 0$: $d_x(s_1 \cdot \ldots \cdot s_n \cdot r' \cdot r^*)$
$$= d_x(s_1) \cdot s_2 \cdot \ldots \cdot s_n \cdot r' \cdot r^* + C(s_1) \cdot d_x(s_2 \cdot \ldots \cdot s_n \cdot r' \cdot r^*)$$
The first summand has the desired form. By induction (on n), each summand of $d_x(s_2 \cdot \ldots \cdot s_n \cdot r' \cdot r^*)$ has the desired form and multiplying with $C(s_1)$ from the left retains this form: By Lemma 19, $C(s_1)$ is still a descendant of $C(s)$ for some descendant s of r.   □

To obtain finiteness it appears that a sufficient condition is to ensure that the subterms s_i are trivial (either ε or ϕ). Via similarity, the explosion of terms derived from Kleene star can then be avoided. To verify this claim we also characterize the descendants of concatenated behaviors.

Lemma 21. *For* $w \in \Sigma^+$, $d_w(r \cdot s)$ *has the form*

$$d_w(r \cdot s) \approx d_w(r) \cdot s + \mathcal{C}(r) \cdot d_w(s) + t$$

where t *is a sum of terms of the form* $r' \cdot s'$ *where* s' *is a descendant of* s *and* r' *is a descendant of* $\mathcal{C}(r'')$ *and* r'' *is a descendant of* r.

Proof. By induction on w.

   Case x: Immediate from the definition of $d_x(r \cdot s)$ with $t = \phi$.
   Case wx: By induction

$$\begin{aligned}
d_x(d_w(r \cdot s)) &\approx d_x(d_w(r) \cdot s + \mathcal{C}(r) \cdot d_w(s) + t) \\
&\approx \underline{d_{wx}(r) \cdot s} + \mathcal{C}(d_w(r)) \cdot d_x(s) \\
&\quad + d_x(\mathcal{C}(r)) \cdot d_w(s) + \underline{\mathcal{C}(r) \cdot d_{wx}(s)} \\
&\quad + d_x(t)
\end{aligned}$$

The underlined summands have the expected forms. The newly created summands have a form corresponding to $r' \cdot s'$. It remains to observe that the derivative of a summand in t has the expected form by Lemmas 7 and 19.

$$d_x(r' \cdot s') = d_x(r') \cdot s' + \mathcal{C}(r') \cdot d_x(s') \qquad \square$$

Definition 22 (Well-behaved Behaviors). *A* behavior t *is* well-behaved *if all subterms of the form* r^* *have the property that* $\mathcal{C}(d_w(r)) \leqq \varepsilon$, *for all* $w \in \Sigma^*$.

   The intuition for this definition is simple: Well-behaved behaviors do not fork processes with non-trivial communication behavior in a loop (i.e., under a star). Indeed, we have a simple decidable sufficient condition for well-behavedness.

Lemma 23. *If* $r \approx r'$ *and* r' *is fork-free, then* $\mathcal{C}(d_w(r)) \leqq \varepsilon$, *for all* $w \in \Sigma^*$.

Thus, a behavior is also well-behaved if, for all subterms of the form r^*, r is similar to a fork-free expression.

   Recall that $d(r)$ is the set of all descendants of r and $d(r)/(\approx)$ denotes the set of equivalence classes of descendants of r. If we pick a representative from each of these equivalence classes, we obtain the dissimilar descendants of r. In a practical implementation, we may want to compute the *canonical* representative of each equivalence class. See the online version for further details. For the purpose of this paper, an *arbitrary* representative is sufficient.

Definition 24 (Dissimilar Descendants). *We define the set of* dissimilar descendants of r, $d_{\approx}(r)$, *as a complete set of arbitrarily chosen representative behaviors for the equivalence classes* $d(r)/(\approx)$.

   We extend the function $\mathcal{C}()$ on behaviors pointwise to sets of behaviors and relation $\approx$ to sets of behaviors by

$$R \approx S \text{ iff } (\forall r \in R. \exists s \in S. r \approx s) \wedge (\forall s \in S. \exists r \in R. r \approx s)$$

Lemma 25. *For any behavior r, $\mathcal{C}(d_{\approx}(\mathcal{C}(d_{\approx}(r)))) \approx d_{\approx}(\mathcal{C}(d_{\approx}(r)))$.*

Proof. Follows from Lemmas 18 and 19.   □

Lemma 26. *For any behavior r, $d(r) \approx d_{\approx}(r)$.*

Proof. We need to verify that for each $t_1 \in d(r)$ there exists $t_2 \in d_{\approx}(r)$ such that $t_1 \approx t_2$. We prove this property by induction on the number of derivative steps.

  Case $w = \varepsilon$: Then, $t_1 = r$. Clearly, there exists $t_2 \in d_{\approx}(r)$ such that $t_1 \approx t_2$.

  Case $w = x \cdot w'$: Then, $t_1 = d_x(d_{w'}(r))$. By the IH, $d_{w'}(r) \approx t_2$ where $t_2 \in d_{\approx}(r)$. By Lemma 18, $t_1 \equiv d_x(t_2)$ where $d_x(t_2) \approx t_3$ for some $t_3 \in d_{\approx}(r)$. Thus, we are done.   □

The next result can be verified via similar reasoning.

Lemma 27. *For any behavior r, $d(d_{\approx}(r)) \approx d_{\approx}(r)$.*

Theorem 28 (Finiteness of Well-Behaved Dissimilar Descendants).
Let t be a well-behaved behavior. Then, $\sharp d_{\approx}(t) < \infty$.

Proof. We need to generalize the statement to obtain the result: If t is well-behaved then $\sharp d_t^i < \infty$ for all $i \geq 0$ where $d_t^0 = d_{\approx}(t)$ and $d_t^{n+1} = d_{\approx}(\mathcal{C}(d_t^n))$.

  Based on Lemmas 25 and 27 we find that $d_t^{n+1} = d_t^n$ for $n \geq 1$. That is, in the induction step it is sufficient to establish that d_t^0 and d_t^1 are finite.

  We proceed by induction on t. For brevity, we only consider the case of concatenation.

Case $r \cdot s$: By the IH, d_r^i and d_s^i are finite for any $i \geq 0$. We first show that $d_{r \cdot s}^0$ is finite.

1. By Lemma 21, the elements of $d(r \cdot s)$ are drawn from the set

$$d(r) \cdot s + \mathcal{C}(r) \cdot d(s) + \sum d(\mathcal{C}(d(r))) \cdot d(s)$$

2. By Lemma 26 the above is similar to

$$d_{\approx}(r) \cdot s + \mathcal{C}(r) \cdot d_{\approx}(s) + \sum d_{\approx}(\mathcal{C}(d_{\approx}(r))) \cdot d_{\approx}(s)$$

3. Immediately, we can conclude that $d_{r \cdot s}^0$ is finite.

Next, we consider $d_{r \cdot s}^1$.

1. From above, the (dissimilar) descendants of $r \cdot s$ are drawn from

$$\underbrace{d(r) \cdot s}_{t_1} + \underbrace{\mathcal{C}(r) \cdot d(s)}_{t_2} + \sum \underbrace{d(\mathcal{C}(d(r))) \cdot d_{\approx}(s)}_{t_3}$$

For each t_i we will show that $d_{t_i}^1$ is finite and thus follows the desired result.

2. By Lemma 21, descendants of $\mathcal{C}(t_1)$ are of the form

$$d(\mathcal{C}(d(r))) \cdot \mathcal{C}(s) + \mathcal{C}(d(r)) \cdot d(\mathcal{C}(s)) + \sum d(\mathcal{C}(d(\mathcal{C}(d(r))))) \cdot d(\mathcal{C}(s))$$

3. As we know by the IH, d_r^i and d_s^i are finite for any $i \geq 0$. Hence, via similar reasoning as above we can conclude that the above is similar to expressions of the form

$$d_r^1 \cdot \mathcal{C}(s) + \mathcal{C}(d_r^0) \cdot d_s^1 + \sum d_r^2 \cdot d_s^1$$

As all sub-components are finite, we conclude that $d_{t_1}^1$ is finite.

4. We consider the descendants of $\mathcal{C}(t_2)$ which are of the following form

$$d(\mathcal{C}(r)) \cdot \mathcal{C}(d(s)) + \mathcal{C}(r) \cdot d(\mathcal{C}(d(s))) + \sum d(\mathcal{C}(d(\mathcal{C}(r)))) \cdot d(\mathcal{C}(d(s)))$$

5. The above is similar to

$$d_r^1 \cdot \mathcal{C}(d_s^0) + \mathcal{C}(r) \cdot d_r^1 + \sum d_r^2 \cdot d_s^1$$

Thus, we find that $d_{t_2}^1$ is finite.

6. Finally, we observe that shape of descendants of $\mathcal{C}(t_3)$

$$d(\mathcal{C}(d(\mathcal{C}(d(r))))) \cdot \mathcal{C}(d(s)) + \mathcal{C}(d(\mathcal{C}(d(r)))) \cdot d(\mathcal{C}(d(s)))$$
$$+ \sum d(\mathcal{C}(d(\mathcal{C}(d(\mathcal{C}(d(r))))))) \cdot d(\mathcal{C}(d(s)))$$

7. The above is similar to

$$d_r^2 \cdot \mathcal{C}(d_s^0) + \mathcal{C}(d_r^1) \cdot d_s^1 + \sum d_r^3 \cdot d_s^1$$

8. Then, $d_{t_3}^1$ is finite which concludes the proof for this case.

$$\square$$

The result no longer holds if we replace the assumption $\mathcal{C}(d_w(r)) \leqq \varepsilon$, for $w \in \Sigma^*$ by a simpler assumption like $\mathcal{C}(r) \leqq \varepsilon$. For example, consider the behavior $(x \cdot Fork(y))^*$ where $\mathcal{C}(x \cdot Fork(y)) = \phi \leqq \varepsilon$. However, the set of dissimilar descendants of $(x \cdot Fork(y))^*$ is infinite as shown by the calculation

$$(x \cdot Fork(y))^*$$
$$\xrightarrow{x} (\varepsilon \cdot Fork(y)) \cdot (x \cdot Fork(y))^*$$
$$\approx Fork(y) \cdot (x \cdot Fork(y))^*$$
$$\xrightarrow{x} Fork(\phi) \cdot (x \cdot Fork(y))^* + Fork(y) \cdot Fork(y) \cdot (x \cdot Fork(y))^*$$
$$\approx Fork(y) \cdot Fork(y) \cdot (x \cdot Fork(y))^*$$
$$\cdots$$

The example also shows that the assumption $\mathcal{C}(d_w(r)) \leqq \varepsilon$, for $w \in \Sigma^*$ is necessary and cannot be weakened to words w of a fixed length.

As an example, consider the behavior $t = (x_1 \cdot \ldots \cdot x_n \cdot x_{n+1} Fork(y))^*$ where for all $w \in \Sigma^*$ with length less or equal n we find that $\mathcal{C}(d_w(x_1 \cdot \ldots \cdot x_n \cdot x_{n+1} Fork(y))) \leqq \varepsilon$. Via a similar calculation as above, we can show that the set of dissimilar descendants of t is infinite.

6 Related Work

Shuffle expressions are regular expressions with operators for shuffle and shuffle closure. Shaw [13] proposes to describe the behavior of software using flow expressions, which extend shuffle expressions with further operators. Gischer [6] shows that shuffle expression generate context-sensitive languages and proposes a connection to Petri net languages.

The latter connection is made precise by Garg and Ragunath [4], who study concurrent regular expressions (CRE), which are shuffle expressions extended with synchronous composition. They show that the class of CRE languages is equal to the class of Petri net languages. The proof requires the presence of synchronous composition. Forkable expressions do not support synchronous composition, but they are equivalent to *unit expressions*, which are also defined by Garg and Ragunath and shown to be strictly less powerful than CREs.

Warmuth and Haussler [14] present more refined complexity results for the languages generated by shuffle expressions. Jedrzejowicz [8] shows that the nesting of iterated closure operators matters.

Acknowledgments. We thank the reviewers for their comments.

References

1. Amtoft, T., Nielson, F., Nielson, H.R.: Type and behaviour reconstruction for higher-order concurrent programs. J. Funct. Programm. **7**(3), 321–347 (1997)
2. Amtoft, T., Nielson, H.R., Nielson, F.: Behavior analysis for validating communication patterns. STTT **2**(1), 13–28 (1998)
3. Brzozowski, J.A.: Derivatives of regular expressions. J. ACM **11**(4), 481–494 (1964)
4. Garg, V.K., Ragunath, M.T.: Concurrent regular expressions and their relationship to Petri nets. Theor. Comput. Sci. **96**(2), 285–304 (1992)
5. Ginsburg, S.: The Mathematical Theory of Context-Free Languages. McGraw-Hill Inc., New York (1966)
6. Gischer, J.: Shuffle languages, Petri nets, and context-sensitive grammars. Commun. ACM **24**(9), 597–605 (1981)
7. Jantzen, M.: Extending regular expressions with iterated shuffle. Theoret. Comput. Sci. **38**, 223–247 (1985)
8. Jedrzejowicz, J.: Nesting of shuffle closure is important. Inf. Process. Lett. **25**(6), 363–368 (1987)
9. Kozen, D.: On Kleene algebras and closed semirings. In: Rovan, B. (ed.) MFCS '90. LNCS, vol. 452, pp. 26–47. Springer, London (1990)
10. Leiß, H.: Towards Kleene algebra with recursion. In: Kleine Büning, H., Jäger, G., Börger, E., Richter, M.M. (eds.) CSL 1991. LNCS, vol. 626, pp. 242–256. Springer, Heidelberg (1992)
11. Riis Nielson, H., Amtoft, T., Nielson, F.: Behaviour analysis and safety conditions: a case study in CML. In: Astesiano, E. (ed.) ETAPS 1998 and FASE 1998. LNCS, vol. 1382, pp. 255–269. Springer, Heidelberg (1998)

12. Nielson, H.R., Nielson, F.: Higher-order concurrent programs with finite communication topology. In: Proceedings of POPL 1994, pp. 84–97. ACM Press, January 1994
13. Shaw, A.C.: Software descriptions with flow expressions. IEEE Trans. Softw. Eng. **4**(3), 242–254 (1978)
14. Warmuth, M.K., Haussler, D.: On the complexity of iterated shuffle. J. Comput. Syst. Sci. **28**(3), 345–358 (1984)

On the Levenshtein Automaton and the Size of the Neighbourhood of a Word

Hélène Touzet[✉]

CRIStAL (UMR CNRS 9189 University of Lille) and Inria,
Villeneuve-d'Ascq, France
helene.touzet@univ-lille1.fr

Abstract. Given a word P and a maximal number of errors k, we address the problem of counting the number of strings whose Levenshtein distance to P does not exceed k. We give an algorithm that scales linearly with the size of P and that is based on a variant of the classical Levenshtein automaton.

1 Introduction

The problem of measuring the similarity between two strings arises in many areas such as computational molecular biology, natural language processing, spelling correction, plagiarism detection, music information retrieval. A common metric for it is the *Levenshtein distance*, also simply called the Edit Distance. This distance is defined as the smallest number of substitutions, insertions, and deletions of symbols required to transform one of the words into the other.

In this paper, we investigate the basic problem of the size of the neighbourhood of a given string: count how many strings are within a bounded distance of a fixed reference string. This problem has been exposed in [2,6], among others. As far as we know, there is no efficient algorithm for solving it. We propose a dynamic programming algorithm that runs linearly in the length of the pattern. This algorithm heavily relies upon the *Universal Levenshtein Automaton*, which is an advanced automaton for the Levenshtein distance problem introduced in [4,5]. This automaton is *universal* in the sense that it does not depend on the two strings that have to be compared.

The paper is organized as follows. In Sect. 2, we present the Universal Levenshtein Automaton. We review its main principles, and revisit them by proposing some new features, such as the introduction of a *Nondeterministic Universal Levenshtein Automaton*. In Sect. 3, we present the algorithm to compute the cardinality of the neighbourhood of a word.

2 Definition and Construction of the Deterministic Universal Levenshtein Automaton

2.1 Preliminaries and Notations

Let Σ be a finite alphabet, and let P and V be two strings over Σ. The *Levenshtein distance* between P and V, denoted $Lev(P,V)$, is the smallest

© Springer International Publishing Switzerland 2016
A.-H. Dediu et al. (Eds.): LATA 2016, LNCS 9618, pp. 207–218, 2016.
DOI: 10.1007/978-3-319-30000-9_16

number of edit operations needed to transform P into V, where the allowed operations are *substitution* of one symbol with another, *deletion* of a symbol and *insertion* of a symbol. A sequence of operations transforming P into V is called an *edit script*. For any pair of strings P and V, the distance between P and V can be computed by dynamic programming in time proportional to the product of the length of P and V with the Wagner-Fischer algorithm [9].

Assume now that we are given a fixed threshold for the number of errors k. We consider the decision problem associated to the Levenshtein distance: for any pair of strings P and V, decide whether the Levenshtein distance is lesser than or equal to k. In this decision problem, we can also introduce the additional hypothesis that the pattern P is fixed, and consider the *neighbourhood* of P, noted $\mathcal{L}ev(P, k)$: $\mathcal{L}ev(P, k) = \{V \in \Sigma^*; \ Lev(P, V) \le k\}$.

When P is fixed, it is highly desirable to preprocess P, so that the decision problem can be solved more efficiently than with the dynamic programming algorithm. There has been an abundance of literature on this subject. A standard approach is to start from the *Levenshtein automaton*, depicted in Fig. 1, which recognizes the set of strings which are at most at distance k to P. This Levenshtein automaton is a Nondeterministic Finite Automaton (NFA). A run can be simulated using dynamic programming or bit parallelism [1,3]. However, it is of little help to count the number of accepted strings, because distinct paths in this NFA can recognize the same string. Another possibility is to transform the Levenshtein automaton into an equivalent Deterministic Finite Automaton (DFA) [7,8]. This is a tedious task, that should be performed for each target string P.

In [4,5], a universal Levenshtein automaton was introduced. The term *universal* conveys its one-time construction and independency of the two input strings. Thus it can be applied to any pair of strings P and V of arbitrary length over any arbitrary alphabet Σ. This new automaton is based upon insightful observations of the nondeterministic Levenshtein automaton for a fixed word, and extends the work done for the determinization algorithm described by the same authors in [7].

The remainder of this section is devoted to a thorough presentation of the Universal Levenshtein Automaton. First, in Sect. 2.2, we explain how to convert the two input strings, P and V, into a single string that contains the required information to determine their distance. Then, in Subsects. 2.3 and 2.4, we explain how to construct the Universal Levenshtein Automaton.

Notation: For a word $V \in \Sigma^*$, $|V|$ is the length of this word, and $|V|_\Sigma$ is the number of distinct symbols of Σ present in V. For each positions i, j in V $(1 \le i \le |V|)$, V_i is the ith letter of V and $V[i..j]$ is the subtring of V starting at position i and ending at position j. ε denotes the empty string.

2.2 Bit Vector Representation

Let P and V be the two strings. In the two-dimensional dynamic programming table of the Wagner-Fischer algorithm, all edit scripts with at most k errors have

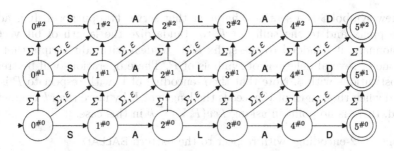

Fig. 1. Levenshtein automaton for the word SALAD. In general, this automaton has $m+1$ columns and $k+1$ levels. The $i^{\#x}$ notation for each state corresponds to i symbols read in the pattern and x errors recorded. Horizontal transitions represent identities, vertical transitions represent insertions, and the two types of diagonal transitions represent substitutions (Σ) and deletions (ε), respectively.

a path that stays around the diagonal. The width of this diagonal is $2k + 1$. To compute the values in this portion of the table, it is sufficient to know which coordinates (i, j) of the cells are such that $P_i = V_j$. This information captures the similarity between the two strings P and V. We show an example in Fig. 2 (left), where grey cells represent coordinates such that $P_i = V_j$, and white cells represent other coordinates. At this point, one can forget the two input words, P and V.

The idea of the Universal Levenshtein Automaton is to work directly on these configurations of white and grey cells. For this, each horizontal line of the diagonal is represented by a bit vector of length $2k + 1$: 0 for white cells, and 1 for grey cells. These bit vectors will serve as the new alphabet, rather than Σ. This encoding captures the local structural properties of the input words, and guarantees alphabet independence.

Definition 1. *Let $P \in \Sigma^*$ and let $s \in \Sigma$. The characteristic vector $\chi(s, P)$ is the bit vector of length $|P|$ such that the ith bit is 1 if $s = P_i$, and 0 otherwise.*

Definition 2. *Let $P \in \Sigma^m$, $V \in \Sigma^n$ such that $n \leq m + k$. Let $P' = \$^k P \$^{2k}$ and $V' = V \$^{m-n+k}$, where $\$$ is a new symbol not present in Σ. The k-encoding of V with respect to P is the sequence of $m + k$ bit vectors of length $2k + 1$ such that the jth element is the characteristic vector $\chi(V'_j, P'[j - k..j + k])$.*

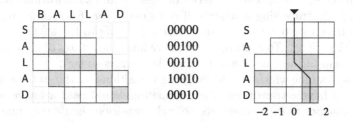

Fig. 2. Bit vector representation for the string SALAD with respect to BALLAD

The new symbol \$ is a sentinel character that serves two purposes. It is added to the prefix and to the suffix of P to standardize the length of bit vectors. Additionnaly, it is added to the suffix of V to deal with edit scripts that end with one or several deletions of symbols of P. These operations will be treated as substitutions with \$. Note that the k-encoding of V with respect to P is not defined when the length of V exceeds the length of P by more than k characters. Indeed, there is no point in asking $Lev(P, V) \leq k$ in this case.

Example 3. 2-encodings with respect to the pattern BALLAD.

	B	A	L	L	A	D		
S A L A D	00000	00100	00110	10010	00010	00011	00111	01111
B A L D	00100	00100	00110	00001	00001	00011	00111	01111
B A L L	00100	00100	00110	00110	00001	00011	00111	01111
B A L L A D S	00100	00100	00110	01100	00100	00100	00000	01111

Finally, we define $\overline{\mathcal{L}\mathrm{ev}}(P, k)$ as the set of bit vector sequences u in $(\{0, 1\}^{2k+1})^*$ such that there exists V in $\mathcal{L}\mathrm{ev}(P, k)$ whose k-encoding wrt P is u.

Remark 4. In [4], the authors use a different encoding. They have bit vectors of length $2k + 2$ bits, instead of $2k + 1$. The last bit is used to identify the transition between non-accepting and accepting states. Here, we get rid of this additional bit by adding \$ symbols to the suffix of V. In Subsect. 2.4, it will allow us to obtain a smaller automaton.

2.3 Construction of the Nondeterministic Universal Levenshtein Automaton

In [4], the authors directly build the *Deterministic Universal Levenshtein Automaton* from the Levenshtein automaton for a fixed word. They consider a symbolic triangular area of a state to simulate multiple active states. Here we take a different approach, and introduce a *Nondeterministic Universal Levenshtein Automaton*. We find this intermediate step useful to facilitate understanding. Moreover it allows us to give an effective algorithm to build the Deterministic Universal Levenshtein Automaton, which is not so easy to infer from [4][1].

Let us come back to the table in Fig. 2. An edit script between P and V can be seen as a path in the white and grey grid. The portions of the path that stay in the same lane, correspond to a series of identity and substitution operations. In this context, traversing a white cell costs one error. Everytime you change lanes, you have to pay either for an insertion (moving to the left) or a deletion (moving to the right). To capture this idea, we introduce states of the form (x, y), meaning "I am in the lane y, and have made x errors so far".

We now examine which are the outgoing transitions for the state (x, y). If we prefer not to have ε-transitions, the automaton should read a bit vector from its input sequence at each time step. So each transition should consume exactly

---

[1] The authors write: "We describe only the basic idea".

one symbol v of V. In a first approach, an edit script can be decomposed into a series of one the two following basic events: an insertion of v, or a series of ℓ deletions in P $(0 \leq \ell \leq k - x)$, followed by either a substitution or an identity with v. In this decomposition, we make the usual hypothesis that a deletion is not followed by an insertion. To reduce the nondeterminism, we add another local condition: a deletion is not followed by a substitution. Indeed, it is always possible to intervert the two operations, and to apply the substitution before the deletion. Doing so, we obtain three types of edit events.

- ins: insertion of v,
- sub: substitution of v,
- ℓ del+id: ℓ deletions in P $(0 \leq \ell \leq k)$, followed by an identity with v.

ins makes the automaton transits from (x, y) to $(x + 1, y - 1)$, sub from (x, y) to $(x + 1, y)$ and ℓ del+id from (x, y) to $(x + \ell, y + \ell)$. Figure 3 shows all these transitions.

We are now ready to formally define NULA(k), the *Nondeterministic Universal Levenshtein Automaton* for k errors.

Definition 5. *Let k be a positive number. The* Nondeterministic Universal Levenshtein Automaton *for k, denoted* NULA(k), *is the NFA represented as follows.*

- *the input alphabet is $\{0, 1\}^{2k+1}$,*
- *the set of states Q_k is $\{(x, y) \in \mathbb{N} \times \mathbb{Z}; \ 0 \leq x \leq k, -x \leq y \leq x\}$,*
- *the transition function $\Delta_k : Q_k \times \{0, 1\}^{2k+1} \to P(Q_k)$ is constituted of three types of transitions.*

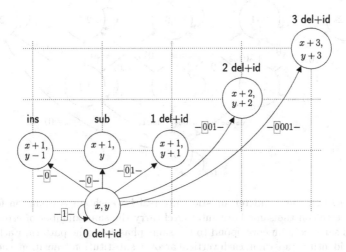

Fig. 3. Outgoing transitions for the state (x, y), $k = 3$. For each bit vector, the bit in position $k + y + 1$ corresponding to the lane y is framed, and $-$ is used to denote any sequence of bits.

insertion transitions: $(x + 1, y - 1) \in \Delta_k((x, y), u)$ for all states $(x, y) \in Q_k$ such that $x < k$ and all $u \in \{0, 1\}^{k+y}0\{0, 1\}^{k-y}$

substitution transitions: $(x + 1, y) \in \Delta_k((x, y), u)$ for all states $(x, y) \in Q_k$, such that $x < k$ and all $u \in \{0, 1\}^{k+y}0\{0, 1\}^{k-y}$

deletion+identity transitions: $(x+\ell, y+\ell) \in \Delta_k((x, y), u)$ for all states $(x, y) \in Q_k$, all ℓ such that $0 \le \ell \le k - x$ and all $u \in \{0, 1\}^{k+y}0^\ell 1\{0, 1\}^{k-y-\ell}$

- the start state is $(0, 0)$,
- all states are accepting.

Figure 4 shows NULA(2). The automaton is very regular and easy to construct. The reader can check that all 2-encodings of Example 3 are accepted. In general, NULA(k) can be seen as an extension of NULA($k-1$): just add $2k+1$ new states of the form (k, y) and incoming transitions. Thus NULA(k) has $(k + 1)^2$ states.

We prove that NULA(k) effectively recognizes the expected language. Recall that the *right language* of a state $q \in Q_k$, denoted $\mathcal{L}(q)$, is the set of all sequences of bit vectors u such that NULA(k) when started in q will accept u.

Proposition 6. Let $(x, y) \in Q_k$. Given $P \in \Sigma^m$ and $V \in \Sigma^n$, we have

- when $y = 0$: $u \in \mathcal{L}(x, y)$ if, and only if, $Lev(P, V) \le k - x$,
- when $y > 0$: $u \in \mathcal{L}(x, y)$ if, and only if, $Lev(P[1 + y..m], V) \le k - x$,
- when $y < 0$: $u \in \mathcal{L}(x, y)$ if, and only if, $Lev(P, V[1 - y..n]) \le k - x$,

where u is the k-encoding of V with respect to P.

Proof. The proof is by induction on the length of u.

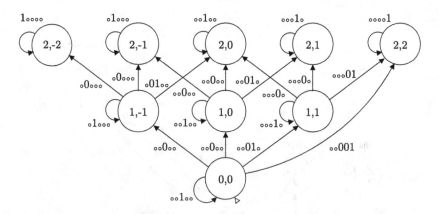

Fig. 4. NULA(2), the Nondeterministic Universal Levenshtein Automaton for $k = 2$ errors. All states on the same horizontal level carry the same number of errors x, and all states in same column correspond to the same phase y in the pattern. Each looping arrow is an identity transition, each vertical arrow a substitution transition, each north-east arrow a deletion transition, and each north-west arrow an insertion transition. For the sake of readability, the symbol o in a bit vector is either 0 or 1. $(0, 0)$ is the start site and all states are accepting.

Corollary 7. *Let $P \in \Sigma^*$, $V \in \Sigma^*$. $Lev(P, V) \leq k$, if, and only if, the k-encoding of V with respect to P is accepted by $NULA(k)$.*

2.4   Construction of the Deterministic Universal Levenshtein Automaton

We now explain how to build the Deterministic Universal Levenshtein Automaton from the NFA introduced in the preceding section. The principle is to use the standard powerset construction that converts a NFA into a DFA. In this construction, we show that it is possible to reduce the number of states by defining *subsumed states*. This notion is already present in [7], where it is defined for the Levenshtein automaton for a fixed word. Here, we adapt it to the states of $NULA(k)$.

Definition 8. *Let (x, y) and (x', y') be two states of Q_k. We say that (x, y) subsumes (x', y'), denoted $(x', y') \sqsubset (x, y)$, if $x < x'$ and $y + x - x' \leq y' \leq y + x' - x$.*

It is clear from the definition that the relation $\sqsubset$ is a well-founded partial order.

Proposition 9. *Let $q \in Q_k$, $q' \in Q_k$, such that $q' \sqsubset q$. Then $\mathcal{L}(q') \subseteq \mathcal{L}(q)$.*

Proof. Consequence of Proposition 6.   □

This proposition implies that all subsumed states can be removed from a subset of Q_k without modifying its right language. In other words, for any subset Q' of Q_k, Q' and $Reduced(Q')$ cannot be distinguished, where $Reduced(Q')$ is defined as the largest subset of Q' such that no two elements of $Reduced(Q')$ are subsumed. It allows us to prune the set of states considered during the construction of $DULA(k)$. Figure 1 gives the corresponding algorithm and Definition 10 the formal definition of $DULA(k)$.

Definition 10. *Let k be a positive number. The Deterministic Universal Levenshtein Automaton for k, denoted $DULA(k)$, is the DFA represented as follows.*

- *the input alphabet is $\{0, 1\}^{2k+1}$,*
- *the set of states is the set of reduced subsets of Q_k,*
- *the transition function δ is given by Algorithm of Fig. 1,*
- *the start site is $\{(0, 0)\}$,*
- *all states are accepting.*

Figure 5 shows $DULA(1)$, that has 8 states. $DULA(2)$, the automaton obtained from $NULA(2)$ visible on Fig. 4, has 50 states, and is not represented here. For each value of k, the number of states of $DULA(k)$ is computed with $R(k)$.

$$R_k(x, y) = \sum_{(x', y') \in Q_k, y < y', (x, y) \not\sqsubset (x', y'), (x', y') \not\sqsubset (x, y)} R_k(x', y')$$
$$R(k) = \sum_{(x, y) \in Q_k} R_k(x, y)$$

Add $\{(0,0)\}$ to $DULA(k)$ as an unmarked state;
while $DULA(k)$ *contains an unmarked state* **do**
    Let T be that unmarked state;
    Mark T;
    for *each bit vector* $u \in \{0,1\}^{2k+1}$ **do**
        $S = \{q' \in Q_k; \exists q \in T \; q' \in \Delta(q,u)\}$;
        $S' = Reduced(S)$;
        Define $\delta(T,u) = S'$;
        if S' *is not in* $DULA(k)$ *already* **then**
          | Add S' to $DULA(k)$ as an unmarked state;
        end
    end
end

Algorithm 1: Construction of DULA(k) from NULA(k)

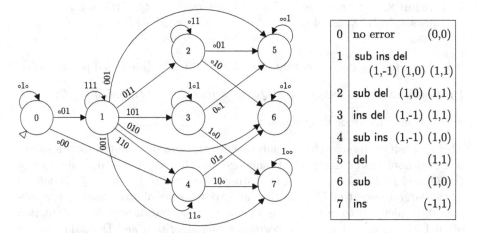

0	no error	(0,0)
1	sub ins del	
		(1,-1) (1,0) (1,1)
2	sub del	(1,0) (1,1)
3	ins del	(1,-1) (1,1)
4	sub ins	(1,-1) (1,0)
5	del	(1,1)
6	sub	(1,0)
7	ins	(-1,1)

Fig. 5. DULA(1), Deterministic Universal Levenshtein Automaton for $k = 1$. It has 8 states, numbered from 0 to 7, that are all accepting. In the table, we report the subset of Q_1 corresponding to each state, as well as the semantics of the state.

$R_k(x,y)$ is the number of reduced subsets of Q_k such that the state (x,y) is the element with the smallest y in the subset. The total number of reduced subsets for Q_k, $R(k)$ is obtained by summing over all possible states of Q_k. For k ranging from 1 to 10, R equals 8, 50, 322, 2187, 15510, 113633, 853466, 6536381, 50852018, 400763222. We conjecture that $R(k)$ is in $O(7^k)$.

3 Application to the Neighbourhood Counting Problem

We now turn to the problem of computing the cardinality of $\mathcal{L}ev(P,k)$. This value depends on the length of the word, the input alphabet and the internal structure of the word. Consider for example the three-letter alphabet {A,B,L}.

$\mathcal{L}$ev(AAA,1) = { AAA, AA, AAB, AAL, ABA, ALA, BAA, LAA, AAAA, BAAA,
            LAAA, ABAA, ALAA, AABA, AALA, AAAB, AAAL}

$\mathcal{L}$ev(LAB,1) = { LAB, LA, AB, LB, AAB, BAB, LBB, LLB, LAA, LAL, ALAB,
            BLAB, LLAB, LAAB, LBAB, LALB, LABA, LABB, LABL}

In the first case, the neighbourhood has 17 elements, and in the latter case 19 elements. The combinatorics is even more complex for greater values of k.

We show how to solve this problem efficiently with the help of DULA(k). Indeed, it is enough to intersect DULA(k) with the set of all sequences of bit vectors that are a valid encoding for some string V with respect to P. We designate by Encod(P, k) this latter language.

3.1   A DFA for Encod(P, k)

Definition 11. *Let* P *be a string of* Σ^**.* *Encod(P, k) is the set* $\{u \in (\{0,1\}^{2k+1})^*; \exists V \in \Sigma^*$ *s.t.* u *is the k-encoding of V wrt to P}.*

From Definition 2, we know that the elements of Encod(P, k) are strings of $k+m$ bit vectors.

Definition 12. *Let* $V \in (\Sigma \cup \{\$\})^*$*. Define* $B(V) = \{\chi(s, V); s \in \Sigma\}$*.*

$B(V)$ is the set of all bit vectors u of length $|V|$ that satisfies the three following properties.

– If $V_i = V_j$, then $u_i = u_j$.
– If $u_i = 1$ and $u_j = 1$, then $V_i = V_j$.
– If $V_i = \$$, then $u_i = 0$.

Example 13. B(ABL)={$000, 001, 010, 100$}, B(ABB)={$000, 011, 100$}, B(AA\$)= {$000, 110$}.

Encod(P, k) is recognized by the following DFA.

– the input alphabet is $(\{0, 1\}^{2k+1})^*$,
– the set of states is $\{0, \ldots, m + k\} \cup \{\$_{m-k+1}, \ldots, \$_{m+k}\}$,
– the transition function γ is defined by

$$\gamma(i - 1, u) = i, \ 1 \leq i \leq m + k, u \in B(P'[i - k..i + k])$$
$$\gamma(i - 1, 0^{k+1+m-i}1^{i+k-m}) = \$_i, \ m - k + 1 \leq i \leq m + k$$
$$\gamma(\$_{i-1}, 0^{k+1+m-i}1^{i+k-m}) = \$_i, \ m - k + 1 < i \leq m + k$$

– the start state is 0,
– there are two accepting states: $m + k$ and $\$_{m+k}$.

Each state i recognizes the encodings of strings of Σ^+ of length i, and each state $\$_i$ recognizes the encodings of strings of $\Sigma^+\$^+$ of length i. The transition from $i - 1$ to $\$_i$ corresponds to the first occurrence of $\$$ in the string. Figure 6 shows the DFAs obtained for AAA, LAB and $k = 1$. We also give the DFA for BALLAD and $k = 2$, for which several encodings were provided in Example 3.

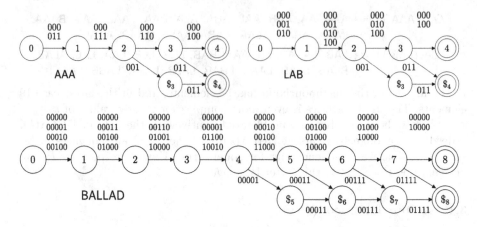

Fig. 6. DFAs for Encod(AAA, 1), Encod(LAB, 1) and Encod(BALLAD, 2)

3.2   Back to the Counting Problem

Considering that $\overline{\mathcal{L}ev}(P,k) = \text{DULA}(k) \cap \text{Encod}(P,k)$, we will exploit the product automaton of $\text{DULA}(k)$ and $\text{Encod}(P,k)$. This product automaton needs not to be constructed explicitly. It will serve us to design the recurrence formula to compute the size of $\mathcal{L}ev(P,k)$.

From the product automaton, we can deduce what is the size of the language recognized by $\overline{\mathcal{L}ev}(P,k)$. This is simply the total number of distinct paths leading from the start state to an accepting state. What we still have to do is to bring the problem back to the initial alphabet Σ. For that, we need a function $\alpha : \{0,1\}^{2k+1} \times \Sigma^{2k+1} \to \mathbb{N}$ that computes the number of symbols s of Σ such that $\chi(s,V) = u$, for each bit vector u and each word V over Σ.

$\alpha(u,V) = 1$, whenever at least one bit of u is 1

$\alpha(u,V) = |\Sigma| - |V|_\Sigma$ otherwise (in this case, $u = 00 \cdots 00$)

α is used to assign a multiplicity to each transition of the product automaton. In this context, the total number of underlying strings of Σ^* is the sum of all multiplicities of all distinct paths. Define S as follows.

$$S : \text{REDUCED}(Q_k) \times (\{0, \ldots, m+k\} \cup \{\$_{m-k+1}, \ldots, \$_{m+k}\}) \to \mathbb{N}$$
$$S(0,0) = 1$$
$$S(q', i+1) = \sum_{u,q,q'=\delta_k(q,u)} \alpha(u, P'[i-k..i+k]) \times S(q,i), \quad 0 \le i < m+k$$
$$S(q', \$_{m-k}) = \sum_{u,q,q'=\delta_k(q,u)} S(q, m-k-1)$$
$$S(q', \$_{i+1}) = \sum_{u,q,q'=\delta_k(q,u)} S(q, i-1) + S(q, \$_i), \quad m-k \le i < m+k$$

m is the length of P. $S(q,i)$ is the number of distinct paths leading from the start site to the state (q,i) in the product automaton. The final result is obtained by summing over the accepting states.

Proposition 14.

$$|\mathcal{L}ev(P,k)| = \sum_{q \in \text{REDUCED}(Q_k)} S(q, m+k) + S(q, \$_{m+k})$$

Figure 7 shows the developments of S for $\mathcal{L}ev(AAA, 1)$ and $\mathcal{L}ev(LAB, 1)$, which were obtained from DULA(1) in Fig. 5 on the one hand, and Encod(AAA,1), Encod(LAB,1) in Fig. 6 on the other hand. S can be implemented by dynamic programming with a table of size $R(k) \times (m + 3k)$. Each element of the table is computed in constant time. So the algorithm has a time complexity of $O(m)$. In practice, for each possible word structure (e.g. AAA, AAB, ABL), the associated transitions in DULA(k) can be extracted during a preprocessing step. As for the space complexity, this is not necessary to store the full dynamic table. At any instant, the algorithm only requires elements from the current row (i) and the previous row ($i-1$). Thus the space complexity is in $O(1)$.

Remark 15. In Sect. 2, we have put a lot of efforts into defining a universal automaton that is able to process any pair of strings. In this Section, we have

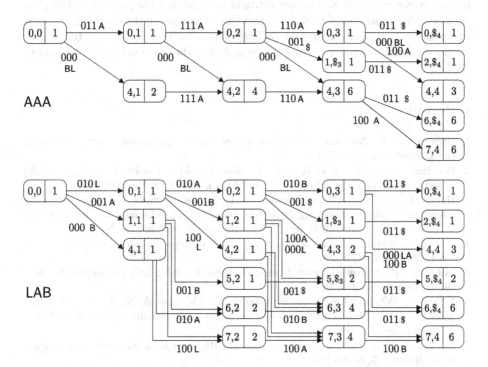

Fig. 7. Computation of S for AAA and LAB, $\Sigma = \{A, B, L\}$ and $k = 1$. For each state (q, i) or $(q, \$_i)$ (left part of the box), we indicate the value of S (right part of the box). The labels on the transitions are the bit vectors. For each bit vector, we also mention the corresponding symbol(s) of Σ. The number of such symbols is α. The result is the sum of S over the last column.

specialized this automaton for a fixed pattern P and have obtained a DFA for $\mathcal{L}ev(P,k)$. Alternatively, we could have directly used the DFA presented in [7]. The construction of this DFA is also in linear time, for a fixed value of k. However it requires a complex table-based preprocessing, that is dependent on each pattern. In our approach, the computational burden is reported to the design of DULA(k), which is performed only once. This is a new route to recover the result established in [7].

4   Conclusion

We have shown how to count the number of strings present in the neighbourhood of some fixed reference word P. The algorithm produces a product automaton, which could also be used to generate the set of all strings in the neighbourhood of P, or to sample it. This generic approach could extend to other target regular languages, instead of the singleton language $\{P\}$. The downside of the method, however, is that it requires the computation of the DFA DULA(k), whose size increases exponentially with the number of errors k. It exceeds one million states with $k = 8$. Another route is to construct the product automaton with NULA(k), instead of DULA(k), and then determinize the resulting automaton with an optimization similar to Proposition 9. This could lead to a lower memory consumption for regular languages that do not need all transitions of DULA(k).

References

1. Baeza-yates, R., Navarro, G.: A faster algorithm for approximate string matching. In: Algorithmica, pp. 1–23 (1996)
2. Becerra-Bonache, L., de la Higuera, C., Janodet, J.C., Tantini, F.: Learning balls of strings from edit corrections. J. Mach. Learn. Res. **9**, 1841–1870 (2008)
3. Holub, J., Melichar, B.: Implementation of nondeterministic finite automata for approximate pattern matching. In: Champarnaud, J.-M., Maurel, D., Ziadi, D. (eds.) WIA 1998. LNCS, vol. 1660, pp. 92–99. Springer, Heidelberg (1999)
4. Mihov, S., Schulz, K.: Fast approximate search in large dictionaries. Comput. Linguist. **30**(4), 451–477 (2004)
5. Mitankin, P.: Universal Levenshtein automata. Building and properties. Masters thesis, University of Sofia (2005)
6. Myers, G.: Whats behind blast. In: Chauve, C., El-Mabrouk, N., Tannier, E. (eds.) Models and Algorithms for Genome Evolution, Computational Biology, vol. 19, pp. 3–15. Springer, London (2013)
7. Schulz, K., Mihov, S.: Fast string correction with levenshtein automata. Int. J. Doc. Anal. Recogn. **5**, 67–85 (2002)
8. Ukkonen, E.: Finding approximate patterns in strings. J. Algorithms **6**(1), 132–137 (1985)
9. Wagner, R.A., Fischer, M.J.: The string-to-string correction problem. J. ACM **21**(1), 168–173 (1974)

Combinatorics on Words

Parallelogram Morphisms and Circular Codes

Alexandre Blondin Massé, Mélodie Lapointe, and Hugo Tremblay[✉]

Laboratoire de Combinatoire et d'informatique Mathématique,
Université du Québec à Montréal, CP 8888, Succ. Centre-ville,
Montréal, Québec H3C 3P8, Canada
blondin_masse.alexandre@uqam.ca, lapointe.melodie@courrier.uqam.ca,
hugo.tremblay@lacim.ca

Abstract. In 2014, it was conjectured that any polyomino can be factorized uniquely as a product of prime polyominoes [7]. In this paper, we present simple tools from words combinatorics and graph topology that seem very useful in solving the conjecture. The main one is called parallelogram network, which is a particular subgraph of $G(\mathbb{Z}^2)$ induced by a parallelogram morphism, i.e. a morphism describing the contour of a polyomino tiling the plane as a parallelogram would. In particular, we show that parallelogram networks are homeomorphic to $G(\mathbb{Z}^2)$. This leads us to show that the image of the letters of parallelogram morphisms is a circular code provided each element is primitive, therefore solving positively a 2013 conjecture [8].

Keywords: Codes · Combinatorics on words · Graphs · Digital geometry · Topological graph theory · Morphisms

1 Introduction

The interaction between combinatorics on words and digital geometry has been extensively studied in the last decades [1,6,9,10]. The most famous example is without doubt the family of Sturmian words, which can be seen as the discrete counterpart of lines having irrational slope [15]. Another remarkable example is about digital convexity: It was recently established that it can be decided very efficiently if some discrete figure is convex by factorizing its boundary in Lyndon and Christoffel words [10]. In the same spirit, one can decide in linear time and space whether some discrete path is self-intersecting, by using combinatorial arguments together with an enriched radix quadtree [9]. Finally, generalizations of discrete lines in 3D have also been proposed, such as in [6].

In parallel, the theory of codes has been developed for more than 50 years. Here, we focus on circular codes, i.e. sets of words that allow unique encoding of words written on a circle. Circular codes were first introduced and studied by Golomb and Gordon [13] and have received a lot of attention from researchers

H. Tremblay—This research is supported by the Natural Sciences and Engineering Research Council of Canada (NSERC).

© Springer International Publishing Switzerland 2016
A.-H. Dediu et al. (Eds.): LATA 2016, LNCS 9618, pp. 221–232, 2016.
DOI: 10.1007/978-3-319-30000-9_17

since then. From an algebraic perspective, Schützenberger has contributed significantly to a better understanding of their structure [17]. His results have been generalized by Bassino who described the generating functions of weighted circular codes [3]. Circular codes have also been extensively studied in bioinformatics. For instance, a remarkable circular code for the protein coding genes of mitochondria has been brought to light by Arques and Michel [2].

More recently, researchers (including both authors) have been interested in the shape of parallelogram tiles (also called *square tiles* in [16]) using words combinatorics formalism [7,8,11,16]. In particular, in 2008, Provençal defined the product (or composition) of a polyomino and a parallelogram polyomino, which consists in substituting each unit square of the first polyomino with a copy of the parallelogram polyomino (see Fig. 1). This leads to the natural definition of prime and composed polyominoes: A polyomino is called *prime* if it cannot be obtained by the composition of two smaller nontrivial polyominoes [16]. Provençal's definition was further studied in [7], where it was proved that every polyomino can be factorized as a product of prime polyominoes, a result in the same spirit than the Fundamental Theorem of Arithmetic. However, the authors were not able to prove that such a factorization is unique and left it as a conjecture:

Conjecture 1. Let U be the unit square polyomino and $P \neq U$ be a polyomino. Then P can be factorized uniquely as a product of a prime polyomino Q and primes parallelogram polyominoes P_1, P_2, ..., P_n, i.e. $P = Q \circ P_1 \circ P_2 \circ \cdots \circ P_n$.

In this paper, we neither prove nor disprove Conjecture 1, but we provide tools that we believe are essential in showing the unicity of the prime factorization. It relies on basic words combinatorics as well as graph topology. In particular, it introduces *parallelogram networks*, i.e. undirected subgraphs of the grid graph $\mathbb{Z}^2$ induced by special morphisms called parallelogram [7]. They turn out to be expressive and easy to manipulate: As a byproduct, we obtain a simple proof that the image of parallelogram morphisms is a circular code under very mild conditions (Theorem 13), thus solving another conjecture stated in [8].

The content is divided as follows. In Sect. 2, we introduce the basic definitions about words and codes. In Sect. 3, we recall basic definitions about graphs and their interaction with words. Section 4 is devoted to the study of the properties

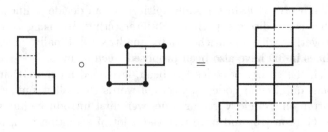

Fig. 1. The composition of a polyomino (left) with a parallelogram polyomino (middle) is a composed polyomino (right).

of parallelogram networks, culminating with Theorem 13 in Sect. 5. We briefly conclude with an open problem.

2   Words and Codes

We recall the basic definitions and notation for words and codes (see [15] for more details). An *alphabet* is a finite set Σ whose elements are called *letters*. A *word* on Σ is a finite sequence $w = w_1 w_2 \cdots w_n$ of letters of Σ. The i-th letter of w is denoted by w_i. The *length* of w, written $|w|$ is the number of elements in the sequence w. The unique word of length 0 is called the *empty word* and is written ε. Whenever $|w| > 0$, we write $\mathrm{FST}(w)$ and $\mathrm{LST}(w)$ for the first and last letter of w. Moreover, for any letter $a \in \Sigma$, $|w|_a$ is the number of occurrences of the letter a in w.

Given two words $u = u_1 u_2 \cdots u_m$ and $v = v_1 v_2 \cdots v_n$, the *concatenation* of u and v, denoted by uv or $u \cdot v$, is the word $u_1 u_2 \cdots u_m v_1 v_2 \cdots v_n$. If u is a word and n is an integer, then $u^n = u \cdot u \cdots u$ (n times). A word w is called *primitive* if there does not exist any word u and integer $n \geq 2$ such that $w = u^n$. A well-known fact is the following:

Proposition 2 [15]. *Let w be a word such that there exist words u and v with $w = uv = vu$. Then w is not primitive.*

The set of all words on Σ having length n is denoted by Σ^n. The *free monoid* is defined by $\Sigma^* = \bigcup_{n \geq 0} \Sigma_n$. Its name comes from the fact that it has a monoid structure when combined with the concatenation operation, and with neutral element ε. A *submonoid* of Σ^* is a subset $M \subseteq \Sigma^*$ which is stable under the concatenation and which includes ε. The submonoid M is *pure* if for all $x \in \Sigma^*$ and $n \geq 1$, $x^n \in M$ implies $x \in M$. Moreover, we say that M is *very pure* if for all $u, v \in \Sigma^*$, the relations $uv \in M$ and $vu \in M$ imply $u, v \in M$. It is straightforward to show that any very pure submonoid is also pure. However, the converse is false: The submonoid of $\{a, b\}^*$ generated by $\{ab, ba\}$ is pure but not very pure.

Let w be some word. Then we say that u is a *factor* of w if there exist words x and y such that $w = xuy$. Moreover, if $x = \varepsilon$ (resp. $y = \varepsilon$), u is called *prefix* (resp. *suffix*) of w. The set of prefixes (resp. suffixes) of a word w is denoted by $\mathrm{Pref}(w)$ (resp. $\mathrm{Suff}(w)$). Also, the unique prefix (resp. suffix) of length ℓ of w is denoted by $\mathrm{Pref}_\ell(w)$ (resp. $\mathrm{Suff}_\ell(w)$), where $0 \leq \ell \leq |w|$.

Given two alphabets A and B, an application $\varphi : A^* \to B^*$ is called *morphism* (resp. *antimorphism*) if $\varphi(uv) = \varphi(u)\varphi(v)$ (resp. $\varphi(uv) = \varphi(v)\varphi(u)$) for all $u, v \in A^*$. Given $w = w_1 w_2 \cdots w_n$, the *reversal of w*, denoted by $\widetilde{w}$, is defined by $\widetilde{w} = w_n w_{n-1} \cdots w_2 w_1$. The operator $\widetilde{\ }$ is an antimorphism. It is easy to see that morphisms and antimorphisms are completely defined by their action on single letters.

Let Σ be an alphabet and $X \subseteq \Sigma^*$. Then X is a *code over Σ* if for all $m, n \geq 1$ and $x_1, x_2, \ldots, x_m, y_1, y_2, \ldots, y_n \in X$, the condition $x_1 x_2 \cdots x_m = y_1 y_2 \cdots y_n$ implies $m = n$ and $x_i = y_i$ for $i = 1, 2, \ldots, n$. Roughly speaking, X is a code if any word in X^* can be written uniquely as a product of words in X. Similarly, we say that X is a *circular code* if for all $m, n \geq 1$ and $x_1, x_2, \ldots, x_m, y_1, y_2, \ldots, y_n \in X$, $p \in \Sigma^*$ and $s \in \Sigma^+$, the relations $sx_2x_3 \cdots x_m p = y_1 y_2 \cdots y_n$ and $x_1 = ps$ imply $m = n$, $p = \varepsilon$ and $x_i = y_i$ for $i = 1, 2, \ldots, n$. In other words, X is a circular code if any circular permutation of a word in X^* can be written uniquely as a product of words in X. It is not hard to prove that any circular code is a code. The reader is referred to [5] for more details about code theory, but one important result for our purpose is the following characterization of circular codes:

Theorem 3 (Proposition 1.1 of [5]). *A submonoid M of A^* is very pure if and only if its minimal set of generators is a circular code.*

3   Discrete Paths and Graphs

An alphabet of particular interest for our purposes is the *Freeman chain code* $\mathcal{F} = \{0, 1, 2, 3\}$, which encodes the four elementary steps on the square grid $\mathbb{Z}^2$ with respect to the bijection

$$0 \mapsto \rightarrow, \quad 1 \mapsto \uparrow, \quad 2 \mapsto \leftarrow, \quad 3 \mapsto \downarrow.$$

Two basic operations on Freeman words have useful geometrical interpretations. The application $\overline{}$ is the morphism defined by

$$\overline{0} = 2, \quad \overline{1} = 3, \quad \overline{2} = 0, \quad \overline{3} = 1,$$

which corresponds geometrically to the application of a rotation of angle π. Also, the antimorphism $\widehat{} = \overline{} \circ \widetilde{}$ corresponds to traveling the sequence of elementary steps in the opposite order.

Given $w \in \mathcal{F}^*$, we write $\overrightarrow{w} = (|w|_0 - |w|_2, |w|_1 - |w|_3)$. Any word $w \in \mathcal{F}^*$ is called *closed* if $\overrightarrow{w}$ is the null vector. Moreover, w is called *simple* if none of its proper factor is closed, and is a *contour word* if it is nonempty, closed and simple.

A *discrete path* is a sequence of connected unit segments whose endpoints are on $\mathbb{Z}^2$. Discrete paths can naturally be represented by an ordered pair $\gamma = (p, w)$, where $p \in \mathbb{Z}^2$ and $w \in \mathcal{F}^*$. Thus, the set of points of $\mathbb{Z}^2$ visited by γ is $\text{Points}(\gamma) = \{p + \overrightarrow{u} \mid u \in \text{Pref}(w)\}$. A discrete path is called *closed* (resp. *simple*) if w is closed (resp. *simple*). Given a closed discrete path γ, the *region of γ*, denoted by $R(\gamma)$, is defined as the closed subset of $\mathbb{R}^2$ whose boundary is exactly described by γ.

Every discrete path yields a unique undirected graph $G(\gamma) = (V, E)$, where $V = \text{Points}(\gamma)$ and $(q, q') \in E$ if and only if there exist two consecutive prefixes u, u' of w such that $q = p + \overrightarrow{u}$ and $q' = p + \overrightarrow{u}'$. Also, the *(graph) distance* between two vertices p and p' is the length of a shortest discrete path γ between p and p'.

The *grid graph* $G(\mathbb{Z}^2)$ is the infinite graph whose set of vertices is $\mathbb{Z}^2$ and whose set of edges E is defined as follows: $\{p, p'\} \in E$ if and only if $\text{dist}(p, p') = 1$, where dist is the usual Euclidean distance. The set of all discrete paths of $G(\mathbb{Z}^2)$ is denoted by $\Gamma(\mathbb{Z}^2)$. Clearly, for any $\gamma \in \Gamma(\mathbb{Z}^2)$, the undirected version of the graph $G(\gamma)$ is a subgraph of $G(\mathbb{Z}^2)$.

We now recall topological graph theoretic definitions. We use the same terminology as in [12]. Let $G = (V, E)$ be a undirected graph. A *subdivision of* G is any graph obtained from G by replacing some edges in E with new paths between their ends such that those paths have no inner vertex in V or in another path. The original vertices of G are then called *branch vertices* and the new vertices are called *inner vertices*. It is clear that inner vertices have degree 2 while branch vertices retain their respective degree from G.

Given two graphs $G = (V, E)$ and $G' = (V', E')$, G and G' are called *isomorphic*, and we write $G \simeq G'$, if there exists a bijection $f : V \to V'$ such that $\{u, v\} \in E$ if and only if $\{f(u), f(v)\} \in E'$. From this, one defines the notion of *graph homeomorphism*: Two graphs G and G' are *homeomorphic* (i.e. topologically isomorphic) if there exist two isomorphic subdivisions T and T' of G and G' respectively. It is easy to show that G and any of its subdivision T are homeomorphic. Also, the notions of graph homeomorphism and standard topological homeomorphism are equivalent when considering the topological representations of graphs (i.e. the topological space obtained by representing vertices as distinct points and edges by homeomorphic images of the closed unit interval $[0, 1]$) [14].

4   Parallelogram Networks

Some morphisms are of particular interest from a geometrical perspective. We recall some definitions from [7].

Definition 4 [7]. *Let $\varphi : \mathcal{F}^* \to \mathcal{F}^*$ be a morphism. Then φ is called*

(i) homologous *if $\varphi(a) = \widehat{\varphi(\overline{a})}$;*
(ii) parallelogram *if it is homologous, $\varphi(\mathbf{0123})$ is a contour word and* $\mathrm{FST}(\varphi(a)) = a$ *for all $a \in \mathcal{F}$.*

Let $\varphi : \mathcal{F}^* \to \mathcal{F}^*$ be a parallelogram morphism. For simplicity of writing, we extend the application φ as follows. For any $p = (x, y) \in \mathbb{Z}^2$, let $\varphi(p) = \varphi(x, y) = (0, 0) + x\overrightarrow{\varphi(0)} + y\overrightarrow{\varphi(1)} \in \mathbb{Z}^2$. Moreover, if $\gamma = (p, w)$ is a discrete path, then $\varphi(\gamma)$ is the discrete path $\varphi(\gamma) = (\varphi(p), \varphi(w))$.

The *graph of* φ is defined by

$$G(\varphi) = \bigcup_{\gamma \in \Gamma(\mathbb{Z}^2)} G(\varphi(\gamma)) = \bigcup_{p \in \mathbb{Z}^2} G(\varphi(p, \mathbf{0123})). \tag{1}$$

Any such graph is called *parallelogram network*. The second equality of Eq. (1) is easy to check: The inclusion $\supseteq$ follows directly from the fact that $(p, \mathbf{0123})$ is a path in $G(\mathbb{Z}^2)$ while the inclusion $\subseteq$ follows from the fact that any path γ in $G(\mathbb{Z}^2)$ can be divided into discrete paths of length 1, each belonging to at least one discrete path of the form $(p, \mathbf{0123})$, for some $p \in \mathbb{Z}^2$.

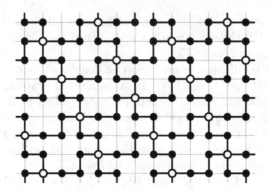

Fig. 2. The parallelogram network $G(\varphi)$ induced by the parallelogram morphism such that $\varphi(0) = 0010$ and $\varphi(1) = 121$. The white dots correspond to branch vertices.

Example 5. The graph $G(\varphi)$ is represented in Fig. 2, where φ is the parallelogram morphism such that $\varphi(0) = 0010$ and $\varphi(1) = 121$.

Clearly, if φ is a parallelogram morphism, then the morphism φ_i defined by $\varphi_i(a) = \varphi(a + i)$ is also a parallelogram morphism for $i = 0, 1, 2, 3$ and $a + i$ is the addition modulo 4. Therefore, for unicity purposes, we assume that $\mathrm{FST}(\varphi(a)) = a$ for all $a \in \mathcal{F}$, and that any discrete path whose associated word is $\varphi(0123)$ is traveled counterclockwise.

The following basic properties of homologous morphisms are useful.

Proposition 6. *Let φ be an homologous morphism and $w \in \mathcal{F}^*$.*

(i) *For any $a \in \mathcal{F}$, $\overrightarrow{\varphi(a)} + \overrightarrow{\varphi(\overline{a})} = \overrightarrow{0}$.*
(ii) *If $\overrightarrow{w} = (x, y)$, then $\overrightarrow{\varphi(w)} = x\overrightarrow{\varphi(0)} + y\overrightarrow{\varphi(1)}$.*

Proof. (i) Since φ is homologous, for any $a \in \mathcal{F}$, we have $\overrightarrow{\varphi(a)} = \overrightarrow{\varphi(\overline{a})} = -\overrightarrow{\varphi(\overline{a})}$.
(ii) Write $w = w_1 w_2 \cdots w_n$. Then

$$\overrightarrow{\varphi(w)} = \sum_{i=1}^{n} \overrightarrow{\varphi(w_i)}$$

$$= \sum_{a \in \mathcal{F}} |w|_a \overrightarrow{\varphi(a)}$$

$$= \sum_{a \in \{0,1\}} \left(|w|_a \overrightarrow{\varphi(a)} + |w|_{\overline{a}} \overrightarrow{\varphi(\overline{a})} \right)$$

$$= \sum_{a \in \{0,1\}} \left(|w|_a - |w|_{\overline{a}} \right) \overrightarrow{\varphi(a)}$$

$$= x\overrightarrow{\varphi(0)} + y\overrightarrow{\varphi(1)},$$

as claimed. □

It is worth noticing that for any parallelogram morphism φ, the graph $G(\varphi)$ is a regular parallelogram tiling of the plane $\mathbb{R}^2$. In other words, it is possible to completely cover the plane by non-overlapping translated copies of $\varphi(\mathbf{0123})$ along the direction of the two vectors $\overrightarrow{\varphi(\mathbf{0})}$ and $\overrightarrow{\varphi(\mathbf{1})}$, i.e.

$$\mathbb{R}^2 = \bigcup_{(a,b)\in\mathbb{Z}^2} \left\{ R((0,0); \varphi(\mathbf{0123})) + a\overrightarrow{\varphi(\mathbf{0})} + b\overrightarrow{\varphi(\mathbf{1})} \right\},$$

Indeed, as shown in [4], a tile admitting a contour word $w \in \mathcal{F}^*$ tiles the plane by translation along the direction of exactly two vectors if and only if w can be factorized as $w = XY\hat{X}\hat{Y}$, where $X, Y \in \mathcal{F}$. Moreover, the authors characterize such regular tiling by describing the surrounding of parallelogram tiles (see Fig. 3). From this, Proposition 7 follows.

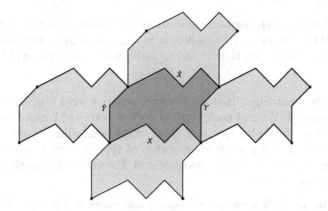

Fig. 3. The surrounding of a tile t coded by $w = XY\hat{X}\hat{Y}$ obtained by taking the four translated copies $t \pm \overrightarrow{\varphi(\mathbf{0})}$ or $\overrightarrow{\varphi(\mathbf{1})}$ and matching the corresponding homologous factors. It induces a regular parallelogram tiling of the plane $\mathbb{R}^2$.

Proposition 7. *Let φ be a parallelogram morphism. Then $\left\{ \overrightarrow{\varphi(\mathbf{0})}, \overrightarrow{\varphi(\mathbf{1})} \right\}$ is a basis of the vector space $\mathbb{R}^2$.*

Proof. Let $\overrightarrow{u} = \overrightarrow{\varphi(\mathbf{0})}$ and $\overrightarrow{v} = \overrightarrow{\varphi(\mathbf{1})}$. It suffices to prove that $\overrightarrow{u}$ and $\overrightarrow{v}$ are linearly independent since $\mathbb{R}^2$ is a vector space of dimension 2. Arguing by contradiction, assume that this is not the case and let

$$T = \bigcup_{(a,b)\in\mathbb{Z}^2} \left\{ R((0,0), \varphi(\mathbf{0123})) + a\overrightarrow{\varphi(\mathbf{0})} + b\overrightarrow{\varphi(\mathbf{1})} \right\}.$$

Now, since the region $R((0,0), \varphi(\mathbf{0123}))$ is bounded, there exist points $p_1, p_2 \in \mathbb{R}^2$ such that $R((0,0), \varphi(\mathbf{0123}))$ lies completely in the region B between the two parallel lines $l_1 = p_1 + t_1 \overrightarrow{u}$ and $l_2 = p_2 + t_2 \overrightarrow{u}$, where $t_1, t_2 \in \mathbb{R}$. Further, the linear dependance of $\overrightarrow{u}$ and $\overrightarrow{v}$ implies that any point of T lies entirely in B, so that T is a subset of B. But then T is a proper subset of $\mathbb{R}^2$, contradicting $T = \mathbb{R}^2$.

A remarkable property of parallelogram morphisms is that they preserve closed and simple paths. The former is an immediate consequence of Proposition 7 while the latter is more complicated to show and we need additional results. First, we recall a result of [18] about tessellation that translates directly to our context:

Theorem 8. *Let φ be a parallelogram morphism, $\overrightarrow{a} = \overrightarrow{\varphi(0)}$, $\overrightarrow{b} = \overrightarrow{\varphi(1)}$, $p, q \in \mathbb{Z}^2$ and P, Q be the regions enclosed inside the discrete paths $(p, \varphi(\mathbf{0123}))$ and $(q, \varphi(\mathbf{0123}))$ respectively. Then exactly one of the following conditions holds:*

  (i) *$P = Q$ and then $p = q$;*
 (ii) *P and Q share a single point and then $\overrightarrow{q - p} = \pm\overrightarrow{a} \pm \overrightarrow{b}$;*
(iii) *P and Q share a chain in $\varphi(\mathcal{F})$ and then $\overrightarrow{q - p} \in \{\pm\overrightarrow{a}, \pm\overrightarrow{b}\}$;*
(iv) *P and Q are disjoint.*

Proof. By definition, the regions enclosed inside the discrete path $\varphi(\mathbf{0123})$ is a polyomino tiling the plane by translation in a parallelogram manner. It follows from Theorem 4.13 of [18] that P and Q verify one and only one of Conditions (i)–(iv). $\qquad\square$

We observe from Fig. 2 that each vertex $x\overrightarrow{\varphi(0)} + y\overrightarrow{\varphi(1)}$, where $x, y \in \mathbb{Z}$ of $G(\varphi)$ has degree 4. We call such vertices *branch vertices*. A non branch vertex p is called *inner vertex of type a* if there exists some discrete path $(p', \varphi(a))$ visiting p. Note that if p is an inner vertex of type a, then it is also an inner vertex of type $\bar{a}$. An immediate consequence of Theorem 8 is a simple description of parallelogram networks.

Corollary 9. *Let φ be some parallelogram morphism and $p \in \mathbb{Z}^2$. Then*

$$\deg(p) = \begin{cases} 4, & \textit{if p is a branch vertex;} \\ 2, & \textit{otherwise.} \end{cases}$$

The remainder of this section is devoted to proving that $G(\mathbb{Z}^2)$ and $G(\varphi)$ are homeomorphic. First, observe that any parallelogram morphism φ induces a subdivision T_φ of $\mathbb{Z}^2$: Subdivide horizontal edges $\{u, v\}$ of $G(\mathbb{Z}^2)$ by adding $|\varphi(0)| - 1$ inner vertices between the two branch vertices u and v. Similarly, vertical edges are subdivided using $|\varphi(1)| - 1$ new inner vertices. Therefore, the new horizontal (*resp.* vertical) chains obtained between two branch vertices adjacent in the original graph have length $|\varphi(0)|$ (*resp.* $|\varphi(1)|$), since $|\varphi(0)| = |\varphi(2)|$ and $|\varphi(1)| = |\varphi(3)|$.

Our first main result follows:

Theorem 10. *Let φ be a parallelogram morphism. Then, $T_\varphi \simeq G(\varphi)$.*

Proof. Let $V(T_\varphi)$ and $V(G(\varphi))$ be the set of vertices of T_φ and $G(\varphi)$ respectively. Also, let (x, y) be a vertex of T_φ. By construction, we have

$$(x, y) \in \left\{ (\lfloor x \rfloor, \lfloor y \rfloor) + \left(\frac{k_1}{|\varphi(0)|}, \frac{k_2}{|\varphi(1)|} \right) \right\}$$

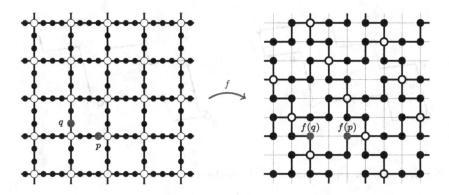

Fig. 4. The effect of f on two vertices of T_φ

with $0 \le k_1 < |\varphi(\mathbf{0})|$, $0 \le k_2 < |\varphi(\mathbf{1})|$ and $k_1 k_2 = 0$. Now consider the function $f : V(T_\varphi) \to V(G(\varphi))$ defined by

$$f(x,y) = \varphi(\lfloor x \rfloor, \lfloor y \rfloor) + \begin{cases} \overrightarrow{\mathrm{Pref}_{k_1}(\varphi(\mathbf{0}))}, & \text{if } k_2 = 0; \\ \overrightarrow{\mathrm{Pref}_{k_2}(\varphi(\mathbf{1}))}, & \text{if } k_1 = 0. \end{cases}$$

Intuitively, the transformation f finds the closest bottom or left branch vertex $(\lfloor x \rfloor, \lfloor y \rfloor)$ of any vertex (x, y), and then consider the k-th vertex in the path $\varphi((x, y), a)$ in $G(\varphi)$, where $k \in \{k_1, k_2\}$ and $a \in \{\mathbf{0}, \mathbf{1}\}$ (see Fig. 4). It is straightforward to check that f is a bijection. It remains to show that $p, q \in \mathbb{Z}^2$ are adjacent in T_φ if and only if $f(p)$ and $f(q)$ are adjacent in $G(\varphi)$.

First, for any $p \in \mathbb{Z}^2$ and $a \in \mathcal{F}$, let $C(p, a)$ be the sequence whose i-th element is $p + i\vec{a}$, for $i = 0, 1, \ldots, |\varphi(a)|$ and consider the sequence $C'(p, a)$ whose i-th element is $f(p + i\vec{a})$, for $i = 0, 1, \ldots, |\varphi(a)|$. Then

$$f(p + i\vec{a}) = \varphi(p) + \overrightarrow{\mathrm{Pref}_i(\varphi(a))}.$$

Consequently, $C(p, a)$ is a chain of T_φ if and only if $C'(p, a)$ is a chain of $G(\varphi)$, since (p, a) is a discrete path of T_φ if and only if $(p, \varphi(a))$ is a discrete path of $G(\varphi)$.

In other words, paths between vertices having integer coordinates in T_φ are isomorphic to path between branch vertices in $G(\varphi)$. By Corollary 9, the degrees of vertices match, so that we have considered all possible neighbors. $\square$

From Theorem 10, we deduce that $G(\mathbb{Z}^2)$ and $G(\varphi)$ have essentially the same structure.

Corollary 11. *Let φ be a parallelogram morphism. Then, $G(\mathbb{Z}^2)$ and $G(\varphi)$ are homeomorphic.*

Finally, from Corollary 11, one deduces that φ preserves both closed and simple paths. In other words, $G(\varphi)$ is a deformed image of $G(\mathbb{Z}^2)$.

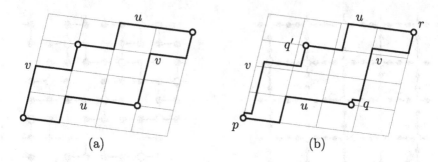

Fig. 5. Geometric representation of the paths u and v. (a) $u, v \in M$. (b) $u, v \notin M$.

5 Main Result

Before proving Theorem 13, we describe the graph distance between particular pairs of vertices in parallelogram networks.

Lemma 12. *Let φ be any parallelogram morphism and p be a vertex of $G(\varphi)$. Moreover, let $q = p + k\overrightarrow{\varphi(a)}$ for some $a \in \mathcal{F}$ and some positive integer k.*

(i) *If p and q are branch vertices, then $(p, \varphi(a)^k)$ is the unique shortest path going from p to q and $\mathrm{dist}_{G(\varphi)}(p, q) = k|\varphi(a)|$.*

(ii) *If p and q are inner vertices of type b, where $b \in \mathcal{F}$ and $b \neq a, \bar{a}$, then $\mathrm{dist}_{G(\varphi)}(p, q) > k|\varphi(a)|$.*

Proof. (i) By definition of parallelogram network, there is a path from p to q in $G(\varphi)$ described by $\varphi(a)^k$. This path is also the shortest: Any other path from p to q must be composed of at least k non-overlapping subpaths of the form $(p_i, \varphi(a))$, where $b = (a + 1) \mod 4$, $p_i = p + i\overrightarrow{\varphi(a)} + j_i\overrightarrow{\varphi(b)}$ and $j_i \in \mathbb{Z}$ for $i = 0, 1, \ldots, k$.

(ii) A shortest path from p to q is obtained by going to the nearest branch vertex, then traveling along $\varphi(b)^k$ and then going to q. Since p and q are inner vertices, the number of edges in this shortest path is more than $k|\varphi(a)|$.   □

We are now ready to solve Conjecture 36 of [8].

Theorem 13. *Let φ be any parallelogram morphism. Then $\varphi(\mathcal{F})$ is a circular code if and only if $\varphi(\mathbf{0})$ and $\varphi(\mathbf{1})$ are both primitive words.*

Proof. ($\Rightarrow$) If $\varphi(\mathcal{F})$ is a circular code, then each of its element must be primitive, in particular $\varphi(\mathbf{0})$ and $\varphi(\mathbf{1})$.

($\Leftarrow$) Let $M = \varphi(\mathcal{F})^*$. We show that M is very pure. Arguing by contradiction, assume the contrary, i.e. there exist $u, v \in \mathcal{F}^*$ such that $uv, vu \in M$ but $u, v \notin M$.

Clearly, $\overrightarrow{uv} = \overrightarrow{vu}$, which implies that the discrete paths (p, uv) and (p, vu) of $G(\varphi)$ end at the same point, for any $p \in \mathbb{Z}^2$. Moreover, there exist branch vertices $p, r \in \mathbb{Z}^2$ of $G(\varphi)$ and inner vertices q, q' of type a, a' of $G(\varphi)$ such that the discrete paths (p, u) and $(r, \widehat{v})$ both end at q and the discrete paths (p, v) and $(r, \widehat{u})$ both end at q' (the situation is depicted in Fig. 5). There are two cases to consider.

First, suppose that $uv = \varphi(b)^k$ for some $b \in \mathcal{F}$. Since $|uv| = |vu|$ and since $(p, \varphi(b)^k)$ is the unique shortest path from p to r (Lemma 12(i)), we deduce that $uv = \varphi(b)^k = vu$. Write $u = u'u''$ and $v = v = v'v''$, where $u', v'' \in M$ and $u''v' = \varphi(b)$ (such a decomposition exists and is unique since $uv \in M$ but $u, v \notin M$). Then $\varphi(b)^k = uv = vu = v'v''u'u''$, which implies that v' is a prefix of $\varphi(b)$ and u'' is a suffix of $\varphi(b)$. Hence, $u''v' = \varphi(b) = v'u''$, so that, by Proposition 2, $\varphi(b)$ is not primitive, contradicting the theorem assumption.

Otherwise, let u' and v' be the maximal words of $\mathcal{F}^*$ such that $\varphi(u')$ is a prefix of u and $\varphi(v')$ is a suffix of v. Let

$$Q = \{q' + \overrightarrow{\varphi(u'')} \mid u'' \in \mathrm{Pref}(u')\} \cup \{q' + \overrightarrow{\varphi(\widehat{v''})} \mid v'' \in \mathrm{Suff}(v')\}.$$

Since $u, v \notin M$, all elements of Q are inner vertices. Moreover, they all are of type a' (the same type as q'). However, there must exist at least two distinct $s, s' \in Q$ such that $s' = s + \overrightarrow{\varphi(b)}$, where $b \neq a', \overline{a'}$: Otherwise, we would have $uv = \varphi(a') = vu$ which was considered in the previous paragraph. But then Lemma 12 applies to s and s', so that $\mathrm{dist}_{G(\varphi)}(s, s') > |\varphi(b)|$, contradicting the fact that s' can be reached from s through the path $(s, \varphi(b))$.   □

6   Concluding Remarks

Theorem 13 might be seen as a first important step in solving Conjecture 1. Indeed, as mentioned in Sect. 4, parallelogram networks are not uniquely represented by a parallelogram morphism φ, since its circular permutations also yield the same parallelogram network. Moreover, there exist examples of parallelogram morphisms having a circular permutation which induces a distinct parallelogram network. In fact, there are infinitely many of them, and their structure has been described in [8].

For instance, it is easy to verify that for any $p \in \mathbb{Z}^2$, $(p, \varphi(\mathbf{0123}))$ is a discrete path of both $G(\varphi)$ and $G(\varphi')$ defined by

$$\varphi(\mathbf{0}) = \mathbf{01010}, \quad \varphi(\mathbf{1}) = \mathbf{121}, \quad \varphi'(\mathbf{0}) = \mathbf{030}, \quad \varphi'(\mathbf{1}) = \mathbf{10101}.$$

However, it seems that no other closed discrete path can exist in both parallelogram networks.

References

1. Anselmo, M., Giammarresi, D., Madonia, M.: Two dimensional prefix codes of pictures. In: Béal, M.-P., Carton, O. (eds.) DLT 2013. LNCS, vol. 7907, pp. 46–57. Springer, Heidelberg (2013)

2. Arqus, D.G., Michel, C.J.: A circular code in the protein coding genes of mito-chondria. J. Theor. Biol. **189**(3), 273–290 (1997). http://www.sciencedirect.com/science/article/pii/S0022519397905130
3. Bassino, F.: Generating functions of circular codes. Adv. Appl. Math. **22**(1), 1–24 (1999)
4. Beauquier, D., Nivat, M.: On translating one polyomino to tile the plane. Discrete Comput. Geom. **6**, 575–592 (1991)
5. Berstel, J., Perrin, D., Reutenauer, C.: Codes and Automata, Encyclopedia of Mathematics and Its Applications, vol. 129. Cambridge University Press, Cambridge (2009)
6. Berthé, V., Labbé, S.: An arithmetic and combinatorial approach to three-dimensional discrete lines. In: Debled-Rennesson, I., Domenjoud, E., Kerautret, B., Even, P. (eds.) DGCI 2011. LNCS, vol. 6607, pp. 47–58. Springer, Heidelberg (2011)
7. Blondin Massé, A., Tall, A.M., Tremblay, H.: On the arithmetics of discrete figures. In: Dediu, A.-H., Martín-Vide, C., Sierra-Rodríguez, J.-L., Truthe, B. (eds.) LATA 2014. LNCS, vol. 8370, pp. 198–209. Springer, Heidelberg (2014)
8. Massé, A.B., Garon, A., Labbé, S.: Combinatorial properties of double square tiles. Theor. Comput. Sci. **502**, 98–117 (2013)
9. Brlek, S., Koskas, M., Provençal, X.: A linear time and space algorithm for detect-ing path intersection in $\mathbb{Z}^d$. Theor. Comput. Sci. **412**(36), 4841–4850 (2011)
10. Brlek, S., Lachaud, J., Provençal, X., Reutenauer, C.: Lyndon + Christoffel = digitally convex. Pattern Recogn. **42**(10), 2239–2246 (2009)
11. Cousineau, G.: Characterization of some periodic tiles by contour words. Oligomer-ization of Chemical and Biological Compounds (2014)
12. Diestel, R.: Graph Theory, 4th edn. Springer, New York (2010)
13. Golomb, S.W., Gordon, B.: Codes with bounded synchronization delay. Inf. Control **8**(4), 355–372 (1965)
14. Gross, J.L., Tucker, T.W.: Topological Graph Theory. Wiley, New York (1987)
15. Lothaire, M.: Combinatorics on Words. Cambridge University Press, Cambridge (1997)
16. Provençal, X.: Combinatoire des mots, géométrie discrète et pavages. Ph.D. thesis, D1715, Université du Québec à Montréal (2008)
17. Schtzenberger, M.: On a factorization of free monoids. Proc. Amer. Math. Soc. **16**, 21–24 (1965)
18. Wijshoff, H., van Leeuwen, J.: Arbitrary versus periodic storage schemes and tes-sellations of the plane using one type of polyomino. Inf. Control **62**(1), 1–25 (1984)

On Del-Robust Primitive Partial Words
with One Hole

Ananda Chandra Nayak[1](✉) and Amit K. Srivastava[2]

[1] Department of Mathematics, Indian Institute of Technology Guwahati,
Guwahati, India
n.ananda@iitg.ernet.in
[2] Department of Computer Science and Engineering, Indian Institute of Technology
Guwahati, Guwahati, India
amit.srivastava@iitg.ernet.in

Abstract. A partial word is a string over a finite alphabet with some
undefined places which are known as holes or "do not know" symbols. A
partial word w is said to be primitive if there does not exist any word
v such that w is contained in v^n with $n \geq 2$. We investigate the effect
of a point mutation on primitive partial words with a single hole. We
characterize a special class of such words, del-robust primitive partial
words with one hole, that remains primitive on deletion of any symbol
or the hole. We identify some important properties of such words and
prove that the language of non-del-robust primitive partial words with
one hole is not context-free. Finally we approximate the counting of del-
robust primitive partial words with one hole for a fixed length.

Keywords: Combinatorics on words · Primitive words · Partial words ·
Del-robust · Context-free language · Reflective

1 Introduction

Let Σ be a finite alphabet. A word is a sequence of symbols from the alphabet Σ.
Words are natural objects of interest in several research areas including formal
language and automata theory [8], coding theory [1], computational biology and
DNA computing [12]. The research in the area of combinatorics on words has
been actively pursued in last two decades. Several aspects of words such as,
algebraic [15], applied [16] and algorithmic [10], have been extensively explored.

A word $w = a_1 a_2 \cdots a_n$ of length n is a sequence of symbols over a finite
alphabet Σ where $a_i \in \Sigma$. A partial word $w = a_1 a_2 \cdots a_n$ of length n over the
finite alphabet where $a_i \in \Sigma \cup \{\Diamond\}$ and $\Diamond$ is referred as hole. Equivalently, a
partial word is a string that contain some "do not know symbols" and a total
word is a string without holes.

In the context of combinatorics on words, primitive words are of special
interest where a word is said to be primitive if it cannot be represented as an
integer power of a smaller word [14]. A partial word is said to be primitive if it
is not contained in an integer(≥ 2) power of a word. The relation between the

© Springer International Publishing Switzerland 2016
A.-H. Dediu et al. (Eds.): LATA 2016, LNCS 9618, pp. 233–244, 2016.
DOI: 10.1007/978-3-319-30000-9_18

language of primitive words and the language of primitive partial words with conventional formal language classes has been extensively studied [9,13,17,19]. It is still an open problem that whether the language of primitive words is context-free [9,19].

In [18], G.Păun et al. have studied the robustness of the language of primitive words with respect to various point mutation operations such as insertion, deletion or substitution of a symbol by another symbol. In [7], Dassow et al. considered the word ww' where w is a primitive word and w' is a modification of w and studied whether ww' is primitive. Similarly, in [6], Sadri et al. extended the work of Dassow et al. to partial words. In this paper we discuss the robustness property of primitive partial words with one hole with respect to deletion of a symbol or a hole from a partial word that preserves the primitivity. This special class of primitive partial words with one hole is referred to as del-robust primitive partial words with one hole.

The rest of the paper is organized as follows. In Sect. 2 we review some basic concepts and preliminaries which are used in the rest of the paper. In Sect. 3, we characterize del-robust primitive partial words with one hole and identify their properties. In Sect. 4, we prove that the language of non-del-robust primitive partial words with one hole is not context-free. We give a lower bound on the number of del-robust primitive partial words with one hole for a given length in Sect. 5. Finally, the conclusions and some open problems are discussed in Sect. 6.

2   Preliminaries

Let Σ be a finite set of symbols known as alphabet. We assume that Σ is a nontrivial alphabet, which means that it has at least two distinct symbols. A total word (referred to as simply a word) $u = a_0 a_1 a_2 \cdots a_{n-1}$ of length n can be defined by a total function $u : \{0, \ldots, n-1\} \mapsto \Sigma$ where each $a_i \in \Sigma$ [2]. We use string and word interchangeably. The set Σ^* is the free monoid generated by Σ which contains all the strings. The length of a string u is the number of symbols contained in it and is denoted by $|u|$, and $\alpha(u)$ is the set of symbols appearing in u from Σ. The empty word, λ, is a word that does not contain any letter and therefore $|\lambda| = 0$. The notation $|u|_a$ denotes the number of times letter a appears in the word u. The set of all words of length n over Σ is denoted by Σ^n. We define $\Sigma^* = \bigcup_{n \geq 0} \Sigma^n$ where $\Sigma^0 = \{\lambda\}$, and $\Sigma^+ = \Sigma^* \setminus \{\lambda\}$ is the free semigroup generated by Σ. A language L over Σ is a subset of Σ^*.

A partial word u of length n over alphabet Σ can be defined by a partial function $u : \{0, \ldots, n-1\} \mapsto \Sigma$. The partial word u contains some do not know symbols known as holes along with the usual symbols. For $0 \leq i < n$, if $u(i)$ is defined, then we say $i \in D(u)$ (the domain of u), otherwise $i \in H(u)$ (the set of holes) [2]. A word is a partial word without any hole. If u and v are two partial words of equal length, then u is said to be *contained* in v, if all elements in $D(u)$ are also in the set $D(v)$ and $u(i) = v(i)$ for all $i \in D(u)$ and u is said to be *compatible* to v if there exists a partial word w such that $u \subset w$ and $v \subset w$. The containment and compatibility are denoted as $\subset$ and $\uparrow$ respectively. Two partial words x and y are conjugate if there exist partial words u and v such that $x \subset uv$ and $y \subset vu$ [3].

A word is said to be primitive if it cannot be expressed as a nontrivial power of another word. Formally, a word w is primitive if there does not exist any word v such that $w = v^n$ with $n \geq 2$. The language of primitive and nonprimitive words over an alphabet Σ are denoted as Q and Z respectively. Similarly, a partial word u is said to be primitive if there does not exist any word v such that $u \subset v^n$, $n \geq 2$. Note that if u is primitive and $u \subset v$, then v is primitive as well [2]. The language of primitive partial words and nonprimitive partial words are denoted as Q_p and Z_p respectively. The language of primitive partial words with i holes is represented as Q_p^i.

Theorem 1 [20]. *Every nonempty word w can be expressed uniquely in the form $w = x^n$, where $n \geq 1$ and x is primitive.*

Observe that the above result is not true for partial words, that is, the uniqueness does not hold in case of partial word. For example, $u = a\Diamond$, we have $u \subset a^2$ and $u \subset ab$ for $\Sigma = \{a, b\}$.

Definition 2 (Reflective Language [18]). *A language L is called reflective if $uv \in L$ implies $vu \in L$, for all $u, v \in \Sigma^*$.*

It is easy to observe that a language L is reflective if it is closed under conjugacy relation. Several facts are known about the language of primitive words Q and the language of primitive partial words Q_p. Let us recall some of them which will be useful later in the paper.

Lemma 3 [20]. *The languages Q and Z are reflective.*

Theorem 4 [2]. *Let u and v be partial words. If there exists a primitive word x such that $uv \subset x^n$ for some positive integer n, then there exists a primitive word y such that $vu \subset y^n$. Moreover, if uv is primitive then vu is primitive.*

Corollary 5. *The language Q_p is reflective.*

Proof. This follows from the Theorem 4.   □

Corollary 6. *The language Z_p is reflective.*

Proof. Let $w = uv$ be a non-primitive partial word. Let $w = uv \subset x^n$ for $n \geq 2$ but $vu \notin Z_p$. If vu is primitive then by Corollary 5 uv is also primitive which is a contradiction. Hence, $vu \in Z_p$.   □

Let $w = uv$ be a nonempty partial word. Then, the partial words u and v are said to be prefix and suffix of w, respectively. A partial word y is said to be a factor of a word w if w can be written as xyz, where $x, z \in (\Sigma \cup \{\Diamond\})^*$ and $y \in (\Sigma \cup \{\Diamond\})^+$. The partial word y is said to be proper factor if $x \neq \lambda$ or $z \neq \lambda$. A prefix (suffix) of length k of a partial word w is denoted as $\mathrm{pref}(w, k)$ ($\mathrm{suff}(w, k)$), respectively, where $k \in \{0, 1, \ldots, |w|\}$ and $\mathrm{pref}(w, 0) = \mathrm{suff}(w, 0) = \lambda$.

The robustness of primitive words has been defined in [18]. Given a primitive word, w, the robustness of w is considered with respect to insertion of a symbol from Σ, deletion of a symbol from w, substitution of a symbol in w by another symbol from Σ. We state some of the properties about robustness of primitive words and primitive partial words that will be useful later.

The next result shows the possibility of obtaining primitive partial words by appending a symbol or removing the last symbol in any nonempty partial word. Specifically, if u is nonempty partial word with one hole, then at least one of the u or ua is primitive for $a \in \Sigma$.

Lemma 7 [2].

(i) Let u be a partial word with one hole such that $|\alpha(u)| \geq 2$. If a is any letter then u or ua is primitive.

(ii) Let u_1, u_2 be nonempty partial words such that $u_1 u_2$ has one hole such that $|\alpha(u_1 u_2)| \geq 2$. Then for any letter a, $u_1 u_2$ or $u_1 a u_2$ is primitive.

The next result is an extension of Lemma 7 in total words.

Lemma 8 [18]. For every word $u \in \Sigma^+$ and all symbols $a, b \in \Sigma$, where $a \neq b$, at least one of the words ua, ub is primitive.

The next proposition holds for partial words with exactly one hole.

Proposition 9 [2]. Let u be a partial word with one hole which is not of the form $x \Diamond x$ for any word x. If a and b are distinct letters, then ua or ub is primitive.

The following proposition shows the possibility of obtaining primitive word by deletion of a symbol in a primitive word.

Proposition 10 [18]. Every word $w \in Q, |w| \geq 2$, can be written in the form $w = u_1 a u_2$, for some u_1, $u_2 \in \Sigma^*$, $a \in \Sigma$, such that $u_1 u_2 \in Q$.

We have the following result for partial primitive word with one hole that gives an alternative to obtain new primitive partial words by deleting a symbol $a \in \Sigma \cup \{\Diamond\}$.

Lemma 11. Every primitive partial word $w \in Q_p^1$, $|w| \geq 3$ can be written as $w = u_1 a u_2$ for some $a \in \Sigma \cup \{\Diamond\}$ such that $u_1 u_2 \in Q_p$.

Proof. Take $w \in Q_p^1$ such that $|w| \geq 3$. Since $w \in Q_p^1$, the word w can be written as either $w = v_1 a \Diamond v_2$ or $w = v_1 \Diamond a v_2$ for some v_1, $v_2 \in \Sigma^*$, $a \in \Sigma$. From Lemma 7, one of the words $u_1 u_2$ or $u_1 a u_2$ is primitive. The partial word $u_1 u_2$ is obtained by deleting the symbol a from $w = v_1 a \Diamond v_2$. Hence, $u_1 = v_1$ and $u_2 = \Diamond v_2$. The partial word $u_1 a u_2$ is obtained from $w = v_1 a \Diamond v_2$ by erasing $\Diamond$ and therefore $u_1 = v_1 a$ and $u_2 = v_2$. The argument is similar for $w = v_1 \Diamond a v_2$. $\qquad\square$

3   Del-Robust Primitive Partial Words with One Hole

In this section, we study a special class of primitive partial words having one hole which remain primitive after deletion of any one symbol. We refer to this special class as del-robust primitive partial words with one hole that are formally defined as follows.

Definition 12 (Del-Robust Primitive Partial Words). *A primitive partial word w of length n with one hole is said to be del-robust if and only if the partial word*

$$\text{pref}(w, i) \cdot \text{suff}(w, n - i - 1)$$

is a primitive partial word for all $i \in \{0, 1, \ldots, n - 1\}$.

Note that a del-robust primitive partial word must remain primitive on deletion of any symbol or the hole. The number of such words is infinite. For example, $a \Diamond bba$ and $a^m \Diamond b^n$ for m, $n \geq 2$ are del-robust primitive partial words with one hole.

   We denote the set of all del-robust primitive partial words with one hole by Q_p^{1D} over an alphabet $\Sigma \cup \{\Diamond\}$. It is obvious that the language of del-robust primitive partial words with one hole is a subset of Q_p^1 and hence $w \in Q_p^1$ for all $w \in Q_p^{1D}$ where Q_p^1 be the set of all primitive partial words with one hole. Let $Q_p^1(1) = Q_p^1 \cup \{\lambda\}$, and for all $n \geq 2$ let $Q_p^1(n) = \{u^n \mid u \in Q_p^1\}$.

Proposition 13. *Let m and n be two distinct positive integers. Then $Q_p^1(m) \cap Q_p^1(n) = \phi$ (ϕ is the usual set symbol denote an empty set).*

   Next we give complete structural characterization of those primitive partial words with one hole which are in the set Q_p^1 but not in Q_p^{1D}, that is, deletion of a symbol or hole from such words will result in non-primitive partial words.

Theorem 14. *A primitive partial word w with one hole is not del-robust if and only if w is contained in any word which is of the form $u^{k_1} u_1 a u_2 u^{k_2}$ where $u = u_1 u_2$, $a \in \Sigma$, k_1, $k_2 \geq 0$ and $k_1 + k_2 \geq 1$.*

Proof. The necessary and sufficient parts are proved as follows.

($\Leftarrow$) The sufficient part is easy. Let us consider a primitive partial word with one hole $w \subset u^{k_1} u_1 a u_2 u^{k_2}$ where $u_1 u_2 = u$ and $a \in \Sigma$. If the symbol in w for which a is replaced in the word $u^{k_1} u_1 a u_2 u^{k_2}$ is deleted then the obtained partial word will be contained in exact integer power of u which is non-primitive. Hence, w is not a del-robust primitive partial word.

($\Rightarrow$) Let w be a primitive partial word with one hole which is not del-robust. Therefore w can be written as $w = w_1 c w_2$ for some $c \in \Sigma \cup \{\Diamond\}$ such that $w_1 w_2$ is not a primitive partial word. Thus $w_1 w_2 \subset u^n$ for $u \in Q$ and $n \geq 2$. Hence, $w_1 \subset u^r u_1$ and $w_2 \subset u_2 u^s$ for r, $s \geq 0$ and $r + s \geq 1$ such that $u_1 u_2 = u$. Therefore, $w \subset u^{k_1} u_1 a u_2 u^{k_2}$ where $c \subset a$. $\square$

It is easy to observe that the language of primitive partial words with one hole Q_p^1 is closed under cyclic permutation. Next we prove that similar result also holds for del-robust primitive partial words with one hole Q_p^{1D}.

Lemma 15. Q_p^{1D} *is reflective.*

Proof. We prove it by contradiction. Let a partial word $w = xy \in Q_p^{1D}$ such that $yx \notin Q_p^{1D}$. Since $w \in Q_p^{1D}$, hence $w \in Q_p^1$. So $yx \subset u^{k_1} u_1 a u_2 u^{k_2}$ where $u = u_1 u_2$, $a \in \Sigma$ and $k_1 + k_2 \geq 1$. We consider two cases depending upon the inclusion of a either in y or in x.

Case 1. If a is included in y, we consider two cases.

    Case 1.1. If $u_1 a u_2$ is completely in y, then $y \subset u^{k_1} u_1 a u_2 u^r u_1'$ and $x \subset u_2' u^s$ where $u_1' u_2' = u$, $r + s + 1 = k_2$. Now $xy \subset u_2' u^s u^{k_1} u_1 a u_2 u^r u_1'$ which will not be del-robust after deletion of the symbol for which a is replaced and the obtained partial word will be non-primitive. Hence, this is a contradiction to the assumption that $xy \in Q_p^{1D}$.

    Case 1.2. If $u_1 a u_2$ is not completely in y, that is, some portion of u_2 is in y, then $y \subset u^{k_1} u_1 a u_2'$ and $x \subset u_2'' u^{k_2}$ where $u = u_1 u_2$ and $u_2 = u_2' u_2''$. Now $xy \subset u_2'' u^{k_2} u^{k_1} u_1 a u_2'$ which will not result in a del-robust primitive partial word after deletion of the symbol for which a is replaced. Moreover, the partial word will be contained in $(u_2'' u_1 u_2')^{k_1 + k_2 + 1}$ and xy is a non-primitive partial word. Hence it is a contradiction.

Case 2. If a is included in x, we need to consider two cases with similar proofs as in the previous case.

Hence the language of del-robust primitive partial words with one hole Q_p^{1D} is reflective.   □

Next we study the subset of primitive partial words with one hole in which deletion of a symbol results in a non-primitive partial word. We call such partial words as non-del-robust primitive partial words with one hole and the set is denoted by $Q_p^{1\overline{D}}$.

Definition 16 (Non-Del-Robust Primitive Partial Words). *A primitive partial word w with one hole is said to be non-del-robust if and only if deletion of a symbol $a \in \Sigma \cup \{\Diamond\}$ from w makes it non-primitive.*

Thus $Q_p^{1\overline{D}} = Q_p^1 \setminus Q_p^{1D}$ where '\' is the set difference operator. The number of such non-del-robust words is infinite. For example, $a \Diamond b$, $ba^n \Diamond$ for $n \geq 1$ are non-del-robust words.

Corollary 17. $Q_p^{1\overline{D}}$ *is reflective.*

Theorem 18. *A primitive partial word w with one hole is non-del-robust if and only if w is contained in $u^n a$ or is contained in its cyclic permutation for some for some $u \in Q_p$ where $a \in \Sigma$, $n \geq 2$, and $|\alpha(u)| \geq 2$.*

Proof. We prove the necessary and sufficient conditions as follows:

($\Rightarrow$) Let $w \in Q_p^1$ be non-del-robust, that is, $w \in Q_p^{1\overline{D}}$. So w is contained in the word which of the form $u^p u_1 c u_2 u^q$ for some $u = u_1 u_2 \in Q$ and $c \in \Sigma$. Since $Q_p^{1\overline{D}}$ is reflective, so $u_2 u^q u^p u_1 c = (u_2 u_1)^{p+q+1} c$ is also in $Q_p^{1\overline{D}}$. Hence $w \subset (u_2 u_1)^{p+q+1} c$.

($\Leftarrow$) Let w be either contained in the word $u^n a$ or its cyclic permutation. Deletion of that symbol in w for which a is replaced from $u^n a$ gives a non-primitive partial word $w' \subset u^n$ which is non-primitive (Z_p is reflective). Hence, $w \in Q_p^{1\overline{D}}$.   □

Let us prove a simple observation which shows that if a primitive partial word w with one hole is del-robust then the reverse of w denoted by $rev(w)$ is also del-robust.

Lemma 19. *If $w \in Q_p^{1D}$ then $rev(w) \in Q_p^{1D}$.*

Proof. We prove this by contradiction. Let w be a primitive partial word with one hole which is del-robust, that is, $w \in Q_p^{1D}$ but $rev(w) \notin Q_p^{1D}$. Using the structural characterization of non-del-robust words, we have $rev(w) \subset u^m u_1 a u_2 u^n$ where $u = u_1 u_2$, $m + n \geq 1$. Now,

$$rev(rev(w)) \quad = w \subset rev(u^m u_1 a u_2 u^n)$$
$$rev(u^m u_1 a u_2 u^n) = (rev(u))^n rev(u_2) rev(a) rev(u_1)(rev(u))^m$$

Since $u = u_1 u_2$, we have $rev(u) = rev(u_2) rev(u_1)$. Thus,

$$w \subset (rev(u_2) rev(u_1))^n rev(u_2) a rev(u_1)(rev(u_2) rev(u_1))^m.$$

It is clear that w is non-del-robust which is a contradiction to the assumption.   □

An existing algorithm [4] that recognizes if a given partial word, w, with at most one hole is primitive uses the fact that if w is primitive and $ww \uparrow xwy$ then it implies that either $x = \lambda$ or $y = \lambda$. Observe that if a partial word with one hole w is in set $Q_p^{1\overline{D}}$ then there exists a cyclic permutation of w that contains a non-primitive partial word factor of length $|w| - 1$. Also note that ww contains all the cyclic permutations of w.

Theorem 20. *Let w be a primitive partial word with one hole. Then w is non-del-robust if and only if ww contains at least one non-primitive partial word factor of length $|w| - 1$.*

Proof. The necessary and sufficient conditions are proved below.

($\Rightarrow$) Let w be a primitive partial word with one hole. Let $w \in Q_p^{1\overline{D}}$. Since w is non-del-robust then $w \subset u^p u_1 a u_2 u^q$ for some $u \in Q$, $u_1 u_2 = u$, $a \in \Sigma$ and $p + q \geq 1$. So, $ww \subset u^p u_1 a u_2 u^q u^p u_1 a u_2 u^q$ which has a factor $u_2 u^q u^p u_1 = (u_2 u_1)^{p+q+1}$ of length $|w| - 1$. Since $p + q + 1 \geq 2$, so $(u_2 u_1)^{p+q+1}$ is a non-primitive word. Hence a factor of ww contains one non-primitive partial word of length $|w| - 1$.

($\Leftarrow$) Let the partial word ww has a non-primitive factor of length $|w| - 1$ where $w \in Q_p^1$. So, $ww \subset u_1 v^n u_2$, where $u_1, u_2 \in \Sigma^*$, $|v^n| = |w| - 1$, $v \in Q$ and $n \geq 2$. There are two cases depending upon whether v^n entirely lies in w or partly in w.

Case 1. Let v^n, $n \geq 2$ is entirely lies in w. Since $|v^n| = |w| - 1$, so $w \subset v^n a$. The deletion of that particular symbol $c \in \Sigma \cup \{\Diamond\}$ for which a is replaced will gives a non-primitive partial word. Hence, w is non-del-robust.

Case 2. Let v^n be partially in w and $ww \subset u_1 v^n u_2$. Since Z is reflective, so $u_2 u_1 v^n \in Z$. So there must exists a partial word x such that $xx \subset u_2 u_1 v^n$ where x is a cyclic permutation of w. If v^n is entirely in x, then either $x \subset a v^n$ or $x \subset v^n a$. In both cases the word x is a non-del-robust primitive partial word. As w is a cyclic permutation of x and Q_p is reflective, we conclude that w is a non-del-robust partial word. $\square$

4   Context-Freeness of $Q_p^{1\overline{D}}$

In this section we investigate the relation between the language of non-del-robust primitive partial words with one hole $Q_p^{1\overline{D}}$ and the conventional language classes in Chomsky hierarchy. In particular, we show that $Q_p^{1\overline{D}}$ is not a Context-Free Language (CFL) over a nontrivial alphabet $\Sigma \cup \{\Diamond\}$. Let us recall the pumping lemma for context-free language which is required to prove this result.

Lemma 21 (Pumping Lemma for Context-Free Languages [11]). *Let L be a CFL. Then there exists an integer $n > 0$ such that for every $u \in L$ with $|u| \geq n$, u can be decomposed into $vwxyz$ such that the following conditions hold:*

(a) $|wxy| \leq n$.
(b) $|wy| > 0$.
(c) $vw^i xy^i z \in L$ for all $i \geq 0$.

Theorem 22. *$Q_p^{1\overline{D}}$ is not a context-free language.*

Proof. We prove it by contradiction. Let us assume that $Q_p^{1\overline{D}}$ is a CFL. Let $n > 0$ be an integer which is the pumping length that is guaranteed to exist by pumping lemma. Since $Q_p^{1\overline{D}}$ is context-free, then it satisfies all the conditions of Lemma 21. Consider a string $w = a^n b^n a^n b^n a^n b^n \Diamond$ where $a, b \in \Sigma$ and are distinct. It is clear that $w \in Q_p^{1\overline{D}}$ and of length at least n.

Hence, by the pumping lemma for CFL, w can be factorized into $uvxyz$ such that $|vy| \geq 1$, $|vxy| \leq n$ and for all $i \geq 0$, $uv^i xy^i z \in Q_p^{1\overline{D}}$. There are several possibilities, that we consider below, depending on whether the substrings v and y contain more than one alphabet symbol or hole.

Case 1. When both v and y contain one type of symbol, that is v does not contain both a's and b's, and same holds for y. Consider one such case. Let v and y contain only a's from the first set of a's. Let $vy = a^k$ for some $k > 0$. Let $u = a^j$, $vxy = a^p$ and $z = a^q b^n a^n b^n a^n b^n \Diamond$ such that $j \geq 0$ and

$j + p + q = n$. Now for $i = 0$, we have $uv^i x y^i z = a^j a^{p-k} a^q b^n a^n b^n a^n b^n \Diamond = a^{j+p+q-k} b^n a^n b^n a^n b^n \Diamond = a^{n-k} b^n a^n b^n a^n b^n \Diamond \notin Q_p^{1\overline{D}}$ for $k > 1$ and for $k = 1$, we have $a^{n-1} b^n a^n b^n a^n b^n \Diamond \subset (a^{n-1} b^n a)^3 \notin Q_p^{1\overline{D}}$.

Similar cases can be handled if both v and y contain only symbol b.

Case 2. If v and y contain more than one type of symbol. There will be several cases depending upon whether v contains combinations of a's and b's and y contains only one type of symbol or v contains one type of symbol and y contains the combination of symbols or x contains combinations of a's and b's. Let us consider one such case.

Let $vxy = a^j b^k$ for some j and k such that $0 < j + k \leq n$. Observe that $j, k > 0$ otherwise it will fall into Case 1. Suppose $u = a^l$, $v = a^{j_1}$, $x = a^{j_2}$, $y = a^{j_3} b^k$ and $z = b^p a^n b^n a^n b^n \Diamond$ such that $j_1 + j_2 + j_3 = j$, $l + j = n$, and $k + p = n$. For $i = 0$, the string $uv^i x y^i z = a^{l+j_2} b^p a^n b^n a^n b^n \Diamond \notin Q_p^{1\overline{D}}$ as $l + j_2 < n$ and $p < n$.

Similarly other cases in which v and y contain more than one symbol can be handled.

Case 3. Let us consider the last case. If $vxy = b^p \Diamond$ then there are following possibilities:

(a) If the symbol $\Diamond$ is in vy then $vy = b^l \Diamond$ and $x = b^{p-l}$. For $i = 0$, $uv^i x y^i z = a^n b^n a^n b^n a^n b^{n-p} \notin Q_p^{1\overline{D}}$.

(b) If the symbol $\Diamond$ is in x then $v = b^l$, $y = \lambda$ and $x = b^{p-l} \Diamond$ and $l \geq 1$. Now, $uv^i x y^i z = a^n b^n a^n b^n a^n b^{n-l} \Diamond \notin Q_p^{1\overline{D}}$ for $i = 0$.

Observe that one of the above cases will occur. Since all the above cases result in a contradiction, the assumption that the language of non-del-robust primitive partial words with one hole $Q_p^{1\overline{D}}$ is context-free is not true. □

Next we prove that the language of non-del-robust primitive partial words is not context-free in general.

Corollary 23. *The language $Q_p^{1\overline{D}}$ is not context-free over the alphabet $\Sigma \cup \{\Diamond\}$ where Σ has at least two distinct letters.*

Proof. This is a direct consequence of Theorem 22. □

5   Counting n-Length Words in Q_p^{1D}

In this section we give a lower bound on the number of del-robust primitive partial words with one hole of length n. We begin the counting with some notation. Denote by $Q_p^{1D}(n)$ (respectively, $Q_p^{1\overline{D}}(n)$) the set of del-robust (respectively, non-del-robust) primitive partial words with one hole of length n over an alphabet Σ.

We use earlier results [4] on counting of primitive words as well as primitive partial words. Let $\mathcal{P}_{h,k}(n)$ (respectively, $\mathcal{N}_{h,k}(n)$) denote the set of primitive (respectively, non-primitive) partial words with h holes of length n over an alphabet of size k. Also, denote by $P_{h,k}(n)$ (respectively, $N_{h,k}(n)$) the number of primitive (respectively, non-primitive) partial words with h holes of length n over an alphabet of size k. Let $T_{h,k}(n)$ denote the total number of partial words of length n with h holes over Σ. We have, from [5],

$$T_{h,k}(n) = P_{h,k}(n) + N_{h,k}(n), \text{ and}$$

$$T_{h,k}(n) = \binom{n}{h} k^{n-h} = \frac{n!}{h!(n-h)!} k^{n-h}$$

Theorem 24 [3]. $N_{1,k}(n) = nN_{0,k}(n)$.

A consequence of Theorem 24 can be observed as follows.

Corollary 25 [3]. *The equality* $P_{1,k}(n) = n(P_{0,k}(n) + k^{n-1} - k^n)$ *holds.*

Next we give an upper bound on the number of non-del-robust primitive partial words of length n over an alphabet of size k.

Theorem 26. *The following inequality holds:*

$$|Q_p^{1\overline{D}}(n)| \leq \begin{cases} [nk - (n-2)]\, N_{1,k}(n-1) - 2k(n-1) & \text{if } 2 \nmid (n-2), \\ [nk - (n-2)]\, N_{1,k}(n-1) - 2k(n-1) - 2k^2 & \text{otherwise.} \end{cases}$$

Proof. We give a constructive proof of the theorem. A non-del-robust primitive partial word with one hole of length n can be obtained in two ways either by inserting a hole $\Diamond$ in a non-primitive word of length $(n-1)$ or inserting a symbol $a \in \Sigma$ in a non-primitive partial word with one hole of length $(n-1)$. But the first case results in repetition because all those partial words that will be generated by inserting $\Diamond$ in a word $w \in Z_{n-1}$ are already in the second case because the partial words are obtained from words by replacing holes. Thus we will consider only the second case, that is, inserting a symbol $a \in \Sigma$ in a partial word $w \in N_{1,k}(n-1)$ to obtain non-del-robust primitive partial words with one hole of length n.

Take a partial word $w \in N_{1,k}(n-1)$. The number of new partial words that will be generated by inserting a symbol $a \in \Sigma$ in w is

$$|\{w_1 a w_2 \mid w = w_1 w_2, \; w_1, w_2 \in (\Sigma \cup \{\Diamond\})^*\}| = n - |w|_a$$

Now for a partial word $w \in N_{1,k}(n-1)$, the total number of partial words generated by inserting any symbol $a \in \Sigma$ is

$$\sum_{a \in \Sigma} |\{w_1 a w_2 \mid w = w_1 w_2, \; w_1, w_2 \in (\Sigma \cup \{\Diamond\})^*\}| = nk - (n-2)$$

where k is the size of the alphabet.

As we can observe there are non-primitive partial words with one hole which remains non-primitive by inserting a symbol $a \in \Sigma$ and such partial words are of the form $\{w \mid w \subset a^n\}$ or $\{x \Diamond x \mid x \in \Sigma^+\}$. Since, in case of partial words, position of a hole gives different partial words, so there are total $k(n-1)$ non-primitive partial words. Inserting a symbol $a \in \Sigma$ before or after $\Diamond$ gives two different partial words. Hence, the total number of partial words is $2k(n-1)$.

Also, there are some non-primitive partial words which are contained in a^{n-1} but inserting a symbol b where $b \neq a$ gives a non-primitive partial word. For example, $aa\Diamond aa$. Inserting $b \neq a$ either in suffix (respectively, prefix) gives us $aa\Diamond aab$ (respectively, $baa\Diamond aa$) which are non-primitive. Those case in which $(n-2)$ is divisible by 2, we have to extract some extra partial words. Hence,

$$|Q_p^{1\overline{D}}(n)| \leq \begin{cases} [nk - (n-2)] N_{1,k}(n-1) - 2k(n-1) & \text{if } 2 \nmid (n-2), \\ [nk - (n-2)] N_{1,k}(n-1) - 2k(n-1) - 2k^2 & \text{otherwise.} \end{cases}$$

□

Now the lower bound on the count of number of del-robust primitive partial words with one hole of length n can be obtained by subtracting the number of non-del-robust primitive words with one hole of length n from the total number of primitive partial words of length n with one hole. This gives us

$$|Q_p^{1D}(n)| \geq \begin{cases} [nk - (n-2)] N_{1,k}(n-1) - 2k(n-1) & \text{if } 2 \nmid (n-2), \\ [nk - (n-2)] N_{1,k}(n-1) - 2k(n-1) - 2k^2 & \text{otherwise.} \end{cases}$$

6 Conclusions

In this paper, we have discussed a special subclass of primitive partial words with one hole referred to as del-robust primitive partial words. We have characterized such words and identified several properties. We have also proved that the language of non-del-robust primitive partial words with one hole $Q_p^{1\overline{D}}$ over a nontrivial alphabet is not context-free. We have provided a lower bound on the number of such partial words of a given length.

Several interesting questions for the language of del-robust primitive partial words need further exploration. We mentioned a few of them: (1) Generalizing the del-robustness properties for primitive partial words with at least two holes. (2) Is the language of del-robust primitive partial words with one hole Q_p^{1D} context-free? We believe that proving Q_p^{1D} is not context-free will help to solve the long standing conjecture that the language of primitive words, Q, is not context-free.

References

1. Berstel, J., Perrin, D., Reutenauer, C.: Codes and Automata. Cambridge University Press, Cambridge (2009)
2. Blanchet-Sadri, F.: Primitive partial words. Discrete Appl. Math. **148**(3), 195–213 (2005)
3. Blanchet-Sadri, F.: Primitive Partial Words. In: Algorithmic Combinatorics on Partial Words, p. 171. CRC Press, Boca Raton(2007)

4. Blanchet-Sadri, F.: Algorithmic Combinatorics on Partial Words. CRC Press, Boca Raton (2007)
5. Blanchet-Sadri, F., Cucuringu, M.: Counting primitive partial words. J. Automata Lang. Comb. **15**(3/4), 199–227 (2010)
6. Blanchet-Sadri, F., Nelson, S., Tebbe, A.: On operations preserving primitivity of partial words with one hole. In: AFL, pp. 93–107 (2011)
7. Dassow, J., Martin, G.M., Vico, F.J.: Some operations preserving primitivity of words. Theor. Comput. Sci. **410**(30), 2910–2919 (2009)
8. Dömösi, P., Horváth, S., Ito, M.: On the connection between formal languages and primitive words. In: First Session on Scientific Communication, Univ. of Oradea, Oradea, Romania, pp. 59–67, June 1991
9. Dömösi, P., Horváth, S., Ito, M., Kászonyi, L., Katsura, M.: Formal languages consisting of primitive words. In: Ésik, Z. (ed.) FCT 1993. LNCS, vol. 710, pp. 194–203. Springer, Heidelberg (1993)
10. Gusfield, D.: Algorithms on Strings, Trees and Sequences: Computer Science and Computational Biology. Cambridge University Press, Cambridge (1997)
11. Hopcroft, J.E., Motwani, R., Ullman, J.D.: Introduction to Automata Theory, Languages, and Computation. Addison-Wesley, Reading (2003)
12. Leupold, P.: Partial words for DNA coding. In: Ferretti, C., Mauri, G., Zandron, C. (eds.) DNA 2004. LNCS, vol. 3384, pp. 224–234. Springer, Heidelberg (2005)
13. Lischke, G.: Primitive words and roots of words. Acta Univ. Sapientiae Informatica **3**(1), 5–34 (2011)
14. Lothaire, M.: Combinatorics on Words. Cambridge University Press, Cambridge (1997)
15. Lothaire, M.: Algebraic Combinatorics on Words, vol. 90. Cambridge University Press, Cambridge (2002)
16. Lothaire, M.: Applied Combinatorics on Words. Cambridge University Press, Cambridge (2005)
17. Nayak, A.C., Kapoor, K.: On the language of primitive partial words. In: Dediu, A.-H., Formenti, E., Martín-Vide, C., Truthe, B. (eds.) LATA 2015. LNCS, vol. 8977, pp. 436–445. Springer, Heidelberg (2015)
18. Păun, G., Santean, N., Thierrin, G., Yu, S.: On the robustness of primitive words. Discrete Appl. Math. **117**(1), 239–252 (2002)
19. Petersen, H.: On the language of primitive words. Theor. Comput. Sci. **161**(1), 141–156 (1996)
20. Shallit, J.: A Second Course in Formal Languages and Automata Theory. Cambridge University Press, Cambridge (2008)

Optimal Bounds for Computing α-gapped Repeats

Maxime Crochemore[1], Roman Kolpakov[2], and Gregory Kucherov[3(✉)]

[1] King's College London,
London WC2R 2LS, UK
Maxime.Crochemore@kcl.ac.uk

[2] Lomonosov Moscow State University, Leninskie Gory, Moscow 119992, Russia
foroman@mail.ru

[3] LIGM/CNRS, Université Paris-Est, 77454 Marne-la-vallée, France
Gregory.Kucherov@univ-mlv.fr

Abstract. Following (Kolpakov et al., 2013; Gawrychowski and Manea, 2015), we continue the study of α-*gapped repeats* in strings, defined as factors uvu with $|uv| \leq \alpha|u|$. Our main result is the $O(\alpha n)$ bound on the number of *maximal* α-gapped repeats in a string of length n, previously proved to be $O(\alpha^2 n)$ in (Kolpakov et al., 2013). For a closely related notion of maximal δ-subrepetition (maximal factors of exponent between $1 + \delta$ and 2), our result implies the $O(n/\delta)$ bound on their number, which improves the bound of (Kolpakov et al., 2010) by a $\log n$ factor.

We also prove an algorithmic time bound $O(\alpha n + S)$ (S size of the output) for computing all maximal α-gapped repeats. Our solution, inspired by (Gawrychowski and Manea, 2015), is different from the recently published proof by (Tanimura et al., 2015) of the same bound. Together with our bound on S, this implies an $O(\alpha n)$-time algorithm for computing all maximal α-gapped repeats.

1 Introduction

Notation and basic definitions. Let $w = w[1]w[2]\ldots w[n] = w[1\,..\,n]$ be an arbitrary word. The length n of w is denoted by $|w|$. For any $1 \leq i \leq j \leq n$, word $w[i]\ldots w[j]$ is called a *factor* of w and is denoted by $w[i\,..\,j]$. Note that notation $w[i\,..\,j]$ denotes two entities: a word and its occurrence starting at position i in w. To underline the second meaning, we will sometimes use the term *segment*. Speaking about the equality between factors can also be ambiguous, as it may mean that the factors are identical words or identical segments. If two factors u, v are identical words, we call them *equal* and denote this by $u = v$. To express that u and v are the same segment, we use the notation $u \equiv v$. For any $i = 1\ldots n$, factor $w[1\,..\,i]$ (resp. $w[i\,..\,n]$) is a *prefix* (resp. *suffix*) of w. By *positions* on w we mean indices $1, 2, \ldots, n$ of letters in w. For any factor $v \equiv w[i\,..\,j]$ of w, positions

---

R. Kolpakov—The author was partially supported by Russian Foundation for Fundamental Research (Grant 15-07-03102).

A.-H. Dediu et al. (Eds.): LATA 2016, LNCS 9618, pp. 245–255, 2016.
DOI: 10.1007/978-3-319-30000-9_19

i and j are called respectively *start position* and *end position* of v and denoted by $beg(v)$ and $end(v)$ respectively. Let u, v be two factors of w. Factor u *is contained* in v iff $beg(v) \leq beg(u)$ and $end(u) \leq end(v)$. Letter $w[i]$ *is contained* in v iff $beg(v) \leq i \leq end(v)$.

A positive integer p is called a *period* of w if $w[i] = w[i + p]$ for each $i = 1, \ldots, n-p$. We denote by $per(w)$ the *smallest period* of w and define the *exponent* of w as $exp(w) = |w|/per(w)$. A word is called *periodic* if its exponent is at least 2. Occurrences of periodic words are called *repetitions*.

Repetitions, squares, runs. Patterns in strings formed by repeated factors are of primary importance in word combinatorics [24] as well as in various applications such as string matching algorithms [10,13], molecular biology [16], or text compression [26]. The simplest and best known example of such patterns is a factor of the form uu, where u is a nonempty word. Such repetitions are called *squares*. Squares have been extensively studied. While the number of all square occurrences can be quadratic (consider word a^n), it is known that the number of *primitively-rooted* squares is $O(n \log n)$ [10], where a square uu is primitively-rooted if the exponent of u is not an integer greater than 1. An optimal $O(n \log n)$-time algorithm for finding all primitively-rooted squares was proposed in [5].

Repetitions can be seen as a natural generalization of squares. A repetition in a given word is called *maximal* if it cannot be extended by at least one letter to the left nor to the right without changing (increasing) its minimal period. More precisely, a repetition $r \equiv w[i \mathbin{..} j]$ in w is called *maximal* if it satisfies the following conditions:

1. $w[i - 1] \neq w[i - 1 + per(r)]$ if $i > 1$,
2. $w[j + 1 - per(r)] \neq w[j + 1]$ if $j < n$.

For example, word $\mathsf{cababaaa}$ has two maximal repetitions: ababa and aaa. Maximal repetitions are usually called *runs* in the literature. Since any repetition is contained in some run, the set of all runs can be considered as a compact encoding of all repetitions in the word, and can then be used to efficiently infer various useful properties related to repetitions [7]. For any word w, we denote by $\mathcal{R}(w)$ the number of maximal repetitions in w and by $\mathcal{E}(w)$ the sum of exponents of all maximal repetitions in w. Let $\mathcal{R}(n) = \max_{|w|=n} \mathcal{R}(w)$ and $\mathcal{E}(n) = \max_{|w|=n} \mathcal{E}(w)$. The following statements are proved in [18].

Theorem 1. $\mathcal{E}(n) = O(n)$.

Corollary 1. $\mathcal{R}(n) = O(n)$.

A series of papers (e.g., [6,9]) focused on more precise upper bounds on $\mathcal{E}(n)$ and $\mathcal{R}(n)$ trying to obtain the best possible constant factor behind the O-notation. A breakthrough in this direction was recently made in [2] where the so-called "runs conjecture" $\mathcal{R}(n) < n$ was proved. To the best of our knowledge, the currently best upper bound $\mathcal{R}(n) \leq \frac{22}{23}n$ on $\mathcal{R}(n)$ is shown in [12].

On the algorithmic side, an $O(n)$-time algorithm for finding all runs in a word of length n was proposed in [18] for the case of constant-size alphabet. Another $O(n)$-time algorithm, based on a different approach, has been proposed in [2]. The $O(n)$ time bound holds for the (polynomially-bounded) integer alphabet as well, see, e.g., [2]. However, for the case of unbounded-size alphabet where characters can only be tested for equality, the lower bound $\Omega(n \log n)$ on computing all runs has been known for a long time [25]. It is an interesting open question (raised over 20 years ago in [3]) whether the $O(n)$ bound holds for an unbounded linearly-ordered alphabet. Some results related to this question have recently been obtained in [23].

Gapped repeats and subrepetitions. Another natural generalization of squares are factors of the form uvu where u and v are nonempty words. We call such factors *gapped repeats*. For a gapped repeat uvu, the left (resp. right) occurrence of u is called the *left* (resp. *right*) *copy*, and v is called the *gap*. The *period* of this gapped repeat is $|u| + |v|$. For a gapped repeat π, we denote the length of copies of π by $c(\pi)$ and the period of π by $p(\pi)$. Note that a gapped repeat $\pi = uvu$ may have different periods, and $per(\pi) \leq p(\pi)$. For example, in string cabacaabaa, segment abacaaba corresponds to two gapped repeats having copies a and aba and periods 7 and 5 respectively. Gapped repeats forming the same segment but having different periods are considered distinct. This means that to specify a gapped repeat it is generally not sufficient to specify its segment. If u', u'' are equal non-overlapping factors and u' occurs to the left of u'', then by (u', u'') we denote the gapped repeat with left copy u' and right copy u''. For a given gapped repeat (u', u''), equal factors $u'[i..j]$ and $u''[i..j]$, for $1 \leq i \leq j \leq |u'|$, of the copies u', u'' are called *corresponding factors* of repeat (u', u'').

For any real $\alpha > 1$, a gapped repeat π is called *α-gapped* if $p(\pi) \leq \alpha c(\pi)$. Maximality of gapped repeats is defined similarly to repetitions. A gapped repeat $(w[i'..j'], w[i''..j''])$ in w is called *maximal* if it satisfies the following conditions:

1. $w[i'-1] \neq w[i''-1]$ if $i' > 1$,
2. $w[j'+1] \neq w[j''+1]$ if $j'' < n$.

In other words, a gapped repeat π is maximal if its copies cannot be extended to the left nor to the right by at least one letter without breaking its period $p(\pi)$. As observed in [21], any α-gapped repeat is contained either in a (unique) maximal α-gapped repeat with the same period, or in a (unique) maximal repetition with a period which is a divisor of the repeat's period. For example, in the above string cabacaabaa, gapped repeat (ab)aca(ab) is contained in maximal repeat (aba)ca(aba) with the same period 5. In string cabaaabaaa, gapped repeat (ab)aa(ab) with period 4 is contained in maximal repetition abaaabaaa with period 4. Since all maximal repetitions can be computed efficiently in $O(n)$ time (see above), the problem of computing all α-gapped repeats in a word can be reduced to the problem of finding all maximal α-gapped repeats.

Several variants of the problem of computing gapped repeats have been studied earlier. In [4], it was shown that all maximal gapped repeats with a gap length belonging to a specified interval can be found in time $O(n \log n + S)$, where n is the word length and S is output size. In [22], an algorithm was proposed for finding all gapped repeats with a fixed gap length d running in time $O(n \log d + S)$. In [21], it was proved that the number of maximal α-gapped repeats in a word of length n is bounded by $O(\alpha^2 n)$ and all maximal α-gapped repeats can be found in $O(\alpha^2 n)$ time for the case of integer alphabet. A new approach to computing gapped repeats was recently proposed in [11,15]. In particular, in [15] it is shown that the longest α-gapped repeat in a word of length n over an integer alphabet can be found in $O(\alpha n)$ time. Finally, in a recent paper [27], an algorithm is proposed for finding all maximal α-gapped repeats in $O(\alpha n + S)$ time where S is the output size, for a constant-size alphabet. The algorithm uses an approach previously introduced in [1].

Recall that repetitions are segments with exponent at least 2. Another way to approach gapped repeats is to consider segments with exponent smaller than 2, but strictly greater than 1. Clearly, such a segment corresponds to a gapped repeat $\pi = uvu$ with $per(\pi) = p(\pi) = |u| + |v|$. We will call such factors (segments) *subrepetitions*. More precisely, for any δ, $0 < \delta < 1$, by a δ-subrepetition we mean a factor v that satisfies $1 + \delta \leq exp(v) < 2$. Again, the notion of maximality straightforwardly applies to subrepetitions as well: maximal subrepetitions are defined exactly in the same way as maximal repetitions. The relationship between maximal subrepetitions and maximal gapped repeats was clarified in [21]. Directly from the definitions, a maximal subrepetition π in a string w corresponds to a maximal gapped repeat with $p(\pi) = per(\pi)$. Futhermore, a maximal δ-subrepetition corresponds to a maximal $\frac{1}{\delta}$-gapped repeat. However, there may be more maximal $\frac{1}{\delta}$-gapped repeats than maximal δ-subrepetitions, as not every maximal $\frac{1}{\delta}$-gapped repeat corresponds to a maximal δ-subrepetition.

Some combinatorial results on the number of maximal subrepetitions in a string were obtained in [20]. In particular, it was proved that the number of maximal δ-subrepetitions in a word of length n is bounded by $O(\frac{n}{\delta} \log n)$. In [21], an $O(n/\delta^2)$ bound on the number of maximal δ-subrepetitions in a word of length n was obtained. Moreover, in [21], two algorithms were proposed for finding all maximal δ-subrepetitions in the word running respectively in $O(\frac{n \log \log n}{\delta^2})$ time and in $O(n \log n + \frac{n}{\delta^2} \log \frac{1}{\delta})$ expected time, over the integer alphabet. In [1], it is shown that all subrepetitions with the largest exponent (over all subrepetitions) can be found in an overlap-free string in time $O(n)$, for a constant-size alphabet.

Our results. In the present work we improve the results of [21] on maximal gapped repeats: we prove an asymptotically tight bound of $O(\alpha n)$ on the number of maximal α-gapped repeats in a word of length n (Sect. 2). From our bound, we also derive an $O(n/\delta)$ bound on the number of maximal δ-subrepetitions occurring in a word, which improves the bound of [20] by a $\log n$ factor. Then, based on the algorithm of [15], we obtain an asymptotically optimal $O(\alpha n)$ time bound for computing all maximal α-gapped repeats in a word (Sect. 3). Note that this bound follows from the recently published paper [27] that presents an $O(\alpha n + S)$ algorithm for computing all maximal α-gapped repeats. In this work, we present an alternative algorithm with the same bound that we obtained independently.

2   Number of Maximal Repeats and Subrepetitions

In this section, we obtain an improved upper bound on the number of maximal gapped repeats and subrepetitions in a string w. Following the general approach of [21], we split all maximal gapped repeats into three categories according to periodicity properties of repeat's copy: periodic, semiperiodic and ordinary repeats. Bounds for periodic and semiperiodic repeats are directly borrowed from [21], while for ordinary repeats, we obtain a better bound.

Periodic repeats. We say that a maximal gapped repeat is *periodic* if its copies are periodic strings (i.e. of exponent at least 2). The set of all periodic maximal α-gapped repeats in w is denoted by $\mathcal{PP}_\alpha$. The following bound on the size of $\mathcal{PP}_\alpha$ was been obtained in [21, Corollary 6].

Lemma 1. $|\mathcal{PP}_k| = O(kn)$ *for any natural* $k > 1$.

Semiperiodic repeats. A maximal gapped repeat is called *prefix (suffix) semiperiodic* if the copies of this repeat are not periodic, but have a prefix (suffix) which is periodic and its length is at least half of the copy length. A maximal gapped repeat is *semiperiodic* if it is either prefix or suffix semiperiodic. The set of all semiperiodic α-gapped maximal repeats is denoted by $\mathcal{SP}_\alpha$. In [21, Corollary 8], the following bound was obtained on the number of semiperiodic maximal α-gapped repeats.

Lemma 2 ([21]). $|\mathcal{SP}_k| = O(kn)$ *for any natural* $k > 1$.

Ordinary repeats. Maximal gapped repeats which are neither periodic nor semiperiodic are called *ordinary*. The set of all ordinary maximal α-gapped repeats in the word w is denoted by $\mathcal{OP}_\alpha$. In the rest of this section, we prove that the cardinality of $\mathcal{OP}_\alpha$ is $O(\alpha n)$. For simplicity, assume that α is an integer number k.

To estimate the number of ordinary maximal k-gapped repeats, we use the following idea from [17]. We represent a maximal repeat $\pi \equiv (u', u'')$ from $\mathcal{OP}_k$ by a triple (i, j, c) where $i = beg(u')$, $j = beg(u'')$ and $c = c(\pi) = |u'| = |u''|$. Such triples will be called *points*. Obviously, π is uniquely defined by values i, j and c, therefore two different repeats from $\mathcal{OP}_k$ can not be represented by the same point.

For any two points (i', j', c'), (i'', j'', c'') we say that point (i', j', c') *covers* point (i'', j'', c'') if $i' \le i'' \le i' + c'/6$, $j' \le j'' \le j' + c'/6$, $c' \ge c'' \ge \frac{2c'}{3}$. A point is *covered* by a repeat π if it is covered by the point representing π. By $V[\pi]$ we denote the set of all points covered by a repeat π. We show that any point can not be covered by two different repeats from $\mathcal{OP}_k$.

Lemma 3. *Two different repeats from* $\mathcal{OP}_k$ *cannot cover the same point.*

Proof. Let $\pi_1 \equiv (u_1', u_1'')$, $\pi_2 \equiv (u_2', u_2'')$ be two different repeats from $\mathcal{OP}_k$ covering the same point (i, j, c). Denote $c_1 = c(\pi_1)$, $c_2 = c(\pi_2)$, $p_1 = per(\pi_1)$,

$p_2 = per(\pi_2)$. Note that $beg(u_1'') - beg(u_1') = p_1$ and $beg(u_2'') - beg(u_2') = p_2$. Without loss of generality we assume $c_1 \geq c_2$. From $c_1 \geq c \geq \frac{2c_1}{3}$, $c_2 \geq c \geq \frac{2c_2}{3}$ we have $c_1 \geq c_2 \geq \frac{2c_1}{3}$, i.e. $c_2 \leq c_1 \leq \frac{3c_2}{2}$. Note that $w[i]$ is contained in both left copies u_1', u_2', i.e. these copies overlap. If $p_1 = p_2$, then repeats π_1 and π_2 must coincide due to the maximality of these repeats. Thus, $p_1 \neq p_2$. Denote $\Delta = |p_1 - p_2| > 0$. From $beg(u_1') \leq i \leq beg(u_1') + c_1/6$ and $beg(u_1'') \leq j \leq beg(u_1'') + c_1/6$ we have

$$(j - i) - c_1/6 \leq p_1 \leq (j - i) + c_1/6.$$

Analogously, we have

$$(j - i) - c_2/6 \leq p_2 \leq (j - i) + c_2/6.$$

Thus $\Delta \leq (c_1 + c_2)/6$ which, together with inequality $c_1 \leq \frac{3c_2}{2}$, implies $\Delta \leq \frac{5c_2}{12}$.

First consider the case when one of the copies u_1', u_2' is contained in the other, i.e. u_2' is contained in u_1'. In this case, u_1'' contains some factor $\widehat{u_2''}$ corresponding to the factor u_2' in u_1'. Since $beg(u_2'') - beg(u_2') = p_2$, $beg(\widehat{u_2''}) - beg(u_2') = p_1$ and $u_2'' = \widehat{u_2''} = u_2'$, we have

$$|beg(u_2'') - beg(\widehat{u_2''})| = \Delta,$$

so Δ is a period of u_2'' such that $\Delta \leq \frac{5}{12}c_2 = \frac{5}{12}|u_2''|$. Thus, u_2'' is periodic which contradicts that π_2 is not periodic.

Now consider the case when u_1', u_2' are not contained in one another. Denote by z' the overlap of u_1' and u_2'. Let z' be a suffix of u_l' and a prefix of u_r' where $l, r = 1, 2$, $l \neq r$. Then u_l'' contains a suffix z'' corresponding to the suffix z' in u_l', and u_r'' contains a prefix $\widehat{z}''$ corresponding to the prefix z' in u_r'. Since $beg(z'') - beg(z') = p_l$ and $beg(\widehat{z}'') - beg(z') = p_r$ and $z'' = \widehat{z}'' = z'$, we have

$$|beg(z'') - beg(\widehat{z}'')| = |p_l - p_r| = \Delta,$$

therefore Δ is a period of z'. Note that in this case

$$beg(u_l') < beg(u_r') \leq i \leq beg(u_l') + c_l/6,$$

therefore $0 < beg(u_r') - beg(u_l') \leq c_l/6$. Thus

$$|z'| = c_l - (beg(u_r') - beg(u_l')) \geq \frac{5}{6}c_l \geq \frac{5}{6}c_2.$$

From $\Delta \leq \frac{5}{12}c_2$ and $c_2 \leq \frac{6}{5}|z'|$ we obtain $\Delta \leq |z'|/2$. Thus, z' is a periodic suffix of u_l' such that $|z'| \geq \frac{5}{6}|u_l'|$, i.e. π_l is either suffix semiperiodic or periodic which contradicts $\pi_l \in \mathcal{OP}_k$. $\qquad \square$

Denote by $\mathcal{Q}_k$ the set of all points (i, j, c) such that $1 \leq i, j, c \leq n$ and $i < j \leq i + (\frac{3}{2}k + \frac{1}{4})c$.

Lemma 4. *Any point covered by a repeat from $\mathcal{OP}_k$ belongs to $\mathcal{Q}_k$.*

Proof. Let a point (i, j, c) be covered by some repeat $\pi \equiv (u', u'')$ from $\mathcal{OP}_k$. Denote $c' = c(\pi)$. Note that $w[i]$ and $w[j]$ are contained respectively in u' and u'' and $n > c' \geq c \geq \frac{2c'}{3} > 0$, so inequalities $1 \leq i, j, c \leq n$ and $i < j$ are obvious. Note also that

$$j \leq beg(u'') + c'/6 = beg(u') + per(\pi) + c'/6 \leq i + kc' + c'/6,$$

therefore, taking into account $c' \leq \frac{3c}{2}$, we have $j \leq i + (\frac{3}{2}k + \frac{1}{4})c.$ $\qquad\square$

From Lemmas 3 and 4, we obtain

Lemma 5. $|\mathcal{OP}_k| = O(nk).$

Proof. Assign to each point (i, j, c) the weight $\rho(i, j, c) = 1/c^3$. For any finite set A of points, we define

$$\rho(A) = \sum_{(i,j,c)\in A} \rho(i, j, c) = \sum_{(i,j,c)\in A} \frac{1}{c^3}.$$

Let π be an arbitrary repeat from $\mathcal{OP}_k$ represented by a point (i', j', c'). Then

$$\rho(V[\pi]) = \sum_{i' \leq i \leq i'+c'/6} \sum_{j' \leq j \leq j'+c'/6} \sum_{2c'/3 \leq c \leq c'} \frac{1}{c^3}$$

$$> \frac{c'^2}{36} \sum_{2c'/3 \leq c \leq c'} \frac{1}{c^3}.$$

Using a standard estimation of sums by integrals, one can deduce that $\sum_{2c'/3 \leq c \leq c'} \frac{1}{c^3} \geq \frac{5}{32} \frac{1}{c'^2}$ for any c'. Thus, for any π from $\mathcal{OP}_k$

$$\rho(V[\pi]) > \frac{1}{36} \frac{5}{36} = \Omega(1).$$

Therefore,

$$\sum_{\pi \in \mathcal{OP}_k} \rho(V[\pi]) = \Omega(|\mathcal{OP}_k|). \tag{1}$$

Note also that

$$\rho(\mathcal{Q}_k) \leq \sum_{i=1}^{n} \sum_{i < j \leq i+(\frac{3}{2}k+\frac{1}{4})c} \sum_{c=1}^{n} \frac{1}{c^3}$$

$$< n(\frac{3}{2}k + \frac{1}{4})c \sum_{c=1}^{n} \frac{1}{c^3} < 2nk \sum_{c=1}^{n} \frac{1}{c^2} < 2nk \sum_{c=1}^{\infty} \frac{1}{c^2} = \frac{nk\pi^2}{3}.$$

Thus,

$$\rho(\mathcal{Q}_k) = O(nk). \tag{2}$$

By Lemma 4, any point covered by repeats from $\mathcal{OP}_k$ belongs to $\mathcal{Q}_k$. On the other hand, by Lemma 3, each point of $\mathcal{Q}_k$ can not be covered by two repeats from $\mathcal{OP}_k$. Therefore,

$$\sum_{\pi \in \mathcal{OP}_k} \rho(V[\pi]) \leq \rho(\mathcal{Q}_k).$$

Thus, using 1 and 2, we conclude that $|\mathcal{OP}_k| = O(nk)$.   □

Putting together Lemmas 1, 2, and 5, we obtain that for any integer $k \geq 2$, the number of maximal k-gapped repeats in w is $O(nk)$. The bound straightforwardly generalizes to the case of real $\alpha > 1$. Thus, we conclude with

Theorem 2. *For any $\alpha > 1$, the number of maximal α-gapped repeats in w is $O(\alpha n)$.*

Note that the bound of Theorem 2 is asymptotically tight. To see this, it is enough to consider word $w_k = (0110)^k$. It is easy to check that for a big enough α and $k = \Omega(\alpha)$, w_k contains $\Theta(\alpha|w_k|)$ maximal α-gapped repeats whose copies are single-letter words.

We now use Theorem 2 to obtain an upper bound on the number of maximal δ-subrepetitions. The following proposition, shown in [21, Proposition 3], follows from the fact that each maximal δ-subrepetition defines at least one maximal $1/\delta$-gapped repeat (cf. Introduction).

Proposition 1 ([21]). *For $0 < \delta < 1$, the number of maximal δ-subrepetitions in a string is no more then the number of maximal $1/\delta$-gapped repeats.*

Theorem 2 combined with Proposition 1 immediately imply the following upper bound for maximal δ-subrepetitions that improves the bound of [20] by a $\log n$ factor.

Theorem 3. *For $0 < \delta < 1$, the number of maximal δ-subrepetitions in w is $O(n/\delta)$.*

The $O(n/\delta)$ bound on the number of maximal δ-subrepetitions is asymptotically tight, at least on an unbounded alphabet : word $\mathtt{ab_1ab_2 \ldots ab_k}$ contains $\Omega(n/\delta)$ maximal δ-subrepetitions for $\delta \leq 1/2$.

3   Computing All Maximal α-gapped Repeats

We now turn to the algorithmic question how to efficiently compute all maximal α-gapped repeats in a given word. Recall (cf Introduction) that an algorithm with running time $O(\alpha^2 n + S)$ has been proposed in [21] for this problem, which becomes $O(\alpha^2 n)$-time taken into account the bound on S. On the other hand, it was shown in [15] that computing the *longest* α-gapped repeat can be done in time $O(\alpha n)$. It is therefore a natural question whether all maximal α-gapped repeats can be computed in time $O(\alpha n + S)$. Here we answer this question positively. Together with the $S = O(\alpha n)$ bound of Theorem 2, this implies the following result.

Theorem 4. *For a fixed* $\alpha > 1$, *all maximal* α-*gapped repeats in a word of length* n *over a constant alphabet can be computed in* $O(\alpha n)$ *time.*

The proof of Theorem 4 can be found in the full version of this work [8]. It is based on a case analysis and uses ideas of [15].

We note that independently of our work, another $O(\alpha n + S)$-time algorithm for computing all maximal α-gapped repeats has been recently announced in [27].

Note that, as mentioned earlier, a word can contain $\Theta(\alpha n)$ maximal α-gapped repeats, and therefore the $O(\alpha n)$ time bound stated in Theorem 4 is asymptotically optimal.

4   Concluding Remarks

In this work, we proved the tight $O(\alpha n)$ bound on the number of maximal α-gapped repeats in a word. We note that while submitting this paper, manuscript [14] appeared that proves that the number of maximal α-gapped repeats is bounded by $18\alpha n$. From our bound, we obtain an $O(n/\delta)$ bound on the number of maximal δ-subrepetitions in a word, which improves the bound of [20] by a $\log n$ factor. We also presented an $O(\alpha n)$-time algorithm (obtained independently from [27]) for computing all maximal α-gapped repeat in a word.

Besides gapped repeats we can also consider gapped palindromes which are factors of the form uvu^R, where u and v are nonempty words and u^R is the reversal of u [19]. A gapped palindrome uvu^R in a word w is called *maximal* if $w[end(u) + 1] \neq w[beg(u^R) - 1]$ and $w[beg(u) - 1] \neq w[end(u^R) + 1]$ for $beg(u) > 1$ and $end(u^R) < |w|$. A maximal gapped palindrome uvu^R is α-gapped if $|u| + |v| \leq \alpha|u|$ [15]. It can be shown analogously to the results of this paper that for $\alpha > 1$ the number of maximal α-gapped palindromes in a word of length n is bounded by $O(\alpha n)$ and for the case of constant alphabet, all these palindromes can be found in $O(\alpha n)$ time[1].

In this paper, we consider maximal α-gapped repeats with $\alpha > 1$. However, this notion can be formally generalized to the case of $\alpha \leq 1$. In particular, maximal 1-gapped repeats are maximal repeats whose copies are adjacent or overlapping. It is easy to see that such repeats form runs whose minimal periods are divisors of the periods of these repeats. Moreover, each run in a word is formed by at least one maximal 1-gapped repeat, therefore the number of runs in a word is not greater than the number of maximal 1-gapped repeats. More precisely, each run r is formed by $\lfloor exp(r)/2 \rfloor$ distinct maximal 1-gapped repeats. Thus, if a word contains runs with exponent greater than or equal to 4 then the number of maximal 1-gapped repeats is strictly greater than the number of runs. However, using an easy modification of the proof of "runs conjecture" from [2], it can be also proved the number of maximal 1-gapped repeats in a word is strictly less than the length of the word. Moreover, denoting by $\mathcal{R}_1(n)$ the maximal possible number of maximal 1-gapped repeats in words of length n,

---

[1] Note that in [15], the number of maximal α-gapped palindromes was conjectured to be $O(\alpha^2 n)$.

we conjecture that $\mathcal{R}(n) = \mathcal{R}_1(n)$ since known words with a large number of runs have no runs with big exponents. We can also consider the case of $\alpha < 1$ for repeats with overlapping copies and, in particular, the case of maximal $1/k$-gapped repeats where k is integer greater than 1. It is easy to see that such repeats form runs with exponents greater than or equal to $k + 1$. It is known from [2, Theorem 11] that the number of such runs in a word of length n is less than n/k, and it seems to be possible to modify the proof of this fact to prove that the number of maximal $1/k$-gapped repeats in the word is also less than $n/k = \alpha n$. These observations together with results of computer experiments for the case of $\alpha > 1$ leads to a conjecture that for any $\alpha > 0$, the number of maximal α-gapped repeats in a word of length n is actually less than αn. This generalization of the "runs conjecture" constitutes an interesting open problem. Another interesting open question is whether the obtained $O(n/\delta)$ bound on the number of maximal δ-subrepetitions is asymptotically tight for the case of constant alphabet.

References

1. Badkobeh, G., Crochemore, M., Toopsuwan, C.: Computing the maximal-exponent repeats of an overlap-free string in linear time. In: Calderón-Benavides, L., González-Caro, C., Chávez, E., Ziviani, N. (eds.) SPIRE 2012. LNCS, vol. 7608, pp. 61–72. Springer, Heidelberg (2012)
2. Bannai, H., Tomohiro, I., Inenaga, S., Nakashima, Y., Takeda, M., Tsuruta, K.: A new characterization of maximal repetitions by Lyndon trees, intermediate version presented to SODA'2015 (2014). CoRR abs/1406.0263
3. Breslauer, D.: Efficient string algorithmics. Ph.D. thesis, Columbia University (1992)
4. Brodal, G.S., Lyngs, R.B., Pedersen, C.N.S., Stoye, J.: Finding maximal pairs withbounded gap. J. Discrete Algorithms 1(1), 77–104 (2000)
5. Crochemore, M.: An optimal algorithm for computing the repetitions in a word. Inf. Process. Lett. 12(5), 244–250 (1981)
6. Crochemore, M., Ilie, L., Tinta, L.: Towards a solution to the "Runs" conjecture. In: Ferragina, P., Landau, G.M. (eds.) CPM 2008. LNCS, vol. 5029, pp. 290–302. Springer, Heidelberg (2008)
7. Crochemore, M., Iliopoulos, C., Kubica, M., Radoszewski, J., Rytter, W., Waleń, T.: Extracting powers and periods in a string from its runs structure. In: Chavez, E., Lonardi, S. (eds.) SPIRE 2010. LNCS, vol. 6393, pp. 258–269. Springer, Heidelberg (2010)
8. Crochemore, M., Kolpakov, R., Kucherov, G.: Optimal searching of gapped repeats in a word (2015). CoRR abs/1509.01221
9. Crochemore, M., Kubica, M., Radoszewski, J., Rytter, W., Walen, T.: On the maximal sum of exponents of runs in a string. J. Discrete Algorithms 14, 29–36 (2012)
10. Crochemore, M., Rytter, W.: Sqares, cubes, and time-space efficient string searching. Algorithmica 13(5), 405–425 (1995)
11. Dumitran, M., Manea, F.: Longest gapped repeats and palindromes. In: Italiano, G.F., Pighizzini, G., Sannella, D.T. (eds.) MFCS 2015. LNCS, vol. 9234, pp. 205–217. Springer, Heidelberg (2015)

12. Fischer, J., Holub, S., Tomohiro, I., Lewenstein, M.: Beyond the runs theorem. CoRR abs/1502.04644 (2015)
13. Galil, Z., Seiferas, J.I.: Time-space-optimal string matching. J. Comput. Syst. Sci. **26**(3), 280–294 (1983)
14. Gawrychowski, P.I.T., Inenaga, S., Köppl, D., Manea, F.: Efficiently finding all maximal α-gapped repeats. CoRR abs/1509.09237 (2015)
15. Gawrychowski, P., Manea, F.: Longest α-gapped repeat and palindrome. In: Kosowski, A., Walukiewicz, I. (eds.) FCT 2015. LNCS, vol. 9210, pp. 27–40. Springer, Heidelberg (2015)
16. Gusfield, D.: Algorithms on Strings, Trees, and Sequences - Computer Science and Computational Biology. Cambridge University Press, New York (1997)
17. Kolpakov, R.: On primary and secondary repetitions in words. Theor. Comput. Sci. **418**, 71–81 (2012)
18. Kolpakov, R., Kucherov, G.: On maximal repetitions in words. J. Discrete Algorithms **1**(1), 159–186 (2000)
19. Kolpakov, R., Kucherov, G.: Searching for gapped palindromes. Theor. Comput. Sci. **410**(51), 5365–5373 (2009)
20. Kolpakov, R., Kucherov, G., Ochem, P.: On maximal repetitions of arbitrary exponent. Inf. Process. Lett. **110**(7), 252–256 (2010)
21. Kolpakov, R., Podolskiy, M., Posypkin, M., Khrapov, N.: Searching of gapped repeats and subrepetitions in a word. In: Kulikov, A.S., Kuznetsov, S.O., Pevzner, P. (eds.) CPM 2014. LNCS, vol. 8486, pp. 212–221. Springer, Heidelberg (2014)
22. Kolpakov, R.M., Kucherov, G.: Finding repeats with fixed gap. In: SPIRE, pp. 162–168 (2000)
23. Kosolobov, D.: Lempel-Ziv factorization may be harder than computing all runs. In: Mayr, E.W., Ollinger, N. (eds.) 32nd International Symposium on Theoretical Aspects of Computer Science, STACS 4–7, 2015, Garching, Germany. LIPIcs, vol. 30, pp. 582–593. Schloss Dagstuhl - Leibniz-Zentrum fuer Informatik, March 2015
24. Lothaire, M.: Combinatorics on Words. Addison Wesley, Reading (1983)
25. Main, M., Lorentz, R.: An $O(n \log n)$ algorithm for finding all repetitions in a string. J. Algorithms **5**(3), 422–432 (1984)
26. Storer, J.A.: Data Compression: Methods and Theory. Computer Science Press, Rockville (1988)
27. Tanimura, Y., Fujishige, Y., I, T., Inenaga, S., Bannai, H., Takeda, M.: A faster algorithm for computing maximal α-gapped repeats in a string. In: Iliopoulos, C., Puglisi, S., Yilmaz, E. (eds.) SPIRE 2015. LNCS, vol. 9309, pp. 124–136. Springer, Heidelberg (2015)

Complexity

On XOR Lemma for Polynomial Threshold Weight and Length

Kazuyuki Amano[✉]

Department of CS, Gunma University, Tenjin 1-5-1,
Kiryu, Gunma 376-8515, Japan
amano@gunma-u.ac.jp

Abstract. Let $f : \{-1, 1\}^n \to \{-1, 1\}$ be a Boolean function. We say that a multilinear polynomial p sign-represents f if $f(x) = sgn(p(x))$ for all $x \in \{-1, 1\}^n$. In this paper, we consider the length and weight of polynomials sign-representing Boolean functions of the form $f \oplus f \oplus \cdots \oplus f$ where each f is on a disjoint set of variables. Obviously, if p sign-represents f, then $p(x)p(y)$ sign-represents $f(x) \oplus f(y)$. We give a constructive proof that there is a shorter polynomial when f is AND on n variables for every $n \geq 3$. In addition, we introduce a parameter v_f^* of a Boolean function and show that the k-th root of the minimum weight of a polynomial sign-representing $f \oplus f \oplus \cdots \oplus f$ (k times) converges between v_f^* and $(v_f^*)^2$ as k goes to infinity.

Keywords: Computational complexity · Boolean functions · PTF · Integer programming

1 Introduction

Throughout the paper, we consider a Boolean function as a mapping from $\{-1, 1\}^n$ to $\{-1, 1\}$; -1 denotes True and 1 denotes False. We discuss the representation of Boolean functions by polynomial threshold functions (PTF, in short).

Let p be a multilinear polynomial on n variables. If $f(x) = sgn(p(x))$ for every $x \in \{-1, 1\}^n$, we say that f is computed by a polynomial threshold function p, or p *sign-represents* f. The PTF representation of Boolean functions have been extensively studied especially in complexity theory and learning theory (see e.g., [1–3, 6–8, 15]).

The most well-investigated measure in the study of PTF representation is its *degree*, which is defined as the minimum degree over all polynomials sign-representing f. Other important but less understood measures are the *weight* and *length*. The weight of a Boolean function f is the minimum value of the sum of the absolute values of integer coefficients of a polynomial that sign-represents f. The length of f is the minimum number of monomials in a polynomial that sign-represents f.

In this paper, we focus on these two measures for a certain class Boolean functions. For Boolean functions f and g, let $f \oplus g$ denote the XOR of f and g

© Springer International Publishing Switzerland 2016
A.-H. Dediu et al. (Eds.): LATA 2016, LNCS 9618, pp. 259–269, 2016.
DOI: 10.1007/978-3-319-30000-9_20

on disjoint sets of variables. If p sign-represents f and q sign-represents g, then pq sign-represents $f \oplus g$. It is natural to ask whether this gives a most economical polynomial for $f \oplus g$.

O'Donnell and Servedio [11] proved that the answer is "yes" for PTF degree. In fact, they proved the "XOR Lemma" saying that the degree of $f \oplus g$ is equal to the sum of the degrees of f and g. It would be natural to expect that such a property holds for PTF length. The problem to verify this is presented, e.g., in a list of open problems compiled by Filmus et al. [5].

Recently, Sezener and Oztop [14] gave a heuristic algorithm to find a short PTF and computationally verified that every 6-variable Boolean function can be sign-represented using at most 26 monomials. This is surprising because it implies that the length of the function $x_1 x_2 \oplus x_3 x_4 \oplus x_5 x_6$ is at most 26, which is strictly smaller than 3^3; the cube of the length of $x_1 x_2 = sgn(x_1 + x_2 + 1)$. This shows that the "XOR Lemma" does not hold for PTF length in its "ideal" form, and gives us a strong motivation for further research. What functions admit such a saving? How much saving can be possible?

In the first part of this paper (in Sect. 3), we see that such a saving is possible for a wider class of functions. Namely, we give an explicit polynomial that sign-represents the XOR of two ANDs on an arbitrary number of variables whose length is less than the square of the length of a single AND function (Theorem 6).

In the second part of this paper (in Sect. 4), we analyze the length and weight of the function of the form $\oplus_k f$ which denotes the XOR of k copies of f on disjoint sets of variables. We introduce a parameter of a Boolean function and obtain an "XOR Lemma" for PTF weight. This parameter v_f^* (the definition will be given in Sect. 4) is given by the value of a certain linear programming problem. Interestingly, this LP problem is a relaxation of two LP problems. The first one is a problem whose value is the weight of f, and the second one is a problem whose value is the spectral norm (a.k.a. Fourier L_1-norm) of f. We show that the k-th root of the weight of $\oplus_k f$ converges between v_f^* and $(v_f^*)^2$ as k goes to infinity by analyzing the tensor product of these LP problems (Theorem 12).

The organization of the paper is as follows: In Sect. 2, we give notations and definitions. In Sect. 3, we exhibit an explicit construction of short polynomials sign-representing the XOR of two ANDs. In Sect. 4, we show that the weight and length of Boolean functions of the form $\oplus_k f$ is strongly related to a parameter based on a certain LP problem. Finally, in Sect. 5, we give some open questions.

2   Preliminaries

Let $[n]$ denote the set $\{1, 2, \ldots, n\}$. Let $\mathbb{Z}$ denote the set of integers and $\mathbb{Z}^+$ denote the set of non-negative integers. For a set $S \subseteq [n]$, we write x^S to denote the monomial $\prod_{i \in S} x_i$. Note that $x^\emptyset$ is the constant 1. A multilinear polynomial is the sum of the monomials

$$p(x_1, \ldots, x_n) = \sum_{S \subseteq [n]} p_S x^S.$$

Throughout the paper, we assume that all the coefficients p_S are integers.

Definition 1. *Let* $f : \{-1,1\}^n \to \{-1,1\}$ *be a Boolean function and let* $p : \{-1,1\}^n \to \mathbb{Z}$ *be a multilinear polynomial with integer coefficients. We say that* p *sign-represents* f *if* $f(x) = sgn(p(x))$ *and* $p(x) \neq 0$ *for every* $x \in \{-1,1\}^n$, *where* $sgn(y) = 1$ *if* $y > 0$ *and* $sgn(y) = -1$ *if* $y < 0$.

Definition 2. *The weight of a polynomial* $p(x) = \sum_{S \subseteq [n]} p_S x^S$ *is the sum of the absolute values of the coefficients* $\sum_{S \subseteq [n]} |p_S|$. *Let* $f : \{-1,1\}^n \to \{-1,1\}$ *be a Boolean function. The weight of* f, *denoted by* $wt(f)$, *is defined as the minimum weight of a polynomial that sign-represents* f. *The length of* f, *denoted by* $len(f)$, *is defined as the minimum number of monomials in a polynomial that sign-represents* f.

Definition 3. *Let* f *be a Boolean function on* $\{x_1, \ldots, x_n\}$ *and* g *be a Boolean function on* $\{y_1, \ldots, y_n\}$. *Let* $f \oplus g$ *denote the Boolean function on* $\{x_1, \ldots, x_n, y_1, \ldots, y_n\}$ *whose value is the XOR of* $f(x_1, \ldots, x_n)$ *and* $g(y_1, \ldots, y_n)$. *Let* $\oplus_k f$ *denote the XOR of* k *copies of* f *on disjoint sets of variables.*

3   Upper Bounds on the Length of XOR of ANDs

Let AND_n denote the AND of n variables. Let $\mathsf{IP}_n : \{-1,1\}^{2n} \to \{-1,1\}$ denote the *inner product* function defined as

$$\mathsf{IP}_n(x_1, \ldots, x_n, y_1, \ldots, y_n) = \oplus_{i \in [n]} x_i y_i.$$

Equivalently, $\mathsf{IP}_n = \oplus_n \mathsf{AND}_2$.

Fact 4. $len(\mathsf{AND}_2) = 3$, $len(\oplus_2 \mathsf{AND}_2) = 9$ *and* $len(\oplus_3 \mathsf{AND}_2) \leq 26$.

The first equality is obvious by observing $\mathsf{AND}_2(x_1, x_2) = sgn(x_1 + x_2 + 1)$ is (one of) the shortest polynomial for AND_2. The second equality can easily be verified by using a computer. The last inequality is proved by Sezener and Oztop [14, Sect.7.2] who gave a polynomial of length 26 and total weight 686 that sign-represents $\oplus_3 \mathsf{AND}_2$.

Below we describe another polynomial $p(x)$ of the same length that sign-represents $x_1 x_2 \oplus x_3 x_4 \oplus x_5 x_6$.

$$p(x) = 2(x^{\{3,5\}} - x^{\{1,2,3,5\}} + x^{\{4,5\}} + x^{\{2,4,5\}} - x^{\{2,3,4,5\}} + x^{\{1,2,3,4,5\}} + x^{\{1,6\}}$$
$$+ x^{\{2,3,6\}} + x^{\{4,6\}} + x^{\{1,2,3,4,6\}} - x^{\{3,5,6\}} - x^{\{2,3,5,6\}} + x^{\{1,2,3,5,6\}}$$
$$- x^{\{2,4,5,6\}} + x^{\{2,3,4,5,6\}})$$
$$+ 3(x^{\{2\}} - x^{\{1,2\}} - x^{\{1,3,4\}} + x^{\{1,5\}} - x^{\{1,2,4,5\}} + x^{\{1,3,6\}}$$
$$- x^{\{1,2,3,6\}} - x^{\{2,3,4,6\}} - x^{\{4,5,6\}} + x^{\{1,2,4,5,6\}} - x^{\{1,2,3,4,5,6\}}).$$

The weight of this polynomial is 63. This is the smallest among all 26-monomial polynomials that we have found. Note that we found this polynomial using an IP solver. We strongly believe that $len(\oplus_3 \mathsf{AND}_2)$ is actually 26, but we have not succeeded in showing this.

Fact 4 shows that $len(\oplus_k AND_2)$ is strictly smaller than $len(AND_2)^k$ for every $k \geq 3$. Below we generalize this to show that $len(\oplus_k AND_n)$ is strictly smaller than $len(AND_n)^k$ for every $k \geq 2$ and $n \geq 3$.

The following fact should be folklore, but we include a proof for completeness.

Fact 5. *For every* n, $len(AND_n) = n + 1$.

Proof. The upper bound is obvious by the representation $AND_n(x_1, \ldots, x_n) = sgn(x_1 + \cdots + x_n + (n-1))$. Below we show the lower bound. For the sake of notational simplicity, we consider the length of NOR_n (Negation of OR) instead of AND_n. This is harmless because $NOR_n(x_1, \ldots, x_n) = AND(-x_1, \ldots, -x_n)$ by the De Morgan's law.

Suppose for a contradiction that a polynomial $p(x) = \sum_{1 \leq i \leq n} p_i x^{S_i}$ sign-represents NOR_n, i.e., $p(x) < 0$ if $x = (1, 1, \ldots, 1)$ and $p(x) > 0$ otherwise.

Let $\mathbf{M}$ be the 0/1-valued $n \times n$ matrix such that its (i, j) entry is 1 iff x_j appears in the i-th monomial in p (i.e., $x_j \in S_i$).

We divide the proof into two cases: (i) the rank of $\mathbf{M}$ is strictly smaller than n, and (ii) $\mathbf{M}$ has full rank.

For the case (i), we consider the system $\mathbf{Mv} = \mathbf{0}$ over $GF(2)$, where $\mathbf{v}$ is a 0/1-valued column vector of length n and $\mathbf{0}$ is the all-zero column vector of length n. Let $\chi : \{0, 1\} \to \{-1, 1\}$ be the mapping such that $\chi(0) = 1$ and $\chi(1) = -1$. Then, for any vector $\mathbf{v} = (v_1, \ldots, v_n)$ satisfying $\mathbf{Mv} = \mathbf{0}$, $x = (x_1, \ldots, x_n) = (\chi(v_1), \ldots, \chi(v_n))$ satisfies $x^{S_i} = 1$ for every i. Since $\mathbf{M}$ is degenerate, there exists a non-zero vector $\mathbf{v}'$ with $\mathbf{Mv}' = \mathbf{0}$. Let $x' \in \{-1, 1\}^n$ be the input vector obtained from $\mathbf{v}'$ by the mapping χ, we have $p(x') = \sum_i p_i$ and hence $\sum_i p_i > 0$ since $x' \neq (1, 1, \ldots, 1)$. This contradicts $p(1, 1, \ldots, 1) = \sum_i p_i < 0$, completing the proof for the case (i).

The proof for the case (ii) is similar. Let $\mathbf{a}$ be any 0/1-valued column vector of length n such that $\mathbf{a} \neq (0, 0, \ldots, 0)^T$ and $\mathbf{a} \neq (1, 1, \ldots, 1)^T$. Let $\bar{\mathbf{a}}$ denote the *complement* of $\mathbf{a}$, that is the vector obtained from $\mathbf{a}$ by flipping 0s and 1s. Let $\mathbf{v}^1$ and $\mathbf{v}^2$ be the solutions to the systems $\mathbf{Mv}^1 = \mathbf{a}$ and $\mathbf{Mv}^2 = \bar{\mathbf{a}}$. Let x^1 and x^2 be the input vectors obtained from $\mathbf{v}^1$ and $\mathbf{v}^2$ by the mapping χ, respectively. Then, we have $p(x^1) = -p(x^2)$, but this contradicts $NOR_n(x^1) = NOR_n(x^2)$ since $x^1, x^2 \neq (1, 1, \ldots, 1)$. $\qquad\square$

Theorem 6. *For every* $n \geq 3$, $len(AND_n \oplus AND_n) \leq n^2 + n + 4$.

Proof. The proof is constructive. Let $U = \{1, \ldots, n\}$ and $V = \{n+1, \ldots, 2n\}$. Consider a polynomial of the form

$$p(x_1, \ldots, x_n, x_{n+1}, \ldots, x_{2n}) = a + b \sum_{i \in U \cup V} x_i + c \sum_{\substack{i \in U, j \in V \\ i \neq j \pmod{n}}} x_i x_j$$

$$+ d(x^U + x^V) + e \cdot x^{U \cup V}.$$

Let $a = (n-1)^2$, $b = (n-1.5)$, $c = 1$, $d = 0.5(-1)^{n+1}$ and $e = 1$. Here we use half integral weights for simplicity. The number of monomials in p is

$$1 + 2n + n(n-1) + 2 + 1 = n^2 + n + 4.$$

Below we show that p sign-represents the XOR of $\mathsf{AND}_n(x_1, \ldots, x_n)$ and $\mathsf{AND}_n(x_{n+1}, \ldots, x_{2n})$.

For any input $x = (x_1, \ldots, x_{2n})$, let u and v denote the number of -1's in the first half and second half of x, respectively, and let $s := |\{i \in [n] \mid x_i \neq x_{n+i}\}|$. Observe that the value of p depends only on u, v and s, which we denote by $p(u, v, s)$. An easy calculation shows that

$$p(u, v, s) = a + b\{2n - 2(u + v)\} + c\{(n - 2u)(n - 2v) + 2s - n\}$$
$$+ d\{(-1)^u + (-1)^v\} + e(-1)^{u+v}. \tag{1}$$

What we should verify is that $p(u, v, s) > 0$ if $u = v = n$ or $u, v < n$ and $p(u, v, s) < 0$ otherwise. Without loss of generality we can assume that $u \geq v$. Since as s increases, $p(u, v, s)$ increases, it is sufficient to verify that (a) $p(n, n, 0) > 0$, (b) $p(u, v, u - v) > 0$ ($\forall v \leq u < n$), and (c) $p(n, v, n - v) < 0$ ($\forall v < n$).

The proof of these is a bit tedious but elementary. First we consider the case $u = v$ which covers (a) and a part of (b). We have that

$$p(u, u, 0) = 4(n - u)(n - u - 1.5) + 1 + 2 \cdot \mathbb{1}[(n - u) \text{ is odd}],$$

here $\mathbb{1}[\cdot]$ is an indicator function. It is easy to check that the value of the right hand side of the above formula is positive for every integer $0 \leq u \leq n$.

Now we consider the case $u \neq v$ which covers the rest of (b) and (c). Let $\tilde{p}(u, v)$ denote the sum of the first three terms of $p(u, v, u - v)$ in Eq. (1). Put $v = u - \alpha$ ($\alpha \geq 1$). We have

$$\tilde{p}(u, u - \alpha) = 4(n - u)(n - u + \alpha - 1.5) - \alpha + 1.$$

What we should show is $p(u, u - \alpha, \alpha)$ is negative when $u = n$, and is positive when $u \leq n - 1$. Since $\tilde{p}(n, n - \alpha) = 1 - \alpha$, we have $p(n, n - \alpha, \alpha) < 0$ for every $\alpha \geq 1$.

Since the derivative $\frac{d\tilde{p}(u, u - \alpha)}{du}$ is negative when $u \leq n - 1 < n + \frac{2\alpha - 3}{4}$, we have

$$\tilde{p}(u, u - \alpha) \geq \tilde{p}(n - 1, n - 1 - \alpha) = 3\alpha - 1 \geq 2,$$

which implies $p(u, u - \alpha, \alpha) \geq 1$ for every $1 \leq \alpha \leq u \leq n - 1$ as desired. $\qquad \square$

By Theorem 6, we have

$$\mathrm{len}(\oplus_2 \mathsf{AND}_n) \leq n^2 + n + 4 = \mathrm{len}(\mathsf{AND}_n)^2 - (n - 3),$$

which is strictly smaller than $\mathrm{len}(\mathsf{AND}_n)^2$ for $n \geq 4$.

For $n = 3$, we can verify that the following polynomial with 15 monomials sign-represents $x_1 x_2 x_3 \oplus x_4 x_5 x_6$.

$$p(x) = 8 + 3x^{\{1\}} + 3x^{\{2\}} + 3x^{\{3\}} + 3x^{\{4\}} + 2x^{\{1,4\}} + 2x^{\{3,4\}} + 3x^{\{5\}} + 2x^{\{1,5\}}$$
$$+ 2x^{\{2,5\}} + 3x^{\{6\}} + 2x^{\{2,6\}} + 2x^{\{3,6\}} + x^{\{4,5,6\}} + 2x^{\{1,2,3,4,5,6\}}.$$

In summary, we have:

Corollary 7. *For $k \geq 2$ and $n \geq 2$, $\mathrm{len}(\oplus_k \mathsf{AND}_n)$ is strictly smaller than $\mathrm{len}(\mathsf{AND}_n)^k$ except for $k = n = 2$.*

4 The XOR Lemma for PTF Weight

In this section, we consider the weight and length of Boolean functions of the form $\oplus_k f$ for large k. We first introduce a parameter v_f^* for a Boolean function f.

The problem to find a minimum weight polynomial $p(x) = \sum_S p_S x^S$ that sign-represents f can be represented by the following integer programming problem, which we call the problem P_f.

$$\text{Minimize:} \quad \sum_{S \subseteq [n]} |p_S|,$$

$$\text{Subject to:} \quad \sum_{S \subseteq [n]} p_S M_{f,(x,S)} \geq 1, \ (\forall x \in \{-1,1\}^n),$$

$$p_S \in \mathbb{Z}, \qquad\qquad (\forall S \subseteq [n]),$$

where $M_{f,(x,S)} := f(x)x^S$ (for $x \in \{-1,1\}^n$ and $S \subseteq [n]$). The value of P_f gives the weight of f and the optimal solution to P_f gives the coefficients of such a polynomial.

It is natural to consider the LP-relaxation of P_f, denoted by P_f^*, in which the integral conditions on p_S's are removed. Let v_f denote the value of P_f and let v_f^* denote the value of P_f^*.

Interestingly, the problem P_f^* is also a relaxation of the one whose value is the *spectral norm* (a.k.a Fourier L_1-norm) of f.

For $S \subseteq [n]$, we define $\chi_S : \{-1,1\}^n \to \{-1,1\}$ by

$$\chi_S(x) = \prod_{i \in S} x_i.$$

In fact, $\chi_S(x) = x^S$ in our notation. Every Boolean function f can be uniquely represented as

$$f(x) = \sum_{S \subseteq [n]} \hat{f}(S)\chi_S(x). \qquad (2)$$

This expression is called the *Fourier expansion* of f, and $\hat{f}(S)$ is called the *Fourier coefficient* of f on S. The spectral norm of f is defined as

$$\|\hat{f}\|_1 = \sum_{S \subseteq [n]} |\hat{f}(S)|.$$

Multiplying $f(x)$ to both sides of Eq. (2) and substituting $\chi_S(x)$ by x^S, we have

$$1 = \sum_{S \subseteq [n]} f(x)\hat{f}(S)x^S.$$

Recalling that $M_{f,(x,S)} = f(x)x^S$, we see that the system

$$\sum_{S \subseteq [n]} p_S M_{f,(x,S)} = 1 \quad (\forall x \in \{-1,1\}^n)$$

has a unique solution $p_S = \hat{f}(S)$ (for $S \subseteq [n]$), and hence $\sum_{S \subseteq [n]} |p_S| = ||\hat{f}||_1$. This means that the value of the following problem, which we denote Q_f,

$$\text{Minimize:} \quad \sum_{S \subseteq [n]} |p_S|,$$

$$\text{Subject to:} \quad \sum_{S \subseteq [n]} p_S M_{f,(x,S)} = 1, \ (\forall x \in \{-1,1\}^n)$$

gives the spectral norm of f. We can immediately see that the problem P_f^* is a relaxation of Q_f by replacing the equality ("=") in the constraints with inequality ("$\geq$").

The following is now obvious.

Fact 8. *For every f, $v_f^* \leq v_f = wt(f)$ and $v_f^* \leq ||\hat{f}||_1$.*

If we wish to remove the absolute symbols in the problem P_f or P_f^*, we can achieve this by splitting each variable p_S into two non-negative variables p_S^+ and p_S^-.

Given the problem P_f, we define the linear programming program P_f' as follows:

$$\text{Minimize:} \quad \sum_{S \subseteq [n]} p_S^+ + \sum_{S \subseteq [n]} p_S^-,$$

$$\text{Subject to:} \quad \sum_{S \subseteq [n]} p_S^+ M_{f,(x,S)}^+ + \sum_{S \subseteq [n]} p_S^- M_{f,(x,S)}^- \geq 1, \ (\forall x \in \{-1,1\}^n),$$

$$p_S^+, p_S^- \in \mathbb{Z}^+, \qquad\qquad (\forall S \subseteq [n]),$$

where $M_{f,(x,S)}^+ := M_{f,(x,S)}$ and $M_{f,(x,S)}^- := -M_{f,(x,S)}$.

The following fact shows that the problems P_f and P_f' are essentially the same, i.e., the value of P_f' is v_f and the value of the LP-relaxation of P_f' is v_f^*.

Fact 9. *The optimal solution to P_f' satisfies that, for every $S \subseteq [n]$, at least one of p_S^+ and p_S^- is 0. This is also true for the LP-relaxation of P_f'.*

Proof. Suppose that $\alpha = \min(p_S^+, p_S^-) > 0$ in the optimal solution to P_f' (or the LP-relaxation of P_f'). Then by changing the value of p_S^+ to $p_S^+ - \alpha$ and p_S^- to $p_S^- - \alpha$, the value of the objective function is decreased but still satisfy all the constraints. $\square$

Fact 9 guarantees that we can interchange the optimal solutions to P_f and P_f' via the mapping

$$p_S := \begin{cases} p_S^+, & (\text{if } p_S^+ > 0), \\ -p_S^-, & (\text{if } p_S^- > 0), \\ 0, & (\text{if } p_S^+ = p_S^- = 0). \end{cases}$$

Below we show that the gap between v_f and v_f^*, which is in fact the integrality gap of P_f^*, is at most quadratic (when v_f^* is sufficiently large).

Theorem 10. *For every Boolean function f on n variables, $v_f = wt(f) \leq 8n(v_f^*)^2$.*

Proof. The proof is analogous to the proof of a weaker bound of $wt(f) = O(n||\hat{f}||_1^2)$ by Bruck and Smolensky [4] (or see [10, Theorem 5.12]).

Let $v(p_S^+)$ and $v(p_S^-)$ be the values of an optimal solution to the problem P_f'. Let D be the distribution on the set of signed monomials $\{\pm x^S \mid S \subseteq [n]\}$ such that the probability of choosing $+x^S$ is $v(p_S^+)/v_f^*$ and $-x^S$ is $v(p_S^-)/v_f^*$. Let Q be a random signed monomial chosen according to D. Then, for every fixed x, we have

$$\Pr[Q(x) = f(x)] \geq \frac{1}{2}\left(1 + (v_f^*)^{-1}\right).$$

Let F be the sum of ℓ random signed monomials $Q_1, \ldots, Q_\ell$ where Q_i is chosen independently from this distribution and the value of ℓ will be chosen later. Let $\mu = \mathbf{E}[|\{i \in [\ell] \mid Q_i(x) = 1\}|] = \frac{\ell}{2}(1 + (v_f^*)^{-1})$ and put $\delta = \frac{(v_f^*)^{-1}}{2}$. An easy calculation shows that $(1 - \delta)\mu > \frac{\ell}{2}$. Then, applying Chernoff Bound (see, e.g., [9, Theorem 4.5-2]), we have that for every fixed x,

$$\Pr[sgn(F(x)) \neq f(x)] \leq \Pr[|\{i \in [\ell] \mid Q_i(x) = 1\}| \leq (1 - \delta)\mu]$$
$$\leq \exp(-\mu\delta^2/2).$$

By setting $\ell = 8n(v_f^*)^2$, this probability is less than 2^{-n}. Then by the union bound, we get

$$\Pr[sgn(F(x)) \neq f(x) \text{ for some } x] < 1.$$

This says that there exists a polynomial F of length $8n(v_f^*)^2$ that sign-represents f, which completes the proof. □

We see next that the quadratic gap in Theorem 10 is tight for almost all functions.

Fact 11. *For almost all functions f, the gap between v_f and v_f^* is quadratic.*

Proof. By Parseval's theorem ($\sum_{S \subseteq [n]} \hat{f}(S)^2 = 1$) and Cauchy-Schwartz inequality, we have $||\hat{f}||_1 \leq 2^{n/2}$ for every f on n variables. This implies $v_f^* \leq 2^{n/2}$ by Fact 8. On the other hand, Saks [13] proved that, for almost all Boolean functions f on n variables, $len(f) \geq (0.11)2^n$. The fact follows from $v_f = wt(f) \geq len(f)$. □

We are now ready to discuss the weight and length of the function $\oplus_k f$. It seems that the parameters

$$wt^\oplus(f) := \lim_{k \to \infty} \sqrt[k]{wt(\oplus_k f)},$$
$$len^\oplus(f) := \lim_{k \to \infty} \sqrt[k]{len(\oplus_k f)}$$

well characterize the complexity of f. Since $wt(\oplus_{i+j}f) \leq wt(\oplus_i f)wt(\oplus_j f)$ and $len(\oplus_{i+j}f) \leq len(\oplus_i f)len(\oplus_j f)$ for all i and j, these limits do exist by Fekete's lemma.

For example, we know

$$2 \leq len^{\oplus}(x_1 \wedge x_2) \leq \log_3 26 < 2.966,$$
$$2 \leq wt^{\oplus}(x_1 \wedge x_2) \leq 3.$$

The lower bounds follow from the bound of $len(\mathsf{IP}_n) \geq 2^n$ by Bruck [3] and the upper bound on $len^{\oplus}(x_1 \wedge x_2)$ follows from Fact 4.

Below we show that $wt^{\oplus}(f)$ is between v_f^* and $(v_f^*)^2$. The key observation in proving this is that the value of $P_{\oplus_k f}$ is given by the *tensor power* of the problem P_f.

We write the following integer linear programming problem as $P = (\mathbf{A}, \mathbf{b}, \mathbf{c})$.

$$\text{Minimize: } \mathbf{c}^T \mathbf{x}$$
$$\text{Subject to: } \mathbf{A}\mathbf{x} \geq \mathbf{b},$$
$$\text{every entry of } \mathbf{x} \text{ is in } \mathbb{Z}^+.$$

where $\mathbf{A}$ is $m \times d$ integer matrix, $\mathbf{b}$ is a non-negative column vector of length m, $\mathbf{c}$ is a non-negative column vector of length d, and $\mathbf{x}$ ranges over the set of non-negative integer column vector of length d.

We define the tensor product of two problems $P_1 = (\mathbf{A}_1, \mathbf{b}_1, \mathbf{c}_1)$ and $P_2 = (\mathbf{A}_2, \mathbf{b}_2, \mathbf{c}_2)$ as $P_1 \otimes P_2 = (\mathbf{A}_1 \otimes \mathbf{A}_2, \mathbf{b}_1 \otimes \mathbf{b}_2, \mathbf{c}_1 \otimes \mathbf{c}_2)$, where $\mathbf{A}_1 \otimes \mathbf{A}_2$ denotes the tensor product of matrices $\mathbf{A}_1$ and $\mathbf{A}_2$ and $\mathbf{b}_1 \otimes \mathbf{b}_2$ ($\mathbf{c}_1 \otimes \mathbf{c}_2$, resp.) denotes the tensor product of vectors $\mathbf{b}_1$ and $\mathbf{b}_2$ ($\mathbf{c}_1$ and $\mathbf{c}_2$, resp.). The k-th power tensor of P is denoted by $P^{\otimes k}$. For an IP problem P, let $v(P)$ denote the value of P and $v^*(P)$ denote the value of the LP-relaxation of P.

Note that $v^*(P_1 \oplus P_2) = v^*(P_1)v^*(P_2)$ and $v(P_1 \otimes P_2) \leq v(P_1)v(P_2)$ always hold, but $v(P_1 \otimes P_2) = v(P_1)v(P_2)$ does not hold in general (see e.g., [12, p. 130] for an example).

Theorem 12. *For every Boolean function f, $v_f^* \leq wt^{\oplus}(f) \leq (v_f^*)^2$.*

Proof. Let n be the number of input variables of f. We first verify that $v_{\oplus_k f}^* = (v_f^*)^k$. Recall that $v_{\oplus_k f}^*$ is the value of the LP relaxation of $P'_{\oplus_k f}$.

Consider two IP problems $P'_{\oplus_k f}$ and $(P'_f)^{\otimes k}$. One may see that these two problems are different because $P'_{\oplus_k f}$ contains 2×2^{nk} variables p_S^+ and p_S^- (for $S = (S_1, \ldots, S_k) \subseteq [n]^k$) and $(P'_f)^{\otimes k}$ contains $2^k \times 2^{nk}$ variables p_S^L (for $L \in \{+, -\}^k$ and $S = (S_1, \ldots, S_k) \subseteq [n]^k$). However, an easy inspection shows that these two problems are equivalent. Namely, if $\{p_S^+, p_S^- \mid S \subseteq [n]^k\}$ is a feasible solution to $P'_{\oplus_k f}$, then any $\{p_S^L \mid L \in \{+, -\}^k, S \subseteq [n]^k\}$ satisfying

$$p_S^+ = \sum_{L:|L|=even} p_S^L,$$
$$p_S^- = \sum_{L:|L|=odd} p_S^L,$$

is a feasible solution to $(P'_f)^{\otimes k}$, and vice versa. Here $|L|$ denotes the number of '$-$' symbols in L. In addition, the value of the objective functions are the same. Since it is known that $v^*(P_1 \otimes P_2) = v^*(P_1)v^*(P_2)$ for every IP problems P_1 and P_2 (see e.g., [12, Proposition 4(iii)]), we have $v^*_{\oplus_k f} = v^*(P'_f)^{\otimes k} = v^*(P'_f)^k = (v^*_f)^k$

Now the proof is completed by observing

$$(v^*_f)^k = v^*_{\oplus_k f} \leq v_{\oplus_k f} = wt(\oplus_k f) \leq 8n(v^*_{\oplus_k f})^2 = 8n(v^*_f)^{2k}.$$

Here we use Theorem 10 to derive the last inequality.   □

As P^*_f is an LP problem with 2^n variables and 2^n constraints, we can compute v^*_f for reasonable size of n, say $n \sim 10$, by an LP-solver. Based on some computer experiments, it is very plausible that, for $f = \mathsf{AND}_n$,

$$3 - 2^{-(n-2)} = ||\hat{f}||_1 = v^*_f < v_f = n + 1, \tag{3}$$

and, for $f = \mathsf{MAJ}_n (= sgn(\sum_{i \in [n]} x_i)$, n is odd),

$$(n+1)/2 = v^*_f < v_f = n,$$

The formal proof of the above values of v^*_f could be provided by considering the dual of P^*_f.

For example, Ineq. (3) implies that $3 - 2^{-(n-2)} \leq wt^{\oplus}(\mathsf{AND}_n) < 9$ and $len^{\oplus}(\mathsf{AND}_n) < 9$, which show that there is a large gap between $wt^{\oplus}(\mathsf{AND}_n)$ and $wt(\mathsf{AND}_n) = 2n - 1$ ($len^{\oplus}(\mathsf{AND}_n)$ and $len(\mathsf{AND}_n) = n + 1$, resp.).

For the lower bound on $len^{\oplus}(\mathsf{AND}_n)$, we only know $2 \leq len^{\oplus}(\mathsf{AND}_n)$ following from $\mathsf{AND}_2(x_1, x_2) = \mathsf{AND}_n(x_1, x_2, -1, \ldots, -1)$. Currently, the relationship between three parameters v_f, v^*_f and $||\hat{f}||_1$ seems mysterious.

5   Concluding Remarks

There are many interesting problems for future research. Below we list some of them.

- Can we obtain a good parameter which gives the lower bound on $len^{\oplus}(f)$? By the result of Bruck [3], we have $1/||\hat{f}||_\infty \leq len^{\oplus}(f)$, where $||\hat{f}||_\infty$ is the Fourier L_∞-norm of f. However, this seems weak for non-bent functions. In fact, we conjecture that $v^*_f \leq len^{\oplus}(f)$.
- We have not found any examples satisfying $wt(f \oplus g) \neq wt(f)wt(g)$. It is interesting to give an explicit construction of such polynomials. As we have seen in Sect. 4, this problem is closely related to the problem to seek the (sufficient/necessary) conditions on an IP problem P satisfying $v(P \otimes P) = v(P)v(P)$, which would be interesting in its own right.

Acknowledgment. The author would like to thank an anonymous referee for pointing out an error in the earlier version of the proof of Theorem 6, and Shoma Tate for helpful comments. This work was partially supported by KAKENHI No. 15K00006, 24106006 and 24500006.

References

1. Amano, K.: New upper bounds on the average PTF density of boolean functions. In: Cheong, O., Chwa, K.-Y., Park, K. (eds.) ISAAC 2010, Part I. LNCS, vol. 6506, pp. 304–315. Springer, Heidelberg (2010)

2. Beigel, R.: The polynomial method in circuit complexity. In: Proceedings of 8th Conference on Structure in Complexity Theory, pp. 82–95 (1993)

3. Bruck, J.: Harmonic analysis of polynomial threshold functions. SIAM J. Disc. Math. $3(2)$, 168–177 (1990)

4. Bruck, J., Smolensky, R.: Polynomial threshold functions, AC^0 functions, and spectral norms. SIAM J. Comput. $21(1)$, 33–42 (1992)

5. Filmus, Y., Hatami, H., Heilman, S., Mossel, E., O'Donnell, R., Sachdeva, S., Wan, A., Wimmer, K.: Real analysis in computer science: a collection of open problems (2014). https://simons.berkeley.edu/sites/default/files/openprobsmerged.pdf

6. Klivans, A.R., O'Donnell, R., Servedio, R.A.: Learning intersections and thresholds of halfspaces. J. Comput. Syst. Sci. $68(4)$, 808–840 (2004)

7. Klivans, A.R., Servedio, R.A.: Learning DNF in time $2^{\tilde{o}(n^{1/3})}$ time. J. Comput. Syst. Sci. $68(2)$, 303–318 (2004)

8. Klivans, A.R., Sherstov, A.A.: Unconditional lower bounds for learning intersections of halfspaces. Mach. Learn. $69(2–3)$, 97–114 (2007)

9. Mitzenmacher, M., Upfal, E.: Probability and Computing. Cambridge University Press, New York (2005)

10. O'Donnell, R.: Analysis of Boolean Functions. Cambridge University Press, Cambridge (2014)

11. O'Donnell, R., Servedio, R.A.: New degree bounds for polynomial threshold functions. Combinatorica $30(3)$, 327–358 (2010)

12. Pemantle, R., Propp, J.G., Ullman, D.: On tensor powers of integer programs. SIAM J. Disc. Math. $5(1)$, 127–143 (1992)

13. Saks, M.E.: Slicing the hypercubes. In: Surveys in Combinatorics, pp. 211–255 (1993)

14. Sezener, C.E., Oztop, E.: Minimal sign representation of boolean functions: algorithms and exact results for low dimensions. Neural Comput. $27(8)$, 1796–1823 (2015)

15. Sherstov, A.A.: The intersection of two halfspaces has high threshold degree. SIAM J. Comput. $42(6)$, 2329–2374 (2013)

The Beachcombers' Problem: Walking and Searching from an Inner Point of a Line

Yu Chen[1], Xiaotie Deng[2], Ziwei Ji[1(✉)], and Chao Liao[1]

[1] Zhiyuan College, Shanghai Jiao Tong University, 800 Dong Chuan Road,
Shanghai, People's Republic of China
{chenyuccc,dengxiaotie,kidjiziwei,zeonsgtr}@sjtu.edu.cn
[2] Department of Computer Science, Shanghai Jiao Tong University,
800 Dong Chuan Road, Shanghai, People's Republic of China

Abstract. We consider n beachcombers who are set to search a line segment whose length can be any real number. Each beachcomber has a high *walking speed* and a lower *searching speed* of its own. The problem is to find the optimal schedule such that the line segment can be searched with the minimum makespan.

We assume that the length of the segment is known in advance and beachcombers all start from an arbitrary inner point of the line segment. We show that the problem is NP-hard even if all beachcombers have *the same walking speed*. Then we give an efficient algorithm for the case where all beachcombers are identical.

Keywords: Computational complexity · Mobile agents · Algorithms · Schedule · Searching · Walking · Speed · Partitioning

1 Introduction

In many applications, we are to search a large domain with mobile agents in the minimum makespan. A mobile agent usually needs to first *walk* towards this area, and then *search* it. The walking speed of an agent is normally considered faster than its searching speed since the latter is a more demanding task. The Beachcombers' Problem is introduced in [6] to model such a setting: We have n mobile agents, the ith of whom has a *walking speed* w_i and a *searching speed* s_i with $w_i > s_i > 0$. The agents are requested to collectively search a given line segment whose length can be any real number. The goal is to schedule them such that the makespan is minimized.

In [6], all mobile agents are required to start from one endpoint of the line segment. If the length of the line segment is known at the beginning, an optimal solution is known to solve the scheduling problem in polynomial time [6]. In [2,7], the mobile agents can be divided into groups and the algorithm can choose the starting point for each group.

This work is supported by the National Science Foundation of China (Grant No. 61173011) and a Project 985 grant of Shanghai Jiao Tong University.

© Springer International Publishing Switzerland 2016
A.-H. Dediu et al. (Eds.): LATA 2016, LNCS 9618, pp. 270–282, 2016.
DOI: 10.1007/978-3-319-30000-9_21

In this work, we consider the problem where agents may all initially start from an arbitrary inner point of the line segment. It will be demonstrated that the problem is NP-hard even if all mobile agents have *the same walking speed* (but possibly different searching speeds). An algorithm will also be given under the condition that all agents are identical. A similar problem is considered in [2,7]. However, in their work an algorithm can choose several starting points and assign each agent to one of them. If we do not have efficient tools to move agents to their starting points, the model in [2,7] is not so intuitive.

1.1 Related Work

Agent exploration has been an extensively studied topic. Some papers consider topological model (i.e., the real world is abstracted as an unknown graph and for each edge we want to figure out where it leads to), such as [1,4,10,11,15]. In [1,10], all the nodes and edges are distinguishable and the competitive ratio is considered (i.e., we compare the effort needed by the algorithm to search the whole graph with the effort needed if the graph is known in advance). In [4,11,15], the nodes and edges are indistinguishable and some pebbles or markers need to be used to mark nodes or edges.

Geometric explorations are studied in other works (c.f. [5,9,12]). The goal is to record geometric information of the real world. While our task is to record more information of the geometric world, the acquired geometric information can also be used as a feedback to help us in the process of exploration and mapping.

Mobile agents with different speeds have been considered in recent studies. In [8,13], mobile agents with different speeds are used for boundary patrolling. In [3], agents' different speeds are used to devise protocols which converge quickly. Different speeds of mobile agents can make the search more efficient but the structure of the optimal schedule also becomes more complicated in this way.

1.2 Outline of the Paper

We introduce notations, terminologies, and techniques we use in Sect. 2.

In Sect. 3, we consider the problem in which all agents start from an inner point of the line segment. This problem is NP-hard even if all agents have the same walking speed. For the case where all agents are the same, we give an efficient algorithm.

2 Preliminaries

We are given a set A of n mobile agents, r_1, r_2, ..., r_n. Without loss of generality, we assume $n \geq 2$. Each agent either takes the *walking mode* where the *maximal walking speed* is denoted by w_i, or the *searching mode* where the *maximal searching speed* is denoted by s_i, with $w_i > s_i > 0$. Using the terminology in [7], if all w_i's are identical, we call this input a *W-uniform* instance.

We need to search a line segment (i.e., each point of the line segment needs to be searched at least once) using these mobile agents. Agents can change their modes, speeds and directions, and cross over each other on the line segment. The length of the segment is denoted by L. In our work, L can be any real number. We want to compute the optimal schedule which minimizes the makespan (i.e., the time by which each point of the line segment is searched at least once).

More formally, a schedule $\mathcal{A}$ contains a series of time points $0 = \tau_0 < \tau_1 < \tau_2 < \ldots < \tau_z$. In each time interval $[\tau_i, \tau_{i+1}]$, $0 \le i < z$, each agent either walks or searches without changing its direction. The mode switching and direction change only happen at τ_1, or τ_2, $\ldots$, or τ_{z-1}. At time τ_z, each point of the line segment must be searched at least once. τ_z is the makespan of this schedule, which we want to minimize.

The area searched by agent r_i is denoted by $\sigma(r_i)$. Note that by definition, $\sigma(r_i)$ may have zero length or contain a large number of subintervals of the segment. Later more properties of $\sigma(r_i)$ will be introduced.

For any possible schedule, we make the following assumptions in our work:

Assumption 1. *For any two agents r_i and r_j, $\sigma(r_i) \cap \sigma(r_j)$ has zero length.*

Assumption 2. *For agent r_i, it either walks with a speed of w_i, or searches with a speed of s_i.*

It should be noted that no generality is sacrificed under these two assumptions since given any schedule we can modify it and make Assumptions 1 and 2 correct without increasing the makespan. Actually, in the following it will be shown that Assumptions 1 and 2 are satisfied by *any optimal schedule*.

Due to Assumption 2, we will just call w_i the walking speed of r_i and s_i the searching speed of r_i, leaving out the adjective "maximal".

2.1   Starting from an Endpoint

If all agents start from the same endpoint of the segment, we call this problem EBP ("E" refers to "endpoint"). The minimum makespan in this case is denoted by T_{EBP}. Note that in EBP, it is unnecessary for an agent to change its direction.

EBP is wholly solved in [6]. First the following lemma is proved there under Assumptions 1 and 2:

Lemma 1. *For an EBP problem, in any optimal schedule,*

- *All agents stop at the same time.*
- *For each agent r_i, $\sigma(r_i)$ has a positive length and contains only one subinterval of the line segment (i.e., each agent must search a continuous area of positive length).*
- *If for some r_i and r_j, $w_i < w_j$, then $\sigma(r_i)$ is closer to the starting point than $\sigma(r_j)$. If $w_i = w_j$, then the order of $\sigma(r_i)$ and $\sigma(r_j)$ does not affect the makespan of the schedule.*

In fact, with Lemma 1 we can prove that every optimal schedule of an EBP problem must satisfy Assumptions 1 and 2. Actually for any optimal schedule $\mathcal{O}$ we can change it into an optimal schedule $\mathcal{O}'$ satisfying Assumptions 1 and 2 where all agents stop at the same time, which is impossible if Assumptions 1 and 2 do not hold in $\mathcal{O}$. Thus Lemma 1 is valid regardless of Assumptions 1 and 2.

What's more, the following crucial notion of *search power* is introduced.

Definition 2 (Search Power). *Consider a set A of mobile agents r_1, r_2, $\ldots$, r_n. Sort the agents such that $w_1 \leq w_2 \leq \ldots \leq w_n$. The search power of A is defined as:*

$$\mathrm{SP}(A) = \sum_{i=1}^{n} s_i \prod_{j=i+1}^{n} \left(1 - \frac{s_j}{w_j}\right) \tag{1}$$

One can easily verify that if some agents have identical walking speed, then the order of them does not matter. Therefore the search power is well-defined.

For later reference, we also give some other forms of the above formula. We set

$$e_i = \prod_{j=i+1}^{n} \left(1 - \frac{s_j}{w_j}\right) \tag{2}$$

and $e_n = 1$. Then

$$\mathrm{SP}(A) = \sum_{i=1}^{n} w_i(e_i - e_{i-1}) \tag{3}$$

Note that we always have $\mathrm{SP}(A) < w_n$. If all w_i's are equal to w, then

$$\mathrm{SP}(A) = w \left(1 - \prod_{j=1}^{n} \left(1 - \frac{s_j}{w}\right)\right) \tag{4}$$

What is the meaning of search power? It is proved in [6] that

Lemma 3.

$$\mathrm{SP}(A) = L/T_{\mathrm{EBP}} \tag{5}$$

To understand this result, we can consider some special cases. For example, as w_i's tend to infinity, the search power tends to $\sum_{i=1}^{n} s_i$, which is consistent with our intuition. Another example is to let s_n tend to w_n and consequently the search power will tend to s_n. This is also intuitive: In this way $s_n \approx w_n \geq w_{n-1} \geq \ldots \geq w_1$ and we will tend to use only r_n to search the whole segment.

2.2 Starting from an Inner Point

In our work, we consider the problem where the starting point can be an arbitrary inner point of the line segment. We call this problem IBP ("I" refers to "inner point"). In IBP agents may need to change their directions.

We assume that the starting point is L_1 away from the left endpoint and L_2 away from the right endpoint. We use $[-L_1, L_2]$ to denote the whole line segment, with the starting point at position 0. By Assumption 1, we can have $-L_1 = c_{-p} < c_{-p+1} < \ldots < c_{-1} < c_0 = 0 < c_1 < \ldots < c_{q-1} < c_q = L_2$, such that for each k, $-p \leq k < q$, the subinterval $[c_k, c_{k+1}]$ is searched wholly by a single agent and no two consecutive subintervals are searched by the same agent except for $[c_{-1}, c_0]$ and $[c_0, c_1]$ (since they are on different sides of the starting point and may be searched by the same agent). Denote $[c_k, c_{k+1}]$ by σ_k and the label of the agent searching σ_k by $a(\sigma_k)$, $-p \leq k < q$. Thus for each agent r_i, $\sigma(r_i) = \cup_{k \in H_i} \sigma_k$, where $H_i = \{k | a(\sigma_k) = i\}$.

Our main result is that IBP, even W-uniform IBP, is NP-hard. To prove this result, we consider two modified versions of IBP:

- NOBACKIBP: Agents are not allowed to change their directions after setting out from the starting point.
- RETURNTOSTARTIBP: Agents can return to the starting point instantaneously (i.e., without spending any time) from any point on the line segment.

For example, suppose an agent with a walking speed of 1 starts from point 0 and reaches point 1. In RETURNTOSTARTIBP, it can return to point 0 instantaneously. However, in IBP it must spend at least 1 time unit to return to 0.

The above two modified models are not intuitive but important in our proof. Our main idea is the following: First, we show the NP-hardness of W-uniform NOBACKIBP; more specifically, we show the NP-hardness of deciding whether we can divide the agents into two groups having the same search power. Then we show that if we can divide the agents into two groups of identical search power, then in any optimal schedule of W-uniform IBP, no agent changes its direction. In the proof, RETURNTOSTARTIBP serves as a bridge between IBP and EBP with which we can invoke previous results.

3   Searching a Line Segment Starting from an Inner Point

3.1   Mobile Agents Which Cannot Turn Back

In this subsection, we consider the NOBACKIBP model in which the mobile agents can choose a direction at the beginning but cannot change the direction. This model is not so natural but to show the NP-hardness of IBP, it is helpful to first understand the NP-hardness of NOBACKIBP.

In such a situation we need to divide the set A of mobile agents into two disjoint sets A_1 and A_2. Intuitively, we want the ratio of their search powers to be close to L_1/L_2.

Next we show that such a problem is NP-hard even if the input is a W-uniform instance. We mention that this proof is actually equivalent to the proof of the NP-hardness of 2-SBP in [7]. However, since the model is a little different, we still put the proof here.

Theorem 4. *W-uniform* NOBACKIBP *is NP-hard.*

Proof. In [14] it is shown that PRODUCTPARTITION (where we are given a set of positive integers and asked whether we can partition them into two groups having the same product) is NP-hard. We prove the theorem by reducing PRODUCTPARTITION to W-uniform NOBACKIBP.

Suppose we are given a set X of integers $x_1, x_2, \ldots, x_n$. Without loss of generality, we assume there is no 1 in X. Take $w_i = w = \prod_{i=1}^{n} x_i$ and $s_i = w(1 - 1/x_i)$. What's more, put all the mobile agents at the middle of a line segment of length 2 (i.e., $L_1 = L_2 = 1$). In this way, we need to divide the set A of mobile agents into two disjoint subsets A_1 and A_2 which maximizes $\min\{\mathrm{SP}(A_1), \mathrm{SP}(A_2)\}$. Since all w_i's are equal, (4) shows that we need to maximize

$$\min\left\{1 - \prod_{i \in A_1}(1 - \frac{s_i}{w}), 1 - \prod_{i \in A_2}(1 - \frac{s_i}{w})\right\} \tag{6}$$

Since $\prod_{i \in A}(1 - s_i/w) = 1/(\prod_{i \in A} x_i)$ is a constant, X can be divided into two disjoint subsets having identical product if and only if the maximum of (6) is $1 - \sqrt{\prod_{i \in A}(1 - s_i/w)} = 1 - \sqrt{1/(\prod_{i \in A} x_i)}$. $\qquad\square$

3.2 Mobile Agents Which Can Return to the Starting Point Instantaneously

In this section we consider the RETURNTOSTARTIBP model where agents can return to the starting point instantaneously from any point on the segment. This model is also significant in proving the NP-hardness of IBP.

Like Lemma 1, we have the following lemma for RETURNTOSTARTIBP:

Lemma 5. *In any optimal schedule of a* RETURNTOSTARTIBP *problem:*

- *All agents stop at the same time.*
- *For each agent r_i, $\sigma(r_i)$ has a positive length and either $\sigma(r_i) = \{\sigma_k\}$, $-p \leq k < q$, or $\sigma(r_i) = \{\sigma_{k_1}, \sigma_{k_2}\}$, $-p \leq k_1 < 0 \leq k_2 < q$.*

The proof is very similar to the proof of Lemma 1 in [6] so we leave it out.

Here is the another lemma, which is about mobile agents that search in both directions:

Lemma 6. *Suppose $r_i, r_j \in A$ both search in two directions. r_i searches σ_u and σ_x while r_j searches σ_v and σ_y, $-p \leq u, v < 0 \leq x, y \leq q - 1$. Then in any optimal schedule, $u < v$ and $x < y$ cannot both hold.*

Proof. We prove the above lemma by contradiction. Assume in an optimal schedule $\mathcal{O}$, $u < v$ and $x < y$. Furthermore assume the length of σ_v is smaller than or equal to σ_x. Then we can change our schedule and let r_i search σ_u, σ_v and $[c_x, c_{x+1} - (c_{v+1} - c_v)]$ while let r_j search σ_y and $[c_{x+1} - (c_{v+1} - c_v), c_{x+1}]$. Their searching distances remain unchanged but their walking distances decrease by at least $(c_{v+1} - c_v)$. Therefore they spend less time in the new schedule, which contradicts the optimality of $\mathcal{O}$ due to Lemma 5. $\qquad\square$

Another observation of the RETURNTOSTARTIBP problem is the following: Suppose mobile agent $r_i \in A$ searches in both directions and spends x time units in the positive direction and y time units in the negative direction. Then the work of r_i can be seen as done by two agents r_{i_1} and r_{i_2}. r_{i_1} has a walking speed of $w_i x/(x+y)$ and a searching speed of $s_i x/(x+y)$ while r_{i_2} has a walking speed of $w_i y/(x+y)$ and a searching speed of $s_i y/(x+y)$. Both r_{i_1} and r_{i_2} use $(x+y)$ time units. Inspired by this observation, we define the *directional speed*:

Definition 7 (directional speed). *Suppose in a given schedule the line segment $[-L_1, L_2]$ is searched using t time units. For each agent $r_i \in A$ which searches a subinterval in the positive direction, if it spends xt $(0 < x \leq 1)$ time units in the positive direction, then its positive directional speed is*

$$w_{i+} = xw_i, \quad s_{i+} = xs_i \tag{7}$$

The negative directional speed (i.e., w_{i-} and s_{i-}) can be defined similarly.

With the notion of directional speed, we can see a RETURNTOSTARTIBP problem as two EBP problem in different directions. With Lemma 1, we have:

Lemma 8. *In any optimal schedule of a RETURNTOSTARTIBP problem, $w_{a(\sigma_0)+} \leq w_{a(\sigma_1)+} \leq \cdots \leq w_{a(\sigma_{q-1})+}$ and $w_{a(\sigma_{-1})-} \leq w_{a(\sigma_{-2})-} \leq \cdots \leq w_{a(\sigma_{-p})-}$. What's more, if in the positive (negative) direction some agents have identical positive (negative) directional walking speed, then the order of them does not affect the makespan of the schedule.*

Combining Lemmas 6 and 8, we have the following crucial lemma:

Lemma 9. *In a W-uniform RETURNTOSTARTIBP problem, if mobile agents all start from the middle of the segment and the set A of agents can be partitioned into two disjoint subsets A_1 and A_2 such that $SP(A_1) = SP(A_2)$, then (A_1, A_2) is one of the optimal schedules of the problem (i.e., for an optimal schedule we can let agents in A_1 search in the positive direction and agents in A_2 search in the negative direction).*

Proof. Suppose $\mathcal{O}$ is an optimal schedule that is strictly better than (A_1, A_2). There must be some agents which search in both directions in $\mathcal{O}$; otherwise $\mathcal{O}$ cannot be strictly better than (A_1, A_2) since we assume $SP(A_1) = SP(A_2)$ (recalling the proof of Theorem 4). Suppose there are k agents searching in both directions. Notice that all the walking speeds are identical. Due to Lemma 8, $\sigma_{-k}, \sigma_{-k+1}, \ldots, \sigma_{k-1}$ are searched by these k agents. We prove k must be 0 case by case and thus achieve a contradiction.

Case 1: $k \geq 2$ and $w_{a(\sigma_0)+} < w_{a(\sigma_{k-1})+}$. Denote $a(\sigma_0)$ by i and $a(\sigma_{k-1})$ by j. Since $w_{i-} + w_{i+} = w_{j-} + w_{j+}$, we have $w_{i-} > w_{j-}$. Because of Lemma 8, in the negative direction the interval searched by r_j must be closer to the starting point than that searched by r_i. This contradicts Lemma 6.

Case 2: $k \geq 2$ and $w_{a(\sigma_0)+} = w_{a(\sigma_{k-1})+}$. Still denote $a(\sigma_0)$ by i and $a(\sigma_{k-1})$ by j. Note that we also have $w_{a(\sigma_{-1})-} = w_{a(\sigma_{-2})-} = \cdots = w_{a(\sigma_{-k})-}$. Due to

Lemma 8, in the negative direction the order of these k agents does not affect the makespan. Thus in the negative direction we can always make the interval searched by r_j closer to the starting point than that searched by r_i, which contradicts Lemma 6.

Case 3: $k = 1$. Suppose σ_{-1} and σ_0 are searched by agent r_i. According to the observation above Definition 7, the work of r_i can be seen as done by r_{i_1} in the positive direction and r_{i_2} in the negative direction. Let X and Y denote the set of agents which search in the positive direction and negative direction respectively, except for r_i. We set (w is the common walking speed):

$$T(X) = \prod_{i \in X} (1 - \frac{s_i}{w})$$

$$T(Y) = \prod_{i \in Y} (1 - \frac{s_i}{w})$$

By Definition 2, we have $\mathrm{SP}(X \cup r_{i_1}) = \mathrm{SP}(X) + s_{i+} \cdot T(X)$ and $\mathrm{SP}(Y \cup r_{i_2}) = \mathrm{SP}(Y) + s_{i-} \cdot T(Y)$. Suppose $T(X) \geq T(Y)$. Then we have

$$\min\{\mathrm{SP}(X \cup r_{i_1}), \mathrm{SP}(Y \cup r_{i_2})\}$$
$$\leq \frac{1}{2}(\mathrm{SP}(X) + s_{i+} \cdot T(X) + \mathrm{SP}(Y) + s_{i-} \cdot T(Y))$$
$$\leq \frac{1}{2}(\mathrm{SP}(X) + s_i \cdot T(X) + \mathrm{SP}(Y))$$
$$\leq \frac{1}{2}(\mathrm{SP}(X \cup r_i) + \mathrm{SP}(Y))$$
$$\leq \frac{1}{2}(\mathrm{SP}(A_1) + \mathrm{SP}(A_2))$$
$$= \mathrm{SP}(A_1)$$

$\square$

3.3   Mobile Agents Which Can Turn Back

Now we consider the IBP problem where agents can turn back. We will show that W-uniform IBP is NP-hard.

We can show that the conclusions of Lemma 5 also hold in the IBP problem (i.e., in any optimal schedule of an IBP problem, all agents stop at the same time and for each agent and each direction, it either does not walk or search in this direction, or search a continuous subinterval in this direction). The only difference is that if an agent wants to turn back and search in the other direction, it must walk across the subinterval between the turning point and the starting point. It is also easy to see that each agent changes its direction at most once.

Lemma 10. *Consider a W-uniform IBP instance. Suppose agents all start from the middle of the line segment. If we can divide the set A of agents into two disjoint subsets A_1 and A_2 such that $\mathrm{SP}(A_1) = \mathrm{SP}(A_2)$, then in any optimal schedule of this instance, no mobile agent turns back.*

Proof. Suppose the conditions of Lemma 10 hold and in an optimal schedule $\mathcal{O}$ of the IBP problem, some agents turn back. The makespan of $\mathcal{O}$ is denoted by $T_I(\mathcal{O})$.

$\mathcal{O}$ can be changed easily to a schedule $\mathcal{O}'$ for the W-uniform RETURN-TOSTARTIBP model without increasing the makespan: We just allow agents which turn back in $\mathcal{O}$ to return (or "jump") to the starting point instantaneously. The makespan of $\mathcal{O}'$ is denoted by $T_{RTS}(\mathcal{O}')$. What's more, (A_1, A_2) is a schedule for both models and the makespan of it is denoted by $T_I((A_1, A_2)) = T_{RTS}((A_1, A_2))$.

If in $\mathcal{O}$ there exists an agent which does not turn back, then in $\mathcal{O}'$ not all agents stop at the same time and thus $\mathcal{O}'$ *is not an optimal schedule* for the W-uniform RETURNTOSTARTIBP model due to Lemma 5. However, by Lemma 9 (A_1, A_2) *is an optimal schedule* for the W-uniform RETURNTOSTART-IBP. Therefore $T_I(\mathcal{O}) \geq T_{RTS}(\mathcal{O}') > T_{RTS}((A_1, A_2)) = T_I((A_1, A_2))$, which contradicts the optimality of $\mathcal{O}$.

If in $\mathcal{O}$ all agents change directions, then we have $T_I(\mathcal{O}) > T_{RTS}(\mathcal{O}') \geq T_{RTS}((A_1, A_2)) = T_I((A_1, A_2))$, which also contradicts the optimality of $\mathcal{O}$. $\quad\square$

Combining Lemma 10 and the proof of Theorem 4, we have

Theorem 11. *W-uniform* IBP *is NP-hard.*

Proof. The construction is the same as that in the proof of Theorem 4. X can be partitioned into two subgroups having identical product if and only if in any optimal schedule of the constructed IBP problem, no agent changes its direction and A is partitioned into two subsets with the same search power. $\quad\square$

3.4   An Algorithm for Identical Agents Which Can Turn Back

In this section, we consider the problem with n identical agents which can turn back. We have the following theorem.

Theorem 12. *If all agents are the same, there exists an optimal schedule such that at most one agent turns back and if there is one (say r_i), it searches σ_0 and σ_{-1}.*

Based on this theorem, we have an algorithm computing the optimal schedule. In fact, we only need to consider:

- whether there exists an agent that turns back and if there exists one, its initial direction and how long are σ_0 and σ_{-1} (in other words, what are c_1 and $(-c_{-1})$);
- the number of agents in each direction.

We just need to calculate the makespan each situation and choose the best one. A formal algorithm is given in the next page.

Now we prove Theorem 12. For agent r_i, let $-u_i(u_i > 0)$ and $v_i(v_i > 0)$ be the leftmost and rightmost point it has ever touched respectively.

---

1 For n identical agents, assume each of them has searching speed s and walking speed w. Let $a = 1 - \frac{s}{w}$ and then the search power is $w(1 - a^n)$. At most one agent needs to turn back.

2 Suppose no agent turns back. Let m be the number of agents walking left. Enumerate m from 1 to $n-1$, and check $\frac{L_1}{w(1-a^m)} = \frac{L_2}{w(1-a^{n-m})}$. If some m (say m_0) satisfies this equation, let $ans = \frac{L_1}{w(1-a^{m_0})}$. Otherwise let $ans = +\infty$.

3 Suppose exactly one agent turns back. For brevity, suppose this agent first walks in negative direction and then turns back. The other case needs to be discussed similarly. Let m be the number of agents that walk in negative direction first and do not turn back. There are three possibilities (c_{-1} and c_1 are unknown):

   1. $m = 0$: Solve $\frac{L_1+c_1}{s} + \frac{L_1}{w} = \frac{L_2-c_1}{w(1-a^{n-1})} + \frac{c_1}{w}$. If $0 \le c_1 < L_2$, update ans with $\frac{L_1+c_1}{s} + \frac{L_1}{w}$.

   2. $0 < m < n-1$: Enumerate m from 1 to $n-2$, and solve

$$\begin{cases} \frac{L_1+c_{-1}}{w(1-a^m)} + \frac{-c_{-1}}{w} = \frac{c_1-c_{-1}}{s} + \frac{-c_{-1}}{w} \\ \frac{L_2-c_1}{w(1-a^{n-m-1})} + \frac{c_1}{w} = \frac{c_1-c_{-1}}{s} + \frac{-c_{-1}}{w} \end{cases}$$

     If $-L_1 < c_{-1} < 0$ and $0 < c_1 < L_2$, update ans with $\frac{c_1-c_{-1}}{s} + \frac{-c_{-1}}{w}$.

   3. $m = n-1$: Solve $\frac{L_2-c_{-1}}{s} + \frac{-c_{-1}}{w} = \frac{L_1+c_{-1}}{w(1-a^{n-1})} + \frac{-c_{-1}}{w}$. If $-L_1 < c_{-1} \le 0$, update ans with $\frac{L_2-c_{-1}}{s} + \frac{-c_{-1}}{w}$.

Algorithm 1: n-identical IBP

Lemma 13. *There exists an optimal schedule where at most one agent turns back.*

Proof. If there are at least two agents that turn back, arbitrarily choose two from them (say r_i, r_j). Supposing $u_i \le u_j$, there are several cases:

1. $v_i \le v_j$
      (a) r_i and r_j take the same direction at the beginning. Suppose they both take the negative direction first.

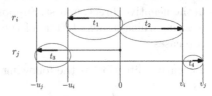

Thick lines in the picture represent search and normal lines represent walk. t_1 is the time of r_i going from 0 to $-u_i$ and returning to 0. t_2 is the time of r_i going from 0 to v_i after returning to 0. t_3 and t_4 are similarly defined as the figure shows. In the remaining proof of this lemma, we use the same notations.

Since they stop at the same time, we know that $t_1 + t_2 > t_3 + t_4 \Rightarrow t_1 > t_4 \lor t_2 > t_3$. If $t_1 > t_4$, let r_i do as r_j in $[v_i, v_j]$. Since r_i and r_j are identical, it takes r_i t_4 units of time additionally. Then let r_i retrace at some point $x \in [-u_i, 0]$ such that r_i spends t_4 units of time in $[-u_i, x]$ originally. Thus the time used by r_i does not change. However, at the same time, let r_j get rid of the work in $[v_i, v_j]$ (which is done by r_i now) and do the work in $[-u_i, x]$ (which is done by r_i before). r_j does not need to turn back now since it does not search some area in $[0, L_2]$ and the ending time does not increase. So this case can be avoided in an optimal schedule. If $t_2 > t_3$, we can make similar arguments.

(b) r_i and r_j take different directions at the beginning. Suppose r_i takes the positive direction first.

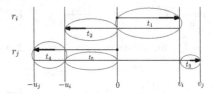

Because they stop at the same time, $t_1 + t_2 > t_3 + t_4 + t_5 \Rightarrow t_1 > t_4 \vee t_2 > t_3 + t_5$. If $t_1 > t_4$, we can make similar argument as 1a. Otherwise $t_2 > t_3 + t_5$, let r_i take the negative direction first and do the work in $[v_i, v_j]$. In this way r_i spends at most $(t_3 + t_5)$ units of time additionally. In order to keep the ending time of r_i unchanged, let r_i get rid of some work in $[-u_i, 0]$ which takes $(t_3 + t_5)$ units of time. This work can be done by r_j because r_j does not need to turn back and do its original work in $[v_i, v_j]$. Now, the ending time does not increase. So this case can be avoided in an optimal schedule.

2. $v_i > v_j$ The proof of this case is essentially the same as before so we leave out it because of page limit. □

Lemma 14. *There exists an optimal schedule where there is either no agent that turns back or exactly one agent that turns back and searches $[-u_i, v_i]$.*

Proof. By Lemma 13, there exists an optimal schedule in which at most one agent turns back. If there is such an agent, denote it by r_i. Without loss of generality, suppose r_i walks in the negative direction first. Suppose agent r_j $(j \neq i)$ searches a subinterval of $[-u_i, v_i]$ which is next to one of the two subintervals searched by r_i. There are two cases:

1. r_j walks in the negative direction as the following figure shows:

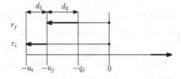

Now we let r_j search $[-u_i, -u_i + x]$ and r_i search $[-u_i + x, -q_j]$. In order to make the ending time of r_j same as before, x need to satisfy $\frac{x}{s} + \frac{d_1 + d_2 - x}{w} = \frac{d_2}{s}$, where s is the searching speed and w is the walking speed. Solving this equation, we have $x = d_2 - \frac{s}{w-s}d_1$. Since r_i and r_j stop at the same time, we have $\frac{d_2}{s} - \frac{d_1+d_2}{w} > 0$, thus $d_2 - \frac{s}{w-s}d_1 > 0$ and $0 < x < d_2$. We also have $\frac{d_2}{s} > \frac{2d_2}{w}$ and then $w > 2s$. What's more, we have $\frac{d_2}{s} > \frac{d_1}{w} + \frac{d_1}{s} + \frac{2d_2}{w} \Leftrightarrow \frac{d_2}{d_1} > \frac{w+s}{w-2s}$. And the ending time of r_i does not increase: $(\frac{d_1+d_2-x}{s} + \frac{d_1+d_2-x}{w}) - (\frac{d_1}{s} + \frac{d_1+2d_2}{w}) < 0 \Leftrightarrow \frac{d_2}{d_1} > \frac{w+s}{2(w-s)} \Leftarrow \frac{d_2}{d_1} > \frac{w+s}{w-2s} > \frac{w+s}{2(w-s)}$.

2. r_j walks in the positive direction:

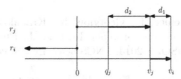

Using the same skill as before, let r_j search $[v_i - x, v_i]$ and r_i search $[q_j, v_i - x]$, and x need to satisfy $\frac{x}{s} + \frac{d_1+d_2-x}{w} = \frac{d_2}{s} \Leftrightarrow x = d_2 - \frac{s}{w-s}d_1$. Since r_i and r_j stop at the same time, we have $\frac{d_2}{s} > \frac{d_2}{w} + \frac{d_1}{s} \Leftrightarrow \frac{d_2}{d_1} > \frac{w}{w-s}$. And the time used by r_i does not increase: $\frac{d_1+d_2-x}{s} - (\frac{d_1}{s} + \frac{d_2}{w}) < 0 \Leftrightarrow \frac{d_2}{d_1} > \frac{w}{w-s}$. $\qquad\square$

4 Conclusions and Open Problems

In this paper, we consider the Beachcombers' Problem where agents start from an inner point. The problem is NP-hard even in W-uniform instances. It would be interesting to find efficient algorithms for some special cases and approximation or randomized algorithms for the NP-hard instances. Also note that many of our discussion can be extended to the star-shaped structure. If we further extend the star-shaped structure, we can get the tree structure and graph structure. Algorithms in these situations would be interesting and useful.

Another interesting problem is to search each point by at least k distinct agents since some agents may be faulty. At first glance, we may still need to partition the agents into k parts and try to maximize the minimum search power. However, the optimal schedule may not follow this pattern. For example, consider searching a segment whose length is 3 using 3 mobile agents with walking speed 2 and searching speed 1 and set $k = 2$. This problem is also of much interest.

References

1. Albers, S., Henzinger, M.R.: Exploring unknown environments. SIAM J. Comput. **29**(4), 1164–1188 (2000)
2. Bampas, E., Czyzowicz, J., Ilcinkas, D., Klasing, R.: Beachcombing on strips and islands. In: Bose, P., Gąsieniec, L.A., Römer, K., Wattenhofer, R. (eds.) Algorithms for Sensor Systems. Lecture Notes in Computer Science, vol. 9536, pp. 155–168. Springer, Heidelberg (2016)
3. Beauquier, J., Burman, J., Clement, J., Kutten, S.: On utilizing speed in networks of mobile agents. In: Proceedings of the 29th ACM SIGACT-SIGOPS Symposium on Principles of Distributed Computing, pp. 305–314. ACM (2010)

4. Bender, M.A., Fernández, A., Ron, D., Sahai, A., Vadhan, S.: The power of a pebble: exploring and mapping directed graphs. In: Proceedings of the Thirtieth Annual ACM Symposium on Theory of Computing, pp. 269–278. ACM (1998)
5. Canny, J., Reif, J.: New lower bound techniques for robot motion planning problems. In: 28th Annual Symposium on Foundations of Computer Science, pp. 49–60. IEEE (1987)
6. Czyzowicz, J., Gąsieniec, L., Georgiou, K., Kranakis, E., MacQuarrie, F.: The Beachcombers' problem: walking and searching with mobile robots. In: Halldórsson, M.M. (ed.) SIROCCO 2014. LNCS, vol. 8576, pp. 23–36. Springer, Heidelberg (2014)
7. Czyzowicz, J., Gasieniec, L., Georgiou, K., Kranakis, E., MacQuarrie, F.: The multi-source Beachcombers' problem. In: Gao, J., Efrat, A., Fekete, S.P., Zhang, Y. (eds.) ALGOSENSORS 2014, LNCS 8847. LNCS, vol. 8847, pp. 3–21. Springer, Heidelberg (2015)
8. Czyzowicz, J., Gąsieniec, L., Kosowski, A., Kranakis, E.: Boundary patrolling by mobile agents with distinct maximal speeds. In: Demetrescu, C., Halldórsson, M.M. (eds.) ESA 2011. LNCS, vol. 6942, pp. 701–712. Springer, Heidelberg (2011)
9. Deng, X., Kameda, T., Papadimitriou, C.H.: How to learn an unknown environment. In: 1991 Proceedings of 32nd Annual Symposium on Foundations of Computer Science, pp. 298–303. IEEE (1991)
10. Deng, X., Papadimitriou, C.H.: Exploring an unknown graph. In: 1990 Proceedings of 31st Annual Symposium on Foundations of Computer Science, pp. 355–361. IEEE (1990)
11. Dudek, G., Jenkin, M., Milios, E., Wilkes, D.: Robotic exploration as graph construction. IEEE Trans. Rob. Autom. 7(6), 859–865 (1991)
12. Guibas, L.J., Motwani, R., Raghavan, P.: The robot localization problem in two dimensions. In: Proceedings of the Third Annual ACM-SIAM Symposium on Discrete Algorithms, pp. 259–268. Society for Industrial and Applied Mathematics (1992)
13. Kawamura, A., Kobayashi, Y.: Fence patrolling by mobile agents with distinct speeds. In: Chao, K.-M., Hsu, T., Lee, D.-T. (eds.) ISAAC 2012. LNCS, vol. 7676, pp. 598–608. Springer, Heidelberg (2012)
14. Ng, C., Barketau, M., Cheng, T.E., Kovalyov, M.Y.: "Product Partition" and related problems of scheduling and systems reliability: Computational complexity and approximation. Eur. J. Oper. Res. 207(2), 601–604 (2010)
15. Wang, H., Jenkin, M., Dymond, P.: The relative power of immovable markers in topological mapping. In: 2011 IEEE International Conference on Robotics and Automation (ICRA), pp. 1050–1057. IEEE (2011)

Using Duality in Circuit Complexity

Silke Czarnetzki[1]([⊠]) and Andreas Krebs[1]

Wilhelm-Schickard-Institut, Universität Tübingen,
Sand 13, 72076 Tübingen, Germany
{czarnetz,krebs}@informatik.uni-tuebingen.de

Abstract. We investigate in a method for proving lower bounds for abstract circuit classes. A well established method to characterize varieties of regular languages are identities. We use a recently established generalization of these identities to non-regular languages by Gehrke, Grigorieff, and Pin: so called equations, which are capable of describing arbitrary Boolean algebras of languages. While the main concern of their result is the existence of these equations, we investigate in a general method that could allow to find equations for circuit classes in an inductive manner. Thereto we extend an important tool – the block product or substitution principle – known from logic and algebra, to non-regular language classes. Furthermore, we abstract this concept by defining it directly as an operation on (non-regular) language classes. We show that this principle can be used to obtain equations for certain circuit classes, given equations for the gate types. Concretely, we demonstrate the applicability of this method by obtaining a description via equations for all languages recognized by circuit families that contain a constant number of (inner) gates, given a description of the gate types via equations.

1 Introduction

In Boolean circuit complexity, deriving lower bounds on circuit size and depth has up to now shown to generally be difficult. While there have been results proving lower bounds, we still lack methods that are applicable in general. Algebraic methods have improved our understanding of circuit complexity. Here we are especially interested in the constant depth circuit complexity classes $\mathbf{AC}^0, \mathbf{CC}^0$, and $\mathbf{ACC}^0$ that have tight connections to algebra via programs. For instance the class $\mathbf{AC}^0$ is equal to the class of languages recognized by polynomial-length programs over finite aperiodic monoids [6]. Using these connections allowed the usage of algebraic methods in circuit complexity [3–5,14,21]. For an overview see the book of Straubing [19].

It is a well known method from algebra to characterize regular language classes by identities and has successfully been applied to describe varieties of regular languages stemming from various logic classes (see for example the book of Pin [15]). Recently, Gehrke, Grigorieff and Pin generalized the approach to work with non-regular language classes [9].

© Springer International Publishing Switzerland 2016
A.-H. Dediu et al. (Eds.): LATA 2016, LNCS 9618, pp. 283–294, 2016.
DOI: 10.1007/978-3-319-30000-9_22

While many concrete characterizations via identities or equations exist for classes of regular languages (see for example the book of Almeida [1]), only few concrete examples are known for non-regular classes [10]. One of the main difficulties is, that these equations hold for all languages in a circuit class and not only the regular ones, for which we have other manageable descriptions. Furthermore, the question arises how to achieve an abstract method to obtain equations for circuit classes, instead of calculating them concretely for each class.

As the result of [9] shows the existence of equational descriptions for arbitrary Boolean algebras, circuit classes form suitable candidates. However, it is not clear how to obtain these equations in a constructive way. The method presented in the paper allows us to obtain equations for more complex classes of circuits, starting with equations from simple classes. In this paper, we would like to dare a first step towards an approach to derive circuit lower bounds for abstract classes of circuits. Even though the classes described are fairly simple and separation results could be proven by using combinatorical arguments, this is the first effort made towards a procedure to compute equations for more general circuit classes.

In order to reach our goal we abstract another powerful technique: the block product or substitution principle [20]. The idea of computing the defining equations for a more complex variety constructed by some principle from simpler varieties has been successfully used in the regular case [2,13]. While all these previous results rely on regular language classes, we extended it to work on non-regular classes by defining an operation purely on language classes, not relying on monoids or automata, which reflects a decomposition of the computation of the circuit.

As our main contribution, we show that in principle it is possible to systematically construct equations for the block product under certain restrictions. To demonstrate that our method can be applied, we concretely compute the equations for languages recognized by constant size circuit families, given equations that describe the gate types allowed in the circuit family.

Organization of the Paper. We organized this paper in a way that all the definitions are introduced along with our demonstration of how to compute the equations for constant size circuit families. Further background on the not (yet) well known theory of topology and duality is put in [18].

As a first step, circuit classes whose gates are defined by a variety of languages are introduced in Sect. 2. In Sect. 3, we define an abstract version of the block product. Then in Sect. 4 we introduce basic definitions and results from Stone duality as far as needed to formulate the main theorem in Sect. 5. We conclude and give hints for further research in Sect. 6. The proof of soundness and completeness of the equations provided in the main theorem can be found in [18]. For a more in depth presentation of the background, we refer to [18].

2   Constant Size Circuits Families

In this paper we consider circuits over arbitrary alphabets. In contrast to the usual notion of size we do not count input gates, and hence only call them inputs. Thus, the size of a circuit is the number of inner nodes. This is necessary to allow circuit families with gates of unbounded fan-in to access inputs of arbitrary length and still have constant size. A formal definition of circuits can be found in [18].

Definition 1 (Family of Boolean functions defined by a language). *A language $L \subseteq \{0,1\}^*$ in a natural way defines a family of Boolean functions, denoted by $f^L = (f_i^L)_{i \in \mathbb{N}}$ where $f_i^L(x_0, \ldots, x_{i-1}) = 1$ iff $x_0 \ldots x_{i-1} \in L$.*

Definition 2. *A* variety of languages *is a class of languages $\mathcal{V}$ such that*

1. *for each alphabet A, $\mathcal{V}_A$ is a Boolean algebra over A^**
2. *for each morphism $\varphi : A^* \to B^*$, the condition $L \in \mathcal{V}_B$ implies $\varphi^{-1}(L) \in \mathcal{V}_A$*
3. *for each $L \in \mathcal{V}_A$ and $a \in A$, we have $a^{-1}L := \{w \in A^* \mid aw \in L\} \in \mathcal{V}_A$ and $La^{-1} := \{w \in A^* \mid wa \in L\} \in \mathcal{V}_A$*

Recall that a base is a set containing Boolean functions and families of Boolean functions. To treat circuits in a more general way we will define a base defined by a variety of languages. This allows us to describe different constant size circuit families over arbitrary alphabets simply by considering different varieties. The definition will allow a base to consist of an infinite number of Boolean families, but a circuit family over an (infinite) base is only allowed to use a finite subset of the elements in the base.

Definition 3 (Bases defined by a variety of languages). *Given a variety $\mathcal{V}$ of regular languages, $\mathcal{V}_{\{0,1\}}$ is a collection of languages in $\{0,1\}^*$ and each of these languages defines a family of Boolean functions. We call the set $\{f^L \mid L \in \mathcal{V}_{\{0,1\}}\}$ the* base defined by $\mathcal{V}$.

For our purpose, it suffices to consider bases generated by varieties of languages, where the languages are regular and commutative. This is not much of a limitation as many gate types correspond to commutative regular languages (see table on the left of Fig. 1).

Gate Type	Language
$\wedge$	1^*
$\vee$	$\{0,1\}^*1\{0,1\}^*$
mod_p	$\{0,1\}^*((10^*)^p)^*$

Circuit class	Base
$\mathbf{AC^0}$	$\{\wedge, \vee\}$
$\mathbf{CC^0}$	$\{\text{mod}_p \mid p \in \mathbb{N}\}$
$\mathbf{ACC^0}$	$\{\wedge, \vee, \text{mod}_p \mid p \in \mathbb{N}\}$

Fig. 1. On the left: Typical gate types and the languages they are defined by. On the right: Typical circuit classes

The following lemma implies that for each circuit over a base generated by a variety of regular commutative languages $\mathcal{V}$, there exists a circuit that accepts exactly the same languages and can be written as one layer of gates from $\mathcal{V}$ accessing the inputs and below an $\mathbf{NC^0}$ circuit.

Lemma 4. *Let* **B** *be a base generated by a variety* $\mathcal{V}$ *of regular commutative languages. A language* L *is recognized by a constant size circuit family over* **B** *if and only if it is recognized by a constant size circuit family where the gates only accessing inputs are from* **B**, *and all other gates are labeled by* $\wedge_2$ *or* $\vee_2$.

The proof can be found in [18].

3   The Block Product for Varieties of Languages

In the last section, we introduced bases for circuits that were defined by a variety of languages. Here we will define an unary operation $\cdot \boxdot \mathcal{P}_{arb}$ on varieties mapping a variety of commutative regular languages $\mathcal{V}$ to the variety of languages $\mathcal{V} \boxdot \mathcal{P}_{arb}$ recognized by constant size circuits over the base $\mathcal{V}$.

This rather strange looking notation comes from the algebraic background where similar ideas have been used on the algebraic side in [7]. Using the algebraic tools from that paper one could show that constant size circuit families recognize the same languages as the finitely typed groups in the block product $\mathbf{V} \boxdot \mathbf{P}_{arb}$, where $\mathbf{V}$ are the (typed) monoids corresponding to the gate types and $\mathbf{P}_{arb}$ are the typed monoids corresponding to arbitrary predicates and hence to the non-uniform wiring of the circuit family. In this paper however, we omit the algebraic definition of the block product of (typed) monoids but rather define a mechanism, that provides us with the same languages as recognized by the block product, and that is purely defined on the language side. For more details on the algebraic and logic side for varieties of regular languages, we refer to a survey about the block product principle [20] or for the non-regular case to [12].

Here we will restrict to the unary operation $\cdot \boxdot \mathcal{P}_{arb}$ suitable for our constant size circuit classes.

There is a natural morphism $|\cdot| : A^* \to \mathbb{N}$ that maps each word to its length. We say a mapping $f : A^* \to B^*$ is length preserving, if $|f(u)| = |u|$ for all $u \in A^*$.

Definition 5 (ℕ-transduction). *Let* $\mathcal{D}$ *be a finite partition of* $\mathbb{N}^2$. *By* $[(i, j)]_{\mathcal{D}}$ *denote the equivalence class that* (i, j) *belongs to. Then a* ℕ-*transduction is a length preserving map* $\tau_{\mathcal{D}} : A^* \to (A \times \mathcal{D})^*$, *where* $(\tau_{\mathcal{D}}(w))_i = (w_i, [(|w_{<i}|, |w_{>i}|)]_{\mathcal{D}}) = (w_i, [(i, |w| - i - 1)]_{\mathcal{D}})$.

Finally we can use these transductions to define the block product. We only define a unary operation that maps a variety $\mathcal{V}$ to a variety $\mathcal{V} \boxdot \mathcal{P}_{arb}$. This notation stems from the strong connection of $\mathcal{P}_{arb}$ with ℕ-transducers.

Definition 6 ($\mathcal{V} \boxdot \mathcal{P}_{arb}$). *Let* $\mathcal{V}$ *be a variety of languages. We define* $\mathcal{V} \boxdot \mathcal{P}_{arb}$ *as the variety of languages, where* $(\mathcal{V} \boxdot \mathcal{P}_{arb})_A$ *is generated by the languages* $\tau_{\mathcal{D}}^{-1}(L)$ *for all partitions* $\mathcal{D}$ *of* $\mathbb{N}^2$, *and* $L \in \mathcal{V}_{A \times \mathcal{D}}$.

Because of the connection to the block product we call the languages $\mathcal{V} \boxdot \mathcal{P}_{arb}$.

Lemma 7. *The languages in constant size circuits over a base defined by the variety $\mathcal{V}$, are exactly the languages in $\mathcal{V} \boxdot \mathcal{P}_{arb}$.*

Proof. Let $L \subseteq A^*$ be a language recognized by a constant size circuit. Then L is recognized by Lemma 4 by a Boolean combination of depth 1 circuits with gates from $\mathcal{V}_{\{0,1\}}$. As $(\mathcal{V} \boxdot \mathcal{P}_{arb})_A$ is a Boolean algebra it suffices to show that every language recognized by a depth 1 circuit with gates from $\mathcal{V}$ is in $(\mathcal{V} \boxdot \mathcal{P}_{arb})_A$. So we assume that L is recognized by a circuit of depth 1 which is just a single gate. Let L' be the language corresponding to the function computed by this single gate. As the functions of the gates are symmetric we can assume that the gate queries the inputs in order. Also as L' is regular the multiplicity of edges querying the same input position to the gate is limited by some constant. Let c be this constant. Hence we can upper bound the different ways an input position is wired to this gate by $(2^{|A|})^c$. Let $\mathcal{D}$ be a partition where the equivalence classes correspond to the different ways the input can be wired. We define a morphism $h \colon (A \times \mathcal{D}) \to \{0,1\}^*$ where each (a, P) is mapped to the way an input in the equivalence class P reading the letter a as input would influence the gate. As $\mathcal{V}$ is a variety $L'' = h^{-1}(L')$ is in $\mathcal{V}_{(A \times \mathcal{D})}$. But then $\tau_{\mathcal{D}}^{-1}(L'') = L$.

For the other direction as constant size circuits over the base generated by $\mathcal{V}$ are closed under Boolean combinations it suffices to show that any language $L \subseteq A^*$ with $L = \tau_{\mathcal{D}}^{-1}(L')$ and $L' \in \mathcal{V}_{(A \times \mathcal{D})}$ is recognized by a constant size circuit. As $L' \in \mathcal{V}_{(A \times \mathcal{D})}$ is a symmetric regular language it is a Boolean combination of languages $L'_1, \ldots, L'_k \in \mathcal{V}_{(A \times \mathcal{D})}$, such that there exists a morphism $h_i \colon (A \times \mathcal{D})^* \to \{0,1\}^*$ and $L'_i = h_i^{-1}(L''_i)$, where $L''_i \in \mathcal{V}_{\{0,1\}}$. Fix an input length n. We construct a circuit that consists of exactly this Boolean combination of gates $g_1, \ldots, g_k$ computing the functions corresponding to $L''_1, \ldots, L''_k$. We wire each input to the gates such that for a word $w \in A^n$ each input position j with $w_j = a$ contributes to the gate g_i the value $h_i((\tau_{\mathcal{D}}(w))_j)$ and some neutral string otherwise. Please note that this definition of the wires is only well defined as $(\tau_{\mathcal{D}}(w))_j$ does only depend on the w_j, the length of w and the position j, but not on the other letters. This completes our construction of the circuit.

4   Duality and the Block Product

We briefly introduce some theory from Stone Duality, which will be used to characterize the classes of languages we are interested in and to obtain separation results for them. A short introduction on the idea behind the theory is stated here. For a more detailed and possibly apprehensible introduction, see [18].

By duality, each Boolean algebra $\mathcal{B}$ has an associated compact space, called its Stone Space $\mathcal{S}(\mathcal{B})$. For any two Boolean algebras $\mathcal{B}$ and $\mathcal{C}$, if $\mathcal{C}$ is a subalgebra of $\mathcal{B}$, a relation between the Stone spaces exists, namely $\mathcal{S}(\mathcal{C})$ is a quotient of $\mathcal{S}(\mathcal{B})$. Since every Boolean algebra of languages is a subalgebra of the powerset of A^*, there always exists a canonical projection from the Stone space of $P(A^*)$ to the Stone space of any Boolean algebra of languages over A^*. The idea is to characterize the Boolean algebra of languages $\mathcal{B}$ by the kernel of said projection, that is finding all pairs of elements in the Stone space of $P(A^*)$ that get identified in $\mathcal{S}(\mathcal{B})$.

In order to define the points of the Stone space we need to define the notion of an ultrafilter.

Definition 8 ((Ultra)Filter). *Let $\mathcal{B}$ be a Boolean algebra. A proper filter of $\mathcal{B}$ is a non-empty subset γ of $\mathcal{B}$ that satisfies*

1. $\emptyset \notin \gamma$,
2. *if $L \in \gamma$ and $K \supseteq L$, then $K \in \gamma$,* 　　　　　(γ *is closed under extension*)
3. *if $L, K \in \gamma$ then $K \cap L \in \gamma$,* 　　(γ *is closed under finite intersections*)

A proper filter is called ultrafilter *if it additionally satisfies*

4. *for each $L \in \mathcal{B}$, either $L \in \gamma$ or $L^c \in \gamma$.* 　　　　(*ultrafilter condition*)

Definition 9 (Stone Space). *Let $\mathcal{B}$ be a Boolean algebra. The Stone space $\mathcal{S}(\mathcal{B})$ of $\mathcal{B}$ is the space of all ultrafilters of $\mathcal{B}$ equipped with the topology generated by the sets $\widehat{L} = \{\gamma \in \mathcal{S}(\mathcal{B}) \mid L \in \gamma\}$ for $L \in \mathcal{B}$.*

The topology that the Stone spaces are equipped with is of importance, since it holds informations about the languages in the underlying Boolean algebra. For those familiar with topology: The clopen sets of $\mathcal{S}(\mathcal{B})$ are exactly the topological closures of the sets $L \in \mathcal{B}$.

The Stone Space of the full Boolean algebra $P(A^*)$ is a special case, also known as the Stone-Čech compactification of A^*, which is denoted by $\beta(A^*) = \mathcal{S}(\mathcal{P}(A^*))$. For a Boolean algebra $\mathcal{B} \subseteq P(A^*)$, we denote the canonical projection from $\beta(A^*)$ onto $\mathcal{S}(\mathcal{B})$ by $\pi_{\mathcal{B}}$. Let A^* and B^* be two free monoids and $f : A^* \to B^*$ be a function. Then there exists a unique continuous extension $\beta f : \beta A^* \to \beta B^*$, which is defined by

$$L \in \beta f(\gamma) \Leftrightarrow f^{-1}(L) \in \gamma.$$

See [11].

Definition 10 (Equation). *An ultrafilter equation is a tuple $(\mu, \nu) \in \beta A^* \times \beta A^*$. Let $\mathcal{B}$ be a boolean algebra. We say that $\mathcal{B}$ satisfies the equation (μ, ν) if $\pi_{\mathcal{B}}(\mu) = \pi_{\mathcal{B}}(\nu)$. With respect to some Boolean algebra $\mathcal{B}$ we say that $[\mu \leftrightarrow \nu]$ holds.*

Lemma 11. *Let $\mathcal{B}$ be a subalgebra of $P(A^*)$. For $\mu, \nu \in \beta A^*$ we have $\pi_{\mathcal{B}}(\mu) = \pi_{\mathcal{B}}(\nu)$ iff for all $L \in \mathcal{B}$ the equivalence $L \in \mu \Leftrightarrow L \in \nu$ holds.*

Proof. The projection $\pi_{\mathcal{B}}$ is given by $\pi_{\mathcal{B}}(\mu) = \{L \in \mu \mid L \in \mathcal{B}\}$ and thus the equivalence holds.

Recently, Gehrke, Grigorieff and Pin [9] were able to show that any Boolean algebra of languages can be defined by a set of equations of the form $[\mu \leftrightarrow \nu]$, where μ and ν are ultrafilters on the set of words. That is $L \in \mathcal{B}$ if and only if for all equations $[\mu \leftrightarrow \nu]$ of $\mathcal{B}$ the equivalence $L \in \mu \Leftrightarrow L \in \nu$ holds. We say a set of equations is sound, if all L in $\mathcal{B}$ satisfy the equivalence above and complete, if a language in A^* satisfying all equations is in $\mathcal{B}$.

This theorem provides us with the existence of ultrafilter equations for $(\mathcal{V} \Box \mathcal{P}_{arb})_A$. However, it does not answer the question on how to obtain them. The following lemma provides us with a set of equations that define precisely the kernel of the projection $\pi_{(\mathcal{V} \Box \mathcal{P}_{arb})_A}$. It builds on the knowledge, that $(\mathcal{V} \Box \mathcal{P}_{arb})_A$ was defined by the functions $\tau_{\mathcal{D}}$.

Lemma 12. *Let $\mu, \nu \in \beta A^*$. Then for each partition $\mathcal{D}$ of $\mathbb{N}^2$ the Boolean algebra $\mathcal{V}_{(A \times \mathcal{D})}$ satisfies the equation $[\beta \tau_{\mathcal{D}}(\mu) \leftrightarrow \beta \tau_{\mathcal{D}}(\nu)]$ if and only if $[\mu \leftrightarrow \nu]$ is an equation of $(\mathcal{V} \Box \mathcal{P}_{arb})_A$.*

Proof. Let $\mu, \nu \in \beta A^*$ such that $[\beta \tau_{\mathcal{D}}(\mu) \leftrightarrow \beta \tau_{\mathcal{D}}(\nu)]$ holds for all partitions $\mathcal{D}$ of $\mathbb{N}^2$ and let $L \in (\mathcal{V} \Box \mathcal{P}_{arb})_A$ be a generator of the Boolean algebra. Recall that by definition there exists a partition $\mathcal{D}$ of $\mathbb{N}^2$ and a language $S \in \mathcal{V}_{(A \times \mathcal{D})}$ such that $L = \tau_{\mathcal{D}}^{-1}(S)$. Then

$$L \in \mu \Leftrightarrow \tau_{\mathcal{D}}^{-1}(S) \in \mu \Leftrightarrow S \in \beta \tau_{\mathcal{D}}(\mu) \Leftrightarrow S \in \beta \tau_{\mathcal{D}}(\nu) \Leftrightarrow \tau_{\mathcal{D}}^{-1}(S) \in \mu \Leftrightarrow L \in \nu.$$

This proves both directions of the claim.

This set of equations already provides us with a full characterization of $(\mathcal{V} \Box \mathcal{P}_{arb})_A$, but we are interested in a set that satisfies conditions that are easier to check and still is sound and complete.

5   Equations for the Block Product

In this chapter we find a set of equations that holds for $(\mathcal{V} \Box \mathcal{P}_{arb})_A$, depending on the equations that define the variety $\mathcal{V}$ of the gate types, which in our case is regular and commutative. As a corollary, we expose separation results for a selection of classes, to demonstrate the applicability of the equations.

As we describe the base of the circuits by a variety $\mathcal{V}$ of regular languages, we use a description that has already been applied in the regular case. Thereto, we introduce the notion of identities of profinite words. Here, we define a profinite word as an ultrafilter on the regular languages. Commonly a different but equivalent [16] definition is used. For the interested reader, we refer to [18]. The combined results of Reiterman [17] and Eilenberg [8] state that for each variety of regular languages $\mathcal{V}$, there is a set of profinite identities, defining the variety. Informally speaking, an identity is an equation that holds not only for a language, but also for all quotients of a language.

As such, we can define the notion of profinite identities in the following way: A Boolean algebra of regular languages $\mathcal{B}$ satisfies the profinite identity $[u = v]$, where $u, v \in S(\mathbf{Reg})$ instead of $u, v \in \beta(A^*)$, if for all $L \in \mathcal{B}$ the equivalence $x^{-1} L y^{-1} \in u \Leftrightarrow x^{-1} L y^{-1} \in v$ for all $x, y \in A^*$ holds. As varieties are closed under quotients, it suffices to consider $L \in u \Leftrightarrow L \in v$ for all $L \in \mathcal{B}$.

To define the equations that hold for $(\mathcal{V} \square \mathcal{P}_{arb})_A$, we define a function, that gets as arguments a word w, another word s and a vector of positions p, such that the positions of p in w are substituted by the letters of s. Naturally, such a substitution only makes sense, if the input is restricted to be reasonable. For instance the positions in p should not exceed the length of the word w. For technical reasons, each element of the vector of positions will be a tuple containing the distance of the position from the beginning and the end of the word.

For an element $p \in \mathbb{N}^2$ denote by p^1 the first and by p^2 the second component of p, i.e. $p = (p^1, p^2)$. We define the set of correct substitutions

$$\mathfrak{D} = \left\{ (w, s, p) \in A^* \times A^* \times (\mathbb{N}^2)^* \middle| \begin{array}{l} |s| = |p| \\ \forall i \colon |w| - 1 = p_i^1 + p_i^2 \\ p_0^1 < \ldots < p_{|p|-1}^1 < |w| \end{array} \right\}.$$

Given a word $w = w_0 \ldots w_{m-1} \in A^*$ of length m and k, l with $0 \leq k \leq l < m$, we define $w_{k,l} = w_k \ldots w_l$. Let n be the length of s, then the function $f \colon A^* \times A^* \times (\mathbb{N}^2)^* \to A^*$ is defined as

$$f(w, s, p) = \begin{cases} w_{0, p_0^1 - 1} s_0 w_{p_0^1 + 1, p_1^1 - 1} s_1 \ldots s_{n-1} w_{p_{n-1}^1 + 1, m-1} & \text{if } (w, s, p) \in \mathfrak{D}, \\ w & \text{otherwise.} \end{cases}$$

Furthermore, define the function that maps the second component to its length as

$$\lambda \colon A^* \times A^* \times (\mathbb{N}^2)^* \to A^* \times \mathbb{N} \times \mathbb{N}^*$$
$$(w, s, p) \mapsto (w, |s|, p)$$

and let $\pi_2 \colon A^* \times A^* \times (\mathbb{N}^2)^* \to A^*$ with $\pi_2(w, s, p) = s$ be the projection on the second component.

As the function f substitutes letters in certain positions, we need the following definition in order to define which positions of p are "indistinguishable" by a language in $(\mathcal{V} \square \mathcal{P}_{arb})_A$.

Define the mapping $\pi_c \colon A^* \times A^* \times (\mathbb{N}^2)^* \to \mathcal{P}(\mathbb{N}^2)$ that maps the third component onto its content, given by $\pi_c(w, s, p) = \{p_0, \ldots, p_{|p|}\}$. Note that any finite subset of $\mathbb{N}^2$ is a finite subset of $\beta(\mathbb{N}^2)$ and thus π_c can be interpreted as a mapping into the space $\mathcal{F}(\mathbb{N}^2)$ of all filters of $\mathcal{P}(\mathbb{N}^2)$, by sending it to the intersection of all ultrafilters containing the set. Furthermore, $\beta(\mathbb{N}^2)$, which contains all ultrafilters of $\mathcal{P}(\mathbb{N}^2)$ can be seen as a subspace of $\mathcal{F}(\mathbb{N}^2)$, which is homeomorphic to Vietoris of $\beta(\mathbb{N}^2)$. Then there exists and extension of π_c denoted by $\beta \pi_c$, known as the Stone-Čech extension $\beta \pi_c \colon \beta(A^* \times A^* \times (\mathbb{N}^2)^*) \to \mathcal{F}(\mathbb{N}^2)$.

Together with these definitions we can formulate the theorem that provides us with a set of equations for $(\mathcal{V} \square \mathcal{P}_{arb})_A$.

Theorem 13. *The variety* $(\mathcal{V} \square \mathcal{P}_{arb})_A$ *is defined by the equations*

$$[\beta f(\gamma_u) \leftrightarrow \beta f(\gamma_v)].$$

where $[u = v]$ *is a profinite equation that holds on* $\mathcal{V}$ *and* $\gamma_u, \gamma_v \in \beta(A^* \times A^* \times (\mathbb{N}^2)^*)$ *satisfying*

(1) $\beta\lambda(\gamma_u) = \beta\lambda(\gamma_v)$
(2) $u \subseteq \beta\pi_2(\gamma_u)$ and $v \subseteq \beta\pi_2(\gamma_v)$
(3) $\beta\pi_c(\gamma_u) = \beta\pi_c(\gamma_v) \in \beta(\mathbb{N}^2)$

Proof. For the proof of soundness and completeness of these equations in [18].

While the following separation result itself is not surprising, and strong separations are known, the proof method has the advantage that there is no need for probabilistic methods to find specific inputs for the circuits to be fixed or swapped.

For a fixed $x \in A^*$ define the profinite word, also denoted by x as

$$x = \{L \in \mathbf{Reg} \mid x \in L\}$$

and x^ω as

$$x^\omega = \{L \in \mathbf{Reg} \mid \exists n_0 \in \mathbb{N} \quad \forall n \geq n_0 \quad : \quad x^{n!} \in L\}$$

The following varieties are used for characterization of $\mathcal{V} \boxdot \mathcal{P}_{arb}$ by equations (Fig. 2).

Gates	Profinite Identities		Circuit Class (constant size)
$\{\wedge, \vee\}$	$xy = yx$	$x^2 y = xy^2$	$\mathbf{AC^0}$
$\{\wedge, \vee, \mathrm{mod}_p \mid p \in \mathbb{N}\}$	$xy = yx$		$\mathbf{ACC^0}$
$\{\mathrm{mod}_p \mid p \in \mathbb{N}\}$	$xy = yx$	$x^\omega = y^\omega$	$\mathbf{CC^0}$

Fig. 2. Varieties defining the gate types and the defining profinite identities.

Corollary 14. *Constant size* $\mathbf{CC^0}$ *is strictly contained in constant size* $\mathbf{ACC^0}$ *and constant size* $\mathbf{AC^0}$ *is strictly contained in constant size* $\mathbf{ACC^0}$. *Also constant size* $\mathbf{CC^0}$ *and constant size* $\mathbf{AC^0}$ *are not comparable.*

Proof. We show this by proving $L_{\mathrm{AND}} = 1^*$ is not contained in constant size $\mathbf{CC^0}$ and parity $L_{\mathrm{PARITY}} = (0^*10^*1)^*0^*$ is not contained in constant size $\mathbf{AC^0}$.

For that we construct ultrafilters from filterbases, using Lemmata 27, 28 and Theorem 29 from [18], such that they satisfy conditions 1.-3. from Theorem 13.

Take the identity $[0^\omega = 1^\omega]$, that holds for the variety providing us with the gate types of $\mathbf{CC^0}$. By Theorem 13, we know that for any two ultrafilters γ_{0^ω} and γ_{1^ω} satisfying the conditions of the Theorem, provide us with an equation $[\beta f(\gamma_{0^\omega}) \leftrightarrow \beta f(\gamma_{1^\omega})]$. In constructing two such filters, such that L_{AND} is contained in $\beta f(\gamma_{0^\omega})$, but not in $\beta f(\gamma_{0^\omega})$, we prove that it is not an element of constant size $\mathbf{CC^0}$. Consider the filter base

$$\mathcal{F}_1 = \{A^* \times \{n! \mid n \geq N\} \times (\mathbb{N}^2)^* \mid N \in \mathbb{N}\}$$

and the second base

$$\mathcal{F}_2 = \{\bigcup_{i=0}^{n} A^* \times \mathbb{N} \times P_i^* \mid \{P_0, \ldots, P_n\} \text{ is a partition of } \mathbb{N}^2\}.$$

Adding the two together yields another filterbase, denoted by $\mathcal{F}$, as none of the elements have empty intersection. Let $\mu \in \beta(A^* \times \mathbb{N} \times \mathbb{N}^*)$ be an ultrafilter containing the filter base $\mathcal{F}$. Next, consider the set

$$1^* \times \{1^{n!} \mid n \in \mathbb{N}\} \times (\mathbb{N}^2)^*.$$

By definition of f, we obtain $f(1^* \times \{1^{n!} \mid n \in \mathbb{N}\} \times (\mathbb{N}^2)^*) \subseteq L_{\text{AND}}$ and thus $1^* \times \{1^{n!} \mid n \in \mathbb{N}\} \times (\mathbb{N}^2)^* \subseteq f^{-1}(L_{\text{AND}})$. Adding this to the pullback by $\lambda^{-1}(\mu)$ yields another filter base, denoted by $\mathcal{F}_{1^\omega}$. By $\mathcal{F}_{0^\omega}$ we denote the base $\lambda^{-1}(\mu)$ when adding the set

$$1^* \times \{0^{n!} \mid n \in \mathbb{N}\} \times (\mathbb{N}^2)^*.$$

Let γ_{0^ω} be an ultrafilter containing $\mathcal{F}_{0^\omega}$ and γ_{1^ω} be an ultrafilter containing $\mathcal{F}_{1^\omega}$. Then both ultrafilters satisfy conditions 1.-3. of the Theorem, such that $[\beta f(\gamma_{0^\omega}) \leftrightarrow \beta f(\gamma_{1^\omega})]$ holds for $\mathbf{CC^0}$. But $L_{\text{AND}} \in \beta f(\gamma_{1^\omega})$ and $L_{\text{AND}} \notin \beta f(\gamma_{0^\omega})$. Hence L_{AND} is not in constant size $\mathbf{CC^0}$.

Equivalently, we use the identity $[110 = 100]$ satisfied by the variety corresponding to the gate types of $\mathbf{AC^0}$. Again let

$$\mathcal{F}_2 = \{\bigcup_{i=0}^{n} A^* \times \mathbb{N} \times P_i^* \mid \{P_0, \ldots, P_n\} \text{ is a partition of } \mathbb{N}^2\}.$$

Adding the set $A^* \times \{3\} \times \mathbb{N}^*$ yields a filter base $\mathcal{F}'$. Let $\nu \in \beta(A^* \times \mathbb{N} \times \mathbb{N}^*)$ be an ultrafilter containing $\mathcal{F}'$. Consider the sets

$$S_{110} = 0^* \times \{110\} \times (\mathbb{N}^2)^* \text{ and } S_{100} = 0^* \times \{100\} \times (\mathbb{N}^2)^*.$$

Adding the set S_{110} to the pullback $\lambda^{-1}(\nu)$ provides us with a new filter base $\mathcal{F}_{110}$ and respectively adding S_{100} to $\lambda^{-1}(\nu)$ with a filter base $\mathcal{F}_{100}$. Let γ_{110} be an ultrafilter containing $\mathcal{F}_{110}$ and γ_{100} be an ultrafilter containing $\mathcal{F}_{100}$. Then both ultrafilters satisfy conditions 1.-3. of the main theorem and thus $[\beta f(\gamma_{110}) \leftrightarrow \beta f(\gamma_{100})]$ is an equation satisfied by constant size $\mathbf{AC^0}$. Since $f(S_{110}) \subseteq L_{\text{PARITY}}$ and $f(S_{100}) \subseteq L_{\text{PARITY}}^c$, we obtain $L_{\text{PARITY}} \in \beta f(\gamma_{110})$ but $L_{\text{PARITY}} \notin \beta f(\gamma_{100})$ and thus it is not in constant size $\mathbf{AC^0}$.

6   Conclusion

We have presented a method to describe circuit classes by equations. The tools and techniques used originate from algebra and topology and have previously been used on regular language classes. Due to recent developments in generalizing these methods to non-regular classes, they are now powerful enough to describe circuit classes. But the knowledge that they are powerful enough itself is not sufficient, as we require a constructive mechanism behind these descriptions.

Since non-uniform circuit classes are by definition not finitely presentable, this seemed to be impossible.

Nevertheless, we were able to find a description of small but natural circuit classes via equations. This description seems helpful as it easily allows to prove non-membership of a language to some circuit class. Another advantage is the possibility of using Zorn's Lemma for the extension of filter bases to ultrafilters, which prevents us from having to use probabilistic arguments in many places. Also in [18], Lemma 34, we use purely topological arguments of convergence to prove completeness, for which it is unclear how this could be achieved purely combinatorially.

The results we acquired are not so different from the results about equations for varieties of regular languages by Almeida and Weil [2]. This gives hope that their results can be used as a roadmap for further research.

In [7] it was shown that a certain restricted version of the block product of our constant size circuit classes would actually yield linear size circuit classes (over the same base). Here having equations for all languages captured by this circuit class, not just the regular ones, would pay off greatly. By showing that a padded version of a language is not in a linear circuit class we could already prove that PARITY is not in a polynomial size circuit class. Equations for non-regular language classes could be used to overcome previous bounds. The separation results in the corollary can easily be extended to show that a padded version of those languages is not contained in these circuit classes.

A different approach would be to examine the way the block product was used here. The evaluation of a circuit is equivalent to a program over finite monoids. While the program itself has little computational power, it allows non-uniform operations like our N-transducers. The finite monoid itself corresponds loosely speaking to the computational power of the gates of the circuit, which was handled by our variety $\mathcal{V}$. For general circuit classes one would need to consider larger varieties containing also non-commutative monoids. While the methods here seem to be extendable to non-commutative varieties, the more complicating problem remaining is to find an extension of the block product that corresponds to polynomial programs over these monoids.

References

1. Almeida, J.: Finite Semigroups and Universal Algebra. World Scientific Publishing Co., Inc., River Edge (1994). Translated from the 1992 Portuguese original and revised by the author
2. Almeida, J., Weil, P.: Profinite categories and semidirect products. J. Pure Appl. Algebra **123**(13), 1–50 (1998)
3. Mix Barrington, D.A., Straubing, H.: Lower bounds for modular counting by circuits with modular gates. In: Proceedings of Theoretical Informatics, Second Latin American Symposium, LATIN 1995, 3–7 April 1995, Valparaíso, Chile, pp. 60–71 (1995)
4. Mix Barrington, D.A., Straubing, H.: Superlinear lower bounds for bounded-width branching programs. J. Comput. Syst. Sci. **50**(3), 374–381 (1995)

5. Mix Barrington, D.A., Thérien, D.: Non-uniform automata over groups. In: Proceedings of 14th International Colloquium Automata, Languages and Programming, ICALP 1987, 13–17 July 1987, Karlsruhe, Germany, pp. 163–173 (1987)
6. Mix Barrington, D.A., Thérien, D.: Finite monoids and the fine structure of NC^1. J. ACM **35**(4), 941–952 (1988)
7. Behle, C., Krebs, A., Mercer, M.: Linear circuits, two-variable logic and weakly blocked monoids. Theor. Comput. Sci. **501**, 20–33 (2013)
8. Eilenberg, S.: Automata, Languages, and Machines. Academic Press Inc., Orlando (1976)
9. Gehrke, M., Grigorieff, S., Pin, J.É.: Duality and equational theory of regular languages. In: Aceto, L., Damgård, I., Goldberg, L.A., Halldórsson, M.M., Ingólfsdóttir, A., Walukiewicz, I. (eds.) ICALP 2008, Part II. LNCS, vol. 5126, pp. 246–257. Springer, Heidelberg (2008)
10. Gehrke, M., Krebs, A., Pin, J.É.: From ultrafilters on words to the expressive power of a fragment of logic. In: Jürgensen, H., Karhumäki, J., Okhotin, A. (eds.) DCFS 2014. LNCS, vol. 8614, pp. 138–149. Springer, Heidelberg (2014)
11. Johnstone, P.T.: Stone Space. Cambridge University Press, Cambridge (1983)
12. Krebs, A., Lange, K.-J., Reifferscheid, S.: Characterizing TC^0 in terms of infinite groups. Theory Comput. Syst. **40**(4), 303–325 (2007)
13. Krebs, A., Straubing, H.: An effective characterization of the alternation hierarchy in two-variable logic. In: IARCS Annual Conference on Foundations of Software Technology and Theoretical Computer Science, FSTTCS, 15–17 2012, Hyderabad, India, pp. 86–98, December 2012
14. McKenzie, P., Péladeau, P., Thérien, D.: NC^1: the automata-theoretic viewpoint. Comput. Complex. **1**, 330–359 (1991)
15. Pin, J.É.: Equational descriptions of languages. Int. J. Found. Comput. Sci. **23**(6), 1227–1240 (2012)
16. Pippenger, N.: Regular languages and stone duality. Theory Comput. Syst. **30**(2), 121–134 (1997)
17. Reiterman, J.: The Birkhoff theorem for finite algebras. Algebra Universalis **14**(1), 1–10 (1982)
18. Krebs, A., Czarnetzki, S.: Using Duality in Circuit Complexity, ArXiv e-prints. October 2015
19. Straubing, H.: Finite Automata, Formal Logic, and Circuit Complexity. Birkhauser Verlag, Basel (1994)
20. Tesson, P., Thérien, D.: Logic meets algebra: the case of regular languages. Logical Methods Comput. Sci. **3**(1), 1–37 (2007)
21. Thérien, D.: Circuits constructed with Mod_q gates cannot compute "And" in sublinear size. Comput. Complex. **4**, 383–388 (1994)

The Minimum Entropy Submodular Set Cover Problem

Gabriel Istrate[1,2]([⊠]), Cosmin Bonchiş[1,2], and Liviu P. Dinu[3]

[1] Department of Computer Science, West University of Timişoara,
Bd. V. Pârvan 4, Timişoara, Romania
gabrielistrate@acm.org
[2] e-Austria Research Institute, Bd. V. Pârvan 4, cam. 045 B, Timişoara, Romania
[3] Faculty of Mathematics and Computer Science, University of Bucharest,
Bucharest, Romania

Abstract. We study *Minimum Entropy Submodular Set Cover*, a variant of the Submodular Set Cover problem (Wolsey [21], Fujito [8], etc.) that generalizes the Minimum Entropy Set Cover problem (Halperin and Karp [11], Cardinal et al. [4]) We give a general bound on the approximation performance of the greedy algorithm using an approach that can be interpreted in terms of a particular type of biased network flows. As an application we rederive known results for the Minimum Entropy Set Cover and Minimum Entropy Orientation problems, and obtain a nontrivial bound for a new problem called the Minimum Entropy Spanning Tree problem. The problem can be applied to (and is partly motivated by) a worst-case approach to fairness in concave cooperative games.

Keywords: Submodular set cover · Minimum entropy · Approximation algorithms

1 Introduction

Submodularity encodes the notion of *diminishing returns* and plays a crucial role in many problems in combinatorial optimization [7], cooperative game theory [5,19], information theory [16] and in applications like clustering, learning, natural language and signal processing or constraint satisfaction. Submodular optimization is well-understood: minimization has polynomial time algorithms [12,18]; maximization is intractable but has efficient approximation algorithms. Minimizing the cost is not the only possible objective for submodular optimization: a problem in computational biology led Halperin and Karp [11] to study a *Minimum Entropy* version of the Set Cover problem (MESC). MESC is NP-hard, but the GREEDY algorithm produces [11] an approximate solution with an additive approximation guarantee. The optimal constant is $\log_2(e)$ [4].

G. Istrate and C. Bonchiş were supported by IDEI Grant PN-II-ID-PCE-2011-3-0981. L.P. Dinu was supported by UEFISCDI, PNII-ID-PCE-2011-3-0959.

© Springer International Publishing Switzerland 2016
A.-H. Dediu et al. (Eds.): LATA 2016, LNCS 9618, pp. 295–306, 2016.
DOI: 10.1007/978-3-319-30000-9_23

It must be stressed that minimizing entropy is a reasonably common scenario: the authors of [4] subsequently studied other combinatorial problems under minimum entropy objectives [13,14]. Minimal Entropy Graph Coloring is relevant in coding and information theory [1]. Entropy minimization has been applied e.g. to *word segmentation* [20] or (for the non-extensive entropy) to maximum parsimony *haplotype inference* [10].

In this paper we join these two directions, submodularity and combinatorial optimization under a minimum entropy objective, by investigating an *extension of MESC* we call *Minimum Entropy Submodular Set Cover* (MESSC). While the problem is clearly NP-hard (as a generalization of MESC), our main result shows that the approximation guarantees of the GREEDY algorithm for MESC extend to MESSC, with the additional appearance of a certain covering parameter that has an interpretation in terms of a type of certain "biased" network flows. This interpretation allows a fairly illuminating rederivation of results in [13,14] and applications to several new problems, which are special cases of MESSC.

Besides the conceptual integration, the framework we investigate was developed with several applications in mind. The most important of them (developed in a companion paper [3]) concerns the development of a **worst-case approach to fairness in concave cooperative games** similar in spirit to the *price of anarchy* from noncooperative game theory. We measure unfairness of an allocation in the core by the entropy of the associated distribution, and seek allocations in the core minimizing entropy. Here we analyze a concrete example of such a game, the *Minimum Entropy Spanning Tree* (MEST) problem.

The organisation of the paper is as follows: in Sect. 2 we briefly review some relevant concepts and notions. In Sect. 3 define the problems we are interested in and show that they are NP-hard; next we introduce a greedy approach to Minimum Entropy Submodular Set Cover. Section 4 contains our main result: we quantify the performance of the GREEDY algorithm in terms of an instance-specific "covering constant". We then rederive (in Sect. 6) existing results on the performance of the GREEDY algorithm for the Minimum Entropy Orientations and Set Cover problems [13,14]. Section 7 contains an interpretation of the covering constant using network flows that allows us to tighten up our main theorem using a "multi-level" version of our covering constant. As an application we obtain in Sect. 8 a result on the approximability of the Minimum Entropy Spanning Tree problem that matches the $\log_2(e)$ bound for MESC from [4].

Because of space constraints, some proofs are omitted from this extended abstract. We refer the reader to an extended version available online [9].

2   Preliminaries

We will use *the Shannon entropy* of a distribution $P = (p_i)_{i \in I}$, defined as $Ent(P) = -\sum_{i \in I} p_i \log_2(p_i)$. We will assume general familiarity with submodular optimization, see e.g. [7]. In particular a set function $f : \mathcal{P}(U) \to \mathbf{R}_+$ will be called **integer** if $range(f) \subseteq \mathbf{Z}$, *monotone* if $f(S) \le f(T)$ whenever $S \subseteq T \subseteq U$, *submodular* if $f(S) + f(T) \ge f(S \cup T) + f(S \cap T)$ for all $S, T \subseteq U$, *modular* if

$f(S) + f(T) = f(S \cup T) + f(S \cap T)$ for all $S, T \subseteq U$, and *polymatroid* if f is monotone, submodular and satisfies $f(\emptyset) = 0$.

An instance of the *(Minimum Cost) Set Cover* (SC) is specified by a universe Z and a family $\mathcal{P} = \{\mathcal{P}_1, \ldots, \mathcal{P}_m\}$ of parts of Z. Each set $\mathcal{P}_i$ comes with a nonnegative *cost* $c(i)$. Given set $X \subseteq Z$, a *cover of X* is a function $g : X \to [m]$ such that for every $x \in X$, $x \in P_{g(x)}$ ("x is covered by $P_{g(x)}$"). The goal is to find a cover g of Z whose cost, defined as $cost[g] = \sum\limits_{j \in range(g)} c(P_j)$, is minimized.

The following classical generalization called *submodular set cover (SSC)* [8, 21] shares many properties with SC: we are given an integer polymatroid f and a *cost function* $c : U \to \mathbf{R}_+$. The cost of a set $S \subseteq U$, denoted $cost(S)$, is simply the sum of costs of its elements. Feasible solutions to SSC are subsets $S \subseteq U$ with $f(S) = f(U)$. The goal is to find a feasible subset $S \subseteq U$ of minimum cost.

An equivalent restatement of SSC relies on the notion of matroids and related concepts (such as basis and flats) we will assume known (see [17]):

Proposition 1. *The following problem is equivalent to SSC: given matroid $M = (U, \mathcal{I})$ and a covering $\mathcal{P} = \{\mathcal{P}_1, \ldots, \mathcal{P}_m\}$ of the universe U find a basis $\mathcal{B}$ of M and a cover $g : \mathcal{B} \to [m]$ of $\mathcal{B}$ such that $c[g]$ is minimized.*

Proof. We use the well-known representation of polymatroids as the rank function of a system of flats in a matroid (Theorem 12.1.9 in [17]): for every $f : [m] \to \mathbf{Z}$ there exists a matroid M and a set of flats of M, $\mathcal{P} = \{\mathcal{P}_1, \ldots, \mathcal{P}_m\}$ such that for all $S \subseteq M$, $f(S) = rank_M(\cup_{i \in S} P_i)$. See the extended online version [9] for more details and a self-contained proof. $\square$

On the other hand, Halperin and Karp introduced [11] the following variation on SC called *Minimum Entropy Set Cover (MESC)*: Consider an instance of *SC*, $Z = \{u_1, u_2, \ldots, u_n\}$, $n \geq 1$, $\mathcal{P} = \{P_1, P_2, \ldots, P_m\}$. Given $X \subseteq Z$ and cover $g : X \to [m]$ of X, the **entropy of g** is $Ent[g] = -\sum\limits_{i \in [m]} \frac{|g^{-1}(\{i\})|}{|X|} \ln\left(\frac{|g^{-1}(\{i\})|}{|X|}\right)$.
The goal is to find a cover g of Z of minimum entropy.

3   Minimum Entropy Submodular Set Cover: Definition, Special Cases and Applications

The following definition states the problem we are interested in this paper:

Definition 2 [Min-Entropy Submodular Set Cover] (MESSC): *Given a matroid $M = (U, \mathcal{I})$ of rank N, find a basis $\mathcal{B}$ and a cover g of $\mathcal{B}$ minimizing $Ent[g]$.*

MESSC indeed generalizes MESC: if M is the transversal matroid of the bipartite graph $G = (U, \mathcal{P})$ naturally associated to instance $X = (U, \mathcal{P})$ of MESC then solving instance M of MESSC is equivalent to solving instance X of MESC.

Problem MESSC has applications to fuzzy set theory: a fuzzy measure (*Choquet capacity*) is an extension of a probability measure. Submodular Choquet capacities coincide with polymatroids. Various definitions of the notion of the entropy of a Choquet capacity have been proposed in the literature. One such definition, due to Dukhovny [6], is of special interest: for submodular capacities, it is equivalent to our definition of Minimum Entropy Cover (proof details could be funded in [9]). Thus solving MESSC is equivalently stated as computing the Dukhovny entropy of a submodular Choquet capacity.

Polymatroids are just a different name for *concave games* in theory of cooperative games [5]. If $(M, \mathcal{P})$ is the rank representation of an integer polymatroid f, then the convex hull of the incidence vectors of bases in M (the matroid polytope) coincides with the core of the cooperative game defined by f. Since entropy is a concave function, its minimum over the *core(f)* is obtained at an extremal point. That is, finding a basis of minimum entropy is equivalent to finding a minimum entropy imputation in the core. MESSC can, therefore, be restated as follows:

Definition 3. *Given an integer-valued polymatroid f, find a vector $(x_1, \ldots, x_m)$ of nonnegative reals satisfying $\sum_{i=1}^{m} x_i = f([m])$ and, for all $S \subseteq [m]$, $\sum_{i \in S} x_i \leq f([S])$ minimizing the entropy of distribution $\left(\frac{x_i}{f([m])} \right)_{i=1,\ldots,m}$.*

We will freely switch between the two equivalent definitions of MESSC.

A class of matroids that yields a particular interesting class of cooperative games is that of cycle matroids of a connected graph. We will call the corresponding particularization of MESSC the *Minimum Entropy Spanning Tree (MEST)* problem; it can be specified as follows: we define a *cover of a spanning tree T* in a graph G as an orientation of its edges. The *entropy of a cover* is the entropy of the distribution of indegrees (the number of edges oriented to the vertex). The objective is to find a spanning tree T of G and a cover of T of minimum entropy. Intuitively in MEST players correspond to graph nodes, each of which may contribute the edges it is adjacent to, each at a unit cost. The submodular (cost) function is $f(S) = |\{v \in V : v \in S \text{ or } \exists w \in S, (v, w) \in E\}|$ for $S \subseteq V$. The goal of the players is to form a spanning tree with the contributed edges. We seek the "most unfair" spanning tree.

Theorem 4 *(proved in the full version [9]). The decision problem associated to MEST is NP-hard.*[1]

Finally, we would like to mention a potential practical application of MESSC to *Web Search Diversification*: Matroids are a natural way to encode diversity in web search results [22]. One could formalize this by following special case of MESSC, a generalization of MESC, called *Minimum Entropy Diverse Multicover* (MEDM). We are given a bipartite graph of *queries* and *pages*, and an integer $k \geq 1$. Each web page $p \in P$ has a *type* $t(p)$ from a set of types T. We assume that each query is adjacent to pages of at least k types. Feasible solutions are

---

[1] The problem is NP-hard, rather than NP-complete, since its specification involves general real numbers that may put it outside class NP.

k-*covers*, i.e. a set of edges covering the n queries, each query by exactly k pages (a *top-k answer*) and *all the types of pages that cover a given query are distinct.* We seek a partial cover with minimal entropy. A justification for this objective function comes from adapting the maximum likelihood approach developed by Halperin and Karp [11] for MESC to our problem.

3.1   The Greedy Algorithm

We will denote by X^{OPT} an optimal solution for an instance of MESSC. Also, given a permutation $\sigma \in S_m$ define vector X^σ as follows: for $1 \le i \le m$, $X^\sigma(\sigma(i)) = f(\{\sigma(1), \sigma(2), \ldots, \sigma(i)\}) - f(\{\sigma(1), \sigma(2), \ldots, \sigma(i-1)\})$. It is easy to see that, for every $\sigma \in S_m$, X^σ is a feasible (but perhaps not optimal) solution to instance f of MESSC, and that one of X^σ, $\sigma \in S_n$, is an optimal solution. This is easy to see using the language of cooperative games: in concave games the core is non-empty, a polytope whose extremal points are those produced greedily on permutations of U, that is X^σ. Since the entropy is a concave function, the optimum is reached at some extremal point X^σ.

A GREEDY approximation algorithm is presented in Algorithm 1. Note that it is well known that the resulting vector is also one of the vectors X^σ. We will use, throughout the rest of the paper, the following notations: $i_1, i_2, \ldots, i_m$ will be the indices chosen by the GREEDY algorithm, in this order. Furthermore, we define for $1 \le r \le m$ the *greedy rank function* by $rank(i_r) = r$. For $1 \le r \le m$, $W_r = \{i_1, \ldots, i_r\}$ is the set of first r elements added by the GREEDY algorithm; also $W_0 = \emptyset$. $X_r^{GR} = f(W_r) - f(W_{r-1})$ is the increase in the objective function caused by the choice of the r'th element.

---

input : An instance (U, f) of MESSC
output: Vector $X^{GR} = (X_r^{GR})_{r \in [m]}$
$S := \emptyset$, $r := 1$;
while $S \ne U$ **do**
    choose $i \in U \setminus S$ that maximizes $f(S \cup \{i\}) - f(S)$ (amount may be 0);
    $X_r^{GR} := f(S \cup \{i\}) - f(S)$;
    $S := S \cup \{i\}; r+ = 1$
end

Algorithm 1. Greedy algorithm for Minimum Entropy Submodular Set Cover.

4   Main Result and Definition of the Covering Coefficient α

By the analogy with MESC, we would expect to upper bound the entropy of the cover produced by the GREEDY algorithm by the entropy of the optimal cover plus $\log_2(e)$. We almost accomplish this: our upper bound further depends on a covering constant α. It can be stated as follows:

Theorem 5. *Given a polymatroid $G = (U, f)$, the greedy algorithm produces a solution X^{GR} to the instance G of MESSC related to the optimal cover X^{OPT} by the relation:*

$$Ent[X^{GR}] \leq \frac{1}{\alpha} \cdot [Ent[X^{OPT}] + \log_2(e)] + [1 - \frac{1}{\alpha}] \log_2(n) \tag{1}$$

The rest of the section is devoted to precisely defining α. First we define a quantity that will play a fundamental role in our results: for any $1 \leq r, j \leq m$ let $a_r^j = f(W_r) - f(W_{r-1}) - (f(W_r \cup \{j\}) - f(W_{r-1} \cup \{j\}))$. The best way to make sense of the (admittedly unintuitive) definition of the a_r^j coefficients is to particularize them in the case of the set cover problem. In this case coefficients a_r^j have a very intuitive description: they represent the size of the intersection of the j'th set P_j to the r'th set in the GREEDY solution. Indeed, $f(W_r) - f(W_{r-1})$ is the number of elements newly added by GREEDY at step r, whereas the subtracted term $f(W_r \cup \{j\}) - f(W_{r-1} \cup \{j\})$ is the number of elements that would still be newly added *if P_j were present too.*

Proposition 6. *For any $1 \leq r, j \leq m$ we have $a_r^j \geq 0$.*

When $j \in W_r$ this follows directly from the monotonicity of f. Assume now that $j \notin W_r$, and employ the submodularity of f with $S = W_r$, $T = W_{r-1} \cup \{j\}$.   □

To define the covering coefficient α we introduce a large number of apparently superfluous variables Z_r^j. Intuitively Z_r^j is the portion of optimal solution X_j^{OPT} that can be assigned to cover X_r^{GR}. This explains the equations below: first, one has to allocate all of X_j^{OPT} and no more than that. Second, one cannot allocate to any "set X_r^{GR}" more than "its intersection with X_j^{OPT}". The quoted statements above make full sense, of course, only for regular set cover.

Definition 7. *Given a polymatroid G, let $\alpha = \alpha_G$ the smallest positive value such that there exist $Z_r^j \in \mathbb{Z}, Z_r^j \geq 1$ so that the following inequalities hold*

$$\sum_{r=1}^{m} Z_r^j = X_j^{OPT}, 1 \leq j \leq m \tag{2}$$

$$\sum_{j=1}^{m} Z_r^j \leq \alpha \cdot X_r^{GR}, 1 \leq r \leq m \tag{3}$$

Given the discussion above, α can be seen as a covering coefficient. It measures the amount of "redundancy" inherent into "assembling" the GREEDY solution from pieces obtained by breaking up the optimal solution.

Proposition 8. *The coefficient α is always greater than or equal to 1.*

Proof. Sum all Eq. (3) for all $r = 1, \ldots, m$. The left-hand side is

$$\sum_{r=1}^{m} \left(\sum_{j=1}^{m} Z_r^j \right) = \sum_{j=1}^{m} \left(\sum_{r=1}^{m} Z_r^j \right) = \sum_{j=1}^{m} X_j^{OPT} = N.$$

On the other hand the right-hand side is $\alpha \cdot \sum_{r=1}^{m} X_r^{GR} \leq \alpha \cdot N$, by the GREEDY algorithm. The result follows.   □

5 Proof of Main Result

By the greedy choice we infer $X_r^{GR} = f(W_{r-1} \cup \{i_r\}) - f(W_{r-1})$ for any $1 \leq r \leq m$. We first prove several auxiliary results:

Lemma 9. *For any $j \in [m]$ we have $\sum_{r=1}^{m} a_r^j = f(\{j\})$.*

Proof: By the definition of a_k^j and using the equalities $f(W_0) = 0, f(W_m \cup \{j\}) = f(W_m) = N$ we have:

$$\sum_{r=1}^{m} a_r^j = \sum_{r=1}^{m} (f(W_r) - f(W_{r-1}) - (f(W_r \cup \{j\}) - f(W_{r-1} \cup \{j\}))) = f(W_m)$$
$$- f(W_0) - (f(W_m \cup \{j\}) - f(W_0 \cup \{j\})) = N - (N - f(\{j\})) = f(\{j\}). \quad \square$$

Lemma 10. *Given $r, j \in [m]$ we have $f(\{j\}) - \sum_{k=1}^{r} a_k^j = f(W_r \cup \{j\}) - f(W_r)$.*

Proof: Similarly with the previous proof, based on the coefficients definition and by using the equations $f(W_0) = 0, f(W_0 \cup \{j\}) = f(\{j\})$ is obtaining the proof. $\sum_{k=1}^{r} a_k^j = f(W_r) - f(W_0) - (f(W_r \cup \{j\}) - f(W_0 \cup \{j\})) = f(W_r) - f(W_r \cup \{j\}) + f(\{j\})$. $\quad \square$

Lemma 11. *We have $\prod_{r=1}^{m} (X_r^{GR})^{\alpha X_r^{GR}} \geq \prod_{j=1}^{m} (X_j^{OPT})!$.*

Proof: By the greedy choice, Lemmas (9), (10) and $Z_r^j \leq a_r^j$: $X_r^{GR} = f(W_{r-1} \cup \{i_r\}) - f(W_{r-1}) \geq f(W_{r-1} \cup \{j\}) - f(W_{r-1}) = f(\{j\}) - \sum_{k=1}^{r-1} a_k^j = \sum_{k=r}^{m} a_k^j \geq \sum_{k=r}^{m} Z_k^j = X_j^{OPT} - \sum_{k=1}^{r-1} Z_k^j$.

Thus $\prod_{j=1}^{m} \left(\prod_{r=1}^{m} (X_r^{GR})^{Z_k^j} \right) \geq \prod_{j=1}^{m} \left(\prod_{r=1}^{m} \left(X_j^{OPT} - \sum_{k=1}^{r-1} Z_k^j \right)^{Z_k^j} \right) \geq \prod_{j=1}^{m} (X_j^{OPT})!$

By Definition (7): $\prod_{j=1}^{m} \left(\prod_{r=1}^{m} (X_r^{GR})^{Z_k^j} \right) = \prod_{r=1}^{m} (X_r^{GR})^{\sum_{j=1}^{m} Z_k^j} \leq \prod_{r=1}^{m} (X_r^{GR})^{\alpha X_r^{GR}}$.

$\quad \square$

With Lemma (11) in hand, we get

$$ENT[X^{GR}] = -\sum_{r=1}^{m} \frac{X_r^{GR}}{n} \log_2 \left(\frac{X_r^{GR}}{n} \right) = -\sum_{r=1}^{m} \frac{X_r^{GR}}{n} \log_2(X_r^{GR}) + \log_2(n)$$

$$= -\frac{1}{n\alpha} \log_2 \prod_{r=1}^{m} (X_r^{GR})^{\alpha X_r^{GR}} + \log_2 n \leq -\frac{1}{n\alpha} \log_2 \prod_{j \in OPT} (X_j^{OPT})! + \log_2 n$$

Using inequality $n! \geq \left(\frac{n}{e}\right)^n$ we infer:

$$\mathbf{ENT[X^{GREEDY}]} \leq -\frac{1}{n\alpha} \log_2 \prod_{j \in OPT} \left(\frac{X_j^{OPT}}{e}\right)^{X_j^{OPT}} + \log_2 n$$

$$= -\frac{1}{n\alpha} \sum_{j \in OPT} X_j^{OPT} \left(\log_2 X_j^{OPT} - \log_2 e\right) + \log_2 n$$

$$= \frac{1}{\alpha}\left(-\sum_{j \in OPT} \frac{X_j}{n} \log_2 \frac{X_j}{n} - \log_2 n\right) + \frac{1}{\alpha} \log_2 e + \log_2 n$$

$$= \frac{1}{\alpha}\left(\mathbf{ENT[X^{OPT}]} + \log_2 e\right) + \left(1 - \frac{1}{\alpha}\right) \log_2 \mathbf{n}. \qquad \square$$

6   Applications: Special Cases with $\alpha = 1$

A simple problem where one can determine the value of α is the Minimum Entropy Orientation (MEO) problem [13,14]. The input to MEO is a graph $G = (V, E)$. An *orientation of* G is a function $u : E \to V$ such that for all $e \in E$, $u(e)$ is one of the vertices of e. The entropy of orientation u is defined in an obvious way, as the entropy of the distribution of indegrees. The objective is to find an orientation u of G that minimizes the entropy.

MEO is a special case of MESC: each instance $G = (V, E)$ of MEO can be regarded as an instance of MESC with submodular cost function $f : V \to \mathbf{Z}, f(S) = |\{e \in E : e \cap S \neq \emptyset\}|$. We first recover (using a different method) the upper bound on the performance of the GREEDY algorithm for MEO (an algorithm that is, however, not optimal [13]).

Proposition 12. *For any instance G of MEO, $\alpha_G = 1$.*

Proof. A simple application of the definition of f yields $a_r^j = 1$, if $i_r \neq j, (i_r, j) \in E, j \notin W_r; a_r^j = X_r^{GR}$, if $i_r = j; a_r^j = 0$, otherwise. This allows us to turn an orientation of minimum entropy (corresponding to an optimal solution) into the greedy orientation (and define coefficients Z_r^j) as follows:

- At each stage r, after the choice of i_r we reorient the edges (j, i_r), $j \notin W_r$ that have different orientations in the optimal and greedy solution. Correspondingly, we define $Z_r^j = 1$ for such edges.
- Also let $Z_r^{i_r}$ be the number of edges (j, i_r) that are oriented towards i_r in both the greedy and the optimal orientation. Note that there are at most $a_r^{i_r} = X_r^{GR}$ such edges.
- Note that an edge that is reoriented at stage r is not reoriented again at a later stage (because of the restriction $j \notin W_r$). Hence the process ends up with the greedy solution. In other words $\sum_{j=1}^{m} Z_r^j = X_r^{GR}$ (as we add one unit for each edge counted by X_r^{GR}). $\qquad \square$

We can also rederive the results of Cardinal et al. on MESC using our approach:

Proposition 13. *For any instance G of MESC $\alpha_G = 1$.*

Proof. For MESC the submodular function is $f(S) = |\cup_{i \in S} P_i|$, for $S \subseteq [m]$. By a direct application of their definition $a_r^j = |(P_{i_r} \setminus \cup_{k=1}^{r-1} P_{i_k}) \cap P_j|$.

Let $u : [N] \to [m]$ be an optimal solution to MESC, i.e. for any $1 \leq i \leq N$, $i \in P_{u(i)}$ and the cover specified by u has minimum entropy. Denote, for $j = 1, \ldots, m$, $U_j = \{x \in [N] : u(x) = j\}$. $U_j \subseteq P_j$ is the set of elements assigned by cover u to set P_j. Define, for $1 \leq r \leq l$

$$Z_r^j = |U_j \cap (P_{i_r} \setminus \cup_{k=1}^{r-1} P_{i_k})|. \tag{4}$$

Then $0 \leq Z_r^j \leq a_r^j$. Moreover

$$\sum_{r=1}^{l} Z_r^j = \sum_{r=1}^{l} |U_j \cap (P_{i_r} \setminus \cup_{k=1}^{r-1} P_{i_k})| = |U_j|$$

$$\sum_{j=1}^{m} Z_r^j = \sum_{j=1}^{m} |U_j \cap (P_{i_r} \setminus \cup_{k=1}^{r-1} P_{i_k})| = |(P_{i_r} \setminus \cup_{k=1}^{r-1} P_{i_k})|,$$

(as each of the two set systems $(U_j)_{j \in [m]}$ and $(P_{i_r} \setminus \cup_{k=1}^{r-1} P_{i_k})_{r=1}^{m}$ consists of disjoint sets), hence $X_j = |U_j|$ and Z_r^j satisfy the conditions for $\alpha = 1$. $\square$

7   Network Flow Interpretation of α and a Multistage Approach

Theorem 5 is, of course, most interesting when $\alpha_G = 1$, matching the $\log_2(e)$ additive guarantee of MESC. However, there exist instances G of MESSC for which the associated constant α_G is *strictly greater than 1*.

To circumvent this problem we will develop a more powerful technique: we first reinterpret the constant α_G using one-stage network flows. This will allow us to generalize our method to multistage flows, characterized by a related constant β_G. A variant of Theorem 5 holds for the constant β_G as well. The extension allows us to prove that *the $\log_2(e)$ additive guarantee is valid for* **all** *instances of MEST*; the result follows from a multistage flow construction witnessing that for any instance G of MEST $\beta_G = 1$.

Consider the flow network in Fig. 1 (a). In addition to source/sink nodes s, t, F has two layers of nodes; the first layer of nodes corresponding to the optimal solution, the second layer of nodes corresponding to the greedy one. In each layer we have a node for every player in the game. Edges appear between nodes of type k and i_r, with capacity equal to a_k^r. The fact that the first layer of nodes corresponds to the optimal solution is reflected by setting capacity X_j^{OPT} on the edge between node s and node j. Similarly, capacities between node i_r of the second layer and node t are set to value X_r^{GR}. These values are seen as *requests*

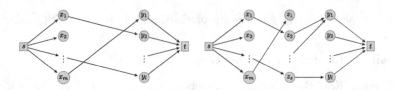

Fig. 1. (a) Network flow interpretation of the constant α. (b) Multistage flow network between solutions.

of node Y_r that may be satisfied by the flow (which in general might send an amount larger than X_r^{GR} to this node).

It follows that $\alpha_G = 1$ amounts to the existence of an integer flow fl of value N in the flow network of Fig. 1 (a) (that is, fl satisfies the request of each node Y_r exactly). More generally, α *is the minimum amount needed to multiply the capacities on the edges entering t in order to accommodate a $s - t$ integer flow with value N*.

Our solution to MEO could be easily recast in terms of flows: we construct the flow iteratively, by considering the paths between a node in the first layer and a node in the second layer. The induction starts determined by the order of second-layer nodes (corresponding to the GREEDY algorithm) and then continue by the order of the nodes in the first layer (according to a fixed ordering).

There are lessons to be learned from the construction this flow and our proof of Theorem 5: The key point was that when we had to reorient an edge towards a node in the greedy solution, *we could do so without overflowing this node*. Similarly, the general proof depended on the following invariant we maintained (*): $X_r^{GR} \geq \sum_{k=r}^{l} Z_k^j$. Condition (*) does not have a direct flow interpretation, since X_r^{GR} is the request, rather than the actual flow value at the given node. However, its relaxation involving α does: the actual flow into node y_r is at most αX_r^{GR}, so requiring that the total flow into node y_r is at least $\sum_{k=r}^{l} Z_k^j$ guarantees the following relaxed version of equation (*): $\alpha \cdot X_r^{GR} \geq \sum_{k=r}^{l} Z_k^j$. We will see (in Theorem 16) that the relaxed condition can be applied as well.

We generalize the setting of Theorem 5 by considering flow networks with $q \geq 1$ levels (see Fig. 1(b) for $q = 2$). The nodes in each level are ordered according to a fixed ordering, e.g. the ordering induced by the GREEDY algorithm, with nodes not chosen by this algorithm coming after all chosen nodes in a fixed, arbitrary sequence. Capacities correspond either to values a_j^r (if the chosen indices are j and i_r, respectively) or ∞, for edges between nodes with the same index j but on different levels. Note that each path ending in a greedy node with index i_r has finite capacity, at most the capacity of its last edge. We use the notation $P : [j \ldots k]$ to indicate the fact that path P starts at node j on the first level and ends at node k on the last. Also write $P \sim v$ to indicate the fact that path P is adjacent to node v. We will also need to consider a total ordering $<$ on paths (explicitly constructed when analyzing particular problems, e.g. MEST):

Definition 14. *A flow fl is admissible with respect to total path ordering $<$ if for any path P between, say, node X_j and Y_r, the remaining flow into node X_j just before path P is considered is at most the final value of the final total flow into node Y_r. Formally $\sum_{Q \sim X_j, P \leq Q} fl(Q) \leq \sum_{W \sim Y_r} fl(W)$.*

Even in a multiple-layer flow network it may not be possible to obtain an admissible flow of value N. As before, the solution is to multiply the capacities of edges into node t by some fixed $\beta \geq 1$.

Definition 15. *Define β_G as the infimum (over all multi-level flow networks corresponding to the optimal and greedy solution) of all values $\beta > 0$ for which there exists a path ordering $<$ and a flow fl admissible w.r.t. $<$ such that for every pair of nodes j and r, $\sum_j \left(\sum_{P:[j...i_r]} fl_P \right) \leq \beta \cdot X_r^{GR}$. The reader is requested to compare Definitions 7 and this definition.*

Similarly to the proof of Proposition 8, we obtain $\beta = \beta_G \geq 1$. On the other hand, admissibility will guarantee in general only a weaker version of Theorem 5 (though no weaker for the main setting we have in mind, $\beta = 1$):

Theorem 16. *Given an instance $G = (N, f)$ of MESSC the greedy algorithm produces a solution X^{GR} satisfying $\beta \cdot Ent[X^{GR}] \leq Ent[X^{OPT}] + \log_2(e) + \beta \log_2(\beta) + (\beta - 1) \log_2(n)$.*

The proof is almost identical to that of Theorem 5, and detailed in [9].

8   Application to MEST

Theorem 16 applies to MEST, yielding a nontrivial special case of MESSC with the same additive constant as that of Minimum Entropy Set Cover:

Theorem 17. *For any instance G of MEST, $\beta_G = 1$. Therefore*

$$Ent[X^{GR}] \leq Ent[X^{OPT}] + \log_2(e).$$

Proof. A rather construction of an admissible multistage flow, given in [9].   □

We conjecture that a similar result holds for MEDM, and that it can be proved using multilevel flows (single level ones do not seem powerful enough).

9   Conclusions and Open Problems

The most important open question raised by our work is whether $\log_2(e)$ is an additive approximation guarantee for **all** instances of MESSC. A deeper matroid-theoretic study of MESSC would be useful in this respect. Finding the optimal additive approximation guarantee for MEST is an open problem as well.

Second, a problem in information theory called the *Minimum Entropy Coupling* [15] problem led us to consider an extension of the framework from this paper to *string submodular* functions. Variations on "worst-case fairness" are a topic for further study, given the large variety of interesting combinatorial games [2]. Finally, MESSC may have many potential practical applications, including MEDM. It would be interesting to study more realistic version of this problem.

References

1. Alon, N., Orlitsky, A.: Source coding and graph entropies. IEEE Trans. Inform. Theory **42**(5), 1329–1339 (1996)
2. Bilbao, J.M.: Cooperative Games on Combinatorial Structures. Kluwer, Boston (2000)
3. Bonchiş, C., Istrate, G.: A parametric worst-case approach to fairness in tu-cooperative games. arXiv.org:1208.0283
4. Cardinal, J., Fiorini, S., Joraet, G.: Tight results on minimum entropy set cover. Algorithmica **51**(1), 49–60 (2008)
5. Driessen, T.: Cooperative Games, Solutions and Applications. Kluwer, Boston (1988)
6. Dukhovny, A.: General entropy of general measures. Int. J. Uncertainty Fuzziness Knowl. Based Syst. **10**(03), 213–225 (2002)
7. Fujishige, S.: Submodular Functions and Optimization. Elsevier, Amsterdam (2005)
8. Fujita, T.: Approximation algorithms for submodular set cover with applications. IEICE Trans. Inf. Syst. **E83–D**(3), 480–487 (2000)
9. Bonchiş, C., Istrate, G., Dinu, L.P.: The minimum entropy submodular set cover problem. Manuscript. http://tcs.ieat.ro/wp-content/uploads/2015/10/lata.pdf
10. Wang, X., Jajamovich, G.: Maximum-parsimony haplotype inference based on sparse representations of genotypes. IEEE Trans. Sign. Proc. **60**, 2013–2023 (2012)
11. Halperin, E., Karp, R.: The minimum entropy set cover problem. Theoret. Comput. Sci. **348**(2–3), 340–350 (2005)
12. Iwata, S., Orlin, J.B.: A simple combinatorial algorithm for submodular function minimization. J. Comb. Theory Ser. B **84**, 1230–1237 (2009)
13. Fiorini, S., Cardinal, J., Joret, G.: Minimum entropy orientations. Oper. Res. Lett. **36**(6), 680–683 (2008)
14. Fiorini, S., Cardinal, J., Joret, G.: Minimum entropy combinatorial optimization problems. Theory Comput. Syst. **51**(1), 4–21 (2012)
15. Stanojević, I., Kovačević, M., Šenk, V.: On the entropy of couplings. Inf. Comput. **242**, 369–382 (2015)
16. Madiman, M., Tetali, P.: Information inequalities for joint distributions, with interpretations and applications. IEEE Trans. Inf. Theory **56**, 2699–2713 (2010)
17. Oxley, J.G.: Matroid Theory. Oxford University Press, Oxford (2006)
18. Schrijver, A.: A combinatorial algorithm minimizing submodular functions in strongly polynomial time. J. Comb. Theory Ser. B **80**, 346–355 (2000)
19. Shapley, L.: Cores of convex games. Int. J. Game Theory **1**, 11–26 (1971)
20. Wang, B.: Minimum entropy approach to word segmentation problems. Phys. A Stat. Mech. Appl. **293**, 583–591 (2001)
21. Wolsey, L.: An analysis of the greedy algorithm for the submodular set covering problem. Combinatorica **2**, 385–393 (1982)
22. Mirrokni, V., Abbassi, Z., Thakur, M.: Diversity maximization under matroid constraints. In: Proceedings of the 19th ACM SIGKDD International Conference on Knowledge Discovery and Data Mining, KDD 2013, pp. 32–40. ACM (2013)

On the Capacity of Capacitated Automata

Orna Kupferman[1]([⊠]) and Sarai Sheinvald[2]

[1] School of Computer Science and Engineering, The Hebrew University,
Jerusalem, Israel
orna@cs.huji.ac.il
[2] Department of Software Engineering, Ort Braude Academic College,
Karmiel, Israel
sarai@braude.ac.il

Abstract. *Capacitated automata* (CAs) have been recently introduced in [8] as a variant of finite-state automata in which each transition is associated with a (possibly infinite) *capacity*. The capacity bounds the number of times the transition may be traversed in a single run.

The study in [8] includes preliminary results about the expressive power of CAs, their succinctness, and the complexity of basic decision problems for them. We continue the investigation of the theoretical properties of CAs and solve problems that have been left open in [8]. In particular, we show that union and intersection of CAs involve an exponential blow-up and that their determinization involves a doubly-exponential blow up. This blow-up is carried over to complementation and to the complexity of the universality and containment problems, which we show to be EXPSPACE-complete. On the positive side, capacities do not increase the complexity when used in the deterministic setting. Also, the containment problem for nondeterministic CAs is PSPACE-complete when capacities are used only in the left-hand side automaton. Our results suggest that while the succinctness of CAs leads to a corresponding increase in the complexity of some of their decision problems, there are also settings in which succinctness comes at no price.

1 Introduction

Finite automata (FAs) are used in the modeling and design of finite-state systems and their behaviors, with applications in engineering, databases, linguistics, biology, and many more. The traditional definition of an automaton does not refer to its transitions as consumable resources. Indeed, a run of an automaton is a sequence of successive transitions, and there is no bound whatsoever on the number of times that a transition may be traversed. In practice, the use of a transition may correspond to the use of some resource. For example, it may be associated with filling a buffer, consumption of bandwidth, or a usage of some energy-consuming machine. In [8], the authors introduced *capacitated automata* (CAs) – a variant of finite automata in which each transition is associated with a (possibly infinite) *capacity*, which limits the number of times the transition may be traversed in a single run.

© Springer International Publishing Switzerland 2016
A.-H. Dediu et al. (Eds.): LATA 2016, LNCS 9618, pp. 307–319, 2016.
DOI: 10.1007/978-3-319-30000-9_24

The study in [8] examines CAs from two points of view. The first views CAs as recognizers of formal languages. The interesting questions that arise in this view are the classical questions about automata: their expressive power, succinctness, closure properties, determinization, etc. The second view, more related to traditional resource-allocation problems, views CAs as labeled flow networks. The interesting questions then have to do with maximum utilization of the system modeled by CA and can be viewed as a generalization of the max-flow problem in networks [6].

Our work here continues the study of the first approach taken in [8]. There, the authors study the expressive power of CAs, their succinctness with respect to FA, and the basic decision problems of non-emptiness and membership. The study in [8] also relates CAs with extensions of FAs of similar flavor. For completeness, we briefly repeat the main findings here. For a full description, see [8]. Augmenting the transitions of automata by numerical values is common in the quantitative setting, for example in probabilistic automata and weighted automata [5,10]. There, the values are used for modeling probabilities, costs, rewards, certainty, and many more. The semantics of these models is multi-valued. CAs, on the other hand, maintain the Boolean nature of regular languages and only augment the way in which acceptance is defined. In this family of extensions, we can find *Parikh automata* [7], whose semantics involves counting of the number of occurrences of each letter in the word, and their variants, in particular the *constrained automata* of [3]. The expressive power of Parikh automata and their variants goes beyond regular languages, and the type of questions studied for them is different than these studied for CAs. Additional strictly more expressive models include *multiple counters automata* [4], where transitions can be taken only if guards referring to traversals so far are satisfied, and *queue-content decision diagrams*, which are used to represent queue content of FIFO-channel systems [1,2]. Finally, a model with the same name – *finite capacity automata* – is used in [9] for modeling the control of an automated manufacturing system. This model is different from the CAs studied here and is more related to Petri nets.

In order to describe the results in [8], we first need some definitions. Let Δ be the set of transitions of a CA $\mathcal{A}$ and let $c : \Delta \rightarrow \mathbb{N} \cup \{\infty\}$ be its capacity function. That is, for every transition $\tau \in \Delta$, a run of $\mathcal{A}$ can traverse τ at most $c(\tau)$ times. A naive translation of $\mathcal{A}$ to an FA involves a blow-up that is polynomial in the number γ of "capacity configurations". Formally, $\gamma = \Pi_{\tau \in \Delta : c(\tau) \neq \infty}(c(\tau) + 1)$. That is, for every transition τ with a bounded capacity, the FA has to remember how many more times τ can be traversed. It is not hard to see that by attributing the states of $\mathcal{A}$ by the current capacity configuration, we can obtain an equivalent FA. Note that γ is exponential in the number of transitions with finite capacities. The above implies that CAs are not more expressive than FAs, but may be exponentially more succinct. Indeed, as shown in [8], the blow-up in γ may not be avoided in some cases. Consider, for example, the language $L_{n,m}$ over the alphabet $\Sigma_n = \{1, \ldots, n\}$ that contains exactly all words in which each letter in Σ_n appears at most m times. It is not hard to see that a traditional, possibly nondeterministic, automaton for $L_{n,m}$ needs at least m^n states. On the other hand, a deterministic CA for $L_{n,m}$ consists of a single state with n self loops, each labeled with a different letter and having capacity m.

A key question in the theory of CAs is the role that γ plays in constructions and decision problems. For example, it is shown in [8] that while determinization of nondeterministic CAs (NCAs) does involve a blow up in γ, the non-emptiness problem for NCAs can be solved in linear time and its complexity is independent of γ. The study of succinctness and complexity in [8] is preliminary. In particular, even for determinization, only an exponential lower bound has been shown, with no matching upper bound, leaving open also the blow-up involved in other basic constructions like union, intersection, and complementation. Decision problems whose solution makes use of these constructions, most notably universality and containment, have been left open too.

We solve the problems left open in [8]. Our news is mainly bad: the exponential lower bound for determinization that is proven there is not tight, and determinization of NCAs actually involves a doubly-exponential blow-up. This is an interesting phenomenon, implying that the power of nondeterminism in the capacitated model goes beyond the subset construction. Even though the deterministic automaton we end up with is capacitated, it sometimes cannot make use of its capacities and has to maintain not only sets of states but also sets of capacity configurations. The doubly-exponential blow-up in determinization is carried over to complementation of NCAs. Also there, even though the complementing NCA is both nondeterministic and capacitated, it may be doubly-exponentially bigger, as an exponential blow up in γ cannot be avoided. Moreover, even the constructions of union and intersection involve a blow-up in γ. Essentially, it follows from the inability to merge the capacity functions of the underlying CAs to a single capacity function in the product CA. In fact, the blow-up applies even when one of the automata is not capacitated. An exception is union of NCAs, which can be defined by putting the NCAs "side by side", and thus involves no blow up.

We continue to the decision problems of universality (given a CA $\mathcal{A}$, decide whether $L(\mathcal{A}) = \Sigma^*$) and containment (given two CAs $\mathcal{A}$ and $\mathcal{A}'$, decide whether $L(\mathcal{A}) \subseteq L(\mathcal{A}')$) and show that the bad news is carried over also to their complexity, but only in the nondeterministic model: While the universality problem for DCAs is NLOGSPACE-complete, thus is not more complex than the problem for DFAs, it is EXPSPACE-complete for NCAs. The EXPSPACE complexity of universality immediately implies an EXPSPACE lower bound also for containment, in fact already containment of DFAs in NCAs. We study the various cases according to whether each of $\mathcal{A}$ and $\mathcal{A}'$ is capacitated and/or nondeterministic. Here, we are also able to come up good news: the containment problem for NCAs in NFAs is PSPACE-complete, thus is not harder than containment for NFAs. We conclude that while the succinctness of CAs often leads to a corresponding increase in the complexity of their decision problems, capacities come at no price when they model systems or when used in a deterministic automaton.

Due to the lack of space, some of the proofs are omitted from this version and can be found in the full version, in the authors' URLs.

2   Preliminaries

A *nondeterministic finite automaton* (NFA, for short) is a tuple $\mathcal{A} = \langle \Sigma, Q, Q_0, \Delta, F \rangle$, where Σ is a finite alphabet, Q is a finite set of states, $Q_0 \subseteq Q$ is a set of initial states, $\Delta \subseteq Q \times \Sigma \times Q$ is a transition relation, and $F \subseteq Q$ is a set of final states. Given a word $w = \sigma_1 \cdot \sigma_2 \cdots \sigma_l$, a *run* of $\mathcal{A}$ on w is a sequence r of successive transitions in Δ that reads w and starts in a transition from the set of initial states. Thus, $r = \langle q_0, \sigma_1, q_1 \rangle, \langle q_1, \sigma_2, q_2 \rangle, \ldots, \langle q_{l-1}, \sigma_l, q_l \rangle$, for $q_0 \in Q_0$. The run is accepting if $q_l \in F$. The NFA $\mathcal{A}$ *accepts* the word w iff it has an accepting run on it. Otherwise, $\mathcal{A}$ *rejects* w. The language of $\mathcal{A}$, denoted $L(\mathcal{A})$, is the set of words that $\mathcal{A}$ accepts. If $|Q_0| = 1$ and for all $q \in Q$ and $\sigma \in \Sigma$ there is at most one $q' \in Q$ with $\Delta(q, \sigma, q')$, then $\mathcal{A}$ is *deterministic*. Note that a deterministic finite automaton (DFA) has at most one run on each word.

A *nondeterministic capacitated automaton* (NCA, for short) is an NFA in which each transition has a *capacity*, bounding the number of times it may be traversed. A transition may not be bounded, in which case its capacity is ∞. Let $\mathbb{N}^\infty = \mathbb{N} \cup \{\infty\}$ and $\mathbb{N}^+ = \mathbb{N} \setminus \{0\}$. Formally, an NCA is a tuple $\mathcal{A} = \langle \Sigma, Q, Q_0, \Delta, F, c \rangle$, where Σ, Q, Q_0, Δ, and F are as in NFAs and $c : \Delta \to \mathbb{N}^\infty$ is a *capacity function* that maps each transition in Δ to its capacity. A legal run of $\mathcal{A}$ is defined as for NFA, with the additional condition that the number of occurrences of each transition $e \in \Delta$ is at most $c(e)$. When the underlying NFA is deterministic, then so is $\mathcal{A}$. The *width* of an NCA $\mathcal{A}$ is the maximal finite capacity that c assigns to a transition in Δ.

For a capacity function $c : \Delta \to \mathbb{N}^\infty$, let $c_\downarrow$ be the set of capacity functions obtained by closing c downwards. Formally, a function $c' : \Delta \to \mathbb{N}^\infty$ is in $c_\downarrow$ if for all transitions $e \in \Delta$ with $c(e) = \infty$, we have $c'(e) = \infty$, and for all transitions $e \in \Delta$ with $c(e) \in \mathbb{N}$, we have $0 \leq c'(e) \leq c(e)$. It is easy to see that the size of $c_\downarrow$, denoted $|c_\downarrow|$, is $\Pi_{e:c(e) \in \mathbb{N}^+}(c(e) + 1)$. Thus, $|c_\downarrow|$ is exponential in the number of transitions with finite capacities and is bounded by $w^{|\Delta|}$, where w is the width of $\mathcal{A}$ plus 1.

Example 1. For a word $w \in \{0,1\}^*$, let $\tilde{w}$ be the word obtained by flipping all the letters in w. For example, if $w = 0010$, then $\tilde{w} = 1101$. We refer to $\tilde{w}$ as the *negative* of w.

We use the NCA $\mathcal{A}_k$ appearing in Fig. 1 as a "box" in some of our constructions. As a warm-up example, consider the NCA $\mathcal{U}_k$ appearing in Fig. 1. It

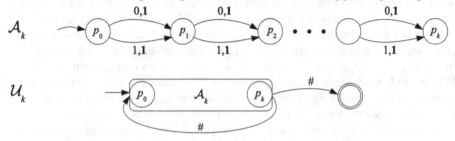

Fig. 1. The NCAs $\mathcal{A}_k$ and $\mathcal{U}_k$, with $L(\mathcal{U}_k) = \{v\# : v \in (0+1)^k\} \cup \{v\#\tilde{v}\# : v \in (0+1)^k\}$.

is easy to see that $\mathcal{U}_k$ accepts only words in $(0+1)^k\# + (0+1)^k\#(0+1)^k\#$. Moreover, since the capacity of all the transitions labeled 0 or 1 is 1, the only way to traverse the sequence of states $p_0,\ldots,p_k$ twice is by reading a word $v \in (0+1)^k$ during the first traversal, then $\#$, and then the negative $\tilde{v}$ of v. Hence $L(\mathcal{U}_k) = \{v\# : v \in (0+1)^k\} \cup \{v\#\tilde{v}\# : v \in (0+1)^k\}$. Note that an NFA for $L(\mathcal{U}_k)$ requires at least 2^k states.

3 Closure Constructions

Since CAs are as expressive as FAs, the closure of regular languages under Boolean operations implies a similar closure for NCAs and DCAs. In this section we study the blow up that is involved in the corresponding constructions. We start with NCA determinization. In [8], the authors show an exponential blow-up for determinization. We tighten this result and show that determinization is tightly doubly exponential.

Theorem 1. *Determinization of NCAs is tightly doubly-exponential.*

Proof: For the upper bound, consider an NCA $\mathcal{A} = \langle \Sigma, Q, Q_0, \Delta, F, c \rangle$. By [8], $\mathcal{A}$ has an equivalent NFA $\mathcal{A}'$ with $|Q| \cdot |c_\downarrow|$ states, which is exponential in the size of $\mathcal{A}$. By determinizing $\mathcal{A}'$, we get an equivalent DFA that is exponential in $\mathcal{A}'$ and hence doubly-exponential in $\mathcal{A}$.

For the lower bound, we define a sequence of languages $L_1, L_2, \ldots$ over the alphabet $\Sigma = \{0, 1, \#, \$\}$ as follows. We define $L^1 = (0+1+\#)^*$, and for $n \geq 1$, we define $L_n^2 = \{(0+1+\#)^*\#w\#(0+1+\#)^*\$(0+1+\#)^*\#\tilde{w}\#(0+1+\#)^* : w \in (0+1)^n\}$. That is, each word in L_n^2 is of the form $u\$v$, for words $u, v \in (0+1+\#)^*$. Inside u, there should be some word $w \in (0+1)^n$ between two $\#$'s, and v includes $\tilde{w}$ between $\#$'s. Now, $L_n = L^1 \cup L_n^2$.

The language L_n can be recognized by an NCA $\mathcal{U}_n$ with $O(n)$ states. The NCA $\mathcal{U}_n$, described in Fig. 2, guesses when the $\#$ before w starts, then reads w and traverses the NCA $\mathcal{A}_n$ from Example 1 once. Reading the $\#$ after w, it guesses whether the input word belongs to L^1, in which case it goes with $\#$ to a state that accepts all words in $(0+1+\#)^*$, or belongs to L_n^2, in which case

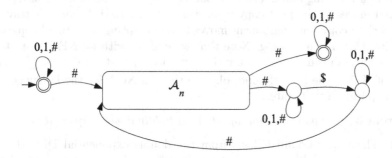

Fig. 2. the NCA $\mathcal{U}_n$.

it moves to a state that waits for a $ and then waits for the # before $\tilde{w}$, with which it returns to the initial state of $\mathcal{A}_n$.

A DFA for L_n must have at least 2^{2^n} states. Indeed, in order to recognize words in L_n^2, upon reaching the $, the DFA must remember the subset of words in $(0+1)^n$ that have appeared between #'s before the $.

The same arguments imply that the language $S_n = \Sigma^* \cdot L_n$ can be recognized by an NCA with $O(n)$ states but requires a DFA with at least 2^{2^n} states. As shown in [8], a minimal DCA for a language of the form $\Sigma^* \cdot L$ is not smaller than a minimal DFA for it. It follows that a minimal DCA for S_n requires 2^{2^n} states, hence the lower bound for determinization.   □

We continue to study complementation. For DFAs, complementation is easy, as one only has to dualize the set of accepting states. In the case of DCAs, we also need to dualize the upper bound on the number of traversals of transitions that is imposed by capacities. Such a dualization amounts to adding lower bounds on the number of traversals, which is not supported in CAs. As we show, this makes complementation of DCAs exponential. When, however, we complement the DCA into an NCA, we can "implement" the lower bound with a polynomial blow-up. Starting, however, with an NCA, a doubly-exponential blow-up in complementation cannot be avoided.

Theorem 2. *Complementing a DCA to a DCA is tightly exponential. Complementing a DCA to an NCA is polynomial.*

Proof: Consider a DCA $\mathcal{A}$. By translating $\mathcal{A}$ to a DFA and complementing the latter, we obtain a complementing DCA (in fact, DFA) of exponential size. For the lower bound, consider the language $L_{m,n}$ described in Sect. 1. A DCA for $L_{n,m}$ consists of a single state with n self loops, one for each letter, with capacity m. Let $\tilde{L}_{n,m}$ be the complement of $L_{n,m}$. That is, $\tilde{L}_{n,m}$ contains all words in which at least one letter appears more than m times. Note that $\tilde{L}_{n,m} = \Sigma_n^* \cdot \tilde{L}_{n,m}$. Hence, by [8], a minimal DCA for $\tilde{L}_{n,m}$ is not smaller than a minimal DFA for it, which needs at least m^n states.

Now, let $\mathcal{A} = \langle \Sigma, Q, Q_0, \Delta, F, c \rangle$. Intuitively, the complementing NCA $\mathcal{A}'$ has $|\Delta|$ components, one for each transition $\tau \in \Delta$. The component $\mathcal{A}_\tau$ accepts exactly all words w such that the run of $\mathcal{A}$ on w traverses τ more than $c(\tau)$ times or ends in a rejecting state. Thus, $\mathcal{A}_\tau$ has $c(\tau)$ copies of $\mathcal{A}$, starts from the first copy and moves to the next copy whenever τ is traversed. When τ is traversed in the $c(\tau)$-th copy, the component moves to an accepting sink. In all copies, the states in $Q \setminus F$ are accepting. Note that we end up with an NFA (in fact, the only nondeterminism is in choosing the component) that consists of $\sum_{\tau \in \Delta} c(\tau)$ copies of $\mathcal{A}$. For the lower bound, observe that an NCA for the language $\tilde{L}_{n,m}$ above requires $O(mn)$ states.   □

Theorem 3. *Complementation of NCAs is tightly doubly-exponential.*

Proof. The upper bound follows from the doubly-exponential DFA that the determinization in Theorem 1 results in, which can be complemented with no blow-up.

For the lower bound, consider the sequence of languages $L_1, L_2, \ldots$ defined in the proof of Theorem 1. An NCA for the complement $\tilde{L}_n = \Sigma^* \setminus L_n$ needs at least 2^{2^n} states. Upon reaching the \$, the NCA must remember the subset of words in $(0+1)^n$ that have appeared between \#'s before the \$. Formally, let $\tilde{\mathcal{U}}_n$ be an NCA for $\tilde{L}_n$. For a word $u \in (0+1+\#)^*$, let S_u be the set of words $w \in (0+1)^n$ in which w appears in u between \#'s, and let $C_u \in 2^{(0+1+\#)^*}$ be the set of words u' such that a word $x \in (0+1)^n$ is in $S_{u'}$ iff $\tilde{x}$ is not in S_u. For example, if $n = 3$ and $u = 0\#000\#10\#010\#110\#$, then $S_u = \{000, 010, 110\}$. Thus, $(0+1)^3 \setminus S_u = \{001, 011, 100, 101, 111\}$, so a possible $u' \in C_u$ is $\#110\#100\#011\#010\#000\#$. Indeed, $S_{u'} = \{110, 100, 011, 010, 000\}$, which contains exactly all the negatives of words not in S_u.

It is easy to see that for all $u \in (0+1+\#)^*$, a word $u\$u'$ such that $u' \in C_u$ is not in L_n and thus should be accepted by $\tilde{\mathcal{U}}_n$. Let Q_u be the set of states that all accepting runs of $\mathcal{U}_n$ over $u\$u'$ reach after reading the \$. In the full version, we prove that if $u_1, u_2 \in (0+1+\#)^n$ are such that $S_{u_1} \neq S_{u_2}$, then it must be that $Q_{u_1} \cap Q_{u_2} = \emptyset$. $\qquad\square$

We now turn to intersection of NCAs. We show that a blow-up in the capacity factor of the underlying automata cannot be avoided. Moreover, it applies even when we construct an NCA for the intersection of a DCA and a DFA.

Theorem 4. *Intersection of NCAs and DCAs is tightly exponential.*

Proof: For the upper bound, given two NCAs (or DCAs), we can construct their intersection by first removing capacities, obtaining exponentially bigger NFAs (or DFAs), and then constructing the product of the latter. Note that this leads to a DFA or an NFA with no capacities.

For the lower bound, we define two sequences of languages $L_1^1, L_2^1, L_3^1, \ldots$ and $L_1^2, L_2^2, L_3^2, \ldots$ such that for all $n \geq 1$, the language L_n^1 can be recognized by a DCA with one state, the language L_n^2 can be recognized by a DFA with $n+1$ states, and the minimal NCA for their intersection requires at least 2^n states.

Let $n \geq 1$, and let $\Sigma_n = \{1, \ldots, n\}$. We define L_n^1 as the set of all words over Σ_n in which every letter $i \in \Sigma_n$ occurs at most once. We define L_n^2 to be the set of all words over Σ_n whose length is n. It is easy to see that indeed L_n^1 can be recognized by a DCA with one state (it has a self-loop transition with capacity 1 for each letter in Σ_n), and L_n^2 can be recognized by a DFA with $n+1$ states. We prove that an NCA for the language $L_n = L_n^1 \cap L_n^2$ needs at least 2^n states. Note that L_n includes exactly all words that form a permutation of Σ_n.

For a set $S \subseteq \Sigma_n$, let w_S be the word obtained by concatenating the letters in S in ascending order, and let $\bar{S} = \Sigma_n \setminus S$. Note that $w_S \cdot w_{\bar{S}} \in L_n$

Consider an NCA $\mathcal{A}$ for L_n. Assume by way of contradiction that $\mathcal{A}$ has less than 2^n states. Then, there are two sets $S, T \subseteq \Sigma_n$ such that $S \neq T$ and there is a state q that is visited by both an accepting run r of $\mathcal{A}$ on $w_S \cdot w_{\bar{S}}$ after it reads w_S and by an accepting run r' of $\mathcal{A}$ on $w_T \cdot w_{\bar{T}}$ after it reads w_T. Recall that r and r' are sequences of transitions. Let $r_S \cdot r_{\bar{S}}$ be a partition of r to the parts that traverse w_S and $w_{\bar{S}}$, respectively, and similarly for $r_T \cdot r_{\bar{T}}$ and r'.

In the full version, we show that the existence of two sets S and T, and a state q as above leads to a contradiction.   □

The union of two NFAs can be easily constructed in linear time by putting the two NFAs "side by side". The product construction can be used to construct a DFA for the union of two DFAs in polynomial time. We now show that while the union of two NCAs can be constructed linearly, the construction of a DCA for the union of two DCAs involves an exponential blow up.

Theorem 5. *Union of NCAs is linear. Union of DCAs is tightly exponential.*

Proof: The union of two NCAs can be constructed linearly, similarly to the union construction for NFAs.

We proceed to DCAs. For the upper bound, given two DCAs, we can construct their union by removing capacities, obtaining exponentially bigger DFAs, and then constructing the product of the latter. As in the case of intersection, this construction ends up with a DFA with no capacities. As we show in the lower bound below, however, a blow-up in the capacity factor of the underlying automata cannot be avoided. Moreover, it applies even when we construct a DCA for the union of a DCA and a DFA.

Consider the two sequences of languages $L_1^1, L_2^1, L_3^1, \ldots$ and $L_1^2, L_2^2, L_3^2, \ldots$ used in the proof of Theorem 4. We prove that a DCA for the language $L_n = L_n^1 \cup L_n^2$ needs at least 2^{n-2} states.

Consider a DCA $\mathcal{A}$ for L_n. Assume by way of contradiction that $\mathcal{A}$ has less than 2^{n-2} states. Then, there are two sets $S, T \subseteq \Sigma_n$ of size at most $n - 2$ such that $S \neq T$, and there is a state q such that both the run of $\mathcal{A}$ on w_S and the run of $\mathcal{A}$ on w_T reach q. Assume w.l.o.g. that there is a letter $i \in S \setminus T$. Consider the transition $\tau = \langle q, i, q' \rangle$. That is, when $\mathcal{A}$ is in state q and reads the letter i, it moves to state q'. Clearly, $w_T \cdot i \in L_n$, so q' is accepting. Also, since $|S| \neq n - 1$, then $w_S \cdot i \notin L_n$. Hence, either q' is not accepting, and we have reached a contradiction, or τ was taken $c(\tau)$ times while w_S is read. Then, however, as $|S| \leq n - 2$, the run of $\mathcal{A}$ on $w_S \cdot i^{n-|S|}$ has to traverse τ more than $c(\tau)$ times, and is thus rejecting. But $w_S \cdot i^{n-|S|}$ is of length n, a contradiction.   □

Note that in all lower-bound proofs we have used NCAs and DCAs of width 1.

4   Decision Procedures

The *universality problem* is to decide, given an automaton $\mathcal{A}$ and an alphabet Σ, whether $L(\mathcal{A}) = \Sigma^*$. The problem is known to be NLOGSPACE-complete for DFAs, and PSPACE-complete for NFAs. The *containment problem* is to decide, given two automata $\mathcal{A}_1$ and $\mathcal{A}_2$, whether $L(\mathcal{A}_1) \subseteq L(\mathcal{A}_2)$. The problem is known to be PSPACE-complete when $\mathcal{A}_2$ is an NFA and NLOGSPACE-complete when $\mathcal{A}_2$ is a DFA.

In this section we study the universality and containment problems for the different classes of CA. We start with the universality problem. In the case of FA, the solution involves a construction of a complementing automaton and checking its emptiness. By [8], the emptiness problem for NCAs is NLOGSPACE complete, as it is for NFAs. Complementation of CAs, however, is expensive (and in fact results in FAs). We show that the blow-up in complementation is carried over to the complexity of the universality problem. Still, in the deterministic setting it is possible to reason directly on the structure of the DCA and circumvent the construction of a complementing automaton.

Theorem 6. *The universality problem for DCA is NLOGSPACE-complete.*

Proof: Since DFAs, for which universality is known to be NLOGSPACE-hard, are a special case of DCAs, the lower bound is immediate.

For the upper bound, we claim that a DCA $\mathcal{D} = \langle \Sigma, Q, q_0, \Delta, F, c \rangle$ is not universal iff there exists a path from q_0 to a state $q \notin F$, or to a state q from which there is no transition labeled σ for some $\sigma \in \Sigma$, or a path from q_0 to a transition $\langle q, a, q' \rangle$ that is part of a cycle and such that $c(\langle q, a, q' \rangle) = k$ for some $k \in \mathbb{N}$. In the first two cases, a path to a rejecting state or to a state from which some letter cannot be read induces a word that is rejected by $\mathcal{D}$. In the third case, a path to a cycle on which $\langle q, a, q' \rangle$ occurs induces a word that gets stuck after at most k traversals through the cycle. Checking all cases can be performed in NLOGSPACE by guessing the path to a rejecting state in the former case, or a path to a finitely capacitated transition $\langle q, a, q' \rangle$ followed by a path from q' to q in the latter case.   □

Theorem 7. *The universality problem for NCAs is EXPSPACE-complete.*

Proof: An NCA can be translated to a DFA with a doubly-exponential blow-up. The upper bound then follows from the NLOGSPACE complexity for universality of DFAs.

For the lower bound, we show a reduction from an exponent version of the *tiling problem*, defined as follows. We are given a finite set T of tiles, two relations $V \subseteq T \times T$ and $H \subseteq T \times T$, an initial tile t_0, a final tile t_f, and a bound $n > 0$. We have to decide whether there is some $m > 0$ and a tiling of a $2^n \times m$-grid such that (1) The tile t_0 is in the bottom left corner and the tile t_f is in the top left corner, (2) A horizontal condition: every pair of horizontal neighbors is in H, and (3) A vertical condition: every pair of vertical neighbors is in V. Formally, we have to decide whether there exist $m \in \mathbb{N}$ and a function $f : \{0, \ldots, 2^n - 1\} \times \{0, \ldots, m-1\} \to T$ such that (1) $f(0,0) = t_0$ and $f(0, m-1) = t_f$, (2) For every $0 \le i \le 2^n - 2$ and $0 \le j \le m-1$, we have that $(f(i,j), f(i+1,j)) \in H$, and (3) For every $0 \le i \le 2^n - 1$ and $0 \le j \le m-2$, we have that $(f(i,j), f(i,j+1)) \in V$. When n is given in unary, the problem is known to be EXPSPACE-complete.

We encode a tiling as a word over $T \cup \{0, 1, \#, \$\}$, consisting of a sequence of rows (each row is of length 2^n). Each row is a sequence of tile blocks, where each tile block contains the tile and its index in the row, as an n-bit counter. Each row starts with $\#$, and the entire grid ends with $\#\$$. Such a word represents a proper tiling if it starts with t_0, has a last row that starts with t_f, every pair of adjacent tiles in a row are in H, and every pair of tiles that are 2^n tiles apart are in V.

We reduce the tiling problem to the universality problem for NCAs. Given a tiling problem $\rho = \langle T, H, V, t_0, t_f, n \rangle$, we construct an NCA $\mathcal{A}_\rho$ that accepts a word w iff w is not an encoding of a legal tiling for ρ. Thus, $\mathcal{A}_\rho$ rejects a word w iff w encodes a legal tiling ρ, and is universal iff no such tiling exists.

The difficulty lies in relating vertical neighbors – tiles that are 2^n far apart. Detecting a violation of the vertical condition amounts to the existence of two tiles t and t' such that $(t, t') \notin V$, and there is a row with tile t in position i, such that the tile in position i in the successive row is t'. Recall the NCA $\mathcal{A}_k$ from Example 1. In order to use it for detecting when the position i repeats in the successive row, we maintain both the counter and its negative. That is, each tile block is of length $1 + 2n$, and consists of the tile, its index i in the row, and the negative $\tilde{i}$. Then, the property we need in order to find a violation of the condition V becomes "there is a row with tile t in position i, such that the tile whose negative position is $\tilde{i}$ in the successive row is t'".

We define $\mathcal{A}_\rho$ as the union of several NCAs, each guessing a different type of violation of an encoding of a legal tiling for ρ. We begin with a violation of the pattern and counters.

- The pattern is not of the form $(\# \cdot (T \cdot (0 + 1)^{2n})^*)^* \# \$$.
- It is not the case that every $\#$ precedes $\$$ or a block with counter value 0.
- It is not the case that $\#$ follows all blocks with counter value $2^n - 1$.
- It is not the case that in every tile block the last n bits are the negation of the n bits before them.
- The counters are not increased properly. We check this separately for even and odd counter values. For even values, there exists a block whose counter value i ends with 0 and either the first $n - 1$ bits of the counter i' in the next block do not agree with the first $n - 1$ bits of i, or the last bit of i' is not 1. For odd counters, there exists a block with counter value i that ends with $0(1^k)$ for some $0 < k < n$ such that the first $n - k - 1$ bits of the counter i' in the next block do not agree with the first $n - k - 1$ bits of i, or the last $k + 1$ bits in i' are not $1(0^k)$.

Secondly, we have violations of the tiling conditions:

- The tiling does not start with t_0.
- The first tile in the last row is not t_f. That is, there is a row that starts with a tile that is different from t_f and the row ends with $\#\$$.
- For a violation of the horizontal condition, namely that there is $0 \le i \le 2^n - 2$ and $0 \le j \le m - 1$ for which $(f(i, j), f(i + 1, j)) \notin H$, we use the union of at most $|T|^2$ NFAs, one for each pair $(t, t') \notin H$. The NFA $\mathcal{A}_{(t,t')}^H$ accepts a word if it has a block with tile t such that the next block in the row (that is, the next letter in T in the word) is t'. Formally, $L(\mathcal{A}_{(t,t')}^H) = \Sigma^* t (0 + 1)^{2n} t' \Sigma^*$.
- For a violation of the vertical condition, namely that there is $0 \le i \le 2^n - 1$ and $0 \le j \le m - 2$ for which $(f(i, j), f(i, j + 1)) \notin V$, we use a union of NCAs, one for each pair $(t, t') \notin V$.

The NCA $\mathcal{A}_{(t,t')}^V$ accepts a word if it has a tile block with t and the tile block with the same counter value in the next row is t'. To check that the counters

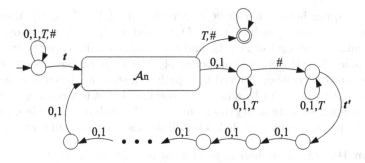

Fig. 3. the NCA $\mathcal{A}^V_{(t,t')}$.

are identical, we check that the negation in the next counter is the negation of the current counter. This can be achieved by using a component similar to $\mathcal{A}_k$ of Example 1, with $k = n$. As described in Fig. 3, the NCA $\mathcal{A}^V_{(t,t')}$ guesses a tile block with t and the counter traverses $\mathcal{A}_n$. After a single occurence of $\#$, it guesses a tile block with t' and the negation of the counter in the tile block traverses $\mathcal{A}_n$ again. If the counter of the two tile blocks is identical, the second traversal of $\mathcal{A}_n$ is successful, and we have indeed identified a violation. Thus, $\mathcal{A}^V_{(t,t')}$ now accepts the rest of the word. Notice that $\mathcal{A}_n$ must be traversed twice in order to accept, as the first letter after the first traversal is the first bit of the negative, causing the run to take the cycle that leads back to $\mathcal{A}_n$.   □

Theorem 8. *The containment problem of a DFA or an NFA in a DCA is NLOGSPACE-complete.*

Proof: The lower bound directly follows from the NLOGSPACE-hardness of the containment problem for DFAs.

For the upper bound, intuitively, we can guess a run on a word w such that $w \in L(A_1)$ and $w \notin L(A_2)$. We do so by running simultaneously, one transition at a time, on A_1 and on the complement of A_2, on the fly. It holds that $w \notin L(A_2)$ iff we reach a rejecting state, or traverse some finitely capacitated transition too many times. For the latter case, we can guess this transition in advance, and track the number of times it is traversed. The full details are in the full version.   □

Theorem 9. *The containment problem of a DCA or an NCA in a DCA or a DFA is co-NP-complete.*

Proof: For the lower bound, we reduce from the Hamiltonian cycle problem for directed graphs. Given a directed graph $G = \langle V, E \rangle$, we construct a DCA $\mathcal{A}_1$ and a DFA $\mathcal{A}_2$ as follows. The DFA $\mathcal{A}_2$ accepts all words over V whose length is at most $2n - 1$. The DCA $\mathcal{A}_1$ is the DCA described in Theorem 6 in [8], which accepts a word of length $2n$ iff G has a Hamiltonian cycle, and in any case, accepts only words whose length is at most $2n$. Therefore, we have that $L(\mathcal{A}_1) \subseteq L(\mathcal{A}_2)$ iff G does not have a Hamiltonian cycle.

For the upper bound, it can be shown that if $L(\mathcal{A}_1) \nsubseteq L(\mathcal{A}_2)$, then there exists a witness word w such that $w \in L(\mathcal{A}_1)$ and $w \notin L(\mathcal{A}_2)$, and whose length is polynomial in the sizes of $\mathcal{A}_1, \mathcal{A}_2$, and the maximal finite capacity in $\mathcal{A}_2$.

Indeed, let $\mathcal{A}$ be the product of $\mathcal{A}_1$ and $\mathcal{A}_2$ which ignores the capacities of $\mathcal{A}_1$. A witness word w is either induced by a path to a state in $\mathcal{A}$ that is rejecting in $\mathcal{A}_1$ and accepting in $\mathcal{A}_2$, or by a path to a state in $\mathcal{A}$ that is accepting in $\mathcal{A}_1$, and traverses some transition in $\mathcal{A}_2$ too many times. In both cases, the length of the path can be polynomially bounded. The full details are in the full version.   □

Theorem 10. *The containment problem of a DCA or an NCA in an NFA is PSPACE-complete.*

Proof: The lower bound follows from the containment problem of a DFA in an NFA.

For the upper bound, intuitively, we can guess a run on a word w such that $w \in L(A_1)$ and $w \notin L(A_2)$. We do so by running simultaneously, one transition at a time, on $\mathcal{A}_1$ and on the complement of $\mathcal{A}_2$. To do so in PSPACE, we run on the latter on-the-fly. The full details are in the full version.   □

Theorem 11. *The containment problem of a DFA, an NFA, a DCA, or an NCA in an NCA is EXPSPACE-complete.*

Proof: The upper bound follows from the doubly-exponential translation of NCA to DFA. Since Σ^* can be recognized by a one-state DFA, the lower bound follows from Theorem 7.   □

5   Conclusion and Future Work

We have shown that while the succinctness that capacities offer often comes with a corresponding increase in the complexity of decision problems for them, there are some cases where it comes for free. In the future, we plan to study the idea of capacities in the formalism of temporal logic. The analogue of a capacitated transition would be *bounding operators* in the logic, which bound the number of prefixes along which a sub-specification may hold. Such operators can conveniently and succinctly bound the number of occurrences of events in a computation. We believe that the classical translation of LTL into automata [11] can be generalized so that LTL with bounded operators are translated to CAs.

References

1. Boigelot, B., Godefroid, P.: Symbolic verification of communication protocols with infinite state spaces using qdds. In: Alur, R., Henzinger, T.A. (eds.) CAV 1996. LNCS, vol. 1102, pp. 1–12. Springer, Heidelberg (1996)
2. Bouajjani, A., Habermehl, P., Vojnar, T.: Verification of parametric concurrent systems with prioritised FIFO resource management. Formal Meth. Syst. Des. **32**(2), 129–172 (2008)

3. Cadilhac, M., Finkel, A., McKenzie, P.: On the expressiveness of parikh automata and related models. In: NCMA, pp. 103–119 (2011)
4. Comon, H., Jurski, Y.: Multiple counters automata, safety analysis and presburger arithmetic. In: Vardi, M.Y. (ed.) CAV 1998. LNCS, vol. 1427, pp. 268–279. Springer, Heidelberg (1998)
5. Droste, M., Kuich, W., Vogler, H. (eds.): Handbook of Weighted Automata. Springer, Berlin (2009)
6. Ford, L.R., Fulkerson, D.R.: Flows in Networks. Princeton University Press, Princeton (1962)
7. Klaedtke, F., Rueß, H.: Monadic second-order logics with cardinalities. In: Baeten, J.C.M., Lenstra, J.K., Parrow, J., Woeginger, G.J. (eds.) ICALP 2003. LNCS, vol. 2719. Springer, Heidelberg (2003)
8. Kupferman, O., Tamir, T.: Properties and utilization of capacitated automata. In: Raman, V., Suresh, S.P. (eds.) FSTTCS 2014. LIPIcs, vol. 29, pp. 33–44. Schloss Dagstuhl-Leibniz-Zentrum fuer Informatik, Dagstuhl (2014)
9. Qiu, R.G., Joshi, S.B.: Deterministic finite capacity automata: a solution to reduce the complexity of modeling and control of automated manufacturing systems. In: CACSD, pp. 218–223 (1996)
10. Rabin, M.O.: Probabilistic automata. Inf. Control 6, 230–245 (1963)
11. Vardi, M.Y., Wolper, P.: Reasoning about infinite computations. Inf. Comput. 115(1), 1–37 (1994)

On Limited Nondeterminism and ACC Circuit Lower Bounds

Holger Spakowski[✉]

Department of Mathematics and Applied Mathematics,
University of Cape Town, Rondebosch 7701, South Africa
Holger.Spakowski@uct.ac.za

Abstract. Williams's celebrated circuit lower bound technique works by showing that the existence of certain small enough nonuniform circuits implies that nondeterministic exponential time can be speeded up in such a way that it implies a contradiction with the nondeterministic time hierarchy.

We apply Williams's technique by speeding up instead (i) *deterministic* exponential-time computations and (ii) nondeterministic exponential-time computations that use only a *limited number of nondeterministic bits*. From (i), we obtain that EXP $\subseteq$ ACC^0 has a consequence that might seem unlikely, while (ii) yields an exponential ACC^0 size-depth tradeoff for $\mathrm{E}^{\mathrm{NP}[2^{n^{c\delta}}]}$, which is the class of exponential-time computation with access to an NP oracle where the number of oracle queries is bounded.

Keywords: Circuit lower bounds · ACC · EXP

1 Introduction

Williams's breaktrough circuit lower bound technique [10,11] can be described as working as follows. For the sake of a contradiction, assume that the desired circuit lower bound does not hold. Show that this implies that we can speed up NTIME$(2^n/n^{10})$ computations in such a way that we get a contradiction with the nondeterministic time hierarchy.

This technique requires that there is a non-trivial satisfiability algorithm for the circuit class that we want to prove lower bounds against. For the case of ACC^0 circuits, Williams [11] provided such an algorithm and thus solves the long-standing open problem of showing NEXP $\not\subseteq$ ACC^0:

Theorem 1 ([11]). *For every depth d and modulus m, there is a $\delta = \delta(d, m)$ such that satisfiability of depth-d ACC circuits with MOD_m gates, n inputs, and 2^{n^δ} size can be determined deterministically in 2^{n-n^δ} time.*

The contribution of this paper is to investigate what happens if we use this approach to speed up instead *deterministic* exponential-time computations or nondeterministic exponential-time computations that use only a *limited number of nondeterministic bits*.

© Springer International Publishing Switzerland 2016
A.-H. Dediu et al. (Eds.): LATA 2016, LNCS 9618, pp. 320–329, 2016.
DOI: 10.1007/978-3-319-30000-9_25

First, we describe a few more details on how Williams achieved the speed-up for $\mathrm{NTIME}(2^n/n^{10})$ computations (assuming the existence of small-enough circuits). Williams's algorithm starts with reducing any problem L in $\mathrm{NTIME}(2^n/n^{10})$ to SUCCINCT-3SAT.[1] SUCCINCT 3SAT is an NEXP-complete version of the classical NP-complete problem 3SAT, where the formulas are succinctly represented by circuits that have these formulas as truth table.

Throughout this paper, we often follow Arora and Barak's presentation [1] of Williams's proof.

Decision Problem SUCCINCT-3SAT [7] (see also [1])
Input: A 3CNF formula ϕ with 2^n clauses and 2^n variables that is described in succinct form by a circuit C of size n^5 as follows.

The circuit C has n inputs and $3n + 3$ outputs. For any $j \in \{0,1\}^n$, $C(j)$ outputs the description $(\mathrm{neg}_1(j), \mathrm{var}_1(j), \mathrm{neg}_2(j), \mathrm{var}_2(j), \mathrm{neg}_3(j), \mathrm{var}_3(j))$ of the jth clause in ϕ. Here, $\mathrm{var}_1(j), \ldots, \mathrm{var}_3(j)$ are the indices of the variables appearing in clause j and $\mathrm{neg}_1(j), \ldots, \mathrm{neg}_3(j)$ are three bits that are 1 iff the corresponding variable appears negated in clause j.
Output: Is the formula ϕ described by C satisfiable?

It has been known for a long time that SUCCINCT-3SAT is NEXP-complete:

Theorem 2 ([7]). SUCCINCT-3SAT *is* NEXP-*complete under polynomial-time many-one reduction.*

However, in order to lose not too much time in the speed up of $\mathrm{NTIME}(2^n/n^{10})$ computations, Williams needs the following stronger reduction to SUCCINCT-3SAT.

Theorem 3 ([10]). *Let* $L \in \mathrm{NTIME}(2^n/n^{10})$. *Then there is a reduction* f *computable in time* $O(n^5)$ *such that for all* $x \in \{0,1\}^n$, $f(x) = C_x$ *is a circuit of size at most* n^5 *with* n *inputs such that*

$$x \in L \iff C_x \in \text{SUCCINCT-3SAT}$$

The *crucial* new property of this stronger reduction is that the number of input bits of circuit C_x equals n. This reduction is obtained as an exponentially scaled-up version of a more efficient Cook-Levin reduction that reduces $\mathrm{NTIME}(t(n))$ languages to 3CNF formulas of length $O(t(n)\log(t(n)))$ [2], which in turn can be shown using the efficient simulation of Turing machines by oblivious 2-tape Turing machines [8]. A key feature of the efficient Cook-Levin reduction, which is needed to obtain that the circuits C_x are of polynomial size, is that each bit of the Cook-Levin reduction can be computed in polylogarithmic time (assuming random access to the input). That this is the case has been observed for instance in [3]. (Note that the size of C_x depends on the time needed to compute each

---

[1] Williams actually takes $L \in \mathrm{NTIME}(2^n)$. As in [1], we instead use $\mathrm{NTIME}(2^n/n^{10})$ so that the circuit C in SUCCINCT-3SAT has exactly n inputs instead of $n + c\log n$ inputs.

bit of the 3CNF formula represented by C_x, while the number of inputs of C_x depends on the length of of the 3CNF formula represented by C_x.)

The next step in Williams algorithm to solve NTIME$(2^n/n^{10})$ faster than allowed by the nondeterministic time hierarchy, is to use the fast ACC SAT algorithm stated in Theorem 1 on a circuit composed of a circuit that is equivalent to C_x and three circuits that succinctly describe satisfying assignments (which are guessed nondeterministically). As the running time of this algorithm is exponential in the number of input bits of the circuit, it is very important that the number of input bits of C_x in this reduction is not larger than $|x|$.

The circuits succinctly describing satisfying assignments needed above exist for instance under the assumption NEXP $\subseteq$ P/*poly*:

Definition 4 [5]**.** SUCCINCT-3SAT *has succinct witnesses if there exists* $c > 0$ *such that for every circuit* $C \in$ SUCCINCT-3SAT *with* n *inputs, there exists a circuit* D *of size at most* n^c *such that the assignment given by the sequence* $D(0^n), \ldots, D(1^n)$ *satisfies the 3CNF formula represented by* C.

Theorem 5 (**[10], follows from** [5])**.** *If* NEXP $\subseteq$ P/*poly then* SUCCINCT-3SAT *has succinct witnesses.*

To make this whole approach work, Williams found a solution to (nondeterministically) convert C_x into an ACC^0 circuit. For this we need only the assumption P $\subseteq$ ACC^0.

This ends the outline of Williams's approach to solve NTIME$(2^n/n^{10})$ problems in NTIME$(o(2^n/n^{10}))$ (under the assumption of the existence of small ACC^0 circuits).

In this paper, we apply the same idea to speed up DTIME$(2^n/n^{10})$ computations (see Sect. 2) and NTIME$(2^n/n^{10})$ computations with bounded number of nondeterministic bits (see Sect. 3).

Of course, since any problem in NTIME$(2^n/n^{10})$ can be efficiently reduced to SUCCINCT-3SAT, also any problem in DTIME$(2^n/n^{10})$ can be efficiently reduced to SUCCINCT-3SAT. Note that the efficient Cook-Levin reduction applied to a problem L in the deterministic class DTIME$(t(n))$ yields a 3CNF formula ϕ of length $O(t(n)\log(t(n)))$ that has a unique satisfying assignment. This unique satisfying assignment can be interpreted as transcript of the deterministic computation and can therefore be computed deterministically in time $poly(t(n))$. This is not very interesting in the case of polynomial-time computation. However, in the context of exponential time computation, in particular in the reduction from DTIME$(2^n/n^{10})$ to SUCCINCT-3SAT, this means that the succinct witness circuits exist under the weaker assumption EXP $\subseteq$ P/*poly* (see Theorem 10). We get as main theorem of Sect. 2 the following theorem:

Theorem 6. *If* EXP $\subseteq$ *ACC^0 then there exist constants* $\delta > 0$ *and* c *such that* DTIME$(2^n/n^{10}) \subseteq$ NTIMEGUESS$(2^{n-n^\delta}, n^c)$.[2]

---

[2] If nothing else is said then ACC^0 stands for the class of languages that have ACC^0 circuits of polynomial size.

The notation NTIMEGUESS($t(n), g(n)$) was introduced by Fortnow and Santhanam [4] for nondeterministic classes with bounded numbers of nondeterministic bits.

Definition 7 [4]. *NTIMEGUESS($t(n), g(n)$) is the class of all languages accepted by nondeterministic Turing machines running in time $O(t(n))$ and using at most $O(g(n))$ nondeterministic bits.*

Theorem 6 is an attempt to improve Williams result $\text{NEXP} \not\subseteq \text{ACC}^0$ to $\text{EXP} \not\subseteq \text{ACC}^0$. It seems unlikely that deterministic exponential time computation can be speeded up this much at the expense of only a polynomial number of nondeterministic bits. It is not clear if it is possible to resolve this question by (for instance) diagonalization or any other techniques.

In Sect. 3, we apply the same construction to speed up the limited nondeterministic class NTIMEGUESS($2^n/n^{10}, 2^{n^{c\delta}}$). Williams showed the following exponential size-depth tradeoff for ACC^0:

Theorem 8 [11]. *For every d and m, there is a $\delta > 0$ and a language in E^{NP} that does not have ACC^0 circuits of depth d and size 2^{n^δ} with MOD_m gates.*

We obtain the following result:

Theorem 9. *There exists a universal constant c such that the following is true. For every d and m, there is a $\delta > 0$ and a language in $\text{E}^{\text{NP}[2^{n^{c\delta}}]}$ that does not have ACC^0 circuits of depth d and size 2^{n^δ} with MOD_m gates.*

This means that we get a tradeoff behavior for classes $\text{E}^{\text{NP}[2^{n^{c\delta}}]}$, which is the class E^{NP} where the number of oracle queries is bounded by $2^{n^{c\delta}}$.

The intuition behind the proof of Theorem 9 is that for the limited nondeterminism class NTIMEGUESS($2^n/n^{10}, 2^{n^{c\delta}}$), the largest accepting computation can be found with only $2^{n^{c\delta}}$ calls to an NP oracle.

2   A Consequence of the Inclusion $\text{EXP} \subseteq \text{ACC}^0$

In this section, we prove the following theorem.

Reminder of Theorem 6. *If $\text{EXP} \subseteq \text{ACC}^0$ then there exist constants $\delta > 0$ and c such that $\text{DTIME}(2^n/n^{10}) \subseteq \text{NTIMEGUESS}(2^{n-n^\delta}, n^c)$.*

Theorem 5 says that $\text{NEXP} \subseteq \text{P}/poly$ implies that SUCCINCT-3SAT has succinct witnesses. It is not known if this follows also from the weaker assumption $\text{EXP} \subseteq \text{P}/poly$, but we can show that $\text{EXP} \subseteq \text{P}/poly$ implies that all SUCCINCT-3SAT instances *obtained by reduction from problems in EXP* have succinct witnesses, and this is all need to prove Theorem 6. More precisely, we can show the following theorem.

Theorem 10. *Suppose $\text{EXP} \subseteq \text{P}/poly$. Let $L \in \text{DTIME}(2^n/n^{10})$. Then there exists a reduction f computable in deterministic time $O(n^5)$ and $c_1 > 0$ such that for any $x \in \{0,1\}^n$, $f(x) = C_x$ is a circuit of size at most n^5 with n inputs and*

324   H. Spakowski

- If $x \in L$ then $C_x \in$ SUCCINCT-3SAT *and there exists a circuit D of size at most n^{c_1} such that the assignment given by the sequence $D(0^n), \ldots, D(1^n)$ satisfies the 3CNF formula represented by C_x.*
- If $x \notin L$ then $C_x \notin$ SUCCINCT-3SAT.

Proof. The reduction f is basically the same as the one in Theorem 3. The difference is that L is now in a deterministic class. Hence we now consider the efficient Cook-Levin reduction of [2] when applied to languages in the deterministic class DTIME($t(n)$). We still get a formula ψ of length $O(t(n) \log(t(n)))$. However, all inputs of ψ originate only from the Tseitin transformation converting boolean circuits to 3CNF formulas. That is, the formula ψ obtained by reduction from DTIME($t(n)$) does not contain any inputs that represent nondeterministic bits of nondeterministic Turing machines. (See the construction in [2]. The second binary input I_2 is not needed here.) Hence the formulas ψ have a unique satisfying assignment that is given by the values of the gates of the circuit that is constructed as intermediate step in the efficient Cook-Levin reduction [2]. Therefore, this assignment can be computed deterministically in time $O(t(n))$.

Now we obtain the succinct witness circuit D for C_x as follows. Define a deterministic exponential-time Turing machine $M(x, i)$ as follows. $M(x, i)$ determines the satisfying assignment of the 3CNF formula represented by C_x and accepts if and only if the variable with index i is set to true. Because of our assumption EXP $\subseteq$ P/*poly*, M can be simulated by a circuit C of size *poly*($|x|$). By hardwiring x into the inputs of C, we obtain a circuit D of size $|x|^{c_1}$ that describes the satisfying assignment. □

Lemma 11 (Folklore, see [1,11]). *Suppose P $\subseteq ACC^0$. Then there exist constants d_0, c_2, and m such that for every (unrestricted) circuit C of size s there exists an equivalent ACC^0 circuit C' of depth at most d_0 and size s^{c_2} with MOD_m gates.*

Proof. If P $\subseteq ACC^0$ then in particular the problem CIRCUIT-EVAL, which given C and x, determines the output of circuit C on input x, is in ACC^0. The circuit C' equivalent to C is obtained by hardwiring the description of C in the ACC^0 circuit for CIRCUIT-EVAL. □

The following lemma is implicit in [11].

Lemma 12. *Suppose P $\subseteq ACC^0$. Then there exists a constant δ such that the following is true. There exists a nondeterministic algorithm that given two (unrestricted) circuits C and D of size at most s with n inputs, checks if they are equivalent, runs in $O(2^{n-n^\delta} s)$ time and uses at most poly(s) nondeterministic bits.*

Proof. We assume P $\subseteq ACC^0$. For every gate i of C, we guess an ACC^0 circuit C'_i with n inputs of depth d_0 and size s^c with MOD_m gates that is equivalent to the value of the ith gate of C, i.e., for all $x \in \{0,1\}^n$, $C'_i(x)$ equals the value of the ith gate in $C(x)$ (needs poly(s) nondeterministic bits). These circuits exist by Lemma 11. If j is the output gate of C then $C' := C'_j$ is an ACC^0 circuit that is equivalent to C. The correctness of the guessed circuits can be checked gate by gate with the help of Williams's ACC^0 satisfiability algorithm stated in Theorem 1 as follows.

Let i be any gate of C. If gate i is an AND of gates j and k then we have to check if for all $x \in \{0,1\}^*$, $C'_i(x) = C'_j(x) \wedge C'_k(x)$. Construct an ACC^0 circuit E such that for all $x \in \{0,1\}^n$, $E(x) = C'_i(x) \leftrightarrow C'_j(x) \wedge C'_j(x)$. Then we have $\neg E$ is unsatisfiable if and only if for all $x \in \{0,1\}^n$, $C'_i(x) = C'_j(x) \wedge C'_k(x)$. To check that $\neg E$ is unsatisfiable, we can use Williams's ACC^0 satisfiability algorithm, which needs 2^{n-n^δ} deterministic time. The case that gate i is an OR gate or a negation gate is treated analogously. Repeating this for all s gates in C needs $O(2^{n-n^\delta} s)$ determinstic time, where $\delta = \delta(d_0, m)$ from Theorem 1.

Next, we apply the same procedure to circuit D to obtain an equivalent ACC^0 circuit D'.

Finally, we construct an ACC^0 circuit F that is equivalent to $C' \leftrightarrow D'$. Then we have that C' and D' are equivalent if and only if $\neg F$ is unsatisfiable. Checking that $\neg F$ is unsatisfiable can again be done with Williams's ACC^0 satisfiability algorithm in time $O(2^{n-n^\delta})$. $\qquad\square$

Proof of Theorem 6. Suppose $EXP \subseteq ACC^0$. Let $L \in DTIME(2^n/n^{10})$. We will show $L \in NTIMEGUESS(2^{n-n^\delta}, n^c)$ for some δ and c. Let $x \in \{0,1\}^n$ be any input for L. Let C_x be the circuit of size at most n^5 constructed according to Theorem 10. Guess a circuit D of size at most n^{c_1} describing a satifying assignment for C_x. It remains to check that D indeed describes a satisfying assignment. To this end we use the construction by Williams [10] (we follow [1]). Construct an (unrestricted) circuit G such that for each clause index $j \in \{0,1\}^n$,

$$G(j) = [D(\mathrm{var}_1(j)) \oplus \mathrm{neg}_1(j)] \vee [D(\mathrm{var}_2(j)) \oplus \mathrm{neg}_2(j)] \vee [D(\mathrm{var}_3(j)) \oplus \mathrm{neg}_3(j)],$$

where $(\mathrm{neg}_1(j), \mathrm{var}_1(j), \mathrm{neg}_2(j), \mathrm{var}_2(j), \mathrm{neg}_3(j), \mathrm{var}_3(j))$ is the output of $C_x(j)$ as described in the definition of SUCCINCT-3SAT.

Circuit G constists of 3 copies of D and one copy of C_x. Hence G has size at most $3n^{c_1} + n^5 + O(1)$. By Lemma 11, there exist constants c_2 and d_0 such that there is an ACC^0 circuit G' of depth d_0 and size n^{c_2} with MOD_m gates that is equivalent to G. Guess this circuit G'. (This takes no more than n^{c_2+1} nondeterministic bits.) Use Lemma 12 to verify that G' is indeed equivalent to G. This takes $O(2^{n-n^\delta} n^{c_2})$ time and uses at most $poly(n^{c_2})$ nondeterministic bits. Finally, use Williams's ACC^0 satisfiability algorithm to check that $\neg G'$ is unsatisfiable, which can be done in deterministic time $O(2^{n-n^\delta} n^{c_2})$.

Altogether, we obtain a nondeterministic algorithm that determines if $x \in L$, runs in time $O(2^{n-n^\delta})$, and uses at most n^c nondeterministic bits for some constants c and δ. Hence we have shown that $L \in NTIMEGUESS(2^{n-n^\delta}, n^c)$. $\square$

3   An Exponential Size-Depth Tradeoff for $E^{NP[2^{n^{c\delta}}]}$

In this section, we prove the following theorem.

Reminder of Theorem 9. *There exists a universal constant c such that the following is true. For every d and m, there is a $\delta > 0$ and a language in $E^{NP[2^{n^{c\delta}}]}$ that does not have ACC^0 circuits of depth d and size 2^{n^δ} with MOD_m gates.*

The following theorem is an analogon to Theorem 10 for the case of bounded nondeterminism, and where the assumption on the circuit size is 2^{n^δ} instead of polynomial. Note however that in this case we assume the existence of ACC^0 circuits and obtain witness circuits D that are already ACC^0 circuits.

Theorem 13. *For every $c > 0$, the following is true.*

Suppose $E^{NP[2^{n^{c\delta}}]}$ has ACC^0 circuits of depth d and size 2^{n^δ} with MOD_m gates. Let $L \in NTIMEGUESS(2^n/n^{10}, 2^{n^{c\delta}})$. Then there exists a reduction f computable in deterministic time $O(n^5)$ such that for any $x \in \{0,1\}^n$, $f(x) = C_x$ is a circuit of size at most n^5 with n inputs and

- *If $x \in L$ then $C_x \in SUCCINCT\text{-}3SAT$ and there exists an ACC^0 circuit D of depth d and size at most 2^{n^δ} with MOD_m gates such that the assignment given by the sequence $D(0^n), \ldots, D(1^n)$ satisfies the 3CNF formula represented by C_x.*
- *If $x \notin L$ then $C_x \notin SUCCINCT\text{-}3SAT$.*

Proof. The proof uses the same ideas as the one for Theorem 10. The difference is that now instead of zero, the computation has $2^{n^{c\delta}}$ nondeterministic bits. Hence the formulas ψ obtained from the efficient Cook-Levin reduction contain not only variables originating from the Tseitin transformation, but also $2^{n^{c\delta}}$ variables corresponding to the nondeterministic bits in $NTIMEGUESS(2^n/n^{10}, 2^{n^{c\delta}})$. However, for any satisfying assignment of ψ, once the variable assignment for the $2^{n^{c\delta}}$ variables corresponding to the nondeterministic bits are determined, the remaining variables (i.e., the variables from the Tseitin transformation) are uniquely determined and can be computed in deterministic exponential time.

Now we obtain the succinct witness circuit D for C_x as follows. Define a deterministic exponential-time Turing machine $M(x, i)$ that makes $2^{n^{c\delta}}$ queries to an NP oracle as follows. As in [11, Fact 2], $M(x, i)$ determines the lexicographically smallest satisfying assignment of the 3CNF formula represented by C_x and accepts if and only if the variable with index i is set to true. Because the number of variables corresponding to nondeterministic bits is limited to $2^{n^{c\delta}}$ (and the remaining variables can be computed deterministically once these are determined), $M(x, i)$ needs in our case only $2^{n^{c\delta}}$ queries to an NP oracle.

Because of our assumption that $E^{NP[2^{n^{c\delta}}]}$ has ACC^0 circuits of depth d and size 2^{n^δ} with MOD_m gates, M can be simulated by such an ACC^0 circuit C. By hardwiring x into the inputs of C, we obtain an ACC^0 circuit D of depth d and size 2^{n^δ} with MOD_m gates that describes (the smallest) satisfying assignment.   □

Lemma 14 (Folklore, See [1,11]). *Suppose* P *has* ACC^0 *circuits of depth* d *and size* 2^{n^δ} *with* MOD_m *gates. Then every (unrestricted) circuit* C *of size* s *there exists an equivalent* ACC^0 *circuit* C' *of depth* d *and size at most* $2^{(s^2)^\delta} = 2^{s^{2\delta}}$ *with* MOD_m *gates.*

Proof. The proof is similar to the proof of Theorem 11.

If P has ACC^0 circuits of depth d and size 2^{n^δ} with MOD_m gates then in particular the problem CIRCUIT-EVAL, which given C and x, determines the output of circuit C on input x has such circuits. The circuit C' equivalent to C is obtained by hardwiring the description of C in the ACC^0 circuit for CIRCUIT-EVAL. Note that the description of a circuit of size s together with the input x needs no more than $O(s \log s) + n = O(s^2)$ bits. $\qquad\square$

Proof of Theorem 9. We prove the theorem for $c = 12$. Fix any depth d and modulus m. Let $\delta = \delta(2d + 5, m)/11$, where $\delta(\cdot, \cdot)$ is the function from Williams's satisfiability algorithm (Theorem 1). To get a contradiction, suppose $\mathrm{E}^{\mathrm{NP}[2^{n^{12\delta}}]}$ has ACC^0 circuits of depth d and size 2^{n^δ} with MOD_m gates. Let L be any language in $\mathrm{NTIMEGUESS}(2^n/n^{10}, 2^{n^{12\delta}})$. We will now show $L \in \mathrm{NTIMEGUESS}(2^{n-n^\delta} n^5, 2^{n^{11\delta}})$. Let $x \in \{0,1\}^n$ be any input for L. Let C_x be the circuit constructed according to Theorem 13. Guess an ACC^0 circuit D of depth d and size at most 2^{n^δ} with MOD_m gates such that the assignment given by the sequence $D(0^n), \ldots, D(1^n)$ satisfies the 3CNF formula represented by C_x. (Such a circuit exists by Theorem 13.) It remains to describe how to verify that D indeed describes a satisfying assignment for C_x.

First, we need to find an ACC^0 circuit C'_x that is equivalent to C_x.[3] Since by our assumption, $\mathrm{E}^{\mathrm{NP}[2^{n^{12\delta}}]}$ has ACC^0 circuits of depth d and size 2^{n^δ} with MOD_m gates, P also has such circuits. Hence by Lemma 14, for each gate i of C_x there exists an ACC^0 circuit $C_x^{i'}$ of depth d and size $\leq 2^{(n^5)^{2\delta}} = 2^{n^{10\delta}}$ with MOD_m gates that is equivalent to the value of the ith gate of C_x, i.e., for each $y \in \{0,1\}^n$, $C_x^{i'}(y)$ equals the ith gate of $C_x(y)$. (Note that C_x has size at most n^5.) We obtain C'_x as follows.

For each gate i of C_x, we guess the corresponding above-described ACC^0 circuit $C_x^{i'}$. This needs no more than $2^{n^{11\delta}}$ nondeterministic bits. The correctness of the guessed circuits can be verified by going through the gates of C_x gate by gate with the help of Williams's ACC^0 circuit satisfiability algorithm stated in Theorem 1 as follows.

Let i be any gate of C_x. If gate i is an AND of gates j and k then we have to check if for each $y \in \{0,1\}^n$, $C_x^{i'}(y) = C_x^{j'}(y) \wedge C_x^{k'}(y)$. To this end, construct an ACC^0 circuit E such that for all $y \in \{0,1\}^n$, $E(y) = C_x^{i'}(y) \leftrightarrow C_x^{j'}(y) \wedge C_x^{j'}(y)$. Then we have $\neg E$ is unsatisfiable if and only if for all $y \in \{0,1\}^n$, $C_x^{i'}(y) = C_x^{j'}(y) \wedge C_x^{k'}(y)$. To check that $\neg E$ is unsatisfiable, we can use Williams's ACC^0

---

[3] Alternatively, C'_x could have been constructed to be ACC^0 right away using the recent result by Jahanjou, Miles, and Viola [6], but we don't know if the required witness circuit can be shown to exist also in this case.

circuit satisfiability algorithm, which needs no more than 2^{n-n^δ} deterministic time. (Note that E has size $\leq 2^{n^{11\delta}}$.) The case that gate i is an OR gate or a negation gate is treated analogously. Repeating this for all n^5 gates of C_x needs $O(2^{n-n^\delta}n^5)$ deterministic time. The desired circuit C'_x is composed of all those circuits $C_x^{j'}$ such that j is an output gate of C_x. Hence C'_x is an ACC^0 circuit with n inputs, $3n+3$ outputs, depth d, size at most $(3n+3)2^{n^{10\delta}}$ with MOD_m gates.

It remains to verify that circuit D guessed above describes a satisfying assignment for C'_x. We use the construction by Williams [10] (we follow [1]). We construct an ACC^0 circuit G such that for each clause index $j \in \{0,1\}^n$,

$$G(j) = [D(\mathrm{var}_1(j)) \oplus \mathrm{neg}_1(j)] \vee [D(\mathrm{var}_2(j)) \oplus \mathrm{neg}_2(j)] \vee [D(\mathrm{var}_3(j)) \oplus \mathrm{neg}_3(j)],$$

where $(\mathrm{neg}_1(j), \mathrm{var}_1(j), \mathrm{neg}_2(j), \mathrm{var}_2(j), \mathrm{neg}_3(j), \mathrm{var}_3(j))$ is the output of $C'_x(j)$ as described in the definition of SUCCINCT-3SAT.

Circuit G consists of 3 copies of D and one copy of C'_x. Hence G has size at most $3 \cdot 2^{n^\delta} + (3n+3)2^{n^{10\delta}} + O(1) \leq 2^{n^{11\delta}}$ and depth at most $2d+5$. Finally, use Williams's ACC^0 satisfiability algorithm to check that $\neg G$ is unsatisfiable. This takes deterministic time $O(2^{n-n^\delta})$.

Altogether, we obtain a nondeterministic algorithm that determines if $x \in L$, runs in time $O(2^{n-n^\delta}n^5)$, and uses no more than $2^{n^{11\delta}}$ nondeterministic bits. This means $L \in \mathrm{NTIMEGUESS}(2^{n-n^\delta}n^5, 2^{n^{11\delta}})$.

We have therefore shown

$$\mathrm{NTIMEGUESS}(2^n/n^{10}, 2^{n^{12\delta}}) \subseteq \mathrm{NTIMEGUESS}(2^{n-n^\delta}n^5, 2^{n^{11\delta}}),$$

which is a contradiction because

$$\mathrm{NTIMEGUESS}(2^n/n^{10}, 2^{n^{12\delta}}) \not\subseteq \mathrm{NTIMEGUESS}(2^{n-n^\delta}n^5, 2^{n^{11\delta}})$$

can be shown by delayed diagonalization [9,12]. □

References

1. Arora, S., Barak, B.: Computational complexity: a modern approach. Web addendum. http://theory.cs.princeton.edu/uploads/Compbook/accnexp.pdf
2. Cook, S.: Short propositional formulas represent nondeterministic computations. Inf. Process. Lett. **26**(5), 269–270 (1988)
3. Fortnow, L., Lipton, R., van Melkebeek, D., Viglas, A.: Time-space lower bounds for satisfiability. J. ACM **52**(6), 835–865 (2005)
4. Fortnow, L., Santhanam, R.: Robust simulations and significant separations. In: Aceto, L., Henzinger, M., Sgall, J. (eds.) ICALP 2011, Part I. LNCS, vol. 6755, pp. 569–580. Springer, Heidelberg (2011)
5. Impagliazzo, R., Kabanets, V., Wigderson, A.: In search of an easy witness: exponential time vs. probabilistic polynomial time. J. Comput. Syst. Sci. **65**(4), 672–694 (2002)

6. Jahanjou, H., Miles, E., Viola, E.: Local reductions. In: Halldórsson, M.M., Iwama, K., Kobayashi, N., Speckmann, B. (eds.) ICALP 2015. LNCS, vol. 9134, pp. 749–760. Springer, Heidelberg (2015)
7. Papadimitriou, C., Yannakakis, M.: A note on succinct representations of graphs. Inf. Control **71**(3), 181–185 (1986)
8. Pippenger, N., Fischer, M.: Relations among complexity measures. J. ACM **26**(2), 361–381 (1979)
9. Seiferas, J., Fischer, M., Meyer, A.: Separating nondeterministic time complexity classes. J. ACM **25**(1), 146–167 (1978)
10. Williams, R.: Improving exhaustive search implies superpolynomial lower bounds. SIAM J. Comput. **42**(3), 1218–1244 (2013)
11. Williams, R.: Nonuniform ACC circuit lower bounds. J. ACM **61**(1), 2:1–2:32 (2014)
12. Žák, S.: A Turing machine time hierarchy. Theoret. Comput. Sci. **26**(3), 327–333 (1983)

The Complexity of Induced Tree Reconfiguration Problems

Kunihiro Wasa[1(✉)], Katsuhisa Yamanaka[2], and Hiroki Arimura[1]

[1] Graduate School of Information Science and Technology, Hokkaido University,
Hokkaido, Japan
{wasa,arim}@ist.hokudai.ac.jp
[2] Faculty of Engineering, Iwate University, Iwate, Japan
yamanaka@cis.iwate-u.ac.jp

Abstract. A *reconfiguration problem* asks when we are given two feasible solutions A and B, whether there exists a *reconfiguration sequence* $(A_0 = A, A_1, \ldots, A_\ell = B)$ such that (i) $A_0, \ldots, A_\ell$ are feasible solutions and (ii) we can obtain A_i from A_{i-1} under the prescribed rule (the *reconfiguration rule*) for each $i = 1, \ldots, \ell$. In this paper, we address the reconfiguration problem for induced trees, where an induced tree is a connected and acyclic induced graph of an input graph. This paper treats the following two rules as the prescribed rules: TOKEN JUMPING; removing u from an induced tree and adding v to the tree, and TOKEN SLIDING; removing u from an induced tree and adding v adjacent to u to the tree, where u and v are vertices in an input graph. As the main results, we show (I) the reconfiguration problem is PSPACE-complete, (II) the reconfiguration problem is W[1]-hard when parameterized by both the size of induced trees and the length of the reconfiguration sequence, and (III) there exists an FPT algorithm when parameterized by both the size of induced trees and the maximum degree of an input graph, under each of TOKEN JUMPING and TOKEN SLIDING.

Keywords: Reconfiguration problem · Induced tree · PSPACE-complete · W[1]-hard · FPT

1 Introduction

A *reconfiguration problem* is the following problem: given two feasible solutions A and B for a search problem $\mathscr{P}$, asking whether there exists a *reconfiguration sequence* $(A_0 = A, A_1, \ldots, A_\ell = B)$ such that (i) $A_0, \ldots, A_\ell$ are feasible solutions and (ii) we can obtain A_i from A_{i-1} under the prescribed rule (*reconfiguration rule*) for each $i \in \{1, \ldots, \ell\}$. For reconfiguration problems of vertex-subset problems on graphs, in which feasible solutions are subsets of vertex set of input graphs, the following three reconfiguration rules are usually considered. For the

This work was supported by JSPS Grant-in-Aid for Scientific Research on Innovative Areas 24106007, Scientific Research(C) 25330001, and JSPS Fellows 25 · 1149.

A.-H. Dediu et al. (Eds.): LATA 2016, LNCS 9618, pp. 330–342, 2016.
DOI: 10.1007/978-3-319-30000-9_26

current feasible solution F and two vertices $u \in F$ and $v \notin F$, (1) TOKEN SLIDING(TS [7]): Removing u from F and adding v adjacent to u to F. (2) TOKEN JUMPING(TJ [14]): Removing u from F and adding v to F. (3) TOKEN ADDITION/REMOVAL(TAR [9]): Removing u from F or adding v to F.

In this decade, for a large number of NP-complete problems, their reconfiguration problems have been shown to be PSPACE-complete [8,10,12,14]. However, interestingly, we can not determine the computational complexity of the reconfiguration version of $\mathscr{P}$ by the complexity of $\mathscr{P}$. For example, the 3-coloring problem for a general graph is NP-complete, however, it is known that the 3-coloring reconfiguration problem [2] is in P. On the other hand, the shortest path problem is in P, however, the shortest path reconfiguration problem [1] is PSPACE-complete. Kamiński et al. [14] showed that given a perfect graph, the independent set reconfiguration problem is PSPACE-complete under TS, TJ, or TAR. On the other hand, they also developed polynomial-time algorithm for the reconfiguration problem when an input graph is even-hole-free or P_4-free.

Some notable theorems of the computational complexity of the reconfiguration problems are shown. Gopalan et al. [6] investigated the st-connectivity problem of 3SAT. This problem asks, given two feasible solutions $\mathbf{s}$ and $\mathbf{t}$ of a Boolean formula W with n variables, whether or not $\mathbf{s}$ and $\mathbf{t}$ are connected. Roughly speaking, $\mathbf{s}$ and $\mathbf{t}$ are connected if $\mathbf{t}$ can be obtained from $\mathbf{s}$ by repeatedly flipping the value of a variable such that the resulting assignment also satisfies W. Their dichotomy theorem states that the st-connectivity problem is in P if W is tight or Schaefer [17]; otherwise is PSPACE-complete. Here, W is tight if (1) every connected component in the n-dimensional hypercube of all tuples in W is bijunctive (componentwise bijunctive), (2) $(x \vee y)$ is not definable from W by fixing $n - 2$ variables (OR-free), or (3) $(\overline{x} \vee \overline{y})$ is not definable from W by fixing $n - 2$ variables (NAND-free), where x and y are any variables in W. W is Schaefer if W is a 2CNF, HORN, DUAL-HORN, or AFFINE formula. Mouawad et al. [15] showed the trichotomy theorem of the shortest st-reconfiguration problem asking whether or not there exists a reconfiguration sequence with length less than a given value. The theorem says that the shortest st-reconfiguration problem is in P if a given formula W is navigable, is NP-complete if W is tight but not navigable, and is PSPACE-complete otherwise, where W is navigable if (1) W is OR-free and $(x \vee \overline{y} \vee \overline{z})$ is not definable from W by fixing $n - 3$ variables (Horn-free), (2) W is NAND-free and $(\overline{x} \vee y \vee z)$ is not definable from W by fixing $n-3$ variables (Dual-Horn-free), or (3) W is componentwise bijunctive, where x, y, and z are any variables in W. For some problems, the characteristics/restrictions on the instances with which the reconfiguration problem becomes polynomial-time solvable have been studied.

There are some other known results for the complexity of the reconfiguration problems. Hearn and Demaine [7] showed that sliding-block puzzles such as Klotski puzzles are PSPACE-complete. They solved the outstanding problem by Martin Gardner [5] since 1964. Kamiński et al. [13] showed that finding the shortest reconfiguration sequence for the shortest path is NP-hard even when we know the sequence has polynomial length. Mouawad et al. [16] showed the meta-theorem of the hardness of reconfiguration problems for graphs with hereditary

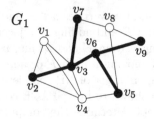

Fig. 1. An example of an induced tree $S_1 = \{v_2, v_3, v_5, v_6, v_7, v_9\}$ in a graph G_1. Bold solid edges indicate that they belong to S_1. Moreover, for any vertex v in $\{v_1, v_4, v_8\}$, $S_1 \cup \{v\}$ has a cycle. Thus, S_1 is also a maximal induced tree.

properties under TAR. In addition, reconfiguration problems for cliques [12], list $L(2,1)$-labelings [11], subset sums [8], list edge-colorings [10], and swapping labeled tokens [18] are studied. However, there are few results for reconfiguration problems for graphs having a *connected* hereditary property.

In this paper, we address the induced tree reconfiguration problem ITRE-CONF under various settings. An *induced tree* is a connected and acyclic induced subgraph in an undirected graph (See Fig. 1) and is well known as a vertex-subset with a connected hereditary property. An informal description of ITRECONF is as follows: Suppose that we are given two distinct induced trees S and T of an input graph and each vertex of S has a token. Then, we obtain T from S by changing the positions of some tokens of S according to the given reconfiguration rule. In our contributions, we first show that ITRECONF is PSPACE-complete under TS and TJ. This is the first hardness result for the induced tree recon-figuration problem. Next, we investigate the problem from the viewpoint of the parameterized complexity. We show that ITRECONF is W[1]-hard when para-meterized by both the size of induced trees and the length of reconfiguration sequences under TJ, and fixed parameter tractable when parameterized by both the size of induced trees and the maximum degree of an input graph under TS and TJ.

Our paper is organized as follows: In Sect. 2, we will give the definitions of terms used in our paper and the definition of ITRECONF. In Sect. 3, we will show the PSPACE-completeness of ITRECONF. In Sect. 4, we will show that ITRECONF is W[1]-hard when parameterized by both the size of induced trees and the length of reconfiguration sequences. In Sect. 5, we will give an FPT algorithm for ITRECONF when parameterized by both the size of induced trees and the maximum degree of an input graph.

2   Preliminaries

2.1   Graphs

An *undirected graph* $G = (V(G), E(G))$ is a pair of a *vertex set* $V(G)$ and an *edge set* $E(G) \subseteq V(G)^2$. In this paper, we assume that G is simple and finite.

For any two vertices u and v in $V(G)$, u and v are *adjacent* in G if $(u, v) \in E(G)$. $N_G(u) = \{v \in V(G) \mid (u, v) \in E(G)\}$ denotes the set of the adjacent vertices of u in G. We define the *degree* $d_G(u)$ of $u \in V(G)$ as the number of vertices adjacent to u. For any vertex subset $S \subseteq V(G)$, $N_G(S) = \left(\bigcup_{u \in S} N_G(u) \right) \setminus S$. In what follows, we omit the subscript G and fix a graph $G = (V(G), E(G))$ if it is clear from the context.

Suppose that, for any two vertices u and v in $V(G)$, $\pi(u, v) = (v_1 = u, \ldots, v_j = v)$ is a sequence of vertices. $\pi(u, v)$ is a *path* from u to v if vertices in $\pi(u, v)$ are distinct and for every $i = 1, \ldots, j - 1$, $(v_i, v_{i+1}) \in E(G)$. The *length* $|\pi(u, v)|$ of a path $\pi(u, v)$ is the number of edges in $\pi(u, v)$. $\pi(u, v)$ is called a *cycle* if $|\pi(u, v)| \geq 3$, $u = v$, vertices in $\pi(u, v)$ are distinct other than u and v, and $(v_i, v_{i+1}) \in E(G)$ for every $i = 1, \ldots, j - 1$. We say that G is *acyclic* if G has no cycle. G is *connected* if for any pair of vertices in G, there exists a path between them. G is a *tree* if G is acyclic and connected.

Let S be a subset of $V(G)$. $G[S] = (S, E[S])$ denotes the graph *induced* by S, where $E[S] = \{(u, v) \in E(G) \mid u, v \in S\}$. We call $G[S]$ the *induced subgraph* of G. If no confusion, we identify S with $G[S]$. We say that S is an *induced tree* if S forms a tree. Moreover, an induced tree S is *maximal* if there exists no induced tree S' such that $S \subseteq S'$ (See Fig. 1). A *connected component* of G is a maximal connected induced subgraph of G.

2.2 Reconfiguration Problem

We firstly consider the following function $f : 2^{V(G)} \times V(G) \times V(G) \to 2^{V(G)}$: $f(S, u, v) = (S \setminus \{u\}) \cup \{v\}$, where $u \in S$ and $v \notin S$. For any two induced trees S and T in G, S and T are *adjacent* each other if there exist two vertices u and v satisfying $f(S, u, v) = T$. That is, for any induced tree S, T is an induced tree adjacent to S if we can obtain T by removing $u \in S$ from S and adding $v \notin S$ to S. We refer f as TOKEN JUMPING (TJ) and f is one of *reconfiguration rules* (we will introduce another rule, TOKEN SLIDING (TS) in Sect. 3.3). Next, we construct a reconfiguration graph under TJ. Let $\mathcal{IT}(G)$ be the set of induced trees of G. A *reconfiguration graph* $R_{\mathrm{IT}}(G) = (\mathcal{IT}(G), \mathcal{E}(G, f))$ is a pair of the set $\mathcal{IT}(G)$ and the set $\mathcal{E}(G, f)$. Here, $\mathcal{E}(G, f) = \{(S, T) \in \mathcal{IT}(G)^2 \mid \exists u \in S, \exists v \notin S \, (f(S, u, v) = T)\}$, that is, each edge in $\mathcal{E}(G, f)$ is a pair of two induced trees that are adjacent. For any two induced trees S and T, a *reconfiguration sequence* from S to T is a path $\pi(S, T)$ on $R_{\mathrm{IT}}(G)$. Figure 2 shows an example of a reconfiguration sequence from S_1 to T_1 on G_1. Now, we give the definition of the induced tree reconfiguration problem ITRECONF.

Problem 1 (Induced Tree Reconfiguration Problem). Given a graph G and two induced trees S and T in G, ITRECONF(G, S, T) asks whether there exists a reconfiguration sequence from S to T on $R_{\mathrm{IT}}(G)$ under TJ.

3 PSPACE-Completeness

In this section, we show ITRECONF is PSPACE-complete. To prove the completeness, we demonstrate that we reduce from 3SAT reconfiguration problem [6]

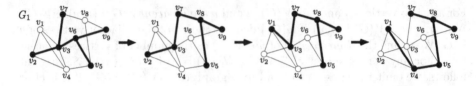

Fig. 2. An example of a reconfiguration sequence. The above figure shows that there exists a reconfiguration sequence $\pi(S_1, T_1) = (S_1, S_2, S_3, S_4 = T_1)$ between two induced trees $S_1 = \{v_2, v_3, v_5, v_6, v_7, v_9\}$ and $T_1 = \{v_1, v_4, v_5, v_7, v_8, v_9\}$ on an input graph G_1, where $S_2 = f(S_1, v_6, v_8)$, $S_3 = f(S_2, v_2, v_1)$, and $S_4 = f(S_3, v_3, v_4)$.

to ITRECONF in polynomial time. In Sect. 3.1, at first, we give the definition of 3SAT reconfiguration problem. Next, in Sect. 3.2, we give a polynomial time reduction. In Sects. 3.3 and 3.4, we show that some variations of ITRECONF are also PSPACE-complete.

3.1 3SAT Reconfiguration Problem

Let $W = C_1 \wedge \cdots \wedge C_m$ be an instance formula of 3SAT on a variable set $X = \{x_1, \ldots, x_n\}$, where C_i is a disjunctive clause consisting of three literals for each $i = 1, \ldots, m$. We say that $\mathbf{s} = (s_1, \ldots, s_n)$ is a *satisfying assignment* if for each $j = 1, \ldots, n$, s_j is a truth value for x_j, and for each $i = 1, \ldots, m$, at least one literal in C_i is true, that is, W is true.

Now we consider the following function $g : \{0,1\}^n \times [n] \rightarrow \{0,1\}^n$: $g(\mathbf{s}, j) = (s_1, \ldots, s_{j-1}, \overline{s_j}, s_{j+1}, \ldots, s_n)$, where $[n] = \{1, \ldots, n\}$. We say that two satisfying assignments $\mathbf{s}$ and $\mathbf{t}$ are *adjacent* if there exists a positive integer j such that $g(\mathbf{s}, j) = \mathbf{t}$. That is, $\mathbf{s}$ is adjacent to $\mathbf{t}$ if the hamming distance between $\mathbf{s}$ and $\mathbf{t}$ is exactly one. In a similar way of a reconfiguration graph of induced trees of G, we define a reconfiguration graph $R_{3SAT}(W) = (\mathcal{SA}(W), \mathcal{E}(W, g))$ of W, where $\mathcal{SA}(W)$ is the set of satisfying assignments of W and $\mathcal{E}(W, g) = \{(\mathbf{s}, \mathbf{t}) \in \mathcal{SA}(W)^2 \mid \exists j \in [n](g(\mathbf{s}, j) = \mathbf{t})\}$. That is, for any two satisfying assignments $\mathbf{s}$ and $\mathbf{t}$ of W, an edge $(\mathbf{s}, \mathbf{t})$ belongs to $\mathcal{E}(W, g)$ if and only if $\mathbf{s}$ and $\mathbf{t}$ are adjacent to each other. We also say that a path between $\mathbf{s}$ and $\mathbf{t}$ on $R_{3SAT}(W)$ is a reconfiguration sequence between them. In the following, we show the definition and the hardness result of the reconfiguration problem 3SATRECONF.

Problem 2 (3SAT Reconfiguration Problem). Given an instance W of 3SAT and two satisfying assignments $\mathbf{s}$ and $\mathbf{t}$ of W, 3SATRECONF$(W, \mathbf{s}, \mathbf{t})$ asks whether there exists a reconfiguration sequence from $\mathbf{s}$ to $\mathbf{t}$ on $R_{3SAT}(W)$.

Theorem 3 (Gopalan *et al.* [6]). 3SATRECONF *is PSPACE-complete.*

3.2 Polynomial-Time Reduction

In this subsection, we will give a polynomial-time reduction from 3SATRECONF to ITRECONF. In what follows, we fix $W = C_1 \wedge \cdots \wedge C_m$ to an instance of 3SAT

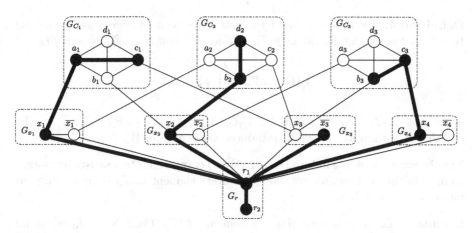

Fig. 3. An example of $H(W_1)$ for a formula $W_1 = C_1 \wedge C_2 \wedge C_3$ on a variable set $\{x_1, \ldots, x_4\}$. Here, $C_1 = (x_1 \vee x_2 \vee x_3)$, $C_2 = (\overline{x_1} \vee x_2 \vee x_3)$, and $C_3 = (\overline{x_2} \vee x_3 \vee x_4)$. Bold vertices and edges indicate a satisfying induced tree in $\mathscr{T}_{W_1}(\mathbf{s_1})$ consisting of $V(G_r) \cup V_{\mathbf{s_1}} \cup \{a_1, c_1, b_2, d_2, b_3, c_3\}$, where $\mathbf{s_1} = (1, 1, 0, 1)$.

on a variable set $X = \{x_1, \ldots, x_n\}$ and $\mathbf{s}$ and $\mathbf{t}$ to two satisfying assignments for W. In addition, let $L(X) = \{x_1, \ldots, x_n, \overline{x_1}, \ldots, \overline{x_n}\}$ be the set of literals of W and $C(W) = \{C_1, \ldots, C_m\}$ be the set of clauses of W.

By combining the following three gadgets, we construct an input graph $H(W)$ of ITRECONF (See Fig. 3). (1) *Variable gadget* G_{x_j}: For each variable $x_j \in X$, $G_{x_j} = (V(G_{x_j}), E(G_{x_j}))$ consists of a vertex set $V(G_{x_j}) = \{x_j, \overline{x_j}\}$ and an edge set $E(G_{x_j}) = \{(x_j, \overline{x_j})\}$. (2) *Clause gadget* G_{C_i}: For each clause $C_i \in C(W)$, $G_{C_i} = (V(G_{C_i}), E(G_{C_i}))$ is a K_4 consisting of a vertex set $V(G_{C_i}) = \{a_i, b_i, c_i, d_i\}$. (3) *Root gadget* G_r: $G_r = (V(G_r), E(G_r))$ consists of a vertex set $V(G_r) = \{r_1, r_2\}$ and an edge set $E(G_r) = \{(r_1, r_2)\}$. Next, by joining the three gadgets by some edges, we construct $H(W) = (V(W), E(W))$ as follows:

$$V(W) = V(G_r) \cup \left(\bigcup_{j=1}^{n} V(G_{x_j}) \right) \cup \left(\bigcup_{i=1}^{m} V(G_{C_i}) \right) \text{ and}$$

$$E(W) = E(G_r) \cup E_{X,r} \cup \left(\bigcup_{i=1}^{m} E_{C_i, X} \right) \cup \left(\bigcup_{j=1}^{n} E(G_{x_j}) \right) \cup \left(\bigcup_{i=1}^{m} E(G_{C_i}) \right),$$

where $E_{X,r} = \{(r_1, x) \mid x \in L(X)\}$ and $E_{C_i, X} = \{(a_i, u), (b_i, v), (c_i, w)\}$ for each clause $C_i = (u \vee v \vee w)$.

In the remaining of this section, we show that there exists a reconfiguration sequence on $R_{3\mathrm{SAT}}(W)$ if and only if there exists a reconfiguration sequence on $R_{\mathrm{IT}}(H(W))$. We denote by $V_{\mathbf{s}} = \{x_j \mid s_j = 1\} \cup \{\overline{x_j} \mid s_j = 0\}$ a *satisfying vertex set*. Then, we consider a family of induced subgraphs corresponding to $\mathbf{s}$.

Definition 4. *Let W be an input formula and* **s** *be a satisfying assignment of W. We define the set $\mathscr{T}_W(\mathbf{s})$ as follows: Each element S of $\mathscr{T}_W(\mathbf{s})$ satisfies*

$$S = V(G_r) \cup V_\mathbf{s} \cup \left(\bigcup_{i=1}^{m} \{\alpha_i, \beta_i\} \right),$$

where $\alpha_i \in V(G_{C_i}) \cap N(V_\mathbf{s})$ and $\beta_i \in V(G_{C_i}) \setminus N(V_\mathbf{s})$ for each $i = 1, \ldots, m$. We say $\mathscr{T}_W(\mathbf{s})$ the family of satisfying induced trees of **s** *for W.*

That is, every satisfying induced tree in $\mathscr{T}_W(\mathbf{s})$ corresponds to a satisfying assignment **s**. In the next lemma, we show that every element in $\mathscr{T}_W(\mathbf{s})$ is actually an induced tree of $H(W)$.

Lemma 5. *Let* **s** *be a satisfying assignment of W. Then, $S \in \mathscr{T}_W(\mathbf{s})$ is an induced tree of $H(W)$.*

Proof. We first show S is connected. By the construction of S, r_1 is adjacent to r_2 and all vertices in $V_\mathbf{s}$. For each $i = 1, \ldots, m$, there exists a vertex in $V_\mathbf{s}$ that is adjacent to α_i since **s** is a satisfying assignment. In addition, α_i and β_i are adjacent. Thus, S is connected. Next, we show S is acyclic. No vertex in G_r is adjacent to vertices in clause gadgets. Moreover, x_j and $\overline{x_j}$ are not adjacent to vertices in other variable gadgets. The same can be said for clause gadgets. □

Example 6. For a given formula W_1, Fig. 3 shows an example of $H(W_1)$ and a satisfying induced tree S_1 belonging to $\mathscr{T}_{W_1}(\mathbf{s_1})$. The vertex set of S_1 is $V(S_1) = V(G_r) \cup V_{\mathbf{s_1}} \cup \{a_1, c_1, b_2, d_2, b_3, c_3\}$. Since S_1 is connected and acyclic, S_1 is an induced tree of $H(W_1)$.

Note that sometimes there are many satisfying induced trees for an instance formula. Actually, in Fig. 3, the tree induced by $V(G_r) \cup V_{\mathbf{s_1}} \cup \{a_1, d_1, b_2, d_2, c_3, d_3\}$ is another satisfying induced tree for W_1 and the tree is in $\mathscr{T}_{W_1}(\mathbf{s_1})$. Lemma 7 implies that if S and T are distinct satisfying induced trees of the same satisfying assignment, then there always exists a reconfiguration sequence from S to T.

Lemma 7. *Let* **s** *be a satisfying assignment of W. Then, $R_{\mathrm{IT}}(H(W))[\mathscr{T}_W(\mathbf{s})]$ is connected.*

Proof. For any two distinct induced trees S and S' in $\mathscr{T}_W(\mathbf{s})$, $S \triangle S'$ consists of vertices in clause gadgets, where $S \triangle S' = (S \setminus S') \cup (S' \setminus S)$. For convenience of explanation, we first assume that $S \triangle S'$ only includes vertices in $V(G_{C_i})$ and $D_i = V(G_{C_i}) \cap (S \triangle S') \neq \emptyset$. Since $|S \cap V(G_{C_i})| = |S' \cap V(G_{C_i})| = 2$, there exist two cases: (I) $|D_i| = 2$ and (II) $|D_i| = 4$. (I) Suppose that $D_i = \{\alpha, \alpha'\}$, $\alpha \in S$, and $\alpha' \in S'$. In this case, both α and α' are adjacent to vertices in variable gadgets, or not adjacent to vertices in variable gadgets. Thus, $f(S, \alpha, \alpha') = S'$. (II) Suppose that $D_i = \{\alpha, \beta, \alpha', \beta'\}$, α and β in S, and α' and β' in S'. Without loss of generality, we can assume that α and α' are adjacent to vertices in variable gadgets. From (I), there exists a satisfying tree $T = f(S, \alpha, \alpha')$.

Then, $T \triangle S' = \{\beta, \beta'\}$. Thus, S and S' are connected in $\mathcal{T}_W(\mathbf{s})$. When $S \triangle S'$ includes vertices of more than one clause gadgets, we can obtain S' from S by iteratively applying the above operation to each clause gadget.   □

From Lemma 8, we can see that for any satisfying induced tree $S \in \mathcal{T}_W(\mathbf{s})$, if an induced tree S' is adjacent to S, S' is a satisfying induced tree of either $\mathbf{s}$ or of another satisfying assignment $\mathbf{t}$ of W.

Lemma 8. *Let S be an induced tree in $\mathcal{T}_W(\mathbf{s})$. For any two vertices $u \in S$ and $v \notin S$, if $S' = f(S, u, v)$ is an induced tree of $H(W)$, then either (I) $u \in V_\mathbf{s}$ and $v = \bar{u}$ or (II) there exists $i \in \{1, \ldots, m\}$ such that $u, v \in G_{C_i}$.*

Proof. The lemma says that if we pick u and v from the different gadgets of $H(W)$, S' is not an induced tree of $H(W)$. Let $i = 1, \ldots, m$ and $j = 1, \ldots, n$ be arbitrary integers. (1) Suppose $u \in G_{C_i}$. If $v \in G_{C_k}$ such that $k \neq i$, $G_{C_k} \cap S'$ is K_3. On the other hand, if $v \in G_{x_j}$, $\{r_1, x_j, \overline{x_j}\}$ is also K_3. (2) Suppose $u \in G_{x_j}$. If $v \in G_{C_i}$, $G_{C_i} \cap S'$ is K_3. On the other hand, if $v \in G_{x_k}$ such that $k \neq j$, $\{r_1, x_k, \overline{x_k}\}$ is also K_3. (3) Suppose $u \in G_r$. If $v \in G_{C_i}$, $G_{C_i} \cap S'$ is K_3. On the other hand, if $v \in G_{x_j}$, $\{r_1, x_j, \overline{x_j}\}$ is K_3. Thus, the statement holds.   □

Then, we show that, in the following lemma, $\mathbf{s}$ and $\mathbf{t}$ are also adjacent to each other on $R_{3SAT}(W)$.

Lemma 9. *Let $\mathbf{s}$ and $\mathbf{t}$ be two distinct satisfying assignments of W. The conditions (I) and (II) are equivalent: (I) $\mathbf{s}$ and $\mathbf{t}$ are adjacent to each other on $R_{3SAT}(W)$. (II) There exist $S \in \mathcal{T}_W(\mathbf{s})$ and $T \in \mathcal{T}_W(\mathbf{t})$ such that S and T are adjacent to each other on $R_{IT}(H(W))$.*

Proof. From Lemma 5, both S and T are induced trees of $H(W)$. Firstly, we show (I) $\rightarrow$ (II). Suppose that the jth variable of $\mathbf{s}$ and $\mathbf{t}$ are different. We can assume that, without loss of generality, $s_j = 1$. For any clause C including x_j, there exist $S \in \mathcal{T}_W(\mathbf{s})$ and $T \in \mathcal{T}_W(\mathbf{t} = g(\mathbf{s}, j))$ such that $S \cap G_C = T \cap G_C$ since C is satisfied by another variable $x_k \neq x_j$ and $s_k = t_k$. Thus, S and T are adjacent to each other. Next, we show (II) $\rightarrow$ (I). Since S and T are adjacent to each other and $\mathbf{s} \neq \mathbf{t}$, there are two vertices $u \in V_\mathbf{s}$ and $v = \bar{u}$ such that $f(S, u, v) = T$ from Lemma 8. Therefore, the hamming distance between $\mathbf{s}$ and $\mathbf{t}$ is exactly one. Thus, $\mathbf{s}$ is adjacent to $\mathbf{t}$ and the statement holds.   □

From Lemma 9, for any connected component $\mathcal{C}$ of $R_{IT}(H(W))$ including a satisfying induced tree, each induced tree belonging to $\mathcal{C}$ has a corresponding satisfying assignment. From the above discussion, we have the main theorem.

Theorem 10. ITRECONF *is PSPACE-complete under TJ.*

Proof. At first, we show ITRECONF is in PSPACE. Ito *et al.* [9] showed that for any problem $\mathcal{P}$ in NP, a reconfiguration version of $\mathcal{P}$ belongs to PSPACE. Thus, ITRECONF is in PSPACE since a problem for finding an induced tree of a graph is in NP. Next, we give a polynomial-time reduction from 3SATRECONF to ITRECONF. Firstly, given an instance $(W, \mathbf{s}, \mathbf{t})$ of 3SATRECONF, we construct $H(W)$, $S \in \mathscr{T}_W(\mathbf{s})$, and $T \in \mathscr{T}_W(\mathbf{t})$. This can be done in polynomial time. From Lemma 5, S and T are induced trees of $H(W)$. From Lemmas 7 and 9, there exists a reconfiguration sequence $\pi(S, T)$ on $R_{\mathrm{IT}}(H(W))$ if and only if there exists a reconfiguration sequence $\pi(\mathbf{s}, \mathbf{t})$ on $R_{3\mathrm{SAT}}(W)$. Since 3SATRECONF is PSPACE-complete by Theorem 3, ITRECONF is also PSPACE-complete.   □

3.3 Token Sliding Case

Until now, we assume that the reconfiguration rule is TJ. Next, we consider another reconfiguration rule $f'(S, u, v) = (S \setminus \{u\}) \cup \{v\}$ such that $u \in S$, $v \notin S$, and $v \in N(u)$. We say f' TOKEN SLIDING (TS). Now, we immediately obtain the following corollary, since Lemma 8 indicates that we always select such u and v that are adjacent to each other on $H(W)$.

Corollary 11. ITRECONF *is PSPACE-complete under* TS.

3.4 Maximal Induced Tree Reconfiguration Problem

In the previous subsections, we demonstrated that ITRECONF is PSPACE-complete. Now, we show that the maximal induced tree version MITRECONF remains PSPACE-complete. $R_{\mathrm{MIT}}(G)$ denotes the reconfiguration graph for maximal induced trees of G. The maximal induced tree reconfiguration problem is defined as follows:

Problem 12 (Maximal Induced Tree Reconfiguration Problem). Let G be a graph and S and T be two maximal induced trees of G. MITRECONF(G, S, T) asks whether there exists a reconfiguration sequence from S to T on $R_{\mathrm{MIT}}(G)$.

Theorem 13. MITRECONF *is PSPACE-complete under both* TS *and* TJ.

Proof. Let $\mathbf{s}$ be a satisfying assignment of a formula W. Firstly, for any $S \in \mathscr{T}_W(\mathbf{s})$, we show that S is maximal. (I) For each clause gadget G_{C_i}, if $u \in G_{C_i} \setminus S$, then $(S \cup \{u\}) \cap G_{C_i}$ is a cycle. (II) For a literal $x_j \in V_{\mathbf{s}}$, $\{x_j, \overline{x_j}, r_1\}$ induces a cycle. From (I) and (II), S is maximal. Thus, the component of $R_{\mathrm{IT}}(G)$ that includes S consists only of maximal ones. Hence, the statement holds.   □

4   W[1]-Hardness

In this section, we show that ITRECONF is W[1]-hard when parameterized by $k + \ell$, where k is the size of induced trees and ℓ is the number of steps in the reconfiguration sequences. To show the hardness, we use the following fact:

Theorem 14 (Downey and Fellows [4]). *Let* $G = (V(G), E(G))$ *be a graph and* k *be a positive integer. The following question is* W[1]-*complete: does there exist a vertex subset* $S \subseteq V$ *such that* S *is an independent set of* G *and* $|S| = k$?

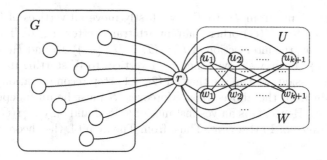

Fig. 4. An example of $I(G)$. The edges in G are omitted. The vertex r is adjacent to all vertices. The induced subgraph $I(G)[U \cup W]$ forms a biclique $K_{k+1,k+1}$.

We refer to the independent set problem as K-IS. Now, we consider an FPT-reduction from K-IS to ITRECONF. To demonstrate the FPT-reduction, firstly, we construct the graph $I(G) = (V(I(G)), E(I(G)))$ from an instance $G = (V(G), E(G))$ of K-IS as follows:

$$V(I(G)) = V(G) \cup \{r\} \cup U \cup W \text{ and}$$
$$E(I(G)) = E(G) \cup \{(r, v) \mid v \in V \cup U \cup W\} \cup \{(u_i, w_j) \mid i, j = 1, \ldots, k+1\},$$

where $U = \{u_1, \ldots, u_{k+1}\}$ and $W = \{w_1, \ldots, w_{k+1}\}$. Note that $T_U = I(G)[\{r\} \cup U]$ and $T_W = I(G)[\{r\} \cup W]$ form induced trees (See Fig. 4). The next theorem shows that ITRECONF is W[1]-hard when parameterized $k + \ell$.

Theorem 15. ITRECONF *is W[1]-hard when parameterized by* $k + \ell$ *under TJ, where* k *is the size of induced trees and* ℓ *is the length of reconfiguration sequences.*

Proof. Let $G = (V(G), E(G))$ be a graph and k be a positive integer. Then, we show that the following (I) and (II) are equivalent: (I) G has an independent set whose size is k. (II) There exists a reconfiguration sequence from T_U to T_W such that the number of steps in the sequence is $2k + 1$. (I) $\rightarrow$ (II): Let S be an independent set whose size is k. For any two vertices u and v in S, $(u, v) \notin E(I(G))$ from the definition of an independent set. Thus, $I(G)[S \cup \{r\}]$ is an induced tree since there is the edge between u and r for any vertex u in S. Hence, we can actually construct the reconfiguration sequence of length $2k + 1$ from T_U to T_W as follows: Firstly, we move $u_1, \ldots, u_k$ to S one by one. Secondly, we move u_{k+1} to w_{k+1}. Finally, we move the vertices in S to W one by one. (II) $\rightarrow$ (I): We assume that there exists a reconfiguration sequence from T_U to T_W of length $2k + 1$. Note that the following two facts: (a) for any distinct four vertices $u, u' \in U$ and $w, w' \in W$, $I(G)[\{u, u', w, w'\}]$ forms a cycle, and (b) for any two vertices $u \in U$ and $w \in W$, $I(G)[\{r, u, w\}]$ forms a cycle. That is, for any two vertices $u \in U$ and $w \in W$, $(T_U \setminus \{u\}) \cup \{w\}$ has a cycle. On the other hand, for any vertex $w \in W$, $(T_U \setminus \{r\}) \cup \{w\}$ is an induced tree. However, for any two vertices $u \in U$ and $w' \in W$, $(T_U \setminus \{r, u\}) \cup \{w, w'\}$ has a cycle. Thus, when we reconfigure from T_U to T_W, we firstly move all vertices of $U \setminus \{u\}$ to a vertex subset $S \subseteq V(G)$ other than an arbitrary vertex $u \in U$. This needs k steps. After that, the intersection of $T' = (T_U \setminus (U \setminus \{u\})) \cup S$ and T_W consists of a singleton $\{r\}$. Then, we move other $k + 1$ vertices to W starting from u. From the above discussion, we can see that all $(2k + 1)$-step reconfiguration sequences can be obtain from the above steps. Moreover, S must be an independent set with k vertices. Hence, G is an yes-instance of K-IS if and only if $(I(G), T_U, T_W)$ is an yes-instance of ITRECONF. Thus, from Theorem 14, the theorem holds. $\square$

5   Fixed Parameter Tractability

As shown in the previous sections, ITRECONF is a computationally hard problem. ITRECONF is PSPACE-complete and W[1]-hard when parameterized by both the size of induced tree and the length of the reconfiguration sequence. In this section, we show that ITRECONF is fixed parameter tractable when parameterized by the size of induced trees and the maximum degree of an input graph.

Theorem 16. ITRECONF *is fixed parameter tractable when parameterized by* $k + \Delta$, *where* k *is the size of induced trees and* Δ *is the maximum degree of an input graph.*

Proof. Let $G = (V(G), E(G))$ be an input graph. If the size of $\mathcal{IT}(G)$ is polynomial, then we can solve ITRECONF in polynomial time since the reachability of the graph can be solved in polynomial time. Thus, to prove this theorem, all we have to do is to show that the number of induced trees is polynomial. For any vertex v in G, the number of vertices whose distance from v is at most k is fewer than $N = \frac{\Delta}{\Delta - 2}(\Delta - 1)^k$ [3], and then the number of induced trees including v is at most $\binom{N}{k}$. Hence, the number of induced trees of G is $O(|V(G)|\binom{N}{k})$, linear in the number of vertices. $\square$

We can easily see that the above theorem also holds for MITRECONF.

Corollary 17. MITRECONF *is fixed parameter tractable when parameterized by* $k + \Delta$, *where* k *is the size of induced trees and* Δ *is the maximum degree of an input graph.*

There exists some variations of ITRECONF such as SHORTESTITRECONF; outputting the shortest length reconfiguration sequence between two induced trees, and CONITRECONF; answering whether the reconfiguration graph is connected or not. From Theorem 16, the size of a reconfiguration graph is linear in the number of the vertices of an input graph when $k + \Delta$ is constant, where k is the size of induced trees and Δ is the maximum degree of an input graph. Hence, not only ITRECONF and MITRECONF but also SHORTESTITRECONF and CONITRECONF are fixed parameter tractable when parameterized by $k + \Delta$.

6 Conclusion

In this paper, we addressed the reconfiguration problem for induced trees. We showed that ITRECONF and MITRECONF are PSPACE-complete under TJ and TS. Moreover, ITRECONF is W[1]-hard when parameterized by $k + \ell$ under TJ. On the other hand, we also showed ITRECONF is FPT when parameterized by $k + \Delta$. Future work includes to consider whether ITRECONF can be solved in polynomial time when input graphs are restricted. In addition, it is still open to determine the hardness of ITRECONF when parameterized by either k or Δ.

Acknowledgments. The authors would like to thank Akira Suzuki, Ryuhei Uehara, and Yukiko Yamauchi for helpful discussions and comments. The authors also thank the anonymous referees for their detailed comments and helpful suggestions.

References

1. Bonsma, P.: The complexity of rerouting shortest paths. Theor. Comput. Sci. **510**, 1–12 (2013)
2. Cereceda, L., van den Heuvel, J., Johnson, M.: Connectedness of the graph of vertex-colourings. Discrete Math. **308**(5–6), 913–919 (2008)
3. Diestel, R.: Graph Theory, 4th edn. Springer, Heidelberg (2012)
4. Downey, R.G., Fellows, M.R.: Fixed-parameter tractability and completeness II: on completeness for W[1]. Theor. Comput. Sci. **141**(1–2), 109–131 (1995)
5. Gardner, M.: The hypnotic fascination of sliding-block puzzles. Sci. Am. **210**, 122–130 (1964)
6. Gopalan, P., Kolaitis, P.G., Maneva, E., Papadimitriou, C.H.: The connectivity of boolean satisfiability: computational and structural dichotomies. SIAM J. Comput. **38**(6), 2330–2355 (2009)
7. Hearn, R.A., Demaine, E.D.: PSPACE-completeness of sliding-block puzzles and other problems through the nondeterministic constraint logic model of computation. Theor. Comput. Sci. **343**(1–2), 72–96 (2005)
8. Ito, T., Demaine, E.D.: Approximability of the subset sum reconfiguration problem. J. Comb. Optim. **28**(3), 639–654 (2012)
9. Ito, T., Demaine, E.D., Harvey, N.J., Papadimitriou, C.H., Sideri, M., Uehara, R., Uno, Y.: On the complexity of reconfiguration problems. Theor. Comput. Sci. **412**(12–14), 1054–1065 (2011)
10. Ito, T., Kamiński, M., Demaine, E.D.: Reconfiguration of list edge-colorings in a graph. Discrete Appl. Math. **160**(15), 2199–2207 (2012)
11. Ito, T., Kawamura, K., Ono, H., Zhou, X.: Reconfiguration of list $L(2, 1)$-labelings in a graph. Theor. Comput. Sci. **544**, 84–97 (2014)
12. Ito, T., Ono, H., Otachi, Y.: Reconfiguration of cliques in a graph. In: Jain, R., Jain, S., Stephan, F. (eds.) TAMC 2015. LNCS, vol. 9076, pp. 212–223. Springer, Heidelberg (2015)
13. Kamiński, M., Medvedev, P., Milanič, M.: Shortest paths between shortest paths. Theor. Comput. Sci. **412**(39), 5205–5210 (2011)
14. Kamiński, M., Medvedev, P., Milanič, M.: Complexity of independent set reconfigurability problems. Theor. Comput. Sci. **439**, 9–15 (2012)

15. Mouawad, A.E., Nishimura, N., Pathak, V., Raman, V.: Shortest reconfiguration paths in the solution space of Boolean formulas. In: Halldórsson, M.M., Iwama, K., Kobayashi, N., Speckmann, B. (eds.) ICALP 2015. LNCS, vol. 9134, pp. 985–996. Springer, Heidelberg (2015)

16. Mouawad, A.E., Nishimura, N., Raman, V., Simjour, N., Suzuki, A.: On the parameterized complexity of reconfiguration problems. In: Gutin, G., Szeider, S. (eds.) IPEC 2013. LNCS, vol. 8246, pp. 281–294. Springer, Heidelberg (2013)

17. Schaefer, T.J.: The complexity of satisfiability problems. STOC **1978**, 216–226 (1978)

18. Yamanaka, K., Demaine, E.D., Ito, T., Kawahara, J., Kiyomi, M., Okamoto, Y., Saitoh, T., Suzuki, A., Uchizawa, K., Uno, T.: Swapping labeled tokens on graphs. Theor. Comput. Sci. **586**, 81–94 (2015)

Grammars

The Missing Case in Chomsky-Schützenberger Theorem

Stefano Crespi Reghizzi[1] and Pierluigi San Pietro[1(✉)]

Dipartimento di Elettronica, Informazione e Bioingegneria (DEIB),
Politecnico di Milano, Piazza Leonardo da Vinci 32, 20133 Milano, Italy
{stefano.crespireghizzi,pierluigi.sanpietro}@polimi.it

Abstract. The theorem by Chomsky and Schützenberger (CST) says
that every context-free language L over alphabet Σ is representable as
$h(D_k \cap R)$ where D_k is the Dyck language over k pairs of brackets, R
is a local (i.e., 2-strictly-locally-testable language) regular language, and
h is an alphabetic homomorphism that may erase symbols; the Dyck
alphabet size depends on the size of the grammar generating L. In the
Stanley variant, the Dyck alphabet size only depends on the size of Σ,
but the homomorphism has to erase many more symbols than in the previous version. Berstel found that the number of erasures in CST can be
linearly limited if the grammar is in Greibach normal form, and recently
Okhotin proved a non-erasing variant of CST for grammars in Double
Greibach normal form. In both statements the Dyck alphabet depends
on the grammar size. We present a new non-erasing variant of CST that
uses a Dyck alphabet independent from the grammar size and a regular
language that is strictly-locally-testable, similarly to a recent generalization of Medvedev theorem for regular languages.

1 Introduction

The theorem by Chomsky and Schützenberger (CST) [2] says that every context-
free language L over an alphabet Σ is representable as $h(D_k \cap R)$, where D_k is
the Dyck language over $k \geq 1$ pairs of brackets, R is a regular language and h is
an erasing alphabetic homomorphism; the Dyck alphabet size is $2k$ and depends
on the size of the grammar G that generates L. In the later variant by Stanley
[11], also presented in Ginsburg [5], the Dyck alphabet size only depends on
the size of Σ, but the homomorphism has to erase many more symbols than
in the original version. Berstel [1] found that the number of erasures in CST
can be linearly limited if G is in Greibach normal form. Recently, Okhotin [9]
proved a non-erasing variant of CST, by using grammars in Double Greibach
normal form (see e.g. [4]). In both Berstel's and Okhotin's statements, however,
the Dyck alphabet depends on the grammar size. Other formal language books
(cited in [9]) includ essentially equivalent statements and proofs of the CST.

Work partially supported by MIUR project PRIN 2010LYA9RH and by CNR-IEIIT.

To sum up, we may classify the existing versions of CST with respect to two parameters: the erasing vs. nonerasing homomorphism, and the grammar-dependence versus grammar-independence of the Dyck alphabet, as shown in the following table:

	grammar-dependent alphabet	grammar-independent alphabet
erasing homomorphism	Chomsky and Schützenberger [2], Berstel [1]	Stanley [11]
nonerasing homomorphism	Okhotin [9]	

We fill the empty case of the table by presenting a new non-erasing version of CST that uses a Dyck alphabet polynomially dependent on the terminal alphabet size, but not on the grammar size (measured by the number of nonterminals or of rules).

Since the interest of our contribution is purely theoretical, we rely on the standard constructions for pushdown automata, grammars and sequential transductions, without any optimization effort. As a consequence, the size of the Dyck alphabet, though independent from the grammar size, is rather large. At the end, we observe that a substantial size reduction is easy in special cases. We quantify this for the linear grammars, by exploiting our recent result [3] that reduces the alphabet needed to homomorphically characterize a regular language by means of Medvedev theorem [8,10].

Paper organization: Section 2 lists the basic definitions and recalls a relevant CST formulation. Section 3 proves the CST variant based on a grammar-independent alphabet and a non-erasing homomorphism, first for the case of languages containing strings of even length, then for the general case. Section 4 exploits the extended Medvedev theorem to show that a much smaller alphabet suffices for linear context-free languages, and concludes. An Appendix shows an example of our version of the CST.

2 Preliminaries

For brevity, we omit the classical definitions of context-free grammars and their normal forms, pushdown and finite automata, and finite transducers, for which we refer primarily to [6]. Let Σ denote a finite terminal alphabet and ε the empty word. For a word x, $|x|$ denotes the length of x; the i-th letter of x is $x(i)$, $1 \le i \le |x|$, i.e., $x = x(1)x(2)\ldots x(|x|)$. The mirror image of word x is denoted as $x^R = x(|x|)\ldots x(2)x(1)$. For finite alphabets Δ, Γ, a *homomorphism* is a mapping $h : \Delta \to \Gamma^*$; if for some $d \in \Delta$, $h(d) = \varepsilon$, then h is called *erasing*, while it is called *letter-to-letter* if for every $d \in \Delta$, $h(d)$ is in Γ.

A *finite automaton* (FA) is specified by a 5-tuple $(\Sigma, Q, \delta, q_0, F)$ where Q is the set of states, δ the state-transition function, q_0 the initial state, and F is the set of final states. A *sequential transducer* [6] is denoted by a 5-tuple

$S_m = (Q, \Sigma, \Delta, \delta, q_0)$ where Δ is the output alphabet and δ is a finite subset of $Q \times \Sigma^* \times \Delta^* \times Q$, i.e., every transition may read a string $u \in \Sigma^*$ outputting a string $v \in \Delta^*$.

A *context-free grammar* is a 4-tuple $G = (\Sigma, N, P, S)$ where N is the non-terminal alphabet, P the rule set, and S the axiom. Since we only deal with context-free grammars and languages, we often omit the term "context-free". A *pushdown automaton* (PDA) [6] is denoted by a 7-tuple $(Q, \Sigma, \Gamma, \delta, Z_0, F)$, where Γ is the stack alphabet and Z_0 the initial stack symbol. The languages recognized by a PDA A, using the traditional accepting conditions, are denoted: $T(A)$ for acceptance by final state and $N(A)$ for acceptance by empty stack.

A function $f : \mathbb{N} \to \mathbb{N}$ is said to be in $POLY(n)$ if $f(n)$ is upper bounded by a polynomial expression in n, i.e., there exists $\alpha > 1$ such that $f(n) < n^\alpha$ for all $n > 1$.

The family SLT of *strictly locally testable* languages [7] is next defined dealing only with ε-free languages for simplicity. For every word $w \in \Sigma^+$, for all $k \geq 2$, let $i_k(w)$ and $t_k(w)$ denote the prefix and, resp., the suffix of w of length k if $|w| \geq k$, or w itself if $|w| < k$. Let $f_k(w)$ denote the set of words of w of length k. Extend i_k, t_k, f_k to languages as usual.

Definition 1. *A language L is k-strictly locally testable (k-SLT), if there exist finite sets $W \subseteq \Sigma \cup \Sigma^2 \cup \cdots \cup \Sigma^{k-1}$, $I_{k-1}, T_{k-1} \subseteq \Sigma^{k-1}$, and $F_k \subseteq \Sigma^k$ such that, for every $x \in \Sigma^+$, $x \in L$ if, and only if,*

$$x \in W \lor (i_{k-1}(x) \in I_{k-1} \land t_{k-1}(x) \in T_{k-1} \land f_k(x) \subseteq F_k).$$

A language is strictly locally testable (SLT) *if it is k-SLT for some k, called its width.*

Value $k = 2$ yields the well known family of *local languages*. The SLT family is strictly included in the family of *regular languages* and forms a hierarchy w.r.t. the width.

The following notation for Dyck alphabets and languages is from [9]. For any finite set X, the set, denoted Ω_X, of brackets labeled with elements of X is

$$\Omega_X = \{ {}_{[x} \mid x \in X \} \cup \{ {}_{]x} \mid x \in X \}.$$ The Dyck language $D_X \subset \Omega_X^*$ is generated by the following grammar:

$S \to {}_{[x} S {}_{]x}$ for each $x \in X$, $S \to SS$, $S \to \varepsilon$ Let $k = |X|$. Clearly, each Dyck language D_X is isomorphic to $D_{\{1,\dots,k\}}$, For brevity we write Ω_k and D_k instead of $\Omega_{\{1,\dots,k\}}$ and $D_{\{1,\dots,k\}}$, respectively.

We need the following statement of CST

Theorem 1 (Theorem 1 of Okhotin [9]). *A language $L \subseteq (\Sigma^2)^*$ is context-free if, and only if, there exists a number k, a regular language $R \subseteq \Omega_k^*$ and a letter-to-letter homomorphism $h : \Omega_k \to \Sigma$, such that $L = h(D_k \cap R)$.*

The proof in [9] assumes that L is generated by a grammar $G = (\Sigma, N, P, S)$ in Chomsky normal form and first converts it into a grammar $G' = (\Sigma, N', P', S')$ in Double Greibach normal form. It then labels each bracket with (essentially) a pair of rules of P'. Therefore the size of the Dyck alphabet Ω_k is in $O(|P'|^2)$, which is in $POLY(|N|)$.

3   Homomorphic Characterization

The first, more complex part shows that every language containing just sentences of even length is homomorphically characterized by means of a letter-to-letter homomorphism of the intersection of a Dyck and a regular language over an alphabet that only depends on Σ. Then, the second easy part extends the property to any language, using a Dyck alphabet completed with neutral symbols.

Definition 2 (Tuple alphabet and homomorphism). *For an alphabet Σ, let $\Delta_r = \{\langle a_1, \ldots, a_r \rangle \mid a_1, \ldots, a_r \in \Sigma\}$ for all $r \geq 2$. An element of alphabet Δ_r is called an r-tuple or simply a tuple. Let $m \geq 2$ be an even number and define the alphabet*

$$\Delta_{[m,3m)} = \bigcup_{r=m}^{3m-1} \Delta_r.$$

The tuple homomorphism $\pi : \Delta_{[m,3m)} \to \Sigma^+$ is: $\pi(\langle a_1, \ldots, a_r \rangle) = a_1 \ldots a_r$.

Mapping by sequential transducer and relation between grammar sizes. In our construction we shall use a mapping $\mathcal{S}_m$ from a language L over alphabet Σ to another language $L' = \mathcal{S}_m(L)$ over alphabet $\Delta_{[m,3m)}$, computed by a finite sequential transducer. We are going to compute a relation between the sizes of the grammars G for L and G' for L'; for that we have to analyze the sizes of the PDAs accepting L and L'; we shall do that following Harrison [6] Theorem 6.4.3 and some standard conversion algorithms from and to grammars and PDAs.

First, we specify the sequential transduction, which complies with Harrison definition ([6], p. 198), but for convenience is equipped with a final state.

Definition 3. *Let $m \geq 2$ be an even number and let $\mathcal{S}_m : \Sigma^{2m}\Sigma^* \to (\Delta_{[m,3m)})^+$ be the mapping defined for all integer $j \geq 0$ and for all $v_1, \ldots, v_{2j}, v_{2j+1} \in \Sigma^m$, for all $w \in \Sigma^+$ with $m \leq |w| \leq 3m - 1$ as:*

$$\mathcal{S}_m(v_1 \ldots v_{2j+1}w) = \pi^{-1}(v_1) \cdots \pi^{-1}(v_{2j+1}) \cdot \pi^{-1}(w).$$

Property 1. For all $x \in \Sigma^+$ such that $|x| \geq 2m$, mapping $\mathcal{S}_m$ is defined and single-valued, i.e., there exists one, and only one, decomposition of x as $x = v_1 \ldots v_{2j+1}w$, for $v_1, \ldots, v_{2j}, v_{2j+1} \in \Sigma^m$ and $w \in \bigcup_{r=m}^{3m-1} \Sigma^r$ such that $\mathcal{S}_m(x) = \mathcal{S}_m(v_1 \ldots v_{2j+1}w)$. Moreover, $\mathcal{S}_m(x) \in \pi^{-1}(x)$ has even length; if $|x|$ is even, then also $|w|$ is even.

Proof. We show that $|x|$ can be written as $m(2j+1)+t$, for some $j \geq 0$, $m \leq t \leq 3m-1$: x can be factorized into $2j+1$ consecutive words of length m and a suffix of length at least m and at most $3m-1$, as in Definition 3. In fact, the length of a word $x \in \Sigma^{2m}\Sigma^*$ can be written as $|x| = hm + r$, for some integer $h \geq 2$, and number r with $0 \leq r \leq m-1$. If h is even, then $h = 2j+2$ for some $j \geq 0$, hence $|x|$ can be represented as $m(2j+1)+(m+r)$. If h is odd, then $h = 2i+1$ for some $i \geq 1$ (since $h \geq 2$ when h is odd it must also be greater than 3); hence, $|x|$ can be represented as $m(2i-1)+(2m+r)$: let $j = i-1$, i.e., $2i-1 = 2j+1$, hence $|x|$

can be represented as $m(2j+1)+(2m+r)$. To show that the decomposition is unique, assume by contradiction that there exist $j, j' \geq 0, j' \neq j$ and $t' \neq t$, with $m \leq t', t \leq 3m-1$, such that both $|x| = m(2j+1)+t$ and $|x| = m(2j'+1)+t'$. By symmetry, it is enough to consider $j' > j$, i.e., $j' - j = p > 0$. Then, $|x| = m(2j'+1)+t' = m(2(j+p)+1)+t' = m(2j+1)+2pm+t'$. Clearly, $t = 2pm+t'$, but $t' \geq m$ and $p > 0$ hence $t \geq 3m$, a contradiction. Finally, if $|x|$ is even, then also r must be even, therefore $|w|$, which is equal either to $m+r$ or $2m+r$, is also even. $\square$

Mapping $\mathcal{S}_m$ is computed by the following (nondeterministic) sequential transducer, also named $\mathcal{S}_m$, specified by the 5-tuple

$$\mathcal{S}_m = \left(Q_{\mathcal{S}_m}, \Sigma, \Delta_{[m,3m)}, \delta_2, q_0, \{q_F\} \right),$$

where $Q_{\mathcal{S}_m} = \{q_0, q_1, q_2, q_F\}$ and :

$$\forall v \in \Sigma^m : \left(q_0, v, \pi^{-1}(v), q_1 \right), \left(q_1, v, \pi^{-1}(v), q_2 \right), \left(q_2, v, \pi^{-1}(v), q_1 \right) \in \delta_2;$$
$$\forall w \in \Sigma^+, m \leq |w| \leq 3m-1 : \left(q_1, w, \pi^{-1}(w), q_F \right) \in \delta_2.$$

It may help to look at the following state-transition graph:

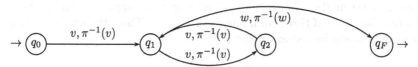

Notice for later reference that, in each move, the length of the input word (v or w) is at most $3m-1$ and the length of the output word is at most one (tuple).

The next lemma states the relation between the grammar sizes.

Lemma 1. *Let* $L = L(G)$, *where* $G = (\Sigma, N, P, S)$ *is in Chomsky normal form. Let* $m \geq 2$ *be an even number and let* $\mathcal{S}_m$ *be the transduction of Definition 3. Then, the language* $L' = \mathcal{S}_m(L)$ *is generated by a grammar* $G' = (\Delta_{[m,3m)}, N', P', S')$ *such that* $|N'|$ *and* $|P'|$ *are in* $POLY(|\Sigma|^m|N|)$.

Proof. The proof tracks the size of the language descriptions corresponding to the PDA $\mathcal{A}_N$ such that $L = N(\mathcal{A})$, the PDA $\mathcal{A}_T$ such that $L = T(\mathcal{A}_T)$, the PDA $\mathcal{A}_S$ such that $L' = \mathcal{S}_m(L) = T(\mathcal{A}_S)$, and the grammar G' equivalent to $\mathcal{A}_S$. $\square$

Claim. The size of the transition function of the PDA $\mathcal{A}_N$ such that $L = N(\mathcal{A}_N)$ is $|\mathcal{A}_N| = O(|P|)$.

This follows from the standard construction of the one-state PDA that accepts by empty stack. The machine pushes onto the stack at most two symbols since G is in Chomsky normal form. The size of the stack alphabet is $|N|+1$.

Converting $\mathcal{A}_N$ to an equivalent machine $\mathcal{A}_T$ that accepts by final state, we claim:

Claim. The size of the transition function of the PDA $\mathcal{A}_T$ such that $T(\mathcal{A}_T) = N(\mathcal{A})$ is $|\mathcal{A}_T| = O(|\mathcal{A}_N|)$ therefore also $|\mathcal{A}_T| = O(|P|)$.

This follows from the standard conversion from acceptance by empty store to acceptance by final state. The conversion adds a constant number of states, of stack symbols, and of moves. Machine $\mathcal{A}_T$, as $\mathcal{A}_N$, pushes onto the stack at most two symbols.

Building the PDA $\mathcal{A}_S$ that recognizes $\mathcal{S}_m\,(T(\mathcal{A}_T))$ by final state, we claim:

Claim. Let $\mathcal{A}_S$ be the PDA such that $T(\mathcal{A}_S) = \mathcal{S}_m\,(T(\mathcal{A}_T)) = L'$. Then, the size of the state set $Q_{\mathcal{A}_S}$ of $\mathcal{A}_S$ is in $POLY(|\Sigma|^m)$ and the size of the transition function of $\mathcal{A}_S$ is $O\left(|Q_{\mathcal{A}_S}|^2\,|P|\right)$.

According to the construction in [6] (Theorem 6.4.3), producing a PDA simulating both $\mathcal{A}_T$ and $\mathcal{S}_m$, the set of states $Q_{\mathcal{A}_S}$ is:

$$Q_{\mathcal{A}_T} \times Q_{\mathcal{S}_m} \times \bigcup_{i=0}^{\lambda} \Sigma^i \times \bigcup_{i=0}^{\lambda}(\Delta_{[m,3m)})^i \tag{1}$$

where $\lambda = 3m - 1$ (see earlier motivation). But, in our case, we do not need that many states: the fourth component in Eq. (1) is meant to record all prefixes of an output word to be emitted by the sequential transducer. Our output word is a single character in $\Delta_{[m,3m)}$, uniquely defined by the mapping $\pi^{-1}(v)$ or $\pi^{-1}(w)$: it is a function only of three components of Eq. (1). Thus, the set $Q_{\mathcal{A}_S}$ of states of $\mathcal{A}_S$ can actually be reduced to:

$$Q_{\mathcal{A}_T} \times Q_{\mathcal{S}_m} \times \bigcup_{i=0}^{3m-1} \Sigma^i.$$

Since the value $|Q_{\mathcal{A}_T}||Q_{\mathcal{S}_m}|$ is a constant, it follows that the number $|Q_{\mathcal{A}_S}|$ is in $POLY(|\Sigma|^m)$. Moreover, since each move may push at most two symbols onto the stack (as in machine $\mathcal{A}_T$), the number of moves in the transition function of $\mathcal{A}_S$ is: $O\left(|Q_{\mathcal{A}_S}|^2\,|P|\right)$.

Claim. There exists a grammar $G' = (\Delta_{[m,3m)}, N', P', S')$, defining the same language of $\mathcal{A}_S$, such that both $|P'|$ and $|N'|$ are in $POLY\,(|\Sigma|^m|N|)$.

This follows from the standard construction of grammar G': every nonterminal of N' is a tuple of the form $\langle q_i, B_1, q_j\rangle$ where $q_i, q_j \in Q_{\mathcal{A}_S}$ and B_1 is a pushdown stack symbol of $\mathcal{A}_S$, which is $O(|N|)$. Hence, $|N'|$ is in $O(|Q_{\mathcal{A}_S}|^2)|N|)$, i.e., in $POLY\,(|\Sigma|^m|N|)$. G' is in Chomsky normal form, therefore $|P'|$ is in $O(|N'|^3)$ and thus also $|P'|$ is in $POLY\,(|\Sigma|^m|N|)$. Since $|P|$ for Chomsky normal form grammar is in $O(|N|^3)$, it follows immediately that also $|P'|$ is in $POLY\,(|\Sigma|^m|N|)$.

The next theorem states the main result. The proof applies Okhotin's Theorem 1 to language $L' = \mathcal{S}_m(L)$ over the tuple alphabet: thus, L' may be represented as $h\,(D_k \cap R)$. Homomorphism h maps every bracket w of D_k into a tuple symbol $h(w)$, corresponding to a word $\pi(h(w))$ in Σ^m. Then the idea is to map, by means of another homomorphism ρ, each open bracket w of D_k into a word $\rho(\omega)$, made by m open brackets of a new Dyck alphabet Ω_n; similarly, every closed bracket ω' is mapped into a sequence $\rho(\omega')$ of closing brackets of Ω_n. Moreover, the opening brackets in $\rho(\omega)$ are exactly matched by the closing brackets of $\pi(h(\omega'))$. Homomorphism ρ is such that the word $\rho(\omega)$ represents two pieces of information: the word $\pi(h(\omega))$ and an identifier of the original bracket ω. The identifier is a number, encoded as m digits in a suitably chosen base. For the closing brackets, the word $\rho(\omega')$ represents $\pi(h(\omega'))$ and the same identifier of the original bracket ω, but in reverse order.

Theorem 2. *Let $L \subseteq \left(\Sigma^2\right)^*$ be a language generated by a Chomsky normal form grammar $G = (\Sigma, N, P, S)$. Then, there exists a number q in $POLY(|\Sigma|)$ (independent of $|N|$), a Dyck alphabet Ω_q, a regular language $T \subseteq \Omega_q^+$ and a letter-to-letter homomorphism ν, such that $L = \nu\,(D_q \cap T)$.*

Proof. Let $m \geq 2$ be an even number, to be bounded later in the proof. The proof for words longer than $2m - 1$ is next outlined and afterwards expanded. At the end, the proof deals with the simple case of short sentences. First, by Lemma 1 we transform L into the language $L' = \mathcal{S}_m(L)$ over the tuple alphabet, which can be characterized using Okhotin's Theorem 1, by means of a letter-to-letter homomorphism h, a Dyck language $D_k \subseteq \Omega_k{}^*$, and a regular language R. Clearly it holds: $L = \pi\,(h\,(D_k \cap R))$.

Second, we define a Dyck alphabet Ω_n, made by matching open and closed brackets, ζ, ζ', each one represented by a 4-tuple carrying the following information:

1. whether the element is an open or closed bracket;
2. the character of Σ to which ζ will be mapped by homomorphism ρ;
3. the character of Σ to which ζ' will be mapped by homomorphism ρ;
4. a digit i in a given constant base $j \geq 2$, as next explained. In any two matching elements ζ, ζ', value i is identical.

We recall that each element $\omega \in \Omega_k$ is identified by a number ι in $1 \ldots k$. We represent the value of ι in base j using m digits. We show that it is possible to choose a base j in $POLY(|\Sigma|)$ as long as m is chosen in $\Omega(\log|N|)$ (as customary, here $\Omega(.)$ denotes a lower bound of a complexity function). As said, each digit in base j occurs as fourth component of 4-tuples ζ and ζ'.

Third, we define a new homomorphism $\tau : \Omega_k \to \Omega_n^+$ such that the image of D_k is a subset of the Dyck language D_n, i.e., $\tau(D_k) \subset D_n$. Such subset will be obtained, by means of the regular language $\tau(R)$, as $\tau(D_k) = D_n \cap \tau(R)$.

Fourth, we define the letter-to-letter homomorphism $\rho : \Omega_n \to \Sigma$ that extracts the second component from each 4-tuple, proving that $\rho\,(D_n \cap \tau(R))$ is exactly $\pi\,(h\,(D_k \cap R))$.

Details of the Proof. Let $L' = \mathcal{S}_m(L)$. By Definition 3 it is $L' \subseteq \Delta_m^* \cdot \Delta_{[m,3m]}$. By Property 1, L' contains only even-length sentences. By Lemma 1, $L' = L(G')$, where grammar $G' = \left(\Delta_{[m,3m]}, N', P', S'\right)$ has $|N'|$ in $POLY\left(|\Sigma|^m |N|\right)$.

By Theorem 1, there exist $k > 0$, a Dyck alphabet Ω_k and a letter-to-letter homomorphism $h : \Omega_k \to \Delta_{[m,3m)}$ such that $L(G') = h\,(D_k \cap R)$, where k is in $POLY\,(|N'|)$.

For a number $j \geq 2$, let $n = j|\Sigma|^2$ and define the new Dyck alphabet:

$$\Omega_n = \{\,`[\,'\,,\,`]\,'\,\} \times \Sigma \times \Sigma \times (\{0,\ldots,j-1\}) \tag{2}$$

and the matching open/closed elements ζ, ζ' in Ω_n:

$$\zeta = \langle\,`[\,',a,b,o\rangle \text{ matches } \zeta' = \langle\,`]\,',b,a,o\rangle \tag{3}$$

Note: the 2nd and 3d components are interchanged; component o is in $0,\ldots,j-1$. Let D_n be the Dyck language over Ω_n.

We want to represent each one of the k open parentheses in Ω_k with a distinct string, composed of m digits in base $j \geq 2$. Therefore, we must choose a number j such that $\log_j k \leq m$. Denoting with log the base 2 logarithm, we require that $\frac{\log k}{\log j} \leq m$, that is the inequality: $\log j \geq \frac{\log k}{m}$. Although j can be chosen arbitrarily, we notice that $n = j|\Sigma|^2$: therefore, n is in $POLY(|\Sigma|)$ if, and only if, also j is in $POLY(|\Sigma|)$.

By Theorem 1, $k \in POLY(|N'|)$, with $|N'| \in POLY\,(|\Sigma|^m|N|)$ by Claim 3. Hence, k is in $POLY\,(|\Sigma|^m|N|)$. Therefore, there exist $\alpha, \beta > 1$ such that $k < |\Sigma|^{\alpha m}|N|^\beta$ for every $k > 1$, i.e., $k^{1/m} < \left(|\Sigma|^{\alpha m}|N|^\beta\right)^{1/m}$, which entails:

$$\log k^{1/m} < \log\left(|\Sigma|^\alpha|N|^{\frac{\beta}{m}}\right), \text{ i.e., } \frac{\log k}{m} < \log|\Sigma|^\alpha + \frac{\log|N|^\beta}{m}$$
$$\text{If } m > \log(|N|^\beta), \text{ then } \frac{\log k}{m} < \log|\Sigma|^\alpha + 1 = \log\left(2|\Sigma|^\alpha\right).$$

Hence, the condition that $\log j \geq \frac{\log k}{m}$ can be verified, when m is in $\Omega(\log|N|)$, by choosing j such that $\log j > \log\left(2|\Sigma|^\alpha\right)$, i.e., it suffices to choose a suitable j in $POLY(|\Sigma|)$. Hence, under the above assumption on j and m, every open parenthesis $\omega \in \Omega_k$ can be represented in base j by a distinct string with m digits, to be denoted in the following as $[\omega]_j$. The closed parenthesis ω' matching ω has no encoding of its own, but it is just represented with the reversal of the encoding of ω, i.e., $\left([\omega]_j\right)^R$ (as we will see, no confusion can arise).

As noticed, if j is in $POLY(|\Sigma|)$, then n is in $POLY(|\Sigma|)$: asymptotically, the cardinality of Ω_n is polynomial in the terminal alphabet Σ and does not depend on the number $|P|$ of productions of G, as long as m is in $\Omega(\log|N|)$.

To define the homomorphism $\tau : \Omega_k \to \Omega_n^+$, we first need the partial mapping:

$$\otimes : (\Sigma_1)^+ \times (\Sigma_2)^+ \times (\Sigma_3)^+ \times (\Sigma_4)^+ \to (\Sigma_1 \times \Sigma_2 \times \Sigma_3 \times \Sigma_4)^+$$

where each Σ_i is a finite alphabet, to combine four words of identical length into one word of the same length over the alphabet of 4-tuples. This *combinator* $\otimes$ is defined for all $l \geq 1$, $x_i \in (\Sigma_i)^l$, $1 \leq i \leq 4$, as:

$$\otimes (x_1, x_2, x_3, x_4) = \langle x_1(1), x_2(1), x_3(1), x_4(1)\rangle \ldots \langle x_1(l), x_2(l), x_3(l), x_4(l)\rangle.$$

For instance, let $x_1 = ab, x_2 = cd, x_3 = ef, x_4 = ca$; then $\otimes (x_1, x_2, x_3, x_4) = \langle a, c, e, c\rangle \langle b, d, f, a\rangle$.

Since L' is a subset of $\Delta_m^* \cdot \Delta_{[m,3m)}$, the image $h(\omega)$ of an open parenthesis $\omega \in \Omega_k$ is in Δ_m, while the image of a closing parentheses ω' may be in $\Delta_{[m,3m)} \supseteq \Delta_m$. The definition of τ is then split into two cases. The first case is when $h(\omega') \in \Delta_m$:

$$\tau(\omega) = \otimes \left(`[^m, \pi\,(h(\omega)), (\pi\,(h(\omega')))^R, [\omega]_j \right)$$
$$\tau(\omega') = \otimes \left(`]^m, \pi\,(h(\omega')), (\pi\,(h(\omega)))^R, ([\omega]_j)^R \right)$$
(4)

All four arguments of $\otimes$ are words of length m, therefore the combinator $\otimes$ returns a word of length m over the alphabet of 4-tuples. For instance, if $h(\omega) = \langle a_1, \ldots, a_m \rangle \in \Delta_m$, $h(\omega') = \langle b_m, \ldots, b_1 \rangle \in \Delta_m$, and $[\omega]_j = o_1 o_2 \ldots o_m$, with $o_1, \ldots, o_m \in \{0, \ldots, j-1\}$, then $([\omega]_j)^R = o_m o_{m-1} \ldots o_1$ and:

$$\tau(\omega) = \langle `[', a_1, b_1, o_1 \rangle \langle `[', a_2, b_2, o_2 \rangle \ldots \langle `[', a_m, b_m, o_m \rangle$$
$$\tau(\omega') = \langle `]', b_m, a_m, o_m \rangle \langle `]', b_{m-1}, a_{m-1}, o_{m-1} \rangle \ldots \langle `]', b_1, a_1, o_1 \rangle$$

The second case for τ is when $h(\omega') \notin \Delta_m$ (still, $h(\omega') \in \Delta_{[m,3m)}$). Since m is even and both L', L only contain even length sentences, then $h(\omega') \in \Delta_{m+2t}$, for some $1 \leq t \leq (3m-2)/2$, i.e., $h(\omega') = \langle a_1, \ldots a_m, b_1, \ldots, b_t, c_t, \ldots, c_1 \rangle$, for some $a_i, b_i, c_i \in \Sigma$. The definition of τ in this case is:

$$\tau(\omega) = \otimes (`[^m, \pi\,(h(\omega)), \pi\,(\langle a_m, \ldots, a_1 \rangle)), [\omega]_j)$$
$$\tau(\omega') = \otimes \left(`]^m, \pi\,(\langle a_1, \ldots, a_m \rangle), (\pi\,(h(\omega)))^R, ([\omega]_j)^R \right) \cdot$$
$$\cdot \langle `[', b_1, c_1, 0 \rangle \ldots \langle `[', b_t, c_t, 0 \rangle \langle `]', c_t, b_t, 0 \rangle \ldots \langle `]', c_1, b_1, 0 \rangle$$
(5)

Claim. (a) Let $\omega, \omega' \in \Omega_k$ be a matching pair. Then $\tau(\omega) = \zeta_1 \ldots \zeta_m$ and either (case (4)) $\tau(\omega') = \zeta'_m \ldots \zeta'_1$, or (case (5)) $\tau(\omega') = \zeta'_m \ldots \zeta'_1 \beta_1 \ldots \beta_t \beta'_t \ldots \beta'_1$ where for all i both the pair ζ_i, ζ'_i and the pair β_i, β'_i are matching in Ω_n.
(b) $\tau(D_k) \subseteq D_n$.

Proof. (a) The fact that the above elements match according to formula (3), follows immediately from the definition of τ. (b) Since, for every $w \in \Omega_k^+$, $\tau(w)$ preserves the parenthetization of w, if $w \in D_k$, $\tau(w) \in D_n$. We show that mapping τ is one-to-one:

Claim. For all $w, w' \in (\Omega_k)^+$, if $\tau(w) = \tau(w')$, then $w = w'$.

Proof. Let $\omega_1, \omega_2 \in \Omega_k$; if $\omega_1 \neq \omega_2$, then $[\omega_1]_j \neq [\omega_2]_j$ by definition of $[\ldots]_j$. Therefore $\tau(\omega_1) \neq \tau(\omega_2)$ because at least one position differs.   □

We define the letter-to-letter homomorphism $\rho : \Omega_n \to \Sigma$ as the projection on the second component of each 4-tuple: $\rho(\langle x_1, x_2, x_3, x_4 \rangle) = x_2 \in \Sigma$.

Claim. For all $w \in (\Omega_k)^+$, $\rho(\tau(w)) = \pi(h(w))$.

Proof. By the definitions of τ and ρ, for every $\chi \in \Omega_k$ it is $\rho(\tau(\chi)) = \pi(h(\chi))$. This is true for both cases (4) and (5).

Claim. $\tau^{-1}(D_n) \subseteq D_k$.

Proof. Although τ^{-1} is not defined for every word in D_n, mapping τ is defined so that, if a word $w \notin D_k$, then $\tau(w) \notin D_n$; hence if $\tau(w) \in D_n$, then also $w \in D_k$.

Then we prove the identity: $\rho\left(D_n \cap \tau(R)\right) = \pi\left(h\left(D_k \cap R\right)\right)$ (6)

By Claim 3, $\tau(D_k) \cap \tau(R) = \tau(D_k \cap R)$; hence, by Claim 3, part (b), $\rho\left(\tau(D_k \cap R)\right)$ is equal to $\rho\left(\tau(D_k) \cap \tau(R)\right) \subseteq \rho\left(D_n \cap \tau(R)\right)$. The inclusion $\pi\left(h\left(D_k \cap R\right)\right) \subseteq \rho\left(\tau(D_k \cap R)\right)$ then follows: if $z \in \pi(h\left(D_k \cap R\right))$, then there exists a word $w \in D_k \cap R$ such that $\pi(h(w)) = z$, hence $z = \rho(\tau(w))$ by Claim 3. Since $w \in D_k \cap R$, then $\tau(w) \in \tau(D_k \cap R)$, hence $z \in \rho\left(\tau(D_k \cap R)\right) \subseteq \rho\left(D_n \cap \tau(R)\right)$.

The opposite inclusion $\rho\left(D_n \cap \tau(R)\right) \subseteq \pi\left(h\left(D_k \cap R\right)\right)$ also follows: if $z \in \rho\left(D_n \cap \tau(R)\right)$, then there exists $w \in R$ such that $\tau(w) \in D_n$ and $\rho(\tau(w)) = z$. By Claim 3, if $\tau(w) \in D_n$, then also $w \in D_k$. Since $z = \pi(h(w))$ by Claim 3, it follows that $z \in \pi\left(h\left(D_k \cap R\right)\right)$.

It remains to consider the "short" words in $F_m = \{x \in L \mid |x|$ is even and $0 \le |x| \le 2m - 1\}$. Let $p = |\Sigma|^2$ and let $\Omega_p = \{[,]\} \times \Sigma \times \Sigma$, where a triple of the form $\langle [, a, b \rangle$ is an open parenthesis and $\langle], b, a \rangle$ is the corresponding closed one, and the Dyck language is D_p. Let σ be the projection $\sigma : \Omega_p \to \Sigma$ defined as: $\sigma\left(\langle [, a, b \rangle\right) = a, \sigma\left(\langle], b, a \rangle\right) = b$, and let R_F be the regular language $\sigma^{-1}(F_m)$. It is then obvious that $F_m = \sigma(D_p \cap R_F)$, i.e., the CST holds for language F_m.

To finish, define the homomorphism $\nu : (\Omega_n \cup \Omega_p) \to \Sigma$ as $\forall \omega \in \Omega_n, \nu(\omega) = \rho(\omega)$ and $\forall \omega \in \Omega_p, \nu(\omega) = \sigma(\omega)$; let $T = R \cup R_F$. The thesis then follows with $q = n + p$, i.e., $\Omega_q = \Omega_n \cup \Omega_p$. (An example is in the Appendix.)   □

Theorem 2 homomorphically characterizes any language having sentences of even length, by means of a regular and a Dyck language over an alphabet with size not depending on the complexity of the original grammar, but only on its terminal alphabet.

We observe that the regular language constructed in the proof of Theorem 2 is no longer local, i.e., 2-SLT (Definition 1). Yet, it would be straightforward to modify our construction to obtain an SLT language having width greater than two: it suffices to modify homomorphism τ (Eqs. (4) and (5)) so that the first bracket of $\tau(\omega)$ and the last one of $\tau(\omega')$ are typographically different from the remaining $m - 1$ brackets.

We also remark that Theorem 2 leaves the value of m unspecified, provided that it is larger than $\beta \log |N|$, for a constant β. In this paper we do not analyze the relation between m and the alphabet size $|\Delta_n|$, but in Sect. 4 we discuss the tradeoff between the alphabet size, $|N|$ and m in a special case.

Homomorphic Characterization for Languages of Words of Arbitrary Length. It is simple to remove the restriction to even-length sentences, thus obtaining a homomorphic characterization that holds for any context-free language. For dealing also with the special case of singleton sentences of the form $a \in \Sigma$, it suffices to consider a variant of the Dyck language equipped with neutral symbols, as in Okhotin's Theorem 3.

Let $\Omega_{n,l}$ be an alphabet with n pairs of parentheses and $l \geq 1$ new symbols called *neutral* [9]; let $D_{n,l}$ denote the corresponding Dyck language with neutral symbols.

In our treatment, $l = |\Sigma|$ and $\Omega_{n,l} = \Omega_n \cup \{\langle -, a, a, 0\rangle \mid a \in \Sigma\}$, where "$-$" is a new symbol. If a word $x \in L$ has odd length, then $\mathcal{S}_m(x)$ is composed of $2j + 1$ tuples in Δ_m (of even total length) and one tuple $h(\omega') \in \Delta_{m+2t+1}$, for some number t, $1 \leq t \leq (3m - 2)/2$, and $\omega' \in \Omega_k$ (i.e., $|\pi(h(\omega')|$ must be odd). Therefore, there exist letters $a, a_i, b_i, c_i \in \Sigma$ such that $h(\omega') = \langle a_1, \ldots a_m, b_1, \ldots, b_t, c_t, \ldots, c_1, a\rangle$. Redefine Case (5) for $\tau(\omega')$ in the proof of Theorem 2 as follows:

$$\tau(\omega') = \otimes \left(`]'^m, \pi\left(\langle a_1, \ldots, a_m\rangle\right), \left(\pi\left(h(\omega)\right)\right)^R, \left([\omega]_j\right)^R\right) \cdot$$
$$\cdot\langle `[', b_1, c_1, 0\rangle \ldots \langle `[', b_t, c_t, 0\rangle \cdot \langle `]', c_t, b_t, 0\rangle \ldots \langle `]', c_1, b_1, 0\rangle \cdot \langle -, a, a, 0\rangle$$

Also, we extend the definition of ρ by setting $\rho(\langle -, a, a, 0\rangle) = a$ for all $a \in \Sigma$.

Theorem 3. *A language $L \subseteq \Sigma^*$ is context-free if, and only if, there exist a number n that only depends, polynomially, on the size of Σ, a regular language $R \subseteq \left(\Omega_{n,|\Sigma|}\right)^*$ and a letter-to-letter homomorphism $h : \Omega_{n,|\Sigma|} \to \Sigma$, such that $L = h\left(D_{n,|\Sigma|} \cap R\right)$.*

4 Relation with Medvedev Theorem and Conclusion

It would be possible but tedious to compute the precise size of the Dyck alphabet founding the proof in Theorem 2, by analyzing the standard transformations involving grammars and PDAs used, but such computation would likely yield a large overestimation, because of the generality of such transformations. For the particular case of linear context-free grammars, we show that a small alphabet suffices.

Since the following remark does not aim to generality, for brevity we consider a language L devoid of odd-length sentences, generated by a linear grammar $G = (\Sigma, P, N, S)$ with rules of the form $A \to aBb, A \to ab, a, b \in \Sigma, A, B \in N$.

It is known that, for every such linear language L, there exist a regular language $R_L \subseteq \Gamma^*$, and two letter-to-letter homomorphisms $h_1, h_2 : \Gamma \to \Sigma$, such that $L = \{h_1(v) \cdot h_2(v^R) \mid v \in R_L\}$.

More precisely, let $\Gamma = \{`['\} \times \Sigma \times \Sigma$ and $R_L \subseteq \Gamma^+$ be the language recognized by the following FA. The state set is $N \cup \{q_F\}$, with S and q_F respectively initial and final; the state transition function is:

$$A \xrightarrow{[\langle a,b\rangle} B \text{ if } A \to aBb \in P, \qquad A \xrightarrow{[\langle a,b\rangle} q_F \text{ if } A \to ab \in P.$$

Define the similar alphabet $\Gamma' = \{`]'\} \times \Sigma \times \Sigma$ and the Dyck alphabet $\Delta' = \Gamma \cup \Gamma'$ of size $2|\Sigma|^2$. Positing that $[_{\langle a,b\rangle}$ and $]_{\langle a,b\rangle}$ are matching, denote the Dyck language as D' and define the regular language $R' = R_L \cdot (\Gamma')^+$. Then, it is straightforward that language L can be represented as

$$L = h'(D' \cap R'), \text{ where } h' : \Delta' \to \Sigma \text{ is : } h'\left([\langle a,b \rangle\right) = a, \ h'\left(]\langle a,b \rangle\right) = b. \quad (7)$$

Notice that the regular language R', in general, is not SLT (Definition 1); e.g., for the language $\{(a^2)^n(b^2)^n \mid n \geq 1\}$, language R_L (and R') is not non-counting hence not SLT [7]. However, we can derive from (7) a representation that uses an SLT language, if we accept to pay a small extra cost in terms of alphabet size.

In [3] we have extended the historical Medvedev theorem [8,10], which states that every regular language R can be represented as a letter-to-letter homomorphism of a 2-SLT language over a (much) larger alphabet. Moving to higher width, we proved the following relation between the alphabet sizes, the complexity of language R (measured by the number of states of its nondeterministic FA), and the SLT width parameter.

Theorem 4 [3]. *Given an alphabet Σ, if a regular language $R \subseteq \Sigma^*$ is accepted by an FA with $|Q|$ states, then for every $c, 2 \leq c \leq |Q|$, there exists a letter-to-letter homomorphism f and an s-SLT language T over an alphabet of size $c|\Sigma|$, such that $R = f(T)$. The width parameter s is in $\Theta(\frac{\log |Q|}{\log c})$.*

By applying Theorem 4 to language R' of (7) and setting parameter c to its minimum value 2, we obtain that every linear language $L(G)$, where G has the form considered, can be homomorphically characterized using a Dyck alphabet of size $4|\Sigma^2|$ and an s-SLT language, where s logarithmically depends on $|N|$. The c parameter of Theorem 4 expresses a tradeoff between the alphabet size and the SLT width.

Returning to the case of unrestricted context-free grammars, to the best of our knowledge the past research on CST has not dealt with the relationship between the Dyck alphabet size and the class of the regular language, in particular, whether it is SLT and its width. The present work may provide a starting point for such investigation.

Appendix: An Example

The example illustrates the crucial part of our constructions, namely the homomorphism τ defined by formulas (4) and (5). Consider language $L = \{a^{2n+4}b^{6n} \mid n \geq 0\}$ (generated, e.g., by grammar $\{S \to aaSb^6 \mid a^4\}$), and choose the value $m = 2$ for Definition 3, meaning that the substrings of length two occurring in the language are mapped on the 2-tuples $\langle a,a \rangle, \langle a,b \rangle, \langle b,b \rangle$, shortened as $\langle aa \rangle$, etc. The following grammar in Double Greibach normal form, though constructed by hand, takes the place of grammar G' of Lemma 1:

$1 : S \to \langle aa \rangle \, S \, B \, \langle bb \rangle$, $2 : S \to \langle aa \rangle \, \langle aa \rangle$, $3 : B \to \langle bb \rangle \, \langle bb \rangle$.
The sentence $a^8b^{12} \in L$ becomes $\langle aa \rangle^4 \langle bb \rangle^6 \in L(G')$, with the syntax tree:

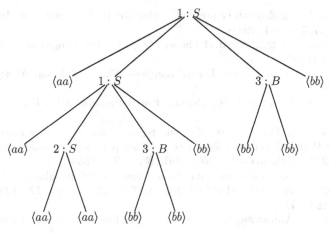

For Okhotin Theorem 1, this sentence is the image by homomorphism h of the following sequence γ of labeled parentheses, where the numbers identify the rules and the dash marks the root:

$$\gamma = (_1^- \quad (_1^1 \quad (_2^1 \quad)_2^1 \quad (_3^1 \quad)_3^1 \quad)_1^1 \quad (_3^1 \quad)_3^1 \quad)_1^-$$

We choose to represent the labeled parentheses with $m = 2$ binary digits, defining τ as:

ω	ω'	$\tau(\omega)$	$\tau(\omega')$
$(_1^-$	$)_1^-$	$[_{a,b,0}\ [_{a,b,0}$	$]_{b,a,0}\ ]_{b,a,0}$
$(_1^1$	$)_1^1$	$[_{a,b,0}\ [_{a,b,1}$	$]_{b,a,1}\ ]_{b,a,0}$
$(_2^1$	$)_2^1$	$[_{a,a,1}\ [_{a,a,0}$	$]_{a,a,0}\ ]_{a,a,1}$
$(_3^1$	$)_3^1$	$[_{b,b,1}\ [_{b,b,1}$	$]_{b,b,1}\ ]_{b,b,1}$

Hence $\tau\left(\pi(h(\gamma))\right)$ is

$$\overbrace{}^{(_1^-} \quad \overbrace{}^{(_1^1} \quad \overbrace{}^{(_2^1} \quad \overbrace{}^{)_2^1} \quad \overbrace{}^{(_3^1} \quad \overbrace{}^{)_3^1}$$

$\overbrace{[_{a,b,0}\ [_{a,b,0}}\ \overbrace{[_{a,b,0}\ [_{a,b,1}}\ \overbrace{[_{a,a,1}\ [_{a,a,0}}\ \overbrace{]_{a,a,0}\ ]_{a,a,1}}\ \overbrace{[_{b,b,1}\ [_{b,b,1}}\ \overbrace{]_{b,b,1}\ ]_{b,b,1}}$

$$\overbrace{}^{)_1^1} \quad \overbrace{}^{(_3^1} \quad \overbrace{}^{)_3^1} \quad \overbrace{}^{)_1^-}$$

$\overbrace{]_{b,a,1}\ ]_{b,a,0}}\ \overbrace{[_{b,b,1}\ [_{b,b,1}}\ \overbrace{]_{b,b,1}\ ]_{b,b,1}}\ \overbrace{]_{b,a,0}\ ]_{b,a,0}}$

Notice that the 2-SLT language of the classical CST (applied to language L) is now replaced by an SLT language of higher width.

References

1. Berstel, J.: Transductions and Context-Free Languages. Teubner, Stuttgart (1979)
2. Chomsky, N., Schützenberger, M.: The algebraic theory of context-free languages. In: Brafford, H. (ed.) Computer Programming and Formal Systems, pp. 118–161. North-Holland, Amsterdam (1963)
3. Crespi Reghizzi, S., San Pietro, P.: From regular to strictly locally testable languages. Int. J. Found. Comput. Sci. **23**(8), 1711–1728 (2012)

4. Engelfriet, J.: An elementary proof of double Greibach normal form. Inf. Process. Lett. **44**(6), 291–293 (1992)
5. Ginsburg, S.: The Mathematical Theory of Context-free Languages. McGraw-Hill, New York (1966)
6. Harrison, M.: Introduction to Formal Language Theory. Addison Wesley, Reading (1978)
7. McNaughton, R., Papert, S.: Counter-free Automata. MIT Press, Cambridge (1971)
8. Medvedev, Y.T.: On the class of events representable in a finite automaton. In: Moore, E.F. (ed.) Sequential machines - Selected papers (translated from Russian), pp. 215–227. Addison-Wesley, New York, NY, USA (1964)
9. Okhotin, A.: Non-erasing variants of the Chomsky–Schützenberger theorem. In: Yen, H.-C., Ibarra, O.H. (eds.) DLT 2012. LNCS, vol. 7410, pp. 121–129. Springer, Heidelberg (2012)
10. Eilenberg, S.: Automata, Languages, and Machines. Academic Press, Orlando (1974)
11. Stanley, R.J.: Finite state representations of context-free languages. M.I.T. Res. Lab. Electron. Quart. Progr. Rept. **76**(1), 276–279 (1965)

Homomorphic Characterizations of Indexed Languages

Séverine Fratani and El Makki Voundy[(⊠)]

Aix-Marseille Université, CNRS, LIF UMR 7279, 13000 Marseille, France
makki.voundy@lif.univ-mrs.fr

Abstract. We study a family of context-free languages that reduce to ε in the free group and give several homomorphic characterizations of indexed languages relevant to that family.

Keywords: Grammars · Homomorphic characterizations · Transductions · Indexed languages

1 Introduction

The well known Chomsky–Schützenberger theorem [6] states that every context-free language L can be represented as $L = h(R \cap \mathcal{D}_k)$, for some integer k, regular set R and homomorphism h. The set $\mathcal{D}_k$ used in this expression, called Dyck language, is the set of well-bracketed words over k pairs of brackets. Combined with Nivat's characterization of rational transductions, this means that any context-free language can be defined as a set $L = h(g^{-1}(\mathcal{D}_2) \cap R)$, for some regular set R, and homomorphisms h and g.

Let us consider wider families of languages, Maslov defines in [13] an infinite hierarchy of languages included in recursively enumerable languages. The level 1 consists of context-free languages, the level 2 of indexed languages (initially defined by Aho [1]). Known as higher order languages since the last decades, the languages of the hierachy and derived objects as higher order trees [12], higher order schemes [9], or higher order graphs [5], are used to model programming languages and have been subject to recent researches in program verification [18].

It is stated in [14] and proved in [8] that each level $\mathcal{L}_k$ of the hierarchy is a principal rational cone generated by a language $M_k \in \mathcal{L}_k$. This means that each language in $\mathcal{L}_k$ is the image of M_k by a rational transduction. Roughly speaking, the language M_k consists of words composed by k embedded Dyck words and can be viewed as a generalization of the Dyck language. Indeed it gives a description of derivations of an indexed grammar of level k, in the same way that the Dyck language encodes derivations of a context-free grammar.

This latter characterization describes $\mathcal{L}_k$ from a single language M_k, but this one is very complicated as soon as $k \geq 2$, as the majority of higher order languages. To better understand higher order languages, we think that it is necessary to characterize them using more simple objects. So, we may wonder

A.-H. Dediu et al. (Eds.): LATA 2016, LNCS 9618, pp. 359–370, 2016.
DOI: 10.1007/978-3-319-30000-9_28

whether it is possible to give versions of the Chomsky–Schützenberger theorem and a characterization by transduction of the level $k + 1$ of the hierarchy, using only the level k of the hierarchy. The fundamental point is then to identify mechanisms that bridge the level k to the level $k + 1$.

In this paper, we solve the problem for the class IL of Indexed Languages (the level 2 of the hierarchy). In order to identify a difficulty, let us remark that from [10], recursively enumerable languages are sets that can be written as $L = h(K \cap \mathcal{D}_k)$ where K is a context-free language, and h a homomorphism. So if we want a homomorphic characterization of IL using only context-free or regular sets, we would have to consider a restricted class of context-free languages. We then introduce the class of ε-Reducible Context-Free Languages (ε-CFLs), which is a strict subclass of context-free languages that reduce to ε by the bilateral reduction $S = \{a\bar{a} \rightarrow \varepsilon, \bar{a}a \rightarrow \varepsilon\}_{a \in \Gamma}$ (these languages are thus defined over an alphabet Γ and its copy $\bar{\Gamma}$). We extend this definition to transductions: an ε-Reducible Context-Free Transduction (ε-CFT) is a context-free transduction whose domain is an ε-CFL. Using these objects, we obtain simple generalizations of the Chomsky–Schützenberger theorem. Indexed languages are:

– the images of $\mathcal{D}_2$ by ε-reducible context-free transductions. (Theorem 15);
– sets $h(Z \cap \mathcal{D}_k)$; where k is an integer, Z an ε-CFL, and h a homomorphism (Theorem 18).

Beyond these two results, we study the classes of ε-CFLs and ε-CFTs defined by means of context-free grammars and context-free transduction grammars.

First we express them using *symmetric homomorphisms* which are homomorphisms under which there are closed. We establish a Chomsky–Schützenberger-like Theorem for ε-CFLs, and a Nivat-like characterization for ε-CFTs: every ε-CFL L can be represented as $L = g(R \cap \mathcal{D}_k)$ for some integer k, regular set R, and symmetric homomorphism g; and ε-CFTs are relations that can be represented as $\{(g(x), h(x)) \mid x \in R \cap \mathcal{D}_k\}$ for some integer k, regular language R, homomorphism h and symmetric homomorphism g. This leads to a third characterization: indexed languages are languages that can be described as $L = h(g^{-1}(\mathcal{D}_2) \cap R \cap \mathcal{D}_k)$, for some integer k, regular set R, homomorphism h and symmetric homomorphism g (Corollary 17).

Similar characterizations have been given for subclasses of indexed languages, by Weir [20] for linear indexed languages, by Kanazawa [11] and Sorokin [19] for yields of tree languages generated by simple context-free grammars. The main difference is that in their cases, the homomorphism g is not symmetric, but is *fixed* in function of k.

Overview. Section 1 is devoted to the study of ε-CFLs. After introducing necessary notions as free groups and Dyck languages, we define the class of grammars generating ε-CFLs. We then study their closure properties, and conclude the section by giving a Chomsky–Schützenberger-like characterization of the class of ε-CFLs. In Sect. 2, we extend our definition to transductions and define the class of ε-CFTs. After a subsection giving background on transductions, we give a Nivat-like characterization of ε-CFTs.

The last section is devoted to indexed languages. After introducing indexed grammars, we prove that indexed languages are images of the Dyck language by ε-CFTs and deduce from this result several homomorphic characterizations.

2   Epsilon-Reducible Context-Free Languages

In this section, we study a family of context-free languages defined over a union of an alphabet and its opposite-disjoint copy and that reduce to the neutral element ε when projected into the free group. The main result here is a Chomsky–Schützenberger-like homomorphic characterization of these languages. We assume the reader to be familiar with context-free grammars and languages (see [3] for example), and present below a few necessary notions on free groups.

2.1   Free Groups and Dyck Languages

Given an alphabet Γ, we denote by $\overline{\Gamma}$ a disjoint copy $\overline{\Gamma} = \{\bar{a} \mid a \in \Gamma\}$ of it, and by $\widehat{\Gamma}$ the set $\Gamma \cup \overline{\Gamma}$. We adopt the following conventions: $\bar{\bar{a}} = a$ for all $a \in \Gamma$, $\bar{\varepsilon} = \varepsilon$ and for any word $u = \alpha_1 \cdots \alpha_n \in \widehat{\Gamma}^*$, $\bar{u} = \bar{\alpha}_n \cdots \bar{\alpha}_1$.

Let us consider the reduction system $S = \{(a\bar{a}, \varepsilon), (\bar{a}a, \varepsilon)\}_{a \in \widehat{\Gamma}}$. A word in $\widehat{\Gamma}^*$ is said to be **reduced** if it is S-reduced, i.e. it does not contain occurrences of $a\bar{a}$, $\bar{a}a$, for $a \in \Gamma$. As S is confluent, each word w is equivalent (mod $\leftrightarrow_S^*$) to a unique reduced word denoted $\rho(w)$. Note that for all $u \in \widehat{\Gamma}^*$, $\rho(u\bar{u}) = \rho(\bar{u}u) = \varepsilon$. Given a set X, we denote by $\rho(X)$ the set $\{\rho(x) \mid x \in X\}$.

The free group $\mathrm{F}(\Gamma)$ consists of reduced words over $\widehat{\Gamma}$. Its neutral element is the empty word and its product $\bullet$ is defined as $u \bullet v = \rho(uv)$.

The set of all words $u \in \widehat{\Gamma}^*$ such that $\rho(u) = \varepsilon$ is denoted $\mathcal{T}_\Gamma$. The Dyck language over Γ, denoted $\mathcal{D}_\Gamma$, is the set of all $u \in \mathcal{T}_\Gamma$, such that for every prefix $v \preccurlyeq u$: $\rho(v) \in \Gamma^*$. We will also write $\mathcal{D}_k$, $k \geq 1$, to refer to the set of Dyck words over any alphabet of size k.

2.2 ε-Reducible Context-Free Languages and Grammars

Definition 1. *An ε-Reducible Context-Free Grammar (ε-CFG) is a context free grammar $G = (N, T, S, P)$ (N is the set of nonterminal symbols, $\widehat{\Gamma}$ the terminal alphabet, $S \in N$ is the start symbol, and P is the set of productions) such that $T = \widehat{\Gamma}$ for some alphabet Γ and every production is in the form:*

$$X \longrightarrow \omega \Omega \bar{\omega}, \qquad for \ \omega \in \widehat{\Gamma}^*, \quad and \ \Omega \in N^*$$

For all $X \in N$, we define $\mathcal{L}_G(X) = \{u \in \widehat{\Gamma}^* \mid X \xrightarrow{*}_G u\}$; the language generated by G is $\mathcal{L}_G = \mathcal{L}_G(S)$.

An ε-Reducible Context-Free Language (ε-CFL) is a context-free language L that can be generated by an ε-CFG.

Example 2. Let $G = (N, \{\alpha, \beta, \bar{\alpha}, \bar{\beta}\}, S, P)$ be the ε-CFG whose productions are:
$$S \longrightarrow \beta X \bar{\beta}, \; X \longrightarrow \alpha X \bar{\alpha} + Y, \; Y \longrightarrow \bar{\alpha} Y Z \alpha + \bar{\beta} \beta, \; Z \longrightarrow \bar{\alpha} Z \alpha + \bar{\beta} \beta.$$

One can easily check that:
$$\mathcal{L}_G(Z) = \bigcup_{n \geq 0} \bar{\alpha}^n \bar{\beta} \beta \alpha^n, \quad \mathcal{L}_G(Y) = \bigcup_{n \geq 0} \bar{\alpha}^n \beta \bar{\beta} (\Pi_{i=1}^n \mathcal{L}_G(Z) \alpha),$$
$$\mathcal{L}_G(S) = \beta \mathcal{L}_G(X) \bar{\beta}, \qquad \mathcal{L}_G(X) = \bigcup_{n \geq 0} \alpha^n \mathcal{L}_G(Y) \bar{\alpha}^n.$$

It follows that: $\mathcal{L}_G = \bigcup_{n,m,r_1,\ldots r_m \geq 0} \beta \alpha^n \bar{\alpha}^m \bar{\beta} \beta (\Pi_{i=1}^m \bar{\alpha}^{r_i} \bar{\beta} \beta \alpha^{r_i+1}) \bar{\alpha}^n \bar{\beta}.$ □

It seems clear that every ε-CFL L satisfies $\rho(L) = \{\varepsilon\}$. One can indeed observe that every terminal word generated from a nonterminal symbol $X \in N$ reduce to ε. However, there are context-free languages that reduce to ε and which are not ε-CFL. We prove this by using a "pumping lemma" for ε-CFLs.

Lemma 3. *If $L \subseteq \widehat{\Gamma}^*$ is an ε-CFL, then there exists some integer $p \geq 1$ such that every word $s \in L$ with $|s| \geq p$ can be written as $s = uvwxy$ with*

1. $\rho(uy) = \varepsilon$, $\rho(vx) = \varepsilon$ and $\rho(w) = \varepsilon$
2. $|vwx| \leq p$,
3. $|vx| \geq 1$, *and*
4. $uv^n wx^n y$ *is in L for all $n \geq 0$.*

Proof (Sketch). Let G be an ε-CFG generated L. The proof of the pumping lemma for context-free languages is based on the fact that if a word $s \in L$ is long enough, there is a non terminal A and terminal words u, v, w, x, y such that $S \xrightarrow{*}_G uAy \xrightarrow{*}_G uvAxy \xrightarrow{*}_G uvwxy$ and $s = uvwxy$. Since G is an ε-CFG, this implies that $\rho(uy) = \varepsilon$, $\rho(vx) = \varepsilon$ and $\rho(w) = \varepsilon$ □

Proposition 4. *There is a context-free language L satisfying $\rho(L) = \varepsilon$ which is not an ε-CFL.*

By applying Lemma 3 to the set $L = \{(\alpha\bar{\alpha})^n \beta (\alpha\bar{\alpha})^n \bar{\beta}, n \geq 0\}$, one can show that L is not an ε-CFL.

Proposition 5. *The class of ε-CFLs is closed under union, intersection with regular sets, concatenation and Kleene star.*

Proof. Obviously, the class of ε-CFLs is closed under union, concatenation and Kleene star. Let us prove the closure under intersection with regular sets. Let L be generated by an ε-CFG $G = (N, \widehat{\Gamma}, P, S)$ and R be a regular language. There is a monoid morphism $\mu : \widehat{\Gamma}^* \to M$, where M is a finite monoid and $H \subseteq M$ such that $R = \mu^{-1}(H)$. We construct the ε-CFG $G' = (N', \widehat{\Gamma}, P', S')$ where $N' = \{X_m \mid X \in N, m \in M\} \cup \{S'\}$ and P' is the set of all productions:

- $X_m \longrightarrow \alpha X_{1,m_1} \cdots X_{n,m_n} \bar{\alpha}$ such that $X \longrightarrow \alpha X_1 \cdots X_n \bar{\alpha} \in P$ and $m = \mu(\alpha) m_1 \cdots m_n \mu(\bar{\alpha})$
- $S' \longrightarrow S_m$ for every $m \in H$

Then for every $u \in \widehat{\Gamma}^*$, for every $X \in N$ and $m \in M$:
$$X_m \xrightarrow{*}_{G'} u \text{ iff } X \xrightarrow{*}_G u \text{ and } u \in \mu^{-1}(m).$$

It follows that $\mathcal{L}_{G'} = L \cap \mu^{-1}(H)$. □

2.3 A Chomsky–Schützenberger-like Theorem for ε-CFLs

The Chomsky–Schützenberger theorem states that a language $L \subseteq \Sigma^*$ is context-free iff there is an alphabet Γ, a regular set $R \subseteq \widehat{\Gamma}^*$, and a homomorphism $h : \widehat{\Gamma}^* \to \Sigma^*$ such that

$$L = h(R \cap \mathcal{D}_\Gamma).$$

This implies that the whole class of context-free languages can be generated as homomorphic images of ε-CFLs since $R \cap \mathcal{D}_B$ is an ε-CFL. To get an homomorphic characterization for ε-CFLs, we introduce a class of homomorphisms under which the family of ε-CFLs is closed.

Definition 6. *A homomorphism* $g : \widehat{\Sigma}^* \to \widehat{\Gamma}^*$ *is said to be* symmetric *if for all* $\alpha \in \widehat{\Sigma}$, $g(\bar{\alpha}) = \overline{g(\alpha)}$.

Proposition 7. *The class of ε-CFLs is closed under symmetric homomorphism.*

Proof. Consider a language L generated by an ε-CFG $G = (N, \widehat{\Gamma}, P, S)$ and $g : \widehat{\Gamma}^* \to \widehat{\Sigma}^*$ be a symmetric homomorphism. We build an ε-CFG $G' = (N, \widehat{\Sigma}, P', S)$ generating $g(L)$ as follows:

$$P' = \{X \longrightarrow g(u)\Omega g(\bar{u}) \mid X \longrightarrow u\Omega\bar{u} \in P, \Omega \in N^*, u \in \widehat{\Gamma}^*\}. \qquad \Box$$

More generally, ε-CFLs are closed under homomorphisms g satisfying
"$\forall u \in \widehat{\Gamma}^* : \rho(u) = \varepsilon \implies \rho(g(u)) = \varepsilon$".
The main result of this section is the following.

Theorem 8. *A set* $L \subseteq \widehat{\Gamma}^*$ *is an ε-CFL iff there is an alphabet* Σ, *a symmetric homomorphism* $g : \widehat{\Sigma}^* \to \widehat{\Gamma}^*$, *and a regular set* $R \subseteq \widehat{\Sigma}^*$ *such that*

$$L = g(R \cap \mathcal{D}_\Sigma).$$

The "if" part of Theorem 8 is a direct consequence of Propositions 5 and 7. The "only if" part is obtained using a slight adaptation of the proof of the non-erasing variant of the Chomsky–Schützenberger theorem given in [17]. Intuitively, in our case, the symmetric homomorphism expresses the opposition of terminals in the productions of an ε-CFG.

We conclude this section by emphasizing that Theorem 8 and Propositions 5 and 7 provide another characterization of the class of ε-CFLs:

Corollary 9. *The family of ε-CFLs is the least family of languages that contains the Dyck language and is closed under union, intersection with regular sets, symmetric homomorphisms, concatenation and Kleene star.*

2.4 Related Works

In [4], the authors define (pure) balanced grammars that are context-free grammars whose set of productions is a (possibly infinite) regular set of rules of the form $X \longrightarrow \alpha m\bar{\alpha}$, where $\alpha \in \Gamma$ and $m \in N^*$. Balanced grammars do not generate all ε-CFLs included in $\mathcal{D}_\Gamma$, for example they cannot generate the set $\{\beta(\alpha\bar{\alpha})^n(\gamma\bar{\gamma})^n\bar{\beta} \mid n \geq 0\}$.

Introduced in [15], input-driven languages–more recently known as Visibly Pushdown Languages (VPLs)–are extensions of balanced languages defined over a structured alphabet: Σ_c is the set of call symbols, Σ_r the set of returns and Σ_ℓ the set of local symbols. They are recognized by pushdown automata that push onto the stack only when reading a call, pop the stack only on returns, and do not use the stack when reading local actions. The input word hence controls the permissible operations on the stack. However, there is no restriction on the symbols that can be pushed or popped. This implies that there are visibly pushdown languages which are not ε-CFLs. However the ε-CFL $\{(\alpha\bar{\alpha})^n(\beta\bar{\beta})^n \mid n \geq 0\}$ is not a VPL when $\Sigma_c = \Gamma$ and $\Sigma_r = \bar{\Gamma}$.

Also note that unlike ε-CFLs, VPLs are closed under intersection. We will see (Theorem 18) that the intersection of an ε-CFL with the Dyck language is an indexed language.

3   Epsilon-Reducible Context-Free Transductions

In this section, we extend the notion of ε-reducibility to transductions. We consider a subclass of context-free transductions such that their domains are ε-CFLs. We give a Nivat-like presentation of those transductions.

3.1   Transductions

We briefly introduce rational and context-free transductions. The reader can refer to [2] for a more detailed presentation.

Let Γ and Σ be two finite alphabets, we consider the monoid $\Gamma^* \times \Sigma^*$ whose product is the product on words, extended to pairs of words: $(u_1, v_1)(u_2, v_2) = (u_1u_2, v_1v_2)$. A subset τ of $\Gamma^* \times \Sigma^*$ is called a (Γ, Σ)-transduction.

Transductions are viewed as (partial) functions from Γ^* toward subsets of Σ^*: for any $u \in \Gamma^*$, $\tau(u) = \{v \in \Sigma^* \mid (u, v) \in \tau\}$. For every $L \subseteq \Gamma^*$, the image (or transduction) of L by τ is $\tau(L) = \bigcup_{u\in L} \tau(u)$. The domain of τ is $\mathrm{Dom}(\tau) = \{u \mid \exists v, (u, v) \in \tau\}$.

Rational Transductions: A rational (Γ, Σ)-transduction is a rational subset of the monoid $\Gamma^* \times \Sigma^*$. Among the different characterizations of rational transductions, let us cite the Nivat theorem [16] stating that rational transductions are relations $\tau = \{(g(u), f(u)) \mid u \in R\}$, for some regular set R and homomorphisms f and g.

Rational transductions are closed by composition and many classes of languages are closed under rational transductions. In particular, $\tau(L)$ is rational if L is rational, and $\tau(L)$ is context-free if L is context-free.

Associated with the Nivat theorem, the Chomsky–Schützenberger theorem establish in a stronger version that a language L is context-free iff there is a rational transduction τ such that $L = \tau(\mathcal{D}_2)$.

Context-Free Transductions: Following [2, page 62], a transduction $\tau \subseteq \Gamma^* \times \Sigma^*$ is context-free if there is an alphabet A, a context-free language $K \subseteq A^*$ and two homomorphisms $f : A^* \to \Sigma^*$ and $g : A^* \to \Gamma^*$ such that $\tau = \{(g(u), f(u)) \mid u \in K\}$. Equivalently, τ is context-free if it is generated by a context-free transduction grammar. This is a context-free grammar whose terminals are pairs of words. Derivations are done as usually but the product used on terminal pairs is the product of the monoid $\Gamma^* \times \Sigma^*$.

Context-free transductions enjoy however fewer good properties, in particular, [2, page 62] they are not closed under composition and classes of languages are usually not closed under them. For example, images of regulars languages are context-free languages and images of context-free languages are recursively enumerable languages.

3.2 ε-Reducible Context-Free Transductions and Transducers

Definition 10. *An ε-**Reducible Context-Free Transduction Grammar** (ε-CFTG) is a context-free transducer $G = (N, \widehat{\Gamma}, \Sigma, S, P)$ in which every production is in the form*

$$X \longrightarrow (\omega, u)\Omega(\bar{\omega}, v), \qquad \text{with } X \in N, \omega \in \widehat{\Gamma}^*, \ u, v \in \Sigma^*, \ \Omega \in N^*.$$

The transduction generated by G is $T_G = \{(u, v) \in \widehat{\Gamma}^* \times \Sigma^* \mid S \xrightarrow{*}_G (u, v)\}$. An ε-reducible context-free transduction (ε-CFT) is a context-free transduction generated by an ε-CFTG.

Example 11. Let $G = (N, \{\alpha, \beta, \bar{\alpha}, \bar{\beta}\}, \{a\}, S, P)$ be the ε-CFTG whose productions are:

$$S \longrightarrow (\beta, \varepsilon)X(\bar{\beta}, \varepsilon) \qquad X \longrightarrow (\alpha, \varepsilon)X(\bar{\alpha}, \varepsilon) \ X \longrightarrow (\varepsilon, \varepsilon)Y(\varepsilon, \varepsilon)$$
$$Y \longrightarrow (\bar{\alpha}, a)YZ(\alpha, \varepsilon) \quad Z \longrightarrow (\bar{\alpha}, a)Z(\alpha, a)Y \longrightarrow (\bar{\beta}, \varepsilon)(\beta, \varepsilon) \ Z \longrightarrow (\bar{\beta}, \varepsilon)(\bar{\beta}, \varepsilon).$$

Let τ be the transduction generated by G. The domain of τ is the ε-CFL given in Example 2 and one can easily check that

$$\tau = \bigcup_{n,m,r_1,\ldots r_m \geq 0} (\beta\alpha^n\bar{\alpha}^m\bar{\beta}\beta(\Pi_{i=1}^m\bar{\alpha}^{r_i}\bar{\beta}\beta\alpha^{r_i+1})\bar{\alpha}^n\bar{\beta}, a^{m+2r_1+\cdots+2r_m}). \qquad \square$$

Theorem 12. *Given a transduction $\tau \subseteq \widehat{\Gamma}^* \times A^*$, the following properties are equivalent:*

1. *τ is an ε-reducible context-free transduction;*
2. *there is an alphabet Δ, an ε-CFL $X \subseteq \widehat{\Delta}^*$, a symmetric homomorphism $g : \widehat{\Delta}^* \to \widehat{\Gamma}^*$ and a homomorphism $h : \widehat{\Delta}^* \to A^*$ such that*

$$\tau = \{(g(u), h(u)) \mid u \in X\};$$

3. *there is an alphabet Δ, a symmetric homomorphism $g : \widehat{\Delta}^* \to \widehat{\Gamma}^*$, a homomorphism $h : \widehat{\Delta}^* \to A^*$ and a regular set $R \subseteq \widehat{\Delta}^*$ such that*

$$\tau = \{(g(u), h(u)) \mid u \in R \cap \mathcal{D}_\Delta\}.$$

Proof. $(1 \Rightarrow 2)$ Suppose τ to be generated by an ε-CFTG $G = (N, \widehat{\Gamma}, \Sigma, S, P)$. We define the ε-CFG $G' = (N, \widehat{\Delta}, S, P')$ where $\Delta = P$, and the set of productions of P' is obtained by transforming every $p : X \longrightarrow (\omega, v)\Omega(\bar{\omega}, w) \in P$ into $X \longrightarrow p\Omega\bar{p}$. Now, let $h : \widehat{\Delta}^* \to A^*$ and $g : \widehat{\Delta}^* \to \widehat{\Gamma}^*$ such that for every $p : X \longrightarrow (\omega, v)\Omega(\bar{\omega}, w) \in P$, $g(p) = \omega$, $g(\bar{p}) = \bar{\omega}$ and $h(p) = v$, $h(\bar{p}) = w$. Clearly we have
$$T_G = \{(g(u), h(u)) \mid u \in \mathcal{L}(G')\}.$$

$(2 \Rightarrow 3)$ Suppose that $\tau = \{(g(u), h(u)) \mid u \in X\}$ where X is an ε-CFL and g symmetric. From Theorem 8, there is an alphabet C, a regular set $R \subseteq \widehat{C}^*$, and a symmetric homomorphism $g' : \widehat{C}^* \to \widehat{\Delta}^*$ such that $X = g'(R \cap \mathcal{D}_\Delta)$. The homomorphism $g \circ g'$ is symmetric as g and g' are both symmetric and $\tau = \{(g(g'(x)), h(g'(x))) \mid x \in R \cap \mathcal{D}_\Delta\}$.

$(3 \Rightarrow 1)$ Let $\tau = \{(g(u), h(u)) \mid u \in R \cap \mathcal{D}_\Delta\}$ where R is a regular language and g is symmetric. From Proposition 5, $R \cap \mathcal{D}_\Delta$ is an ε-CFL. Let us suppose that $R \cap \mathcal{D}_\Delta$ is generated by the ε-CFG $G = (N, \widehat{\Delta}, P, S)$, then τ is generated by the ε-CFTG $G' = (N, \widehat{\Gamma}, \widehat{\Sigma}P', S)$ where
$$P' = \{X \longrightarrow (g(u), f(u))\Omega(g(\bar{u}), h(\bar{u})) \mid X \longrightarrow u\Omega\bar{u} \in P, \Omega \in N^*, u \in \widehat{\Gamma}^*\}. \qquad \square$$

Theorem 12 implies that the image of a set X by an ε-CFT can be represented as $h(g^{-1}(X) \cap R \cap \mathcal{D}_\Delta)$ with R being a regular set, h a morphism and g a symmetric morphism. It is then clear that the family of images of regular sets by ε-CFTs is the family of context-free languages; we will see (Theorem 15) that the family of images of the Dyck language is that of indexed languages, but more generally, images of ε-CFLs by ε-CFTs are recursively enumerable languages.

Proposition 13. *Given a recursively enumerable language E, there is an ε-CFT τ, and an ε-CFL Z such that $E = \tau(Z)$.*

Proof. Let $E \subseteq \Sigma^*$. From [10], there is an alphabet Γ, a homomorphism $h : \widehat{\Gamma}^* \to \Sigma^*$, and a context-free language $K \subseteq \widehat{\Gamma}^*$ such that $E = h(K \cap \mathcal{D}_\Gamma)$. Let $g : \widehat{\Gamma}^* \to \widehat{\Gamma}^*$ be the injective symmetric homomorphism defined by $x \mapsto x\bar{x}$, for all $x \in \widehat{\Gamma}$. Then $E = h(g^{-1}(Z) \cap \mathcal{D}_\Gamma)$, for $Z = g(K)$. Note finally that Z is an ε-CFL: from the grammar in Chomsky normal form generating K, one obtain an ε-CFG generating Z by replacing the terminal productions $X \longrightarrow a$ by $X \longrightarrow g(a)$. Then from Theorem 12, $E = \tau(Z)$, where τ is an ε-CFT. $\qquad \square$

4   Characterizations of Indexed Languages

In this final section, we relate indexed languages to ε-CFTs by showing that indexed language are sets $\tau(\mathcal{D}_2)$, where τ is an ε-CFT. This gives rise to various homomorphic characterizations of indexed languages.

4.1   Indexed Grammars and Languages

Introduced by Aho [1], indexed grammars extend context-free grammars by allowing nonterminals to yield a stack. Derivable elements are then represented by symbols X^ω where X is a nonterminal and ω is a word called *index word*. Index words are accessed by a FIFO process: during a step of derivation of X^ω, it is possible to add a symbol in head ω, or to remove its first letter. Additionally, ω can be duplicated and distributed over other nonterminals.

Formally, an **indexed grammar** is a structure $\mathfrak{I} = (N, I, \Sigma, S, P)$, where N is the set of nonterminals, Σ is the set of terminals, $S \in N$ is the start symbol, I is a finite set of indexes, and P is a finite set of productions of the form

$$X_0{}^{\eta_0} \longrightarrow u_0 X_1{}^{\eta_1} u_1 \cdots X_n{}^{\eta_n} u_n$$

with $u_i \in \Sigma^*, X_i \in N$ and $\eta_i \in I \cup \{\varepsilon\}$ for $i \in \{0, \ldots, n\}$.

Indexes are denoted as *superscript*, and we do not write indexes equal to ε.

Sentences are words $u_1 A_1{}^{\omega_1} \ldots u_n A_n{}^{\omega_n} u_{n+1}$ with $u_i \in \Sigma^*$, $A_i \in N$ and $\omega_i \in I^*$. The derivation rule "$\longrightarrow_\mathfrak{I}$" is a binary relation over sentences defined by

$$\Omega_1 A^{\eta\omega} \Omega_2 \longrightarrow_\mathfrak{I} \Omega_1 u_0 B_1{}^{\eta_1\omega} \cdots B_n{}^{\eta_1\omega} u_n \Omega_2$$

iff there is a production $A^\eta \longrightarrow u_0 B_1{}^{\eta_1} u_1 \ldots B_n{}^{\eta_n} u_n \in P$.

The language generated by $\mathfrak{I}$ is $\mathcal{L}_\mathfrak{I} = \{u \in \Sigma^* \mid S \xrightarrow{*}_\mathfrak{I} u\}$. Languages generated by indexed grammars are called **indexed languages**.

Example 14. Let us consider the following indexed grammar $\mathfrak{I} = (N, I, A, S, P)$ with $N = \{S, X, A, B, C\}$, $I = \{\beta, \alpha\}$, $A = \{a, b, c\}$ and P consists of the following rules:

$p_1 : S \longrightarrow X^\beta,$ $p_2 : S \longrightarrow \varepsilon,$ $p_3 : X \longrightarrow X^\alpha,$ $p_4 : X \longrightarrow ABC,$
$p_5 : A^\alpha \longrightarrow aA,$ $p_6 : A^\beta \longrightarrow \varepsilon,$ $p_7 : B^\alpha \longrightarrow bB,$ $p_8 : B^\beta \longrightarrow \varepsilon,$
$p_9 : C^\alpha \longrightarrow cC,$ $p_{10} : C^\beta \longrightarrow \varepsilon.$

Here is a possible derivation:

$$S \xrightarrow{p_1} X^\beta \xrightarrow{p_3}_\mathfrak{I} X^{\alpha\beta} \xrightarrow{p_3}_\mathfrak{I} X^{\alpha\alpha\beta} \xrightarrow{p_4}_\mathfrak{I} A^{\alpha\alpha\beta} B^{\alpha\alpha\beta} C^{\alpha\alpha\beta} \xrightarrow{p_5}_\mathfrak{I} aA^{\alpha\beta} B^{\alpha\alpha\beta} C^{\alpha\alpha\beta}$$

$$\xrightarrow{p_5}_\mathfrak{I} aaA^\beta B^{\alpha\alpha\beta} C^{\alpha\alpha\beta} \xrightarrow{p_6}_\mathfrak{I} aaB^{\alpha\alpha\beta} C^{\alpha\alpha\beta} \xrightarrow{p_7 p_7 p_8}_\mathfrak{I} aabbC^{\alpha\alpha\beta} \xrightarrow{p_9 p_9 p_{10}}_\mathfrak{I} aabbcc$$

The language generated by $\mathfrak{I}$ is $\{a^n b^n c^n, n \geq 0\}$.   □

4.2   Characterizations of Indexed Languages

We provide now homomorphic characterizations of indexed languages by establishing a strong connexion between indexed languages and ε-CFTs.

Theorem 15. *A language L is indexed iff there is an ε-CFT τ such that*

$$L = \tau(\mathcal{D}_2).$$

Let us informally explain the proof of Theorem 15. First we need to consider normal forms of indexed grammars (which extend the normal form given in [1]) and ε-CFT grammars.

An indexed grammar is said to be *reduced* if its productions are in the forms:

$$X_0 \longrightarrow u X_1{}^\alpha \cdots X_n{}^\alpha v, \qquad \text{or } X_0{}^\alpha \longrightarrow u X_1 \cdots X_n v;$$
$$\text{with } n \geq 0, X_i \in N, u, v \in \Sigma^* \text{ and } \alpha \in I \cup \{\varepsilon\}.$$

An ε-CFTG is said to be *reduced* if its productions are in the form:

$$X_0 \longrightarrow (\alpha, u)\Omega(\bar{\alpha}, v) \text{ with } \Omega \in N^*, u, v \in \Sigma^* \text{ and } \alpha \in \widehat{\Gamma} \cup \{\varepsilon\}.$$

Let us consider the bijective mapping φ that maps a reduced indexed grammar $\mathfrak{I} = (N, I, \Sigma, P, S)$ into a reduced ε-CFTG $\varphi(\mathfrak{I}) = (N, \widehat{I}, \Sigma, \varphi(P), S)$ by transforming every production

$$p: X_0 \longrightarrow u\, X_1{}^\alpha \cdots X_n{}^\alpha v \text{ into } \varphi(p): X_0 \longrightarrow (\alpha, u)\, X_1 \cdots X_n\, (\bar{\alpha}, v), \text{ and}$$
$$p: X_0{}^\alpha \longrightarrow u\, X_1 \cdots X_n\, v \text{ into } \varphi(p): X_0 \longrightarrow (\bar{\alpha}, u)\, X_1 \cdots X_n\, (\alpha, v).$$

The idea behind the construction is to write, into the terminal inputs of the ε-CFTG, the index operations made by the indexed grammar. The transduction grammar thus created is able to capture every index modifications of the initial indexed grammar, but also accepts bad computations. We claim that by restricting the domain to Dyck words, we exactly get derivations equivalent to those of the indexed grammar.

For example, there would be a derivation

$$X \longrightarrow u_1 X_1{}^\alpha v_1 \longrightarrow u_2 Y_1{}^\alpha Y_2{}^\alpha v_2 \longrightarrow u_3 Y_1{}^\alpha w_3 Z v_3$$

in $\mathfrak{I}$ iff there was a derivation of the following form in $\varphi(\mathfrak{I})$:

$$X \longrightarrow (\alpha, u_1) X_1(\bar{\alpha}, v_1) \longrightarrow (\alpha, u_2) Y_1 Y_2(\bar{\alpha}, v_2) \longrightarrow (\alpha, u_3) Y_1(\bar{\alpha}, w_3) Z(\alpha\bar{\alpha}, v_3).$$

Claim: *There is a derivation* $S \xrightarrow{*}_{\mathfrak{I}} v_1 Y_1{}^{w_1} v_2 \cdots Y_n{}^{w_n} v_{n+1}$ *iff there is a derivation* $S \xrightarrow{*}_{\varphi(\mathfrak{I})} (u_1, v_1) Y_1(u_2, v_2) \cdots Y_n(u_{n+1}, v_{n+1})$ *where* $u_1 \cdots u_{n+1}$ *belongs to* $\mathcal{D}_I$ *and* $\rho(u_1 \cdots u_i) = w_i^R$ *for* $i \in \{1, \ldots, n\}$ *(w_i^R is the mirror image of w_i).*

This can be proved by induction over the length of derivations, and implies that $T_{\varphi(\mathfrak{I})}(\mathcal{D}_I) = \mathcal{L}_{\mathfrak{I}}$. Because of the bijectivity of the construction, we obtain:
    "A language L is indexed iff there is an ε-CFT τ and $k \in \mathbb{N}$ s.t. $L = \tau(\mathcal{D}_k)$."

Finally, it is possible to define from every ε-CFT τ, an ε-CFT τ' such that $\tau(\mathcal{D}_\Gamma) = \tau'(\mathcal{D}_2)$, by encoding every $\alpha_i \in \Gamma$ by a word 01^i0 and $\bar{\alpha}_i$ by $\bar{0}\bar{1}^i\bar{0}$.

Example 16. Let $\mathfrak{I} = (N, I, \Sigma, S, P)$ be an indexed grammar with $N = \{S, X, Y, W, Z\}$, $I = \{\beta, \alpha\}$, $A = \{a\}$ and P consists of the rules:

$$S \longrightarrow X^\beta, \quad X \longrightarrow X^\alpha, \quad X \longrightarrow Y, \quad Y^\alpha \longrightarrow aYZ$$
$$Y^\beta \longrightarrow \varepsilon, \quad Z^\alpha \longrightarrow aZa, \quad Z^\beta \longrightarrow \varepsilon.$$

Initially defined in [7], the grammar $\mathfrak{J}$ generates the language $L = \{a^{n^2} \mid n \geq 0\}$.

Applying the bijection φ defined above to $\mathfrak{J}$, we get the ε-CFTG G given in Example 11 and generating the transduction

$$\tau = \bigcup_{n,m,r_1,\dots r_m \geq 0} (\beta\alpha^n\bar{\alpha}^m\bar{\beta}\beta(\Pi_{i=1}^m\bar{\alpha}^{r_i}\bar{\beta}\beta\alpha^{r_i+1})\bar{\alpha}^n\bar{\beta}, a^{m+2r_1+\dots+2r_m}).$$

For every $u = \beta\alpha^n\bar{\alpha}^m\bar{\beta}\beta(\Pi_{i=1}^m\bar{\alpha}^{r_i}\bar{\beta}\beta\alpha^{r_i+1})\bar{\alpha}^n\bar{\beta} \in \mathrm{Dom}(\tau)$,

u is a Dyck word $\implies m = n, r_1 = 0$, and for all $i \in [0, m-1]$, $r_{i+1} = r_i + 1$

$$\implies \tau(u) = a^{n+2(0+1+\dots+n-1)}$$

$$\implies \tau(u) = a^{n^2}.$$

It follows that $\tau(\mathcal{D}_I) = \{a^{n^2}\}_{n \geq 0} = \mathcal{L}_{\mathfrak{J}}$. $\qquad\square$

Corollary 17. *A language L is indexed if there is a homomorphism h, a symmetric homomorphism g, a regular set R and $k \in \mathbb{N}$ such that*

$$L = h(g^{-1}(\mathcal{D}_2) \cap R \cap \mathcal{D}_k).$$

Theorem 18. *A language L is indexed iff there is an ε-CFL K, a morphism h, and an alphabet Γ such that*

$$L = h(K \cap \mathcal{D}_\Gamma).$$

Proof. ($\Rightarrow$) Let $L \subseteq A^*$ be an indexed language. From Theorems 12 and 15, there are alphabets Σ, Γ, an ε-CFL $K \subseteq \widehat{\Gamma}^*$, a homomorphism $h : \widehat{\Gamma}^* \to A^*$ and a symmetric homomorphism $g : \widehat{\Gamma}^* \to \widehat{\Sigma}^*$ such that $L = h(K \cap g^{-1}(\mathcal{D}_\Sigma))$. We suppose that $\Sigma \cap A = \emptyset$ (otherwise, it suffices to work with a copy of Σ), and define the homomorphism $\mu : \widehat{\Gamma}^* \to \widehat{\Delta}^*$, for $\Delta = \Sigma \cup A$, by $\alpha \mapsto g(\alpha)h(\alpha)\overline{h(\alpha)}$. For all $u \in \widehat{\Gamma}^*$, $\mu(u) \in \mathcal{D}_\Delta$ iff $u \in g^{-1}(\mathcal{D}_\Sigma)$; in addition, $\pi_A(\mu(u)) = h(u)$, with π_A being the projection of $\widehat{\Delta}^*$ into A^*. Then we have:

$$\pi_A(\mu(K) \cap \mathcal{D}_\Delta) = h(K \cap g^{-1}(\mathcal{D}_\Sigma)) = L.$$

Now, as the homomorphism μ satisfies "$\rho(u) = \varepsilon \implies \rho(g(u)) = \varepsilon$", since K is an ε-CFL, so is $\mu(K)$.

($\Leftarrow$) Obvious from Theorem 15 and Proposition 12, by choosing g to be the identity mapping. $\qquad\square$

Acknowledgement. The authors would like to thank Pr. Jean-Marc Talbot whose remarks and suggestions greatly improved the development of this paper.

References

1. Aho, A.: Indexed grammars-an extension of context-free grammars. J. ACM **15**, 647–671 (1968)
2. Berstel, J.: Transductions and Context-Free Languages. Teubner Verlag, Wiesbaden (1979)
3. Berstel, J., Boasson, L.: Context-free languages. In: van Leeuwen, J. (ed.) Handbook of Theoretical Computer Science (Vol. B), pp. 59–102. MIT Press, Cambridge (1990)
4. Berstel, J., Boasson, L.: Balanced grammars and their languages. In: Brauer, W., Ehrig, H., Karhumäki, J., Salomaa, A. (eds.) Formal and Natural Computing. LNCS, pp. 3–25. Springer, Heidelberg (2002)
5. Caucal, D.: On infinite transition graphs having a decidable monadic theory. Theor. Comput. Sci. **290**(1), 79–115 (2003)
6. Chomsky, N., Schützenberger, M.P.: The algebraic theory of context-free languages. In: Braffort, P., Hirschberg, D. (eds.) Computer Programming and Formal Systems. Studies in Logic and the Foundations of Mathematics, vol. 35, pp. 118–161. Elsevier, North-Holland (1963). doi:10.1016/S0049-237X(08)72023-8. http://www.sciencedirect.com/science/article/pii/S0049237X08720238. ISSN: 0049-237X
7. Fischer, M.J.: Grammars with macro-like productions. In: 9th Annual Symposium on Switching and Automata Theory. pp. 131–142. IEEE Computer Society (1968)
8. Fratani, S.: Automates à piles de piles.. de piles. Ph.D. thesis, Université Bordeaux 1 (2005)
9. Hague, M., Murawski, A.S., Ong, C.L., Serre, O.: Collapsible pushdown automata and recursion schemes. In: Proceedings LICS, pp. 452–461. IEEE Computer Society (2008)
10. Hirose, S., Nasu, M.: Left universal context-free grammars and homomorphic characterizations of languages. Inf. Control **50**(2), 110–118 (1981)
11. Kanazawa, M.: Multidimensional trees and a Chomsky-Schützenberger Weir representation theorem for simple context-free tree grammars. J. Logic Comput. (2014). doi:10.1093/logcom/exu043. http://logcom.oxfordjournals.org/content/early/2014/06/30/logcom.exu043.abstract
12. Knapik, T., Niwiński, D., Urzyczyn, P.: Higher-order pushdown trees are easy. In: Nielsen, M., Engberg, U. (eds.) FOSSACS 2002. LNCS, vol. 2303, pp. 205–222. Springer, Heidelberg (2002)
13. Maslov, A.N.: Hierarchy of indexed languages of arbitrary level. Sov. Math. Dokl **115**(14), 1170–1174 (1974)
14. Maslov, A.N.: Multilevel stack automata. Prob. Inf. Transm. **12**, 38–43 (1976)
15. Mehlhorn, K.: Pebbling mountain ranges and its application to DCFL-recognition. Automata, Lang. Program. **85**, 422–435 (1980)
16. Nivat, M.: Transductions des langages de Chomsky. Ann. Inst. Fourier **18**(1), 339–455 (1968)
17. Okhotin, A.: Non-erasing variants of the Chomsky–Schützenberger theorem. In: Yen, H.-C., Ibarra, O.H. (eds.) DLT 2012. LNCS, vol. 7410, pp. 121–129. Springer, Heidelberg (2012)
18. Ong, L.: Higher-order model checking: an overview. In: LICS, pp. 1–15. IEEE Computer Society (2015)
19. Sorokin, A.: Monoid automata for displacement context-free languages. CoRR abs/1403.6060 (2014)
20. Weir, D.: Characterizing Mildly Context-Sensitive Grammar Formalisms. Ph.D. thesis, University of Pennsylvania , available as Technical Report MS-CIS-88-74 (1988)

Ogden's Lemma, Multiple Context-Free Grammars, and the Control Language Hierarchy

Makoto Kanazawa(✉)

National Institute of Informatics and SOKENDAI,
2-1-2 Hitotsubashi, Chiyoda-ku, Tokyo 101-8430, Japan
kanazawa@nii.ac.jp

Abstract. I present a simple example of a multiple context-free language for which a very weak variant of generalized Ogden's lemma fails. This language is generated by a non-branching (and hence well-nested) 3-MCFG as well as by a (non-well-nested) binary-branching 2-MCFG; it follows that neither the class of well-nested 3-MCFLs nor the class of 2-MCFLs is included in Weir's control language hierarchy, for which Palis and Shende proved an Ogden-like iteration theorem. I then give a simple sufficient condition for an MCFG to satisfy a natural analogue of Ogden's lemma, and show that the corresponding class of languages is a substitution-closed full AFL which includes Weir's control language hierarchy. My variant of generalized Ogden's lemma is incomparable in strength to Palis and Shende's variant and is arguably a more natural generalization of Ogden's original lemma.

Keywords: Grammars · Ogden's lemma · Multiple context-free grammars · Control languages

1 Introduction

A *multiple context-free grammar* [12] is a context-free grammar on tuples of strings (of varying length). An analogue of the pumping lemma, which asserts the existence of a certain number of substrings that can be simultaneously iterated, has been established for *well-nested* MCFGs and (non-well-nested) MCFGs of dimension 2 [6]. So far, it has been unknown whether an analogue of Ogden's [10] strengthening of the pumping lemma holds of these classes. This paper negatively answers the question for both classes, and moreover proves a generalized Ogden's lemma for the class of MCFGs satisfying a certain simple property. The class of languages generated by the grammars in this class includes Weir's [13] control language hierarchy, the only non-trivial subclass of MCFLs for which an Ogden-style iteration theorem has been proved so far [11].

M. Kanazawa—This work was supported by JSPS KAKENHI Grant Number 25330020.

A.-H. Dediu et al. (Eds.): LATA 2016, LNCS 9618, pp. 371–383, 2016.
DOI: 10.1007/978-3-319-30000-9_29

2   Preliminaries

The set of natural numbers is denoted $\mathbb{N}$. If i and j are natural numbers, we write $[i, j]$ for the set $\{ n \in \mathbb{N} \mid i \leq n \leq j \}$. We write $|w|$ for the length of a string w and $|S|$ for the cardinality of a set S; the context should make it clear which is intended. If u, v, w are strings, we write $(u[v]w)$ for the subinterval $[|u| + 1, |uv|]$ of $[1, |uvw|]$. If w is a string, w^R denotes the reversal of w.

2.1   Multiple Context-Free Grammars

A *multiple context-free grammar* (MCFG) [12] is a quadruple $G = (N, \Sigma, P, S)$, where N is a finite set of *nonterminals*, each with a fixed *dimension* ≥ 1, Σ is a finite alphabet of *terminals*, P is a set of *rules*, and S is the distinguished *initial nonterminal* of dimension 1. We write $N^{(q)}$ for the set of nonterminals in N of dimension q. A nonterminal in $N^{(q)}$ is interpreted as a q-ary predicate over Σ^*. A rule is stated with the help of *variables* interpreted as ranging over Σ^*. Let $\mathcal{X}$ be a denumerable set of variables. We use boldface lower-case letters as elements of $\mathcal{X}$. A rule is a *definite clause* (in the sense of logic programming) constructed with *atoms* of the form $A(\alpha_1, \ldots, \alpha_q)$, with $A \in N^{(q)}$ and $\alpha_1, \ldots, \alpha_q$ *patterns*, i.e., strings over $\Sigma \cup \mathcal{X}$. An MCFG rule is of the form

$$A(\alpha_1, \ldots, \alpha_q) \leftarrow B_1(\boldsymbol{x}_{1,1}, \ldots, \boldsymbol{x}_{1,q_1}), \ldots, B_n(\boldsymbol{x}_{n,1}, \ldots, \boldsymbol{x}_{n,q_n}),$$

where $n \geq 0$, $A, B_1, \ldots, B_n$ are nonterminals of dimensions $q, q_1, \ldots, q_n$, respectively, the $\boldsymbol{x}_{i,j}$ are pairwise distinct variables, and each α_i is a string over $\Sigma \cup \{ \boldsymbol{x}_{i,j} \mid i \in [1, n], j \in [1, q_i] \}$, such that $(\alpha_1, \ldots, \alpha_q)$ contains at most one occurrence of each $\boldsymbol{x}_{i,j}$. An MCFG is an *m-MCFG* if the dimensions of its nonterminals do not exceed m; it is *r-ary branching* if each rule has no more than r occurrences of nonterminals in its *body* (i.e., the part that follows the symbol $\leftarrow$). We call a unary branching grammar *non-branching*.[1]

   An atom $A(\alpha_1, \ldots, \alpha_q)$ is *ground* if $\alpha_1, \ldots, \alpha_q \in \Sigma^*$. A *ground instance* of a rule is the result of substituting a string over Σ for each variable in the rule. Given an MCFG $G = (N, \Sigma, P, S)$, a ground atom $A(w_1, \ldots, w_q)$ *directly follows* from a sequence of ground atoms $B_1(v_{1,1}, \ldots, v_{1,q_1}), \ldots, B_n(v_{n,1}, \ldots, v_{n,q_n})$ if $A(w_1, \ldots, w_q) \leftarrow B_1(v_{1,1}, \ldots, v_{1,q_1}), \ldots, B_n(v_{n,1}, \ldots, v_{n,q_n})$ is a ground instance of some rule in P. A ground atom $A(w_1, \ldots, w_q)$ is *derivable*, written $\vdash_G A(w_1, \ldots, w_q)$, if it directly follows from some sequence of derivable ground atoms. In particular, if $A(w_1, \ldots, w_q) \leftarrow$ is a rule in P, we have $\vdash_G A(w_1, \ldots, w_q)$.

   A derivable ground atom is naturally associated with a *derivation tree*, each of whose nodes is labeled by a derivable ground atom, which directly follows from the sequence of ground atoms labeling its children. The language generated by G is defined as $L(G) = \{ w \in \Sigma^* \mid \vdash_G S(w) \}$, or equivalently, $L(G) = \{ w \in \Sigma^* \mid G \text{ has a derivation tree for } S(w) \}$. The class of languages generated by m-MCFGs is denoted m-MCFL, and the class of languages generated by r-ary branching m-MCFGs is denoted m-MCFL(r).

---

[1] Non-branching MCFGs have been called *linear* in [1].

Example 1. Consider the following 2-MCFG:

$$S(x_1\#x_2) \leftarrow D(x_1, x_2) \qquad D(x_1y_1, y_2x_2) \leftarrow E(x_1, x_2), D(y_1, y_2)$$
$$D(\varepsilon, \varepsilon) \leftarrow \qquad E(ax_1\bar{a}, \bar{a}x_2a) \leftarrow D(x_1, x_2)$$

Here, S is the initial nonterminal and D and E are both nonterminals of dimension 2. This grammar is binary branching and generates the language $\{w\#w^R \mid w \in D_1^*\}$, where D_1^* is the (one-sided) *Dyck language* over the alphabet $\{a, \bar{a}\}$. Figure 1 shows the derivation tree for $aa\bar{a}\bar{a}a\bar{a}\#\bar{a}a\bar{a}\bar{a}aa$.

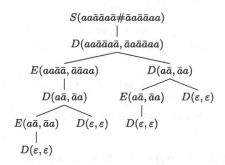

Fig. 1. A derivation tree of a 2-MCFG.

It is also useful to define the notion of a derivation of an atom $A(\alpha_1, \ldots, \alpha_q)$ from an assumption $C(x_1, \ldots, x_r)$, where $x_1, \ldots, x_r$ are pairwise distinct variables. An atom $A(\alpha_1, \ldots, \alpha_q)$ is *derivable from an assumption* $C(x_1, \ldots, x_r)$, written $C(x_1, \ldots, x_r) \vdash_G A(\alpha_1, \ldots, \alpha_q)$, if either

1. $A = C$ and $(\alpha_1, \ldots, \alpha_q) = (x_1, \ldots, x_r)$, or
2. there are some atom $B_i(\beta_1, \ldots, \beta_{q_i})$ and ground atoms $B_j(v_{j,1}, \ldots, v_{j,q_j})$ for each $j \in [1, i-1] \cup [i+1, n]$ such that $C(x_1, \ldots, x_r) \vdash_G B_i(\beta_1, \ldots, \beta_{q_i})$, $\vdash_G B_j(v_{j,1}, \ldots, v_{j,q_j})$, and

$$A(\alpha_1, \ldots, \alpha_q) \leftarrow B_1(v_{1,1}, \ldots, v_{1,q_1}), \ldots, B_{i-1}(v_{i-1,1}, \ldots, v_{i-1,q_{i-1}}),$$
$$B_i(\beta_1, \ldots, \beta_{q_i}), B_{i+1}(v_{i+1,1}, \ldots, v_{i+1,q_{i+1}}), \ldots, B_n(v_{n,1}, \ldots, v_{n,q_n})$$

is an instance of some rule in P.

Let us write $[v_1/x_1, \ldots, v_r/x_r]$ for the simultaneous substitution of strings $v_1, \ldots, v_r$ for variables $x_1, \ldots, x_r$. Evidently, when we have $\vdash_G B(v_1, \ldots, v_r)$ and $B(x_1, \ldots, x_r) \vdash_G A(\alpha_1, \ldots, \alpha_q)$, the two derivations can be combined into one witnessing $\vdash_G A(\alpha_1, \ldots, \alpha_q)[v_1/x_1, \ldots, v_r/x_r]$. The following lemma says that when $B(v_1, \ldots, v_r)$ is derived in the course of a derivation of $A(w_1, \ldots, w_q)$, the derivation can be decomposed into one for $B(v_1, \ldots, v_r)$ and a derivation from an assumption $B(x_1, \ldots, x_r)$:

Lemma 2. *Let τ be a derivation tree of an MCFG G for some ground atom $A(w_1, \ldots, w_q)$, and let $B(v_1, \ldots, v_r)$ be the label of some node of τ. Then there is an atom $A(\alpha_1, \ldots, \alpha_q)$ such that $B(x_1, \ldots, x_r) \vdash_G A(\alpha_1, \ldots, \alpha_q)$ and $(w_1, \ldots, w_q) = (\alpha_1, \ldots, \alpha_q)[v_1/x_1, \ldots, v_r/x_r]$.*

Example 3. Consider the derivation tree in Fig. 1 and the node ν labeled by $E(a a \bar{a} \bar{a}, \bar{a} \bar{a} a a)$. Let τ be the subtree of this derivation tree consisting of ν and the nodes that lie below it. Consider the node ν_1 labeled by $E(a\bar{a}, \bar{a}a)$ in τ. The rules used in the portion of τ that remains after removing the nodes below ν_1 determine a derivation tree for $E(x_1, x_2) \vdash_G E(a x_1 \bar{a}, \bar{a} x_2 a)$, depicted in Fig. 2. Note that substituting $a\bar{a}, \bar{a}a$ for x_1, x_2 in $E(a x_1 \bar{a}, \bar{a} x_2 a)$ gives back $E(a a \bar{a} \bar{a}, \bar{a} \bar{a} a a)$.

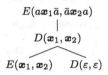

Fig. 2. A derivation of $E(a x_1 \bar{a}, \bar{a} x_2 a)$ from assumption $E(x_1, x_2)$.

An MCFG rule $A(\alpha_1, \ldots, \alpha_q) \leftarrow B_1(x_{1,1}, \ldots, x_{1,q_1}), \ldots, B_n(x_{n,1}, \ldots, x_{n,q_n})$ is said to be

- *non-deleting* if all variables $x_{i,j}$ in its body occur in $(\alpha_1, \ldots, \alpha_q)$;
- *non-permuting* if for each $i \in [1, n]$, the variables $x_{i,1}, \ldots, x_{i,q_i}$ occur in $(\alpha_1, \ldots, \alpha_q)$ in this order;
- *well-nested* if it is non-deleting and non-permuting and there are no $i, j \in [1, n], k \in [1, q_i - 1], l \in [1, q_l - 1]$ such that $x_{i,k}, x_{j,l}, x_{i,k+1}, x_{j,l+1}$ occur in $(\alpha_1, \ldots, \alpha_q)$ in this order.

Every m-MCFG(r) has an equivalent m-MCFG(r) whose rules are all non-deleting and non-permuting, and henceforth we will always assume that these conditions are satisfied. An MCFG whose rules are all well-nested is a *well-nested MCFG* [6]. The 2-MCFG in Example 1 is well-nested. It is known that there is no well-nested MCFG for the language $\{ w \# w \mid w \in D_1^* \}$ [9], although it is easy to write a non-well-nested 2-MCFG for this language.

Every (non-deleting and non-permuting) non-branching MCFG is by definition well-nested. The class $\bigcup_m m$-MCFL(1) coincides with the class of output languages of *deterministic two-way finite-state transducers* (see [1]).

2.2   The Control Language Hierarchy

Weir's [13] *control language hierarchy* is defined in terms of the notion of a *labeled distinguished grammar*, which is a 5-tuple $G = (N, \Sigma, P, S, f)$, where $\overline{G} = (N, \Sigma, P, S)$ is an ordinary context-free grammar and $f : P \to \mathbb{N}$ is a function such that if $\pi \in P$ is a context-free production with n occurrences of nonterminals on its right-hand side, then $f(\pi) \in [0, n]$. We view P as a finite alphabet, and use a language $C \in P^*$ to restrict the derivations of G. The pair (G, C) is a *control grammar*. For each nonterminal $A \in N$, define $R_{(G,C)}(A) \subseteq \Sigma^* \times P^*$ inductively as follows: for each production $\pi = A \to w_0 B_1 w_1 \ldots B_n w_n$ in P,

- if $f(\pi) = 0$ and $(\{v_j\} \times C) \cap R_{(G,C)}(B_j) \neq \varnothing$ for each $j \in [1, n]$, then $(w_0 v_1 w_1 \ldots v_n w_n, \pi) \in R_{(G,C)}(A)$;
- if $f(\pi) = i \in [1, n]$, $(v_i, z) \in R_{(G,C)}(B_i)$, and $(\{v_j\} \times C) \cap R_{(G,C)}(B_j) \neq \varnothing$ for each $j \in [1, i-1] \cup [i+1, n]$, then $(w_0 v_1 w_1 \ldots v_n w_n, \pi z) \in R_{(G,C)}(A)$.

The language of the control grammar (G, C) is $L(G, C) = \{ w \in \Sigma^* \mid (\{w\} \times C) \cap R_{(G,C)}(S) \neq \varnothing \}$.

The first level of the control language hierarchy is $\mathcal{C}_1 = \text{CFL}$, the family of context-free languages, and for $k \geq 1$,

$$\mathcal{C}_{k+1} = \{ L(G, C) \mid (G, C) \text{ is a control grammar and } C \in \mathcal{C}_k \}.$$

The second level $\mathcal{C}_2$ is known to coincide with the family of languages generated by well-nested 2-MCFGs, or equivalently, the family of *tree-adjoining languages* [13].

Example 4. Let $G = (N, \Sigma, P, S, f)$ be a labeled distinguished grammar consisting of the following productions:

$$\pi_1 : S \to a S \bar{a} S, \qquad \pi_2 : S \to b S \bar{b} S, \qquad \pi_3 : S \to \varepsilon,$$

where $f(\pi_1) = 1, f(\pi_2) = 1, f(\pi_3) = 0$. Let $C = \{ \pi_1^n \pi_2^n \pi_3 \mid n \in \mathbb{N} \}$. Then $L(G, C) = D_2^* \cap (\{ a^n b^n \mid n \in \mathbb{N} \} \{ \bar{a}, \bar{b} \}^*)^*$, where D_2^* is the Dyck language over $\{ a, \bar{a}, b, \bar{b} \}$. Since C is a context-free language, this language belongs to $\mathcal{C}_2$.

Palis and Shende [11] proved the following Ogden-like theorem for $\mathcal{C}_k$:

Theorem 5 (Palis and Shende). *If $L \in \mathcal{C}_k$, then there is a number p such that for all $z \in L$ and $D \subseteq [1, |z|]$, if $|D| \geq p$, there are $u_1, \ldots, u_{2^k+1}, v_1, \ldots, v_{2^k} \in \Sigma^*$ that satisfy the following conditions:*

(i) $z = u_1 v_1 u_2 v_2 \ldots u_{2^k} v_{2^k} u_{2^k+1}$.

(ii) *for some $j \in [1, 2^k]$,*

$$D \cap (u_1 v_1 \ldots [u_j] v_j u_{j+1} v_{j+1} \ldots u_{2^k} v_{2^k} u_{2^k+1}) \neq \varnothing,$$
$$D \cap (u_1 v_1 \ldots u_j [v_j] u_{j+1} v_{j+1} \ldots u_{2^k} v_{2^k} u_{2^k+1}) \neq \varnothing,$$
$$D \cap (u_1 v_1 \ldots u_j v_j [u_{j+1}] v_{j+1} \ldots u_{2^k} v_{2^k} u_{2^k+1}) \neq \varnothing.$$

(iii) $|D \cap (u_1 v_1 \ldots u_{2^{k-1}} [v_{2^{k-1}} u_{2^{k-1}+1} v_{2^{k-1}+1}] \ldots u_{2^k} v_{2^k} u_{2^k+1})| \leq p$.

(iv) $u_1 v_1^n u_2 v_2^n \ldots u_{2^k} v_{2^k}^n u_{2^k+1} \in L$ *for all $n \in \mathbb{N}$.*

Kanazawa and Salvati [8] proved the inclusion $\mathcal{C}_k \subseteq 2^{k-1}\text{-MCFL}$, while using Theorem 5 to show that the language $\text{RESP}_{2^{k-1}}$ belongs to $2^{k-1}\text{-MCFL} - \mathcal{C}_k$ for $k \geq 2$, where $\text{RESP}_l = \{ a_1^m a_2^m b_1^n b_2^n \ldots a_{2l-1}^m a_{2l}^m b_{2l-1}^n b_{2l}^n \mid m, n \in \mathbb{N} \}$.

3   The Failure of Ogden's Lemma for Well-Nested MCFGs and 2-MCFGs

Let G be an MCFG, and consider a derivation tree τ for an element z of $L(G)$. When a node of τ and one of its descendants are labeled by ground atoms $B(w_1, \ldots, w_r)$ and $B(v_1, \ldots, v_r)$ sharing the same nonterminal B, the portion of τ consisting of the nodes that are neither above the first node nor below the second node determines a derivation tree σ witnessing $B(\boldsymbol{x}_1, \ldots, \boldsymbol{x}_r) \vdash_G B(\beta_1, \ldots, \beta_r)$ (called a *pump* in [6]), where $(\beta_1, \ldots, \beta_r)[v_1/\boldsymbol{x}_1, \ldots, v_r/\boldsymbol{x}_r] = (w_1, \ldots, w_r)$. This was illustrated by Example 3. When each $\boldsymbol{x}_i$ occurs in β_i, i.e., $\beta_i = v_{2i-1}\boldsymbol{x}_i v_{2i}$ for some $v_{2i-1}, v_{2i} \in \Sigma^*$ (in which case σ is an *even pump* [6]), iterating σ gives a derivation tree for $B(\boldsymbol{x}_1, \ldots, \boldsymbol{x}_r) \vdash_G B(v_1^n \boldsymbol{x}_1 v_2^n, \ldots, v_{2r-1}^n \boldsymbol{x}_r v_{2r}^n)$. Combining this with the rest of τ gives a derivation tree for $z(n) = u_1 v_1^n u_2 v_2^n \ldots u_{2r} v_{2r}^n u_{2r+1} \in L(G)$ for every $n \in \mathbb{N}$, where $z(1) = z$. When some $\boldsymbol{x}_i$ occurs in β_j with $j \neq i$ (σ is an *uneven pump*), however, the result of iterating σ exhibits a complicated pattern that is not easy to describe.

A language L is said to be *k-iterative* if all but finitely many elements of L can be written in the form $u_1 v_1 u_2 v_2 \ldots u_k v_k u_{k+1}$ so that $v_1 \ldots v_k \neq \varepsilon$ and $u_1 v_1^n u_2 v_2^n \ldots u_k v_k^n u_{k+1} \in L$ for all $n \in \mathbb{N}$. A language that is either finite or includes an infinite k-iterative subset is said to be *weakly k-iterative*. (These terms are from [3,4].) The possibility of an uneven pump explains the difficulty of establishing $2m$-iterativity of an m-MCFL. In 1991, Seki et al. [12] proved that every m-MCFL is weakly $2m$-iterative, but whether every m-MCFL is $2m$-iterative remained an open question for a long time, until Kanazawa et al. [7] negatively settled it in 2014 by exhibiting a (non-well-nested) 3-MCFL that is not k-iterative for any k. Earlier, Kanazawa [6] had shown that the language of a well-nested m-MCFG is always $2m$-iterative, and moreover that a 2-MCFL is always 4-iterative. The proof of this last pair of results was much more indirect than the proof of the pumping lemma for the context-free languages, and did not suggest a way of strengthening them to an Ogden-style theorem. Below, we show that there is indeed no reasonable way of doing so.

Let us say that a language L has the *weak Ogden property* if there is a natural number p such that for every $z \in L$ and $D \subseteq [1, |z|]$ with $|D| \geq p$, there are strings $u_1, \ldots, u_{k+1}, v_1, \ldots, v_k$ ($k \geq 1$) satisfying the following conditions:

1. $z = u_1 v_1 \ldots u_k v_k u_{k+1}$,
2. $D \cap (u_1 v_1 \ldots u_i[v_i] \ldots u_k v_k u_{k+1}) \neq \varnothing$ for some $i \in [1, k]$, and
3. $u_1 v_1^n \ldots u_k v_k^n u_{k+1} \in L$ for all $n \geq 0$.

The elements of D are referred to as *distinguished positions* in z.

Theorem 6. *There is an $L \in$ 3-MCFL(1) $\cap$ 2-MCFL(2) that does not satisfy the weak Ogden property.*

$$A(\varepsilon) \leftarrow$$
$$A(b\boldsymbol{x}_1) \leftarrow A(\boldsymbol{x}_1)$$
$$B(\boldsymbol{x}_1, \varepsilon) \leftarrow A(\boldsymbol{x}_1)$$
$$B(a\boldsymbol{x}_1, b\boldsymbol{x}_2) \leftarrow B(\boldsymbol{x}_1, \boldsymbol{x}_2)$$
$$C(\boldsymbol{x}_1, \boldsymbol{x}_2, \varepsilon) \leftarrow B(\boldsymbol{x}_1, \boldsymbol{x}_2)$$
$$C(\boldsymbol{x}_1, a\boldsymbol{x}_2, b\boldsymbol{x}_3) \leftarrow C(\boldsymbol{x}_1, \boldsymbol{x}_2, \boldsymbol{x}_3)$$
$$C(\boldsymbol{x}_1\$\boldsymbol{x}_2, \boldsymbol{x}_3, \varepsilon) \leftarrow C(\boldsymbol{x}_1, \boldsymbol{x}_2, \boldsymbol{x}_3)$$
$$D(\boldsymbol{x}_1\$\boldsymbol{x}_2, \boldsymbol{x}_3) \leftarrow C(\boldsymbol{x}_1, \boldsymbol{x}_2, \boldsymbol{x}_3)$$
$$D(\boldsymbol{x}_1, a\boldsymbol{x}_2) \leftarrow D(\boldsymbol{x}_1, \boldsymbol{x}_2)$$
$$S(\boldsymbol{x}_1\$\boldsymbol{x}_2) \leftarrow D(\boldsymbol{x}_1, \boldsymbol{x}_2)$$

$$A(\varepsilon) \leftarrow$$
$$A(b\boldsymbol{x}_1) \leftarrow A(\boldsymbol{x}_1)$$
$$B(\boldsymbol{x}_1, \varepsilon) \leftarrow A(\boldsymbol{x}_1)$$
$$B(a\boldsymbol{x}_1, b\boldsymbol{x}_2) \leftarrow B(\boldsymbol{x}_1, \boldsymbol{x}_2)$$
$$C(\varepsilon, \varepsilon) \leftarrow$$
$$C(a\boldsymbol{x}_1, b\boldsymbol{x}_2) \leftarrow C(\boldsymbol{x}_1, \boldsymbol{x}_2)$$
$$D(\boldsymbol{x}_1\$\boldsymbol{y}_1\boldsymbol{x}_2, \boldsymbol{y}_2) \leftarrow B(\boldsymbol{x}_1, \boldsymbol{x}_2), C(\boldsymbol{y}_1, \boldsymbol{y}_2)$$
$$D(\boldsymbol{x}_1\$\boldsymbol{y}_1\boldsymbol{x}_2, \boldsymbol{y}_2) \leftarrow D(\boldsymbol{x}_1, \boldsymbol{x}_2), C(\boldsymbol{y}_1, \boldsymbol{y}_2)$$
$$E(\boldsymbol{x}_1, \boldsymbol{x}_2) \leftarrow D(\boldsymbol{x}_1, \boldsymbol{x}_2)$$
$$E(\boldsymbol{x}_1, a\boldsymbol{x}_2) \leftarrow E(\boldsymbol{x}_1, \boldsymbol{x}_2)$$
$$S(\boldsymbol{x}_1\$\boldsymbol{x}_2) \leftarrow E(\boldsymbol{x}_1, \boldsymbol{x}_2)$$

Fig. 3. Two grammars generating the same language.

Proof. Let L be the set of all strings over the alphabet $\{a, b, \$\}$ that are of the form

$$a^{i_1} b^{i_0} \$ a^{i_2} b^{i_1} \$ a^{i_3} b^{i_2} \$ \ldots \$ a^{i_n} b^{i_{n-1}} \qquad (\dagger)$$

for some $n \geq 3$ and $i_0, \ldots, i_n \geq 0$. This language is generated by the non-branching 3-MCFG (left) as well as by the binary branching 2-MCFG (right) in Fig. 3. Now suppose L has the weak Ogden property, and let p be the number satisfying the required conditions. Let

$$z = a\$a^2 b\$a^3 b^2 \$ \ldots \$a^{p+1} b^p,$$

and let D consist of the positions in z occupied by $\$$. Note that $|D| = p$. By the weak Ogden property, there must be strings $u_1, \ldots, u_{k+1}, v_1, \ldots, v_k$ ($k \geq 1$) such that $z = u_1 v_1 \ldots u_k v_k u_{k+1}$, at least one of $v_1, \ldots, v_k$ contains an occurrence of $\$$, and $u_1 v_1^n \ldots u_k v_k^n u_{k+1} \in L$ for all n. Without loss of generality, we may assume that $v_1, \ldots, v_k$ are all nonempty strings. Let us write $z(n)$ for $u_1 v_1^n \ldots u_k v_k^n u_{k+1}$. First note that none of $v_1, \ldots, v_k$ can start in a and end in b, since otherwise $z(2)$ would contain ba as a factor and not be of the form $(\dagger)$. Let i be the greatest number such that v_i contains an occurrence of $\$$. Since none of $v_{i+1}, \ldots, v_k$ contains an occurrence of $\$$, it is easy to see that $v_{i+1}, \ldots, v_k$ are all in $a^+ \cup b^+$. We consider two cases, depending on the number of occurrences of $\$$ in v_i. Each case leads to a contradiction.

Case 1. v_i contains just one occurrence of $\$$. Then $v_i = x\$y$, where x is a suffix of $a^{j+1} b^j$ and y is a prefix of $a^{j+2} b^{j+1}$ for some $j \in [0, p-1]$. Note that $z(3)$ contains $\$yx\$yx\$$ as a factor. Since $z(3)$ is of the form $(\dagger)$, this means that $yx = a^l b^l$ for some $l \geq 0$.

Case 1.1. $l \leq j+1$. Then y must be a prefix of a^{j+1} and since x is a suffix of $a^{j+1} b^j$, it follows that $l \leq j$. Since $yu_{i+1}v_{i+1} \ldots u_k v_k u_{k+1}$ has $a^{j+2} b^{j+1}$ as a prefix and $v_{i+1}, \ldots, v_k \in a^+ \cup b^+$, $\$yx\$yu_{i+1}v_{i+1}^2 \ldots u_k v_k^2 u_{k+1}$ has $\$a^l b^l \$a^q b^r$ as a prefix for some $q \geq j+2$ and $r \geq j+1$. The string $\$a^l b^l \$a^q b^r$ is a factor of $z(2)$ and since $z(2)$ is of the form $(\dagger)$, we must have $l \geq r$, but this contradicts $l \leq j$.

Case 1.2. $l \geq j + 2$. In this case x must be a suffix of b^j and y must have $a^{j+2}b^2$ as a prefix, so $l = j + 2$. Note that

$$\$yx\$yu_{i+1}v_{i+1}^2 \ldots u_k v_k^2 u_{k+1} = \$a^l b^l \$yu_{i+1}v_{i+1}^2 \ldots u_k v_k^2 u_{k+1}$$

is a suffix of $z(2)$, so either $yu_{i+1}v_{i+1}^2 \ldots u_k v_k^2 u_{k+1}$ equals $a^q b^l$ or has $a^q b^l \$$ as a prefix for some q. Since $l = j + 2$ and $yu_{i+1}v_{i+1} \ldots u_k v_k u_{k+1}$ either equals $a^{j+2}b^{j+1}$ or has $a^{j+2}b^{j+1}\$$ as a prefix, it follows that there is some $h > i$ such that $v_h = b$ and $v_{i+1}, \ldots, v_{h-1}$ are all in a^+. But then $z(3)$ will contain

$$\$yx\$yu_{i+1}v_{i+1}^3 \ldots u_k v_k^3 u_{k+1},$$

which must have

$$\$a^{j+2}b^{j+2}\$a^{q'} b^{j+3}$$

as a prefix for some q', contradicting the fact that $z(3)$ is of the form (†).

Case 2. v_i contains at least two occurrences of $\$$. Then we can write

$$v_i = x\$a^{l+1}b^l\$ \ldots \$a^{m+1}b^m\$y,$$

where $1 \leq l \leq m \leq p - 1$, x is a suffix of $a^l b^{l-1}$, and y is a prefix of $a^{m+2}b^{m+1}$. Since

$$\$a^{m+1}b^m\$yx\$a^{l+1}b^l\$$$

is a factor of $z(2)$, we must have

$$yx = a^l b^{m+1}.$$

Since y is a prefix of $a^{m+2}b^{m+1}$ and $l < m + 2$, y must be a prefix of a^l. It follows that x has b^{m+1} as a suffix. But then b^{m+1} must be a suffix of $a^l b^{l-1}$, contradicting the fact that $l - 1 < m + 1$.   □

Since Theorem 5 above implies that every language in Weir's control language hierarchy satisfies the weak Ogden property, we obtain the following corollary:[2]

Corollary 7. *There is a language in* 3-MCFL(1) ∩ 2-MCFL(2) *that lies outside of Weir's control language hierarchy.*

Previously, Kanazawa et al. [7] showed that Weir's control language hiearchy does not include 3-MCFL(2), but left open the question of whether the former includes the languages of well-nested MCFGs. The above corollary settles this question in the negative.

---

[2] The language L in the proof of Theorem 6 was inspired by Lemma 5.4 of Greibach [5], where a much more complicated language was used to show that the range of a deterministic two-way finite-state transducer need not be *strongly iterative*. One can see that the language Greibach used is an 8-MCFL(1). In her proof, Greibach essentially relied on a stronger requirement imposed by her notion of strong iterativity, namely that in the factorization $z = u_1 v_1 \ldots u_k v_k u_{k+1}$, there must be some i such that u_i and u_{i+1} contain at least one distinguished position and v_i contains at least *two* distinguished positions. Strong iterativity is not implied by the condition in Theorem 5, so Greibach's lemma fell short of providing an example of a language in $\bigcup_m$ m-MCFL(1) that does not belong to Weir's hierarchy.

4   A Generalized Ogden's Lemma for a Subclass of the MCFGs

An easy way of ensuring that an m-MCFG G satisfies a generalized Ogden's lemma is to demand that whenever $B(x_1, \ldots, x_r) \vdash_G B(\beta_1, \ldots, \beta_r)$, each x_i occurs in β_i. This is a rather strict requirement, however, and the resulting class of grammars does not seem to cover even the second level C_2 of the control language hierarchy. In this section, we show that a weaker condition implies a natural analogue of Ogden's [10] condition; we prove in the next section that the result covers the entire control language hierarchy.

Let us say that a derivation of $B(\beta_1, \ldots, \beta_r)$ from assumption $A(x_1, \ldots, x_q)$ is *non-decreasing* if it cannot be broken down into two derivations witnessing $A(x_1, \ldots, x_q) \vdash_G C(\gamma_1, \ldots, \gamma_s)$ and $C(z_1, \ldots, z_s) \vdash_G B(\beta'_1, \ldots, \beta'_r)$ such that $s < q$. (If $q > r$, there can be no non-decreasing derivation witnessing $A(x_1, \ldots, x_q) \vdash_G B(\beta_1, \ldots, \beta_r)$.) An m-MCFG $G = (N, \Sigma, P, S)$ is *proper* if for each $A \in N^{(q)}$, whenever $A(x_1, \ldots, x_q) \vdash_G A(\alpha_1, \ldots, \alpha_q)$ with a non-decreasing derivation, each x_i occurs in α_i. It is easy to see that properness is a decidable property of an MCFG.

Theorem 8. *Let L be the language of a proper m-MCFG. There is a natural number p such that for every $z \in L$ and $D \subseteq [1, |z|]$ with $|D| \geq p$, there are strings $u_1, \ldots, u_{2m+1}, v_1, \ldots, v_{2m}$ satisfying the following conditions:*

1. $z = u_1 v_1 \ldots u_{2m} v_{2m} u_{2m+1}$,
2. *for some $j \in [1, 2m]$,*

$$D \cap (u_1 v_1 \ldots [u_j] v_j u_{j+1} v_{j+1} \ldots u_{2m} v_{2m} u_{2m+1}) \neq \varnothing,$$
$$D \cap (u_1 v_1 \ldots u_j [v_j] u_{j+1} v_{j+1} \ldots u_{2m} v_{2m} u_{2m+1}) \neq \varnothing,$$
$$D \cap (u_1 v_1 \ldots u_j v_j [u_{j+1}] v_{j+1} \ldots u_{2m} v_{2m} u_{2m+1}) \neq \varnothing,$$

3. $|D \cap \bigcup_{i=1}^{m} (u_1 v_1 \ldots u_{2i-1} [v_{2i-1} u_{2i} v_{2i}] \ldots u_{2m} v_{2m} u_{2m+1})| \leq p$,
4. $u_1 v_1^n u_2 v_2^n \ldots u_{2m} v_{2m}^n u_{2m+1} \in L$ *for all $n \in \mathbb{N}$.*

The case $m = 1$ of Theorem 8 exactly matches the condition in Ogden's [10] original lemma (as does the case $k = 1$ of Theorem 5).

Proof. Let $G = (N, \Sigma, P, S)$ be a proper m-MCFG. For a rule $A(\alpha_1, \ldots, \alpha_q) \leftarrow B_1(x_{1,1}, \ldots, x_{1,q_1}), \ldots, B_n(x_{n,1}, \ldots, x_{n,q_n})$, let its *weight* be the number of occurrences of terminal symbols in $\alpha_1, \ldots, \alpha_q$ plus n, and let d be the maximal weight of a rule in P.

Let $z \in L$, $D \subseteq [1, |z|]$, and τ be a derivation tree for z. We refer to elements of D as *distinguished positions*. Note that it makes sense to ask whether a particular symbol occurrence in the atom $A(w_1, \ldots, w_q)$ labeling a node ν of τ is in a distinguished position or not. This is because by Lemma 2, there are strings $z_1, \ldots, z_{q+1}$ such that ν determines a derivation witnessing $A(x_1, \ldots, x_q) \vdash_G S(z_1 x_1 z_2 x_2 \ldots z_q x_q z_{q+1})$, which tells us where in z each argument of $A(w_1, \ldots, w_q)$ ends up. Henceforth, when the ground atom labeling a node ν contains a symbol occurrence in a distinguished position, we simply say that ν contains a distinguished position. We call a node ν a *B-node* (cf. [10]) if at least one of its children contains a distinguished position and ν contains more distinguished positions than any of its children. The *B-height* of a node ν is defined as the maximal B-height h of its children if ν is not a B-node, and $h + 1$ if ν is a B-node. (When ν has no children, its B-height is 0.) It is easy to see that a node of B-height h can contain no more than d^{h+1} distinguished positions.

Our goal is to find an h such that, when $|D| \geq d^{h+1}$, we can locate four nodes $\mu_1, \mu_2, \mu_3, \mu_4$, all of B-height $\leq h$, on the same path of τ that together decompose τ into five derivations witnessing

$$A(\boldsymbol{x}_1, \ldots, \boldsymbol{x}_q) \vdash_G S(z_1 \boldsymbol{x}_1 z_2 \boldsymbol{x}_2 \ldots z_q \boldsymbol{x}_q z_{q+1}), \tag{1}$$

$$B(\boldsymbol{x}_1, \ldots, \boldsymbol{x}_q) \vdash_G A(y_1 \boldsymbol{x}_1 y_2, \ldots, y_{2q-1} \boldsymbol{x}_q y_{2q}), \tag{2}$$

$$B(\boldsymbol{x}_1, \ldots, \boldsymbol{x}_q) \vdash_G B(v_1 \boldsymbol{x}_1 v_2, \ldots, v_{2q-1} \boldsymbol{x}_q v_{2q}), \tag{3}$$

$$C(\boldsymbol{x}_1, \ldots, \boldsymbol{x}_q) \vdash_G B(x_1 \boldsymbol{x}_1 x_2, \ldots, x_{2q-1} \boldsymbol{x}_q x_{2q}), \tag{4}$$

$$\vdash_G C(w_1, \ldots, w_q), \tag{5}$$

where for some $j \in [1, 2q]$, each of x_j, v_j, y_j contains at least one distinguished position. Since $y_1 v_1 x_1 w_1 x_2 v_2 y_2, \ldots, y_{2q-1} v_{2q-1} x_{2q-1} w_q x_{2q} v_{2q} y_{2q}$ together can contain no more than d^{h+1} distinguished positions, this establishes the theorem, with $p = d^{h+1}$ and $u_1 = z_1 y_1, u_2 = x_1 w_1 x_2, u_3 = y_2 z_2 y_3$, etc.

We let $h = \sum_{q=1}^{m} h(q)$, where $h(0) = 0$ and $h(q) = (2q \cdot (|N|+1)+1) \cdot (h(q-1)+1)$ for $q \in [1, m]$. By the "dimension" of a node, we mean the dimension of the nonterminal in the label of that node. Assume $|D| \geq d^{h+1}$. Then the root of τ has B-height $\geq h$, and τ must have a path that contains a node of each B-height $\leq h$. For each $i = 0, \ldots, h$, from among the nodes of B-height i on that path, pick a node ν_i of the lowest dimension.

By a *q-stretch*, we mean a contiguous subsequence of $\nu_0, \nu_1, \ldots, \nu_h$ consisting entirely of nodes of dimension $\geq q$. We claim that some q-stretch contains more than $2q \cdot (|N|+1)+1$ nodes of dimension q. For, suppose not. Then we can show by induction on q that $\nu_0, \nu_1, \ldots, \nu_h$ contains no more than $h(q)$ nodes of dimension q, which contradicts $h = \sum_{q=1}^{m} h(q)$. Since the entire sequence $\nu_0, \nu_1, \ldots, \nu_h$ is a 1-stretch, the sequence contains at most $2 \cdot (|N|+1)+1 = h(1)$ nodes of dimension 1. If the sequence contains at most $h(q-1)$ nodes of dimension $q-1$, then there are at most $h(q-1)+1$ maximal q-stretches, so the number of nodes of dimension q in the sequence cannot exceed $(2q \cdot (|N|+1)+1) \cdot (h(q-1)+1) = h(q)$.

So we have a q-stretch that contains nodes $\nu_{i_0}, \ldots, \nu_{i_k}$ of dimension q for some $q \in [1, m]$, where $k = 2q \cdot (|N|+1)+1$. Let A_n be the non-terminal in the label of ν_{i_n}. By the definition of a q-stretch and the way the original sequence $\nu_0, \ldots, \nu_h$ is defined, the nodes of τ that are neither below $\nu_{i_{n-1}}$ nor above ν_{i_n} determine a non-decreasing derivation witnessing $A_{n-1}(\boldsymbol{x}_1, \ldots, \boldsymbol{x}_q) \vdash_G A_n(x_{n,1} \boldsymbol{x}_1 x_{n,2}, \ldots, x_{n,2q-1} \boldsymbol{x}_q x_{n,2q})$ for some strings $x_{n,1}, \ldots, x_{n,2q}$. Since there must be a B-node lying above $\nu_{i_{n-1}}$ and below or at ν_{i_n}, at least one of $x_{n,1}, \ldots, x_{n,2q}$ must contain a distinguished position.

By the pigeon-hole principle, there is a $j \in [1, 2q]$ such that $\{n \in [1, k] \mid x_{n,j}$ contains a distinguished position $\}$ has at least $|N| + 2$ elements. This means that we can pick three elements n_1, n_2, n_3 from this set so that $n_1 < n_2 < n_3$ and $A_{n_1} = A_{n_2}$. Letting $\mu_1 = \nu_{i_0}, \mu_1 = \nu_{i_{n_1}}, \mu_2 = \nu_{i_{n_2}}, \mu_3 = \nu_{i_{n_3}}$, we see that (2), (3) and (4) hold with $C = A_{i_0}, B = A_{i_{n_1}} = A_{i_{n_2}}, A = A_{i_{n_3}}$ and x_j, v_j, y_j all containing a distinguished position, as desired.   □

Let us write m-$\mathrm{MCFL}_{\mathrm{prop}}$ for the family of languages generated by proper m-MCFGs. Using standard techniques (cf. Theorem 3.9 of [12]), we can easily show that for each $m \geq 1$, m-$\mathrm{MCFL}_{\mathrm{prop}}$ is a substitution-closed full abstract family of languages.

5   Relation to the Control Language Hierarchy

Kanazawa and Salvati [8] showed $\mathcal{C}_k \subseteq 2^{k-1}$-MCFL for each k through a tree grammar generating the derivation trees of a level k control grammar (G, C). In fact, detour through tree languages is not necessary—a level k control language can be obtained from a level $k - 1$ control language by certain string language operations. It is easy to see that the family $\bigcup_m m$-$\mathrm{MCFL}_{\mathrm{prop}}$ is closed under those operations.

Let us sketch the idea using Example 4. We start by applying a *homomorphic replication* [2,5] $\langle (1, R), h_1, h_2 \rangle$ to the control set $C = \{\pi_1^n \pi_2^n \pi_3 \mid n \in \mathbb{N}\}$, obtaining

$$\langle (1, R), h_1, h_2 \rangle (C) = \{h_1(w) h_2(w^R) \mid w \in C\}, \tag{6}$$

where $h_1(\pi_1) = a, h_1(\pi_2) = b, h_1(\pi_3) = \varepsilon, h_2(\pi_1) = \bar{a}S, h_2(\pi_2)\bar{b}S, h_2(\pi_3) = \varepsilon$. For instance, $\pi_1^2 \pi_2^2 \pi_3$ is mapped to $aabb\bar{b}S\bar{b}S\bar{a}S\bar{a}S$. Iterating the substitution $S \leftarrow \langle (1, R), h_1, h_2 \rangle (C)$ on the resulting language and then throwing away strings that contain S gives the language of the control grammar of this example.

In general, if π is a production $A \rightarrow w_0 B_1 w_1 \ldots B_n w_n$ of a labeled distinguished grammar $G = (N, \Sigma, P, S, f)$ and $f(\pi) = i \in [1, n]$, then we let $h_1(\pi) = w_0 B_1 w_1 \ldots B_{i-1} w_{i-1}$ and $h_2(\pi) = w_i B_{i+1} w_{i+1} \ldots B_n w_n$. In case $f(\pi) = 0$, $h_1(\pi)$ is the entire right-hand side of π and $h_2(\pi) = \varepsilon$. The control set C is first intersected with a local set so as to ensure consistency of nonterminals in adjacent productions, and then partitioned into sets C_A indexed by nonterminals, with C_A holding only those strings whose first symbol is a production that has A on its left-hand side. Let $L_A = \langle (1, R), h_1, h_2 \rangle (C_A)$ for each $A \in N$. The final operation is iterating simultaneous substitution $A \leftarrow L_A$ and throwing away strings containing nonterminals:

$$L_0 = L_S, \qquad L_{n+1} = L_n[A \leftarrow L_A]_{A \in N}, \qquad L = \bigcup_n L_n \cap \Sigma^*. \tag{7}$$

This last step may be thought of as the fixed point computation of a "context-free grammar" with an infinite set of rules $\{A \rightarrow \alpha \mid A \in N, \alpha \in L_A\}$.

Lemma 9. *If $L \in m$-$\mathrm{MCFL}_{\mathrm{prop}}$ and h_1, h_2 are homomorphisms, then the language $\langle (1, R), h_1, h_2 \rangle (L)$ defined by (6) belongs to $2m$-$\mathrm{MCFL}_{\mathrm{prop}}$.*

Example 1 in Sect. 2.1 illustrates Lemma 9 with $m = 1$, $L = D_1^*$, and h_1, h_2 both equal to the identity function.

The proof of the next lemma is similar to that of closure under substitution.

Lemma 10. *If $L_A \subseteq (N \cup \Sigma)^*$ belongs to m-$\mathrm{MCFL}_{\mathrm{prop}}$ for each $A \in N$, then the language L defined by (7) also belongs to m-$\mathrm{MCFL}_{\mathrm{prop}}$.*

Theorem 11. *For each $k \geq 1$, $\mathcal{C}_k \subsetneq 2^{k-1}$-$\mathrm{MCFL}_{\mathrm{prop}}$.*

Again, the language $\mathrm{RESP}_{2^{k-1}}$ separates 2^{k-1}-$\mathrm{MCFL}_{\mathrm{prop}}$ from $\mathcal{C}_k$. For $k = 2$, $\{ w \# w \mid w \in D_1^* \}$ also witnesses the separation. I currently do not see how to settle the question of whether the inclusion of $\bigcup_k \mathcal{C}_k$ in $\bigcup_m m$-$\mathrm{MCFL}_{\mathrm{prop}}$ is strict.

6   Conclusion

Theorems 5 and 8 with $m = 2^{k-1}$ both apply to languages in $\mathcal{C}_k$, but place incomparable requirements on the factorization $z = u_1 v_1 \ldots u_{2^k} v_{2^k} u_{2^k+1}$. Theorem 8 does not require $v_{2^{k-1}} u_{2^{k-1}} v_{2^{k-1}+1}$ to contain $\leq p$ distinguished positions. On the other hand, it does not seem easy to derive additional restrictions on $v_{2i-1} u_{2i} v_{2i}$ from Palis and Shende's [11] proof. From the point of view of MCFGs, the conditions in Theorem 8 are very natural: the substrings that are simultaneously iterated should contain only a small number of distinguished positions.

References

1. Engelfriet, J.: Context-free graph grammars. In: Rozenberg, G., Salomaa, A. (eds.) Handbook of Formal Languages. Beyond Words, vol. 3, pp. 125–213. Springer, Berlin (1997)
2. Ginsburg, S., Spanier, E.H.: AFL with the semilinear property. J. Comput. Syst. Sci. **5**(4), 365–396 (1971)
3. Greibach, S.A.: Hierarchy theorems for two-way finite state transducers. Acta Informatica **11**, 89–101 (1978)
4. Greibach, S.A.: One-way finite visit automata. Theor. Comput. Sci. **6**, 175–221 (1978)
5. Greibach, S.A.: The strong independence of substitution and homomorphic replication. R.A.I.R.O. Informatique théorique **12**(3), 213–234 (1978)
6. Kanazawa, M.: The pumping lemma for well-nested multiple context-free languages. In: Diekert, V., Nowotka, D. (eds.) DLT 2009. LNCS, vol. 5583, pp. 312–325. Springer, Heidelberg (2009)
7. Kanazawa, M., Kobele, G.M., Michaelis, J., Salvati, S., Yoshinaka, R.: The failure of the strong pumping lemma for multiple context-free languages. Theor. Comput. Syst. **55**(1), 250–278 (2014)
8. Kanazawa, M., Salvati, S.: Generating control languages with abstract categorial grammars. In: Preliminary Proceedings of FG-2007: The 12th Conference on Formal Grammar (2007)

9. Kanazawa, M., Salvati, S.: The copying power of well-nested multiple context-free grammars. In: Dediu, A.-H., Fernau, H., Martín-Vide, C. (eds.) LATA 2010. LNCS, vol. 6031, pp. 344–355. Springer, Heidelberg (2010)
10. Ogden, W.: A helpful result for proving inherent ambiguity. Math. Syst. Theor. **2**(3), 191–194 (1968)
11. Palis, M.A., Shende, S.M.: Pumping lemmas for the control language hierarchy. Math. Syst. Theor. **28**(3), 199–213 (1995)
12. Seki, H., Matsumura, T., Fujii, M., Kasami, T.: On multiple context-free grammars. Theor. Comput. Sci. **88**(2), 191–229 (1991)
13. Weir, D.J.: A geometric hierarchy beyond context-free languages. Theor. Comput. Sci. **104**(2), 235–261 (1992)

Grammatical Inference, Algorithmic Learning, and Patterns

Steganography Based on Pattern Languages

Sebastian Berndt[(✉)] and Rüdiger Reischuk

Institute of Theoretical Computer Science, University of Lübeck,
Ratzeburger Allee 160, 23552 Lübeck, Germany
{berndt,reischuk}@tcs.uni-luebeck.de

Abstract. In order to transmit secret messages such that the information exchange itself cannot be detected, steganography needs a channel, a set of strings with some distribution that occur in an ordinary communication. The elements of such a language or concept are called coverdocuments. The question how to design secure stegosystems for natural classes of languages is investigated for pattern languages. We present a randomized modification scheme for strings of a pattern language that can reliably encode arbitrary messages and is almost undetectable.

Keywords: Language-based cryptography · Steganography · Pattern

1 Introduction

Steganography, the art of hiding secret messages in unsuspicious communication, is an interesting topic, in theory as well as for practical applications. While in cryptographic information transfer an observer is aware of the fact that messages are exchanged, but their contents cannot be detected due to encryption, a steganographic system even tries to keep the fact undetected that secret information is transmitted at all. Therefore, the transmission channel itself plays an important role. Such a channel is described by a subset Σ' of a large alphabet Σ with elements called *coverdocuments* that might be sent over the channel, and a probability distribution on the documents. In the simplest case of uniform probabilities, to determine the channel means learning concepts Σ' of universe Σ.

A computational model for steganography was introduced by Hopper, von Ahn, and Langford [7] and independently by Katzenbeisser and Petitcolas [9]. A *stegosystem* consists of an encoding and a decoding algorithm. The encoding algorithm (also called Alice) tries to hide a secret message in a sequence of strings called *stegodocuments* that are transmitted over the channel. The decoding algorithm (Bob) tries to reconstruct the message from these stegodocuments. As the channel is completely monitored by an adversary (Warden), the distribution of stegodocuments should be indistinguishable from the distribution of coverdocuments. In the steganographic setting, learning such a channel distribution can only be done via positive samples. The only thing Alice can do is sampling from the channel to get an idea of typical coverdocuments.

In [7] a stegosystem has been proposed that can embed up to $\log n$ bits of information into documents of length n securely (under cryptographic assumptions). It is *universal* or *black-box* since it works for arbitrary channels as long as

A.-H. Dediu et al. (Eds.): LATA 2016, LNCS 9618, pp. 387–399, 2016.
DOI: 10.1007/978-3-319-30000-9_30

their *min-entropy* is large enough to allow the transmission of $\log n$ bits. Later Dedić et al. [5] have proven under cryptographic assumptions that no universal polynomial time stegosystem can embed more than $O(\log n)$ bits per document by giving a family of channels called *pseudorandom flat h-channels*.

It has been observed that the stegosystems used in practice typically embed up to $O(\sqrt{n})$ bits in documents of length n [10,11], but they are non-universal and tailored to specific types of channels. In order to close this gap between theory and practice, Liśkiewicz, Reischuk and Wölfel [13] have introduced the model of *grey-box stegosystems* that are specialized to certain subsets $\mathcal{F}$ of all possible channels – thus there is some a priori information how the channel may look like. In addition, they have investigated a weaker notion of security called *undetectability*, where both stegoencoder *and* adversary face the same learning problem of determining the actual channel out of the possible elements in $\mathcal{F}$.

In [13] it has been shown that the family of channels described by arbitrary monomials, a family that can be learned easily, possesses a secure stegosystem that can embed up to $\sqrt{n}$ bits in a single document. Monomials are rather simple objects, thus cannot model many real communication channels. It is therefore an interesting question whether secure grey-box stegosystems can be designed for more complex communication channels. Since some common structure is necessary in order to apply embedding techniques for secret messages, channels that can be described by formal languages are of special interest. To construct a good stegosystem two tasks have to be solved efficiently: learning the channel distribution and modifying this distribution in an (almost) undetectable way. Obviously, one cannot allow arbitrary distributions on the document space Σ since for simple information theoretic reasons they cannot be learned efficiently. Recently, progress has been made for the case of k-term DNF-formulas [6]. The goal of this work is to investigate this question for pattern languages, and therefore let us call the corresponding channels *pattern channels*. Learning algorithms for pattern languages have been studied intensively. Thus, here we concentrate on the second issue, the undetectable modification of strings within such a language.

Pattern languages have been introduced by Angluin [1]. It makes a significant difference whether erasing substitutions are allowed or not [14]. Both cases have sparked a huge amount of work both in the fields of formal languages (e.g. [16]) and machine learning (e.g. [3,4,12,14,15,17,19]). Some of these results were also used in the context of molecular biology (e.g. [18]). An important example of communication channels that can be defined by pattern languages is the set of filled out forms (either in paper or digital).

1.1 Our Contributions

We design a method to alter strings of a pattern language that are provided according to some distribution in an almost undetectable way. On this basis we show how a rate-efficient, secure and reliable stegosystem can be constructed for a wide class of pattern channels if the patterns can be learned efficiently or are given explicitly. To the best of our knowledge this is the first stegosystem ever for this class of channels. As a novel technical contribution we analyze the rank of random matrices that are generated by the distribution of random strings when substituting variables in a pattern. We present a generalized form of the *poisson approximation* typically used for randomized processes that may be of independent interest.

2 Basics and Notations

Let $[n]$ denote the set $\{1, 2, \ldots, n\}$, $\mathbb{F}_q$ the finite field on q elements and $A \in \mathbb{F}_q^{\mu \times \sigma}$ and $b \in \mathbb{F}_q^\mu$. The set $\mathrm{Sol}(A, b) = \{x \in \mathbb{F}_q^\sigma \mid Ax = b\}$ denotes the solutions of the linear equation system (LES) $Ax = b$. The rank $\mathrm{rk}(A)$ of a matrix A is the size of the largest subset of rows or columns that are linearly independent. It is a known fact that $\mathrm{Sol}(A, b)$ is either empty or of size $q^{\sigma - \mathrm{rk}(A)}$. For a fixed matrix A, varying over $b \in \mathbb{F}_q^\mu$ defines a partition of $\mathbb{F}_q^\sigma$. Hence, the number of b with $|\mathrm{Sol}(A, b)| > 0$ is exactly $q^{\mathrm{rk}(A)}$.

In the following we assume that the elements of a finite set S can be described by binary strings of length $O(\log |S|)$. Writing $s \in_R S$ we mean that s is a uniformly distributed random element of S. As computational model we use probabilistic Turing machine (PTM) equipped with different *oracles*:

- For a random variable X the PTM M^X gets a sample x distributed according to X. If X is the uniform distribution on a set S we simply write M^S.
- If $f \colon U \to V$ is a function, M^f can provide an element $u \in U$ and gets back the value $f(u)$.

If M can access several oracles $O_1, O_2, \ldots$ we write $M^{O_1, O_2, \ldots}$.

2.1 Steganography

We give a short formal description of the steganographic model. More details can be found in the references cited.

Definition 1 (channel and history). For a set Σ (the set of possible strings that may be sent) the set of all probability distributions on Σ will be denoted by $\mathrm{Prob}(\Sigma)$. A channel $\mathcal{C}$ over Σ is a mapping $\mathcal{C} \colon \Sigma^* \to \mathrm{Prob}(\Sigma)$ that for each *history* $h \in \Sigma^*$ (a sequence of previous strings) defines a probability distribution $\mathcal{C}(h)$, also denoted by $\mathcal{C}^h$. An element in the support of $\mathcal{C}^h$ is called a *document*. A history $h \in \Sigma^* = h_1 h_2 \cdots h_r$ is called *legal for* $\mathcal{C}$ iff $\mathcal{C}^{h_1 h_2 \cdots h_{i-1}}(h_i) > 0$ for every $0 < i < r$, that means each h_i can actually occur within the prefix history.

A *pattern channel* is a channel where for every history h the support of the distribution $\mathcal{C}^h$ equals a subset of all strings generated by some pattern π that may depend on h. In the next section to keep the exposition simple we will only discuss the case where $\mathcal{C}^h$ is identical for all h, that means there is a single pattern π defining the support – the channel is memoryless. Such a channel will be denoted by $\mathcal{C}_\pi$. Our techniques also carry over to the more complex case where every history implies a different pattern.

Definition 2 (stegosystem). Given a *key space* $\mathcal{K}$, a *message space* $\mathcal{M}$ and a family $\mathcal{F}$ of channels over Σ, a *stegosystem* $\mathcal{S} = [SE, SD]$ consists of two probabilistic algorithms: an *encoding* algorithm SE and a *decoding* algorithm SD. Given a key $K \in \mathcal{K}$, an unknown channel $\mathcal{C} \in \mathcal{F}$ and a legal history h, SE has access to a *sampling oracle* for $\mathcal{C}$ (denoted by $SE^{\mathcal{C}(\cdot)}$). It takes a message $m \in \mathcal{M}$ and produces a sequence c of l elements of Σ, the *stegotext* that invisibly should include m. Using the same key K, the decoding algorithm SD, given a sequence $c \in \Sigma^l$, computes an element in $\mathcal{M}$ (hopefully the original m).

Definition 3 (reliability and security). For $\rho \geq 0$ a stegosystem $\mathcal{S} = [SE, SD]$ is ρ-*reliable* on $\mathcal{F}$ if

$$\max_{h \text{ legal}, \ m \in \mathcal{M}, \ \mathcal{C} \in \mathcal{F}} \left\{ \Pr_{K \in_R \mathcal{K}}[SD(K, SE^{\mathcal{C}(\cdot)}(K, m, h)) \neq m]\right\} \leq \rho \, ,$$

where in addition to the random choice of $K \in_R \mathcal{K}$ the probability is taken with respect to the coin flips of SE and SD and the output of the sampling oracle $\mathcal{C}(\cdot)$.

In order to define the security of a stegosystem, we consider an attacker, called a *warden* W. This is a PTM equipped with the sampling oracle $\mathcal{C}(\cdot)$ and in addition a *challenging oracle* $CH(\cdot, \cdot)$ that is either distributed according to $SE^{\mathcal{C}(\cdot)}(\cdot, m, h)$ (the stego case) or distributed according to the channel distribution $\mathrm{EX}_{\mathcal{C}}^l(h)$ (the nonstego case), where

$$\Pr[\mathrm{EX}_{\mathcal{C}}^l(h) = d_1 d_2 \ldots d_l] = \prod_{i=1}^{l} \Pr_{d \leftarrow \mathcal{C}^{h d_1 d_2 \cdots d_{i-1}}}[d = d_i].$$

Warden W can call CH with message m and legal history h and gets a sequence $d_1 d_2 \cdots d_l$ and its goal is to distinguish between the two cases outputting 1 if he believes that the challenging oracle is $SE^{\mathcal{C}}$ and 0 otherwise. A stegosystem $\mathcal{S}$ is (t, ϵ)-*secure* for $\mathcal{F}$ if $\left| \Pr_{K \in_R \mathcal{K}}[W^{\mathcal{C}(\cdot), SE^{\mathcal{C}(\cdot)}(K, \cdot, \cdot)} = 1] - \Pr[W^{\mathcal{C}(\cdot), \mathrm{EX}_{\mathcal{C}}^l(\cdot)} = 1] \right| \leq \epsilon$ for all wardens W with running time at most t and all $\mathcal{C} \in \mathcal{F}$, where the probability is taken over the output of the oracles and the coin flips of the warden.

As W may choose history and message, this security notion is called *security against chosen-message attacks* or *security against chosen-hiddentext attacks*.

2.2 Cryptographic Primitives

For two finite sets U, V, let $\mathrm{Fun}(U, V)$ be the set of all functions from U to V. A function $F : \mathcal{K} \times U \to V$ is called a (t, ϵ)-*secure pseudorandom function (PRF)* with respect to U and V if $\left| \Pr_{f \in_R \mathrm{Fun}(U,V)}[A^{f(\cdot)} = 1] - \Pr_{K \in_R \mathcal{K}}[A^{F_K(\cdot)} = 1] \right| \leq \epsilon$ for every probabilistic algorithm A with running time at most t where $F_K(\cdot) = F(K, \cdot)$. Such a PRF is thus indistinguishable from a random function. We extend this notion to the case of side information since the warden has access to the channel oracle $\mathcal{C}$. The function F is a (t, ϵ)-*secure PRF relative to $\mathcal{C}$* if

$$\left| \Pr_{f \in_R \text{Fun}(U,V)} [A^{\mathcal{C}(\cdot),f(\cdot)} = 1] - \Pr_{K \in_R \mathcal{K}} [A^{\mathcal{C}(\cdot),F_K(\cdot)} = 1] \right| \leq \epsilon .$$

Bellare et al. [2] have shown that the existence of a PRF $F \colon \{0,1\}^\kappa \times \{0,1\}^\mu \to \{0,1\}^\mu$ implies the existence of a secure encryption scheme (the *XOR-scheme*). They designed the so called *random counter mode* working as follows:

Algorithm 1. CTR$\$_F(K, mes)$

Data: secret key K of suitable length κ,
   a binary string $mes = m_1 m_2 \ldots m_l$ of l blocks of length μ
choose $r \in_R \{0,1\}^\mu$;
return
$(r, F_K(r) \oplus m_1, F_K(r + 1 \bmod 2^n) \oplus m_2, \ldots, F_K(r + l - 1 \bmod 2^\mu) \oplus m_l)$

In [2] it has been proven that for every (t, ϵ)-secure PRF $F \colon \{0,1\}^\kappa \times \{0,1\}^\mu \to \{0,1\}^\mu$, probabilistic algorithm A running in time t and $mes \in \{0,1\}^{l\mu}$

$$\left| \Pr_{K \in_R \mathcal{K}} [A^{\text{CTR}\$_F(K,mes)} = 1] - \Pr[A^{\{0,1\}^{(l+1)\mu}} = 1] \right| \leq 2\epsilon + t^2 \cdot (l+1) \, 2^{-\mu} .$$

The output of CTR$\$_F(\cdot, mes)$ is thus indistinguishable from a random element of $\{0,1\}^{(l+1)\mu}$. In particular, each output block $m = F_K(r + j - 1 \bmod 2^\mu) \oplus m_j$ is indistinguishable from a random string of length μ. As the reduction is a black-box reduction, this property also holds if A has side information that is independent of the construction of F. We use this randomization technique for the steganographic transmission of a message mes. Thus, we have reduced the problem to embed a single string m of length μ that looks almost random in a document such that this embedding cannot be detected.

2.3   Pattern Languages

Let Γ be a finite alphabet of size at least 2, $V = \{v_1, v_2, \ldots\}$ be a disjoint set of variables and PAT $:= (\Gamma \cup V)^+$. An element $\pi = \pi_1 \pi_2 \cdots \pi_m$ of PAT is called a *pattern*. Let $\text{Var}(\pi)$ denote the set of variables appearing in π — we may assume $\text{Var}(\pi) = \{v_1, \ldots, v_d\}$ for some $d \in \mathbb{N}$. For $v \in \text{Var}(\pi)$ let $\text{occ}(v, \pi)$ be the number of occurrences of v in π, that is $\text{occ}(v, \pi) = |\{j \in [1..m] : \pi_j = v\}|$.

A (possibly erasing) *substitution* Θ is a string homomorphism $\Gamma \cup V \to \Gamma^*$ such that $\Theta(a) = a$ for all $a \in \Gamma$. By $\pi\Theta$ we denote the application of Θ to π i.e., $\pi\Theta := \Theta(\pi_1)\Theta(\pi_2) \cdots \Theta(\pi_m)$. For $n \in \mathbb{N}$ let $\text{Subs}_n(\pi)$ denote the set of all substitutions that generate strings of length n and $\text{Lang}_n(\pi)$ these strings, i.e., $\text{Lang}_n(\pi) := \{\pi\Theta \mid \Theta \in \text{Subs}_n(\pi)\} \subseteq \Gamma^n$. The set $\text{Lang}(\pi) = \bigcup_n \text{Lang}_n(\pi)$ is the language generated by π.

According to the length of variable substitutions we further partition $\text{Subs}_n(\pi)$ into subsets $\text{Subs}_n^{[\ell]}(\pi)$ where $\ell = (\ell_1, \ldots, \ell_d) \in [0..n]^d$:

$$\text{Subs}_n^{[\ell]}(\pi) := \{\Theta \in \text{Subs}_n(\pi) \mid \forall i \, |\Theta(v_i)| = \ell_i\} .$$

Such a set may be empty for many parameters n, ℓ, but if not, then its size is exactly $|\Gamma|^{\sigma(\ell)}$ where $\sigma(\ell) := \sum_i \ell_i$ denotes the total length of variable substitutions. Let $\text{Lang}_n^{[\ell]}(\pi)$ denote the set of strings generated by $\text{Subs}_n^{[\ell]}(\pi)$.

For steganographic applications it is necessary that a substitution generates enough entropy. This could either be guaranteed by requiring that the pattern contains a sufficient number of different variables with the restriction that erasing substitutions are not allowed. Alternatively, if we do not want to exclude erasing substitutions the number of independent symbols that are generated by all variables substitutions has to be of a certain size – the number $\sigma(\ell)$ as defined above. Otherwise, a pattern like $v_1 v_2 v_1 v_3 \ldots v_1 v_n$ could generate strings c^{n-1} for $c \in \Gamma$ by substituting v_1 by c and erasing all other variables. Such strings are obviously not suitable for embedding secret information, as $\sigma(\ell) = 1$.

For steganography with strings generated by a pattern π we model the application of a substitution Θ to a variable v as generating a sequence of new intermediate variables $u_v^{(1)}, u_v^{(2)}, \ldots, u_v^{(|\Theta(v)|)}$ which later can be replaced by a single letter of Γ. The *intermediate pattern* $[\pi]\Theta$ for π and Θ is thus defined as $[\pi]\Theta := [\pi_1]\Theta[\pi_2]\Theta \cdots [\pi_m]\Theta$ with $[a]\Theta = a$ for all $a \in \Gamma$ and $[v]\Theta = u_v^{(1)} u_v^{(2)} \cdots u_v^{(|\Theta(v)|)}$ for new variables $u_v^{(j)}$. Note that two substitutions Θ, Θ' generate the same intermediate pattern ($[\pi]\Theta = [\pi]\Theta'$) iff they belong to the same subset $\text{Subs}_n^{[\ell]}(\pi)$. Thus, we denote the intermediate pattern also by $[\pi_\ell]$.

Example 4. Let $\pi = v_1 v_2 0 0 v_2 0 v_1 v_1$ and $\ell_1 = |\Theta(v_1)| = 1$, $\ell_2 = |\Theta(v_2)| = 3$, thus $\sigma(\ell) = 4$. Then Θ belongs to $\text{Subs}_{12}^{[(1,3)]}(\pi)$. The intermediate pattern of length $n = 12$ has the form

$$[\pi_{(1,3)}] = [\pi]\Theta = u_{v_1}^{(1)} u_{v_2}^{(1)} u_{v_2}^{(2)} u_{v_2}^{(3)} \, 0 \, 0 \, u_{v_2}^{(1)} u_{v_2}^{(2)} u_{v_2}^{(3)} \, 0 \, u_{v_1}^{(1)} u_{v_1}^{(1)} \ .$$

3   Steganography Using Patterns

This section develops a stegosystem for pattern channels. The general strategy works as follows. Beforehand, Alice and Bob agree on the number μ of bits that should be hidden in a document. When Alice wants to transmit a longer message she splits it into blocks m of length μ. As part of their secret key they choose a pseudorandom partition of the positions i of a string $y = y_1 y_2 \ldots y_i \ldots y_n$ into μ subsets $B_1, \ldots, B_\mu$. The letters at positions in B_j will be used to encode the j-th secret bit. If they want to use strings of different lengths they define a separate partition for each such n.

When Alice has access to a pattern channel $\mathcal{C}_\pi$ in order to transmit stegodocuments she needs information about π. Either this is given to her explicitly, or in case of a grey-box situation she has to learn the pattern by sampling from the channel. It has been shown that this can be done efficiently for certain subclasses of pattern languages. For the moment, let us assume that Alice knows π. To generate a stegotext that encodes the secret m, Alice tries to modify a document slightly into a stegotext y of the same length that can also be generated by π. In order to make this modification undetectable, Alice must ensure that the distribution of these stegodocuments y is (almost) identical to the original distribution of documents generated by $\mathcal{C}_\pi$. In the next section we will show that with high probability a nonempty subset $\mathrm{Lang}_n^{[\ell]}(\pi)$ is able to encode every possible secret m.

3.1   Coding Bits by Random Subsets

In the following we restrict to the case of a binary alphabet $\Gamma = \{0, 1\}$, which turns out to be the most difficult case. Arithmetic in Γ will be done as in the field $\mathbb{F}_2$. For larger alphabets these techniques can be adopted easily. Let π be a pattern and ℓ a vector for the length of variable substitutions that generate an intermediate pattern $[\pi_\ell]$ of length n with variables $\mathrm{Var}([\pi_\ell]) = \{v_1, v_2, \ldots, v_{\sigma(\ell)}\}$. In the following we consider only parameters such that $\mathrm{Subs}_n^{[\ell]}(\pi) \neq \emptyset$. For a partition of $[n]$ into μ subsets specified by a function $f \colon [n] \to [\mu]$ we define a binary $(\mu \times \sigma(\ell))$-matrix $Z_{f,\pi,\ell} = (z_{\nu,i})$. The entry $z_{\nu,i}$ equals the parity of the number of positions in $[\pi_\ell]$ that hold the i-th variable and are mapped to ν, i.e.

$$z_{\nu,i} := |\{j \in [n] : [\pi_\ell]_j = v_i \wedge f(j) = \nu\}| \bmod 2.$$

Example 5. For π and ℓ, resp. Θ used in the previous example and the partition $f(j) = (j \bmod 3) + 1$, the subset B_1 collects the symbols at position $3, 6, 9, 12$ (which are $u_{v_2}^{(2)}, 0, u_{v_2}^{(3)}, u_{v_1}^{(1)}$), the set B_2 those at position $1, 4, 7, 10$ (which are $u_{v_1}^{(1)}, u_{v_2}^{(3)}, u_{v_2}^{(1)}, 0$) and B_3 those at $2, 5, 8, 11$, namely $u_{v_2}^{(1)}, 0, u_{v_2}^{(2)}, u_{v_1}^{(1)}$. Then the matrix $Z_{f,\pi,\ell}$ has rank 3 and looks as follows

$$Z_{f,\pi,\ell} = \begin{array}{c} \\ \\ 1 \\ 2 \\ 3 \end{array} \begin{array}{c} \overset{u_{v_1}^{(1)}\ \ u_{v_2}^{(1)}\ \ u_{v_2}^{(2)}\ \ u_{v_2}^{(3)}}{} \\ \left[\begin{array}{cccc} 1 & 0 & 1 & 1 \\ 1 & 1 & 0 & 1 \\ 1 & 1 & 1 & 0 \end{array} \right] \end{array}.$$

For a reliable embedding of an arbitrary secret message of length μ into a string of $\mathrm{Lang}_n(\pi)$ the matrix $Z_{f,\pi,\ell}$ must have maximal rank μ. As already noted, this implies that the pattern and the substitution must generate enough entropy with respect to μ. In particular, $\sigma(\ell)$ has to be larger than μ.

3.2   Bounding the Rank of Matrices Obtained by Random Assignments of Intermediate Patterns

With high probability a random $(0, 1)$-matrix of dimension $\mu \times \sigma$ has maximal rank μ over $\mathbb{F}_2$ if σ is slightly larger than μ. In $Z_{f,\pi,\ell}$, however, the entries are not independent. In addition, an entry does not necessarily take value 0 and 1 with probability exactly $1/2$. The second problem can be solved by showing that the deviation from the uniform distribution is not too large. To handle the non-independence significantly more technical effort is required.

For this purpose let us define the function

$$\zeta(\sigma,\mu) := \frac{\mu \cdot (4/5)^{\mu}}{1 - 2\exp\left(-2(1 - \frac{2}{e\sqrt[4]{\sigma}} + \frac{1}{e^2}\sqrt{\sigma})\right)}$$

which for larger σ goes exponentially fast to 0 for growing μ.

Theorem 6. *For every* $\pi \in \mathrm{PAT}$, $n \in \mathbb{N}$ *and every vector* $\ell \in [0..n]^{|\mathrm{Var}(\pi)|}$ *with* $\sigma(\ell) \geq \mu^2$ *and* $\max_{v \in \mathrm{Var}(\pi)}\{\mathrm{occ}(v,\pi)\} \leq \sigma(\ell) \cdot e^{-2}$

$$\Pr_{f \in_R \mathrm{Fun}([n],[\mu])}[Z_{f,\pi,\ell} \text{ has maximal rank}\, \mu] \geq 1 - \zeta(\sigma(\ell),\mu) .$$

The main idea of the proof is to show that a random assignment of pattern variables to subsets can be approximated by independent Poisson processes. This follows from the following claims.

Lemma 7. *Let* $C = (c_{\nu,j})_{\nu\in[1..\mu],\ j\in[1..\sigma]}$ *be a matrix of random variables (RV) that are obtained as follows. We are given a sequence of colored balls, where color* j *appears* $a_j \geq 1$ *often for* $j \in [1..\sigma]$. *Let* $\xi := \max_j a_j$. *The balls are thrown uniformly and independently into* μ *bins. Then* $c_{\nu,j}$ *denotes the number of balls of color* j *that fall into the* ν-th *bin.*

Similarly, let $X = (x_{\nu,j})_{\nu\in[1..\mu],\ j\in[1..\sigma]}$ *be a matrix of pairwise independent Poisson-distributed RVs, where* $x_{\nu,j}$ *has mean* $\lambda_j = a_j/\mu$ *(the same as* $c_{\nu,j}$).

Let P *be an arbitrary predicate over* $\mathbb{N}^{\mu \times l}$. *For a* $(\mu \times \sigma)$-matrix X *of RVs define the predicate* $\mathcal{E}_P(X)$ *as the probability that* X *has a subset of* l *columns* $X_{z_1}, \ldots, X_{z_l}$ *such that* $P(X_{z_1}, \ldots, X_{z_l})$ *holds. Then for* $\eta(l,\sigma,\xi) := (\frac{\sigma}{e\cdot\sqrt{\xi}} - l)^2/\sigma$,

$$\Pr[\mathcal{E}_P(C)] \leq \frac{\Pr[\mathcal{E}_P(X)]}{1 - 2\exp(-2\,\eta(l,\sigma,\xi))} .$$

For $l < \frac{\sigma}{e\cdot\sqrt{\xi}} - \sqrt{\sigma}$ and $\sigma > e^2\xi$ the denominator in the inequality above is at least $1 - 2e^{-2} \geq 0.729$, thus the probability for C is at most a constant factor larger than for independent Poison variables.

Lemma 8. *Let* X *be a* $(\mu \times \sigma)$-matrix *of independent Poisson RVs with* $\mathbf{E}[x_{\nu,j}] = a_j/\mu > 0$ *and* $\sigma \geq \mu^2 \geq 6$.

The matrix $M = (m_{\nu,j})$ *with* $m_{\nu,j} = x_{\nu,j} \bmod 2$ *has full rank over* $\mathbb{F}_2$ *with probability at least* $1 - \mu \cdot (4/5)^{\mu}$.

In the steganographic application described below we replace the random function $f \in_R \mathrm{Fun}([n],[\mu])$ by a pseudo-random function F_K. Its seed is determined by the secret key of Alice and Bob. The function F_K may add another super-polynomial small error to the property that $Z_{f,\pi,\ell}$ has maximal rank.

3.3 Modifying Strings of a Pattern Language to Embed Secrets

Note that the equation $Z_{f,\pi,\ell} \cdot x = b$ has a solution $x \in \{0,1\}^{\sigma(\ell)}$ for every $b \in \{0,1\}^{\mu}$ if the matrix $Z_{f,\pi,\ell}$ has full rank.

Example 9. For the matrix in the previous example and $b = (1,1,0)$, the vector $x = (0,0,0,1)$ is a solution to the linear equation $Z_{f,\pi,\ell} \cdot x = b$. The corresponding substitution $\Theta_x(v_1) = 0, \Theta_x(v_2) = 001$ applied to π yields the string $y = 0\,0\,0\,1\,0\,0\,0\,0\,1\,0\,0\,0$.

This example illustrates how we generate a string $y = y(x)$ in $\mathrm{Lang}_n^{[\ell]}$ from a solution $x \in \{0,1\}^{\sigma(\ell)}$ of the equation $Z_{f,\pi,\ell} \cdot x = b$: Simply replace each intermediate variable by the corresponding symbol in x.

To embed a message m into a string of $\mathrm{Lang}_n^{[\ell]}$ we use the following algorithm MODIFY. For a given pattern $\pi \in \mathrm{PAT}$ and length vector ℓ let $\mathrm{Ter}(\pi, \ell)$ be those positions in $[\pi_\ell]$ that are taken by constants.

Algorithm 2. MODIFY

Data: function $f \colon [n] \to [\mu]$, message $m = m_1 \ldots m_\mu \in \{0,1\}^\mu$,
       pattern $\pi \in \mathrm{PAT}$, vector ℓ.

for $\nu = 1, \ldots, \mu$ **do**
    | let $b_\nu \leftarrow m_\nu + \sum_{j \in \mathrm{Ter}(\pi,\ell), f(j)=\nu} [\pi_\ell]_j$

let $b \leftarrow (b_1, b_2, \ldots, b_\mu)$;
if $Z_{f,\pi,\ell}$ has rank μ **then**
    | choose randomly $x \in_{\mathrm{R}} \mathrm{Sol}(Z_{f,\pi,\ell}, b)$;
    | **return** *the string* $y = y(x)$
else
    | **return** $y \in_{\mathrm{R}} \mathrm{Lang}_n^{[\ell]}$

The running time of MODIFY is $\mathcal{O}(\mu \cdot n)$. One can prove the following lemma.

Lemma 10. *For every* $\pi \in \mathrm{PAT}$, $n \in \mathbb{N}$ *and every vector* $\ell \in [0..n]^{|\mathrm{Var}(\pi)|}$ *holds: the output* y *of* MODIFY(f, m, π, ℓ) *is uniformly distributed over* $\mathrm{Lang}_n^{[\ell]}$ *if* $m \in_{\mathrm{R}} \{0,1\}^\mu$ *is chosen at random.*

Furthermore, if $f \in_{\mathrm{R}} \mathrm{Fun}([n], [\mu])$ *is chosen randomly and* $\sigma(\ell) \geq \mu^2$ *and* $\max_{v \in \mathrm{Var}(\pi)} \{ \mathrm{occ}(v, \pi) \} \leq \sigma(\ell) \cdot e^{-2}$, *with probability at least* $1 - \zeta(\sigma(\ell), \mu)$ *the output* y *satisfies: for every* $m \in \{0,1\}^\mu$ *and every* $\nu \in [1..\mu] : \sum_{j \colon f(j)=\nu} y_j = m_\nu$.

The second property shows how the receiver of string y can decrypt a bit m_ν of the hiddentext: Add up all symbols in y whose position is mapped to ν by f.

Proof. If $Z_{f,\pi,\ell}$ has maximal rank, for each vector $b \in \{0,1\}^\mu$ the set $\mathrm{Sol}(Z_{f,\pi,\ell}, b)$ is nonempty. These sets form an equal size partition of $\{0,1\}^{\sigma(\ell)}$. If m is chosen at random the vector b generated in the for-loop is random, too. Thus, MODIFY returns a random element of $\mathrm{Lang}_n^{[\ell]}$. In the other case this property is obvious.

By the previous lemma, with probability at least $1 - \zeta(\sigma(\ell), \mu)$ the rank is maximal. If we take any solution $x \in \mathrm{Sol}(Z_{f,\pi,\ell}, b)$ a simple calculation shows that the string $y(x)$ specifies all bits m_ν correctly.

3.4   Sampling a Pattern Channel

Next we discuss how to select n and ℓ in order to match the distribution of the pattern channel $\mathcal{C}_\pi$. In general, we cannot sample directly from $\mathcal{C}_\pi$ to determine the parameters n and ℓ. From a sample y we obviously get $n = |y|$ for free. But for complex patterns, determining the substitution lengths of the variables might be difficult since this information allows to solve the membership problem for $\mathrm{Lang}(\pi)$ easily and this is already $\mathcal{NP}$-hard in case of arbitrary patterns [1].

We call a distribution on $\mathrm{Lang}(\pi)$ *fixed variable length* if independently to each variable v_i a substitution of length ℓ_i is applied where the value ℓ_i is chosen according to some distribution Δ_i. For fixed ℓ_i each possible substitution by a string in Γ^{ℓ_i} is equally likely. In this case we assume that the Δ_i are known to the stegoencoder. Thus, a typical channel document can be generated by selecting a value ℓ_i for each v_i and then a random string of Γ^{ℓ_i}. For the modification procedure described above it suffices to generate a random vector $\ell = (\ell_1, \ldots, \ell_d)$ that matches the distribution of $\mathcal{C}_\pi$.

We can also handle a second type of distributions that focuses on the length n of the documents. Let us call a distribution D on $\mathrm{Lang}(\pi)$ *total length-uniform* if for every n every nonempty set $\mathrm{Subs}_n^{[\ell]}(\pi)$ has the same probability and within such a set all substitutions are equally likely. Note that this a nontrivial class because the probabilities for generating a specific length n may be very different. In particular, it includes the simple case that there is only a single $\bar{n}$ with positive probability, that means the pattern channel may generate only strings of fixed length $\bar{n}$ that are generated by arbitrary variable substitutions. Let $D_{\pi,n}$ be the marginal distribution of D on $\mathrm{Lang}_n(\pi)$ for nonempty $\mathrm{Lang}_n(\pi)$. If D is total length-uniform and $x \in \mathrm{Lang}_n(\pi)$ we get $D_{\pi,n}(x) = |\{\Theta \in \mathrm{Subs}_n(\pi) \mid \pi\Theta = x\}| / |\mathrm{Subs}_n(\pi)|$. We now describe how to sample such length vectors ℓ uniformly in order to sample strings from a total length-uniform distribution.

For $\mathrm{Var}(\pi) = \{v_1, \ldots, v_d\}$ and $a_i = \mathrm{occ}(v_i, \pi)$ let $\boldsymbol{a} = a(\pi) := (a_1, \ldots, a_d) \in \mathbb{N}^d$. Given n, consider the task to uniformly generate vectors $\ell = (\ell_1, \ldots, \ell_d) \in [0..n]^d$ that satisfy the diophantine equation $\sum_{i=1}^d a_i \ell_i = n$ and let $S_{\boldsymbol{a}}(n)$ denote the set of such vectors ℓ. For $k \in [1..d]$ define

$$F_{\boldsymbol{a}}(n, k) := |\{\ell \in [1..n]^k \mid \sum_{i=1}^k a_i \ell_i = n\}| .$$

The value $|S_{\boldsymbol{a}}(n)| = F_{\boldsymbol{a}}(n, d)$ can be computed by dynamic programming. It holds $F_{\boldsymbol{a}}(n, 1) = 1$ iff a_1 divides n and 0 else. If $F_{\boldsymbol{a}}(n', k)$ is known for all $n' \leq n$ we can compute $F_{\boldsymbol{a}}(n, k+1)$ as $F_{\boldsymbol{a}}(n, k+1) = \sum_{i=0}^{\lfloor n/a_{k+1} \rfloor} F_{\boldsymbol{a}}(n - a_{k+1} \cdot i, k)$.

Thus, $|S_{\boldsymbol{a}}(n)|$ can be obtained in time $\mathcal{O}(n^2 \cdot d)$. Since the problem of computing such diophantine sets is self-reducible, the work of Jerrum, Valiant and Vazirani [8] (Theorem 6.3) implies the existence of a PTM M that generates these elements with arbitrary precision efficiently. For every $\ell \in S_{\boldsymbol{a}}(n)$ and every $\epsilon > 0$

$$(1+\epsilon)^{-1}|S_{\boldsymbol{a}}(n)|^{-1} \leq \Pr[M(\boldsymbol{a}, n, \epsilon) = \ell] \leq (1+\epsilon)|S_{\boldsymbol{a}}(n)|^{-1}$$

and M is polynomially time-bounded with respect to $n, \boldsymbol{a}, \log \epsilon^{-1}$. The statistical distance between the output of $M(\boldsymbol{a}, n, \epsilon)$ and the uniform distribution on $S_a(n)$ is at most ϵ. The statistical distance of the string $\pi\Theta$ generated by the following algorithm and a total length-uniform distribution $\mathcal{C}_\pi$ on $\mathrm{Lang}(\pi)$ is thus at most ϵ:

Algorithm 3. *Samp*

Data: $\pi \in \mathrm{PAT}$ and $\epsilon > 0$
let $\mathrm{Var}(\pi) = \{v_1, \ldots, v_d\}$ and $a_i = \mathrm{occ}(v_i, \pi)$;
sample x from the channel $\mathcal{C}_\pi$;
sample $\ell \in [1..|x|]^d \leftarrow M((a_1, \ldots, a_d), |x|, \epsilon)$;
for $i = 1$ *to* d **do**
    choose $\Theta(v_i) \in_R \{0,1\}^{\ell_i}$;
return $|x|, \ell, \pi\Theta$

3.5 A Secure Stegosystem for Pattern Channels

Let Π be a subset of PAT that restricts the family of pattern channels $\mathcal{C}_\pi$. We consider two cases: either Π is a simple concept like 1-variable patterns or regular patterns with terminal blocks of fixed length that efficiently can be learned probabilistically exact [4]. Alternatively, Π may be more complex, but then we have to assume that the stegoencoder is told the pattern π of the channel to be used. But note that in any case Alice and Bob first have to agree on a stegosystem and a secret key. After that the pattern channel is determined, and this may even be done by an adversary.

In addition, one cannot allow arbitrary distributions on $\mathrm{Lang}(\pi)$ since the stegoencoder needs information on the distribution and such a description in general is at least of exponential size. Above, we have introduced two families of meaningful distributions, *fixed variable length* and *total length-uniform*. In both cases, for an arbitrary π a pattern channel $\mathcal{C}_\pi$ with such a distribution can be sampled efficiently given π.

The new techniques to design a stegosystem for pattern channels have been described above. To get a complete picture we list the main steps of the encoder:

1. Alice and Bob have agreed on a secret key K used as seed for two pseudo-random functions;
2. Alice learns or gets the pattern defining the channel and is informed about the type of the channel distribution;
3. given a message m Alice randomizes it by $\mathrm{CTR\$}_{F_K}$ to a string m';
4. Alice draws a length vector ℓ using $Samp(\pi)$ in case of a total length-uniform distribution, or samples it for each variable individually in case of a fixed variable length distribution;
5. using MODIFY(F_K, m', π, ℓ) Alice generates a stegotext y that encodes m', which is then sent to Bob

Based on the analysis given above, the following theorem can be proved.

Theorem 11. *There exists a stegosystem S for pattern languages. It embeds secret messages of length μ for any number μ and can be applied to arbitrary families $\mathcal{F}$ of pattern channels if the channels can be sampled efficiently and have entropy at least μ^2. The stegosystem S is ρ-reliable and (t, δ)-secure, where the parameters ρ, δ and t depend on the security of the pseudorandom functions for randomizing the message and partitioning the bits of a coverdocument. The values ρ and δ decrease super-polynomially with respect to μ.*

The precise estimation of the error parameters are tedious and skipped due to space limitations.

References

1. Angluin, D.: Finding patterns common to a set of strings. JCSS **21**(1), 46–62 (1980)
2. Bellare, M., Desai, A., Jokipii, E., Rogaway, P.: A concrete security treatment of symmetric encryption. In: Proceedings of 38th Annual Symposium on Foundations of Computer Science, FOCS, pp. 394–403. IEEE (1997)
3. Case, J., Jain, S., Le, T.D., Ong, Y.S., Semukhin, P., Stephan, F.: Automatic learning of subclasses of pattern languages. Inf. Comput. **218**, 17–35 (2012)
4. Case, J., Jain, S., Reischuk, R., Stephan, F., Zeugmann, T.: Learning a subclass of regular patterns in polynomial time. TCS **364**(1), 115–131 (2006)
5. Dedić, N., Itkis, G., Reyzin, L., Russell, S.: Upper and lower bounds on black-box steganography. J. Cryptol. **22**(3), 365–394 (2009)
6. Ernst, M., Liśkiewicz, M., Reischuk, R.: Algorithmic learning for steganography: proper learning of k-term DNF formulas from positive samples. In: Elbassioni, K., Makino, K. (eds.) ISAAC 2015. LNCS, vol. 9472, pp. 151–162. Springer, Heidelberg (2015)
7. Hopper, N., von Ahn, L., Langford, J.: Provably secure steganography. IEEE Trans. Comput. **58**(5), 662–676 (2009)
8. Jerrum, M., Valiant, L.G., Vazirani, V.V.: Random generation of combinatorial structures from a uniform distribution. TCS **43**, 169–188 (1986)
9. Katzenbeisser, S., Petitcolas, F.A.: Defining security in steganographic systems. In: Electronic Imaging 2002, SPIE, pp. 50–56 (2002)
10. Ker, A.D., Bas, P., Böhme, R., Cogranne, R., Craver, S., Filler, T., Fridrich, J., Pevný, T.: Moving steganography and steganalysis from the laboratory into the real world. In: Proceedings 1st ACM WS on Information Hiding and Multimedia Security, pp. 45–58. ACM (2013)
11. Ker, A.D., Pevný, T., Kodovský, J., Fridrich, J.J.: The square root law of steganographic capacity. In: Proceedings of 10th WS Multimedia & Security, pp. 107–116 (2008)
12. Lange, S., Wiehagen, R.: Polynomial-time inference of arbitrary pattern languages. New Gener. Comput. **8**(4), 361–370 (1991)
13. Liśkiewicz, M., Reischuk, R., Wölfel, U.: Grey-box steganography. TCS **505**, 27–41 (2013)
14. Reidenbach, D.: A negative result on inductive inference of extended pattern languages. In: Cesa-Bianchi, N., Numao, M., Reischuk, R. (eds.) ALT 2002. LNCS (LNAI), vol. 2533, pp. 308–320. Springer, Heidelberg (2002)
15. Reischuk, R., Zeugmann, T.: An average-case optimal one-variable pattern language learner. JCSS **60**(2), 302–335 (2000)

16. Salomaa, A.: Patterns. EATCS Bull. **54**, 46–62 (1994)

17. Shinohara, T.: Polynomial time inference of extended regular pattern languages. In: Goto, E., Furukawa, K., Nakajima, R., Nakata, I., Yonezawa, A. (eds.) RIMS Symposia on Software Science and Engineering. LNCS, pp. 115–127. Springer, Heidelberg (1983)

18. Shinohara, T., Arikawa, S.: Pattern inference. In: Lange, S., Jantke, K.P. (eds.) GOSLER 1994. LNCS, vol. 961, pp. 259–291. Springer, Heidelberg (1995)

19. Stephan, F., Yoshinaka, R., Zeugmann, T.: On the parameterised complexity of learning patterns. In: Proceedings of 26th Computer and Information Sciences, pp. 277–281 (2011)

Inferring a Relax NG Schema from XML Documents

Guen-Hae Kim, Sang-Ki Ko, and Yo-Sub Han[(✉)]

Department of Computer Science, Yonsei University, 50, Yonsei-Ro,
Seodaemun-Gu, Seoul 120-749, Republic of Korea
{guenhaekim,narame7,emmous}@cs.yonsei.ac.kr

Abstract. An XML schema specifies the structural properties of XML documents generated from the schema and, thus, is useful to manage XML data efficiently. However, there are often XML documents without a valid schema or with an incorrect schema in practice. This leads us to study the problem of inferring a Relax NG schema from a set of XML documents that are presumably generated from a specific XML schema. Relax NG is an XML schema language developed for the next generation of XML schema languages such as document type definitions (DTDs) and XML Schema Definitions (XSDs). Regular hedge grammars accept regular tree languages and the design of Relax NG is closely related with regular hedge grammars. We develop an XML schema inference system using hedge grammars. We employ a genetic algorithm and state elimination heuristics in the process of retrieving a concise Relax NG schema. We present experimental results using real-world benchmark.

Keywords: XML schema inference · Regular hedge grammar · Relax NG · Genetic algorithm

1 Introduction

Most information in the real-world has structured with linear ordering and hierarchies. Structural information is often represented in XML (Extensible Markup Language) format, which is both human-readable and machine-readable. An XML schema describes properties and constraints about XML documents. We can manipulate XML data efficiently if we know the corresponding XML schema for the input XML data [11,16,20]. Many researchers demonstrated several advantages when the corresponding XML schema exists [14,21].

All valid XML documents should conform to a DTD or a XML schema. However, in practice, we may not have a valid schema or have an incorrect schema or have an incorrect schema for an input XML data [2,17]. For these reasons, there were several attempts to infer a valid XML schema from a given XML data [3–5]. Bex et al. [3,4] presented an idea for learning deterministic regular expressions for inferring *document type definitions* (DTDs) concisely from XML documents. Bex et al. [5] designed an algorithm for inferring *XML Schema Definitions* (XSDs), which are more powerful than DTDs, from XML documents.

© Springer International Publishing Switzerland 2016
A.-H. Dediu et al. (Eds.): LATA 2016, LNCS 9618, pp. 400–411, 2016.
DOI: 10.1007/978-3-319-30000-9_31

We consider a schema language called *Relax NG*, which is more powerful than both DTDs and XSDs [19]—due to its expressive power, researchers proposed several applications based on Relax NG schema language [1,15]. League and Eng [15] proposed a compression technique of XML data with a Relax NG schema and showed its effectiveness, especially, for highly tagged and nested data.

We study the problem of inferring a concise Relax NG schema from XML documents based on the genetic algorithm approach. Note that a XML document can be described as an ordered tree. Then a Relax NG schema is a regular tree language. Therefore, we use normalized regular hedge grammars (NRHGs) [18] as theoretical tools for representing Relax NG schema since NRHGs exactly captures the class of regular tree languages. We employ a genetic algorithm for learning NRHGs from a set of trees and design a conversion algorithm for obtaining a concise Relax NG schema from NRHGs.

The main idea of inferring a Relax NG schema consists of the following three steps:

1. construct an initial NRHG that only generates all positive instances,
2. reduce the size of the NRHG using genetic algorithm while considering all negative examples, and
3. convert the obtained NRHG into a concise Relax NG schema with the help of state elimination algorithm and variable dependency computation.

We present experimental results with three benchmark schemas and show the preciseness and conciseness of our approach. Our Relax NG inference system successfully infers three benchmark schemas with instances randomly generated by a Java library called xmlgen. We measure the accuracy of our inference algorithm by validating XML documents generated from the original schema against the inferred schema. The inferred schema accepts about 90 % of positive examples and rejects about 80 % of negative examples.

We give some basic notations and definitions in Sect. 2, and present a technique for inferring an NRHG from a set of positive examples in Sect. 3. In Sect. 4, we present a conversion algorithm that converts an NRHG into a Relax NG schema. Experimental results are presented in Sect. 5.

2   Preliminaries

Let Σ be a finite alphabet and Σ^* be the set of all strings over the alphabet Σ including the empty string λ. The size $|\Sigma|$ of Σ is the number of characters in Σ. For a string $w \in \Sigma^*$, we denote the length of w by $|w|$ and the ith character of w by w_i. A language over Σ is any subset of Σ^*.

Let t be a tree, t_n be the number of nodes in t and t_e be the number of edges in t. We define the size $|t|$ of a tree t to be $t_n + t_e$, which is the number of nodes and edges in t. The root node of t is denoted by t_{root}. Let v be a node of a tree. Then, we denote the parent node of v by v_{parent}, the left sibling of v by v_{sibling}, and the ith child of v by $v[i]$. We also denote the number of children of the node v by $|v_{\text{child}}|$ and the label of a node v by $\mathsf{label}(v)$.

A *regular expression* over Σ is $\emptyset, \lambda$, or $a \in \Sigma$, or is obtained by applying the following rules finitely many times. For two regular expressions R_1 and R_2, the union $R_1 + R_2$, the catenation $R_1 \cdot R_2$, and the star R_1^* are regular expressions.

A *nondeterministic finite-state automaton* (NFA) A is specified by a tuple $(Q, \Sigma, \delta, s, F)$, where Q is a finite set of states, Σ is an input alphabet, $\delta : Q \times \Sigma \to 2^Q$ is a multi-valued transition function, $s \in Q$ is the initial state and $F \subseteq Q$ is a set of final states. The transition function δ can be extended to a function $Q \times \Sigma^* \to 2^Q$ that reflects sequences of inputs. A string w over Σ is accepted by A if there is a labeled path from s to a state in F such that this path spells out the string w; namely, $\delta(s, w) \cap F \neq \emptyset$. The language $L(A)$ recognized by A is the set of all strings that are spelled out by paths from s to a final state in F: $L(A) = \{w \in \Sigma^* \mid \delta(s, w) \cap F \neq \emptyset\}$.

A *normalized regular hedge grammar* (NRHG) G is specified by a quintuple (Σ, V_T, V_F, P, s), where Σ is an alphabet, V_T is a finite set of tree variables, V_F is a finite set of forest variables, $s \in V_T$ is the starting symbol, and P is a set of production rules, each of which takes one of the following four forms:

(a) $v_t \to x$, where v_t is a tree variable in V_T, and x is a terminal in Σ,
(b) $v_t \to x\langle v_f \rangle$, where v_t is a tree variable in V_T, x is a terminal in Σ and v_f is a forest variable in V_F,
(c) $v_f \to v_t$, where v_f is a forest variable and v_t is a tree variable,
(d) $v_f \to v_t v_f'$, where v_f and v_f' are forest variables and v_t is a tree variable.

We consider a *derivation* of NRHGs. Given a sequence of variables, we repeatedly replace variables with productions on the right-hand side.

1. for a production rule $v_t \to x$, a node labeled by $x \in \Sigma$ is derived from the tree variable v_t,
2. for a production rule $v_t \to x\langle v_f \rangle$, a tree with a root node labeled by x and its child node v_f is derived from v_t,
3. for a production rule $v_f \to v_t$, a node v_t is derived from v_f, and
4. for a production $v_f \to v_t v_f'$, a sequence of nodes v_t and v_f' is derived from v_f.

The language generated by G is the set of trees derived from s. Given an NRHG $G = (\Sigma, V_T, V_F, P, s)$, the size $|G|$ of G to be $|V_T| + |V_F| + |P|$. Note that the NRHGs generate *regular tree languages* of unranked trees. Thus an NRHG can be converted into an unranked tree automaton and vice versa [6].

For more knowledge in automata and formal language theory, the reader may refer to the textbooks [12, 22].

3   Inference of an NRHG from Trees

A XML document is useful to store a structured information and, often, represented as a labeled and ordered tree. Given a set of positive examples (trees) and a set of negative examples, we aim at learning an NRHG that generates all positive examples and does not generate all negative examples.

Let T be a given set of examples, $T_+ \subseteq T$ denote the set of positive examples and $T_- \subseteq T$ be a set of negative examples. First, we construct *primitive NRHGs* from positive examples; For each tree $t \in T_+$, we construct an NRHG G_t that only accepts the tree t. Namely, $L(G_t) = \{t\}$. Then, we take the union of all primitive NRHGs and construct an NRHG G_{T_+} that only accepts all positive examples, that is,

$$L(G_{T_+}) = \bigcup_{t \in T_+} L(G_t).$$

We reduce the size of the resulting NRHG by merging variables using a genetic algorithm. Note that the problem of minimizing the NRHG is at least as hard as any PSPACE-complete problem as the problem of minimizing NFAs is PSPACE-complete [13]. Therefore, we rely on the genetic algorithm approach to reduce the size of very large primitive NRHGs as much as possible.

3.1 Primitive NRHG Construction

We present a linear-time algorithm for constructing a primitive NRHG that only accepts a given tree t. Our algorithm runs recursively from the root node of t to the leaf nodes.

For each node of t, we recursively construct an NRHG according to the number of children of t. If a node v labeled by x has no child, then we create a production rule $T \rightarrow x$. If a node v labeled by x has more than one child, then we create a production rule $T \rightarrow x \langle F \rangle$. Then, we make a forest variable F to generate the sequence of subtrees of v. Let us assume that v has n subtrees from t_1 to t_n and a tree variable T_i generates a subtree t_i. Then, by creating the following variables and production rules, we let F generate the sequence of subtrees: $F \rightarrow T_1 F_1$, $F_1 \rightarrow T_2 F_2$, $\cdots$ $F_{n-2} \rightarrow T_{n-1} F_{n-1}$, $F_{n-1} \rightarrow T_n$. Since each node in t is represented by a tree variable, each edge in t is represented by a forest variable and left side of each production rule is a tree variable or a forest variable, $|V_T| = t_n$, $|V_F| = t_e$ and $|P| = |t| = t_n + t_e$. Therefore, it is easy to verify that the algorithm produces an NRHG $G = (\Sigma, V_T, V_F, P, s)$ in $O(|t|)$ time such that $L(G) = t$, where $|V_T| = t_n$, $|V_F| = t_e$, and $|P| = |t|$.

Suppose that we have n positive examples from t_1 to t_n. We first construct n primitive NRHGs called G_1 to G_n, where $L(G_i) = \{t_i\}, 1 \le i \le n$. From the n primitive NRHGs, we take the union of the NRHGs and obtain a single NRHG G such that $L(G) = \cup_{i=1}^{n} G_i$. Let $G_i = (\Sigma_i, V_{T_i}, V_{F_i}, P_i, s_i)$ be an NRHG that accepts t_i. Then, we construct a new NRHG $G = (\Sigma, V_T, V_F, P, s)$ that accepts all instances as follows :

$$\Sigma = \bigcup_{i=1}^{n} \Sigma_i, \quad V_T = \bigcup_{i=1}^{n} V_{T_i}, \quad V_F = \bigcup_{i=1}^{n} V_{F_i}, \text{ and } P = \bigcup_{i=1}^{n} P_i.$$

Namely, $L(G) = \{t_1, t_2, \ldots, t_n\}$. Then, we merge all occurrences of starting symbols from s_1 to s_n and denote the merge symbol by s.

3.2   Grammar Optimization by Merging Indistinguishable Variables

Now we have an NRHG $G = (\Sigma, V_T, V_F, P, s)$ that only generates all positive examples. Since we aim at retrieving a concise Relax NG schema, we need to reduce the size of G as much as possible. As a first step, we find redundant structures from G by identifying *indistinguishable* variables. Given a variable v, which is either in V_T or V_F, we define $RS(v)$ to be the set of variables that appear in the right-hand side of production rules, where v appears in the left-hand side. Similarly, we define $LS(v)$ to be the set of variables that appear in the left-hand side of production rules, where v appears in the right-hand side.

$$RS(v) = \begin{cases} \{x \mid v \rightarrow x \in P\} \cup \{(x, v_f) \mid v \rightarrow x\langle v_f \rangle \in P\}, & \text{if } v \in V_T, \\ \{v_t \mid v \rightarrow v_t \in P\} \cup \{(v_t, v_f) \mid v \rightarrow v_t v_f \in P\}, & \text{if } v \in V_F. \end{cases}$$

We say that two variables v_1 and v_2, where $v_1, v_2 \in V_T$ or $v_1, v_2 \in V_F$ are *right-indistinguishable* if and only if $RS(v_1) = RS(v_2)$. It is easy to see that we do not change the language of G if we replace the occurrences of v_1 and v_2 with a new variable v' since two variables generate exactly the same structures.

$$LS(v) = \begin{cases} \{v_f \mid v_f \rightarrow v \in P\} \cup \{(v_{f_1}, v_{f_2}) \mid v_{f_1} \rightarrow vv_{f_2} \in P\}, & \text{if } v \in V_T, \\ \{(v_t, x) \mid v_t \rightarrow x\langle v \rangle \in P\}, & \text{if } v \in V_F. \end{cases}$$

We also define two variables v_1 and v_2 to be *left-indistinguishable* if and only if $LS(v_1) = LS(v_2)$. We say that two variables v_1 and v_2 are *indistinguishable* if they are right-indistinguishable or left-indistinguishable. Here we show that the language of the resulting NRHG does not change when we merge indistinguishable variables into a single variable.

Theorem 1. *Given two indistinguishable variables v_1 and v_2 of an NRHG $G = (\Sigma, V_T, V_F, P, s)$, let G' be a new NRHG where all indistinguishable variables are merged according to indistinguishable classes of G. Then, $L(G) = L(G')$.*

We repeat the merging process until there is no pair of indistinguishable variables in the NRHG. Empirically, we have obtained about 80 % size reduction by merging indistinguishable variables in primitive NRHGs. We show the experimental evidence in Sect. 5.

3.3   Genetic Algorithm

We now have a primitive NRHG from a set of positive examples without any indistinguishable variables. Next, we need to make it as small as possible while considering a set of negative examples. Recall that our main goal is to compute a concise NRHG that generates all positive examples and does not generate all negative examples.

We employ a genetic algorithm (GA), which involves an evolutionary process, to find a small NRHG from a given primitive NRHG. In a genetic algorithm, we first make a population of candidate solutions called *individuals* and make it evolve to the population of better solutions by the help of genetic operators such as *structural crossover* and *structural mutation*.

We explain how these genetic operators work in our schema inference algorithm. We assume that every individual in the first population is indeed an NRHG with seven variables and thus encoded as a string of length 7.

- *structural crossover:* In the population, we randomly select two encoded individuals p_1 and p_2 as parents. We use $p_1 = 1324133$ and $p_2 = 2234144$ as a running example for explaining our approach.

$$p_1 = 1324133 \text{ and } p_2 = 2234144.$$

These strings encode two partitions π_{p_1} and π_{p_2} as follows:

$$\pi_{p_1} : \{1,5\}, \{3\}, \{2,6,7\}, \{4\} \text{ and } \pi_{p_2} : \{5\}, \{1,2\}, \{3\}, \{4,6,7\}.$$

Namely, the ith number of p_1 implies the index of the block where the ith variable of p_1 belongs. For example, the third number of p_1 is 2, and therefore, the variable 3 belongs to the second block #2 of π_{p_1}. Now we randomly select two blocks, say #2 and #4. The #2 block is copied from p_1 to p_2 by taking the union of #2 blocks in p_1 and p_2 and #2 block to be moved from p_1, and #4 block is copied from p_2 to p_1 in the same way. We obtain the following results:

$$\pi_{p'_1} : \{1,5\}, \{3\}, \{2\}, \{4,6,7\} \quad \pi_{p'_2} : \{5\}, \{1,2\}, \emptyset, \{3,4,6,7\}$$

In this way, the number of blocks of each partition can diminish by merging randomly selected blocks.

- *structural mutation:* We randomly select an individual $p = 1324133$ and replace a character in the string by some random number. For example, if we replace the second character by 4, then the following offspring is produced:

$$\pi_{p'} : \{1,5\}, \{3\}, \{2,6,7\}, \{2,4\}.$$

We employ the GA approach for inferring a concise NRHG from a set of positive examples and a set of negative examples as follows:

1. Initialize the population of candidate solutions. Here we set the population size to, 1000. The initial candidate solutions are the primitive NRHGs reduced by merging indistinguishable variables.
2. Select some pairs of individuals according to the crossover rate (0.4) and construct new pairs of individuals by applying the crossover operator to the selected pairs.
3. Select some individuals according to the mutation rate (0.03) and modify the selected individuals by applying the mutation operator.
4. Calculate a fitness value $f(p)$ for each individual p by the fitness function. Let p be an individual encoding for an NRHG $G = (\Sigma, V_T, V_F, P, s)$. Then, the fitness value $f(p)$ of p is defined as follows:

$$f(p) = \begin{cases} \dfrac{1}{|V_F| + |V_T|} + \dfrac{1}{|P|} + \dfrac{|\{w \in U_+ \mid w \in L(G)\}|}{|U_+|}, & \text{if } U_- \cap L(G) = \emptyset, \\ 0 & \text{otherwise.} \end{cases}$$

where U_+ is the set of positive examples and U_- is the set of negative examples.

5. Generate a next generation by *roulette-wheel selection* from the current population of solutions. Note that we retain the individuals from the best 10 % of the current population unchanged in the next generation and select only the remaining 90 % by roulette-wheel selection.

6. Iterate 1–5 steps until the fitness value of the best individual reaches the given threshold.

4   Converting NRHG into Relax NG Schema

Here we present a conversion algorithm from an NRHG into a corresponding Relax NG schema. A Relax NG schema uses references to *named pattern* using the **define** elements. Now tree or forest variables in NRHGs can be directly converted into the **define** elements for being used as references.

4.1   Horizontal NFA Construction

Given an NRHG $G = (\Sigma, V_T, V_F, P, s)$, let us consider the starting symbol s. Without loss of generality, we assume that there is a production rule $s \rightarrow x\langle v_f \rangle \in P$, where $x \in \Sigma$ and $v_f \in V_F$. Then, we convert s into the corresponding **define** element in the resulting Relax NG schema. Now it remains to convert the forest variable v_f into the corresponding element in the schema.

We construct an NFA that accepts all possible sequences of tree variables that can be generated by v_f. We call this procedure the *horizontal NFA construction* for v_f. For each forest variable $v_f \in V_F$, where $v_t \rightarrow x\langle v_f \rangle \in P$, we construct a horizontal NFA $A_{v_f} = (Q, \Sigma, \delta, q_0, q_f)$ as follows:

1. $Q = V_F \cup \{q_f\}$ is a finite set of states,
2. $\Sigma = V_T$ is an input alphabet,
3. $q_0 = v_f$ is the initial state, and
4. the transition function δ is defined as follows:
   (a) $q' \in \delta(q, v_t)$ for each $q \rightarrow v_t q' \in P$, and
   (b) $q_f \in \delta(q, v_t)$ for each $q \rightarrow v_t \in P$.

For example, Fig. 1 shows how we construct horizontal NFAs from a simple NRHG. From the first two production rules, we know that F_0 generates a sequences of sub-elements of **records** and F_1 generates a sequences of sub-elements of **car**. Therefore, we generate two horizontal NFAs A_{F_0} and A_{F_1} with the initial states are F_0 and F_1, respectively.

4.2   State Elimination for Obtaining Regular Expressions

Recall that a Relax NG schema can specify a regular tree language when we consider the tree structures of XML documents captured by the schema. When a Relax NG schema describes a set of possible sequences of trees, we use several elements such as **choice**, **group**, **zeroOrMore**, and so on.

$T_0 \rightarrow \texttt{records}\langle F_0 \rangle, \quad T_1 \rightarrow \texttt{car}\langle F_1 \rangle,$

$F_1 \rightarrow T_2 F_2, \quad F_0 \rightarrow T_1 F_0,$

$F_0 \rightarrow T_1, \quad F_2 \rightarrow T_3,$

$T_3 \rightarrow \texttt{record}, \quad T_2 \rightarrow \texttt{country}$

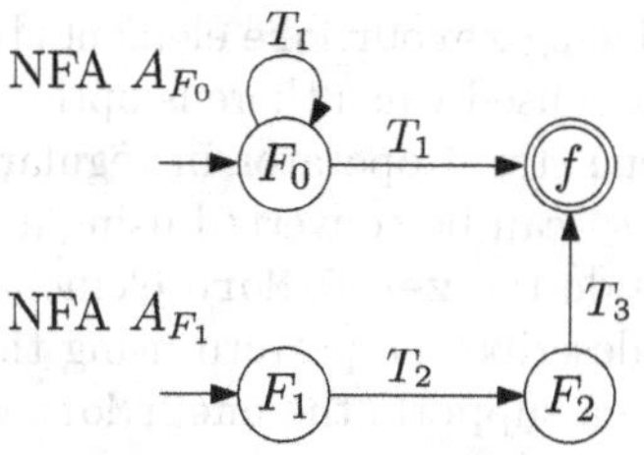

Fig. 1. Since two forest variables F_0 and F_1 are used for generating sequences of sub-elements of **records** and **car**, respectively, we construct two horizontal NFAs A_{F_0} and A_{F_1}. Note that the final states of A_{F_0} and A_{F_1} are merged into a single final state f.

The **choice** element implies that one of the sub-elements inside the element can be chosen. Therefore, we can say that the role of the **choice** element in Relax NG corresponds to the role of the union operator in regular expression. Similarly, the **group** element implies that the sub-elements inside the element should appear in exactly the same order. Thus, the **group** element corresponds to the catenation operator in regular expression. For this reason, we need to convert the obtained horizontal NFAs into the corresponding regular expressions since regular expressions are described in a very similar manner with the Relax NG schema.

State elimination is an intuitive algorithm that computes a regular expression from a finite-state automaton (FA) [9]. There are several heuristics of state elimination for obtaining shorter regular expressions from FAs [7–10]. Delgado et al. [7] observed that an order in eliminating states is crucial for obtaining a shorter regular expression. They defined the weight of a state to be the size of new transition labels that are created as a result of eliminating the state. We borrow their idea and define the weight of a state q in an FA $A = (Q, \Sigma, \delta, s, F)$ as follows:

$$\sum_{i=1}^{\mathsf{IN}} (\mathbb{W}_{\text{in}}(i) \times \mathsf{OUT}) + \sum_{i=1}^{\mathsf{OUT}} (\mathbb{W}_{\text{out}}(i) \times \mathsf{IN}) + \mathbb{W}_{\text{loop}} \times (\mathsf{IN} \times \mathsf{OUT}),$$

where IN is the number of in-transitions excluding self-loops, OUT is the number of out-transitions excluding self-loops, $\mathbb{W}_{\text{in}}(i)$ is the size of the transition label on the ith in-transition, $\mathbb{W}_{\text{out}}(i)$ is the size of the transition label on the ith out-transition, and $\mathbb{W}_{\text{loop}}$ is the size of the self-loop label. After calculating the weights of all states, we eliminate the state with the smallest weight and calculate the state weights again. We repeat this procedure until there are only the initial state and the single final state.

4.3 Schema Refinement

Now we have regular expressions for forest variables of an NRHG and are ready to convert these regular expressions into the form of Relax NG schema. We also use several additional techniques for converting an NRHG into a Relax NG schema.

- Replacing `zeroOrMore` elements by `oneOrMore` elements: The `zeroOrMore` element is used when there is optionally a repetition of a certain pattern like the Kleene star * operator in regular expression. For example, a regular expression a^* can be converted using a `zeroOrMore` element and an element named a inside the `zeroOrMore` element. In some cases such as $a^* \cdot a$ or $a \cdot a^*$, we can describe the pattern using the Kleene plus operator a^+. Since Relax NG schema supports the `oneOrMore` element that corresponds to the Kleene plus in regular expression, we can make the resulting schema smaller in some cases.
- Checking the dependency of forest variables: Consider two forest variables v_1 and v_2, where v_2 can be derived from v_1, but v_1 cannot be derived from v_2. If we construct regular expressions for two forest variables, then there should be some redundancy in two regular expression since the pattern described for v_2 is already contained in the pattern described for v_1. Therefore, we can write the pattern for v_2 by just referring to the pattern for v_1 instead of writing two redundant patterns.

5   Experimental Results

We conduct experiments for inferring a Relax NG schema from a given XML data. For the experiments, we use a Java library xmlgen developed as a part of Sun Multi-Schema XML Validator[1] from randomly generating positive and negative XML instances for an input Relax NG schema.

5.1   Experimental Setup

We use three benchmark Relax NG schemas—XENC, XML-DSig, IBTWSH—to evaluate the performance of our Relax NG inference algorithm.

We aim at inferring only the relationship between elements from XML data. Therefore, we ignore the descriptions for attributes of XML data from benchmark schemas. Moreover, we manually replace the elements defined by the `anyName` name by the elements with the name `anyName` since otherwise randomly generated instances may have too many elements with arbitrary names. We also limit the maximum number of appearing sub-elements in the `zeroOrMore` elements to 2 since otherwise we may have very large XML instances compare to the size of input schema.

5.2   Size Reduction of NRHGs by Optimization

We show that the optimization process helps to reduce the size of primitive NRHGs generated by positive instances.

Table 1 exhibits that the optimization process reduce 84.15 %, 82.13 %, 77.04 % of redundancy from the primitive NRHGs constructed from the positive instances of XENC, XML-DSig and IBTWSH, respectively. Note that the optimization process contributes to the speedup of the genetic process as the size of initial population also decreases substantially.

---

[1] The Oracle Multi-Schema XML Validator (MSV). https://msv.java.net/.

Table 1. The compression ratio ([size of NRHG before optimization]/[size of NRHG before optimization]) achieved by merging indistinguishable variables from primitive NRHGs.

Benchmark schema	XENC		XML-DSig		IBTWSH	
	Before	After	Before	After	Before	After
# of tree variables	755.17	123.67	1048.73	123.20	1222.86	162.57
# of forest variables	705.17	187.17	998.73	176.93	1122.86	208.29
# of production rules	1460.33	359.83	2047.47	349.13	2345.71	467.29
Average size of NRHG	4094.93	649.27	4691.43	838.14	2920.67	670.67
Compression ratio (%)	15.85		17.87		22.96	

5.3   Precision of Inferred Schema

For three benchmark schemas, we generate 50 positive instances and 25 negative instances by xmlgen. Note that we set the error rate to 1/100 when generating negative instances. Then we run our inference algorithm to infer Relax NG schemas for the instances. We repeat the process 100 times and calculate the average value.

We evaluate the precision of our inference system in two directions. First, we validate the inferred schema against 1,000 positive instances generated from the original benchmark schema. Second, we validate the inferred schema against 1,000 negative instances generated from the original benchmark schema. We expect that the inferred schema should generate the positive instances and not generate the negative instances if they are inferred closely to the original schemas.

Table 2. The precision of our Relax NG schema inference system

Benchmark schema	XENC	XML-DSig	IBTWSH (50/25)	IBTWSH (100/50)
Precision for positive instances (%)	91.98	93.50	46.05	96.87
Precision for negative instances (%)	82.43	83.85	83.45	74.67

Table 2 shows the precision of our inference algorithm. Note that the fourth column shows the results for the benchmark IBTWSH with 100 positive and 50 negative instances. Speaking of the first two results, precisions for positive and negative instances are very similar. The inferred schemas generate more than 90 % of positive instances of the original schemas and do not generate more than 80 % of negative instances. The performance is good considering that we infer schemas with only a small number of instances.

For the benchmark IBTWSH, only 46.05 % of positive instances can be generated by the inferred schema. The inference precision improves significantly when we double the number of instances to 100 positive and 50 negative instances. We suspect that the reason why the precision for the benchmark IBTWSH is especially low is because IBTWSH schema has several `interleave` elements that allow the sub-elements to occur in any order. Since our inference system does not support the inference of `interleave` elements, it seems more difficult to infer schemas with `interleave` elements.

6   Conclusions

XML schemas are formal languages that describe structures and constraints about XML instances. They are crucial to maintain and manipulate XML documents efficiently—an XML document should conform a specific XML schema. However, in practice, we may not have a valid schema or have an incorrect schema. This has led many researchers to design an efficient schema inference algorithm.

We have presented an Relax NG schema—one of the most powerful XML schema languages—inference algorithm based on 1) a genetic algorithm for learning process and 2) state elimination heuristics for retrieving a concise Relax NG schema. We have implemented the proposed algorithm and measured the preciseness and conciseness of the algorithm using well-know benchmark schemas. Our experiments have showed that the proposed algorithm has 90 % preciseness for accepting positive examples and 80 % preciseness for rejecting negative examples. In future, we plan to consider other learning approaches for better performance and more Relax NG specifications such as interleave.

References

1. Athan, T., Boley, H.: Design and implementation of highly modular schemas for XML: customization of RuleML in relax NG. In: Palmirani, M. (ed.) RuleML - America 2011. LNCS, vol. 7018, pp. 17–32. Springer, Heidelberg (2011)
2. Barbosa, D., Mignet, L., Veltri, P.: Studying the XML web: gathering statistics from an XML sample. World Wide Web 8(4), 413–438 (2005)
3. Bex, G.J., Gelade, W., Neven, F., Vansummeren, S.: Learning deterministic regular expressions for the inference of schemas from XML data. ACM Trans. Web 4(4), 14 (2010)
4. Bex, G.J., Neven, F., Schwentick, T., Tuyls, K.: Inference of concise DTDs from XML data. In: Proceedings of the 32nd International Conference on Very Large Data Bases, pp. 115–126. VLDB Endowment (2006)
5. Bex, G.J., Neven, F., Vansummeren, S.: Inferring XML schema definitions from XML data. In: Proceedings of the 33rd International Conference on Very Large Data Bases, pp. 998–1009. VLDB Endowment (2007)
6. Comon, H., Dauchet, M., Jacquemard, F., Lugiez, D., Tison, S., Tommasi, M.: Tree Automata Techniques and Applications (2007). http://www.tata.gforge.inria.fr
7. Delgado, M., Morais, J.J.: Approximation to the smallest regular expression for a given regular language. In: Domaratzki, M., Okhotin, A., Salomaa, K., Yu, S. (eds.) CIAA 2004. LNCS, vol. 3317, pp. 312–314. Springer, Heidelberg (2005)

8. Gruber, H., Holzer, M.: Provably shorter regular expressions from deterministic finite automata. In: Ito, M., Toyama, M. (eds.) DLT 2008. LNCS, vol. 5257, pp. 383–395. Springer, Heidelberg (2008)
9. Han, Y.S.: State elimination heuristics for short regular expressions. Fundam. Inf. **128**(4), 445–462 (2013)
10. Han, Y.S., Wood, D.: Obtaining shorter regular expressions from finite-state automata. Theor. Comput. Sci. **370**(1), 110–120 (2007)
11. He, B., Tao, T., Chang, K.C.-C.: Clustering structured web sources: a schema-based, model-differentiation approach. In: Lindner, W., Fischer, F., Türker, C., Tzitzikas, Y., Vakali, A.I. (eds.) EDBT 2004. LNCS, vol. 3268, pp. 536–546. Springer, Heidelberg (2004)
12. Hopcroft, J., Ullman, J.: Introduction to Automata Theory, Languages, and Computation. Addison-Wesley, Boston (1979)
13. Jiang, T., Ravikumar, B.: Minimal nfa problems are hard. SIAM J. Comput. **22**(6), 1117–1141 (1993)
14. Koch, C., Scherzinger, S., Schweikardt, N., Stegmaier, B.: Schema-based scheduling of event processors and buffer minimization for queries on structured data streams. In: Proceedings of the 30th International Conference on Very Large Data Bases, pp. 228–239. VLDB Endowment (2004)
15. League, C., Eng, K.: Schema-based compression of XML data with RELAX NG. J. Comput. **2**(10), 9–17 (2007)
16. Löser, A., Siberski, W., Wolpers, M., Nejdl, W.: Information integration in schema-based peer-to-peer networks. In: Eder, J., Missikoff, M. (eds.) CAiSE 2003. LNCS, vol. 2681, pp. 258–272. Springer, Heidelberg (2003)
17. Mignet, L., Barbosa, D., Veltri, P.: The XML web: a first study. In: Proceedings of the 12th International Conference on World Wide Web, pp. 500–510 (2003)
18. Murata, M.: Hedge automata: a formal model for XML schemata (1999)
19. Paige, R., Tarjan, R.E.: Three partition refinement algorithms. SIAM J. Comput. **16**(6), 973–989 (1987)
20. Shvaiko, P.: A classification of schema-based matching approaches (2004)
21. Wang, G., Liu, M., Yu, G., Sun, B., Yu, G., Lv, J., Lu, H.: Effective schema-based XML query optimization techniques. In: Proceedings of the 7th International Symposium on Database Engineering and Applications, pp. 230–235 (2003)
22. Wood, D.: Theory of Computation. Harper & Row, New York (1987)

Noise Free Multi-armed Bandit Game

Atsuyoshi Nakamura[1]($\boxtimes$), David P. Helmbold[2], and Manfred K. Warmuth[2]

[1] Hokkaido University, Kita 14, Nishi 9, Kita-ku, Sapporo 060-0814, Japan
atsu@main.ist.hokudai.ac.jp
[2] University of California at Santa Cruz, Santa Cruz, USA
{dph,manfred}@cse.ucsc.edu

Abstract. We study the loss version of adversarial multi-armed bandit problems with one lossless arm. We show an adversary's strategy that forces any player to suffer $K - 1 - O(1/T)$ loss where K is the number of arms and T is the number of rounds.

Keywords: Algorithmic learning · Online learning · Bandit problem

1 Introduction

In this paper, we study a kind of the loss version of adversarial multi-armed bandit problems with one lossless arm, that is, the problem in the noise-free case. The gain (reward) version of an adversarial multi-armed bandit problem was studied first by Auer et al. [1] and its loss version has been also studied in some papers such as [2,3]. To the best of our knowledge, however, the problem in the noise-free setting has not been studied yet. This could be because the problem is too trivial to study. In fact, it is easy in the *full-information* case; with $\{0,1\}$ losses and K arms, loss $\sum_{i=2}^{K}(1/i) = \Theta(\ln K)$ is achieved by the minmax strategy: the adversary's maximization strategy sets the loss of an arm with the highest probability of being chosen to 1 in addition to the past lossy arms, and the player's minimization strategy always chooses one of the arms with no loss so far randomly with equal probability. In the bandit case, however, it does not seem so trivial because the arms with no loss so far in player's observation might already have suffered 1-loss. Thus, the adversary may have to stick to its loss assignments, waiting for the player to choose one of the lossy arms.

In this paper, we focus on an adversary's strategy. The adversary's strategy studied in [1] selects a best arm randomly and sets losses of the non-best arms to 1 with probability $1/2$ and sets the loss of the best arm to 1 with probability $1/2 - \epsilon$ at each time for some small ϵ. Their strategy can be modified for the noise free case by changing the probabilities of 1-loss to ϵ and 0. This adversary's strategy is very weak against the player's strategy that sticks to the same random arm until he/she suffers 1-loss, forcing only $(K - 1)/2$ loss in the K-arm case. We show an adversary's strategy that forces any player to suffer $K - 1 - O(1/T)$ loss for K-arm and T-round case.

© Springer International Publishing Switzerland 2016
A.-H. Dediu et al. (Eds.): LATA 2016, LNCS 9618, pp. 412–423, 2016.
DOI: 10.1007/978-3-319-30000-9_32

2   Problem Setting

For any natural numbers i, j with $i \leq j$, $[i..j]$ denotes the set $\{i, \dots, j\}$ and we let $[j]$ denote $[1..j]$. For any sequence $x_1, \dots, x_n$, we let $\mathbf{x}[b..e]$ denote its contiguous subsequence $x_b, \dots, x_e$.

The *noise-free multi-armed bandit problem* we consider here is the loss version of an adversarial multi-armed bandit problem with one lossless arm. It is a T-round game between a player and an adversary. There are K arms (of slot machines): arm $1, \dots,$ arm K. At each time $t = 1, \dots, T$, the adversary picks a loss $\ell_{t,i} \in [0,1]$ for each arm $i \in [K]$. Let $\boldsymbol{\ell}_t \in [0,1]^K$ denote a K-dimensional vector $(\ell_{t,1}, \dots, \ell_{t,K})$. The player, who does not know $\boldsymbol{\ell}_t$, chooses arm I_t and suffers loss ℓ_{t,I_t}. The player's and the adversary's objectives are minimization and maximization, respectively, of player's (expected) cumulative loss $\sum_{t=1}^{T} \ell_{t,I_t}$.

The most popular measure for evaluating player's strategies is *regret*, which is difference between player's and the best arm's cumulative losses. Throughout this paper, we assume that there is an arm whose cumulative loss is zero. In this case, regret coincides with cumulative loss. Note that this assumption constrains the adversary's choices.

We allow the player to use a randomized strategy, so at each time t the player's choice I_t is a random variable. Let i_t denote a realization of random variable I_t. We call (i_t, ℓ_{t,i_t}) a player's *observation* at time t and denote it by o_t. Each player's choice I_t can depend only on his/her past observations $\mathbf{o}[1..t-1]$. The adversary that we consider here is assumed to have infinite computation power; no limitation is set on adversary's computational time and space. The Adversary is also allowed to behave adaptively: the adversary's decision $\boldsymbol{\ell}_t$ can depend on both the player's past choices $\mathbf{i}[1..t-1]$ and the adversary's past decisions $\boldsymbol{\ell}[1..t-1]$.

Example 1. Let $K = 2$. Consider a *randomized consistent conservative player* who chooses each arm i with equal probability at $t = 1$ and continues to choose the same arm i until suffering a non-zero loss, and after that chooses the other arm, which must be a lossless arm by the assumption of one lossless arm. For this player, $\mathbb{E}\left[\sum_{t=1}^{T} \ell_{t,I_t}\right] = 1/2$ if $\boldsymbol{\ell}_1 = \cdots = \boldsymbol{\ell}_T = (1,0)$ or $(0,1)$. The adversary, however, can achieve $\mathbb{E}\left[\sum_{t=1}^{T} \ell_{t,I_t}\right] = 1$ using $\boldsymbol{\ell}_1 = (0,0)$ because the adversary can know I_2 from I_1. For a mere *randomized consistent player* who chooses each arm of no loss so far with equal probability, using $\boldsymbol{\ell}_1 = (0,0)$ only reduces $\mathbb{E}\left[\sum_{t=1}^{T} \ell_{t,I_t}\right]$ and the best strategy of the adversary is to use $\boldsymbol{\ell}_1 = \cdots = \boldsymbol{\ell}_T = (1,0)$ or $(0,1)$ which achieves $\mathbb{E}\left[\sum_{t=1}^{T} \ell_{t,I_t}\right] = 1 - (1/2)^T$. Note that, for this loss sequence, $\mathbb{E}\left[\sum_{t=1}^{T} \ell_{t,I_t}\right] = 1/2$ holds in the full-information setting because the lossless arm can be identified at $t = 1$ regardless of the loss suffered by the player.

Algorithm 1. RepeatW&S$[K, T]$

---

parameter: K: number of arms, T: number of trials
initialize : $t_0, \ldots, t_m, k_1, \ldots, k_m \leftarrow$ the solution of Problem 2, $d \leftarrow 1$
for *time* $t = 1, \ldots, T$ **do**
    if $t \geq t_{d-1}$ **then**
        $b \leftarrow t_{d-1}, e \leftarrow t_d - 1$
        $c \leftarrow$ BestSwitchingTime$(b, e, \mathbf{o}[1..b-1], k_d)$
        $d \leftarrow d + 1$
    end
    $\ell_t \leftarrow$ Wait&Sticking$[b, c, e, \mathbf{o}[1..b-1], k_d](t, \mathbf{o}[b, ..t-1])$
    Observe the player's choice i_t
end

Algorithm 2. BestSwitchingTime$(b, e, \mathbf{o}[1..b-1], k)$

---

input : $b, e \in \mathbb{N}$: beginning and ending times with $1 \leq b \leq e$
      $\mathbf{o}[1..b-1]$: players observations from time 1 to time $b-1$
      k: number of no-loss arms to switch to 1-loss
output: c^*: best time to switch from waiting to sticking

$S \leftarrow$ the set of arms with no loss by time $b - 1$
$p_{\max} = -1$
for $c = b, \ldots, e$ **do**
$$
p_c \leftarrow E_{I[b..e-1]}\left[\max_{s \subseteq S, |s| = k} P\left\{ \sum_{t=c}^{e} \ell_{t, I_t} \geq 1 \,\middle|\, \begin{array}{l} \mathbf{O}[1..b-1] = \mathbf{o}[1..b-1], I[b..c-1], \\ \ell_b = \cdots = \ell_{c-1} = \mathbb{1}_{[K]\setminus S}, \\ \ell_c = \cdots = \ell_e = \mathbb{1}_{([K]\setminus S)\cup s} \end{array} \right\} \right]
$$
    if $p_c > p_{max}$ **then**
        $c^* \leftarrow c$
        $p_{\max} \leftarrow p_c$
    end
end
return c^*

---

3 Adversary's Strategy

For any set $S \subseteq [K]$, define $\mathbb{1}_S$ to be the K-dimensional $\{0, 1\}$-vector whose ith component is 1 if and only if $i \in S$. At any time $t \in [T]$, the decision ℓ_t made by our adversary algorithm RepeatW&S$[K, T]$ (Algorithm 1) is $\mathbb{1}_{S_t}$ for some set $S_t \subseteq [K]$ and will satisfy $\ell_1 \leq \cdots \leq \ell_T$, that is, once the ith component becomes 1, it never becomes 0. Based on the loss analysis in Sect. 4, for some natural number $m \in [K - 1]$, the adversary divides $[T]$ into m parts $[t_0..t_1 - 1], \cdots, [t_{m-1}, t_m - 1]$, and also divides $K - 1$ into m non-negative integers $k_1, \ldots, k_m$ with $\sum_{i=1}^{m} k_i = K - 1$. We explain how to divide $[T]$ and $K - 1$ later in this section.

During the ith period $[t_{i-1}..t_i - 1]$, that is, for times $t \in [t_{i-1}, t_i - 1]$, the adversary switches the loss vector ℓ_t at most once: beginning with $\mathbb{1}_{[K]\setminus S}$, where S is the set of lossless arms so far, it calculates the best time $c \in$

Algorithm 3. Wait&Sticking$[b, c, e, \mathbf{o}[1..b - 1], k](t, \mathbf{o}[b..t - 1])$

---

parameter: $b, c, e \in \mathbb{N}$: beginning, switch, and end times, $1 \le b \le c \le e$
           $\mathbf{o}[1..b - 1]$: players observations from time 1 to time $b - 1$
           k: positive integer at most the number of arms with no loss.
input   : $t \in \mathbb{N}$: $b \le t \le e$
           $\mathbf{o}[b..t - 1]$: players observations from time b to time $t - 1$
output : $\ell_t \in \{0, 1\}^K$: loss vector at time t

$S \leftarrow$ the set of arms with no loss by time $b - 1$
if $t < c$ **then**
  | **return** $\mathbb{1}_{[K] \setminus S}$
else
  | **if** $t = c$ **then**

$$s_* \leftarrow \arg\max_{s \subseteq S, |s| = k} P\left\{ \sum_{t'=c}^{e} \ell_{t', I_{t'}} \ge 1 \;\middle|\; \begin{matrix} \mathbf{O}[1..c - 1] = \mathbf{o}[1..c - 1], \\ \ell_c = \cdots = \ell_e = \mathbb{1}_{([K] \setminus S) \cup s} \end{matrix} \right\}$$

  | **end**
  | **return** $\mathbb{1}_{([K] \setminus S) \cup s_*}$
end

$[t_{i-1}..t_i]$ to switch the loss vector using BestSwitchingTime$(t_{i-1}, t_i, \mathbf{o}[1..t_{i-1} - 1], k_i)$ (Algorithm 2). At this time $t = c$, the adversary calculates the best subset $s \subseteq S$ to add to $[K] \setminus S$ and changes the loss vector ℓ_t to $\mathbb{1}_{([K] \setminus S) \cup s}$ using Wait&Sticking$[t_{i-1}, c, t_i, \mathbf{o}[1..t_{i-1} - 1], k](t, \mathbf{o}[b..t - 1])$ (Algorithm 3).

In RepeatW&S$[K, T]$, $[T]$ and $K - 1$ is divided by the solution of Problem 2 shown below. We introduce notation $F(T', K', k)$ for simple description of the problem: for any three integers $T' \ge 1$, $K' \ge 2$ and $k \ge 0$, define function $F(T', K', k)$ as

$$F(T', K', k) = \frac{T' + \binom{K' - 1}{k - 1} - 1}{T' + \binom{K'}{k} - 1}.$$

Consider the following problem.

Problem 2. Given two integers $T \ge 1$ and $K \ge 2$, find two non-negative integer sequences $t_0, \ldots, t_m$ and $k_1, \ldots, k_m$ that maximize

$$\sum_{i=1}^{m} F\left(t_i - t_{i-1}, K - \sum_{j=1}^{i-1} k_j, k_i\right)$$

subject to

$$1 \le m \le K - 1, \tag{1}$$
$$1 = t_0 < t_1 < \cdots < t_m = T + 1 \text{ and} \tag{2}$$
$$k_1 + \cdots + k_m = K - 1. \tag{3}$$

Example 3. Assume that $K = 2$. Then $m = 1, t_0 = 1, t_1 = T + 1, k_1 = 1$ is the solution of Problem 2. The best time c^* to switch is 2 $(p_1 = 1/2, p_2 = \cdots = p_T = 1)$ for the randomized consistent *conservative* player and 1 $(p_c = 1 - (1/2)^{T+1-c})$ for the randomized consistent player.

4   Lower Loss Bound

Lemma 4. *Let b and T' be arbitrary positive integers and let k be an arbitrary non-negative integer. Let c^* be the returned value from BestSwitchingTime$(b, b + T' - 1, \mathbf{o}[1..b - 1], k)$. Then, for any player algorithm, the following holds with respect to the loss vectors generated by Wait&Sticking$[b, c^*, b + T' - 1, \mathbf{o}[1..b - 1]](t, \mathbf{o}[b..t-1])$:*

$$\mathbb{E}_{I[b..b+T'-1]} \left[\sum_{t=b}^{b+T'-1} \ell_{t,I_t} \right] \geq F(T', |S|, k),$$

where S is the set of arms with no loss by time $b - 1$, that is, $S = \{i \in [K] : \sum_{t=1}^{b-1} \ell_{t,i} = 0\}$.

Remark 5. In the case with $K = 2$, Lemma 4 implies

$$\mathbb{E}_{I[1..T]} \left[\sum_{t=1}^{T} \ell_{t,I_t} \right] \geq F(T, 2, 1) = \frac{T}{T+1}. \tag{4}$$

Equation (4) trivially holds for $T = 1$ because

$$\max_{i \in \{1,2\}} P\left\{\ell_{1,I_1} \geq 1 | \ell_1 = \mathbb{1}_{\{i\}}\right\} = \max_{i \in \{1,2\}} P\left\{I_1 = i\right\} \geq 1/2.$$

Let $p_1 = P\{I_1 = 1\}$, $p_{11} = P\{I_2 = 1 | \mathbf{o}_1 = (1,0)\}$ and $p_{21} = P\{I_2 = 1 | \mathbf{o}_1 = (2,0)\}$. Then, the maximum of the three probabilities

$$P\{\ell_{1,I_1} + \ell_{2,I_2} \geq 1 | \ell_1 = \ell_2 = \mathbb{1}_{\{1\}}\} = P\{I_1 = 1 \text{ or } I_2 = 1\} = p_1 + (1 - p_1)p_{21},$$
$$P\{\ell_{1,I_1} + \ell_{2,I_2} \geq 1 | \ell_1 = \ell_2 = \mathbb{1}_{\{2\}}\} = (1 - p_1) + p_1(1 - p_{11})$$

and

$$\mathbb{E}_{I_1} \left[\max_{i \in \{1,2\}} P\{\ell_{2,I_2} \geq 1 | I_1, \ell_1 = \mathbf{0}, \ell_2 = \mathbb{1}_{\{i\}}\} \right]$$

$$= \mathbb{E}_{I_1} \left[\max_{i \in \{1,2\}} P\{I_2 = i | I_1, \ell_1 = \mathbf{0}\} \right]$$

$$= p_1 \max\{p_{11}, 1 - p_{11}\} + (1 - p_1) \max\{p_{21}, 1 - p_{21}\}$$

is at least 2/3 because the sum of them is at least 2. The above probabilities are lower bounds of $\mathbb{E}[\ell_{1,I_1} + \ell_{2,I_2}]$ for the cases with $\ell_1 = \ell_2 = \mathbb{1}_{\{1\}}, \ell_1 = \ell_2 = \mathbb{1}_{\{2\}}$ and $\ell_1 = \mathbf{0}$ and $\ell_2 = \mathbb{1}_{\{i^*\}}$, respectively, where $i^* = \arg\max_{i \in \{1,2\}} P\{I_2 = i | I_1\}$. Thus, Eq. (4) also holds for $T = 2$. This idea of the proof can be extended to that of Lemma 4. □

Proof of Lemma 4. Let

$$p_{b,s}(\mathbf{o}[1..b-1], T')$$

$$=P\left\{\sum_{t=b}^{b+T'-1} \ell_{t,I_t} \geq 1 \;\middle|\; \begin{array}{l} \mathbf{O}[1..b-1] = \mathbf{o}[1..b-1], \\ \ell_b = \cdots = \ell_{b+T'-1} = \mathbb{1}_{([K]\setminus S)\cup s} \end{array}\right\}$$

and for any positive integer $c \in [b, b+T'-1]$, let $p_c(\mathbf{o}[1..b-1], T')$ denote the value p_c that is set at Line 4 of BestSwitchingTime$(b, b+T'-1, \mathbf{o}[1..b-1], k)$. Note that,

$$p_b(\mathbf{o}[1..b-1], T') = \max_{s\subseteq S, |s|=k} p_{b,s}(\mathbf{o}[1..b-1], T')$$

holds. Then, with respect to the loss vectors generated by Wait&Sticking $[b, c^*, b+T'-1, \mathbf{o}[1..b-1]](t, \mathbf{o}[b..t-1])$,

$$p_{c^*}(\mathbf{o}[1..b-1], T') \leq \mathbb{E}_{I[b..b+T'-1]}\left[\sum_{t=b}^{b+T'-1} \ell_{t,I_t}\right]$$

holds. We prove

$$p_{\max} = \max_{t=b,\ldots,e} p_t(\mathbf{o}[1..b-1], T') \geq F(T', |S|, k),$$

which is implied from the inequality

$$\sum_{s\subseteq S, |s|=k} p_{b,s}(\mathbf{o}[1..b-1], T') + \sum_{c=b+1}^{b+T'-1} p_c(\mathbf{o}[1..b-1], T') \geq T' + \binom{|S|-1}{k-1} - 1$$

$$(5)$$

because the average of the term values in the left hand side of the inequality is at least

$$\left(T' + \binom{|S|-1}{k-1} - 1\right) \middle/ \left(T' + \binom{|S|}{k} - 1\right)$$

if Eq. (5) holds. We prove Eq. (5) for any positive integer b by mathematical induction on T'. When $T' = 1$,

$$p_{b,s}(\mathbf{o}[1..b-1], 1) = P\left\{\ell_{b,I_b} \geq 1 | \mathbf{O}[1..b-1] = \mathbf{o}[1..b-1], \ell_b = \mathbb{1}_{([K]\setminus S)\cup s}\right\}$$
$$= P\left\{I_b \in ([K]\setminus S)\cup s | \mathbf{O}[1..b-1] = \mathbf{o}[1..b-1]\right\}$$

holds. Thus, Eq. (5) holds for any positive integer b because

$$\sum_{s\subseteq S, |s|=k} p_{b,s}(\mathbf{o}[1..b-1], 1) + \sum_{c=b+1}^{b} p_c(\mathbf{o}[1..b-1], 1)$$

$$= \sum_{s\subseteq S, |s|=k} P\left\{I_b \in ([K]\setminus S)\cup s | \mathbf{O}[1..b-1] = \mathbf{o}[1..b-1]\right\}$$

$$\geq \binom{|S|-1}{k-1} = \binom{|S|-1}{k-1} + T' - 1.$$

Here, the above inequality holds because each arm $i \in S$ belongs to at least $\binom{|S|-1}{k-1}$ size-k subsets of S. Assume that Eq. (5) holds when $T' = n$ for any positive integer b. Then,

$$p_{b,s}(\mathbf{o}[1..b-1], n+1)$$

$$=P\left\{\sum_{t=b}^{b+n} \ell_{t,I_t} \geq 1 \middle| \mathbf{O}[1..b-1] = \mathbf{o}[1..b-1], \ell_b = \cdots = \ell_{b+n} = \mathbb{1}_{([K]\setminus S)\cup s}\right\}$$

$$=P\left\{\ell_{b,I_b} \geq 1 \middle| \mathbf{O}[1..b-1]=\mathbf{o}[1..b-1], \ell_b = \mathbb{1}_{([K]\setminus S)\cup s}\right\}$$

$$+ P\left\{\ell_{b,I_b} < 1, \sum_{t=b+1}^{b+n} \ell_{t,I_t} \geq 1 \middle| \begin{array}{l} \mathbf{O}[1..b-1] = \mathbf{o}[1..b-1], \\ \ell_b = \cdots = \ell_{b+n} = \mathbb{1}_{([K]\setminus S)\cup s} \end{array}\right\}$$

$$=P\{I_b \in ([K]\setminus S)\cup s | \mathbf{O}[1..b-1] = \mathbf{o}[1..b-1]\}$$

$$+ P\left\{I_b \notin ([K]\setminus S)\cup s, \sum_{t=b+1}^{b+n} \ell_{t,I_t} \geq 1 \middle| \begin{array}{l} \mathbf{O}[1..b-1] = \mathbf{o}[1..b-1], \\ \ell_b = \mathbb{1}_{[K]\setminus S}, \\ \ell_{b+1} = \cdots = \ell_{b+n} = \mathbb{1}_{([K]\setminus S)\cup s} \end{array}\right\}$$

$$=P\{I_b \in ([K]\setminus S)\cup s | \mathbf{O}[1..b-1] = \mathbf{o}[1..b-1]\}$$

$$+ P\left\{\sum_{t=b+1}^{b+n} \ell_{t,I_t} \geq 1 \middle| \begin{array}{l} \mathbf{O}[1..b-1] = \mathbf{o}[1..b-1], \ell_b = \mathbb{1}_{[K]\setminus S}, \\ \ell_{b+1} = \cdots = \ell_{b+n} = \mathbb{1}_{([K]\setminus S)\cup s} \end{array}\right\}$$

$$- P\left\{I_b \in ([K]\setminus S)\cup s, \sum_{t=b+1}^{b+n} \ell_{t,I_t} \geq 1 \middle| \begin{array}{l} \mathbf{O}[1..b-1] = \mathbf{o}[1..b-1], \\ \ell_b = \mathbb{1}_{[K]\setminus S}, \\ \ell_{b+1} = \cdots = \ell_{b+n} = \mathbb{1}_{([K]\setminus S)\cup s} \end{array}\right\}$$

$$=P\{I_b \in ([K]\setminus S)\cup s | \mathbf{O}[1..b-1] = \mathbf{o}[1..b-1]\}$$

$$+ \mathbb{E}_{I_b}[p_{b+1,s}(\mathbf{o}[1..b-1], (I_b, \mathbb{1}_{[K]\setminus S, I_b}), n)]$$

$$- \sum_{i\in[K]\setminus S\cup s} P\{I_b = i | \mathbf{O}[1..b-1] = \mathbf{o}[1..b-1]\}$$

$$\times p_{b+1,s}(\mathbf{o}[1..b-1], (i, \mathbb{1}_{[K]\setminus S, i}), n)$$

and

$$p_c(\mathbf{o}[1..b-1], n+1) = \mathbb{E}_{I_b}\left[p_c(\mathbf{o}[1..b-1], (I_b, \mathbb{1}_{[K]\setminus S, I_b}), n)\right]$$

holds for $c \geq b+1$. Therefore,

$$\sum_{s\subseteq S, |s|=k} p_{b,s}(\mathbf{o}[1..b-1], n+1) + \sum_{c=b+1}^{b+n} p_c(\mathbf{o}[1..b-1], n+1)$$

$$= \sum_{s\subseteq S, |s|=k} \left(P\{I_b \in ([K]\setminus S)\cup s | \mathbf{O}[1..b-1] = \mathbf{o}[1..b-1]\}\right.$$

$$\left. + \mathbb{E}_{I_b}\left[p_{b+1,s}(\mathbf{o}[1..b-1], (I_b, \mathbb{1}_{[K]\setminus S, I_b}), n)\right]\right)$$

$$- \sum_{i\in[K]\setminus S\cup s} P\{I_b = i | \mathbf{O}[1..b-1] = \mathbf{o}[1..b-1]\}$$

$$\times p_{b+1,s}(\mathbf{o}[1..b-1], (i, \mathbb{1}_{[K]\setminus S, i}), n)$$

$$+ \sum_{c=b+1}^{b+n} \mathbb{E}_{I_b} \left[p_c(\mathbf{o}[1..b-1], (I_b, \mathbb{1}_{[K]\setminus S, I_b}), n) \right]$$

$$\geq \mathbb{E}_{I_b} \left\{ \sum_{s \subseteq S, |s|=k} p_{b+1,s}(\mathbf{o}[1..b-1], (I_b, \mathbb{1}_{[K]\setminus S, I_b}), n) \right.$$

$$\left. + \sum_{c=b+2}^{b+n} p_c(\mathbf{o}[1..b-1], (I_b, \mathbb{1}_{[K]\setminus S, I_b}), n) \right\}$$

$$+ \sum_{i \in S} P\{ I_b = i \mid \mathbf{O}[1..b-1] = \mathbf{o}[1..b-1] \}$$

$$\times \left\{ \sum_{i \in s \subseteq S, |s|=k} (1 - p_{b+1,s}(\mathbf{o}[1..b-1], (i, \mathbb{1}_{[K]\setminus S, i}), n)) \right.$$

$$\left. + \max_{s \subseteq S, |s|=k} p_{b+1,s}(\mathbf{o}[1..b-1], (i, \mathbb{1}_{[K]\setminus S, i}), n) \right\}$$

$$+ \sum_{i \in [K]\setminus S} P\{ I_b = i \mid \mathbf{O}[1..b-1] = \mathbf{o}[1..b-1] \}$$

$$\times \left\{ \sum_{s \subseteq S, |s|=k} (1 - p_{b+1,s}(\mathbf{o}[1..b-1], (i, \mathbb{1}_{[K]\setminus S, i}), n)) \right.$$

$$\left. + \max_{s \subseteq S, |s|=k} p_{b+1,s}(\mathbf{o}[1..b-1], (i, \mathbb{1}_{[K]\setminus S, i}), n) \right\}$$

$$\geq \mathbb{E}_{I_b} \left[\binom{|S|-1}{k-1} + n - 1 \right] + 1 = \binom{|S|-1}{k-1} + (n+1) - 1$$

holds, which means Eq. (5) holds for any positive integer b when $T' = n + 1$. Note that the last inequality holds by the assumption that Eq. (5) holds for any positive integer b when $T' = n$. □

The following theorem is trivial by Lemma 4.

Theorem 6. *RepeatW&S[K, T] forces any player algorithm to suffer the expected loss of at least*

$$\sum_{i=1}^{m} F\left(t_i - t_{i-1}, K - \sum_{j=1}^{i-1} k_j, k_i\right)$$

for any positive integers $t_0, \ldots, t_m$, $k_1, \ldots, k_m$ that satisfies (1),(2) and (3).

Corollary 7. *RepeatW&S[K, T] forces any player algorithm to suffer the expected loss of at least*

$$T(1 - K^{-1/T}) - \frac{1 - K^{(T-1)/T}}{K(1 - K^{1/T})} \tag{6}$$

for $T \leq K - 2$,

$$H_K - H_{h+1} + h - \frac{A^2(h)(B(h) + 4h)}{2B^2(h)} \tag{7}$$

for $T \geq \frac{h(h-1)}{2} + K - 1$ *(h = 1, ..., K-1), where* H_n *is the nth harmonic number for any natural number n and*

$$A(h) = 2\sum_{j=1}^{h} \sqrt{j}$$

$$B(h) = 2T - 2(K-1) + (h+3)h.$$

Proof. For $T \leq K - 2$, consider positive integers $t_0, \ldots, t_m$ that satisfy

$$m = T \text{ and } t_1 = 2, t_2 = 3, ..., t_{m-1} = T, t_m = T + 1$$

Then, in this case,

$$
\sum_{i=1}^{m} F(t_i - t_{i-1}, K - \sum_{j=1}^{i-1} k_j, k_i) = \sum_{i=1}^{T} F(1, K - \sum_{j=1}^{i-1} k_j, k_i)
$$

$$
= \sum_{i=1}^{T} \frac{\binom{K - \sum_{j=1}^{i-1} k_j - 1}{k_i - 1}}{\binom{K - \sum_{j=1}^{i-1} k_j}{k_i}}
$$

$$
= \sum_{i=1}^{T} \frac{k_i}{K - \sum_{j=1}^{i-1} k_j}
$$

$$
= T - \sum_{i=1}^{T} \frac{K - \sum_{j=1}^{i} k_j}{K - \sum_{j=1}^{i-1} k_j}
$$

holds. By the inequality of arithmetic and geometric means, we have

$$
T - \sum_{i=1}^{T} \frac{K - \sum_{j=1}^{i} k_j}{K - \sum_{j=1}^{i-1} k_j} \leq T - TK^{-1/T}
$$

with equality if and only if

$$
\frac{K - \sum_{j=1}^{i} k_j}{K - \sum_{j=1}^{i-1} k_j} = K^{-1/T}
\tag{8}
$$

for all $i = 1, \ldots, T$. Unfortunately, $k_1, \ldots, k_T$ that satisfy (8) are not integers. As an approximate solution, use $k_1, \ldots, k_T$ that satisfy

$$
K - \sum_{j=1}^{i} k_j = \lceil K^{(T-i)/T} \rceil,
$$

then

$$T - \sum_{i=1}^{T} \frac{K - \sum_{j=1}^{i} k_j}{K - \sum_{j=1}^{i-1} k_j} = T - \sum_{i=1}^{T} \frac{\lceil K^{-(T-i)/T} \rceil}{\lceil K^{-(T-i+1)/T} \rceil}$$

$$\geq T - \sum_{i=1}^{T-1} \frac{K^{(T-i)/T} + 1}{K^{(T-i+1)/T}} + \frac{1}{K^{1/T}}$$

$$= T(1 - K^{-1/T}) - \frac{1 - K^{(T-1)/T}}{K(1 - K^{1/T})}.$$

Let h be an arbitrary integer in $[1, K-1]$. Consider the case with $T \geq \frac{h(h-1)}{2} + K - 1$. Let $m = K - 1$ and let

$$t_i = \begin{cases} t_{i-1} + 1 & (i = 1, \ldots, K - h - 1) \\ t_{i-1} + T_i & (i = K - h, \ldots, K - 1) \end{cases}$$

and $k_1 = \cdots = k_{K-1} = 1$, where T_i is non-negative integer with

$$\sum_{i=1}^{h} T_{K-i} = T - (K - 1). \tag{9}$$

Then,

$$\sum_{i=1}^{m} F(t_i - t_{i-1}, K - \sum_{j=1}^{i-1} k_j, k_i)$$

$$= \sum_{i=1}^{K-h-1} F(1, K - i + 1, 1) + \sum_{i=K-h}^{K-1} F(T_i + 1, K - i + 1, 1)$$

$$= \sum_{i=1}^{K-h-1} \frac{1}{K - i + 1} + \sum_{i=K-h}^{K-1} \frac{1 + T_i}{K - i + 1 + T_i}$$

$$= H_K - H_{h+1} + h - \sum_{i=K-h}^{K-1} \frac{K - i}{K - i + 1 + T_i}$$

$$= H_K - H_{h+1} + h - \sum_{i=1}^{h} \frac{i}{i + 1 + T_{K-i}} \tag{10}$$

holds. Let

$$f(T_{K-h}, \ldots, T_{K-1}) = \sum_{i=1}^{h} \frac{i}{i + 1 + T_{K-i}}.$$

By solving the problem of maximizing $f(T_{K-h}, \ldots, T_{K-1})$ subject to Constraint (9) using the method of Lagrange multipliers, we obtain

$$T_{K-i} = \frac{B(h)\sqrt{i}}{A(h)} - (i + 1) \text{ for } i = 1, \ldots, h. \tag{11}$$

All the T_{K-i} are non-negative because

$$\frac{B(h)\sqrt{i}}{A(h)} - (i+1) = \frac{2T - 2(K-1) + h(h+3)}{2\sum_{j=1}^{h}\sqrt{j}}\sqrt{i} - (i+1)$$

$$\geq \frac{2h(h+1)}{2\sum_{j=1}^{h}\sqrt{j}}\sqrt{i} - (i+1)$$

$$\geq \frac{2h(h+1)}{h\sqrt{2(h+1)}}\sqrt{i} - (i+1)$$

$$= \sqrt{2(h+1)i} - (i+1) \geq 0$$

holds for $i = 1, \ldots, h$. Here, the first inequality holds because $T \geq \frac{h(h-1)}{2} + K - 1$ and the second inequality holds by inequality $\sum_{j=1}^{h}\sqrt{j} \leq h\sqrt{\frac{h+1}{2}}$. Due to integer constraint, instead of T_{K-i} defined by Eq. (11), we use T_{K-i} defined as follows:

$$i + 1 + T_{K-i} = \left\lfloor \sum_{j=1}^{i} \frac{B(h)\sqrt{j}}{A(h)} \right\rfloor - \left\lfloor \sum_{j=1}^{i-1} \frac{B(h)\sqrt{j}}{A(h)} \right\rfloor .$$

Then,

$$\sum_{i=1}^{h} \frac{i}{i + 1 + T_{K-i}} < \sum_{i=1}^{h} \frac{i}{\frac{B(h)}{A(h)}\sqrt{i} - 1}$$

$$= A(h) \sum_{i=1}^{h} \frac{i}{B(h)\sqrt{i} - A(h)}$$

$$= \frac{A^2(h)}{2B(h)} + \frac{A^2(h)}{B^2(h)}h + \frac{A^3(h)}{B^2(h)} \sum_{i=1}^{h} \frac{1}{B(h)\sqrt{i} - A(h)}$$

$$\leq \frac{A^2(h)}{2B(h)} + \frac{A^2(h)}{B^2(h)}h + \frac{A^2(h)}{B^2(h)}h$$

$$= \frac{A^2(h)(B(h) + 4h)}{2B^2(h)} \tag{12}$$

holds. Here, the first inequality uses

$$\left\lfloor \sum_{j=1}^{i} \frac{B(h)\sqrt{j}}{A(h)} \right\rfloor - \left\lfloor \sum_{j=1}^{i-1} \frac{B(h)\sqrt{j}}{A(h)} \right\rfloor > \frac{B(h)\sqrt{i}}{A(h)} - 1$$

and the second inequality uses the fact that

$$B(h)\sqrt{i} - A(h) \geq B(h) - A(h) \geq A(h),$$

which can be implied from inequalities

$$A(h) \leq h\sqrt{2(h+1)}$$

and

$$B(h) = 2\left\{ T - (K-1) - \frac{h(h-1)}{2} \right\} + 2h(h+1) \le 2h(h+1).$$

By Eqs. (10) and (12), Bound (7) holds in this case.   □

Corollary 8. *Repeat$W\&S[K,T]$ forces any player algorithm to suffer expected loss at least*

$$K - 1 - \frac{K(K-1)^2(2T + (K-1)(K+4))}{(2T + K(K-1))^2}$$

for $T \ge K(K-1)/2$.

Proof. This corollary can be derived from Bound (7) of Corollary 7 with $h = K - 1$ and the fact that

$$A^2(K-1) = 4\left(\sum_{j=1}^{K-1} \sqrt{j} \right)^2 \le 4(K-1) \sum_{j=1}^{K-1} j = 2K(K-1)^2.$$

□

5   Concluding Remark

In this paper, we focus on the adversary's strategy. Any consistent player that avoids arms with previously observed loss incurs total loss at most $K - 1$. This leaves a small $O(1/T)$ gap between the loss forced by our adversary and this trivial loss bound for consistent players. Finding minimax strategies to close this gap remains an open problem.

Acknowledgment. We would like to thank anonymous reviewers for helpful comments. This work was partially supported by JSPS KAKENHI Grant Number 25280079.

References

1. Auer, P., Cesa-Bianchi, N., Freund, Y., Schapire, R.E.: The nonstochastic multi-armed bandit problem. SIAM J. Comput. **32**(1), 48–77 (2003)
2. Bubeck, S., Cesa-Bianchi, N.: Regret analysis of stochastic and nonstochastic multi-armed bandit problems. Found. Trends Mach. Learn. **5**(1), 1–122 (2012)
3. Cesa-Bianchi, N., Lugosi, G.: Prediction, Learning, and Games. Cambridge University Press, Cambridge (2006)

Graphs, Trees, and Weighted Automata

Properties of Regular DAG Languages

Johannes Blum and Frank Drewes[✉]

Department of Computing Science, Umeå University, 901 87 Umeå, Sweden
{mcs13jbm,drewes}@cs.umu.se

Abstract. A DAG is a directed acyclic graph. We study the properties of DAG automata and their languages, called regular DAG languages. In particular, we prove results resembling pumping lemmas and show that the finiteness problem for regular DAG languages is in P.

1 Introduction

String and tree languages have been the subject of many years of research, with thousands of papers contributing to their theory. Tree languages are especially useful in computational linguistics because trees can represent the syntactic structure of sentences. However, trees are not equally well suited for dealing with meaning representations, which naturally lead to directed acyclic graphs (DAGs). Recently, the question of developing an appropriate theory of DAG languages has gained interest as a basis for the study of so-called *abstract meaning representations* (AMRs, see [2]). Quernheim and Knight [10] have therefore proposed a notion of DAG automata. Chiang et al. [5] propose a mathematically simpler model of *ranked* DAG automata, which coincide with the *edge-marking DAG automata* of [8].[1] They can also be viewed as a slightly simplified ranked version of the DAG automata in [9]. Here, *ranked* means that the vertices of the DAGs in a regular DAG language have bounded degree. While the DAGs that occur in AMRs are unranked, Chiang et al. point out that for almost all purposes it suffices to study ranked DAG languages since unranked ones can be binarized in a way similar to the first-sibling next-child encoding used in the study of unranked tree languages.

In this paper, we study some basic language theoretic and algorithmic properties of these ranked DAG languages, which we call regular DAG languages. To keep the theory closer to the theory of regular tree languages, we deviate slightly from the definition by Chiang et al. in that ordered DAGs are considered, i.e. the ingoing edges of a node form a sequence, and similarly for the outgoing edges. Imagining a ranked and ordered version of Priese's DAG automata, the difference between both types is that his would allow to specify admissible sequences of root states (initial states), and admissible sequences of leaf states (final states), whereas our automata intentionally lack this ability. As shown in [5], one of the

---

This paper is based on the master thesis [3] of the first author.

[1] For further types of DAG automata that, however, are significantly different from those studied here, see [1,4,7].

© Springer International Publishing Switzerland 2016
A.-H. Dediu et al. (Eds.): LATA 2016, LNCS 9618, pp. 427–438, 2016.
DOI: 10.1007/978-3-319-30000-9_33

consequences is that the so-called path languages of our regular DAG languages are regular string languages whereas Priese shows that they are not even in general context-free in his formalism.

In this paper, we first study closure properties of the class of regular DAG languages. Expectedly, it turns out that it is closed under union and intersection. It is, however, not closed under complement. These results, shown in Sect. 3, parallel those of Priese. We then turn to necessary properties of regular DAG languages in Sect. 4, where we show results resembling pumping lemmas. These yield a characterization of the finite regular DAG languages. Finally, we prove in Sect. 5 that useless rules can be removed from DAG automata, and that this can be used to decide finiteness of regular DAG languages in polynomial time, thus answering a problem left open in [5].

2   Regular DAG Languages

In this section we compile some basic notation and introduce DAG automata and their languages.

The set of non-negative integers is denoted by $\mathbb{N}$. For $n \in \mathbb{N}$ we let $[n] = \{1, \ldots, n\}$; in particular, $[0] = \emptyset$. The composition of functions $f \colon A \to B$ and $g \colon B \to C$ is denoted by $g \circ f$. Given a set S, its cardinality is denoted by $|S|$, and the set of all finite sequences (or strings) over S, including the empty string λ, is denoted by S^*. The length of $s \in S^*$ is denoted by $|s|$, and $[s]$ denotes the smallest set S such that $s \in S^*$. If f is a function defined on S, then the canonical extensions of f to functions on the powerset of S and to S^* are denoted by f as well, i.e. $f(S') = \{f(s) \mid s \in S'\}$ for $S' \subseteq S$ and $f(s_1 \cdots s_n) = f(s_1) \cdots f(s_n)$.

An *alphabet* Σ is a finite set of symbols. Given such an alphabet, a Σ-*graph* is a tuple $G = (V, E, lab, in, out)$ with the following components:

- V and E are the finite sets of *vertices* and *edges*, resp.,
- $lab \colon V \to \Sigma$ is the *vertex labelling*, and
- $in, out \colon V \to E^*$ assign to each vertex its ingoing and outgoing edges, in such a way that, for every edge $e \in E$, there is exactly one pair $(u, v) \in V^2$ such that $e \in [out(u)] \cap [in(v)]$, i.e. each edge has unique start and end vertices.

Given a Σ-graph G as above, we may refer to its components by V_G, E_G, lab_G, in_G, and out_G, resp. For an edge $e \in E_G$ we let $nod_G(e) = (u, v)$ where u and v are the unique vertices such that $e \in [out_G(u)] \cap [in_G(v)]$. An alternating sequence $v_0 e_1 v_1 \cdots e_n v_n$ of vertices $v_0, \ldots, v_n \in V_G$ and edges $e_1, \ldots, e_n \in E_G$ is a *path* from v_0 to v_n if $nod_G(e_i) \in \{(v_{i-1}, v_i), (v_i, v_{i-1})\}$ for all $i \in [n]$. The path is *simple* if $v_i \neq v_j$ for all distinct $i, j \in [n]$, *empty* if $n = 0$, *directed* if $nod_G(e_i) = (v_{i-1}, v_i)$ for all $i \in [n]$, and a *cycle* if $n > 0$ and $v_0 = v_n$. G is *connected* if, for all $u, v \in V_G$, there exists a (not necessarily directed) path from u to v in G. Naturally, G is said to be *empty* if $V_G = \emptyset$, in which case we may also denote G by $\emptyset$.

As usual, graphs G, G' are said to be *isomorphic* if there is a pair (g, h) of bijective mappings $g \colon V_G \to V_{G'}$ and $h \colon E_G \to E_{G'}$, called an *isomorphism*, such that $lab_G = lab_{G'} \circ g$, $h \circ in_G = in_{G'} \circ g$, and $h \circ out_G = out_{G'} \circ g$. Throughout this paper we will not distinguish between isomorphic graphs except in a few cases where it is necessary in order to avoid confusion.

Definition 1 (DAG). *Let Σ be an alphabet. A Σ-DAG (or simply DAG) is a Σ-graph that does not contain any nonempty directed cycle. The set of all nonempty connected Σ-DAGs is denoted by $\mathcal{D}_\Sigma$. A DAG language is a subset of $\mathcal{D}_\Sigma$, for some alphabet Σ.*

Sometimes it is convenient to assign explicit ranks to the symbols of an alphabet. We say that an alphabet Σ is *doubly ranked* or simply *ranked* if a rank $rk_\Sigma(\sigma) \in \mathbb{N}^2$ is specified for every symbol $\sigma \in \Sigma$. If Σ is ranked, we restrict the set $\mathcal{D}_\Sigma$ of all Σ-DAGs to those DAGs G over Σ such that every vertex $v \in V_G$ satisfies $rk_\Sigma(lab_G(v)) = (|in_G(v)|, |out_G(v)|)$.

Now we are ready to introduce our notion of DAG automata.

Definition 2 (DAG automaton). *A DAG automaton is a triple $A = (Q, \Sigma, R)$ where Q is a finite set of states, Σ is an alphabet and R is a finite set of rules of the form $\alpha \overset{\sigma}{\leftrightarrow} \beta$ where $\sigma \in \Sigma$ and $\alpha, \beta \in Q^*$.*

A run of A on a DAG $G = (V, E, lab, in, out)$ is a mapping $\rho: E \to Q$ such that R contains the rule $\rho(in(v)) \overset{lab(v)}{\longleftrightarrow} \rho(out(v))$ for every vertex $v \in V$. If such a run exists, we say that A accepts G. The language accepted by A is the set $L(A)$ of all DAGs in $\mathcal{D}_\Sigma$ accepted by A.

The class RDL of regular DAG languages consists of all DAG languages accepted by DAG automata.

If the alphabet Σ of A is ranked, we say that A is ranked and require that all rules respect the ranks of symbols, i.e. every rule $\alpha \overset{\sigma}{\leftrightarrow} \beta$ satisfies $rk_\Sigma(\sigma) = (|\alpha|, |\beta|)$. Thus, both in the general and in the ranked case we have $L(A) \subseteq \mathcal{D}_\Sigma$. For a vertex $v \in V$ with $lab(v) = \sigma$, a run ρ on G is said to *use* the rule $\rho(in(v)) \overset{\sigma}{\leftrightarrow} \rho(out(v))$ in v.

One may note that trees over an alphabet Σ, in the usual sense of tree language theory, are a special case of DAGs. A DAG automaton "is" a finite-state tree automaton if all rules $\alpha \overset{\sigma}{\leftrightarrow} \beta$ satisfy $|\alpha| \leq 1$. Thus, it is straightforward to show that a tree language over Σ is regular (in the sense of tree language theory) if and only if it is a regular DAG language in the sense defined above.

RDL is also closed under "turning DAGs upside down": for a DAG G, let $G^{\updownarrow} = (V_G, E_G, lab_G, out_G, in_G)$, and, for a DAG automaton $A = (Q, \Sigma, R)$, $A^{\updownarrow} = (Q, \Sigma, R^{\updownarrow})$ where $r^{\updownarrow} = (\beta \overset{\sigma}{\leftrightarrow} \alpha)$ for a rule $r = (\alpha \overset{\sigma}{\leftrightarrow} \beta)$. Then $L(A^{\updownarrow}) = L(A)^{\updownarrow}$. This yields a strong duality principle: properties and results regarding DAG automata and their languages also hold if they are "turned upside down".

As a final remark regarding Definition 2, it is worth emphasizing that $L(A)$ only contains nonempty connected DAGs, even though DAG automata can *run* on arbitrary DAGs. This definition of $L(A)$ is motivated by the fact that A accepts a disjoint union of DAGs if and only if it accepts each of its connected components. More precisely, given two disjoint DAGs G_1 and G_2 (where disjointness means that $V_{G_1} \cap V_{G_2} = \emptyset = E_{G_1} \cap E_{G_2}$), let us denote their union by

$G_1 \& G_2$, i.e. $G_1 \& G_2 = (V_{G_1} \cup V_{G_2}, E_{G_1} \cup E_{G_2}, lab_{G_1} \cup lab_{G_2}, in_{G_1} \cup in_{G_2}, out_{G_1} \cup out_{G_2})$, where the union of functions is defined by viewing them as special binary relations. If G_1 and G_2 are not disjoint, we silently take disjoint copies to be able to build $G_1 \& G_2$. For a set L of DAGs, let $L^{\&}$ be the set of all DAGs of the form $G_1 \& \cdots \& G_n$ such that $n \in \mathbb{N}$ and $G_i \in L$ for all $i \in [n]$. Then we have the following as a direct consequence of the definition of runs:

Observation 1. *For every DAG automaton A, a DAG G is in $L(A)^{\&}$ if and only if A accepts G.*

Thus, our definition of $L(A)$ makes it sensible to speak about DAG automata accepting finite, empty, and infinite DAG languages, whereas $L(A)^{\&}$ is never empty (it always contains the empty DAG $\emptyset$), and is finite if and only if it equals $\{\emptyset\}$ if and only if $L(A) = \emptyset$.

Example 3. Consider the DAG automaton $A = (Q, \Sigma, R)$ where $Q = \{p, q\}$, $\Sigma = \{s, a, b, t\}$ and $R = \{\lambda \overset{s}{\leftrightarrow} pq, p \overset{a}{\leftrightarrow} qq, q \overset{b}{\leftrightarrow} pp, qp \overset{t}{\leftrightarrow} \lambda\}$. Thus, A is actually ranked, with $rk_\Sigma(s) = (0, 2)$, $rk_\Sigma(a) = rk_\Sigma(b) = (1, 2)$, and $rk_\Sigma(t) = (2, 0)$. A run of A and a sample DAG G is shown in Fig. 1 on the left. In such drawings ingoing and outgoing edges are ordered from left to right.

The run can be constructed in a top-down manner, as follows. In the root of G the only rule we can use is $\lambda \overset{s}{\leftrightarrow} pq$. Now, since the left child of the root has the label a, we can use the rule $p \overset{a}{\leftrightarrow} qq$ there. In the child with label b we use the rule $q \overset{b}{\leftrightarrow} pp$. In the two leaves the rule $(q, p) \overset{t}{\leftrightarrow} \lambda$ can then be used, which completes the run. Thus, A accepts G.

The right part of Fig. 1 shows a run on another DAG accepted by A. It also illustrates the fact that $L(A)$ is not finite as one can systematically construct a sequence of DAGs of increasing size that are accepted by A.

The DAG automaton in the previous example is in fact *top-down deterministic*, meaning that for all $\alpha \in Q^*$, $\sigma \in \Sigma$, and $n \in \mathbb{N}$ there is at most one

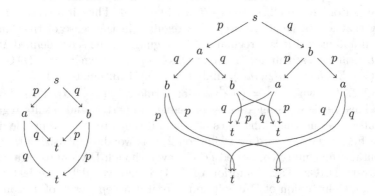

Fig. 1. Runs of the DAG automaton in Example 3

$\beta \in Q^*$ with $|\beta| = n$ such that $\alpha \overset{\sigma}{\hookleftarrow} \beta$ is a rule in A. In such a case, an input DAG permits at most one run, and this run can be constructed deterministically from the roots downwards. A dual notion of bottom-up determinism can be defined in the obvious way. For tree languages it is well known that the class of top-down deterministic regular tree languages is strictly included in the class of regular tree languages, the standard example being the finite tree language $L_0 = \{f(a, b), f(b, a)\}$ (where trees are denoted as terms). Clearly, a top-down deterministic DAG automaton accepting a tree language "is" a top-down deterministic finite-state tree automaton. Consequently, it follows from the fact that a tree language is regular if and only if it is a regular DAG language, that top-down deterministic DAG automata fail to accept all regular DAG languages. By the duality principle mentioned above this holds for bottom-up deterministic DAG automata as well, in contrast to the fact that bottom-up deterministic finite-state tree automata are equally powerful as nondeterministic ones. In fact, looking a bit more closely at the argument, even the union of the classes of top-down deterministic and bottom-up deterministic regular DAG languages turns out to be strictly included in RDL. This is because, for a DAG automaton A, removing all rules $\alpha \overset{\sigma}{\hookleftarrow} \beta$ with $|\alpha| > 1$ yields a DAG automaton that accepts $\{G \in L(A) \mid G \text{ is a tree}\}$. As this construction preserves top-down determinism, it implies (together with its dual) that the regular DAG language $L_0 \cup L_0^{\uparrow}$ is neither top-down nor bottom-up deterministic.

3   Closure Properties

Before discussing more interesting properties of the class RDL, let us note that it is, similarly to the case of tree languages, closed under union and intersection:

Lemma 4. *RDL is closed under union and intersection.*

We omit the proofs, which are direct generalizations of the corresponding proofs for finite-state tree automata.

In view of the lemma above, it may be an interesting observation that $RDL^{\&} = \{L^{\&} \mid L \in RDL\}$ is *not* closed under union. To see this, consider two arbitrary regular DAG languages $L_1 = \{G_1\}$ and $L_2 = \{G_2\}$ where G_1 and G_2 are DAGs that are not isomorphic. We have $L_i^{\&} \in RDL^{\&}$ for $i = 1, 2$. Assume now that there is a DAG automaton A that accepts a DAG G if and only if $G \in L_1^{\&} \cup L_2^{\&}$. Then A accepts $G_1 \& G_2$ as we can mix connected components of graphs accepted by A, but we have $G_1 \& G_2 \notin L_1^{\&} \cup L_2^{\&}$.

We are now going to show that RDL is not closed under complement. For this, we introduce a simple but very useful operation on DAGs.

Definition 5 (edge swap). *Let $G = (V, E, lab, in, out)$ be a DAG. Two edges $e_0, e_1 \in E$ with $nod_G(e_i) = (u_i, v_i)$ are independent if there is no directed path between u_0 and u_1.[2] In this case, the edge swap of e_0 and e_1 is defined and yields the DAG $G[e_0 \bowtie e_1] = (V, E, lab, h \circ in, out)$ given by*

---

[2] A directed path *between* u and v is a directed path from u to v or from v to u.

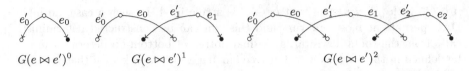

Fig. 2. The repeated edge swap between $k+1$ copies of G; $G(e \bowtie e')^0$ is isomorphic to G and each further copy G_{k+1} is attached to $G(e \bowtie e')^k$ by swapping its edge e' with the edge e of the preceding copy.

$$h(e) = \begin{cases} e_{1-i} & if\ e = e_i\ for\ some\ i \in \{0,1\} \\ e & otherwise. \end{cases}$$

By the requirement of independence $G[e_0 \bowtie e_1]$ is indeed a DAG. Note that the condition is vacuously true in $(G \& G')[e \bowtie e']$, where $e \in E_G$ and $e' \in E_{G'}$.

For $k \in \mathbb{N}$, we can moreover connect $k + 1$ disjoint isomorphic copies of a DAG G by systematically swapping copies of two edges $e, e' \in E_G$ between them. For this, choose disjoint isomorphic copies $G_0, G_1, \ldots$ of G. For $i \in \mathbb{N}$ let e_i and e'_i be the corresponding copies of e and e' in G_i. Then the graph $G(e \bowtie e')^k$ is formally defined as follows, for $k \in \mathbb{N}$ (see also the illustration in Fig. 2):

$$G(e \bowtie e')^0 = G_0 \quad and \quad G(e \bowtie e')^{k+1} = (G(e \bowtie e')^k \& G_{k+1})[e_k \bowtie e'_{k+1}].$$

The usefulness of edge swaps is due to the fact that, given a run, independent edges that are assigned the same state can obviously be swapped without affecting the validity of the run. Thus, we have the following lemma.

Lemma 6. *Let A be a DAG automaton and let $G \in L(A)^\&$. Then for all independent edges $e_1, e_2 \in E_G$, every run ρ of A on G that satisfies $\rho(e_1) = \rho(e_2)$ is also a run on $G[e_1 \bowtie e_2]$. In particular, we have $G[e_1 \bowtie e_2] \in L(A)^\&$ if such a run ρ exists.*

Edge swapping allows us to prove that *RDL* is not closed under complement.

Theorem 7. *There are ranked alphabets Σ and regular DAG languages $L \subseteq \mathcal{D}_\Sigma$ such that $\mathcal{D}_\Sigma \setminus L$ is not regular.*

Proof. Consider the ranked alphabets $\Sigma = \{s, a, b\}$ and $\Sigma_0 = \{s, a\}$, where the rank of s is $(0, 2)$ and a and b have the rank $(2, 0)$. Then $L = \mathcal{D}_{\Sigma_0} \in RDL$ by the top-down deterministic DAG automaton $A = (\{p\}, \Sigma, \{\lambda \overset{s}{\hookrightarrow} (p, p), (p, p) \overset{a}{\hookrightarrow} \lambda\})$.

Now consider $\bar{L} = \mathcal{D}_\Sigma \setminus L$, the set of all DAGs in $\mathcal{D}_\Sigma$ that contain at least one vertex with label b. Assume that there is a DAG automaton $\bar{A}$ with $L(\bar{A}) = \bar{L}$. In particular, $\bar{A}$ accepts, for every $i \geq 1$, the DAG G_i with $V_{G_i} = \{s_1, \ldots, s_i, a_1, \ldots, a_{i-1}, b\}$, built as follows. For $j \in [i]$ we have $lab_{G_i}(s_j) = s$, for $j \in [i-1]$ we have $lab_{G_i}(a_j) = a$ and $lab_{G_i}(b) = b$. Moreover every vertex a_j has two ingoing edges e_j^1 and e_j^2 from $s_{(j-2 \bmod i)+1}$ and s_j,

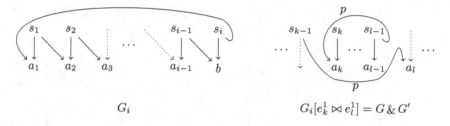

$$G_i \qquad\qquad G_i[e_k^1 \bowtie e_l^1] = G \,\&\, G'$$

Fig. 3. The generic DAG G_i and its decomposition into $G \,\&\, G'$

respectively, and vertex b has two ingoing edges from s_{i-1} and s_i; see the left DAG in Fig. 3.

Since $\bar{A}$ has only a finite number of states, we can now choose i large enough such that for every run ρ of $\bar{A}$ on G_i there are distinct a_k, a_l such that the states of their first ingoing edges are identical, i.e. $\rho(e_k^1) = \rho(e_l^1)$. Now let ρ be a run of $\bar{A}$ on G_i and k, l satisfy $\rho(e_k^1) = \rho(e_l^1) = p$. If we swap the two edges e_k^1 and e_l^1, G_i falls apart into two connected components, i.e. we have $G_i[e_k^1 \bowtie e_l^1] = G \,\&\, G'$ for DAGs G and G', as is illustrated in Fig. 3 on the right.

According to Lemma 6 we have $G \,\&\, G' \in L(\bar{A})^{\&}$ which means that $G, G' \in L(\bar{A})$. But one of these DAGs contains only vertices with labels s and a and is thus in L, a contradiction. □

4 Necessary Properties of Regular DAG Languages

In this section we present two necessary properties of regular DAG languages that exploit the fact that a DAG automaton has to reuse some states when it processes a DAG of a certain size. Using Lemma 6, we can thus pump the DAG up by connecting it to an isomorphic copy through an edge swap without changing the acceptance behaviour of the automaton.

Lemma 8. *For every $L \in RDL$ there is a constant $k \in \mathbb{N}$ such that for all DAGs $G \in L$ with $|E_G| > k$ there are two edges $e, e' \in E_G$ such that for all $n \in \mathbb{N}$ the DAG $G(e \bowtie e')^n$ is in $L^{\&}$. Furthermore, each $G(e \bowtie e')^n$ contains a connected component C such that $|V_C| > n$.*

Proof Sketch. Let $L \in RDL$ be a regular DAG language and $A = (Q, \Sigma, R)$ be a DAG automaton with $L(A) = L$. Let $G \in L$ with $|E_G| > |Q|$. Given a run ρ of A on G, there are two edges $e, e' \in E_G$ such that $\rho(e) = \rho(e')$ as ρ must use one state at least twice. By induction on n, and using Lemma 6 it can thus be shown that $G(e \bowtie e')^n \in L(A)^{\&}$ for all $n \in \mathbb{N}$. This is illustrated in Fig. 4.

But the proof is not finished yet, because the DAGs $G(e \bowtie e')^n$ may fall apart into n connected components C_i with $|V_{C_i}| = |V_G|$. As illustrated in Fig. 5 this can happen if e and e' have different orientations in the sense that there is no path between the end vertex of e' and the start vertex of e that does not contain the edge e or e'. However, if e and e' point into the same direction then

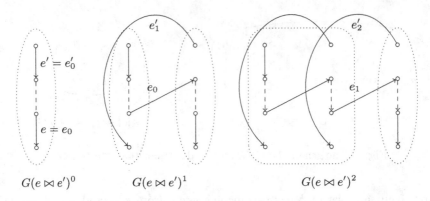

$$G(e \bowtie e')^0 \qquad G(e \bowtie e')^1 \qquad G(e \bowtie e')^2$$

Fig. 4. An example of the graphs $G = G(e \bowtie e')^0, G(e \bowtie e')^1$ and $G(e \bowtie e')^2$

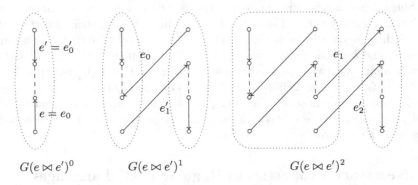

$$G(e \bowtie e')^0 \qquad G(e \bowtie e')^1 \qquad G(e \bowtie e')^2$$

Fig. 5. The components of $G(e \bowtie e')^n$ do not necessarily increase in size

all copies of e in $G(e \bowtie e')^n$ are part of the same connected component, which means that this component contains more than n vertices as two copies of e share at most one vertex. This is the situation depicted in Fig. 4.

Now, for $m \in \mathbb{N}$ let $\mathcal{D}_m$ be the set of all DAGs $G \in L$ such that the maximum length of all shortest paths between two vertices is at most m. As R is finite, the degree of all vertices is bounded and $\mathcal{D}_m$ is finite. Let k be the maximum number of edges of all $G \in \mathcal{D}_{2|Q|}$. Then every DAG $G \in L$ with $|E_G| > k$ contains a simple path ω of length greater than $2|Q|$. Thus, a run of A on such a DAG G assigns the same state to three distinct edges on ω, of which at least two point into the same direction. This completes the proof.   □

The above lemma required a DAG of a certain size. However, one can pump up *every* DAG that contains a simple undirected cycle, regardless of its size.

Lemma 9. *For every DAG language $L \in RDL$ and every DAG $G \in L$ that contains a simple (undirected) cycle there is an edge $e \in E_G$ with $G(e \bowtie e)^n \in L$ for all $n \in \mathbb{N}$.*

Proof. Let e be an edge which belongs to a simple cycle in G. Since $G \in L$, Lemma 6 tells us that $G_n = G(e \bowtie e)^n$ is in $L^\&$ for all $n \in \mathbb{N}$. It remains to

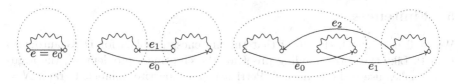

Fig. 6. Cycles in the DAGs G_0, G_1 and G_2

be shown that G_n is connected. For this, one can argue by induction on n that all copies of e in each G_n are part of the same cycle. By the choice of e this is true for G_0. Now, up to isomorphism $G_{n+1} = (G_n \& G)[e' \bowtie e]$, where e' is the relevant copy of e in G_n. Now the cycle in G_n and that in G together form a bigger cycle in G_{n+1} that contains all the copies of e. This is illustrated in Fig. 6. Consequently, for all $n \in \mathbb{N}$ the entire DAG G_n is connected. □

Thus no finite DAG language in RDL contains a DAG with a simple undirected cycle. On the other hand, all finite DAG languages whose DAGs do not contain any such cycle belong to RDL. This yields the following characterization.

Theorem 10. *Let L be a finite DAG language. Then we have $L \in RDL$ if and only if there is no DAG $G \in L$ that contains a simple undirected cycle.*

Proof. By Lemma 9 and the closedness of RDL under union it suffices to show that $\{G\} \in RDL$ for every DAG G that does not contain a simple cycle.

Given such a DAG G, build $A = (Q, \Sigma, R)$ as follows: $Q = E_G$, and for every vertex v of G, R contains the rule $r(v) = in_G(v) \overset{lab(v)}{\longleftrightarrow} out_G(v)$.

We have to show that $L(A) = \{G\}$. For two rules $r(v) = \alpha \overset{\sigma}{\leftrightarrow} \beta$ and $r(v') = \alpha' \overset{\sigma'}{\leftrightarrow} \beta'$ we have $[\beta] \cap [\alpha'] \neq \emptyset$ if and only if there is an edge from v to v' in G. Consider now a DAG $G' \in L(A)$ and a run ρ of A on G'. We show that ρ uses every rule in R exactly once. If ρ uses a rule $r(v_1) \in R$ more than once, there is a $k > 0$ and a sequence of rules $r(v_i) = \alpha_i \overset{\sigma_i}{\leftrightarrow} \beta_i$ ($0 \leq i \leq k$) such that $v_0 = v_k$ and $([\beta_{i-1}] \cap [\alpha_i]) \cup ([\alpha_{i-1}] \cap [\beta_i]) \neq \emptyset$ for all $i \in [k]$. But this means that there is a simple cycle from v_0 to v_0 in G, contradicting the assumption. Thus, ρ uses every rule at most once. Moreover we can observe that for every state $q \in Q$ there is exactly one rule such that q occurs in its left-hand side and exactly one rule such that q occurs in its right-hand side. Therefore the run ρ must use every rule at least once (as $G' \neq \emptyset$ and G is connected).

This means that we have $|V_{G'}| = |V_G|$ and $|E_{G'}| = |E_G|$, and we can define the following isomorphism (g_V, g_E) from G to G'. Let $v \in V_G$ with $r(v) = e_1 \cdots e_m \overset{\sigma}{\leftrightarrow} f_1 \cdots f_n$, and assume that v' is the vertex ρ uses $r(v)$ in, where $in_{G'}(v') = e'_1 \cdots e'_m$ and $out_{G'}(v') = f'_1 \cdots f'_n$. Then we let $g_V(v) = v'$, $g_E(e_i) = e'_i$ and $g_E(f_j) = f'_j$ for all $i \in [m]$ and $j \in [n]$. It should be clear that (g_V, g_E) is an isomorphism, which shows that $L(A) = \{G\}$. □

5   Finiteness

We shall now sketch the proof that the finiteness problem for DAG automata is in P. Like the corresponding proof for the emptiness problem [5], the argument makes use of Petri nets. A Petri net is a bipartite unlabeled graph $N = (V, E, in, out)$, where $V = P \cup T$ for disjoint sets P and T of *places* and *transitions*, and $nod(e) \in (P \times T) \cup (T \times P)$ for all $e \in E$. A *configuration of* N is a function $\phi \colon P \to \mathbb{N}$ that assigns $\phi(p)$ to every place $p \in P$, interpreted as the number of *tokens* on that place. Given such a configuration, a transition t is *enabled* if $\phi(p) \geq \Delta(p, t)$ for all $p \in P$, where $\Delta(u, v) = |\{e \in E \mid nod(e) = (u, v)\}|$. If t is enabled it can *fire* by consuming $\Delta(p, t)$ tokens from each place p and adding $\Delta(t, p)$ tokens to it. Hence the firing of t leads from ϕ to ϕ' defined by $\phi'(p) = \phi(p) - \Delta(p, t) + \Delta(t, p)$ for all $p \in P$. We denote this as $\phi \xrightarrow{t} \phi'$.

As observed in [5], every DAG automaton $A = (Q, \Sigma, R)$ gives rise to a Petri net $N_A = (P \cup T, E, in, out)$ where $P = Q$, $T = R$ and for every rule $r = p_1 \cdots p_m \xleftrightarrow{\sigma} q_1 \cdots q_n$ there is an edge from every p_i to r $(i \in [m])$ and one from r to every q_i $(i \in [n])$. The idea behind this construction is that, if we view a run as a top-down process, using r can be seen as an action that consumes states $p_1, \ldots, p_m$ and produces states $q_1, \ldots, q_n$. Every run ρ of A on a DAG G that uses the rules $r_1, \ldots, r_k$ in a top-down fashion gives thus rise to a firing sequence $\phi_0 \xrightarrow{r_1} \ldots \xrightarrow{r_k} \phi_k$ in which the initial configuration has no states at all, i.e. ϕ_0 is the *null configuration* $\mathbf{0}$ with $\mathbf{0}(p) = 0$ for all $p \in P$. Since the run uses a rule in each vertex of G, all tokens are eventually consumed, i.e. we have $\phi_k = \mathbf{0}$ as well. Below, we call such a (nonempty) firing sequence a *zero cycle*. It is no difficulty to show that also the converse of the above holds: every zero cycle $\mathbf{0} \xrightarrow{r_1} \ldots \xrightarrow{r_k} \mathbf{0}$ of N_A gives rise to a run ρ on a DAG with k vertices such that ρ uses the rules $r_1, \ldots, r_k$ in those vertices. (But note that G is neither uniquely determined and nor necessarily connected!)

It is thus of interest to be able to decide whether a Petri net N admits a zero cycle. This problem has been shown to be in P in [6]. We use this to approach the finiteness problem. In order to solve it, we first show that a DAG automaton can be turned into a *reduced* one, where a DAG automaton is reduced if each of its rules r is *useful* in the sense that there is a run ρ on some DAG G such that ρ uses r in some vertex of G.

Lemma 11. *For every DAG automaton A a reduced DAG automaton A_{red} with $L(A) = L(A_{red})$ can be computed in polynomial time.*

Proof. Let $A = (Q, \Sigma, R)$ be a DAG automaton. As explained above, a rule $r \in R$ is useful if and only if it occurs in a zero cycle of N_A. In other words, A_{red} can be constructed by keeping only those rules r that occur in a zero cycle of N_A. This set is denoted by $\Lambda(N_A)$ in [6], where it was shown to be computable in polynomial time; see [6, Theorem 6.2]. $\square$

Using this result we can now show the finiteness problem to be in P.

Theorem 12. *For DAG automata A it is decidable in polynomial time if $L(A)$ is finite.*

Proof. Owing to space restrictions, we only sketch the proof, omitting some details and illustrations (especially in the second last paragraph).

Let $A = (Q, \Sigma, R)$ be a reduced DAG automaton. We define a *cycle* in A as a nonempty sequence of rules $\alpha_1 \overset{\sigma_1}{\leftrightarrow} \beta_1, \ldots, \alpha_k \overset{\sigma_k}{\leftrightarrow} \beta_k$ for which there exist states $p_1, \ldots, p_k$ such that (a) for $i \in [k]$ we have $p_i \in ([\beta_i] \cap [\alpha_{(i \bmod k)+1}]) \cup ([\alpha_i] \cap [\beta_{(i \bmod k)+1}])$ and (b) if $p_i = p_{(i \bmod k)+1}$ then the state p_i occurs at least twice in the rule $r_{(i \bmod k)+1}$. (State q is said to occur at least twice in $\alpha \overset{\sigma}{\leftrightarrow} \beta$ if it occurs in two or more distinct positions in $\alpha\beta$.) If A contains a cycle it is called *cyclic*. We show that $L(A)$ is infinite if and only if A is cyclic.

Assume first that A is not cyclic, let $G \in L(A)$ be a DAG and ρ be a run of A on G. Consider a simple undirected path $\omega = v_0 e_1 v_1 \cdots e_k v_k$ in G, let $r_i = \alpha_i \overset{\sigma_i}{\leftrightarrow} \beta_i$ be the rule used by ρ in v_i ($i \in \{0, \ldots, k\}$) and let $\rho(e_i) = p_i$ ($i \in [k]$). Then for $i \in [k]$ we have $p_i \in ([\beta_{i-1}] \cap [\alpha_i]) \cup ([\alpha_{i-1}] \cap [\beta_i])$ and if $p_i = p_{i+1}$ for some $i \in [k-1]$ then p_i occurs at least twice in the rule r_i.

We argue that at most two edges along ω can be marked with the same state. Assume that three edges along ω are marked with a state $p \in Q$. Then at least two of them have the same orientation, i.e. there is an undirected path $v_{i-1} e_i v_i \cdots v_{j-1} e_j v_j$ in G such that $\rho(e_i) = \rho(e_j) = p$ and either $e_i \in in(v_{i-1})$ and $e_j \in out(v_j)$ or $e_i \in out(v_{i-1})$ and $e_j \in in(v_j)$. But this means that the sequence $r_i, \ldots, r_j$ is a cycle. Hence every simple undirected path in G contains at most $2|Q|$ edges which means that $L(A)$ is finite.

For the other direction assume that A is cyclic. Let $r_1, \ldots, r_k$ be a cycle in A and $p_1, \ldots, p_k$ a corresponding sequence of states. As A is reduced there is a DAG $G_i \in L(A)$ for every $i \in [k]$ such that there exists an run ρ_i of A on G_i which uses the rule r_i in a vertex $v_i \in V_{G_i}$. Every vertex v_i is incident to two distinct edges e_i and e_i' such that $\rho_i(e_i) = p_i = \rho_{(i \bmod k)+1}(e_{(i \bmod k)+1}')$ and $e_i \in [out_{G_i}(v_i)]$ iff $e_{(i \bmod k)+1}' \in [in_{G_{(i \bmod k)+1}}(v_{(i \bmod k)+1})]$ for all $i \in [k]$. Then $G = (G_1 \& \cdots \& G_k)[e_1 \bowtie e_2'] \cdots [e_{k-1} \bowtie e_k']$ is defined and contains a path that contains the edges e_1' and e_k such that, by Lemma 6, there is a run ρ of A on G with $\rho(e_1') = \rho(e_k)$. Again by Lemma 6 it follows that $G(e_1' \bowtie e_k)^n \in L(A)^\&$ for all $n \in \mathbb{N}$ and as e_1' and e_k have the same orientation every $G(e_1' \bowtie e_k)^n$ contains a connected component C with $|V_C| > nk$ (cf. the proof of Lemma 8). Therefore $L(A)$ is infinite.

The language recognized by a reduced DAG automaton A is therefore infinite if and only if A is cyclic. Given an arbitrary DAG automaton A we can hence decide if $L(A)$ is finite by constructing an equivalent reduced DAG automaton A_{red} and checking if A_{red} is cyclic. According to Lemma 11 the construction of A_{red} can be performed in polynomial time and it should be clear that the same holds for checking whether A_{red} is cyclic. $\square$

6   Conclusion

We have shown that regular DAG languages admit pumping lemmas similar to those of regular string and tree languages, and that the finiteness problem for languages accepted by DAG automata is in P. In almost all of this, edge swapping turned out to be a central technique. Further results that can be found in [3] had to be left out for lack of space. In particular, this includes the connection between regular DAG languages and regular tree languages. Again making use of edge swapping it can be shown that "unfolding" DAGs into trees is an operation that turns regular DAG languages into regular tree languages. In addition to being interesting in its own right, this provides a constructive proof of the fact (known from [5], though by means of a non-constructive proof) that the path languages of regular DAG languages are regular. An even more recent result by the first author is that top-down deterministic DAG automata can be minimized. As in the tree case the minimal top-down deterministic DAG automaton is uniquely determined, and hence the equivalence of top-down deterministic DAG automata is decidable. These results will appear in a forthcoming long version of this paper.

Altogether, we dare to conclude that the notion of DAG automata considered in this paper has desirable properties and is thus worth being studied further.

Acknowledgment. We thank the referees for their careful work, and especially for pointing out a mistake in the original version of the proof of Lemma 8. The second author is grateful to David Chiang, Daniel Gildea, Adam Lopez, and Giorgio Satta for many inspiring discussions regarding DAG automata.

References

1. Anantharaman, S., Narendran, P., Rusinowitch, M.: Closure properties and decision problems of dag automata. Inf. Process. Lett. **94**(5), 231–240 (2005)
2. Banarescu, L., Bonial, C., Cai, S., Georgescu, M., Griffitt, K., Hermjakob, U., Knight, K., Koehn, P., Palmer, M., Schneider, N.: Abstract meaning representation for sembanking. In: Proceedings of 7th Linguistic Annotation Workshop, ACL 2013 Workshop (2013)
3. Blum, J.: DAG Automata - Variants, Languages and Properties. Master thesis, Umeå University (2015)
4. Charatonik, W.: Automata on dag representations of finite trees. Research Report MPI-I-1999-2-001, Max-Planck-Institut für Informatik, Saarbrücken (1999)
5. Chiang, D., Drewes, F., Gildea, D., Lopez, A., Satta, G.: Weighted and extended DAG automata for semantic graphs (2015) (in preparation)
6. Drewes, F., Leroux, J.: Structurally cyclic petri nets. Logical Methods in Computer Science (2015) (to appear)
7. Kamimura, T., Slutzki, G.: Parallel and two-way automata on directed ordered acyclic graphs. Inf. Control **49**, 10–51 (1981)
8. Potthoff, A., Seibert, S., Thomas, W.: Nondeterminism versus determinism of finite automata over directed acyclic graphs. Bull. Belgian Math. Soc. Simon Stevin **1**(2), 285 (1994)
9. Priese, L.: Finite automata on unranked and unordered DAGs. In: Harju, T., Karhumäki, J., Lepistö, A. (eds.) DLT 2007. LNCS, vol. 4588, pp. 346–360. Springer, Heidelberg (2007)
10. Quernheim, D., Knight, K.: Towards probabilistic acceptors and transducers for feature structures. In: Proceedings of 6th Workshop on Syntax, Semantics and Structure in Statistical Translation, pp. 76–85. Association for Computational Linguistics (2012)

Normal Form on Linear Tree-to-Word Transducers

Adrien Boiret[1,2]([✉])

[1] University Lille 1, Villeneuve-d'Ascq, France
[2] Links (Inria Lille & CRIStAL, UMR CNRS 9189),
Villeneuve-d'Ascq, France
adrien.boiret@inria.fr

Abstract. We study a subclass of tree-to-word transducers: linear tree-to-word transducers, that cannot use several copies of the input. We aim to study the equivalence problem on this class, by using minimization and normalization techniques. We identify a Myhill-Nerode characterization. It provides a minimal normal form on our class, computable in EXPTIME. This paper extends an already existing result on tree-to-word transducers without copy or reordering (sequential tree-to-word transducers), by accounting for all the possible reorderings in the output.

Keywords: Transducers · Tree-to-word transducers · Normal form

1 Introduction

Transducers and their properties have long been studied in various domains of computer sciences. The views on transducers that motivate this paper's field of research are mostly the result of the intersection of two approaches.

Language theory sees transducers as the natural extension of automata, with an output. This view extends almost as far back as the study of regular languages, and developed techniques to solve classical problems such as equivalence, type-checking, or even learning problems (e.g. [4,9,10]) on increasingly wide classes of transducers.

Functional programming sees transducers as a formal representation of some programs. In order to study languages such as XSLT, XQuery, or XProc, used to transform XML trees, classes of transducers that acted more and more like functional programs were designed and studied. For example, deterministic top-down tree transducers can be seen as a functional program that transform trees from the root to the leaves, with finite memory. Different classes extend the reach of transducers to encompass more of the functionalities of programming languages.

Concatenation in the output, notably, plays an important role in the way XSLT produces its outputs. Classes like macro-tree transducers [5], tree-to-word transducers, or even word-to-word transducers with copies in the output [1] allow such concatenation, but as this functionality appears to be difficult to combine

© Springer International Publishing Switzerland 2016
A.-H. Dediu et al. (Eds.): LATA 2016, LNCS 9618, pp. 439–451, 2016.
DOI: 10.1007/978-3-319-30000-9_34

with the classical techniques of language theory, this is to the cost of very few results carrying to these classes.

Tree-to-word transducers and Macro-tree transducers are of particular relevance, as they allow concatenation in their output, and are at the current frontier between the language theory approach of transducers and the approach of transducers seen as functional programs.

Many problems are left open in these classes. Notably, in the general case for Macro-tree transducers, the decidability equivalence is a famous long-standing question, that has yet to be resolved. However, some pre-existing results exist for fragments of these classes.

Equivalence for the subclass of linear size increase macro-tree transducers [3] is proven to be decidable. It comes from a *logic* characterization, as if we bound the number of times a transducer can copy the same subtree in the output, then we limit the expressivity of macro-tree transducers into MSO-definable translations, where equivalence is decidable in non-elementary complexity [4].

Equivalence for all tree-to-word transducers has recently been proven to be decidable in randomized polynomial time [11]. Note that this result uses neither classic logic methods nor the classic transducer methods, and does not provide a characterization or Myhill-Nerode theorem.

Equivalence is PTIME for sequential tree-to-word transducers [6], that prevents copying in the output and forces subtrees to produce following the order of the input. Furthermore, using a *Myhill-Nerode* characterization, a normal form computable in EXPTIME is shown to exist. This normal form was later proven to be learnable in PTIME [7].

In this paper, we aim to study the linear tree-to-word transducers (or LTWs), a restriction of deterministic tree-to-word transducers that forbids copying in the output, but allows the image of subtrees to be flipped in any order. This is a more general class than sequential tree-to-word transducers, but still less descriptive than general tree-to-word transductions. In this class, we show the existence of a normal form, computable in EXPTIME.

Note that even if equivalence is already known to be decidable in a reasonable complexity, finding a normal form is of general interest in and of itself. For example, in [7, 8, 10], normal forms on transducers defined using a Myhill-Nerode theorem are used to obtain a learning algorithm.

To define a normal form on LTWs, we start by the methods used for sequential tree-to-words transducers (STWs) in [6]. We consider the notion of *earliest* STWs, which normalizes the output production. We can extend this notion to LTWs and study only earliest LTWs without any loss of expressivity.

In [6], this is enough to obtain a Myhill-Nerode characterization. However, by adding the possibility to flip subtree images to LTWs, we created another way for equivalent transducers to differ. The challenge presented by the extension of the methods of [6] becomes to resolve this new degree of freedom, in order to obtain a good normal form with a Myhill-Nerode characterization.

Outline. After introducing basic notions on words and trees, we will present our class of linear tree-to-word transducers in Sect. 2. Then in Sect. 3 we will

extend the notion of *earliest* production in [6] to the linear case, and find out
that we can also extend the algorithm that takes a transducer and compute and
equivalent earliest one. However, this is no longer sufficient, as transducers can
now also differ in the order they produce their subtrees' output in. Section 4
will detail exactly how two earliest transducers can still differ, by categorizing
all possible flips. Finally, Sect. 5 will compile these results into a Myhill-Nerode
theorem. This will allow us to establish a normal form, computable in EXPTIME.
We will conclude by a brief recap of the result, and propose several possible next
steps for this line of research.

2   Preliminaries

Words and Trees

We begin by fixing notations on standard notions over words and ranked
trees.

Words. For a finite set of symbols Δ, we denote by Δ^* the set of finite words
over Δ with the concatenation operator $\cdot$ and the empty word ε. For a word u,
$|u|$ is its length. For a set of words L, we denote $\text{lcp}(L)$ the longest word u that is
a prefix of every word in L, or *largest common prefix*. Also, $\text{lcs}(L)$ is the largest
common suffix of L. For $w = u \cdot v$, the left quotient of w by u is $u^{-1} \cdot w = v$, and
the right quotient of w by v is $w \cdot v^{-1} = u$.

Ranked Trees. A *ranked alphabet* is a finite set of ranked symbols
$\Sigma = \bigcup_{k \geq 0} \Sigma^{(k)}$, where $\Sigma^{(k)}$ is the set of k-ary symbols. Every symbol has a
unique arity. A *tree* is a ranked ordered term over Σ. For example, $t = f(a, g(b))$
is a tree over Σ if $f \in \Sigma^{(2)}$, $g \in \Sigma^{(1)}$, $a, b \in \Sigma^{(0)}$. The set of all trees on Σ is T_Σ.

Linear Tree-to-Word Transducers

We define linear tree-to-word transducer, that define a function from T_Σ
to Δ^*.

Definition 1. *A   linear   tree-to-word   transducer   (LTW)   is   a   tuple
$M = \{\Sigma, \Delta, Q, ax, \delta\}$ where*

- *Σ is a tree alphabet,*
- *Δ is a finite word alphabet of output symbols,*
- *Q is a finite set of states,*
- *ax is a axiom of form $u_0 q u_1$, where $u_0, u_1 \in \Delta^*$ and $q \in Q$,*
- *δ is a set of rules of the form*

$$q, f \to u_0 q_1(x_{\sigma(1)}) \ldots q_n(x_{\sigma(n)}) u_n$$

*where $q, q_1, \ldots, q_n \in Q$, $f \in \Sigma$ of rank n and $u_0 \ldots u_n \in \Delta^*$; σ is a permuta-
tion on $\{1, \ldots, n\}$. There is at most one rule per pair q, f.*

We define recursively the function $[M]_q$ of a state q. $[M]_q(f(t_1 \ldots t_n))$ is

- $u_0[M]_{q_1}(t_{\sigma(1)})u_1 \ldots [M]_{q_n}(t_{\sigma(n)})u_n$, if $q, f \to u_0 q_1(x_{\sigma(1)}) \ldots q_n(x_{\sigma(n)})u_n \in \delta$
- undefined, if there is no rule for q, f in δ.

The function $[M]$ of a transducer M with axiom $u_0 q u_1$ is defined as $[M](s) = u_0[M]_q(s)u_1$.

Note that to get the definition of STWs as made in [6], we just have to impose that in every rule, σ is the identity.

Example 2. Consider the function $[M] : t \mapsto 0^{|t|}$, that counts the number of nodes in t and writes a 0 in the output for each of them. Our LTW has only one state q, and its axiom is $\text{ax} = q$

$$q(f(x_1, x_2)) \to 0 \cdot q(x_1) \cdot q(x_2)$$
$$q(a) \to 0, \quad q(b) \to 0$$

The image of $f(a, b)$ is $[M](f(a, b)) = [M]_q(f(a, b))$, using the axiom. Then we use the first rule to get $0 \cdot [M]_q(a) \cdot [M]_q(b)$, and finally, $0 \cdot 0 \cdot 0$.

We denote with $\text{dom}([M])$ the domain of a transducer M, i.e. all trees such that $[M](t)$ is defined. Similarly, $\text{dom}([M]_q)$ is the domain of state q.

We define accessibility between states as the transitive closure of appearance in a rule. This means q is accessible from itself, and if there is a rule $q, f \to u_0 q_1(x_{\sigma(1)}) \ldots q_n(x_{\sigma(n)})u_n$, and q accessible from q', then all states q_i, $1 \leqslant i \leqslant n$, are accessible from q'.

We note L_q the set of all productions of q: $L_q = \{[M]_q(t) | t \in \text{dom}([M]_q)\}$. We call a state *periodic* of period $w \in \Delta^*$ if $L_q \subseteq w^*$.

We start the normalization process with a natural notion of trimmed LTWs.

Definition 3. *A* LTW *is* trimmed *if its axiom is* $u_0 q_0 v_0$, *and every state* q *is accessible from* q_0 *and of non-empty domain.*

Note that all LTWs can be made trimmed by deleting all their useless states.

Lemma 4. *For* M *a* LTW, *one can compute an equivalent trimmed* LTW *in linear time.*

3   Earliest Linear Transducers

It is possible for different LTWs to encode the same transformation. To reach a normal form, we start by requiring our LTWs to produce their output "as soon as possible". This method is common for transducers [2,9], and has been adapted to sequential tree-to-word transducers in [6]. In this case, the way an output word is produced by a tree-to-word can be "early" in two fashions: it can be produced sooner in the input rather than later, or it can output letters on the left of a rule rather than on the right. We take the natural extension of this definition for LTWs and find we can reuse the results and algorithms of [6].

Example 5. Consider our previous example (Example 2). The function $[M] : t \mapsto 0^{|t|}$, Our transducer has only one state q, and its axiom is ax $= q$

$$q(f(x_1, x_2)) \to 0 \cdot q(x_1) \cdot q(x_2)$$
$$q(a) \to 0, \quad q(b) \to 0$$

Since all productions of q start with a 0, this LTW does not produce as up in the input as possible. To change this, we form a new state q' that produces one 0 less than q. By removing the 0 at the beginning of each rule of q, and replacing each call $q(x_i)$ by $0q'(x_i)$, we get a new equivalent LTW M' of axiom ax$' = 0 \cdot q'$

$$q'(f(x_1, x_2)) \to 0 \cdot q'(x_1) \cdot 0 \cdot q'(x_2)$$
$$q'(a) \to \varepsilon \quad q'(b) \to \varepsilon$$

Example 6. Consider our previous example (Example 5). We could replace the first rule by $q'(f(x_1, x_2)) \to 0 \cdot 0 \cdot q'(x_1) \cdot q'(x_2)$. This new LTW would produce "more to the left", but still be equivalent to the first M.

In order to eliminate these differences in output strategies, we want transducers to produce the output as up in the input tree as possible, and then as to the left as possible. We formalize these notions in the definition of *earliest* LTWs.

To simplify notations, we note lcp(q) (or lcs(q)) for lcp(L_q) (or lcs(L_q)). By extension, for $u \in \Delta^*$, we note lcp(qu) (or lcs(qu)) for lcp($L_q.u$) (or lcs($L_q.u$)).

Definition 7. *A* LTW *M is* earliest *if it is trimmed, and:*

- *For every state q, lcp(q) = lcs(q) = ε*
- *For each rule $q, f \to u_0q_1(x_{\sigma(1)}) \ldots q_n(x_{\sigma(n)})u_n \in \delta$, for every i from 1 to n, lcp(q_iu_i) = ε*

This definition is a generalization of the one found in [6] from STWs to all LTWs. The first item ensures an earliest LTW outputs as soon as possible, the second that it produces as to the left as possible. Note that this means that $u_0q_1(x_{\sigma(1)}) \ldots q_i(x_{\sigma(i)})u_i$ produces as much of $[M]_q(f(s_1 \ldots s_n))$ by just knowing $s_{\sigma(1)}, \ldots, s_{\sigma(i)}$, i.e. the lcp of all $[M]_q(f(s_1 \ldots s_n))$ for some fixed $s_{\sigma(1)}, \ldots, s_{\sigma(i)}$.

Lemma 8. *For M an earliest* LTW, $q, f \to u_0q_1(x_{\sigma(1)}) \ldots q_n(x_{\sigma(n)})u_n \in \delta$, *for i such that $i \leqslant n$, $t_{\sigma(1)}, \ldots, t_{\sigma(i)}$ respectively in dom($[M]_{q_1}$), ..., dom($[M]_{q_i}$), then $u_0[M]_{q_1}(t_{\sigma(1)}) \ldots [M]_{q_i}(t_{\sigma(i)})u_i$ is the lcp of the set:*

$$\left\{ [M]_q(f(s_1, \ldots s_n)) | s_{\sigma(1)} = t_{\sigma(1)}, \ldots, s_{\sigma(i)} = t_{\sigma(i)} \right\}.$$

In intuition, this comes from the fact that in an earliest, on the right of $u_0[M]_{q_1}(t_{\sigma(1)}) \ldots [M]_{q_i}(t_{\sigma(i)})u_i$, one cannot guess the first letter of $[M]_{q_{i+1}}(t_{\sigma(i+1)}) \ldots [M]_{q_n}(t_{\sigma(n)})u_n$.

Some important properties extend from [6] to earliest LTWs, most notably the fact that all LTWs can be made earliest.

Lemma 9. *For M a* LTW, *one can compute an equivalent earliest* LTW *in exponential time.*

This result is a direct generalization of the construction in Section 3 of [6]. We build the equivalent earliest LTW M' with two kinds of steps:

- If $\text{lcp}(qu) = v$, where v is a prefix of u, we can slide v through state q by creating a new state $[v^{-1}qv]$ such that for all t, $[M']_{[v^{-1}qv]}(t) = v^{-1}[M]_q(t)v$. Every occurrence of $q(x_i)v$ in a rule of M is replaced by $v\left[v^{-1}qv\right](x_i)$.
- If $\text{lcp}(q) = v$, we can produce v outside of q by creating a new state $\left[v^{-1}q\right]$ such that for all t, $[M']_{[v^{-1}q]}(t) = v^{-1}[M]_q(t)$. Every occurrence of $q(x_i)$ in a rule of M is replaced by $v\left[v^{-1}q\right](x_i)$.
  Symmetrically, if $\text{lcs}(q) = v$, we create a state $\left[qv^{-1}\right]$, and every occurrence of $q(x_i)$ in a rule of M is replaced by $\left[qv^{-1}\right](x_i)v$.

Note that the exponential bound is, in fact, an exact bound, as some LTWs gain an exponential number of states through this process.

In [6], earliest STWs are actually enough to make a normal form using a Myhill-Nerode theorem: by minimizing earliest STWs (merging states with the same $[M]_q$), we end up with a normal form with a minimal number of states. However, in the wider case of LTWs, there are still ways for two states to be equivalent and yet not syntactically equal. This impedes the process of minimization. As we will see in the next part, it remains to study how the images of subtrees can be reordered in earliest LTWs while preserving equivalence.

4   Reordering in Earliest Transducers

Syntactically different earliest LTWs may still be equivalent. Indeed, unlike sequential tree transducers [6], which impose the output to follow the order of the input, LTWs permit to flip the order.

The main point of this paper is the observation that it is sufficient to normalize the flips in the output production of earliest LTWs, in order to find a unique normal form for equivalent LTWs. To this end, we will prove that order differences are only possible in very specific cases. We start illustrating such flips in some examples, and then discuss the necessary and sufficient condition that dictates when a flip is possible.

Example 10. We reconsider Example 6. This earliest transducer "counts" the number of nodes in the input tree has only one state q'. It has the axiom $\text{ax}' = 0 \cdot q'$ and the following rules:

$$q'(f(x_1, x_2)) \to 0 \cdot 0 \cdot q'(x_1) \cdot q'(x_2), \qquad q'(a) \to \varepsilon, \qquad q'(b) \to \varepsilon.$$

We can now flip the order of the terms $q'(x_2)$ and $q'(x_1)$ in the first rule, and replace it by:

$$q'(f(x_1, x_2)) \to 0 \cdot 0 \cdot q'(x_2) \cdot q'(x_1).$$

This does not change $[M']$, since just the order is changed in which the nodes of the first and second subtree of the input are counted.

Of course, it is not always possible to flip two occurrences of terms $q_1(x_{\sigma(1)})$ and $q_2(x_{\sigma(2)})$ in LTW rules.

Example 11. Consider an earliest transducer that outputs the frontier of the input tree while replacing a by 0 and b by 1. This transducer has a single state q, the axiom $ax = q$, and the following rules:

$$q(f(x_1, x_2)) \rightarrow q(x_1) \cdot q(x_2), \qquad q(a) \rightarrow 0, \qquad q(b) \rightarrow 1.$$

Clearly, replacing the first rule by a flipped variant $q(f(x_1, x_2)) \rightarrow q(x_2) \cdot q(x_1)$ would not preserve transducer equivalence since $f(a, b)$ would be transformed to 10 instead of 01. More generally, no LTW with rule $q(f(x_1, x_2)) \rightarrow u_0 \cdot q_1(x_2) \cdot u_1 \cdot q_2(x_1) \cdot u_2$ produces the correct output.

Our goal is to understand the conditions when variable flips are possible.

Definition 12. *For M, M' two LTWs, $q \in Q$, $q' \in Q'$,*

$$q, f \rightarrow u_0 q_1(x_{\sigma(1)}) \ldots q_n(x_{\sigma(n)}) u_n \in \delta$$
$$q', f \rightarrow u_0' q_1'(x_{\sigma'(1)}) \ldots q_n'(x_{\sigma'(n)}) u_n' \in \delta'$$

are said to be twin rules *if q and q' are equivalent.*

4.1 Reordering Erasing States

We start the study of possible reordering with the obvious case of states that only produce ε: they can take every position in every rule without changing the semantics of the states. The first step towards normalization would then be to fix the positions of erasing states in the rules, to prevent differences in equivalent earliest LTWs: we put all erasing states at the end of any rule they appear in, in ascending subtree order.

Definition 13. *For M a LTW, a state q is* erasing *if for all $t \in dom([M]_q)$, $[M]_q(t) = \varepsilon$.*

We show that if two states are equivalent, they call erasing states on the same subtrees. We start by this length consideration:

Lemma 14. *For two twin rules of earliest LTWs*

$$q, f \rightarrow u_0 q_1(x_{\sigma(1)}) \ldots q_n(x_{\sigma(n)}) u_n$$
$$q', f \rightarrow u_0' q_1'(x_{\sigma'(1)}) \ldots q_n'(x_{\sigma'(n)}) u_n'$$

For i, j such that $\sigma(i) = \sigma'(j)$, and $t_{\sigma(i)} \in dom([M]_{q_i})$ then $|[M]_{q_i}(t_{\sigma(i)})| = |[M']_{q_i'}(t_{\sigma(i)})|$

Proof. The equivalence of q and q' gives for all $t_1, \ldots, t_n$:

$$u_0[M]_{q_1}(t_{\sigma(1)}) \ldots [M]_{q_n}(t_{\sigma(n)}) u_n = u_0'[M']_{q_1'}(t_{\sigma'(1)}) \ldots [M']_{q_n'}(t_{\sigma'(n)}) u_n'$$

By fixing every t_k except $t_{\sigma(i)}$ we get that for some u, v, u', v', $u[M]_{q_i}(t_{\sigma(i)})v = u'[M']_{q'_j}(t_{\sigma(i)})v'$. If $|[M]_{q_i}(t_{\sigma(i)})| > |[M']_{q'_j}(t_{\sigma(i)})|$ then $|u| < |u'|$, or $|v| < |v'|$. If $|u| < |u'|$, then $u' = uw$. For all $t_{\sigma(i)}$, $[M]_{q_i}(t_{\sigma(i)}) \neq \varepsilon$ (it is longer than $[M']_{q'_j}(t_{\sigma(i)})$), and its first letter is always the first letter of w. This means $\mathrm{lcp}(q_i) \neq \varepsilon$, which is impossible in an earliest LTW. $|v| < |v'|$ leads to $\mathrm{lcs}(q_i) \neq \varepsilon$, another contradiction. By symmetry, $|[M']_{q'_j}(t_{\sigma(i)})| > |[M]_{q_i}(t_{\sigma(i)})|$ also leads to contradiction. Therefore, both are of same size.

Lemma 15. *For two twin rules of earliest* LTW*s*

$$q, f \to u_0 q_1(x_{\sigma(1)}) \ldots q_n(x_{\sigma(n)})u_n$$
$$q', f \to u'_0 q'_1(x_{\sigma'(1)}) \ldots q'_n(x_{\sigma'(n)})u'_n$$

For i, j such that $\sigma(i) = \sigma'(j)$, If q_i is erasing, then q'_j is erasing.

To normalize the order of erasing states in twin rules, we note that since an erasing state produces no output letter, its position in a rule is not important to the semantics or the earliest property. We can thus push them to the right.

Lemma 16. *For M an earliest* LTW, $q, f \to u_0 q_1(x_{\sigma(1)}) \ldots q_n(x_{\sigma(n)})u_n$ *a rule in M, and q_i an erasing state. Then replacing this rule by*

$$q, f \to u_0 q_1(x_{\sigma(1)})\ldots u_{i-1}u_i\ldots q_n(x_{\sigma(n)})u_n q_i(x_{\sigma(i)})$$

does not change $[M]_q$, and M remains earliest.

Note that the earliest property also imposes that if q_i is erasing, $u_i = \varepsilon$.

Given this lemma, we can define a first normalization step where all erasing states appear at the end of the rules in ascending subtree order.

Definition 17. *An earliest* LTW *M is erase-ordered if for every rule $q, f \to u_0 q_1(x_{\sigma(1)}) \ldots q_n(x_{\sigma(n)})u_n \in \delta$, if q_i is erasing, then for all $j > i$, q_j is erasing, and $\sigma(i) < \sigma(j)$.*

Lemma 18. *For M an earliest* LTW, *one can make M erase-ordered in polynomial time without changing the semantic of its states.*

We can detect if a state q is erasing by checking that no accessible rule produces a letter. From there, Lemma 16 ensures that making a LTW erase-ordered is just a matter of pushing all erasing states at the end of the rules and them sorting them in ascending subtree order.

4.2   Reordering Producing States

As we saw in Example 11, some flips between states are not possible. We will now study what makes reordering non-erasing states possible. As we will see, only few differences are possible between twin rules in erase-ordered earliest LTWs. Two states transforming the same subtree are equivalent, and the only order differences are caused by flipping states whose productions commute in Δ^*.

To prove this, we begin by establishing a few preliminary results. We first show that to the left of σ and σ''s first difference, both rules are identical.

Lemma 19. *For two twin rules of erase-ordered earliest* LTW*s* M, M'

$$q, f \rightarrow u_0 q_1(x_{\sigma(1)}) \ldots q_n(x_{\sigma(n)}) u_n$$
$$q', f \rightarrow u'_0 q'_1(x_{\sigma'(1)}) \ldots q'_n(x_{\sigma'(n)}) u'_n$$

For i such that if $k \leqslant i$ then $\sigma(k) = \sigma'(k)$, $[M]_{q_i} = [M']_{q'_i}$, and $u_i = u'_{i'}$

Proof. This results from Lemma 8: if σ and σ' coincide before i, then for all $t_{\sigma(1)}, \ldots, t_{\sigma(i)}$, $u_0[M]_{q_1}(t_{\sigma(1)}) \ldots u_i$ and $u'_0[M']_{q_1}(t_{\sigma'(1)}) \ldots u'_i$ are both equal to the lcp of $\{[M]_q(f(s_1, \ldots, s_n)) | s_{\sigma(1)} = t_{\sigma(1)}, \ldots, s_{\sigma(n)} = t_{\sigma(n)}\}$. This means that:

$$u_0[M]_{q_1}(t_{\sigma(1)}) \ldots [M]_{q_i}(t_{\sigma(i)}) u_i = u_0[M']_{q'_1}(t_{\sigma'(1)}) \ldots [M']_{q'_i}(t_{\sigma'(i)}) u'_i$$

Since this is also true for $i-1$, we can remove everything but the last part for each side of this equation, to obtain that for all $t_{\sigma(i)}$, $[M]_{q_i}(t_{\sigma(i)}) u_i = [M']_{q'_i}(t_{\sigma(i)}) u'_i$. Lemma 14 gives us $|[M]_{q_i}(t_{\sigma(i)})| = |[M']_{q'_i}(t_{\sigma'(i)})|$, and $u_i = u'_i$. This means that q_i and q'_i are equivalent, and $u_i = u'_i$.

It still remains to show what happens when σ and σ' stop coinciding. We study the leftmost order difference between two twin rules in erasing-ordered earliest LTWs, that is to say the smallest i such that $\sigma(i) \neq \sigma'(i)$. Note that Lemma 15 ensures that such a difference occurs before the end of the rule where the erasing states are sorted.

Lemma 20. *For two twin rules of erase-ordered earliest* LTW*s* M, M'

$$q, f \rightarrow u_0 q_1(x_{\sigma(1)}) \ldots q_n(x_{\sigma(n)}) u_n$$
$$q', f \rightarrow u'_0 q'_1(x_{\sigma'(1)}) \ldots q'_n(x_{\sigma'(n)}) u'_n$$

For i such that $\sigma(i) \neq \sigma'(i)$ and for any $k < i$, $\sigma(k) = \sigma'(k)$, for j such that $\sigma'(i) = \sigma(j)$, we have:

(A) For all k from i to $j-1$, $u_k = \varepsilon$ and there exists $t^\varepsilon_{\sigma(k)}$ such that $[M]_{q_k}(t^\varepsilon_{\sigma(k)}) = \varepsilon$
(B) For all k from i to j, for k' such that $\sigma(k) = \sigma'(k')$, q_k is equivalent to $q'_{k'}$
(C) All $q_i, \ldots, q_j$ are periodic of same period.

As a proof intuition, we first prove point (A), then use it to show point (B), then from (A) and (B) we finally show point (C).

For point (A), we use the equivalence of q and q'. For all $t_1, \ldots, t_n$,

$$u_0[M]_{q_1}(t_{\sigma(1)}) \ldots [M]_{q_n}(t_{\sigma(n)}) u_n = u_0[M']_{q'_1}(t_{\sigma'(1)}) \ldots [M']_{q'_n}(t_{\sigma'(n)}) u'_n$$

Lemma 19 gives us that everything up to u_{i-1} and u'_{i-1} coincide. We then get

$$[M]_{q_i}(t_{\sigma(i)}) \ldots [M]_{q_n}(t_{\sigma(n)}) u_n = [M']_{q'_i}(t_{\sigma'(i)}) \ldots [M']_{q'_n}(t_{\sigma'(n)}) u'_n$$

Since q'_i is not erasing, we can fix $t_{\sigma'(i)}$ such that $[M']_{q'_i}(t_{\sigma'(i)}) \neq \varepsilon$. We call its first letter a. All non-ε productions of q_i must begin by a. This is only possible in an earliest if there exists $t^\varepsilon_{\sigma(i)}$ such that $[M]_{q_i}(t^\varepsilon_{\sigma(i)}) = \varepsilon$. We now fix $t_{\sigma(i)} = t^\varepsilon_{\sigma(i)}$.

If $u_i \neq \varepsilon$, its first letter is a. This is impossible in an earliest since it would mean $\mathrm{lcp}(q_i u_i) \neq \varepsilon$. Hence $u_i = \varepsilon$ We can make the same reasoning for q_{i+1} and u_{i+1}, and so on all the way to q_{j-1} and u_{j-1}.

For point (B), we use point (A) to eliminate everything in front of q_k and $q'_{k'}$ by picking all $t^\varepsilon_{\sigma(l)}$ up to $k-1$ and all $t^\varepsilon_{\sigma'(l')}$ up to $k'-1$.

$$[M]_{q_k}(t_{\sigma(k)})...[M]_{q_n}(t_{\sigma(n)})u_n = [M']_{q'_{k'}}(t_{\sigma'(k')})...[M']_{q'_n}(t_{\sigma'(n)})u'_n$$

From Lemma 14, we know that $|[M]_{q_k}(t_{\sigma(k)})| = |[M']_{q'_{k'}}(t_{\sigma(k)})|$. We conclude that q_k and $q'_{k'}$ are equivalent.

For point (C), we take k' such that $\sigma(k) = \sigma'(k')$. We use (A) to erase everything but q_k, q_j, q'_i and $q'_{k'}$ by picking every $t^\varepsilon_{\sigma(l)}$ and $t^\varepsilon_{\sigma'(l')}$ except theirs.

$$[M]_{q_k}(t_{\sigma(k)})[M]_{q_j}(t_{\sigma(j)})...u_n = [M']_{q'_i}(t_{\sigma'(i)})[M']_{q'_{k'}}(t_{\sigma'(k')})...u'_n$$

Point (B) gives q_k is equivalent to $q'_{k'}$ and q_j is equivalent to q'_i. We get that $[M]_{q_k}(t_{\sigma(k)})[M]_{q_j}(t_{\sigma(j)}) = [M]_{q_j}(t_{\sigma(j)})[M]_{q_k}(t_{\sigma(k)})$. This means that the productions of q_k and q_j commute, which in Δ^* is equivalent to say they are words of same period. Therefore, q_j and q_k are periodic of same period.

This result allows us to resolve the first order different between two twin rules by flipping q_j with neighbouring periodic states of same period. We can iterate this method to solve all order differences.

Theorem 21. *For two twin rules of erase-ordered earliest* LTW*s,*

$$q, f \rightarrow u_0 q_1(x_{\sigma(1)}) \ldots q_n(x_{\sigma(n)}) u_n$$
$$q', f \rightarrow u'_0 q'_1(x_{\sigma'(1)}) \ldots q'_n(x_{\sigma'(n)}) u'_n$$

One can replace the rule of q to another rule of same subtree order as the rule of q' only by flipping neighbour states q_k and q_{k+1} of same period where $u_k = \varepsilon$.

We can use Lemma 20 to solve the leftmost difference: for i first index such that $\sigma(i) \neq \sigma'(i)$, and j such that $\sigma(i) = \sigma'(j)$, we have $u_i = ... = u_{j-1} = \varepsilon$ and $q_i, ..., q_j$ commute with each other. This means we can replace the first rule by:

$$q, f \rightarrow u_0...q_j(x_{\sigma(j)})q_i(x_{\sigma(i)})...q_{j-1}(x_{\sigma(j-1)})u_j...u_n$$

where $q_j(x_{\sigma(j)})$ is to the left of $q_i(x_{\sigma(i)})...q_{j-1}(x_{\sigma(j-1)})$ without changing $[M]_q$.

This solves the leftmost order difference: we can iterate this method until both rules have the same order.

Finally, we call Lemma 19 on the rules reordered by Theorem 21 to show that two twin rules use equivalent states and the same constant words:

Theorem 22. *For two twin rules of erase-ordered earliest* LTW*s,*

$$q, f \rightarrow u_0 q_1(x_{\sigma(1)}) \ldots q_n(x_{\sigma(n)}) u_n$$
$$q', f \rightarrow u'_0 q'_1(x_{\sigma'(1)}) \ldots q'_n(x_{\sigma'(n)}) u'_n$$

$u_0 = u'_0, ..., u_n = u'_n$, *and for k, k' such that $\sigma(k) = \sigma'(k')$, $[M]_{q_k} = [M']_{q'_{k'}}$.*

5   Myhill-Nerode Theorem and Normal Form

In Sect. 3, we showed that LTWs can be made earliest. In Sect. 4, we first showed that all earliest LTWs can be made erase-ordered, then we made explicit what reorderings are possible between two rules of two equivalent states. In this section, we use these results to fix a reordering strategy. This will give us a new normal form, *ordered earliest* LTWs. We will show that each LTW in equivalent to a unique minimal ordered earliest LTW, whose size is at worst exponential.

We first use Theorem 21 to define a new normal form: ordered earliest LTWs.

Definition 23. *A* LTW *M is said to be* ordered earliest *if it is earliest, and for each rule $q, f \to u_0 q_1(x_{\sigma(1)}) \ldots q_n(x_{\sigma(n)}) u_n$:*

- *If q_i is erasing, then for any $j > i$, q_j is erasing.*
- *If $u_i = \varepsilon$, and q_i and q_{i+1} are periodic of same period, $\sigma(i) < \sigma(i + 1)$.*

Note that this definition notably implies that any ordered earliest is erase-ordered earliest. On top of that, we impose that if two adjacent states are periodic of same period, and thus could be flipped, they are sorted by ascending subtree.

Lemma 24. *For M an earliest* LTW*, one can make M ordered in polynomial time without changing the semantic of its states.*

We saw in Lemma 18 that one can push and sort erasing states. For this result, sorting periodic states is not more complicated. However, one must test first whether two states are periodic of same period. This can be done in polynomial time. One can prove that the productions of a LTW state q form an algebraic language (described by a context-free grammar). Then, the problem of deciding if two algebraic languages are periodic of same period is known to be polynomial.

Our goal is now to show the existence of a unique minimal normal LTW equivalent to any M. To this end, we first show that two equivalent LTWs will use the same states: any $q \in Q$ has an equivalent $q' \in Q'$.

Lemma 25. *For two equivalent earliest* LTW*s M and M', for q state of M, there exist an equivalent state q' in M'.*

Proof. We start by the axioms: if $ax = u_0 q_0 v_0$ and $ax' = u'_0 q'_0 v'_0$, since M and M' are earliest, $u_0 = \mathrm{lcp}([M]) = \mathrm{lcp}([M']) = u'_0$. Then, $v_0 = \mathrm{lcs}(q_0 v_0) = \mathrm{lcs}(q'_0 v'_0) = v'_0$. We then get that q_0 and q'_0 are equivalent.

We can then call Theorem 22 to twin rules of equivalent states q, q' to get new equivalent pairs $q_k, q'_{k'}$ for $\sigma(k) = \sigma'(k')$. Since M is trimmed, this recursive calls will eventually reach all $q \in Q$ and pair them with an equivalent $q' \in Q'$.

Since all equivalent earliest LTWs use the same states, they have the *minimal* amount of states when they don't have two redundant states q, q' such that $[M]_q = [M]_{q'}$. We show this characterises a *unique minimal normal form*.

Theorem 26. *For M a* LTW*, there exists a unique minimal ordered earliest* LTW *M' equivalent to M (up to state renaming).*

The existence of such a minimal ordered earliest LTW derives directly from Lemma 24. All we need to make an ordered earliest M' minimal is to merge its equivalent states together, which is always possible without changing $[M']$.

The uniqueness derives from several properties we showed in this paper. Imagine M and M' two equivalent minimal ordered earliest LTWs. The fact that they have equivalent states come from Lemma 25. Since both are minimal, neither have redundant state: each q of M is equivalent to exactly one q' of M' and vice-versa. From Theorem 22, we know that two equivalent states call equivalent states in their rules, with only the possibility of reordering periodic states. Since M and M' are ordered, twin rules also have same order.

6   Conclusion and Future Work

This paper's goal was to solve the equivalence problem on linear tree-to-word transducers, by establishing a normal form and a Myhill-Nerode theorem on this class. To do so we naturally extended the notion of earliest transducers that already existed in sequential tree transducers [6]. However it appeared that this was no longer enough to define a normal form: we studied all possible reorderings that could happen in an earliest LTW. We then used this knowledge to define a new normal form, that has both an output strategy (earliest) and an ordering strategy (ordered earliest), computable from any LTW in EXPTIME.

There are several ways to follow up on this result: one would be adapting the learning algorithm presented in [7], accounting for the fact that we now also have to learn the order in which the images appear. It could also be relevant to note that in [6], another algorithm decides equivalence in polynomial time, which is more efficient than computing the normal form. Such an algorithm would be an improvement over the actual randomized polynomial algorithm by [11]. As far as Myhill-Nerode theorems go, the next step would be to consider all tree-to-word transducers. This problem is known to be difficult. Recently, [11] gave a randomized polynomial algorithm to decide equivalence, but did not provide a Myhill-Nerode characterization.

References

1. Alur, R., D'Antoni, L.: Streaming tree transducers. CoRR abs/1104.2599 (2011)
2. Choffrut, C.: Minimizing subsequential transducers: a survey. Theor. Comput. Sci. **292**(1), 131–143 (2003)
3. Engelfriet, J., Maneth, S.: Macro tree translations of linear size increase are MSO definable. SIAM J. Comput. **32**(4), 950–1006 (2003)
4. Engelfriet, J., Maneth, S.: The equivalence problem for deterministic MSO tree transducers is decidable. In: Sarukkai, S., Sen, S. (eds.) FSTTCS 2005. LNCS, vol. 3821, pp. 495–504. Springer, Heidelberg (2005)
5. Engelfriet, J., Vogler, H.: Macro tree transducers. J. Comput. Syst. Sci. **31**(1), 71–146 (1985)

6. Laurence, G., Lemay, A., Niehren, J., Staworko, S., Tommasi, M.: Normalization of sequential top-down tree-to-word transducers. In: Dediu, A.-H., Inenaga, S., Martín-Vide, C. (eds.) LATA 2011. LNCS, vol. 6638, pp. 354–365. Springer, Heidelberg (2011)
7. Laurence, G., Lemay, A., Niehren, J., Staworko, S., Tommasi, M.: Learning sequential tree-to-word transducers. In: Dediu, A.-H., Martín-Vide, C., Sierra-Rodríguez, J.-L., Truthe, B. (eds.) LATA 2014. LNCS, vol. 8370, pp. 490–502. Springer, Heidelberg (2014)
8. Lemay, A., Maneth, S., Niehren, J.: A learning algorithm for top-down XML transformations. In: Proceedings of the Twenty-Ninth ACM SIGMOD-SIGACT-SIGART Symposium on Principles of Database Systems, PODS 2010, June 6–11, 2010, Indianapolis, Indiana, USA, pp. 285–296 (2010)
9. Maneth, S., Seidl, H.: Deciding equivalence of top-down XML transformations in polynomial time. In: PLAN-X , Programming Language Technologies for XML, An ACM SIGPLAN Workshop Colocated with POpPL 2007, Nice, France, January 20, 2007, pp. 73–79 (2007)
10. Oncina, J., Garcia, P., Vidal, E.: Learning subsequential transducers for pattern recognition interpretation tasks. IEEE Trans. Pattern Anal. Mach. Intell. **15**(5), 448–458 (1993)
11. Seidl, H., Maneth, S., Kemper, G.: Equivalence of deterministic top-down tree-to-string transducers is decidable. CoRR abs/1503.09163 (2015)

A Kleene Theorem for Weighted Tree Automata over Tree Valuation Monoids

Manfred Droste[1], Zoltán Fülöp[2], and Doreen Götze[1][(✉)]

[1] Institut für Informatik, Universität Leipzig,
04109 Leipzig, Germany
{droste,goetze}@informatik.uni-leipzig.de
[2] Institute of Informatics, University of Szeged,
Szeged 6701, Hungary
fulop@inf.u-szeged.hu

Abstract. Cauchy unital tree valuation monoids are introduced as weight structures for weighted tree automata. Rational tree series over this kind of monoids are defined and Kleene's classical theorem is proved for this setting: a tree series over a Cauchy unital tree valuation monoid is recognizable if and only if it is rational.

Keywords: Weighted tree automata · Rational expressions · Valuation monoids · Kleene theorem

1 Introduction

Trees or terms are one of the most fundamental concepts both in mathematics and in computer science. Tree automata were introduced in the 1960s and since then the theory of tree automata and tree languages has developed rapidly, see [4,13] for surveys. At the beginning of the 1980s also quantitative aspects gained attention and weighted tree automata (wta) were introduced [1,2]. Since then a wide range of wta models have been considered.

Wta recognize tree series which are mappings from the set of trees into a weight structure D. The semantics of a wta is usually defined in terms of runs and is based on the algebraic structure of D. The weight structure D has a binary operation called addition, and the weight of an input tree is obtained by summing up the weights of all runs over the tree. When D is a semiring (i.e., for s-wta), the weight of a run is the product of the weights of the transitions in the run [8,12]. If D is a multi-operator monoid (i.e., for m-wta), then the weight of each transition is an operation over D and the weight of a run is the evaluation of the weights of the transitions in that run, see [15,16]. In the weighted (word) automaton on real numbers introduced in [3], the weight of a run is determined in a global way, for instance by calculating the average of all weights of the run.

Supported by the German Academic Exchange Service (DAAD) and the Hungarian Scholarship Board Office (MÖB) project "Theory and Applications of Automata" (grant 5567). The second and the third author were partially supported by the NKFI grant no. K 108448 and by the DFG Graduiertenkolleg 1763 (QuantLA), respectively.

A.-H. Dediu et al. (Eds.): LATA 2016, LNCS 9618, pp. 452–463, 2016.
DOI: 10.1007/978-3-319-30000-9_35

This concept was generalized to weighted automata over a more general weight structure called valuation monoid in [6] and to wta over tree valuation monoids (tv-wta) in [5].

In this paper we focus on Kleene-type results for wta. Kleene's classical theorem [14] states that the class of recognizable languages and that of rational languages are the same. This theorem has already been extended to various discrete structures like trees [18] as well as to the weighted settings, cf. e.g. Schützenberger's fundamental paper [17]. In particular, a Kleene-type result was proved for s-wta in [10] and [8], and for m-wta in [11]. The corresponding result for tv-wta was missing up to now. In this paper we fill the gap and provide a Kleene-type characterization of tv-wta (Theorem 12).

The Kleene-type results are based on the fact that the semantics of the involved automaton models can be defined inductively. (For instance, the weight of a run of an s-wta on a tree is, roughly speaking, the product of the weights of the corresponding sub-runs on the direct subtrees and the weight of the transition applied at the root of the tree.) This phenomenon makes it possible to show that, under appropriately defined rational operations, automata and rational expressions are equally powerful. However, the semantics of a tv-wta cannot be defined inductively because the weight of a run on a tree is simply delivered by a global valuation function. Therefore, we enrich the tree valuation monoid by a family of decomposition operations. Such a decomposition operation is parameterized by a tree domain which later determines the shape of the decomposed tree, some incomparable positions of the tree domain which represent the positions at which the decomposition took place, and an additional component for technical reasons. By these decomposition operations we can define the rational operations concatenation and Kleene-star on tree series appropriately. We call this enriched structure a Cauchy tree valuation monoid because our approach is based on the ideas in [7] where Cauchy valuation monoids were introduced and a Kleene-type result was proved for weighted automata over this kind of monoids.

In order to ensure that the concatenation and the Kleene-star of tree series defined by using the decomposition operations preserve recognizability, we follow [11] and use variables as additional labels for leaves of trees. Moreover, we assume that the tree valuation monoid has a unit element and hereby our weight structures will be Cauchy unital tree valuation monoids. This seems to be convenient to show that the expressive power of rational tree series expressions is the same as that of wta. As usual, we represent rational trees series by rational expressions. Then our main result states that weighted rational expressions and wta over Cauchy unital tree valuation monoids are expressively equivalent.

2   Trees and Tree Valuation Monoids

Let $\mathbb{N} = \{1, 2, \ldots\}$ be the set of all natural numbers and $\mathbb{N}_0 = \mathbb{N} \cup \{0\}$. For a set H, we denote by $|H|$ the cardinality of H and by H^* the set of all finite words over H. We denote by $|w|$ the length of a word $w \in H^*$. The empty word is denoted by ε.

A *tree domain* is a finite, non-empty subset $\mathcal{B}$ of $\mathbb{N}^*$ such that for all $u \in \mathbb{N}^*$ and $i \in \mathbb{N}$, if $u.i \in \mathcal{B}$, then $u, u.1, \ldots, u.(i-1) \in \mathcal{B}$. The *subtree domain* $\mathcal{B}|_u$ of $\mathcal{B}$ at u is $\{v \in \mathbb{N}^* \mid u.v \in \mathcal{B}\}$. An *unranked tree* over a set H (of labels) is a mapping $t \colon \mathcal{B} \to H$ such that $\mathrm{dom}(t) = \mathcal{B}$ is a tree domain. The set of all unranked trees over H is denoted by U_H. For every $h \in H$, we denote also by h the particular tree defined by $t \colon \{\varepsilon\} \to H$ and $t(\varepsilon) = h$. Let $t \in U_H$. The elements of $\mathrm{dom}(t)$ are called *positions* of t and the set $\mathrm{im}(t) = \{t(u) \mid u \in \mathrm{dom}(t)\}$ is the *image* of t. For every $G \subseteq H$ and $h \in H$, we define $\mathrm{dom}_G(t) = \{u \in \mathrm{dom}(t) \mid t(u) \in G\}$ and abbreviate $\mathrm{dom}_{\{h\}}(t)$ with $\mathrm{dom}_h(t)$. Let $u \in \mathrm{dom}(t)$. We call $t(u)$ the *label* of t at u. The *rank* $\mathrm{rk}_t(u)$ of u is defined to be $\max\{i \in \mathbb{N} \mid u.i \in \mathrm{dom}(t)\}$. If $\mathrm{rk}(u) = 0$, then u is also called a *leaf of* t. We denote by $\mathrm{leaf}(t)$ the set of all leaves of t. Moreover, for every $G \subseteq H$ and $h \in H$, we define $\mathrm{leaf}_G(t) = \mathrm{leaf}(t) \cap \mathrm{dom}_G(t)$ and we write $\mathrm{leaf}_h(t)$ for $\mathrm{leaf}_{\{h\}}(t)$. The *height* $\mathrm{hg}(t)$ of a tree t is $\max\{|u| \mid u \in \mathrm{dom}(t)\}$ and its *size* $\mathrm{size}(t)$ is $|\mathrm{dom}(t)|$. The *subtree* $t|_u$ of t at position u is defined by $\mathrm{dom}(t|_u) = \mathrm{dom}(t)|_u$ and $t|_u(v) = t(u.v)$ for all $v \in \mathrm{dom}(t|_u)$. Moreover, we denote by $t[u \leftarrow t']$ the tree which is obtained from t by replacing $t|_u$ by t'. Let $\mathcal{B}$ be a tree domain. We denote by $\sqsubseteq_{\mathcal{B}}$ the lexicographic order on $\mathcal{B}$. Note that $\sqsubseteq_{\mathcal{B}}$ is a total order. For a subset $P \subseteq \mathcal{B}$, we denote by $\overrightarrow{P}$ the vector obtained by enumerating the elements of P in the order $\sqsubseteq_{\mathcal{B}}$. Moreover, we define the partial order $<_{\mathcal{B}} = \sqsubseteq_{\mathcal{B}} \setminus \{(u, v) \in \mathcal{B} \times \mathcal{B} \mid v = u.w \text{ for some } w \in \mathbb{N}^*\}$.

A *ranked alphabet* is a pair $(\Sigma, \mathrm{rk}_\Sigma)$, where Σ is an alphabet and $\mathrm{rk}_\Sigma \colon \Sigma \to \mathbb{N}_0$ is a mapping which assigns to each symbol of Σ its rank. We denote by $\Sigma^{(k)}$ the set of all symbols which have rank k. Usually we drop rk_Σ and denote a ranked alphabet simply by Σ. In this paper we assume that $\Sigma^{(0)} \neq \emptyset$. We define $\max_\Sigma = \max\{\mathrm{rk}_\Sigma(\sigma) \mid \sigma \in \Sigma\}$.

Let X be a finite set of variables disjoint with Σ. A *ranked tree* over a ranked alphabet Σ and X is an unranked tree over the set $\Sigma \cup X$ such that for all $u \in \mathrm{dom}(t)$, $\mathrm{rk}_t(u) = k$ whenever $t(u) \in \Sigma^{(k)}$ and $\mathrm{rk}_t(u) = 0$ for $t(u) \in X$. We denote the set of all ranked trees over Σ and X by $T_\Sigma(X)$. If $X = \emptyset$, then $T_\Sigma(X)$ is written as T_Σ. Variables can be seen as symbols with rank zero and thus $T_\Sigma(X) = T_{\Sigma \cup X}$. Let $t \in T_\Sigma(X)$, $x \in X$, $r \in \mathbb{N}_0$ be the number of occurrences of x in t, and $t_1, \ldots, t_r \in T_\Sigma(X)$. We denote by $t[x \leftarrow (t_1, \ldots, t_r)]$ the tree obtained by replacing the i-th occurrence of x in t by t_i (counted from left to right).

A *tree valuation monoid* (*tv-monoid* for short) [5,6] is a quadruple $(D, +, \mathrm{V}, \mathbb{0})$ such that $(D, +, \mathbb{0})$ is a commutative monoid and $\mathrm{V} \colon U_D \to D$ is a function, called *(tree) valuation function*, which satisfies that $\mathrm{V}(d) = d$ for every tree $d \in D$, and $\mathrm{V}(t) = \mathbb{0}$ for every $t \in U_D$ with $\mathbb{0} \in \mathrm{im}(t)$.

Next we generalize unital valuation monoids [9] and define unital tv-monoids. Roughly speaking, the presence of the unit element in a tree $t \in U_D$ does not change $\mathrm{V}(t)$. However, we require the unit element to behave so only if it is in a leaf position and no inner positions of the tree are labeled with it.

Let $t \in U_D$ and $ui \in \mathrm{leaf}(t)$ for some $u \in \mathbb{N}^*$ and $i \in \mathbb{N}$. We denote by $[t \backslash ui]$ the tree obtained by dropping its leaf ui and moving each position $u(j+1)v$ with $j + 1 > i$ to the position ujv. More exactly, we define $\mathrm{dom}([t \backslash ui]) = (\mathrm{dom}(t) \setminus \{ujv \mid j \geq i, v \in \mathbb{N}^*\}) \cup \{ujv \mid j \geq i, u(j+1)v \in \mathrm{dom}(t)\}$ and

$$[t\backslash ui](v) = \begin{cases} t(u(j+1)w) & \text{if } v = ujw \text{ for some } j \geq i \text{ and } w \in \mathbb{N}^*, \\ t(v) & \text{otherwise,} \end{cases}$$

for every $v \in \text{dom}([t\backslash ui])$. Now a *unital tree valuation monoid* is a system $(D, +, \text{V}, 0, 1)$ where $(D, +, \text{V}, 0)$ is a tv-monoid and $1 \in D$ is a unit element satisfying that for every $t \in U_D$ with $\text{dom}_1(t) = \text{leaf}_1(t)$ and $ui \in \text{leaf}_1(t)$ with $i \in \mathbb{N}$, we have $\text{V}([t\backslash ui]) = \text{V}(t)$. Next we generalize Cauchy valuation monoids of [7] and introduce Cauchy unital tree valuation monoids.

Definition 1. *A* Cauchy unital tree valuation monoid *(cutv-monoid for short) is a structure* $\mathbb{D} = (D, +, \text{V}, \Pi, 0, 1)$ *such that* $(D, +, \text{V}, 0, 1)$ *is a unital tree valuation monoid and* Π *is a family of decomposition operations*

$$\Pi_{\mathcal{B}, \mathcal{B}', u_1, \dots, u_r} : D \times D^r \to D$$

where $\mathcal{B}$ *is a tree domain,* $\mathcal{B}'$ *is a subset of the leaves of* $\mathcal{B}$, $r \in \mathbb{N}_0$, *and* $u_1, \dots, u_r \in \mathcal{B}$ *with* $u_1 <_{\mathcal{B}} \cdots <_{\mathcal{B}} u_r$. *The operation* $\Pi_{\mathcal{B}, \mathcal{B}', u_1, \dots, u_r}$ *satisfies:*

1. *For all* $d, d_1, \dots, d_r \in D$, *we have* $\Pi_{\mathcal{B}, \mathcal{B}', u_1, \dots, u_r}(d, d_1, \dots, d_r) = 0$ *if* $0 \in \{d, d_1, \dots, d_r\}$.
2. *For each tree* $t \in U_D$ *with* $\text{dom}(t) = \mathcal{B}$ *and* $\text{leaf}_1(t) \subseteq \mathcal{B}' \subseteq \text{leaf}_{\{0,1\}}(t)$, *we have* $\text{V}(t) = \Pi_{\mathcal{B}, \mathcal{B}', u_1, \dots, u_r}(\text{V}(s), \text{V}(t_1), \dots, \text{V}(t_r))$, *where* $s = t[u_1 \leftarrow 1] \dots [u_r \leftarrow 1]$ *and* $t_i = t|_{u_i}$ *for every* $1 \leq i \leq r$.
3. *For all finite subsets* $A, A_1, \dots, A_r$ *of* D:

$$\Pi_{\mathcal{B}, \mathcal{B}', u_1, \dots, u_r}\left(\sum_{d \in A} d, \sum_{d_1 \in A_1} d_1, \dots, \sum_{d_r \in A_r} d_r\right)$$

$$= \sum_{d \in A, d_1 \in A_1, \dots, d_r \in A_r} \Pi_{\mathcal{B}, \mathcal{B}', u_1, \dots, u_r}(d, d_1, \dots, d_r).$$

Note that $\text{V}(d(1, \dots, 1)) = d$ for any occurrences of 1 and cutv-monoid $\mathbb{D}$.

Example 2.[1] From any semiring $(S, +, \cdot, 0, 1)$ we can derive a cutv-monoid $(S, +, \text{V}, 0, 1)$ by letting $\text{V}(t) = \prod_{u \in \text{dom}(t)} t(u)$ (where the weights are multiplied in the order induced by depth first search) for $t \in U_S$ and $\Pi_{\mathcal{B}, \mathcal{B}', u_1, \dots, u_r}(d, d_1, \dots, d_r) = d \cdot \prod_{1 \leq i \leq r} d_r$ for all tree domains $\mathcal{B}$, set of leaves $\mathcal{B}' \subseteq \mathcal{B}$, $r \in \mathbb{N}_0$, positions $u_1 <_{\mathcal{B}} \cdots <_{\mathcal{B}} u_r$ of $\mathcal{B}$, and $d, d_1, \dots, d_r \in S$.

Example 3. The structure $\mathbb{Q}_{\max} = (\mathbb{Q} \cup \{-\infty\}, \max, \text{avg}, -\infty)$ with $\text{avg}(t) = \frac{\sum_{u \in \text{dom}(t)} t(u)}{\text{size}(t)}$ for all $t \in U_{\mathbb{Q} \cup \{-\infty\}}$ of [5, Example 3.2] is a tv-monoid[2]. The valuation function avg calculates the average of all labels of a tree. The idea for the average calculation was already suggested in [3,6] for words.

---

[1] We are thankful to an anonymous referee for spotting a mistake in a previous version of this example.

[2] Here $\sum$ denotes the ordinary sum of numbers with the natural extension to $-\infty$.

We can extend $\mathbb{Q}_{\max}$ to the Cauchy unital tree valuation monoid

$$\mathbb{Q}_{\max}^{\mathrm{Cu}} = (\mathbb{Q} \cup \{-\infty, \square\}, \max', \mathrm{avg}', \Pi, -\infty, \square)$$

as follows. We define

$$\max'(a, b) = \begin{cases} \max(a, b) & \text{if } a, b \in \mathbb{Q} \cup \{-\infty\}, \\ a & \text{if } (a \geq 0 \wedge b = \square) \text{ or } (a = \square \wedge b < 0), \\ b & \text{if } (b \geq 0 \wedge a = \square) \text{ or } (b = \square \wedge a < 0), \\ \square & \text{if } a = b = \square, \end{cases}$$

(i.e., with respect to the linear order $<$ on $\mathbb{Q}_{\max}$ we insert $\square$ between the negative numbers and 0) and

$$\mathrm{avg}'(t) = \frac{\sum_{u \in \mathrm{dom}_{\mathbb{Q} \cup \{-\infty\}}(t)} t(u)}{|\mathrm{dom}(t) \setminus \mathrm{leaf}_{\square}(t)|},$$

respectively, for all $t \in (U_{\mathbb{Q} \cup \{-\infty, \square\}} \setminus \{\square\})$; and $\mathrm{avg}'(\square) = \square$. Moreover, we define decomposition operators as follows. Let $\mathcal{B}$ be a tree domain, $\mathcal{B}' \subseteq \mathcal{B}$ a set of leaves, $r \in \mathbb{N}_0$, and $u_1 <_\mathcal{B} \cdots <_\mathcal{B} u_r$ positions of $\mathcal{B}$. Let $\overline{\mathcal{B}} = \mathcal{B} \setminus \mathcal{B}'$, $\mathcal{B}_i = \{u_i v \mid v \in \mathbb{N}^*\}$, $\mathcal{B}_i' = \{v \in \mathbb{N}^* \mid u_i v \in \mathcal{B}'\}$ for $1 \leq i \leq r$, and define $n \cdot \square = 0$ for all $n \in \mathbb{N}$ and $0 \cdot (-\infty) = -\infty$. For every $d, d_1, \ldots, d_r \in (\mathbb{Q} \cup \{-\infty, \square\})$ we define

$$\Pi_{\mathcal{B}, \mathcal{B}', u_1, \ldots, u_r}(d, d_1, \ldots, d_r) = \frac{|\overline{\mathcal{B}} \setminus (\bigcup_{1 \leq i \leq r} \mathcal{B}_i)| \cdot d + \sum_{1 \leq i \leq r} |(\mathcal{B}|_{u_i}) \setminus \mathcal{B}_i'| \cdot d_i}{|\overline{\mathcal{B}}|}$$

whenever $\mathcal{B}' \neq \mathcal{B}$; and let $\Pi_{\mathcal{B}, \mathcal{B}, \varepsilon}(d, d_1) = \Pi_{\mathcal{B}, \mathcal{B}, \vec{\emptyset}}(d) = -\infty$ for $\mathcal{B}' = \mathcal{B} = \{\varepsilon\}$. It is clear that $\Pi_{\mathcal{B}, \mathcal{B}', u_1, \ldots, u_r}$ calculates the average of the labels of a tree with domain $\mathcal{B}$ in which all positions "outside" $u_1, \ldots, u_r$ are labeled with d and all positions of the form $u_i v$ ($v \in \mathbb{N}^*$) are labeled with d_i for every $1 \leq i \leq r$, and such that the positions in $\mathcal{B}'$ are not taken into account. One can check that $\mathbb{Q}_{\max}^{\mathrm{Cu}}$ is a cutv-monoid.

For the rest of this paper, let Σ be a ranked alphabet, X a finite set of variables, and $\mathbb{D} = (D, +, \mathrm{V}, \Pi, \mathbb{0}, \mathbb{1})$ a cutv-monoid.

A mapping $S : T_\Sigma(X) \to D$ is called a *tree series* (*over Σ, X, and $\mathbb{D}$*). The set of all tree series is denoted by $\mathbb{D} \langle\!\langle T_\Sigma(X) \rangle\!\rangle$. Let $d \in D$, $t \in T_\Sigma(X)$. The tree series which maps every tree to d is denoted by $\tilde{d}$. Moreover, we denote by $d.t$ the tree series which maps the tree t to d and every other tree to $\mathbb{0}$. Such a tree series is called a *monomial*. For $x \in X$, a tree series S is called *x-proper* if $S(x) = \mathbb{0}$.

3   Rational Operations and Rational Tree Series

In this section we introduce rational operations over tree series. Moreover, we define the concept of rational expressions and rational tree series. In what follows, we often specify a decomposition operator in the form $\Pi_{\mathrm{dom}(t),\mathrm{dom}_L(t),u_1,\ldots,u_r}$, for some $t \in T_\Sigma(X)$, $L \subseteq X$, and $u_1,\ldots,u_r \in \mathrm{dom}(t)$. For the sake of brevity, we shorten this notation to $\Pi_{t,L,u_1,\ldots,u_r}$.

Let $k \geq 0$, $\sigma \in \Sigma^{(k)}$, and $d \in D$. The *top-concatenation with σ and d* is a mapping $\mathrm{top}_\sigma^d : \mathbb{D}\langle\langle T_\Sigma(X)\rangle\rangle^k \to \mathbb{D}\langle\langle T_\Sigma(X)\rangle\rangle$ which is defined for every $t \in T_\Sigma(X)$ such that

$$\mathrm{top}_\sigma^d(S_1,\ldots,S_k)(t) = \begin{cases} \Pi_{t,X,1,\ldots,k}(d, S_1(t_1),\ldots S_k(t_k)) & \text{if } t = \sigma(t_1,\ldots,t_k), \\ \mathbb{0} & \text{otherwise.} \end{cases}$$

Let $x \in X$ and $S, S' \in \mathbb{D}\langle\langle T_\Sigma(X)\rangle\rangle$. The *sum $S + S'$ of S and S'* is defined by $(S + S')(t) = S(t) + S'(t)$ for all trees $t \in T_\Sigma(X)$. The *x-concatenation $S \cdot_x S'$ of S and S'* is the tree series defined by

$$(S \cdot_x S')(t) = \sum_{\substack{s,t_1,\ldots,t_r \in T_\Sigma(X) \\ t=s[x\leftarrow(t_1,\ldots,t_r)]}} \Pi_{t,X,\overrightarrow{\mathrm{dom}_x(s)}}(S(s), S'(t_1),\ldots,S'(t_r))$$

for all $t \in T_\Sigma(X)$. Note that in the index set of the sum we have $r \geq 0$.

Let $x \in X$ and $S \in \mathbb{D}\langle\langle T_\Sigma(X)\rangle\rangle$ be x-proper. For every $n \in \mathbb{N}_0$, we define *n-th x-iteration $S^{n,x}$ of S* by induction: $S^{0,x} = \mathbb{0}$ and $S^{n+1,x} = (S \cdot_x S^{n,x}) + \mathbb{1}.x$.

Lemma 4. (cf. [8, Lemma 3.10]) *Let $x \in X$, $S \in \mathbb{D}\langle\langle T_\Sigma(X)\rangle\rangle$ be x-proper and $t \in T_\Sigma(X)$. Then $S^{n+1,x}(t) = S^{n,x}(t)$ for every $n \geq \mathrm{hg}(t) + 1$.*

Proof. We prove by induction on $\mathrm{hg}(t)$. If $t \in (\Sigma^{(0)} \cup X)$, then three cases are possible. If $t = x$, then for all $n \geq 0$: $S^{n+1,x}(x) = (S \cdot_x S^{n,x})(x) + \mathbb{1}.x(x) = \Pi_{x,X,\varepsilon}(S(x), S^{n,x}(x)) + \mathbb{1}.x(x) = \Pi_{x,X,\varepsilon}(\mathbb{0}, S^{n,x}(x)) + \mathbb{1}.x(x) = \mathbb{0} + \mathbb{1}.x(x) = \mathbb{1}$. Similarly, $S^{n+1,x}(y) = \Pi_{y,X,\overrightarrow{\emptyset}}(S(y))$ and $S^{n+1,x}(\alpha) = S(\alpha)$ for all $x \neq y \in X$, $\alpha \in \Sigma^{(0)}$, and $n \geq 0$. Now we assume that $\mathrm{hg}(t) > 0$ and let $n \geq \mathrm{hg}(t) + 1$. Then

$$S^{n+1,x}(t) = (S \cdot_x S^{n,x})(t) + \mathbb{1}.x(t) = (S \cdot_x S^{n,x})(t)$$

$$= \sum_{t=s[x\leftarrow(t_1,\ldots,t_r)], s\neq x} \Pi_{t,X,\overrightarrow{\mathrm{dom}_x(s)}}(S(s), S^{n,x}(t_1),\ldots,S^{n,x}(t_r))$$

$$\stackrel{(*)}{=} \sum_{t=s[x\leftarrow(t_1,\ldots,t_r)], s\neq x} \Pi_{t,X,\overrightarrow{\mathrm{dom}_x(s)}}(S(s), S^{n-1,x}(t_1),\ldots,S^{n-1,x}(t_r))$$

$$= S^{n,x}(t).$$

We can restrict the summation to $s \neq x$ because $S(x) = \mathbb{0}$. This allows us to apply the induction hypothesis at $(*)$. $\square$

Again, let $x \in X$ and $S \in \mathbb{D}\langle\langle T_\Sigma(X)\rangle\rangle$ be x-proper. The *x-Kleene star $S^{*,x}$ of S* is defined by $S^{*,x}(t) = S^{\mathrm{hg}(t)+1,x}(t)$ for all $t \in T_\Sigma(X)$.

Lemma 5. (cf. [8, Lemma 3.13]) *Let $x \in X$ and $S \in \mathbb{D}\langle\langle T_\Sigma(X)\rangle\rangle$ be x-proper. Then $S^{*,x}$ is the unique solution of the equation $\xi = S \cdot_x \xi + \mathbb{1}.x$.*

Proof. Let $t \in T_\Sigma(X)$ and $n = \text{hg}(t) + 1$. Then

$$(S^{*,x})(t) \stackrel{(*)}{=} (S^{n+1,x})(t) = (S \cdot_x S^{n,x})(t) + 1.x(t)$$

$$= \sum_{t=s[x \leftarrow (t_1,\ldots,t_r)]} \Pi_{t,X,\overline{\text{dom}_x(s)}}(S(s), S^{n,x}(t_1), \ldots, S^{n,x}(t_r)) + 1.x(t)$$

$$\stackrel{(\dagger)}{=} \sum_{t=s[x \leftarrow (t_1,\ldots,t_r)]} \Pi_{t,X,\overline{\text{dom}_x(s)}}(S(s), S^{*,x}(t_1), \ldots, S^{*,x}(t_r)) + 1.x(t)$$

$$= (S \cdot_x S^{*,x} + 1.x)(t)$$

At $(*)$ and $(\dagger)$ we could apply Lemma 4, since $n = \text{hg}(t) + 1 \geq \text{hg}(t_i) + 1$. To show that $S^{*,x}$ is the unique solution, we can prove that $T = S^{*,x}$ for every solution T. The proof is analogous to that of Lemma 4. $\square$

The operations top_σ^d, $+$, x-concatenation, and x-Kleene star are called *rational operations*. The set $\text{Rex}(\Sigma, X, \mathbb{D})$ of *rational tree series expressions* (over Σ, X, and $\mathbb{D}$) is defined to be the smallest set R satisfying the following conditions. For every $E \in \text{Rex}(\Sigma, X, \mathbb{D})$, we define its semantics $[\![E]\!] \in \mathbb{D}\langle\!\langle T_\Sigma(X)\rangle\!\rangle$ simultaneously.

- for all $x \in X$: $x \in R$ and $[\![x]\!] = 1.x$,
- for all $\sigma \in \Sigma^{(k)}$, $d \in (D \setminus \{1\})$, and $E_1, \ldots, E_k \in R$ with $k \geq 0$: $d.\sigma(E_1, \ldots, E_k) \in R$ and $[\![d.\sigma(E_1, \ldots, E_k)]\!] = \text{top}_\sigma^d([\![E_1]\!], \ldots, [\![E_k]\!])$,
- for all $E_1, E_2 \in R$: $E_1 + E_2 \in R$ and $[\![E_1 + E_2]\!] = [\![E_1]\!] + [\![E_2]\!]$,
- for all $E_1, E_2 \in R$ and $x \in X$: $E_1 \cdot_x E_2 \in R$ and $[\![E_1 \cdot_x E_2]\!] = [\![E_1]\!] \cdot_x [\![E_2]\!]$,
- for all $E \in R$ and $x \in X$ s.t. $[\![E]\!]$ is x-proper: $E^{*,x} \in R$ and $[\![E^{*,x}]\!] = [\![E]\!]^{*,x}$.

A tree series $S \in \mathbb{D}\langle\!\langle T_\Sigma(X)\rangle\!\rangle$ is *rational* if there is an $E \in \text{Rex}(\Sigma, X, \mathbb{D})$ such that $[\![E]\!] = S$. The set of all rational tree series is denoted by $\mathbb{D}^{\text{rat}}\langle\!\langle T_\Sigma(X)\rangle\!\rangle$. Note that $\tilde{0}$ is rational, since $\tilde{0} = [\![0.\sigma]\!]$ for each $\sigma \in \Sigma^{(0)}$. Obviously, rational tree series are closed under rational operations.

Example 6. Let $\mathbb{Q}_{\max}^{\text{Cu}}$ be the cutv-monoid from Example 3, $\alpha \in \Sigma^{(0)}$, $\sigma \in \Sigma^{(1)}$, and $x \in X$ a variable. We abbreviate $t = \sigma(\sigma(\ldots \sigma(x) \ldots))$ with n occurrences of σ by $\sigma^n x$. For the rational expression $0.\sigma(x)$ we have $[\![0.\sigma(x)]\!](\sigma x) = \text{top}_\sigma^0([\![x]\!]) = \Pi_{\sigma x, X, 1}(0, [\![x]\!](x)) = \Pi_{\sigma x, X, 1}(0, \square) = 0$ and $[\![0.\sigma(x)]\!](t) = -\infty$ for every other $t \in T_\Sigma(X)$. Let us abbreviate $[\![0.\sigma(x)]\!]$ to T and consider now $0.\sigma(x)^{*,x}$. We have

$$[\![0.\sigma(x)^{*,x}]\!](\sigma^n x) = [\![0.\sigma(x)^{n+1,x}]\!](\sigma^n x) =$$

$$= \Pi_{\sigma^n x, X, 1}(T(\sigma x), \Pi_{\sigma^{n-1}x, X, 1}(T(\sigma x), \ldots \Pi_{\sigma x, X, 1}(T(\sigma x), T^{2,x}(x)) \ldots))$$

$$= \Pi_{\sigma^n x, X, 1}(0, \Pi_{\sigma^{n-1}x, X, 1}(0, \ldots \Pi_{\sigma x, X, 1}(0, \square) \ldots)) = 0$$

and $[\![0.\sigma(x)^{*,x}]\!](t) = -\infty$ for all $t \notin \{\sigma^n x \mid n \geq 1\}$. For the expression $E = 0.\sigma(x)^{*,x} \cdot_x 1.\alpha$ with $S = [\![0.\sigma(x)^{*,x}]\!]$, we get

$$[\![E]\!](\sigma^n \alpha) = \Pi_{\sigma^n \alpha, X, 1^n}(S(\sigma^n x), [\![1.\alpha]\!](\alpha)) = \Pi_{\sigma^n \alpha, X, 1^n}(0, 1) = \frac{1}{n+1}$$

and $[\![E]\!](t) = -\infty$ for all $t \notin \{\sigma^n x \mid n \geq 1\}$. Using similar arguments, we can show that the semantics of the expression

$$\Big(\sum_{\sigma \in \Sigma^{(k)}, k \geq 0} 0.\sigma(x, \dots, x) \Big)^{*, x} \cdot_x \Big(\sum_{\alpha \in \Sigma^{(0)}} 1.\alpha \Big)$$

where $\sum$ denotes the "syntactic sum" of rational expressions, calculates the leaves-to-size ratio of every tree in T_Σ.

4   Weighted Tree Automata over Cutv-Monoids

In this section we introduce weighted tree automata over cutv-monoids. Basically, we follow the definition of a weighted tree automaton over a tv-monoid in [5]. However, our input trees, like input trees in [11], may contain variables.

Definition 7. *A weighted tree automaton (wta for short) (over Σ, X, and $\mathbb{D}$) is a system $\mathcal{M} = (Q, \Sigma, X, \mu, \nu, F)$ where Q is a non-empty, finite set of states, the sets Σ, Q, and X are pairwise disjoint, μ is a family $(\mu_k \mid 0 \leq k \leq \max_\Sigma)$ of transition mappings $\mu_k \colon Q^k \times \Sigma^{(k)} \times Q \to (D \setminus \{\mathbb{1}\})$, $\nu \colon X \times Q \to \{0, \mathbb{1}\}$, and $F \subseteq Q$ is a set of final states.*

In order to be able to use the unital property of $\mathbb{D}$ for the weight of a run (defined below), we do not allow $\mathbb{1}$ to be the weight of a transition. Unfortunately, this yields some limitations which should be explored later.

In the rest of this section, let $\mathcal{M} = (Q, \Sigma, X, \mu, \nu, F)$ be a wta over $\mathbb{D}$. We define the behavior of $\mathcal{M}$ by a run semantics. First we give a general definition of a run which is similar to the one in [8, Definition 4.2]. Let $P \subseteq Q$ and $t \in T_\Sigma(X \cup Q)$. A *run r of $\mathcal{M}$ on t using P* is a mapping $r \colon \mathrm{dom}(t) \to Q$ such that $r(u) \in P$ for all $u \in (\mathrm{dom}(t) \setminus (\mathrm{dom}_Q(t) \cup \{\varepsilon\}))$, and $r(u) = t(u)$ for all $u \in \mathrm{dom}_Q(t)$. Such a run r *reaches* $q \in Q$ if $r(\varepsilon) = q$. The set of all runs on t using P reaching q is denoted by $R_\mathcal{M}^{P,q}(t)$. Moreover, we put $R_\mathcal{M}^q(t) = R_\mathcal{M}^{Q,q}(t)$ and $R_\mathcal{M}(t) = \bigcup_{q \in Q} R_\mathcal{M}^q(t)$. If $\mathcal{M}$ is clear from the context, then we drop the index and write $R^{P,q}(t)$, $R^q(t)$, and $R(t)$ for the sets of the corresponding runs, respectively. Let $r \in R(t)$. We define the *weight mapping* $\mathrm{wt}(t, r) \colon \mathrm{dom}(t) \to D$ by

$$\mathrm{wt}(t, r)(u) = \begin{cases} \mu_k(r(u.1) \dots r(u.k), t(u), r(u)) & \text{if } t(u) \in \Sigma^{(k)}, k \geq 0, \\ \nu(t(u), r(u)) & \text{if } t(u) \in X, \\ \mathbb{1} & \text{if } t(u) \in Q \end{cases}$$

for every $u \in \mathrm{dom}(t)$. We call $\mathrm{wt}(t, r)(u)$ the *weight of r on t at u*. Note that $\mathrm{wt}(t, r)$ is an unranked tree in U_D. The run r is called *valid* if $0 \notin \mathrm{im}(\mathrm{wt}(t, r))$. We call $V(\mathrm{wt}(t, r))$ the *weight of r on t*. The *behavior* of $\mathcal{M}$ is the tree series $\|\mathcal{M}\| \colon T_\Sigma(X) \to D$ defined by

$$\|\mathcal{M}\|(t) = \sum_{r \in R^q(t), q \in F} V(\mathrm{wt}(t, r))$$

for every $t \in T_\Sigma(X)$. Note that if $F = \emptyset$, then $\|\mathcal{M}\| = \tilde{0}$.

For examples of wta see [5, Example 3.2]. The automata given there can be easily adapted to our settings. Note that the first automaton of [5, Example 3.2] computes the leaves-to-size ratio which we described by the rational tree series expression of Example 6.

A tree series S is called *recognizable* if $S = \|\mathcal{M}\|$ for some wta $\mathcal{M}$. Then we say that $\mathcal{M}$ *recognizes* S. We denote by $\mathbb{D}^{\mathrm{rec}}\langle\!\langle T_\Sigma(X)\rangle\!\rangle$ the class all recognizable tree series (over Σ, X, and $\mathbb{D}$). Moreover, we abbreviate the class $\bigcup_{Q \text{ finite set}} \mathbb{D}^{\mathrm{rec}}\langle\!\langle T_\Sigma(X \cup Q)\rangle\!\rangle$ by $\mathbb{D}^{\mathrm{rec}}\langle\!\langle T_\Sigma(X \cup Q_\infty)\rangle\!\rangle$.

5 From Automata to Rational Expressions and Vice Versa

In this section we show that wta and rational expressions are equally powerful. First we prove that the behavior of a wta over a cutv-monoid is rational. Throughout the section $\mathcal{M} = (Q, \Sigma, X, \mu, \nu, F)$ denotes an arbitrary wta over $\mathbb{D}$. For $Q' \subseteq Q$, we define the tree series $\|\mathcal{M}\|^{Q',P,q}$ which describes the behavior of $\mathcal{M}$ on trees in $T_\Sigma(X \cup Q')$ provided that we restrict to runs which use P and reach q. More exactly, for all $P, Q' \subseteq Q$ and $q \in Q$ we define

$$\|\mathcal{M}\|^{Q',P,q}(t) = \begin{cases} \sum_{r \in R^{P,q}(t)} V(wt(t,r)) & \text{if } t \in (T_\Sigma(X \cup Q') \setminus Q'), \\ \mathbb{0} & \text{if } t \in Q' \end{cases}$$

for all $t \in T_\Sigma(X \cup Q')$. Obviously, $\|\mathcal{M}\|^{Q',P,q}$ is p-proper for all $p \in Q'$. Analogously to [8, Lemma 5.1] we can prove the following key lemma.

Lemma 8. *Let $P, Q' \subseteq Q$ and $q \in Q$. Moreover, let $p \in (Q' \setminus P)$. Then*

$$\|\mathcal{M}\|^{Q',P\cup\{p\},q} = \|\mathcal{M}\|^{Q',P,q} \cdot_p (\|\mathcal{M}\|^{Q',P,p})^{*,p}.$$

Let $\mathbb{D}^{\mathrm{rat}}\langle\!\langle T_\Sigma(X \cup Q_\infty)\rangle\!\rangle = \bigcup_{Q \text{ finite set}} \mathbb{D}^{\mathrm{rat}}\langle\!\langle T_\Sigma(X \cup Q)\rangle\!\rangle$.

Theorem 9. $\mathbb{D}^{\mathrm{rec}}\langle\!\langle T_\Sigma(X \cup Q_\infty)\rangle\!\rangle \subseteq \mathbb{D}^{\mathrm{rat}}\langle\!\langle T_\Sigma(X \cup Q_\infty)\rangle\!\rangle$.

Proof. Since $T_\Sigma(X \cup Q)$ can be seen as $T_{\Sigma \cup Q}(X)$, it suffices to show that $\mathbb{D}^{\mathrm{rec}}\langle\!\langle T_\Sigma(X)\rangle\!\rangle \subseteq \mathbb{D}^{\mathrm{rat}}\langle\!\langle T_\Sigma(X \cup Q_\infty)\rangle\!\rangle$. The proof is analogous to [8, Theorem 5.2].

Consider the wta $\mathcal{M}$ and assume that $Q = \{q_1, \ldots, q_m\}$. Let $P \subseteq Q$ and abbreviate $\|\mathcal{M}\|^{Q,P,q}$ by $\|\mathcal{M}\|^{P,q}$. Let us recall that $\|\mathcal{M}\|^{P,q}$ is p-proper for all $p \in Q$ and note that $\|\mathcal{M}\|^{Q,q}(t)$ may not be $\mathbb{0}$ for some $t \in (T_\Sigma(X \cup Q) \setminus T_\Sigma(X))$. However,

$$\|\mathcal{M}\| = \sum_{q \in F} (\cdots ((\|\mathcal{M}\|^{Q,q} \cdot_{q_1} \tilde{0}) \cdot_{q_2} \tilde{0}) \cdots) \cdot_{q_m} \tilde{0}.$$

It means that $\|\mathcal{M}\|$ is rational if $\|\mathcal{M}\|^{Q,q}$ is rational for all $q \in Q$. Therefore, we prove by induction on $|P|$ that $\|\mathcal{M}\|^{P,q}$ is rational for all $P \subseteq Q$ and $q \in Q$. Firstly, let $P = \emptyset$. For every tree $t \in T_\Sigma(X \cup Q)$, if $t \neq \sigma(p_1, \ldots, p_k)$ (for some $k \geq 0$, $\sigma \in \Sigma^{(k)}$, and $p_1, \ldots, p_k \in Q$) or $t \notin X$, then $R^{\emptyset,q}(t) = \emptyset$ and thus $\|\mathcal{M}\|^{\emptyset,q}(t) = 0$. If $t = \sigma(p_1,\ldots,p_k)$, then $R^{\emptyset,q}(t) = \{r^{\sigma,q}_{p_1,\ldots,p_k}\}$, where the run $r^{\sigma,q}_{p_1,\ldots,p_k}\colon \mathrm{dom}(\sigma(p_1,\ldots,p_k)) \to Q$ is defined by $r^{\sigma,q}_{p_1,\ldots,p_k}(\varepsilon) = q$ and $r^{\sigma,q}_{p_1,\ldots,p_k}(i) = p_i$ for all $1 \leq i \leq k$. Thus

$$\|\mathcal{M}\|^{\emptyset,q}(t) = \|\mathcal{M}\|^{\emptyset,q}(\sigma(p_1,\ldots,p_k)) = \mathrm{V}(\mathrm{wt}(\sigma(p_1,\ldots,p_k), r^{\sigma,q}_{p_1,\ldots,p_k}))$$

$$= \mathrm{V}(\mu_k(p_1 \ldots p_k, \sigma, q)(\mathbb{1},\ldots,\mathbb{1})) \stackrel{(*)}{=} \mu_k(p_1 \ldots p_k, \sigma, q)\,.$$

At $(*)$ we used that $\mathbb{D}$ is a unital tv-monoid. Moreover, for $t = x \in X$, $\|\mathcal{M}\|^{\emptyset,q}(x) = \mathrm{V}(\mathrm{wt}(x,q)) = \mathrm{V}(\nu(x,q)) = \nu(x,q)$. Thus, $\|\mathcal{M}\|^{\emptyset,q}$ is a finite sum of monomials of the form $\mu_k(p_1 \ldots p_k, \sigma, q).\sigma(p_1,\ldots,p_k)$ and $\nu(x,q).x$. We can show easily that both kinds of monomials are rational. Since rational tree series are closed under $+$, the tree series $\|\mathcal{M}\|^{\emptyset,q}$ is rational.

Now let $P \subseteq Q$ and $p \in (Q \setminus P)$. By the induction hypothesis, $\|\mathcal{M}\|^{P,q}$ is rational. Since rational tree series are closed under rational operations, by Lemma 8 (with $Q' = Q$), the tree series $\|\mathcal{M}\|^{P\cup\{p\},q} = \|\mathcal{M}\|^{Q,P\cup\{p\},q}$ is also rational. □

Next we will show that rational tree series are recognizable by wta. For this first we prove that recognizable tree series are closed under rational operations.

Theorem 10. *The class $\mathbb{D}^{rec}\langle\!\langle T_\Sigma(X)\rangle\!\rangle$ is closed under rational operations.*

Proof. To show the closedness under weighted top-concatenation we let $k \geq 0$, $\sigma \in \Sigma^{(k)}$, $d \in D$, $S_1,\ldots,S_k \in \mathbb{D}^{rec}\langle\!\langle T_\Sigma(X)\rangle\!\rangle$ such that $S_i = \|\mathcal{M}_i\|$ for the wta $\mathcal{M}_i = (Q_i, \Sigma, X, \mu^i, \nu_i, F_i)$ for $1 \leq i \leq k$. Assume that $Q_i \cap Q_j = \emptyset$ for all $j \neq i$. We will construct an automaton $\mathcal{M}^d_\sigma$ recognizing $\mathrm{top}^d_\sigma(S_1,\ldots,S_k)$ as follows. We build the disjoint union of the $\mathcal{M}_i$ and add a new state f which will be the only final state. Moreover, all transitions $(q_1 \ldots q_k, \sigma, f)$ where $q_i \in F_i$ have weight d. All other transitions containing f have weight zero. Formally, $\mathcal{M}^d_\sigma = (Q, \Sigma, X, \mu, \nu, F)$, where $Q = \{f\} \cup \bigcup_{1\leq i \leq k} Q_i$ with $F = \{f\}$,

$$\mu_n(q_1 \ldots q_n, \gamma, q) = \begin{cases} (\mu^i)_n(q_1 \ldots q_n, \gamma, q) & \text{if } q_1,\ldots,q_n, q \in Q_i, \\ d & \text{if } \gamma = \sigma, q = f, \text{ and } q_i \in F_i, \\ 0 & \text{otherwise,} \end{cases}$$

for all $n \geq 0$, $\gamma \in \Sigma^{(n)}$, $q_1,\ldots,q_n, q \in Q$, and for all $x \in X$ we set $\nu(x,q) = \nu_i(x,q)$ for $q \in Q_i$ and $\nu(x,f) = 0$.

The proof of closedness of $\mathbb{D}^{rec}\langle\!\langle T_\Sigma(X)\rangle\!\rangle$ under sum is analogous to the proof of [5, Theorem 5.12].

Let $x \in X$ and $S_i = \|\mathcal{M}_i\|$ with the wta $\mathcal{M}_i$ defined above for $i = 1, 2$. We will build an automaton $\mathcal{M}_x$ with $\|\mathcal{M}_x\| = S_1 \cdot_x S_2$ by "sticking together" $\mathcal{M}_1$ and $\mathcal{M}_2$. For this we will have glue states (p, p'), where p is a final state of $\mathcal{M}_2$ and p' is a state of $\mathcal{M}_1$ with $\nu(x,p') = \mathbb{1}$. These glue states will later be attached to the concatenation points of the input tree. Above the concatenation points $\mathcal{M}_x$ will behave like $\mathcal{M}_1$ and below like $\mathcal{M}_2$. Let $\mathcal{M}_x = (Q, \Sigma, X, \mu, \nu, F)$ be defined by $Q = Q_1 \cup Q_2 \cup F_2 \times Q_1$, $F = F_1 \cup F_2 \times F_1$;

$$\mu_k(q_1\ldots q_k,\sigma,q) = \begin{cases} (\mu^1)_k(q_1^2\ldots q_k^2,\sigma,q) & \text{if } q_1,\ldots,q_k \in (Q_1 \cup F_2 \times Q_1), \\ & q \in Q_1, \\ (\mu^2)_k(q_1\ldots q_k,\sigma,q^1) & \text{if } q_1,\ldots,q_k \in Q_2 \wedge (q \in Q_2 \vee \\ & q \in F_2 \times Q_1 \text{ with } \nu_1(x,q^2) = \mathbb{1}), \\ \mathbb{0} & \text{otherwise,} \end{cases}$$

for all $\sigma \in \Sigma^{(k)}$, $q_1,\ldots,q_k,q \in Q$ where $q^1 = q^2 = q$ if $q \in (Q_1 \cup Q_2)$, and $q^1 = p$ and $q^2 = p'$ if $q = (p,p') \in F_2 \times Q_1$; and for all $y \in X$:

$$\nu(y,q) = \begin{cases} \nu_1(y,q) & \text{if } y \neq x \text{ and } q \in Q_1, \\ \nu_2(x,q^1) & \text{if } q \in F_2 \times Q_1 \text{ and } \nu_1(x,q^2) = \mathbb{1}, \\ \mathbb{0} & \text{otherwise.} \end{cases}$$

By decomposing the valid runs of $\mathcal{M}_x$ into sub-runs of $\mathcal{M}_1$ and $\mathcal{M}_2$ we can show that $\|\mathcal{M}_x\| = S_1 \cdot_x S_2$.

To show that recognizable tree series are closed under x-Kleene star, basically we will follow up the ideas for $\mathcal{M}_x$. But now a new state f_x will mark the concatenation points. Let $\mathcal{M} = (Q,\Sigma,X,\mu,\nu,F)$ be a wta recognizing a tree series S. We define $\mathcal{M}_* = (Q_*,\Sigma,X,\mu_*,\nu_*,F_*)$ by $Q_* = Q \cup F \times (Q \cup \{f_x\}) \cup \{f_x\}$, $F_* = F \cup \{f_x\}$;

$$(\mu_*)_k(q_1\ldots q_k,\sigma,q) = \begin{cases} \mu_k(q_1^2\ldots q_k^2,\sigma,q^1) & \text{if } q_1,\ldots,q_k \in Q_*, q \in Q, \\ & \text{or } (q \in F \times Q \wedge \nu(x,q^2) = \mathbb{1}), \\ \mathbb{0} & \text{otherwise,} \end{cases}$$

for all $\sigma \in \Sigma^{(k)}$, $q_1,\ldots,q_k,q \in Q_*$ where $q^1 = q^2 = q$ if $q \in Q$ and $q^1 = p$ and $q^2 = p'$ if $q = (p,p') \in F \times (Q \cup \{f_x\})$; and for all $y \in X$:

$$\nu_*(y,q) = \begin{cases} \nu(y,q^1) & \text{if } y \neq x, q \in (Q \cup F \times Q), \text{ and } \nu(x,q^2) = \mathbb{1}, \\ \mathbb{1} & \text{if } y = x, q \in (\{f_x\} \cup F \times \{f_x\}), \\ \mathbb{0} & \text{otherwise.} \end{cases}$$

Similarly to the closedness under $\cdot_x$, we can show that $\|\mathcal{M}_*\| = S^{*,x}$. $\square$

Let $x \in X$. Then $[\![x]\!]$ is recognizable by $\mathcal{M} = (\{q\},\Sigma,X,\mu,\nu,\{q\})$ with $\nu(x,q) = \mathbb{1}$, $\mu_k(q\ldots q,\sigma,q) = \mathbb{0}$ for all $k \geq 0$ and $\sigma \in \Sigma^{(k)}$, and $\nu(y,q) = \mathbb{0}$ for all $y \neq x$. This fact and Theorem 10 justify the following theorem.

Theorem 11. $\mathbb{D}^{rat}\langle\!\langle T_\Sigma(X)\rangle\!\rangle \subseteq \mathbb{D}^{rec}\langle\!\langle T_\Sigma(X)\rangle\!\rangle$.

By Theorems 9 and 11 we can conclude the main result of our paper.

Theorem 12. $\mathbb{D}^{rat}\langle\!\langle T_\Sigma(Q_\infty)\rangle\!\rangle = \mathbb{D}^{rec}\langle\!\langle T_\Sigma(Q_\infty)\rangle\!\rangle$.

References

1. Alexandrakis, A., Bozapalidis, S.: Weighted grammars and Kleene's theorem. Inf. Process. Lett. **24**(1), 1–4 (1987)
2. Berstel, J., Reutenauer, C.: Recognizable formal power series on trees. Theor. Comput. Sci. **18**(2), 115–148 (1982)
3. Chatterjee, K., Doyen, L., Henzinger, T.A.: Quantitative languages. In: Kaminski, M., Martini, S. (eds.) CSL 2008. LNCS, vol. 5213, pp. 385–400. Springer, Heidelberg (2008)
4. Comon, H., Dauchet, M., Gilleron, R., Löding, C., Jacquemard, F., Lugiez, D., Tison, S., Tommasi, M.: Tree automata techniques and applications (2007). http://www.grappa.univ-lille3.fr/tata
5. Droste, M., Götze, D., Märcker, S., Meinecke, I.: Weighted tree automata over valuation monoids and their characterization by weighted logics. In: Kuich, W., Rahonis, G. (eds.) Algebraic Foundations in Computer Science. LNCS, vol. 7020, pp. 30–55. Springer, Heidelberg (2011)
6. Droste, M., Meinecke, I.: Describing average- and longtime-behavior by weighted MSO logics. In: Hliněný, P., Kučera, A. (eds.) MFCS 2010. LNCS, vol. 6281, pp. 537–548. Springer, Heidelberg (2010)
7. Droste, M., Meinecke, I.: Weighted automata and regular expressions over valuation monoids. Int. J. Found. Comput. Sci. **22**, 1829–1844 (2011)
8. Droste, M., Pech, C., Vogler, H.: A Kleene theorem for weighted tree automata. Theory Comput. Syst. **38**(1), 1–38 (2005)
9. Droste, M., Vogler, H.: The Chomsky-Schützenberger theorem for quantitative context-free languages. Int. J. Found. Comput. Sci. Special Issue of DLT **2013**(25), 955–969 (2014)
10. Ésik, Z., Kuich, W.: Formal tree series. J. Automata Lang. Comb. **8**(2), 219–285 (2003)
11. Fülöp, Z., Maletti, A., Vogler, H.: A Kleene theorem for weighted tree automata over distributive multioperator monoids. Theory Comput. Syst. **44**(3), 455–499 (2009)
12. Fülöp, Z., Vogler, H.: Weighted tree automata and tree transducers. In: Droste, M., Kuich, W., Vogler, H. (eds.) Handbook of Weighted Automata, pp. 313–403. Springer, Heidelberg (2009)
13. Gécseg, F., Steinby, M.: Tree Automata. Akadémiai Kiadó (1984). http://www.arxiv.org
14. Kleene, S.: Representation of events in nerve nets and finite automata. In: Shannon, C.E., McCarthy, J. (eds.) Automata Studies, pp. 3–42. Princeton University Press, Princeton (1956)
15. Kuich, W.: Linear systems of equations and automata on distributive multioperator monoids. In: Contributions to General Algebra 12 - Proceedings of the 58th Workshop on General Algebra "58, Arbeitstagung Allgemeine Algebra", Vienna University of Technology, June 3–6, 1999, vol. 12, pp. 1–10. Verlag Johannes Heyn (1999)
16. Maletti, A.: Relating tree series transducers and weighted tree automata. Int. J. Found. Comput. Sci. **16**, 723–741 (2005)
17. Schützenberger, M.P.: On the definition of a family of automata. Inf. Control **4**, 245–270 (1961)
18. Thatcher, J.W., Wright, J.B.: Generalized finite automata theory with an application to a decision problem of second-order logic. Theory Comput. Syst. **2**, 57–81 (1968)

Hankel Matrices for Weighted Visibly Pushdown Automata

Nadia Labai[1]($\boxtimes$) and Johann A. Makowsky[2]

[1] Department of Informatics, Vienna University of Technology,
Vienna, Austria
labai@forsyte.at
[2] Department of Computer Science,
Technion - Israel Institute of Technology, Haifa, Israel
janos@cs.technion.ac.il

Abstract. Hankel matrices (aka connection matrices) of word functions and graph parameters have wide applications in automata theory, graph theory, and machine learning. We give a characterization of real-valued functions on nested words recognized by weighted visibly pushdown automata in terms of Hankel matrices on nested words. This complements C. Mathissen's characterization in terms of weighted monadic second order logic.

1 Introduction and Background

1.1 Weighted Automata for Words and Nested Words

Classical word automata can be extended to *weighted* word automata by assigning weights from some numeric domain to their transitions, thereby having them assign values to their input words rather than accepting or rejecting them. Weighted (word) automata define the class of recognizable word functions, first introduced in the study of stochastic automata by A. Heller [39]. Weighted automata are used in verification, [6,52], in program synthesis, [13,14], in digital image compression, [19], and speech processing, [1,28,53]. For a comprehensive survey, see the Handbook of Weighted Automata [27]. Recognizable word functions over commutative semirings S were characterized using logic through the formalism of Weighted Monadic Second Order Logic (WMSOL), [26], and the formalism of MSOLEVAL[1], [44].

Nested words and nested word automata are generalizations of words and finite automata, introduced by Alur and Madhusudan [2]. A *nested word* $nw \in NW(\Sigma)$ over an alphabet Σ is a sequence of linearly ordered positions, augmented with forward-oriented edges that do not cross, creating a nested structure. In the context of formal verification for software, execution paths in procedural programs are naturally modeled by nested words whose hierarchical

Nadia Labai—Supported by the National Research Network RiSE (S114), and the LogiCS doctoral program (W1255) funded by the Austrian Science Fund (FWF).

Johann A. Makowsky—Partially supported by a grant of Technion Research Authority.

[1] This formalism was originally introduced in [18] for graph parameters.

© Springer International Publishing Switzerland 2016
A.-H. Dediu et al. (Eds.): LATA 2016, LNCS 9618, pp. 464–477, 2016.
DOI: 10.1007/978-3-319-30000-9_36

structure captures calls and returns. Nested words also model annotated linguistic data and tree-structured data which is given by a linear encoding, such as HTML/XML documents. Nested word automata define the class of regular languages of nested words. The key feature of these automata is their ability to propagate hierarchical states along the augmenting edges, in addition to the states propagated along the edges of the linear order. We refer the reader to [2] for details. Nested words $nw \in NW(\Sigma)$ can be (linearly) encoded as words over an extended tagged alphabet $\hat{\Sigma}$, where the letters in $\hat{\Sigma}$ specify whether the position is a call, a return, or neither (internal). Such encodings of regular languages of nested words give the class of *visibly pushdown languages* over the tagged alphabet $\hat{\Sigma}$, which lies between the parenthesis languages and deterministic context-free languages. The accepting pushdown automata for visibly pushdown languages push one symbol when reading a call, pop one symbol when reading a return, and only update their control when reading an internal symbol. Such automata are called *visibly pushdown automata*. Since their introduction, nested words and their automata have found applications in specifications for program analysis [24, 25, 37], XML processing [31, 54], and have motivated several theoretical questions, [3, 20, 55].

Visibly pushdown automata and nested word automata were extended by assigning weights from a commutative semiring $\mathcal{S}$ to their transitions as well. Keifer et al. introduced *weighted visibly pushdown automata*, and their equivalence problem was showed to be logspace reducible to polynomial identity testing, [41]. Mathissen introduced *weighted nested word automata*, and proved a logical characterization of their functions using a modification of WMSOL, [51].

1.2 Hankel Matrices and Weighted Word Automata

Given a word function $f : \Sigma^\star \to \mathcal{F}$, its *Hankel matrix* $\mathsf{H}_f \in \mathcal{F}^{\Sigma^\star \times \Sigma^\star}$ is the infinite matrix whose rows and columns are indexed by words in $\Sigma^\star$ and $\mathsf{H}_f(u, v) = f(uv)$, where uv is the concatenation of u and v. In addition to the logical characterizations, there exists a characterization of recognizable word functions via Hankel matrices, by Carlyle and Paz [12].

Theorem 1 (Carlyle and Paz, 1971). *A real-valued word function f is recognized by a weighted (word) automaton iff H_f has finite rank.*

The theorem was originally stated using the notion of external function rank, but the above formulation is equivalent. Multiplicative words functions were characterized by Cobham [15] as exactly those with a Hankel matrix of rank 1.

Hankel matrices proved useful also in the study of graph parameters. Lovász introduced a kind of Hankel matrices for graph parameters [48] which were used to study real-valued graph parameters and their relation to partition functions, [30, 49]. In [33], the definability of graph parameters in monadic second order logic was related to the rank of their Hankel matrices. Meta-theorems involving logic, such as Courcelle's theorem and generalizations there of [16, 17, 23, 50], were made logic-free by replacing their definability conditions with conditions on Hankel matrices, [43, 45, 46].

1.3   Our Contribution

The goal of this paper is to prove a characterization of the functions recognizable by weighted visibly pushdown automata (WVPA), called here *recognizable nested word functions*, via Hankel matrices. Such a characterization would nicely fill the role of the Carlyle-Paz theorem in the words setting, complementing results that draw parallels between recognizable word functions and nested word functions, such as the attractive properties of closure and decidability the settings share [2], and the similarity between the WMSOL-type formalisms used to give their logical characterizations.

The first challenge is in the choice of the Hankel matrices at hand. A naive straightforward adaptation of the Carlyle-Paz theorem to the setting of nested words would involve Hankel matrices for words over the extended alphabet $\hat{\Sigma}$ with the usual concatenation operation on words. However, then we would have functions recognizable by WVPA with Hankel matrices of infinite rank. Consider the Hankel matrix of the characteristic function of the language of balanced brackets, also known as the Dyck language. This language is not regular, so its characteristic function is not recognized by a weighted word automaton. Hence, by the Carlyle-Paz Theorem 1, its Hankel matrix would have infinite rank despite the fact its encoding over a tagged alphabet is recognizable by VPA, hence also by WVPA.

Main Results. We introduce *nested Hankel matrices* over *well-nested words* (see Sect. 2) to overcome the point described above and prove the following characterization of WVPA-recognizable functions of well-nested words:

Theorem 2 (Main Theorem). *Let $\mathcal{F} = \mathbb{R}$ or $\mathcal{F} = \mathbb{C}$, and let f be an $\mathcal{F}$-valued function on well-nested words. Then f is recognized by a weighted visibly pushdown automaton with n states iff the nested Hankel matrix nH_f has rank $\leq n^2$.*

As opposed to the characterizations of word functions, which allow f to have values over a semiring, we require that f is over $\mathbb{R}$ or $\mathbb{C}$. This is due to the second challenge, which stems from the fact that in our setting of functions of well-nested words, the helpful decomposition properties exhibited by Hankel matrices for word functions are absent. This is because, as opposed to words, well-nested words cannot be split in arbitrary positions and result in two well-nested words. Thus, we use the singular value decomposition (SVD) Theorem, see, e.g., [35], which is valid only over $\mathbb{R}$ and $\mathbb{C}$.

Outline. In Sect. 2 we complete the background on well-nested words and weighted visibly pushdown automata, and introduce nested Hankel matrices. The rather technical proof of Theorem 2 is given in Sect. 5. In Sect. 3 we discuss the applications of Theorem 2 to learning theory. In Sect. 4 we briefly discuss limitations of our methods and possible extensions of our characterization.

2   Preliminaries

For the remainder of the paper, we assume that $\mathcal{F}$ is $\mathbb{R}$ or $\mathbb{C}$. Let Σ be a finite alphabet. For $\ell \in \mathbb{N}$, we denote the set $\{1, \ldots, \ell\}$ by $[\ell]$. For a matrix or vector N, denote its transpose by N^T. Vectors are assumed to be column vectors unless stated otherwise.

2.1   Well-Nested Words

We follow the definitions in [2,51]. A *well-nested word* over Σ is a pair (w, ν) where $w \in \Sigma^\star$ of length ℓ and ν is a matching relation for w. A *matching relation*[2] for a word of length ℓ is a set of edges $\nu \subset [\ell] \times [\ell]$ such that the following holds:

1. If $(i, j) \in \nu$, then $i < j$.
2. Any position appears in an edge of ν at most once: For $1 \le i \le \ell$,
   $|\{j \mid (i, j) \in \nu\}| \le 1$ and $|\{j \mid (j, i) \in \nu\}| \le 1$
3. If $(i, j), (i', j') \in \nu$, then it is not the case that $i < i' \le j < j'$. That is, the edges do not cross.

Denote the set of well-nested words over Σ by $\mathrm{WNW}(\Sigma)$.

   Given positions i, j such that $(i, j) \in \nu$, position i is a *call* position and position j is a *return* position. Denote $\Sigma_{call} = \{\langle s \mid s \in \Sigma\}$, $\Sigma_{ret} = \{s\rangle \mid s \in \Sigma\}$, and $\hat{\Sigma} = \Sigma_{call} \cup \Sigma_{ret} \cup \Sigma_{int}$ where $\Sigma_{int} = \Sigma$ and is disjoint from Σ_{call} and Σ_{ret}. By viewing calls as opening parentheses and returns as closing parentheses, one can define an encoding taking nested words over Σ to words over $\hat{\Sigma}$ by assigning to a position labeled $s \in \Sigma$:

- the letter $\langle s$, if it is a call position,
- the letter $s\rangle$, if it is a return position,
- the same letter s, if it is an internal position.

We denote this encoding by $nw\_w : WNW(\Sigma) \to \hat{\Sigma}^\star$ and give an example in Fig. 1. Note that any parentheses appearing in such encoding will be well-matched (balanced) parentheses. Denote its partial inverse function, defined only for words with well-matched parentheses, by $w\_nw : \hat{\Sigma}^\star \to WNW(\Sigma)$. See [2] for details. We will freely pass between the two forms.

   Given a function $f : WNW(\Sigma) \to \mathcal{F}$ on well-nested words, one can naturally define a corresponding function $f' : \hat{\Sigma}^\star \to \mathcal{F}$ on words with well-matched parentheses by setting $f'(w) = f(w\_nw(w))$. We will denote both functions by f.

2.2   Nested Hankel Matrices

Given a function on well-nested words $f : WNW(\Sigma) \to \mathcal{F}$, define its *nested Hankel matrix* nH_f as the infinite matrix whose rows and columns are indexed by *words* over $\hat{\Sigma}$ with *well-matched parentheses*, and $\mathrm{nH}_f(u, v) = f(uv)$. That

---

[2] The original definition of nested words allowed "dangling" edges. We will only be concerned with nested words that are well-matched.

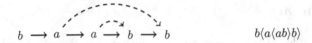

Fig. 1. On the left, a well-nested word, where the successor relation of the linear order is in bold edges, the matching relation is in dashed edges. On the right, its encoding as a word over a tagged alphabet.

is, the entry at the row labeled with u and the column labeled with v is the value $f(uv)$. A nested Hankel matrix nH_f has finite rank if there is a finite set of rows in nH_f that linearly span it. We stress the fact that nH_f is defined over words whose parentheses are well-matched, as this is crucial for the proof of Theorem 2.

As an example, consider the function f which counts the number of pairs of parentheses in a well-nested word over the alphabet $\Sigma = \{a\}$. Then the corresponding word function is on words over the tagged alphabet $\hat{\Sigma} = \{a, \langle a, a\rangle\}$. In Fig. 2 we see (part of) the corresponding nested Hankel matrix nH_f with labels on its columns and rows.

2.3   Weighted Visibly Pushdown Automata

For notational convenience, now let $\Sigma = \Sigma_{call} \cup \Sigma_{ret} \cup \Sigma_{int}$. We follow the definition given in [41]. An $\mathcal{F}$-*weighted visibly pushdown automaton* (WVPA) on Σ is a tuple $A = (n, \boldsymbol{\alpha}, \boldsymbol{\eta}, \Gamma, M)$ where

- n is the number of states,
- $\boldsymbol{\alpha}, \boldsymbol{\eta} \in \mathcal{F}^n$ are initial and final vectors, respectively,
- Γ is a finite stack alphabet, and
- M are matrices in $\mathcal{F}^{n \times n}$ defined as follows.

	ε	a	$\langle aa\rangle$	aa	$\langle aaa\rangle$	$\langle a\langle aa\rangle a\rangle$	$\cdots$
ε	0	0	1	0	1	2	$\cdots$
a	0	0	1	0	1	2	$\cdots$
$\langle aa\rangle$	1	1	2	1	2	3	$\cdots$
aa	0	0	1	0	1	2	$\cdots$
$\langle aaa\rangle$	1	1	2	1	2	3	$\cdots$
$\langle a\langle aa\rangle a\rangle$	2	2	3	2	3	4	$\cdots$
$\vdots$	$\vdots$	$\vdots$	$\vdots$	$\vdots$	$\vdots$	$\vdots$	$\vdots$

Fig. 2. The nested Hankel matrix nH_f. Note that nH_f has rank 2.

For every $\gamma \in \Gamma$ and every $c \in \Sigma_{call}$, the matrix $M_{call}^{(c,\gamma)} \in \mathcal{F}^{n \times n}$ is given by

$$M_{call}^{(c,\gamma)}(i,j) = \text{the weight of a } c\text{-labeled transition from}$$
$$\text{state } i \text{ to state } j \text{ that pushes } \gamma \text{ on the stack.}$$

The matrices $M_{ret}^{(r,\gamma)} \in \mathcal{F}^{n \times n}$ are given similarly for every $r \in \Sigma_{ret}$, and the matrices $M_{int}^{(s)} \in \mathcal{F}^{n \times n}$ are given similarly for every $s \in \Sigma_{int}$.

For each well-nested word $u \in WNW(\Sigma)$, the automaton A inductively computes $M_u^{(A)} \in \mathcal{F}^{n \times n}$ for u in the following way.

- Base cases:

$$M_\varepsilon^{(A)} = I, \quad \text{and} \quad M_s^{(A)} = M_{int}^{(s)} \quad \text{for } s \in \Sigma_{int}.$$

- Closure:

$$M_{uv}^{(A)} = M_u^{(A)} \cdot M_v^{(A)} \quad \text{for } u, v \in WNW(\Sigma),$$
$$M_{cur}^{(A)} = \sum_{\gamma \in \Gamma} M_{call}^{(c,\gamma)} \cdot M_u^{(A)} \cdot M_{ret}^{(r,\gamma)} \quad \text{for } c \in \Sigma_{call} \text{ and } r \in \Sigma_{ret}.$$

The *behavior* of A is the function $f_A : WNW(\Sigma) \to \mathcal{F}$ where

$$f_A(u) = \boldsymbol{\alpha}^T \cdot M_u^{(A)} \cdot \boldsymbol{\eta}$$

A function $f : WNW(\Sigma) \to \mathcal{F}$ is *recognizable by WVPA* if it is the behavior of some WVPA A.

3   Applications in Computational Learning Theory

A passive learning algorithm for classical automata is an algorithm which given a set of strings accepted by the target automaton (positive examples) and a set of strings rejected by the target automaton (negative examples), and is required to output an automaton which is consistent with the set of examples. It is well known that in a variety of passive learning models, such as Valiant's PAC model, [58], and the mistake bound models of Littlestone and Haussler et al., [38,47], it is intractable to learn or even approximate classical automata, [4,34,56]. However, the problem becomes tractable when the learner is allowed to make membership and equivalence queries, as in the active model of learning introduced by Angluin, [4,5]. This approach was extended to weighted automata over fields, [10].

The problem of learning *weighted* automata is of finding a weighted automaton which closely estimates some target function, by considering examples consisting of pairs of strings with their value. The development of efficient learning techniques for weighted automata was immensely motivated by the abundance of their applications, with many of the techniques exploiting the relationship between weighted automata and their Hankel matrices, [9,11,36].

3.1 Learning Weighted Visibly Pushdown Automata

The proof of our Theorem 2 suggests a template of learning algorithms for weighted visibly pushdown automata, with the difficult part being the construction of the matrices that correspond to call and return symbols. The proof of Lemma 5 spells out the construction of these matrices, given an algorithm for finding SVD expansions (see Subsect. 5.2). To the best of our knowledge, learning algorithms for weighted visibly pushdown automata have not been proposed so far.

In recent years, the spectral method of Hsu et al. [40] for learning hidden Markov models, which relies on the SVD of a Hankel matrix, has driven much follow-up research, see the survey [8]. Balle and Mohri combined spectral methods with constrained matrix completion algorithms to learn arbitrary weighted automata, [7]. We believe the possibility of developing spectral learning algorithms for WVPA is worth exploring in more detail.

Lastly, we should note that one could employ existing algorithms to produce a weighted automaton from a nested Hankel matrix, if it is viewed as a partial Hankel matrix for a word function. However, any automaton which is consistent with the matrix will have as many states as the rank of the nested Hankel matrix, [12,29]. This may be less than satisfying when considering how, in contrast, Theorem 2 assures the existence of a weighted visibly pushdown automaton with n states, given a nested Hankel matrix of rank $\leq n^2$. This discrepancy fundamentally depends on the SVD Theorem.

4   Extension to Semirings

The proof of Theorem 2 relies on the SVD Theorem, which, in particular, assumes the existence of an inverse with respect to addition. Furthermore, notions of orthogonality, rank, and norms do not readily transfer to the semiring setting. Thus it is not clear what an analogue to the SVD theorem would be in the context of semirings, nor whether it could exist. Therefore the proof of Theorem 2 cannot be used to characterize nested word functions recognized by WVPA over semirings.

However, in the special case of the tropical semirings, De Schutter and De Moor proposed an extended max algebra corresponding to $\mathbb{R}$, called the *symmetrized max algebra*, and proved an analogue SVD theorem for it, [21]. See also [22] for an extended presentation. These results suggest a similar Hankel matrix based characterization for WVPA-recognizable nested word functions may be possible over the tropical semirings. This would be beneficial in situations where we have a function that has a nested Hankel matrix of infinite rank when interpreted over $\mathbb{R}$, but has finite rank when it is interpreted over a tropical semiring. It is easy to verify that any function on well-nested words which is maximizing or minimizing with respect to concatenation would fall in this category.

5   The Characterization of WVPA-Recognizability

In this section we prove both directions of Theorem 2.

Let $p, q \in [n]$. Define the matrix $A^{(p,q)} \in \mathcal{F}^{n \times n}$ as having the value 1 in the entry (p, q) and zero in all other entries. That is,

$$A^{(p,q)}(i, j) = \begin{cases} 1, & \text{if } (i, j) = (p, q) \\ 0, & \text{otherwise} \end{cases}$$

Obviously, for any matrix $M \in \mathcal{F}^{n \times n}$ with entries $M(i, j) = m_{ij}$ we have $M = \sum_{i,j \in [n]} m_{ij} A^{(i,j)}$.

5.1   Recognizability Implies Finite Rank of Nested Hankel Matrix

Theorem 3. *Let f be recognized by a weighted visibly pushdown automaton A with n states. Then the nested Hankel matrix nH_f has rank $\leq n^2$.*

Proof. We define infinite row vectors $\mathbf{v}^{(i,j)}$ where $i, j \in [n]$, whose entries are indexed by well-nested words $w \in WNW(\Sigma)$, and show they span the rows of nH_f. We define the entries of $\mathbf{v}^{(i,j)}$ to be

$$\mathbf{v}^{(i,j)}(w) = \boldsymbol{\alpha}^T \cdot A^{(i,j)} M_w^{(A)} \cdot \boldsymbol{\eta}$$

Note that there are n^2 such vectors. Now let $u \in WNW(\Sigma)$ and let $M^{(A)}(u)$ be the matrix computed for u by A. By the definition of the behavior for A, the row r_u corresponding to u in nH_f has entry $r_u(w) = \boldsymbol{\alpha}^T \cdot M^{(A)}(u) \cdot M^{(A)}(w) \cdot \boldsymbol{\eta}$. Consider the linear combination

$$\mathbf{v}_u = \sum_{1 \leq i,j \leq n} M_u^{(A)}(i, j) \cdot \mathbf{v}^{(i,j)}.$$

Then

$$\mathbf{v}_u(w) = \sum_{1 \leq i,j \leq n} M_u^{(A)}(i, j) \cdot \mathbf{v}^{(i,j)}(w) = \sum_{1 \leq i,j \leq n} M_u^{(A)}(i, j) \cdot \left(\boldsymbol{\alpha}^T \cdot A^{(i,j)} M_w^{(A)} \cdot \boldsymbol{\eta} \right)$$

$$= \boldsymbol{\alpha}^T \cdot M_u^{(A)} \cdot M_w^{(A)} \cdot \boldsymbol{\eta} = r_u(w)$$

Therefore the rank of nH_f is at most n^2.   □

5.2   Finite Rank of Nested Hankel Matrix Implies Recognizability

Theorem 4 (The SVD Theorem, See [35]). *Let $N \in \mathcal{F}^{m \times n}$ be a non-zero matrix, where $\mathcal{F} = \mathbb{R}$ or $\mathcal{F} = \mathbb{C}$. Then there exist orthogonal matrices*

$$X = [\mathbf{x}_1 \dots \mathbf{x}_m] \in \mathcal{F}^{m \times m}, \quad Y = [\mathbf{y}_1 \dots \mathbf{y}_n] \in \mathcal{F}^{n \times n}$$

such that

$$Y^T N X = diag(\sigma_1, \dots, \sigma_p) \in \mathcal{F}^{m \times n}$$

where $p = \min\{m, n\}$, $diag(\sigma_1, \dots, \sigma_p)$ is a diagonal matrix with the values $\sigma_1, \dots, \sigma_p$, and $\sigma_1 \geq \sigma_2 \geq \dots \geq \sigma_p$.

As a consequence, if we define r by $\sigma_1 \geq \ldots \geq \sigma_r > \sigma_{r+1} = \ldots = 0$, then we have the *SVD expansion* of N:

$$N = \sum_{i=1}^{r} \sigma_i \mathbf{x}_i \mathbf{y}_i^T$$

In particular, if N is of rank $r = 1$, then $N = \mathbf{x}\mathbf{y}^T$.

The SVD is perhaps the most important factorization for real and complex matrices. It is used in matrix approximation techniques, signal processing, computational statistics, and many more areas. See [32, 42, 57] and references therein.

Lemma 5. *Let* $f : WNW(\Sigma) \to \mathcal{F}$ *and let its nested Hankel matrix* nH_f *have finite rank* $r(\mathrm{nH}_f) \leq n^2$ *with spanning rows* $\mathcal{B} = \{w_{1,1}, \ldots, w_{n,n}\}$. *There are*

- *matrices* $M_{w_{i,j}} \in \mathcal{F}^{n \times n}$ *for* $w_{i,j} \in \mathcal{B}$,
- *vectors* $\boldsymbol{\alpha}, \boldsymbol{\eta} \in \mathcal{F}^n$,
- *matrices* M_a *for* $a \in \Sigma_{int}$, *and*
- *matrices* $M_{call}^{(c,\gamma)}$ *and* $M_{ret}^{(r,\gamma)}$ *for* $\gamma \in \Gamma$, *and* $c \in \Sigma_{call}$, $r \in \Sigma_{ret}$,

such that the following equations hold:

$$f(w_{i,j}) = \boldsymbol{\alpha}^T \cdot M_{w_{i,j}} \cdot \boldsymbol{\eta} \tag{1}$$

$$f(a) = \boldsymbol{\alpha}^T \cdot M_a \cdot \boldsymbol{\eta} \tag{2}$$

$$f(cw_{i,j}r) = \boldsymbol{\alpha}^T \left(\sum_{\gamma \in \Gamma} M_{call}^{(c,\gamma)} \cdot M_{w_{i,j}} \cdot M_{ret}^{(r,\gamma)} \right) \boldsymbol{\eta} \tag{3}$$

Proof. Consider the matrix $N \in \mathcal{F}^{n \times n}$ defined as $N(i,j) = f(w_{1,j})$. By Theorem 4, since N has rank 1, there exist vectors $\mathbf{x}, \mathbf{y} \in \mathcal{F}^n$ such that $N = \mathbf{x}\mathbf{y}^T$. Set $\boldsymbol{\eta} = \mathbf{y}$ and $\boldsymbol{\alpha} = \mathbf{x}$, and $M_{w_{i,j}} = \beta_{i,j} \cdot A^{(i,j)}$, where $\beta_{i,j} = f(w_{i,j})f(w_{1,j})^{-1}$. Note that for $w_{1,j}$, we have $\beta_{1,j} = 1$ and $M_{w_{1,j}} = A^{(1,j)}$.

We need to show that $\boldsymbol{\alpha}^T \cdot M_{w_{i,j}} \cdot \boldsymbol{\eta} = f(w_{i,j})$ for $w_{i,j} \in \mathcal{B}$. Since the entries of $M_{w_{i,j}}$ are zero except for entry (i,j), we have

$$\boldsymbol{\alpha}^T \cdot M_{w_{i,j}} \cdot \boldsymbol{\eta} = \boldsymbol{\alpha}(i) \cdot \beta_{i,j} \cdot \boldsymbol{\eta}(j) \tag{4}$$

Since $\boldsymbol{\alpha}(i)\boldsymbol{\eta}(j) = f(w_{1,j})$, we have

$$\boldsymbol{\alpha}^T \cdot M_{w_{i,j}} \cdot \boldsymbol{\eta} = \beta_{i,j} \cdot f(w_{1,j}) = f(w_{i,j}) \cdot f(w_{1,j})^{-1} \cdot f(w_{1,j}) = f(w_{i,j})$$

and Eq. 1 holds.

Let r_a denote the row in nH_f corresponding to some letter $a \in \Sigma_{int}$. $\mathcal{B}$ spans the matrix, so there is a linear combination

$$r_a = \sum_{1 \leq i,j \leq n} z(a)_{i,j} \cdot r_{w_{i,j}}$$

and in particular, $f(a) = r_a(\varepsilon) = \sum_{1 \leq i,j \leq n} z(a)_{i,j} f(w_{i,j})$.
Set

$$M_a = \sum_{1 \leq i,j \leq n} z(a)_{i,j} M_{w_{i,j}}$$

We need to show that $f(a) = \boldsymbol{\alpha}^T \cdot M_a \cdot \boldsymbol{\eta}$. We have

$$\boldsymbol{\alpha}^T \cdot M_a \cdot \boldsymbol{\eta} = \boldsymbol{\alpha}^T \cdot \left(\sum_{1 \leq i,j \leq n} z(a)_{i,j} M_{w_{i,j}} \right) \cdot \boldsymbol{\eta} = \sum_{1 \leq i,j \leq n} z(a)_{i,j} \left(\boldsymbol{\alpha}^T M_{w_{i,j}} \boldsymbol{\eta} \right)$$

$$= \sum_{1 \leq i,j \leq n} z(a)_{i,j} f(w_{i,j}) = f(a)$$

and Eq. 2 holds.

Lastly, we show there exist matrices $M_{call}^{c,\gamma}$ and $M_{ret}^{r,\gamma}$ for $c \in \Sigma_{call}$, $r \in \Sigma_{ret}$ and $\gamma \in \Gamma$ such that Eq. 3 holds. The summation in Eq. 3 can be replaced by multiplication of block matrices as follows. In the sequel, all the defined matrices are $n^2 \times n^2$ block matrices with n blocks of $n \times n$ matrices. We define the following notation, given any matrices $M_{call}^{c,i}$ and $M_{ret}^{r,j}$ for $i = 1, \ldots, n$.

- M_{call}^c is the matrix where the ith block is $M_{call}^{c,i}$.
- M_{ret}^r is the matrix where the jth block is $M_{ret}^{r,j}$.
- $\tilde{M}_{w_{i,j}}$ is the matrix where each block is $M_{w_{i,j}}$. That is, $M_{w_{i,j}}$ is repeated along the diagonal n times.
- $\tilde{\boldsymbol{\alpha}}$ denotes the column vector of length n^2 which is the vertical concatenation of $\boldsymbol{\alpha}$ for n times. $\tilde{\boldsymbol{\eta}}$ is defined similarly for $\boldsymbol{\eta}$.

There exist matrices $M_{call}^{c,\gamma}$ and $M_{ret}^{r,\gamma}$, where $\gamma \in \Gamma$, such that Eq. 3 holds if and only if:

$$f(cw_{i,j}r) = \tilde{\boldsymbol{\alpha}}^T \cdot M_{call}^c \cdot \tilde{M}_{w_{i,j}} \cdot M_{ret}^r \cdot \tilde{\boldsymbol{\eta}} \tag{5}$$

Consider matrices of the following form. For a matrix $M_{call}^{c,\gamma}$, the only row which is not zero is the row associated with γ. We denote this row vector by $\mathbf{q}^{c,\gamma}$. For a matrix $M_{ret}^{r,\gamma}$, the column associated with γ is the only column which is not zero. We denote this column vector by $\mathbf{q}^{r,\gamma}$.

Then there exist matrices of the above form such that Eq. 5 holds if and only if

$$f(cw_{i,j}r) = \sum_{k=1}^n \boldsymbol{\alpha}(i)(\mathbf{q}^{c,k}(i) \cdot \beta_{i,j} \cdot \mathbf{q}^{r,k}(j))\boldsymbol{\eta}(j)$$

$$= (\boldsymbol{\alpha}(i) \cdot \beta_{i,j} \cdot \boldsymbol{\eta}(j)) \sum_{k=1}^n \mathbf{q}^{c,k}(i) \cdot \mathbf{q}^{r,k}(j)$$

if and only if the $n \times n$ matrix $N(i,j) = f(cw_{i,j}r) \cdot (\boldsymbol{\alpha}(i)\beta_{i,j}\boldsymbol{\eta}(j))^{-1}$ has a decomposition $N = \sum_{k=1}^n \mathbf{q}^{c,k} \cdot \mathbf{q}^{r,k}$. Since N has rank $\leq n$, Theorem 4 implies this decomposition exists. Therefore there exist matrices $M_{call}^{(c,i)}$ and $M_{ret}^{(r,j)}$, of the form described above, such that Eq. 3 holds. $\qquad \square$

We are now ready to prove the second direction of Theorem 2.

Theorem 6. *Let $f : WNW(\Sigma) \to \mathcal{F}$ have a nested Hankel matrix nH_f of rank $\leq n^2$. Then f is recognizable by a weighted visibly pushdown automaton A with n states.*

Proof. Use Lemma 5 to build a WVPA A with n states, and set $M_\varepsilon^{(A)} = I$. It remains to show that $M_{ut}^{(A)} = M_u^{(A)} \cdot M_t^{(A)}$, for $u, t \in WNW(\Sigma)$.

Note that we defined the matrices $M_{w_{i,j}}^{(A)}$ such that $r_{w_{i,j}} = \mathbf{v}^{(i,j)}$ up to a constant factor. We show that if $r_u = \sum_{1 \leq i,j \leq n} M_u^{(A)}(i,j) \cdot \mathbf{v}^{(i,j)}$ and $r_t = \sum_{1 \leq i,j \leq n} M_t^{(A)}(i,j) \cdot \mathbf{v}^{(i,j)}$, then

$$r_{ut} = \sum_{1 \leq i,j \leq n} (M_u^{(A)} \cdot M_t^{(A)})(i,j) \cdot \mathbf{v}^{(i,j)}$$

Or, equivalently, that for every $w \in WNW(\Sigma)$,

$$r_{ut}(w) = \boldsymbol{\alpha}^T \cdot M^{(A)}(u) \cdot M^{(A)}(t) \cdot M^{(A)}(w) \cdot \boldsymbol{\eta}$$

Consider the linear combination:

$$\mathbf{v}_{ut} = \sum_{1 \leq i,j \leq n} (M_u^{(A)} \cdot M_t^{(A)})(i,j) \cdot \mathbf{v}^{(i,j)} = \sum_{1 \leq i,k,j \leq n} M_u^{(A)}(i,k) \cdot M_t^{(A)}(k,j) \cdot \mathbf{v}^{(i,j)}$$

Then, for $w \in WNW(\Sigma)$ we have

$$\mathbf{v}_{ut}(w) = \sum_{1 \leq i,k,j \leq n} M_u^{(A)}(i,k) \cdot M_t^{(A)}(k,j) \cdot \mathbf{v}^{(i,j)}(w)$$

$$= \sum_{1 \leq i,k,j \leq n} M_u^{(A)}(i,k) \cdot M_t^{(A)}(k,j) \cdot \left(\boldsymbol{\alpha}^T \cdot A^{(i,j)} M_w^{(A)} \cdot \boldsymbol{\eta}\right)$$

Note that for $N = A^{(i,j)} M_w^{(A)}$, the row i of N is row j of $M_w^{(A)}$ and all other rows are zero. Then

$$\mathbf{v}_{ut}(w) = \sum_{1 \leq i,k,j \leq n} M_u^{(A)}(i,k) \cdot M_t^{(A)}(k,j) \cdot \left(\sum_{l=1}^{n} \boldsymbol{\alpha}(i) M_w^{(A)}(j,l) \cdot \boldsymbol{\eta}(l)\right)$$

$$= \sum_{1 \leq i,k,j,l \leq n} \boldsymbol{\alpha}(i) \cdot M_u^{(A)}(i,k) \cdot M_t^{(A)}(k,j) \cdot M_w^{(A)}(j,l) \cdot \boldsymbol{\eta}(l)$$

$$= \boldsymbol{\alpha}^T \cdot M_u^{(A)} \cdot M_t^{(A)} \cdot M_w^{(A)} \cdot \boldsymbol{\eta} = r_{ut}(w)$$

$\square$

From Theorems 3 and 6 we have our main result, Theorem 2.

Acknowledgments. We thank Boaz Blankrot for helpful discussions on matrix decompositions and the anonymous referees for valuable feedback.

References

1. Allauzen, C., Mohri, M., Riley, M.: Statistical modeling for unit selection in speech synthesis. In: Proceedings of the 42nd Annual Meeting on Association for Computational Linguistics, pp. 55. Association for Computational Linguistics (2004)
2. Alur, R., Madhusudan, P.: Adding nesting structure to words. In: Ibarra, O.H., Dang, Z. (eds.) DLT 2006. LNCS, vol. 4036, pp. 1–13. Springer, Heidelberg (2006)
3. Alur, R., Arenas, M., Barceló, P., Etessami, K., Immerman, N., Libkin, L.: First-order and temporal logics for nested words. In: 22nd Annual IEEE Symposium on Logic in Computer Science, 2007, LICS 2007, pp. 151–160. IEEE (2007)
4. Angluin, D.: On the complexity of minimum inference of regular sets. Inf. Control **39**(3), 337–350 (1978)
5. Angluin, D.: Learning regular sets from queries and counterexamples. Inf. Comput. **75**(2), 87–106 (1987)
6. Arnold, A., Plaice, J.: Finite Transition Systems: Semantics of Communicating Systems. Prentice Hall International (UK) Ltd., Hert- fordshire (1994)
7. Balle, B., Mohri, M.: Spectral learning of general weighted automata via constrained matrix completion. In: Advances in neural information processing systems, pp. 2168–2176 (2012)
8. Balle, B., Mohri, M.: Learning weighted automata. In: Maletti, A. (ed.) Algebraic Informatics. LNCS, vol. 9270, pp. 1–21. Springer, New York (2015)
9. Beimel, A., Bergadano, F., Bshouty, N., Kushilevitz, E., Varricchio, S.: Learning functions represented as multiplicity automata. J. ACM (JACM) **47**(3), 506–530 (2000)
10. Bergadano, F., Varricchio, S.: Learning behaviors of automata from multiplicity and equivalence queries. SIAM J. Comput. **25**(6), 1268–1280 (1996)
11. Bisht, L., Bshouty, N.H., Mazzawi, H.: On optimal learning algorithms for multiplicity automata. In: Lugosi, G., Simon, H.U. (eds.) COLT 2006. LNCS (LNAI), vol. 4005, pp. 184–198. Springer, Heidelberg (2006)
12. Carlyle, J., Paz, A.: Realizations by stochastic finite automata. J. Comput. Syst. Sci. **5**, 26–40 (1971)
13. Chatterjee, K., Doyen, L., Henzinger, T.A.: Probabilistic weighted automata. In: Bravetti, M., Zavattaro, G. (eds.) CONCUR 2009. LNCS, vol. 5710, pp. 244–258. Springer, Heidelberg (2009)
14. Chatterjee, K., Henzinger, T.A., Jobstmann, B., Singh, R.: Measuring and synthesizing systems in probabilistic environments. In: Touili, T., Cook, B., Jackson, P. (eds.) CAV 2010. LNCS, vol. 6174, pp. 380–395. Springer, Heidelberg (2010)
15. Cobham, A.: Representation of a word function as the sum of two functions. Math. Syst. Theory **11**, 373–377 (1978)
16. Courcelle, B., Engelfriet, J.: Graph Structure and Monadic Second-Order Logic: A Language-Theoretic Approach, vol. 138. Cambridge University Press, New York (2012)
17. Courcelle, B., Makowsky, J.A., Rotics, U.: Linear time solvable optimization problems on graphs of bounded clique width. In: Hromkovič, J., Sýkora, O. (eds.) WG 1998. LNCS, vol. 1517, pp. 1–16. Springer, Heidelberg (1998)
18. Courcelle, B., Makowsky, J., Rotics, U.: On the fixed parameter complexity of graph enumeration problems definable in monadic second order logic. Discrete Appl. Math. **108**(1–2), 23–52 (2001)
19. Culik II, K., Kari, J.: Image compression using weighted finite automata. In: Borzyszkowski, A.M., Sokolowski, S. (eds.) MFCS 1993. LNCS, vol. 711, pp. 392–402. Springer, Heidelberg (1993)

20. D'Antoni, L., Alur, R.: Symbolic visibly pushdown automata. In: Biere, A., Bloem, R. (eds.) CAV 2014. LNCS, vol. 8559, pp. 209–225. Springer, Heidelberg (2014)

21. De Schutter, B., De Moor, B.: The singular-value decomposition in the extended max algebra. Linear Algebra Appl. **250**, 143–176 (1997)

22. De Schutter, B., De Moor, B.: The qr decomposition and the singular value decomposition in the symmetrized max-plus algebra revisited. SIAM Rev. **44**(3), 417–454 (2002)

23. Downey, R., Fellows, M.: Parametrized Complexity. Springer, New York (1999)

24. Driscoll, E., Burton, A., Reps, T.: Checking compatibility of a producer and a consumer. Citeseer (2011)

25. Driscoll, E., Thakur, A., Reps, T.: OpenNWA: a nested-word automaton library. In: Madhusudan, P., Seshia, S.A. (eds.) CAV 2012. LNCS, vol. 7358, pp. 665–671. Springer, Heidelberg (2012)

26. Droste, M., Gastin, P.: Weighted automata and weighted logics. In: Caires, L., Italiano, G.F., Monteiro, L., Palamidessi, C., Yung, M. (eds.) ICALP 2005. LNCS, vol. 3580, pp. 513–525. Springer, Heidelberg (2005)

27. Droste, M., Kuich, W., Vogler, H.: Handbook of Weighted Automata. Springer Science & Business Media, Heidelberg (2009)

28. Fernando, C., Pereira, N., Riley, M.: Speech recognition by composition of weighted finite automata. In: Roche, E., Schabes, Y. (eds.) Finite-State Language Processing. MIT Press, Cambridge (1997)

29. Fliess, M.: Matrices de hankel. J. Math. Pures Appl. **53**(9), 197–222 (1974)

30. Freedman, M., Lovász, L., Schrijver, A.: Reflection positivity, rank connectivity, and homomorphism of graphs. J. Am. Math. Soc. **20**(1), 37–51 (2007)

31. Gauwin, O., Niehren, J.: Streamable fragments of forward Xpath. In: Bouchou-Markhoff, B., Caron, P., Champarnaud, J.-M., Maurel, D. (eds.) CIAA 2011. LNCS, vol. 6807, pp. 3–15. Springer, Heidelberg (2011)

32. Gentle, J.: Computational Statistics, vol. 308. Springer, New York (2009)

33. Godlin, B., Kotek, T., Makowsky, J.A.: Evaluations of Graph Polynomials. In: Broersma, H., Erlebach, T., Friedetzky, T., Paulusma, D. (eds.) WG 2008. LNCS, vol. 5344, pp. 183–194. Springer, Heidelberg (2008)

34. Gold, E.: Complexity of automaton identification from given data. Inf. Control **37**(3), 302–320 (1978)

35. Golub, G., Van Loan, C.: Matrix Computations, vol. 3. JHU Press, Baltimore (2012)

36. Habrard, A., Oncina, J.: Learning multiplicity tree automata. In: Sakakibara, Y., Kobayashi, S., Sato, K., Nishino, T., Tomita, E. (eds.) ICGI 2006. LNCS (LNAI), vol. 4201, pp. 268–280. Springer, Heidelberg (2006)

37. Harris, W.R., Jha, S., Reps, T.: Secure programming via visibly pushdown safety games. In: Madhusudan, P., Seshia, S.A. (eds.) CAV 2012. LNCS, vol. 7358, pp. 581–598. Springer, Heidelberg (2012)

38. Haussler, D., Littlestone, N., Warmuth, M.: Predicting {0, 1}-functions on randomly drawn points. In: 29th Annual Symposium on Foundations of Computer Science, 1988, pp. 100–109. IEEE (1988)

39. Heller, A.: Probabilistic automata and stochastic transformations. Theory Comput. Syst. **1**(3), 197–208 (1967)

40. Hsu, D., Kakade, S., Zhang, T.: A spectral algorithm for learning hidden markov models. J. Comput. Syst. Sci. **78**(5), 1460–1480 (2012)

41. Kiefer, S., Murawski, A.S., Ouaknine, J., Wachter, B., Worrell, J.: On the complexity of equivalence and minimisation for Q-weighted automata. Log. Meth. Comput. Sci. (LMCS) **9**(1:8), 1–22 (2013)

42. Klema, V., Laub, A.: The singular value decomposition: its computation and some applications. IEEE Trans. Autom. Control **25**(2), 164–176 (1980)

43. Labai, N.: Definability and Hankel Matrices. Master's thesis, Technion - Israel Institute of Technology, Faculty of Computer Science (2015)

44. Labai, N., Makowsky, J.: Weighted automata and monadic second order logic. In: EPTCS Proceedings of GandALF, vol. 119, pp. 122–135 (2013)

45. Labai, N., Makowsky, J.: Tropical graph parameters. In: DMTCS Proceedings of FPSAC, vol. 01, pp. 357–368 (2014)

46. Labai, N., Makowsky, J.: Meta-theorems using hankel matrices (2015)

47. Littlestone, N.: Learning quickly when irrelevant attributes abound: a new linear-threshold algorithm. Mach. Learn. **2**(4), 285–318 (1988)

48. Lovász, L.: Connection matrices. Oxford Lect. Ser. Math. Appl. **34**, 179 (2007)

49. Lovász, L.: Large Networks and Graph Limits, vol. 60. Colloquium Publications, New York (2012)

50. Makowsky, J.: Algorithmic uses of the Feferman-Vaught theorem. Ann. Pure Appl. Logic **126**(1–3), 159–213 (2004)

51. Mathissen, C.: Weighted logics for nested words and algebraic formal power series. In: Aceto, L., Damgård, I., Goldberg, L.A., Halldórsson, M.M., Ingólfsdóttir, A., Walukiewicz, I. (eds.) ICALP 2008, Part II. LNCS, vol. 5126, pp. 221–232. Springer, Heidelberg (2008)

52. McMillan, K.: Symbolic Model Checking. Springer, New York (1993)

53. Mohri, M.: Finite-state transducers in language and speech processing. Comput. Linguist. **23**(2), 269–311 (1997)

54. Mozafari, B., Zeng, K., Zaniolo, C.: High-performance complex event processing over xml streams. In: Proceedings of the 2012 ACM SIGMOD International Conference on Management of Data, pp. 253–264. ACM (2012)

55. Murawski, A.S., Walukiewicz, I.: Third-order idealized algol with iteration is decidable. In: Sassone, V. (ed.) FOSSACS 2005. LNCS, vol. 3441, pp. 202–218. Springer, Heidelberg (2005)

56. Pitt, L., Warmuth, M.: The minimum consistent dfa problem cannot be approximated within any polynomial. J. ACM (JACM) **40**(1), 95–142 (1993)

57. Poularikas, A.: Transforms and Applications Handbook. CRC Press, London (2010)

58. Valiant, L.: A theory of the learnable. Commun. ACM **27**(11), 1134–1142 (1984)

Linear Context-Free Tree Languages and Inverse Homomorphisms

Johannes Osterholzer$^{(\boxtimes)}$, Toni Dietze, and Luisa Herrmann

Faculty of Computer Science, Technische Universität Dresden,
01062 Dresden, Germany
{johannes.osterholzer,toni.dietze,luisa.herrmann}@tu-dresden.de

Abstract. We prove that the class of linear context-free tree languages
is not closed under inverse linear tree homomorphisms. The proof is by
contradiction: we encode Dyck words into a context-free tree language
and prove that its preimage under a certain linear tree homomorphism
cannot be generated by any context-free tree grammar. However, the
closure can be proved for the linear monadic context-free tree languages.

1 Introduction

Context-free tree grammars (cftg), introduced by Rounds [10], generalize the
concept of context-free rewriting to the realm of tree languages. In general, cftg
can *copy* parts of a sentential form in the application of a production, and a
lot of their complexity in comparison to regular tree grammars is due to the
interplay between copying and nondeterminism (cf. e.g. [5]).

Recently, there has been renewed interest in cftg in natural language processing, where trees are used to express the structure of the processed sentences. In
this area only non-copying, or *linear*, cftg (l-cftg) are considered, as copying adds
too much undesired power when it comes to linguistic applications.

The modular design of syntax-based language processing systems requires
that the used class of tree languages $\mathcal{C}$ possesses certain closure properties. In
particular, for translation tasks it is important that $\mathcal{C}$ is closed under inverse
linear tree homomorphisms. But this closure property does *not* hold when $\mathcal{C}$ is
the class of context-free tree languages [1]. The proof in [1] works by constructing
a *copying* cftg G, and the preimage of the tree language of G under a certain
tree homomorphism is shown to be non-context-free.

But since copying is not required anyway – are maybe the *linear* context-free
tree languages closed under inverse linear tree homomorphisms? In this work,
we answer this question in the negative: there are an l-cftg G_{ex} and a linear tree
homomorphism h such that $L = h^{-1}(\mathcal{L}(G_{\mathrm{ex}}))$ is *not* a context-free tree language
(where $\mathcal{L}(G_{\mathrm{ex}})$ is the tree language of G_{ex}).

The 3rd [1st] author was [partially] supported by DFG RTG 1763 (QuantLA).

A.-H. Dediu et al. (Eds.): LATA 2016, LNCS 9618, pp. 478–489, 2016.
DOI: 10.1007/978-3-319-30000-9_37

The intuition behind our proof is as follows. Every tree t in L is of the form

$$\sigma \underline{\hspace{0.6cm}} \sigma \underline{\hspace{1cm}} \cdots \underline{\hspace{1cm}} \sigma \underline{\hspace{0.6cm}} \sigma - \#$$

$$\#\quad |u_1\ |v_1\ |u_2 \qquad\qquad |v_{n-1}|u_n\ |v_n\ \#$$

$$\#\ \#\ \# \qquad\qquad \#\quad \#\ \#$$

for some $n \geq 1$ and monadic trees $u_1, v_1, \ldots, u_n, v_n$. Here, the root of t is the leftmost symbol σ. The subtrees u_i, v_i, called *chains* in the following, are built up over a parenthesis alphabet, such that the chains u_i contain only opening parentheses, the chains v_i only closing parentheses, and $u_1^R v_1 \cdots u_n^R v_n$ is a well-parenthesized word (w^R denotes the reversal of the word w).

If one were to cut such a tree t into two parts t_1 and t_2, right through an edge between two σs, then one could observe that there are some chains u_j in t_1 which contain opening parentheses which are not closed in t_1, but only in t_2. A similar observation holds of course for some chains v_j in t_2. The "unclosed" parts of the chains u_j and v_j will be called their *defects*.

We assume that there is some (not necessarily linear) cftg G with $\mathcal{L}(G) = L$, and show that if G exists, then it can be assumed to be of a special normal form. We analyze the derivations of such a G in normal form. A derivation of a tree t will be shown to cut t into two pieces as described above, as it must branch at some moment of the derivation. If G exists, it must therefore prepare the defects of t_1 and t_2 such that they "fit together" before the derivation branches. But there are only finitely many nonterminal parameters in which the defects could be prepared. We give a sequence of trees in L such that the number of their defects is strictly increasing, no matter how they are cut apart. Then there is some tree t in this sequence whose defects cannot be prepared fully. Hence it is possible to show by a pumping argument that if $t \in \mathcal{L}(G)$, then there is also a tree $t' \in \mathcal{L}(G)$ whose respective parts do not fit together, and therefore $t' \notin L$. Thus the existence of G is ruled out.

We conclude our work by stating a positive result: the tree languages of linear *monadic* cftg (lm-cftg), i.e. of l-cftg where each nonterminal has at most one parameter, *are* closed under inverse linear tree homomorphisms. The importance of lm-cftg is underscored by their expressive equivalence to the well-known linguistic formalism of *tree-adjoining grammars* [7,8].

Due to space limitations, some proofs had to be omitted. The reader can find an extended version of this paper, with all proofs, on arXiv [9].

2   Preliminaries

The set of natural numbers with zero is denoted by $\mathbb{N}$. For every $m, n \in \mathbb{N}$, the set $\{i \in \mathbb{N} \mid m \leq i \leq n\}$ is denoted by $[m, n]$, and the set $[1, n]$ by $[n]$. We use the standard notions from formal language theory. In particular, the empty word is denoted by ε, the length of a word w by $|w|$, and the reversal of w by w^R. The *Dyck language* over a parenthesis alphabet A is denoted by D_A^* and the Dyck congruence is $\equiv$.

An alphabet Σ equipped with a function $\mathrm{rk}_\Sigma \colon \Sigma \to \mathbb{N}$ is a *ranked alphabet*. Let Σ be a ranked alphabet. When Σ is obvious, we write rk instead of rk_Σ. Let $k \in \mathbb{N}$. Then $\Sigma^{(k)} = \mathrm{rk}^{-1}(k)$. We often write $\sigma^{(k)}$ and mean that $\mathrm{rk}(\sigma) = k$.

Let U be a set and Λ denote $\Sigma \cup U \cup C$, where C consists of the three symbols '(', ')', and ','. The set $T_\Sigma(U)$ of *trees (over Σ indexed by U)* is the smallest set $T \subseteq \Lambda^*$ such that $U \subseteq T$, and for every $k \in \mathbb{N}$, $\sigma \in \Sigma^{(k)}$, and $t_1, \ldots, t_k \in T$, we also have that $\sigma(t_1, \ldots, t_k) \in T$. A tree $\gamma(t)$, $\gamma \in \Sigma^{(1)}$, is abbreviated by γt, and $T_\Sigma(\emptyset)$ by T_Σ. The notation γt suggests a bijection between $\Sigma^* U$ and $T_\Sigma(U)$ for monadic ranked alphabets Σ (i.e. $\Sigma = \Sigma^{(1)}$), and in fact we will often confuse such monadic trees with words.

Let $s, t \in T_\Sigma(U)$. The set of *positions of* t is denoted by $\mathrm{pos}(t) \subseteq \mathbb{N}^*$, and the number of occurrences of a symbol $\sigma \in \Sigma$ in t by $|t|_\sigma$. The *size* of t is $|t| = \sum_{\sigma \in \Sigma} |t|_\sigma$. Denote the *label* of t at its position w by $t(w)$, and the *subtree* of t at w by $t|_w$. The result of *replacing* the subtree $t|_w$ in t by s is $t[s]_w$. Fix the infinite set of *variables* $X = \{x_1, x_2, \ldots\}$. For each $k \in \mathbb{N}$, let $X_k = \{x_i \mid i \in [k]\}$. Given $n, k \in \mathbb{N}$, $t \in T_\Sigma(X_n)$, and $s_1, \ldots, s_n \in T_\Sigma(X_k)$, denote by $t[s_1, \ldots, s_n]$ the result of substituting s_i for each occurrence of x_i in t, where $i \in [n]$. When no other variable is used, we will also write x instead of x_1.

We use notation associated with *magmoids* [2,3]. Let $k, n \in \mathbb{N}$. Then the set $\{\langle k, t_1, \ldots, t_n \rangle \mid t_1, \ldots, t_n \in T_\Sigma(X_k)\}$ is denoted by $T(\Sigma)_k^n$. From now on we omit the component k from such a tuple. We identify the sets $T_\Sigma(X_k)$ and $T(\Sigma)_k^1$ and write t instead of $\langle t \rangle$. The tree $\sigma(x_1, \ldots, x_k) \in T(\Sigma)_k^1$ is identified with the symbol $\sigma \in \Sigma^{(k)}$. In particular, we can write α instead of $\alpha()$, for $\alpha \in \Sigma^{(0)}$. The set of all $u \in T(\Sigma)_k^n$ such that the left-to-right sequence of variables in u is $x_1, \ldots, x_k$ is denoted $\widetilde{T}(\Sigma)_k^n$. The set Θ_k^n of *torsions* is $\{\langle x_{i_1}, \ldots, x_{i_n} \rangle \mid i_1, \ldots, i_n \in [k]\}$. Note that $\Theta_k^n \subseteq T(\Sigma)_k^n$. The torsion $\langle x_1, \ldots, x_n \rangle \in \Theta_n^n$ is denoted by Id_n, and the torsion $\langle x_i \rangle \in T(\Sigma)_k^1$, $i \in [k]$, by π_i^k (when k is clear from the context, we write π_i instead). A tuple $u \in \widetilde{T}(\Sigma)_k^n$ is called *torsion-free*.

Let $n, \ell, k \in \mathbb{N}$, and let $u = \langle u_1, \ldots, u_n \rangle \in T(\Sigma)_\ell^n$, $v = \langle v_1, \ldots, v_\ell \rangle \in T(\Sigma)_k^\ell$. We define $u \cdot v \in T(\Sigma)_k^n$ by $u \cdot v = \langle u_1[v_1, \ldots, v_\ell], \ldots, u_n[v_1, \ldots, v_\ell] \rangle$. Note that the operation $\cdot$ is associative. If $u \in T(\Sigma)_n^n$, then let $u^0 = \mathrm{Id}_n$ and $u^{(j+1)} = u \cdot u^j$ for every $j \in \mathbb{N}$. Moreover, if $u \in T(\Sigma)_k^n$ and $v \in T(\Sigma)_k^\ell$, define $[u, v] \in T(\Sigma)_k^{n+\ell}$ by $[u, v] = \langle u_1, \ldots, u_n, v_1, \ldots, v_\ell \rangle$. Clearly, this operation is associative, so we will write, e.g., $[u, v, t]$ instead of $[[u, v], t]$.

Let us introduce the following auxiliary notation. Let $n, k \in \mathbb{N}$. Then $\mathrm{S}(\Sigma)_k^n = \{[u, x_{k+1}] \mid u \in T(\Sigma)_k^n\}$, $\widetilde{\mathrm{S}}(\Sigma)_k^n = \widetilde{T}(\Sigma)_{k+1}^{n+1} \cap \mathrm{S}(\Sigma)_k^n$, and $\widehat{\Theta}_k^n = \Theta_{k+1}^{n+1} \cap \mathrm{S}(\Sigma)_k^n$. Moreover, for every $s \in T(\Sigma)_{n+1}^1$ and $t \in T(\Sigma)_k^1$, let $s \multimap t = s \cdot [\mathrm{Id}_n, t]$. For example, let $t = \sigma(b(x_2), x_1, x_5)$, then $t \multimap t = \sigma(b(x_2), x_1, \sigma(b(x_2), x_1, x_5))$. We assume $\cdot$ to bind stronger than $\multimap$. So, for instance, $t \cdot u \multimap s \cdot v$ means $(t \cdot u) \multimap (s \cdot v)$.

A *context-free tree grammar (cftg)* over Σ is a tuple $G = (N, \Sigma, \eta_0, P)$ such that Σ and N are disjoint ranked alphabets (of *terminal* resp. *nonterminal* symbols), $\eta_0 \in T(N \cup \Sigma)_0^1$ (the *axiom*) and P is a finite set of *productions* of the form $A(x_1, \ldots, x_k) \to t$ for some $k \in \mathbb{N}$, $A \in N^{(k)}$, and $t \in T(N \cup \Sigma)_k^1$. By the above convention, such a production can be abbreviated by $A \to t$.

The cftg G is said to be *linear* if in every production $A \to t$ in P the right-hand side t contains each variable x_i, $i \in [k]$, at most once. Further, G is *monadic* if $N = N^{(0)} \cup N^{(1)}$. Linear (and monadic) cftg are abbreviated *l-cftg* (*lm-cftg*).

Let $G = (N, \Sigma, \eta_0, P)$ be a cftg, and let n, $k \in \mathbb{N}$. Given η, $\zeta \in T(N \cup \Sigma)^1_n$, we write $\eta \Rightarrow_G \zeta$ if there are $A \to t$ in P, $A \in N^{(k)}$, and $\kappa \in T(N \cup \Sigma)^1_{n+1}$, $\tau \in T(N \cup \Sigma)^k_n$ such that κ contains x_{n+1} exactly once, $\eta = \kappa \cdot [\mathrm{Id}_n, A \cdot \tau]$ and $\zeta = \kappa \cdot [\mathrm{Id}_n, t \cdot \tau]$. Let $\eta \in T(N \cup \Sigma)^1_n$. Then the set $\{t \in T(\Sigma)^1_n \mid \eta \Rightarrow^*_G t\}$ is denoted by $\mathcal{L}(G, \eta)$, and the *tree language of* G, denoted by $\mathcal{L}(G)$, is $\mathcal{L}(G, \eta_0)$. We call $L \subseteq T(\Sigma)^1_0$ a *(linear) (monadic) context-free tree language* if there is a (linear) (monadic) cftg G with $L = \mathcal{L}(G)$. Two cftg G and G' are *equivalent* if $\mathcal{L}(G) = \mathcal{L}(G')$.

Let Σ and Δ be ranked alphabets. A mapping $h\colon \Sigma \to T_\Delta(X)$ is said to be a *tree homomorphism* if $h(\Sigma^{(k)}) \subseteq T_\Delta(X_k)$ for every $k \in \mathbb{N}$. We extend h to a mapping $\widehat{h}\colon T_\Sigma(X) \to T_\Delta(X)$ by setting $\widehat{h}(x_i) = x_i$ for every $i \in \mathbb{N}$ and

$$\widehat{h}(\sigma(t_1, \ldots, t_k)) = h(\sigma)[\widehat{h}(t_1), \ldots, \widehat{h}(t_k)]$$

for every $k \in \mathbb{N}$, $\sigma \in \Sigma^{(k)}$, and t_1, ..., $t_k \in T_\Sigma(X)$. In the following, we will no longer distinguish between h and $\widehat{h}$. Recall the following properties of tree homomorphisms (e.g. from [4]). Let $h\colon T_\Sigma(X) \to T_\Delta(X)$ be a tree homomorphism. We say that h is *linear* (resp. *nondeleting*) if for every $k \in \mathbb{N}$, $\sigma \in \Sigma^{(k)}$, and $i \in [k]$, x_i occurs at most (resp. at least) once in $h(\sigma)$. A linear and nondeleting tree homomorphism is called *simple*. Finally, h is *strict* if $h(\Sigma) \cap X = \emptyset$.

3   The Tree Language L

We start out by introducing the l-cftg G_{ex}. The preimage L of $\mathcal{L}(G_{\mathrm{ex}})$ under a simple tree homomorphism h, introduced afterwards, will be shown to be non-context-free later on.

Let $\Delta = \{\delta^{(2)}_1, \delta^{(2)}_2, \#^{(0)}\} \cup \Gamma$, where $\Gamma = \{a^{(1)}, b^{(1)}, c^{(1)}, d^{(1)}\}$. Consider the linear cftg $G_{\mathrm{ex}} = (N_{\mathrm{ex}}, \Delta, \eta_{\mathrm{ex}}, P_{\mathrm{ex}})$ with nonterminal set $N_{\mathrm{ex}} = \{A^{(3)}\}$, axiom $\eta_{\mathrm{ex}} = \delta_1(\#, A(c\#, d\#, \delta_2(\#, \#)))$, and productions in P_{ex} given by

$$A \to A(ax_1, bx_2, x_3) + A(ccx_1, d\#, A(c\#, ddx_2, x_3)) + \delta_2(cx_1, \delta_1(dx_2, x_3)).$$

Example 1. The following is an example derivation of a tree in $\mathcal{L}(G_{\mathrm{ex}})$.

$$\eta_{\mathrm{ex}} \Rightarrow^*_{G_{\mathrm{ex}}} \delta_1(\#, x) \multimap A(c^2a^2c\#, d\#, x) \multimap A(c\#, d^2b^2d\#, x) \multimap \delta_2(\#, \#)$$
$$\Rightarrow^*_{G_{\mathrm{ex}}} \delta_1(\#, x) \multimap \delta_2(cac^2a^2c\#, x) \multimap \delta_1(dbd\#, x)$$
$$\multimap \delta_2(ca^2c\#, x) \multimap \delta_1(db^2d^2b^2d\#, x) \multimap \delta_2(\#, \#).$$

Let $\Sigma = \{\sigma^{(3)}, \#^{(0)}\} \cup \Gamma$ and let $h\colon T_\Sigma(X) \to T_\Delta(X)$ be such that

$$h(\sigma(x_1, x_2, x_3)) = \delta_1(x_1, \delta_2(x_2, x_3)) \quad \text{and} \quad h(\omega) = \omega \text{ for each } \omega \in \Sigma \setminus \{\sigma\}.$$

Note that h is injective, strict, and simple. In the following, we will analyse the tree language $L = h^{-1}(\mathcal{L}(G_{\mathrm{ex}}))$. It is easy to see that every $t \in L$ is of the form

$$\sigma(\#, u_1\#, x) \multimap \sigma(v_1\#, u_2\#, x) \multimap \cdots \multimap \sigma(v_{n-1}\#, u_n\#, x) \multimap \sigma(v_n\#, \#, \#)$$

for some $n \geq 1$, and $u_i \in (ca^*c)^+$, $v_i \in (db^*d)^+$, for each $i \in [n]$. In general, given a tree t of the form

$$\sigma(v_1\#, u_1\#, x) \circ\!\!-\ \cdots\ \circ\!\!-\ \sigma(v_n\#, u_n\#, \zeta) \qquad \text{with } n \geq 1, \quad \zeta \in \{\#\} \cup X, \quad (1)$$

where $v_i \in (db^*d)^*$ and $u_i \in (ca^*c)^*$, $i \in [n]$, we will call the monadic subtrees u_j (resp. v_j) of t the a-chains (resp. the b-chains) of t. A chain is either an a- or a b-chain. The rightmost root-to-leaf path in t (that is labeled $\sigma \cdots \sigma\zeta$) will be referred to as t's spine.

For every tree t of the form as in (1), we let $\iota(t) = v_1 u_1^R v_2 u_2^R \cdots v_n u_n^R$. We view Γ as a parenthesis alphabet, such that b acts as right inverse to a, and d to c. Then $\iota(t)$ is a Dyck word, for every $t \in L$.

Proposition 2. *For every $t \in L$, $\iota(t) \in D_\Gamma^*$.*

There is the following relation between the numbers of symbols in $t \in L$.

Proposition 3. *For every $t \in L$, $|t|_c = |t|_d = 4 \cdot |t|_\sigma - 6$.*

Both propositions are proved by induction on the derivations in G_{ex}. Each chain of $t \in L$ is determined by the other chains of t, because $\iota(t)$ is a Dyck word, and every chain contains either only symbols from $\{a, c\}$, or from $\{b, d\}$.

Observation 4. Let $t \in L$, let $w \in \text{pos}(t)$ with $t(w) \in \Gamma \cup \{\#\}$, and let $s = t[x_1]_w$. There is exactly one $u \in \text{T}(\Gamma \cup \{\#\})_0^1$ such that $s \cdot u \in L$.

Example 5. The preimage under h of the tree from Example 1 is

$$t = \sigma(\#, cac^2a^2c\#, x) \circ\!\!-\ \sigma(dbd\#, ca^2c\#, x) \circ\!\!-\ \sigma(db^2d^2b^2d\#, \#, \#).$$

We have $\iota(t) = ca^2c^2acdbdca^2cdb^2d^2b^2d$, and one can verify that $\iota(t) \in D_\Gamma^*$.

In the following sections, we prove the following theorem.

Theorem 6. *The preimage of the linear context-free tree language $\mathcal{L}(G_{\text{ex}})$ under the injective, strict, and simple tree homomorphism h is not context-free.*

Corollary 7. *The class of linear context-free tree languages is not closed under inverse linear tree homomorphisms.*

The theorem might seem surprising, as L and $\mathcal{L}(G_{\text{ex}})$ are nearly the same: their only difference is that σ is split up into δ_1 and δ_2. However, this separation gives G_{ex} the power to create the chains under δ_1 and δ_2 independently, while a cftg generating L would have to derive them simultaneously. As described in the introduction, and proved further on, this would require nonterminals of unbounded rank and is therefore impossible.

Assume that there is a cftg G with $\mathcal{L}(G) = L$. The following lemma shows that then G may be chosen to be of a very constrained form. Due to space restrictions, we cannot reproduce its proof (but see [9]). In a nutshell, the lemma follows from two properties which hold for every $t \in L$: *(i)* that t has a spine, and *(ii)* that each chain of t is determined by its siblings (see Observation 4). This means that a derivation's nondeterminism is confined to the spine of t, and the productions of G are quite close to those of context-free word grammars.

Lemma 8. *We may assume that G is of the form $G = (N, \Sigma, \eta_0, P)$, such that $N = N^{(p+1)}$ for some $p \in \mathbb{N}$, $\eta_0 = S(\#, \ldots, \#)$ for some $S \in N$ and every production in P is of the form*

$$A \to B \cdot u, \quad A \to B \cdot \vartheta_1 \circ\!\!- C \cdot \vartheta_2, \quad \text{or} \quad A \to \sigma(x_i, x_j, x_{p+1}),$$

where $u \in \widetilde{S}(\Gamma)_p^p$, $\vartheta_1, \vartheta_2 \in \widehat{\Theta}_p^p$, $i, j \in [p]$, and $A, B, C \in N$.

Assume until contradiction that there is a cftg G of the form in Lemma 8 such that $\mathcal{L}(G) = L$. Let χ denote the tuple $\langle \#, \ldots, \# \rangle$. Then $\eta_0 = S \cdot \chi$.

4 Derivation Trees

A derivation of a tree $t \in \mathcal{L}(G)$ can be described faithfully by a binary tree κ.[1] These *derivation trees* will help us analyze the structure of the derivations in G.

Formally, let κ be a binary tree such that each position $\delta \in \text{pos}(\kappa)$ is equipped with two nonterminal symbols A_δ and $B_\delta \in N$, a torsion-free tuple $s_\delta \in \widetilde{S}(\Gamma)_p^p$, a torsion $\vartheta_\delta \in \widehat{\Theta}_p^p$, and two numbers i_δ and $j_\delta \in [p]$. Then κ is an $(A_\varepsilon, \vartheta_\varepsilon)$-*derivation tree* if for every $\delta \in \text{pos}(\kappa)$,

(i) $A_\delta \Rightarrow_G^* B_\delta \cdot s_\delta$,
(ii) if δ is a leaf of κ, then the production $B_\delta \to \sigma(x_{i_\delta}, x_{j_\delta}, x_{p+1})$ is in P,
(iii) otherwise, $B_\delta \to A_{\delta 1} \cdot \vartheta_{\delta 1} \circ\!\!- A_{\delta 2} \cdot \vartheta_{\delta 2}$ is a production in P.

Let $t \in T(\Sigma)_{p+1}^1$. Then κ is an $(A_\varepsilon, \vartheta_\varepsilon)$-*derivation tree of t* (or: κ *derives* t) if either κ has only one node and $t = \sigma(x_{i_\varepsilon}, x_{j_\varepsilon}, x_{p+1}) \cdot s_\varepsilon \cdot \vartheta_\varepsilon$, or, otherwise, there are $t_1, t_2 \in T(\Sigma)_{p+1}^1$ such that $\kappa|_1$ derives t_1, $\kappa|_2$ derives t_2, and $t = (t_1 \circ\!\!- t_2) \cdot s_\varepsilon \cdot \vartheta_\varepsilon$. An (S, Id_{p+1})-*derivation tree* (of t) will simply be called a *derivation tree* (of t). There is the following relation between derivations and derivation trees.

Proposition 9. *Let $t \in T(\Sigma)_{p+1}^1$, let $A \in N$, and $\vartheta \in \widehat{\Theta}_p^p$. Then $A \cdot \vartheta \Rightarrow_G^* t$ if and only if there is an (A, ϑ)-derivation tree of t.*

As a direct corollary, $t \cdot \chi \in L$ if and only if there is a derivation tree of t. We close our discussion of derivation trees with the following pumping lemma. It states that if there is some s_δ in κ which has a sufficiently large component, then an iterable pair of nonterminals occurs in the derivation of s_δ.

In the sequel, fix the pumping number $H = |N| \cdot h_{\max}$, where $h_{\max}$ is the maximal size of a component of u in a production of G of form $A \to B \cdot u$.

Lemma 10. *Let κ be a derivation tree and $\delta \in \text{pos}(\kappa)$. If there are $i \in [p]$ and $w, w' \in \Gamma^*$ such that $\pi_i \cdot s_\delta = w'wx_i$ and $|w| > H$, then there exist $v, y, z \in \widetilde{S}(\Sigma)_p^p$ such that (i) $s_\delta = v \cdot y \cdot z$, (ii) $\pi_i \cdot y \cdot z$ is a suffix of wx_i, (iii) $|\pi_i \cdot y| > 0$, and (iv) for each $j \in \mathbb{N}$, $A_\delta \Rightarrow_G^* B_\delta \cdot v \cdot y^j \cdot z$.*

---

[1] As a prefix-closed subset $\text{pos}(\kappa) \subseteq \{1, 2\}^*$ and such that $w1 \in \text{pos}(\kappa)$ iff $w2 \in \text{pos}(\kappa)$.

5 Dyck Words and Sequences of Chains

This section prepares some necessary notions for the upcoming counterexample. We introduce a sequence U_1, U_2, ... of Dyck words. Later, an element of this sequence will contribute the chains to the tree t used in the counterexample. As described in the introduction, the proof revolves around the factorization of t into trees t_1 and t_2 that is induced by a derivation of t. So we will analyze the corresponding factorizations of the Dyck words U_i.

Moreover, we will introduce here the notion of *defects*, which can be understood as the "unclosed parentheses" in t_1, resp. t_2. Finally, a lemma on *perturbations* is given, which will be used to show that if the defects in t_1 are modified (or: perturbed), then the word formed by the chains of the resulting tree lies in another Dyck congruence class. This implies that the resulting tree does not "fit together" with t_2 any longer.

Let $q = 2p$, and let $m = 2^{q-1} + 1$. For every $i \in \mathbb{N}$, let $\alpha_i = ca^{imH}c$ and $\beta_i = db^{imH}d$. Note that $\alpha_i^R = \alpha_i$, and $\beta_i^R = \beta_i$. Define the sequence U_1, U_2, ... of words over Γ by $U_1 = \alpha_1\beta_1$ and $U_{i+1} = \alpha_{i+1}U_iU_i\beta_{i+1}$ for every $i \geq 1$.

Observation 11. For every $i \geq 1$,

(1) $U_i \in D_\Gamma^*$, and
(2) $U_i = u_1v_1 \cdots u_nv_n$, where $n = 2^{i-1}$, $u_j \in (ca^+c)^+$, $v_j \in (db^+d)^+$, $j \in [n]$.

For each U_i of the above form, let $Z_i = \langle u_1^R, v_1, \ldots, u_n^R, v_n \rangle$. The components u_ℓ^R and v_ℓ of Z_i will also be called *chains*, as later on they will end up as the chains of some $t \in L$. For every factorization of Z_i into

$$Z_i' = \langle u_1^R, v_1, u_2^R, v_2, \ldots, u_j^R \rangle \quad \text{and} \quad Z_i'' = \langle v_j, u_{j+1}^R, v_{j+1}, \ldots, u_n^R, v_n \rangle, \quad j \in [n],$$

consider the factors $P_{i,j} = u_1v_1u_2v_2 \cdots u_j$ and $S_{i,j} = v_ju_{j+1}v_{j+1} \cdots u_nv_n$ of U_i.

Proposition 12. *The factors $P_{i,j}$ and $S_{i,j}$ can be written as*

$$P_{i,j} = \alpha_iV_{i-1}\alpha_{i-1} \cdots V_1\alpha_1 \quad and \quad S_{i,j} = \beta_1W_1 \cdots \beta_{i-1}W_{i-1}\beta_i, \quad (2)$$

such that V_ℓ, $W_\ell \in \{\varepsilon, U_\ell\}$ and $V_\ell \neq W_\ell$ for every $\ell \in [i-1]$.

Proof. By induction on i. The base case $U_1 = \alpha_1\beta_1$ has only one factorization, $P_{1,1} = \alpha_1$ and $S_{1,1} = \beta_1$, which fulfills the property. Let $i \geq 1$ and consider $U_{i+1} = \alpha_{i+1}U_iU_i\beta_{i+1}$. A factorization $P_{i+1,j}S_{i+1,j}$ of U_{i+1} induces a factorization of either the first or the second occurrence of U_i into, say, $P_{i,j'}$ and $S_{i,j'}$ for some $j' \in [2^{i-1}]$. Therefore, $U_{i+1} = \alpha_{i+1}V_iP_{i,j'}S_{i,j'}W_i\beta_{i+1}$ for V_i, $W_i \in \{\varepsilon, U_i\}$ with $V_i \neq W_i$. By induction, $P_{i,j'} = \alpha_iV_{i-1}\alpha_{i-1} \cdots V_1\alpha_1$, and therefore $P_{i+1,j} = \alpha_{i+1}V_i\alpha_iV_{i-1}\alpha_{i-1} \cdots V_1\alpha_1$, for V_i, ..., V_1 as given above. The same kind of argument works for $S_{i+1,j}$. □

Let U_i be factorized into $P_{i,j}$ and $S_{i,j}$ as in (2). We denote by $D_{i,j}$ the word

$$\$\alpha_iV_{i-1}'\alpha_{i-1} \cdots V_1'\alpha_1\$\beta_1W_1' \cdots \beta_{i-1}W_{i-1}'\beta_i\$$$

over $\Gamma \cup \{\$\}$, where for every $\ell \in [i-1]$, $V_\ell' = \$$ if $V_\ell = U_\ell$, and $V_\ell' = \varepsilon$ if $V_\ell = \varepsilon$, and, analogously, $W_\ell' = \$$ if $W_\ell = U_\ell$, and $W_\ell' = \varepsilon$ if $W_\ell = \varepsilon$. Let $\ell, k \in \mathbb{N}$ with $\ell \le k$. We say that a word $\gamma = \alpha_\ell \cdots \alpha_k$ (resp. $\gamma = \beta_\ell \cdots \beta_k$) is an a-*defect* (resp. a b-*defect*) in $D_{i,j}$ if $\$\gamma^R\$$ (resp. $\$\gamma\$$) occurs in $D_{i,j}$. When the factorization is clear, the reference to $D_{i,j}$ is omitted. Both a-defects and b-defects will be called *defects*. A chain in Z_i whose suffix is a defect is called a *critical chain*.

Proposition 13. *Consider a factorization of U_i into $P_{i,j}$ and $S_{i,j}$.*

(1) There is no $\ell \in [i]$ such that α_ℓ (or β_ℓ) occurs in two distinct defects.
(2) The number of defects in $D_{i,j}$ is $i+1$.
(3) Each a-defect (resp. b-defect) is the suffix of some chain u_n (resp. v_n) in Z_i, with $n \in [2^{i-1}]$.

Proof. For *(1)*, observe that the a-defects in $D_{i,j}$ are disjoint (non-overlapping) factors of the word $\alpha_1 \cdots \alpha_i$, analogously for the b-defects in $D_{i,j}$. For *(2)*, it is easy to see from Proposition 12 that there are exactly $i+2$ occurrences of the symbol $\$$ in $D_{i,j}$. So there are $i+1$ factors of the form $\$\gamma\$$ in $D_{i,j}$, for $\gamma \in \Gamma^*$. By *(1)*, the defects are pairwise distinct, so $D_{i,j}$ contains precisely $i+1$ defects. Regarding *(3)*, let $\gamma = \alpha_\ell \cdots \alpha_k$, $\ell \le k$, be an a-defect in $D_{i,j}$ and let

$$D_{i,j} = D'\$\underbrace{\alpha_k \cdots \alpha_\ell}_{\gamma^R}\$D'' \qquad \text{for some } D', D'' \in (\Gamma \cup \{\$\})^*.$$

By definition of $D_{i,j}$, $P_{i,j}$ is of the form $P_{i,j} = P'U_k\alpha_k \cdots U_\ell\alpha_\ell P''$ for some $P', P'' \in \Gamma^*$ if $k < i$, and $P_{i,j} = \alpha_k \cdots U_\ell\alpha_\ell P''$ if $k = i$. As U_k ends with β_k, γ is the suffix of some chain u_n in Z_i. A similar argument can be made if γ is a b-defect. □

Let $P, P' \in (ca^*c)^*$. We say that P' is a *perturbation* of P if it results from P by modifying the exponents of a in P. More precisely, let P be of the form $P = w_0 a^{f_1} w_1 \cdots w_{\ell-1} a^{f_\ell} w_\ell$, such that $\ell \in \mathbb{N}$, $w_0, \ldots, w_\ell \in c^*$, and for each $i \in [\ell]$, $f_i > 0$. Then $P' \in \Gamma^*$ is called a *perturbation* of P if $P' = w_0 a^{f_1'} w_1 \cdots w_{\ell-1} a^{f_\ell'} w_\ell$, for some $f_1', \ldots, f_\ell' \in \mathbb{N}$. The only perturbation of ε is ε itself.

Lemma 14. *Consider a factorization of U_i into $P_{i,j}$ and $S_{i,j}$, and let $P_{i,j}'$ be a perturbation of $P_{i,j}$, i.e.*

$$P_{i,j} = \alpha_i V_{i-1}\alpha_{i-1} \cdots V_1\alpha_1 \qquad \text{and} \qquad P_{i,j}' = \alpha_i' V_{i-1}'\alpha_{i-1}' \cdots V_1'\alpha_1'. \qquad (3)$$

Then $P_{i,j}' \equiv P_{i,j}$ if and only if $V_\ell' \equiv \varepsilon$ for every $\ell \in [i-1]$ and $\alpha_\ell' = \alpha_\ell$ for every $\ell \in [i]$.

Proof. The direction "if" is trivial. For the other direction, one can prove by induction that for every $i > 0$ and every perturbation U_i' of U_i, either $U_i' \equiv \varepsilon$ or $U_i' = cXd$ for some $X \not\equiv \varepsilon$. Let $P_{i,j}' \equiv P_{i,j}$. As $V_\ell \in \{U_\ell, \varepsilon\}$ for every $\ell \in [i-1]$, $P_{i,j}$ reduces to $\alpha_i \cdots \alpha_1$. Assume that there is some $\ell \in [i-1]$ with $V_\ell' \not\equiv \varepsilon$. Then the reduction of $P_{i,j}'$ would contain an occurrence of d, by the property shown above. But this is in contradiction to the assumption that $P_{i,j}' \equiv P_{i,j}$. Hence, $V_1', \ldots, V_{i-1}' \equiv \varepsilon$. Then clearly also $\alpha_\ell' = \alpha_\ell$ for every $\ell \in [i]$. □

Let us remark that an analogous lemma can be formulated for perturbations of $S_{i,j}$. However, we will only consider perturbations of $P_{i,j}$ afterwards.

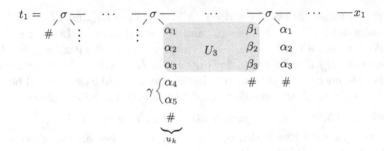

Fig. 1. Occurrence of a defect γ in the critical chain u_k of t_1

6   A Witness for $\mathcal{L}(G) \neq L$

In this section, we choose a tree $t \in L$ whose chains form a sufficiently large word U_i. By viewing a derivation tree κ of t, which induces a factorization $t = t_1 \circ\!\!- t_2$, we will see that the pumping lemma from Sect. 4 can be applied, and this leads to a perturbation in the defects of t_1. By Lemma 14, we receive the desired contradiction.

Let $Z_q = \langle u_1, v_1, \ldots, u_{m-1}, v_{m-1} \rangle$, recalling that $m = 2^{q-1} + 1$, and denote

$$\sigma(\#, u_1\#, x) \circ\!\!- \sigma(v_1\#, u_2\#, x) \circ\!\!- \cdots \circ\!\!- \sigma(v_{m-2}\#, u_{m-1}\#, x) \circ\!\!- \sigma(v_{m-1}\#, \#, \#)$$

by t. Observe that t contains m occurrences of σ, and that $\iota(t) = U_q$. Moreover, it is easy to see that $t \in L$. The chains of t are of the form $\alpha_1 \cdots \alpha_\ell$, resp. $\beta_1 \cdots \beta_\ell$, for some $\ell \in [q]$. As $m > 1$, there are $t_1, t_2 \in T(\Sigma)^1_1$ such that

$$A_\varepsilon \cdot \chi \Rightarrow^*_G B_\varepsilon \cdot s_\varepsilon \cdot \chi \Rightarrow_G (A_1 \cdot \vartheta_1 \cdot s_\varepsilon \circ\!\!- A_2 \cdot \vartheta_2 \cdot s_\varepsilon) \cdot \chi \Rightarrow^*_G t_1 \circ\!\!- t_2 = t.$$

Since both t_1 and t_2 contain at least one occurrence of σ, there is a $j \in [m-1]$ such that

$$t_1 = \sigma(\#, u_1\#, x) \circ\!\!- \sigma(v_1\#, u_2\#, x) \circ\!\!- \cdots \circ\!\!- \sigma(v_{j-1}\#, u_j\#, x) \quad \text{and}$$
$$t_2 = \sigma(v_j\#, u_{j+1}\#, x) \circ\!\!- \cdots \circ\!\!- \sigma(v_{m-2}\#, u_{m-1}\#, x) \circ\!\!- \sigma(v_{m-1}\#, \#, \#),$$

and this factorization of t induces a factorization of Z_q into Z' and Z'' with

$$Z' = \langle u_1, v_1, \ldots, u_j \rangle \quad \text{and} \quad Z'' = \langle v_j, \ldots, u_{m-1}, v_{m-1} \rangle.$$

Example 15. Let us consider an example which relates the introduced concepts. Figure 1 displays the critical chain u_k in t_1, whose defect is $\gamma = \alpha_4 \alpha_5$. As u_k is critical, every a-chain $u_{k'}$ in t_1 to its right (i.e., with $k' > k$) is of the form $\alpha_1 \cdots \alpha_\ell$, for some $\ell \leq 3$. In our intuition, γ is a sequence of opening parentheses which have no corresponding closing parenthesis in t_1. Therefore, t_2 must contain a suitable sequence of closing parentheses. Formally, γ^R occurs in $P_{i,j}$ as $P_{i,j} = P' U_5 \alpha_5 \alpha_4 U_3 P''$, so $D_{i,j} = D' \$ \gamma^R \$ D''$, for some $P', P'' \in \Gamma^*$ and $D', D'' \in (\Gamma \cup \{\$\})^*$. Therefore, γ is indeed a defect by definition.

By Proposition 13(2), the number of defects in $D_{q,j}$ is $q + 1 = 2p + 1$. Thus either t_1 contains at least $p + 1$ critical chains, or t_2 does. For the rest of this work, assume that t_1 contains at least $p + 1$ critical chains. The proofs for the other case are obtained mainly by substituting b for a and β for α.

By Proposition 9, there are a $\hat{t} \in T(\Sigma)_{p+1}^1$ with $t = \hat{t} \cdot \chi$, and a derivation tree κ of $\hat{t}$. Note that the height of κ is at most m. Therefore $|\delta| < m$ for every $\delta \in \text{pos}(\kappa)$. If $\delta = i_1 \cdots i_d$, then we denote the prefix $i_1 \cdots i_{d-\ell}$ of δ by δ_ℓ, for every $\ell \in [0, d]$. In particular, $\delta_0 = \delta$ and $\delta_d = \varepsilon$.

Let $s \in S(\Gamma)_p^p$ and $w \in \Gamma^*$. If there is no possibility of confusion, we will briefly say that w is a component of s if s has a component of the form wx_i, for some $i \in [p]$.

Proposition 16. *Let u_i be an a-chain of t_1, with $i \in [j]$. There is a leaf δ of κ such that $u_i = w_0 \cdots w_d$, where $d = |\delta|$, and w_ℓ is a component of s_{δ_ℓ}, for $\ell \in [0, d]$. Moreover, $\delta_{d-1} = 1$.*

Proof. Every leaf node of κ contributes exactly one occurrence of σ to t. So the chain u_i is contributed to t by κ's i-th leaf node δ, when enumerated from left to right. Let $d = |\delta|$. By tracing the path from δ to the root of κ, we see that

$$u_i \# = \pi_{j_{\delta_0}} \cdot s_{\delta_0} \cdot \vartheta_{\delta_0} \cdots s_{\delta_{d-1}} \cdot \vartheta_{\delta_{d-1}} \cdot s_\varepsilon \cdot \chi.$$

Therefore $u_i = w_0 \cdots w_d$, where w_ℓ is a component of s_{δ_ℓ}, for each $\ell \in [0, d]$. $\square$

In particular, w_d is a component of s_ε. The next lemma is a consequence of the fact that s_ε has only p components apart from x_{p+1}.

Lemma 17. *There is an a-defect γ whose critical chain is of the form $w'w$ for some $w', w \in \Gamma^*$ such that w is a component of s_ε, and $|\gamma| > |w| + mH$.*

Proof. Since t_1 contains more than p critical chains, by Proposition 16 and the pigeonhole principle, there must be two critical chains, say $u\gamma\alpha_i$ and $u'\gamma'\alpha_j$, where $\gamma\alpha_i$ and $\gamma'\alpha_j$ are distinct a-defects with $i < j$, such that

$$u\gamma\alpha_i = w'w \quad \text{and} \quad u'\gamma'\alpha_j = w''w \quad \text{for some } w', w'' \in \Gamma^*,$$

and some component w of s_ε. Observe that α_i is not a suffix of w, as otherwise α_i would be a suffix of α_j. Therefore $|w| < |\alpha_i|$, and hence

$$|w| + mH < |\alpha_i| + mH = |\alpha_{i+1}| \leq |\alpha_j| \leq |\gamma'\alpha_j|.$$

So the a-defect $\gamma'\alpha_j$ satisfies the properties in the lemma. $\square$

Theorem 18. *There is some $t' \in \mathcal{L}(G) \setminus L$.*

Proof. Let γ be the a-defect from Lemma 17. Assume that γ's critical chain in t_1 is u_k, where $k \in [j]$. Then $u_k = w_0 \cdots w_d$, where w_ℓ is a component of s_ℓ, for each $\ell \in [0, d]$. Moreover, $|\gamma| > |w_d| + mH$. Let f be the largest number such that $w_f \cdots w_d$ has γ as suffix, then $f \in [0, \ldots, d-1]$, and there are $w, w' \in \Gamma^*$ such that $w_f = w'w$ and $\gamma = ww_{f+1} \cdots w_d$.

Since $d < m$ and $|ww_{f+1} \cdots w_{d-1}| > mH$, there is a $\tilde{w} \in \{w, w_{f+1}, \ldots, w_{d-1}\}$ such that $|\tilde{w}| > H$. In other words, there is an $\ell \in [f, d-1]$ such that $A_{\delta_\ell} \Rightarrow_G^* B_{\delta_\ell} \cdot s_{\delta_\ell}$, and there is some $i \in [p]$ such that either (i) $\ell = f$ and $\pi_i \cdot s_{\delta_\ell} = w'\tilde{w}x_i$, or (ii) $\ell \neq f$ and $\pi_i \cdot s_{\delta_\ell} = \tilde{w}x_i$. In both cases Lemma 10 can be applied, and we receive that $s_{\delta_\ell} = v \cdot y \cdot z$, and by pumping zero times, also $A_{\delta_\ell} \Rightarrow_G^* B_{\delta_\ell} \cdot v \cdot z$. Therefore a derivation tree κ' can be constructed from κ by replacing the tuple s_{δ_ℓ} by $v \cdot z$. As δ_ℓ begins with the symbol 1, this alteration does only concern t_1, thus κ' derives a tree $\hat{t}' \in T(\Sigma)^1_{p+1}$ such that $\hat{t}' \cdot \chi = t'_1 \multimap t_2$, for some $t'_1 \in T(\Sigma)^1_1$. Denote $\hat{t}' \cdot \chi$ by t'.

Let us compare the k-th a-chain u'_k of t'_1 to u_k. Assume that the i-th components of v, y, and z are, respectively, $v'x_i$, $y'x_i$ and $z'x_i$. Then in case (i), there is a $w'' \in \Gamma^*$ such that $v' = w'w''$, as $y'z'$ is a suffix of w. Therefore,

$$u_k = w_1 \cdots w' \underbrace{w''y'z'w_{f+1} \cdots w_d}_{\gamma} , \quad u'_k = w_1 \cdots w'w''z'w_{f+1} \cdots w_d .$$

In case (ii),

$$u_k = w_1 \cdots \underbrace{ww_{f+1} \cdots w_{\ell-1}v'y'z'w_{\ell+1} \cdots w_d}_{\gamma} , \quad u'_k = w_1 \cdots w_{\ell-1}v'z'w_{\ell+1} \cdots w_d .$$

It is easy to see that $|t'|_\sigma = |t|_\sigma$, as the shape of κ was not modified. Thus Proposition 3 implies that if $t' \in L$, then also $|t'|_c = |t|_c$ and $|t'|_d = |t|_d$. In particular, $y' \in a^*$. Therefore, both in case (i) and (ii), $P'_{i,j} = \iota(t'_1)$ is a perturbation of $P_{i,j}$. Say that $P_{i,j}$ and $P'_{i,j}$ are of the form as in (3). Since $|y'| > 0$ by Lemma 10, at least one a was removed from the occurrence of γ^R in $P_{i,j}$. Therefore, there is some $e \in [q]$ such that $\alpha_e \neq \alpha'_e$. By Lemma 14, therefore $P'_{i,j} \not\equiv P_{i,j}$, and hence

$$\iota(t') \equiv \iota(t'_1)\,\iota(t_2) \not\equiv \iota(t_1)\,\iota(t_2) \equiv \varepsilon .$$

So $\iota(t') \notin D^*_\Gamma$, and by Proposition 2, $t' \notin L$.   □

Therefore, there is no cftg G with $\mathcal{L}(G) = h^{-1}(\mathcal{L}(G_{ex}))$, and we have proven Theorem 6. We close this paper with the positive result announced in the introduction.

Theorem 19. *The class of linear monadic context-free tree languages is closed under inverse linear tree homomorphisms.*

The proof of this theorem, omitted due to space constraints and to be found in [9], is based on the Greibach normal form of lm-cftg [6]. In fact, the closure of Greibach cftg under inverse linear tree homomorphisms was already proven [4], but this construction results in a copying cftg of higher nonterminal rank. Our proof follows the same line as the one in [4]. As every linear tree homomorphism can be decomposed into a linear alphabetic and a finite number of elementary ordered tree homomorphisms, it suffices to prove the closure of lm-cftg under the inverse of these restricted types. The respective proofs rely heavily on the Greibach normal form and are akin to those in [4].

7   Conclusion

In this work, we proved that the class of linear context-free tree languages is not closed under inverse linear tree homomorphisms. However, the tree languages of linear monadic context-free tree grammars, which are employed in praxis under the pseudonym of tree-adjoining grammars, are closed under this operation.

In applications which require nonmonadicity and closure under inverse homomorphisms, it may prove beneficial to revisit the formalism of k-*algebraic grammars*, i.e. context-free tree grammars over magmoids, where a nonterminal may derive a tuple of trees [2, Chapter V]. The class of languages defined by this type of grammar is indeed closed under inverse linear tree homomorphisms.

Acknowledgements. We are grateful for the helpful comments of an anonymous reviewer. Most importantly, we want to thank André Arnold for his help in this work. In our email conversations, which we enjoyed very much, he showed us the flaws in our first proof attempts, and encouraged us to try on. Moreover, the idea for an intermediate normal form of G is due to him, and he showed us how to significantly improve the presentation of the results in Sects. 5 and 6.

References

1. Arnold, A., Dauchet, M.: Forêts Algébriques et Homomorphismes Inverses. Inf. Control **37**, 182–196 (1978)
2. Arnold, A., Dauchet, M.: Théorie des Magmoïdes. RAIRO Inform. Théor. Appl. **12**(3), 235–257 (1978) and **13**(2), 135–154 (1979)
3. Arnold, A., Dauchet, M.: Morphismes et Bimorphismes d'Arbres. Theor. Comput. Sci. **20**, 33–93 (1982)
4. Arnold, A., Leguy, B.: Une Propriété des Forêts Algébriques «de Greibach». Inf. Control **46**(2), 108–134 (1980)
5. Engelfriet, J., Schmidt, E.M.: IO and OI. J. Comput. Syst. Sci. **15**(3), 328–353 (1977) and **16**(1), 67–99 (1978)
6. Fujiyoshi, A.: Analogical conception of Chomsky normal form and Greibach normal form for linear, monadic context-free tree grammars. IEICE Trans. Inf. Syst. **E89–D**(12), 2933–2938 (2006)
7. Gebhardt, K., Osterholzer, J.: A direct link between tree adjoining and context-free tree grammars. In: Proceedings of FSMNLP (2015)
8. Kepser, S., Rogers, J.: The equivalence of tree adjoining grammars and monadic linear context-free tree grammars. J. Log. Lang. Inf. **20**(3), 361–384 (2011)
9. Osterholzer, J., Dietze, T., Herrmann, L.: Linear context-free tree languages and inverse homomorphisms. http://arxiv.org/abs/1510.04881
10. Rounds, W.C.: Mappings and grammars on trees. Theor. Comput. Syst. **4**(3), 257–287 (1970)

Language Varieties and Semigroups

Scalar Ambiguity and Freeness in Matrix Semigroups over Bounded Languages

Paul C. Bell[1][(✉)], Shang Chen[1], and Lisa Jackson[2]

[1] Department of Computer Science, Loughborough University,
Loughborough LE11-3TU, UK
{P.Bell,S.Chen3}@lboro.ac.uk
[2] Department of Aeronautical and Automobile Engineering,
Loughborough University, Loughborough LE11-3TU, UK
L.M.Jackson@lboro.ac.uk

Abstract. There has been much research into freeness properties of finitely generated matrix semigroups under various constraints, mainly related to the dimensions of the generator matrices and the semiring over which the matrices are defined. A recent paper has also investigated freeness properties of matrices within a bounded language of matrices, which are of the form $M_1 M_2 \cdots M_k \subseteq \mathbb{F}^{n \times n}$ for some semiring $\mathbb{F}$ [9]. Most freeness problems have been shown to be undecidable starting from dimension three, even for upper-triangular matrices over the natural numbers. There are many open problems still remaining in dimension two.

We introduce a notion of freeness and ambiguity for scalar reachability problems in matrix semigroups and bounded languages of matrices. Scalar reachability concerns the set $\{\rho^T M \tau | M \in \mathcal{S}\}$, where $\rho, \tau \in \mathbb{F}^n$ are vectors and $\mathcal{S}$ is a finitely generated matrix semigroup. Ambiguity and freeness problems are defined in terms of uniqueness of factorizations leading to each scalar. We show various undecidability results.

Keywords: Matrix semigroup freeness · Scalar ambiguity · Bounded languages · Undecidability

1 Introduction

We start with some general notations and motivation.

Let $A = \{x_1, x_2, \ldots, x_k\}$ be a finite set of *letters* called an *alphabet*. A word w is a finite sequence of letters from A, the set of all words over A is denoted A^* and the set of nonempty words is denoted A^+. The *empty word* is denoted by ε. For two words $u = u_1 u_2 \cdots u_i$ and $v = v_1 v_2 \cdots v_j$, where $u, v \in A^*$, the concatenation of u and v is denoted by $u \cdot v$ (or by uv for brevity) such that $u \cdot v = u_1 u_2 \cdots u_i v_1 v_2 \cdots v_j$. Given a word $u = u_1 u_2 \cdots u_i$, a prefix of u is any word $u = u_1 u_2 \cdots u_j$, where $j \leq i$. A subset L of A^* is called a *language*. A language $L \subseteq A^*$ is called a *bounded language* if and only if there exist words $w_1, w_2 \ldots, w_m \in A^+$ such that $L \subseteq w_1^* w_2^* \cdots w_m^*$.

© Springer International Publishing Switzerland 2016
A.-H. Dediu et al. (Eds.): LATA 2016, LNCS 9618, pp. 493–505, 2016.
DOI: 10.1007/978-3-319-30000-9_38

We denote by $\mathbb{F}^{n \times n}$ the set of all $n \times n$ matrices over a semiring $\mathbb{F}$. Given $M \in \mathbb{F}^{m \times m}$ and $N \in \mathbb{F}^{n \times n}$, we define the direct sum $M \oplus N$ of M and N by:

$$M \oplus N = \begin{pmatrix} M & \overline{\varnothing} \\ \overline{\varnothing} & N \end{pmatrix},$$

where $\overline{\varnothing}$ is the zero matrix of appropriate dimension. Given a finite set of matrices $\mathcal{G} \subseteq \mathbb{F}^{n \times n}$, $\langle \mathcal{G} \rangle$ is the semigroup generated by $\mathcal{G}$.

For a semigroup $\mathcal{S}$, and a subset $\mathcal{G}' \subseteq \mathcal{S}$, we say that $\mathcal{G}'$ is a *code* if $x_1 \cdots x_{k_1} = y_1 \cdots y_{k_2}$, where $x_i, y_i \in \mathcal{G}'$ implies that $k_1 = k_2$ and $x_i = y_i$ for $1 \leq i \leq k_1$. Alternatively stated, $\mathcal{G}'$ is not a code if and only if some element of $\mathcal{S}$ has more than one factorization over $\mathcal{G}'$. We call $\mathcal{G}'$ a *prefix code* if no $w_1 \in \mathcal{G}'$ is a prefix of another word $w_2 \in \mathcal{G}'$.

Given a set $\mathcal{G} \subseteq \mathbb{F}^{n \times n}$, the *freeness problem* is to determine if $\mathcal{G}$ is a code for $\mathcal{S} = \langle \mathcal{G} \rangle$. It was proven by Klarner et al. that the freeness problem is undecidable over $\mathbb{N}^{3 \times 3}$ in [12] and this result was improved by Cassaigne et al. to hold even for upper-triangular matrices over $\mathbb{N}^{3 \times 3}$ in [6].

There are many open problems related to freeness in 2×2 matrices, see [8–10] for good surveys. The freeness problem over $\mathbb{H}^{2 \times 2}$ is undecidable [4], where $\mathbb{H}$ is the skew-field of quaternions (in fact the result even holds when all entries of the quaternions are rationals). The freeness problem for two upper-triangular 2×2 rational matrices remains open, despite many partial results being known [9].

The freeness problem for matrix semigroups defined by a bounded language was recently studied. Given a finite set of matrices $\{M_1, \ldots, M_k\} \subseteq \mathbb{Q}^{n \times n}$, we define a bounded language of matrices to be of the form:

$$\{M_1^{j_1} \cdots M_k^{j_k} | j_i \geq 0 \text{ where } 1 \leq i \leq k\}.$$

The freeness problem for a bounded language of matrices asks if there exists $j_1, \ldots, j_k, j_1', \ldots, j_k' \geq 0$, where at least one $j_i \neq j_i'$ such that $M_1^{j_1} \cdots M_k^{j_k} = M_1^{j_1'} \cdots M_k^{j_k'}$ in which case the bounded language of matrices is not free. This problem was shown to be decidable when $n = 2$, but undecidable in general [9].

In this paper we will introduce two notions of freeness in matrix semigroups called Scalar Ambiguity and Scalar Freeness problems. These are related to the uniqueness of factorizations of a set of scalar values of the form $\{\rho^T M \tau | M \in \mathcal{S}\}$, where $\mathcal{S}$ is a finitely generated matrix semigroup (see Sect. 2 for details). Such a set of scalars can be used to represent computations in many models. Related problems for *vector ambiguity* were studied in [3], where we were interested in the uniqueness of factorizations of a set of *vectors* $\{M\tau | M \in \mathcal{S}\}$.

In Sect. 3, we also study a related ambiguity problem for *Probabilistic Finite Automata* (PFA), defined in Sect. 1.1. The reachability problem for PFA (or emptiness problem) is known to be *undecidable* [14], even in a fixed dimension [5,11]. The reachability problem for PFA defined on a bounded language (i.e. where input words are from a bounded language which is given as part of the input), was recently shown to be undecidable [2].

Associated with each input word is the probability of that word being accepted by the PFA. In this paper, we show that determining whether every probability is unique is undecidable over a bounded language. In other words, to determine if there exists two input words which have the same probability of being accepted is undecidable. This is a similar concept to the *threshold isolation problem* shown in [5] to be undecidable, where we ask if each probability can be approximated arbitrarily closely.

1.1 Probabilistic Finite Automata

A vector $y \in \mathbb{Q}^n$ is a *probability distribution* if its elements are nonnegative and sum to 1 (y has an L_1 norm of 1). Matrix M is called a *column stochastic matrix* if each column is a probability distribution, a *row stochastic matrix* if each row is a probability distribution and it is called a *doubly stochastic matrix* if it is both row and column stochastic. For any row stochastic matrix M, if y is a probability distribution, then so is $y^T M$, since M preserves the L_1 norm on vectors and is nonnegative. The product of two row/column/doubly stochastic matrices is also row/column/doubly stochastic (respectively) as is not difficult to verify.

A *Probabilistic Finite Automaton* (PFA, see [5,14] for further details) over an alphabet A is a triplet (u, φ, v), where $u \in \mathbb{Q}^n$ is the *initial probability distribution*, $\varphi : A^* \to \mathbb{Q}^{n \times n}$ is a monoid homomorphism whose range is the set of n-dimensional row stochastic matrices and $v \in \mathbb{Q}^n$ is the *final state vector* whose ith coordinate is 1, if state i is final, and 0 otherwise.[1]

For a given PFA denoted $R = (u, \varphi, v)$ and a word $w \in A^*$, we can define a function $f_R : A^* \to [0, 1]$, where:

$$f_R(w) = u^T \varphi(w) v \in [0, 1]; \quad w \in A^*.$$

This is the probability of R being in a final state after reading word $w \in A^*$.

We will require the following undecidable problem for proving later results, which is a variant of the famous *Post's Correspondence Problem* (PCP).

Problem 1 (Mixed Modification PCP (MMPCP)). *Given a finite set of letters Σ, a binary alphabet Δ, and a pair of homomorphisms $h, g : \Sigma^* \to \Delta^*$, the MMPCP asks to decide whether there exists a word $w = x_1 \ldots x_k \in \Sigma^+, x_i \in \Sigma$ such that*

$$h_1(x_1)h_2(x_2) \ldots h_k(x_k) = g_1(x_1)g_2(x_2) \ldots g_k(x_k),$$

where $h_i, g_i \in \{h, g\}$, and there exists at least one j such that $h_j \neq g_j$.

Theorem 1 [7]. - *The Mixed Modification PCP is undecidable.*

2   Scalar Ambiguity and Freeness for Matrices

Consider a finite set $\mathcal{G} = \{G_1, G_2, \ldots, G_k\} \subset \mathbb{F}^{n \times n}$, generating a semigroup of matrices $\mathcal{S} = \langle \mathcal{G} \rangle$ and two column vectors $\rho, \tau \in \mathbb{F}^n$. Let $\Lambda(\mathcal{G})$ be the set of scalars

---

[1] The definition of a PFA in the literature often interchanges the roles of u and v from our definition and requires column stochastic matrices, but the two can easily be seen to be equivalent by transposing all matrices and interchanging u and v.

such that $\Lambda(\mathcal{G}) = \{\lambda : \lambda = \rho^T M \tau | M \in \mathcal{S}\}$. If for $\lambda \in \Lambda(\mathcal{G})$ there exists a unique matrix $M \in \mathcal{S}$ such that $\lambda = \rho^T M \tau$, then we say that λ is *unambiguous* with respect to $\mathcal{G}, \rho, \tau$. $\Lambda(\mathcal{G})$ is called unambiguous if every $\lambda \in \Lambda(\mathcal{G})$ is unambiguous. If for $\lambda \in \Lambda(\mathcal{G})$ there exists a unique product $G_{i_1} G_{i_2} \cdots G_{i_m} \in \mathcal{S}$, with each $G_{i_l} \in \mathcal{G}$ such that $\lambda = \rho^T G_{i_1} G_{i_2} \cdots G_{i_m} \tau$, then we say that λ is *free* with respect to $\mathcal{G}, \rho, \tau$. $\Lambda(\mathcal{G})$ is called free if every $\lambda \in \Lambda(\mathcal{G})$ is free.

Problem 2 (Scalar Ambiguity). *Is $\Lambda(\mathcal{G})$ unambiguous with respect to $\mathcal{G}, \rho, \tau$?*

Problem 3 (Scalar Freeness). *Is $\Lambda(\mathcal{G})$ free with respect to $\mathcal{G}, \rho, \tau$?*

Problems 2 and 3 look similar at first glance. However, the scalar ambiguity problem concentrates more on the properties of the semigroup $\mathcal{S}$ while the scalar freeness problem cares more about the properties of the set $\mathcal{G}$. A fact one can see from the definitions is that if the identity matrix I is contained in set $\mathcal{G}$, then the corresponding scalar set $\Lambda(\mathcal{G})$ is not free, but the same property does not hold for the scalar ambiguity problem. Also, we define the scalar freeness problem in a similar way of the matrix semigroup freeness problem. The links between the scalar ambiguity problem, scalar freeness problem and matrix semigroup freeness problem are illustrated in the following theorem.

Proposition 1. *Given a semigroup of matrices $\mathcal{S}$ generated by a finite set $\mathcal{G}$, and two column vectors ρ and τ, let $\Lambda(\mathcal{G})$ be a set of scalars generated by $\mathcal{G}, \rho$ and τ. Then the following relations hold:*

(1) If $\Lambda(\mathcal{G})$ is ambiguous, then $\Lambda(\mathcal{G})$ is not free.
(2) if $\Lambda(\mathcal{G})$ is free, then $\mathcal{S}$ is free.

Proof. (1) Suppose $\Lambda(\mathcal{G})$ is ambiguous, then by definition there exist two matrices $M_1, M_2 \in \mathcal{S}, M_1 \neq M_2$ such that $\rho^T M_1 \tau = \rho^T M_2 \tau$. Thus, there exists factorizations $M_1 = G_{i_1} G_{i_2} \ldots G_{i_{m_1}} \neq G_{j_1} G_{j_2} \ldots G_{j_{m_2}} = M_2$, where each $G_i, G_j \in \mathcal{G}$ so $\Lambda(\mathcal{G})$ is not free.

(2) We proceed by contradiction. Suppose $\Lambda(\mathcal{G})$ is free but $\mathcal{S}$ is not. If $\mathcal{S}$ is not free, there exists $G_{i_1} G_{i_2} \ldots G_{i_{m_1}} = G_{j_1} G_{j_2} \ldots G_{j_{m_2}} \in \mathcal{S}$, where $G_i, G_j \in \mathcal{G}$, and for at least one k, $G_{i_k} \neq G_{j_k}$, or $m_1 \neq m_2$. Thus, clearly it also holds that $\rho^T G_{i_1} G_{i_2} \ldots G_{i_{m_1}} \tau = \rho^T G_{j_1} G_{j_2} \ldots G_{j_{m_2}} \tau$, which contradicts the definition of scalar freeness. $\qquad \square$

It can be seen that by answering the scalar freeness problem, one can 'partly' answer the scalar ambiguity problem and the matrix semigroup freeness problem. However, neither problem is a sub-problem of the other, and it seems there is no direct connection between the scalar ambiguity problem and the matrix semigroup freeness problem. We are now ready to prove the main result of this section.

Theorem 2. *The Scalar Freeness Problem is undecidable over $\mathbb{Z}^{3 \times 3}$ and the Scalar Ambiguity Problem is undecidable over $\mathbb{Z}^{4 \times 4}$.*

Proof. We prove the result by encoding an instance of the MMPCP problem. The basic idea is inspired by [7]. We start by showing the undecidability of the scalar freeness problem. We construct a finite set of matrices $\mathcal{G}$, generating a matrix semigroup $\mathcal{S}$ and two fixed vectors ρ and τ such that the encoded MMPCP instance has a solution if and only if the scalar set $\Lambda(\mathcal{G})$ is free. In other words, there exists a scalar $\lambda \in \Lambda(\mathcal{G})$ such that $\lambda = \rho^T G_{i_1} G_{i_2} \cdots G_{i_{m_1}} \tau = \rho^T G_{j_1} G_{j_2} \cdots G_{j_{m_2}} \tau$, where $G_i, G_j \in \mathcal{G}$ and some $G_{i_k} \neq G_{j_k}$ or $m_1 \neq m_2$.

Let $\Sigma = \{x_1, x_2, \ldots, x_{n-2}\}$ and $\Delta = \{x_{n-1}, x_n\}$ be distinct alphabets and $h, g : \Sigma^* \to \Delta^*$ be an instance of the mixed modification PCP. The naming convention will become apparent below. We define two homomorphisms $\alpha, \beta :$ $(\Sigma \cup \Delta)^* \to \mathbb{Q}$ by:

$$\alpha(x_{i_1} x_{i_2} \cdots x_{i_m}) = \Sigma_{j=1}^m i_j (n+1)^{m-j},$$
$$\beta(x_{i_1} x_{i_2} \cdots x_{i_m}) = \Sigma_{j=1}^m i_j (n+1)^{j-m-1},$$

and $\alpha(\varepsilon) = \beta(\varepsilon) = 0$. Thus α represents $x_{i_1} x_{i_2} \cdots x_{i_m}$ as an $(n+1)$-adic number and β represents $x_{i_1} x_{i_2} \cdots x_{i_m}$ as a fractional number $(0.x_{i_m} \cdots x_{i_2} x_{i_1})_{(n+1)}$ (e.g. the number 123 may be represented as 0.321, base 10). Note that $\forall w \in (\Sigma \cup \Delta)^*, \alpha(w) \in \mathbb{N}$ and $\beta(w) \in (0,1) \cap \mathbb{Q}$. It is not difficult to see that $\forall w_1, w_2 \in (\Sigma \cup \Delta)^*, (n+1)^{|w_2|} \alpha(w_1) + \alpha(w_2) = \alpha(w_1 w_2)$ and $(n+1)^{-|w_2|} \beta(w_1) + \beta(w_2) = \beta(w_1 w_2)$.

Define $\gamma' : (\Sigma \cup \Delta)^* \times (\Sigma \cup \Delta)^* \to \mathbb{Q}^{3 \times 3}$ by

$$\gamma'(u, v) = \begin{pmatrix} (n+1)^{|u|} & 0 & \alpha(u) \\ 0 & (n+1)^{-|v|} & \beta(v) \\ 0 & 0 & 1 \end{pmatrix}.$$

It is easy to verify that $\gamma'(u_1, v_1)\gamma'(u_2, v_2) = \gamma'(u_1 u_2, v_1 v_2)$, i.e., γ' is a homomorphism. Define two more matrices T and T^{-1} :

$$T = \begin{pmatrix} 1 & 1 & 0 \\ 0 & 1 & 0 \\ 0 & 0 & 1 \end{pmatrix}, \quad T^{-1} = \begin{pmatrix} 1 & -1 & 0 \\ 0 & 1 & 0 \\ 0 & 0 & 1 \end{pmatrix}.$$

We now define $\gamma : (\Sigma \cup \Delta)^* \times (\Sigma \cup \Delta)^* \to \mathbb{Q}^{3 \times 3}$:

$$\gamma(u, v) = T\gamma'(u, v)T^{-1} = \begin{pmatrix} (n+1)^{|u|} & (n+1)^{-|v|} - (n+1)^{|u|} & \alpha(u) + \beta(v) \\ 0 & (n+1)^{-|v|} & \beta(v) \\ 0 & 0 & 1 \end{pmatrix}.$$

We can now verify that, $\gamma(u_1, v_1)\gamma(u_2, v_2) = T\gamma'(u_1, v_1)TT^{-1}\gamma'(u_2, v_2)T^{-1} = T\gamma'(u_1 u_2, v_1 v_2)T^{-1} = \gamma(u_1 u_2, v_1 v_2)$, hence γ is a homomorphism.

Let $\mathcal{G} = \{\gamma(x_i, g(x_i)), \gamma(x_i, h(x_i)) | x_i \in \Sigma, 1 \leq i \leq n-2\}$, $\mathcal{S} = \langle \mathcal{G} \rangle$, $\rho = (1, 0, 0)^T$ and $\tau = (0, 0, 1)^T$. Assume that there exists $M_1 = G_{i_1} G_{i_2} \cdots G_{i_t} \in \langle \mathcal{G} \rangle$ and $M_2 = G_{j_1} G_{j_2} \cdots G_{j_{t'}} \in \langle \mathcal{G} \rangle$ such that $t \neq t'$ or else at least one $G_{i_p} \neq G_{j_p}$ where $1 \leq p \leq t$ and $\lambda = \rho^T M_1 \tau = \rho^T M_2 \tau$. We see that:

$$\lambda = \rho^T M_1 \tau = (M_1)_{[1,3]} = \alpha(x_{i_1} x_{i_2} \cdots x_{i_t}) + \beta(f_1(x_{i_1})f_2(x_{i_2}) \cdots f_t(x_{i_t})),$$
$$\lambda = \rho^T M_2 \tau = (M_2)_{[1,3]} = \alpha(x_{j_1} x_{j_2} \cdots x_{j_{t'}}) + \beta(f_1'(x_{j_1})f_2'(x_{j_2}) \cdots f_{t'}'(x_{j_{t'}})),$$

where each $f_i, f'_i \in \{g, h\}$. Since $\alpha(w) \in \mathbb{N}$ and $\beta(w) \in (0, 1) \cap \mathbb{Q}$, $\forall w \in (\Sigma \cup \Delta)^*$, injectivity of α and β implies that if $\rho^T M_1 \tau = \rho^T M_2 \tau$, then $t = t'$ and $i_k = j_k$ for $1 \leq k \leq t$. Furthermore, if $\rho^T M_1 \tau = \rho^T M_2 \tau$, we have that $\beta(f_1(x_{i_1}) f_2(x_{i_2}) \cdots f_t(x_{i_t})) = \beta(f'_1(x_{i_1}) f'_2(x_{i_2}) \cdots f'_t(x_{i_t}))$ and since at least one $f_p \neq f'_p$ for $1 \leq p \leq t$ by our above assumption, then this corresponds to a correct solution to the mixed modification PCP instance (h, g). On the other hand, if there does not exist a solution to (h, g), then $\beta(f_1(x_{i_1}) f_2(x_{i_2}) \cdots f_t(x_{i_t})) \neq \beta(f'_1(x_{i_1}) f'_2(x_{i_2}) \cdots f'_t(x_{i_t}))$, and injectivity of β implies that $\rho^T M_1 \tau \neq \rho^T M_2 \tau$.

Since set $\mathcal{G} \subseteq \mathbb{Q}^{3 \times 3}$ is finite and has a finite description, there exists a computable constant $c \in \mathbb{N}$ such that $c \cdot \mathcal{G} \subseteq \mathbb{Z}^{3 \times 3}$ (based on the least common multiple of the denominators of elements of the matrices of $\mathcal{G}$). This completes the proof of the scalar freeness problem.

For the scalar ambiguity problem, we sketch the proof technique. The above encoding has the property that if some $\lambda = \rho^T M_1 \tau = (M_1)_{[1,3]} = \rho^T M_2 \tau = (M_2)_{[1,3]}$, then it implies that $M_1 = M_2$. If there exists a solution to the PCP instance, then some matrix $M \in \mathcal{S}$ has two distinct factorizations as above, each using a different sequence of morphisms f, g. We increase the dimension of γ by 1 to store an additional word, using mapping α, which is unique for each matrix. For example $x_1^i x_2$ for matrices corresponding to $h(x_i)$ and $x_3^i x_4$ for matrices corresponding to $g(x_i)$. Any two different matrix products will now have a distinct word stored in this element since $\{x_1^i x_2, x_3^i x_4 | 1 \leq i \leq n - 2\}$ is clearly a code. We modify ρ and τ to have an additional dimension which does not select this new word (i.e. they have zeros in the corresponding elements), and therefore its inclusion does not affect the set Λ.   □

3   Ambiguity and Freeness over a Bounded Language

We now study the concept of scalar ambiguity and scalar freeness for a *bounded language* of matrices, showing that these problems are undecidable. We start with the definition of Hilbert's tenth problem, which was shown to be undecidable by Matiyasevich. The following problem was stated as part of 23 open problems for the 20th century by David Hilbert in his 1900 address:

Hilbert's Tenth Problem (HTP) - "Given a Diophantine equation with any number of unknown quantities and with rational integral numerical coefficients: To devise a process according to which it can be determined by a finite number of operations whether the equation is solvable in rational integers".

To use a more modern terminology, Hilbert's tenth problem is to determine if there exists $n_1, n_2, \ldots n_k \in \mathbb{N}$ such that $P(n_1, n_2, \ldots, n_k) = 0$ is a Diophantine equation (i.e. P is a polynomial with integer coefficients). The undecidability of Hilbert's tenth problem was shown in 1970 by Yu. Matiyasevich building upon earlier work of many mathematicians, including M. Davis, H. Putman and J. Robinson. For more details of the history of the problem as well as the full proof of its undecidability, see the excellent reference [13]. We may restrict all the variables of the problem to be natural numbers without loss of generality, see [13, p.6].

The following corollary can be found in [2], or from the proof construction shown in [1].

Corollary 1 [2]. *Given an integer polynomial $P(n_1, n_2, \ldots, n_k)$, one can construct two vectors $\rho = (1, 0, \ldots, 0)^T \in \mathbb{N}^n$ and $\tau = (0, \ldots, 0, 1)^T \in \mathbb{N}^n$, an alphabet $\Sigma = \{x_1, x_2, \ldots, x_k\}$ and a homomorphism $\mu : \Sigma^* \to \mathbb{Z}^{n \times n}$, such that for any word of the form $w = x_1^{y_1} x_2^{y_2} \ldots x_k^{y_k} \in \Sigma^+$:*

$$\rho^T \mu(w) \tau = P(y_1, y_2, \ldots, y_k)^2,$$

and $\rho^T \mu(\varepsilon) \tau = 0$ for the empty word ε. The triple (ρ, μ, τ) is a linear representation of a $\mathbb{Z}$-regular formal power series $Z \in \mathbb{N}\langle\langle \Sigma \rangle\rangle$.

We will require the following lemma.

Lemma 1. *Given two integer polynomials P_1 and P_2 over variables $(x_1, \ldots, x_k)$ and with integer coefficients. It is undecidable to decide whether there exist integers $(y_1, \ldots, y_k)$ such that $P_1^2(y_1, \ldots, y_k) = P_2^2(y_1, \ldots, y_k)$.*

Proof. Let $P(x_2, \ldots, x_k)$ be an instance of Hilbert's tenth problem, i.e. a polynomial with integer coefficients and variables. Define $P_1(x_1, x_2, \ldots, x_k) = (x_1^2 + 1)P$ and $P_2(x_1, x_2, \ldots, x_k) = (x_1^2 + 2)P$. Since $0 < x_1^2 + 1 < x_1^2 + 2$, we see that $P_1^2(x_1, x_2, \ldots, x_k) = P_2^2(x_1, x_2, \ldots, x_k) \Leftrightarrow P_1 = P_2 = 0$, which implies that $P(x_2, \ldots, x_k) = 0$, which is undecidable to determine. This result holds for any value of x_1 since $x_1^2 + 1 \neq x_1^2 + 2$. We will use this property in the later proof. □

Now we show the main result of this section.

Theorem 3. *The Scalar Freeness Problem over a bounded language is undecidable. In other words, given k matrices $M_1, M_2, \ldots, M_k \in \mathbb{Q}^{n \times n}$, generating bounded language $M = M_1^* M_2^* \cdots M_k^*$, and two vectors $\rho, \tau \in \mathbb{Z}^n$, it is undecidable to decide if there exist $l_1, l_2, \ldots, l_k, r_1, r_2, \ldots, r_k \in \mathbb{N}$ such that*

$$\rho^T M_1^{l_1} M_2^{l_2} \ldots M_k^{l_k} \tau = \rho^T M_1^{r_1} M_2^{r_2} \ldots M_k^{r_k} \tau,$$

where $l_j \neq r_j$ for at least one j.

Proof. We prove this theorem by 4 steps. We will define a set of matrices $\{M_i, N_i | 0 \leq i \leq k+1\}$ for some $k+1 > 0$, which will define the bounded language of matrices $M = M_0^* M_1^* M_2^* \cdots M_k^* M_{k+1}^* N_0^* N_1^* N_2^* \cdots N_k^* N_{k+1}^*$. The matrices $\{M_i\}$ encode a polynomial P_1 and matrices $\{N_i\}$ will encode a separate polynomial P_2. The proof will show that if we have $\rho^T A_1 \tau = \rho^T A_2 \tau$, where $A_1, A_2 \in M$ and A_1, A_2 have different factorizations, then $A_1 = M_0^{j_0} M_1^{j_1} M_2^{j_2} \cdots M_k^{j_k} M_{k+1}^{j_{k+1}}$ and $A_2 = N_0^{j_0'} N_1^{j_1} N_2^{j_2} \cdots N_k^{j_k} N_{k+1}^{j_{k+1}'}$ (or vice versa). We will show that this implies that $P_1^2(j_1, \cdots, j_k) = P_2^2(j_1, \cdots, j_k)$, the determination of which was shown to be undecidable in Lemma 1.

Step 1. Given two integer coefficient polynomials P_1 and P_2 of same number of variables, from Corollary 1, we can construct an alphabet $\Sigma = \{x_1, x_2, \ldots, x_k\}$, two vectors $\rho' = (1, 0, \ldots, 0)^T, \tau' = (0, \ldots, 0, 1)^T \in \mathbb{N}^n$, and two homomorphisms $\mu_1, \mu_2 : \Sigma^* \to \mathbb{Z}^{n \times n}$ such that:

$$\rho'^T \mu_i(w) \tau' = \begin{cases} P_i(y_1, y_2, \ldots, y_k)^2, & \text{if} \quad w \in L \setminus \{\varepsilon\}; \\ 0, & \text{if} \quad w = \varepsilon; \end{cases}$$

where $i \in \{1, 2\}$ and L is the bounded language $L = x_1^* x_2^* \ldots x_k^* \subset \Sigma^*$.

Step 2. Given alphabets $K = \{0, 1, \ldots, k, k+1\}$ and $\Omega = K \cup \{\#, *\}$, define left and right desynchronizing morphisms l and $r : K^* \to \Omega^*$ by

$$l(0) = \#0, \quad l(1) = *1, \quad l(i) = \#i, \quad l(k+1) = \#(k+1)\#,$$
$$r(0) = \#0*, \quad r(1) = 1\#, \quad r(i) = i\#, \quad r(k+1) = (k+1)\#,$$

where $2 \leq i \leq k$. In the sequel, by abuse of notation, we use l_j, r_j to represent the words derived from the morphisms $l(j), r(j), 0 \leq j \leq k+1$. We define a word $u \in \Omega^*$ as 'free' if there is a unique factorization of u over $\{l_j, r_j\}$.

Let $L' = l_0^* l_1^* \cdots l_{k+1}^* r_0^* r_1^* \cdots r_{k+1}^* \in \Omega^*$. We shall now prove that any word $u = l_0^{j_0} l_1^{j_1} \cdots l_{k+1}^{j_{k+1}} r_0^{j_0'} r_1^{j_1'} \cdots r_{k+1}^{j_{k+1}'} \in L'$ is *not* free if and only if all $j_i = 0$ or all $j_i' = 0$ where $1 \leq i \leq k$.

Note that no element of $\Gamma = \{l_t, r_t | 0 \leq t \leq (k+1)\}$ is a prefix of any other word from the set, except for l_0 which is a prefix of r_0. Thus, $\Gamma \setminus \{l_0\}$ is a prefix code. If u does not begin with l_0 to some nonzero power, then by the definition of L', word u thus has a unique factorization.

If u has a prefix $\#0$, but not $\#0*$, then the prefix only matches with l_0, not r_0 and this prefix can be extracted from u since it has only a single possible factorization. We can continue this argument iteratively, until we reach u which begins with $\#0*$. Thus assume that u begins with $\#0*$. Let $u = l_0 u_1 = r_0 v_1$ be the two possible factorizations. Since u_1 must start with $*$, then $u_1 = l_1 u_2$. This implies that v_1 starts with symbol '1', which implies $v_1 = r_1 v_2$ since r_1 is the only word with prefix 1. Now, u_2 must be of the form $l_p u_3$ for some $2 \leq p \leq k$. Then v_2 must be of the form $r_p v_3$. This matching continues iteratively, until eventually we reach $(k+1)$, at which point we must use l_{k+1} for the u-word and r_{k+1} for the v-word.

At this point we have the two factorizations $u = l_0^* l_0 l_1 l_2^{j_2} \cdots l_k^{j_k} l_{k+1} r_{k+1}^*$ and $u = l_0^* r_0 r_1 r_2^{j_2} \cdots r_k^{j_k} r_{k+1} r_{k+1}^*$ as the only possibilities. An example of this follows:

$$u = \#0 * 1\#3\#5\#(k+1)\# = l_0 l_1 l_3 l_5 l_{k+1} \quad = \#0 \cdot *1 \cdot \#3 \cdot \#5 \cdot \#(k+1)\#$$
$$= r_0 r_1 r_3 r_5 r_{k+1} = \#0 * \cdot 1\# \cdot 3\# \cdot 5\# \cdot (k+1)\#$$

Step 3. We now encode the words l_i and r_j ($0 \leq i, j \leq k+1$) into rational numbers in the interval $(0, 1)$. For simplicity we first define a mapping $\sigma : \Omega \to X$, where $X = \{x_0, x_1, \ldots, x_{k+3}\}$ such that

$$\sigma(z) = \begin{cases} x_z & \text{if} \quad z \in \{0, 1, \ldots, k+1\}; \\ x_{k+2} & \text{if} \quad z = \#; \\ x_{k+3} & \text{if} \quad z = *. \end{cases}$$

We can extend σ to be a homomorphism $\sigma : \Omega^* \rightarrow X^*$. We then define a homomorphism $\beta : X^* \rightarrow (0, 1) \cap \mathbb{Q}$ in a similar way as in the proof of Theorem 2:

$$\beta(x_{i_1} x_{i_2} \cdots x_{i_m}) = \Sigma_{j=1}^m i_j (n + 1)^{j-m-1},$$

and $\beta(\varepsilon) = 0$, where $n = |X| = k + 4$. Moreover, we use a similar definition as in the proof of Theorem 2 for γ, but only on a single word $v \in X^*$, such that $\gamma : X^* \rightarrow \mathbb{Q}^{2 \times 2}$:

$$\gamma(v) = \begin{pmatrix} (n+1)^{-|v|} & \beta(v) \\ 0 & 1 \end{pmatrix}.$$

It can be verified that $\gamma(v_1 v_2) = \gamma(v_1)\gamma(v_2)$, and thus γ is a homomorphism.

Finally, we define $\gamma_l, \gamma_r : K^* \rightarrow \mathbb{Q}^{2 \times 2}$ by $\gamma_l(i) = \gamma(\sigma(l_i))$ and $\gamma_r(i) = \gamma(\sigma(r_i))$, where $0 \leq i \leq k + 1$. It can be seen that $\rho''^T \gamma_l \tau''$ and $\rho''^T \gamma_r \tau''$ are two homomorphisms from K^* to $(0, 1)$, where $\rho'' = (1, 0)^T$ and $\tau'' = (0, 1)^T$, mapping the words derived from left and right desynchronizing morphisms l and r to $(0, 1) \cap \mathbb{Q}$.

Step 4. In step 1 we showed how to encode an integer polynomial into a matrix. In step 2 and 3 we defined left and right desynchronizing morphisms and wrote them into matrix form. We now combine these steps together by defining a set of matrices $\{M_i, N_i\} \subset \mathbb{Q}^{(n+2) \times (n+2)}$:

$$M_0 = I \oplus \gamma_l(0), \quad M_i = \mu_1(x_i) \oplus \gamma_l(i), \quad M_{k+1} = I \oplus \gamma_l(k + 1),$$
$$N_0 = I \oplus \gamma_r(0), \quad N_i = \mu_2(x_i) \oplus \gamma_r(i), \quad N_{k+1} = I \oplus \gamma_r(k + 1),$$

where $1 \leq i \leq k$, and I is the $n \times n$ identity matrix. Then we let a scalar λ be written as:

$$\lambda = \rho^T M_0^{p_0} M_1^{p_1} \ldots M_{k+1}^{p_{k+1}} N_0^{q_0} N_1^{q_1} \ldots N_{k+1}^{q_{k+1}} \tau$$
$$= \rho'^T \mu_1(w_1)\mu_2(w_2)\tau' + \rho''^T \gamma_l(v_1)\gamma_r(v_2)\tau'',$$

where $\rho = (\rho'^T, \rho''^T)^T, \tau = (\tau'^T, \tau''^T)^T, w_1, w_2 \in L, v_1, v_2 = 0^*1^* \ldots (k + 1)^* \in K^*$. It can be seen that scalar λ contains two parts, one part consists of the homomorphisms μ_1, μ_2 we constructed in step 1 related to the polynomials, which is the integer part; the other part consists of the homomorphisms γ_l, γ_r we constructed in step 3 related to the desynchronizing morphisms, which is the fractional part. We now show that scalar λ is *not* free if and only if there exists some nonzero integer variables $(y_1, \ldots, y_k)$ such that $P_1^2(y_1, \ldots, y_k) = P_2^2(y_1, \ldots, y_k)$.

If λ is not free, by definition there must be integers $p_0, \ldots, p_{k+1}, q_0, \ldots, q_{k+1}$ and $p_0', \ldots, p_{k+1}', q_0', \ldots, q_{k+1}'$ such that

$$\lambda = \rho^T M_0^{p_0} \ldots M_{k+1}^{p_{k+1}} N_0^{q_0} \ldots N_{k+1}^{q_{k+1}} \tau = \rho^T M_0^{p_0'} \ldots M_{k+1}^{p_{k+1}'} N_0^{q_0'} \ldots N_{k+1}^{q_{k+1}'} \tau,$$

where $p_t \neq p_t'$ or $q_t \neq q_t'$ for at least one $0 \leq t \leq k + 1$. Since the value of the fractional part of λ only depends on the desynchronizing morphisms, l, r, and the fractional parts are identical in both factorizations, from step 2 we have

$$p_i = q_j' \text{ and } q_i = p_j' = 0, \quad \text{for } 1 \leq i, j \leq k, \text{ or}$$
$$p_i = q_i' = 0 \text{ and } q_j = p_j', \quad \text{for } 1 \leq i, j \leq k.$$

We only consider the first case, the second case can be analysed in a similar way. As the integer parts of λ in both factorizations are also identical, and $M_0, M_{k+1}, N_0, N_{k+1}$ are defined in a way that the value of $p_0, p_{k+1}, q_0, q_{k+1}$ and $p_0', p_{k+1}', q_0', q_{k+1}'$ do not affect the value of the integer part, we have

$$\rho'^T \mu_1^{p_1}(x_1) \ldots \mu_1^{p_k}(x_k)\tau' = \rho'^T \mu_2^{p_1}(x_1) \ldots \mu_2^{p_k}(x_k)\tau',$$

which implies that $P_1^2(p_1, \ldots, p_k) = P_2^2(p_1, \ldots, p_k)$. So $(p_1, \ldots, p_k)$ is a solution.

If λ is free, we show there is no solution such that $P_1^2 = P_2^2$ by contradiction. Assume there is a nonzero solution $(y_1, \ldots, y_k)$, such that $P_1^2(y_1, \ldots, y_k) = P_2^2(y_1, \ldots, y_k)$. From the way we construct P_1 and P_2 in Lemma 1, we know the value of y_1 can be any integer value without changing the equality. Thus it must be true that $P_1^2(1, y_2, \ldots, y_k) = P_2^2(1, y_2, \ldots, y_k)$, and there exists a word $w = x_1 x_2^{y_2} \ldots x_k^{y_k} \in L^*$ such that

$$\rho'^T \mu_1(w)\tau' = \rho'^T \mu_2(w)\tau',$$

which implies that

$$\rho'^T \mu_1(x_1)\mu_2^{y_2}(x_2) \ldots \mu_k^{y_k}(x_k)\tau' = \rho'^T \mu_1(x_1)\mu_2^{y_2}(x_2) \ldots \mu_k^{y_k}(x_k)\tau'.$$

Since

$$M_i = \mu_1(x_i) \oplus \gamma_l(i),$$
$$N_i = \mu_2(x_i) \oplus \gamma_r(i),$$

for $1 \leq i \leq k$, we can set $v = 0 \cdot 1 \cdot 2^{y_2} \ldots k^{y_k} \cdot (k+1)$, and scalar λ can be written as

$$\lambda = \rho'^T \mu_1(w)\tau' + \rho''^T \gamma_l(v)\tau'' = \rho^T M_0 M_1 M_2^{y_2} \cdots M_k^{y_k} M_{k+1}\tau$$
$$= \rho'^T \mu_2(w)\tau' + \rho''^T \gamma_r(v)\tau'' = \rho^T N_0 N_1 N_2^{y_2} \cdots N_k^{y_k} N_{k+1}\tau.$$

This shows that λ has two different factorizations, which is a contradiction. Thus we showed that scalar freeness problem can be reduced to the problem stated in Lemma 1, which is undecidable.   □

Theorem 4. *The Scalar Ambiguity Problem over a bounded language is undecidable.*

Proof. We can use the same idea as in the proof of Theorem 2, increasing the dimension of matrices M_i, N_i constructed in the proof of Theorem 3 to store an additional word which is unique for each matrix. Vectors ρ, τ are modified with an additional zero-value dimension such that the value of scalar λ is not affected. Hence in the case $\lambda = \rho^T M_1 \tau = \rho^T M_2 \tau$, we must have $M_1 \neq M_2$.   □

Corollary 2. *Vector ambiguity over a bounded language is undecidable.*

Proof. Immediately from Theorem 4 in the case when only one vector τ is considered.   □

Finally, we show a result related to Probabilistic Finite Automata (PFA).

Problem 4 (PFA Ambiguity Problem). *Given a PFA $R = (u, \varphi, v)$ over a bounded language $L \in A^*$, do there exist two different words $w_1, w_2 \in L$ such that $u^T \varphi(w_1) v = u^T \varphi(w_2) v$?*

Corollary 3. *Ambiguity for PFA over a bounded language is undecidable.*

Proof. This proof follows the construction of [15]; see also [2,11].

Let $M_i, N_i \in \mathbb{Q}^{(t-2)\times(t-2)}$ be matrices of dimension $(t-2)$ defined in the proof of Theorem 3, where $0 \le i \le k+1$. First, define a morphism $\zeta : A^* = \{a_0, a_1, \ldots, a_{2k+3}\}^* \to \{M_i, N_i\}$:

$$\zeta(a_j) = \begin{cases} M_j & \text{if} \quad 0 \le j \le k+1; \\ N_{j-(k+2)} & \text{if} \quad k+2 \le j \le 2k+3. \end{cases}$$

We then extend the dimension of the matrix $\zeta(a_j)$ to t by defining $\zeta' \to \mathbb{Q}^{t\times t}$:

$$\zeta'(a_j) = \begin{pmatrix} 0 & 0 & 0 \\ p_j & \zeta(a_j) & 0 \\ r_j & q_j & 0 \end{pmatrix},$$

where $p_j, q_j \in \mathbb{Q}^{(t-2)\times(t-2)}$ and $r_j \in \mathbb{Q}$ are properly chosen such that, for each $\zeta'(a_j)$, the row and column sums of $\zeta'(a_j)$ are all 0.

We now modify $\zeta'(a_j)$ so that every entry is positive. To do this we let Δ be the matrix of dimension t with all elements being 1. Assume b_i is in the set of entries of all $\zeta'(a_j)$, let $c > \max\{|b_i|\} \in \mathbb{Q}$. Define $\hat{\zeta} : A^* \to \mathbb{Q}_+^{t\times t}$ as

$$\hat{\zeta}(a_j) = \zeta'(a_j) + c\Delta.$$

It can be seen that all entries of the matrices $\hat{\zeta}(a_j)$ are positive. Finally, let $\varphi : A^* \to [0,1]^{t\times t}$ be

$$\varphi(a_j) = \frac{1}{ct}\hat{\zeta}(a_j) = \frac{1}{ct}\zeta'(a_j) + \frac{1}{t}\Delta.$$

Since row and column sums of $\zeta'(a_j)$ are all 0, and Δ is a matrix of dimension t with all elements being 1, it can be verified that all $\varphi(a_j)$ are stochastic matrices.

Then let $u = (0, \frac{1}{2}\rho^T, 0)^T$ and $v = (0, \frac{1}{2}\tau^T, 0)^T$, where $\rho, \tau \in \mathbb{Z}^{(t-2)\times(t-2)}$ are defined the same as in the proof of Theorem 3, we have constructed a PFA (u, φ, v) over a bounded language $w = a_0^* a_1^* \ldots a_{2k+3}^* \in L \subset A^*$.

To see that ambiguity for PFA (u, φ, v) is undecidable, we notice that $\Delta^n = t^{n-1}\Delta$ (as $\Delta^2 = t\Delta$), and by the definition of $\zeta'(a_j)$, it holds that $\zeta'(a_j) \cdot \Delta = \Delta \cdot \zeta'(a_j) = \overline{\varnothing}$ (the zero matrix). Thus,

$$u^T \varphi(w) v = u^T \left(\left(\frac{1}{ct}\right)^{|w|} \zeta'(w) + \left(\frac{1}{t}\right)^{|w|} \Delta^{|w|} \right) v$$

$$= \left(\frac{1}{ct}\right)^{|w|} (\rho^T \zeta(w)\tau) + u^T \left(\frac{\Delta}{t}\right) v$$

$$= \left(\frac{1}{ct}\right)^{|w|} (\rho^T M_0^{p_0} \cdots M_{k+1}^{p_{k+1}} N_0^{q_0} \cdots N_{k+1}^{q_{k+1}} \tau) + \frac{1}{t}$$

$$= \rho^T \left(\frac{M_0}{ct}\right)^{p_0} \cdots \left(\frac{M_{k+1}}{ct}\right)^{p_{k+1}} \left(\frac{N_0}{ct}\right)^{q_0} \cdots \left(\frac{N_{k+1}}{ct}\right)^{q_{k+1}} \tau + \frac{1}{t}$$

Since c and t are all fixed, the question of whether there exist two different words $w_1, w_2 \in L$ such that $u^T \varphi(w_1) v = u^T \varphi(w_2) v$, can be reduced to the scalar ambiguity problem over bounded languages, hence is undecidable.   □

4   Conclusion

We defined two related problems for matrix semigroups: the scalar ambiguity problem and the scalar freeness problem. We discussed the relations between these two problems and the matrix semigroup freeness problem. We showed that both problems are undecidable in low dimensions, three for ambiguity and four for freeness. These two problems remain undecidable even over bounded languages, but require higher dimensions. Using these results, we showed the ambiguity problem for probabilistic finite automata is also undecidable.

References

1. Bell, P.C., Halava, V., Harju, T., Karhumäki, J., Potapov, I.: Matrix equations and Hilbert's tenth problem. Int. J. Algebra Comput. **18**, 1231–1241 (2008)
2. Bell, P.C., Halava, V., Hirvensalo, M.: Decision problems for probabilistic finite automata on bounded languages. Fundamenta Informaticae **123**(1), 1–14 (2012)
3. Bell, P., Potapov, I.: Periodic and infinite traces in matrix semigroups. In: Geffert, V., Karhumäki, J., Bertoni, A., Preneel, B., Návrat, P., Bieliková, M. (eds.) SOFSEM 2008. LNCS, vol. 4910, pp. 148–161. Springer, Heidelberg (2008)
4. Bell, P.C., Potapov, I.: Reachability problems in quaternion matrix and rotation semigroups. Inf. Comput. **206**(11), 1353–1361 (2008)
5. Blondel, V., Canterini, V.: Undecidable problems for probabilistic automata of fixed dimension. Theor. Comput. Syst. **36**, 231–245 (2003)
6. Cassaigne, J., Harju, T., Karhumäki, J.: On the undecidability of freeness of matrix semigroups. Int. J. Algebra Comput. **9**(3–4), 295–305 (1999)
7. Cassaigne, J., Karhumäki, J., Harju, T.: On the decidability of the freeness of matrix semigroups. Technical report, Turku Centre for Computer Science (1996)
8. Cassaigne, J., Nicolas, F.: On the decidability of semigroup freeness. RAIRO - Theor. Inf. Appl. **46**(3), 355–399 (2012)
9. Charlier, E., Honkala, J.: The freeness problem over matrix semigroups and bounded languages. Inf. Comput. **237**, 243–256 (2014)
10. Choffrut, C., Karhumäki, J.: Some decision problems on integer matrices. Inf. Appl. **39**, 125–131 (2005)
11. Hirvensalo, M.: Improved undecidability results on the emptiness problem of probabilistic and quantum cut-point languages. In: van Leeuwen, J., Italiano, G.F., van der Hoek, W., Meinel, C., Sack, H., Plášil, F. (eds.) SOFSEM 2007. LNCS, vol. 4362, pp. 309–319. Springer, Heidelberg (2007)

12. Klarner, D., Birget, J.C., Satterfield, W.: On the undecidability of the freeness of integer matrix semigroups. Int. J. Algebra Comput. **1**(2), 223–226 (1991)
13. Matiyasevich, Y.: Hilbert's Tenth Problem. MIT Press, Cambridge (1993)
14. Paz, A.: Introduction to Probabilistic Automata. Academic Press, Orlando (1971)
15. Turakainen, P.: Generalized automata and stochastic languages. Proc. Am. Math. Soc. **21**, 303–309 (1969)

The Word Problem for HNN-extensions of Free Inverse Semigroups

Tatiana Baginová Jajcayová[⊠]

Department of Applied Informatics, Comenius University,
Mlynská Dolina, Bratislava, Slovakia
jajcayova@fmph.uniba.sk

Abstract. In our work [4], we studied the structure of HNN-extensions of inverse semigroups by investigating properties of their Schützenberger graphs. We introduced certain classes of HNN-extensions for which we were able to provide an iterative procedure for building Schützenberger automata. In the present paper, we show that in some cases this procedure yields an effective construction of the Schützenberger automata and thus provides a solution for the word problem. We analyze conditions under which the procedure is effective and show that the word problem is solvable in particular for HNN-extensions of free inverse semigroups.

Keywords: Semigroups · Monoids · Automata · Decidability · Word-problem

1 Preliminaries

The concept of HNN-extensions of groups was introduced by Higman, Neumann and Neumann [3] in their study of embedability of groups in 1949. In combinatorial group theory, HNN-extensions play an important role in applications to algorithmic problems ([3,6]). Work of Yamamura [14] shows the usefulness of HNN-extensions also in the category of inverse semigroups by proving the undecidability of any Markov and several non-Markov properties for finitely presented inverse semigroups. Similar questions for amalgams of inverse semigroups have been addressed in [1,2].

An inverse semigroup is a semigroup S which fulfills the following "weak invertibility" property: for each element $a \in S$ there is a unique element $a^{-1} \in S$ such that $a = aa^{-1}a$ and $a^{-1} = a^{-1}aa^{-1}$; the element a^{-1} is called the inverse of a. As a consequence of the definition, the set of the idempotents $E(S)$ is a semilattice. One may also define a natural partial order on S by setting $a \leq b$ if and only if $a = eb$ for some $e \in E(S)$.

Inverse semigroups may be regarded as semigroups of partial one-to-one transformations, so they arise very naturally in several areas of mathematics and more recently also in computer science, mainly since the inverse of an element can be seen as an "undo with a trace" of the action represented by that element. We refer the reader to the book of Petrich [11] for basic results and

© Springer International Publishing Switzerland 2016
A.-H. Dediu et al. (Eds.): LATA 2016, LNCS 9618, pp. 506–517, 2016.
DOI: 10.1007/978-3-319-30000-9_39

notation about inverse semigroups and to the more recent books of Lawson [5] and Paterson [10] for many references to the connections between inverse semi-groups and other branches of mathematics and computer science.

The free object on a set X in the category of inverse semigroups is denoted by $FIS(X)$. It is the quotient of the free semigroup $(X \cup X^{-1})^+$ by the least congruence ν that makes the resulting quotient semigroup inverse (see [11]). The inverse semigroup S presented by a set X of generators and a set R of relations is denoted by $S = Inv\langle X|R\rangle$. This is the quotient of the free semigroup $(X \cup X^{-1})^+$ by the least congruence τ that contains ν and the relations in R. The structure of $FIS(X)$ was studied via graphical methods by Munn [9]. Munn's work was greatly extended by Stephen [13] who introduced the notion of Schützenberger graphs associated with presentations of inverse semigroups.

The HNN-extension operation, is a classical operation over groups that builds, from a "base" group G, and two isomorphic subgroups A, B, another group H which is an extension of G. Although the general idea of HNN-extensions of semi-groups is similar to the idea behind HNN-extensions of groups, generalizing this operation over semigroups, in such a way that the resulting semigroup is an exten-sion of the base, is not trivial. In our paper, we will use the definition of HNN-extension of inverse semigroups introduced by Yamamura in [14]:

Definition 1 (A.Yamamura). Let S be an inverse semigroup. Let $\varphi : A \longrightarrow B$ be an isomorphism of an inverse subsemigroup A onto an inverse subsemigroup B where $e \in A \subseteq eSe$ and $f \in B \subseteq fSf$ (or $e \notin A \subseteq eSe$ and $f \notin B \subseteq fSf$ for some $e, f \in E(S)$). Then the inverse semigroup presented by

$$S^* = Inv\langle S, t \mid t^{-1}at = a\varphi, t^{-1}t = f, tt^{-1} = e, \forall a \in A\rangle$$

is called the *HNN-extension* of S associated with $\varphi : A \longrightarrow B$ and is denoted by $[S; A, B; \varphi]$.

We denote the set of all t-rules $\{t^{-1}at = a\varphi, \forall a \in A, t^{-1}t = f, tt^{-1} = e,\}$ by R_{HNN}. Then $S^* = Inv\langle X, t \mid R \cup R_{HNN}\rangle$. HNN-extensions have a natural universal mapping property which is described by the next proposition.

Proposition 2. *Let $S^* = [S; A, B; \varphi]$ be an HNN-extension of an inverse semi-group S. Then S^* has the following universal mapping property:*

(i) There is a unique homomorphism $\sigma : S \to S^*$ such that

$$t^{-1}\sigma(a)t = \sigma(a\varphi)\forall a \in A \quad and \quad t^{-1}t = \sigma(f), \ tt^{-1} = \sigma(e).$$

(ii) For each inverse semigroup T and homomorphism $\rho : S \to T$ such that there exists $p \in T$ such that $p^{-1}\rho(a)p = \rho(a\varphi)\forall a \in A$, $pp^{-1} = \rho(e)$ and $p^{-1}p = \rho(f)$, there is a unique homomorphism $\rho' : S \to T$ such that $\rho'(t) = p$ and $\rho' \circ \sigma = \rho$.

Yamamura in [14] showed that S embeds into S^*, provided that $e \in A \subseteq eSe$ and $f \in B \subseteq fSf$ (or $e \notin A \subseteq eSe$ and $f \notin B \subseteq fSf$) for some $e, f \in E(S)$.

One of the main purposes for studying HNN-extensions is to employ HNN-extensions to study algorithmic problems in inverse semigroups.

The Schützenberger automata are crucial in the study of inverse semigroup presentations (See for instance [7,8,13]).

Definition 3 (Schützenberger graph). Let w be a word in $(X \cup X^{-1})^+$. The Schützenberger graph of w relative to the presentation $Inv\langle X|R\rangle$ is the graph $S\Gamma(X, R, w\tau)$ whose vertices are the elements of the $\mathcal{R}$-class $\mathcal{R}_{w\tau}$ of $w\tau$ in S, and whose set of edges is

$$\{(v_1, x, v_2) \mid v_1, v_2 \in \mathcal{R}_{w\tau} \text{ and } v_1(x\,\tau) = v_2\}.$$

It is clear that for any two words w and w' representing the same element s in S, i.e., $w = s = w'$ in S (in other words, $w = w' \pmod{\tau}$), the Schützenberger graphs $S\Gamma(X, R, w\tau)$ and $S\Gamma(X, R, w'\tau)$ are isomorphic. Therefore the Schützenberger graph depends only on the element s represented by the word w and we speak of Schützenberger graphs of elements of S. It should be also noted that any two elements from the same $\mathcal{R}$-class determine the same Schützenberger graph.

The *Schützenberger automaton* of w relative $Inv\langle X|R\rangle$ can now be defined to be the inverse automaton $\mathcal{A}(X, R, w\tau) = (ww^{-1}\tau, S\Gamma(X, R, w\tau), w\tau)$. The language of $\mathcal{A}(X, R, w\tau)$ is the set $\{v \in (X \cup X^{-1})^+ \mid v\tau \geq w\tau\}$ which we denote by $[w]_\tau \uparrow$.

In particular, for the word problem, we have the result of Stephen [13] stating that for any two words u and v in $(X \cup X^{-1})^+$, $u = v$ in $S = Inv\langle X|R\rangle$ if and only if their corresponding Schützenberger automata are isomorphic, i.e. $\mathcal{A}(X, R, u) \cong \mathcal{A}(X, R, v)$. Equivalently, $u = v$ in S if and only if $u \in L[\mathcal{A}(X, R, v)]$ and $v \in L[\mathcal{A}(X, R, u)]$. Thus, an effective construction of the automaton $\mathcal{A}(X, R, w)$ for each $w \in S$ provides a solution to the word problem for S. We say that an *automaton $\mathcal{A}$ is effectively constructible* if there exists an algorithm that decides for any word $w \in (X \cup X^{-1})^+$ whether or not w belongs to the language of $\mathcal{A}$.

Conversely, if an inverse semigroup S has a decidable word problem and $w \in (X \cup X^{-1})^+$, then there is an algorithm to decide for each word $u \in (X \cup X^{-1})^+$ whether $uu^{-1} = ww^{-1}$ in S and thereby to decide whether $u\mathcal{R}w$ in S. Thus, for each letter $x \in X \cup X^{-1}$ and each word $u \in (X \cup X^{-1})^+$, there is an algorithm to decide whether or not the edge that starts at u, ends at ux, and is labelled by x belongs to the Schützenberger automaton $\mathcal{A}(X, R, w)$. Thus the word problem for an inverse semigroup S is decidable if and only if each Schützenberger automaton $\mathcal{A}(X, R, w)$ for S is effectively constructible.

In [4], we introduced an iterative procedure for constructing Schützenberger automata corresponding to, so called, lower bounded HNN-extensions. In the present paper, we first analyze this procedure and later apply this analysis to study HNN-extensions of the free inverse semigroups.

2   Effectiveness Analysis

In [4], we showed that if one starts with a lower bounded HNN-extension and a word $w \in (X \cup X^{-1} \cup \{t, t^{-1}\})^+$, then it is possible to carefully organize repeated

Stephen's [13] elementary expansions and determinations into Constructions 1–6 ([4], Chap. 3). In turn, repeated applications of Constructions 1–6 yields the Schützenberger automaton $\mathcal{A}(X \cup \{t\}, R \cup R_{HNN}, w)$. Recall the defintion of a lower bounded HNN-extension ([4]):

Definition 4. An HNN-extension $S^* = [S; A, B; \varphi]$ of an inverse semigroup S is called lower bounded if it satisfies the following conditions:

1. For all idempotents $e \in E(S)$, each of the sets $U_A(e) = \{u \in A | e \leq u\}$ and $U_B(e) = \{u \in B | e \leq u\}$ is either empty or has a least element.
2. For all idempotents $e \in E(S)$, there does not exist an infinite sequence $\{u_i\}_{i=1}^\infty$, where $u_i \in E(A)$, such that $u_i > f_A(eu_i) > u_{i+1}$ for all i; and for all idempotents $e \in E(S)$, there does not exist an infinite sequence $\{u_i\}_{i=1}^\infty$, where $u_i \in E(B)$, such that $u_i > f_B(eu_i) > u_{i+1}$ for all i.

A motivation for defining the class of lower bounded extensions comes from the fact that t-free subgraphs (lobes) of a Schützenberger graph of an HNN-extension S^* relative to the usual presentation are not, in general, Schützenberger graphs relative to the presentation of the original semigroup. For the class of lower bounded HNN-extensions all lobes of Schützenberger graphs relative to $Inv\langle X, t \mid R \cup R_{HNN}\rangle$ will always stay Schützenberger graphs relative to $Inv\langle X \mid R\rangle$. Furthermore, those graphs which can be Schützenberger graphs of lower bounded HNN-extensions have nice lobe structure: they are characterized as complete T-automata that possess a host – a T-subgraph with only finitely many lobes (lobes themselves can be infinite) such that the whole automaton may be obtained from any automaton containing the host by repeated applications of Construction 6 - that is a finite-lobe host already contains all "vital" information about the whole automaton with possibly infinitely many lobes. In [4], we showed that the class of lower bounded HNN-extensions is quite rich, and contains many well known inverse semigroups.

In the group case, thanks to Britton's Lemma the word problem for HNN-extension is decidable. The situation for inverse semigroups is more complicated. In the case of another extension, the free product with amalgamation, the group case is decidable, but the inverse semigroup case turned out to be undecidable in general (see E. Rodaro and P.V. Silva [12]).

In certain cases the iterative procedure of building the Schützenberger automaton $\mathcal{A}(X \cup \{t\}, R \cup R_{HNN}, w)$, gives an effective construction of the Schützenberger automata corresponding to lower bounded HNN-extensions and thus provides a solution for the word problem for the associated presentations. We will now examine the conditions under which our construction is effective.

First observe that if an HNN-extension $S^* = [S; A, B; \varphi]$ of an inverse semigroup $S = Inv\langle X | R\rangle$ satisfies Condition 1 of the definition of a lower bounded HNN-extension (Definition 4), then it is possible to perform all the steps of the iterative construction of the Schützenberger automaton relative to $Inv\langle X, t \mid R \cup R_{HNN}\rangle$ as described in [4]. Condition (2) of Definition 4 guarantees that only finitely many applications of Construction 3 are needed in the process of building the Schützenberger automaton relative to this presentation.

Note that for a given arbitrary HNN-extension $S^* = [S; A, B; \varphi]$, it may be hard to decide whether or not S^* is lower bounded. We believe that in general this problem may be undecidable. Consequently, in our analysis, we will assume that the HNN-extension $S^* = [S; A, B; \varphi]$ considered is lower bounded.

Before we examine the procedure from [4] step by step, let us first make some general observations. In order for the iterative procedure for building a Schützenberger graph corresponding to a lower bounded HNN-extension $S^* = [S; A, B; \varphi]$ to be effective, the following conditions must clearly be satisfied:

(C1) *The word problem for the original inverse semigroup $S = Inv\langle X|R\rangle$ is decidable.*

This requirement, as we have pointed out above, is equivalent to the condition that each Schützenberger automaton $\mathcal{A}(X, R, w)$ is effectively constructible.

(C2) *The isomorphisms φ and φ^{-1} from the definition of HNN-extension (Definition 1) are effectively calculable.*

(C3) *The membership problem for both inverse subsemigroups A and B in S is decidable.*

We now analyze Constructions 1–6 as they are used in the iterative procedure that begins with a linear automaton, $lin\mathcal{A}(w)$.

Construction 1. Is used to close our graphs with respect to $Inv\langle X|R\rangle$. One step of the construction consists of either a folding of a single pair of t-edges or a replacement of a single lobe by the Schützenberger graph of the element of S approximated by the lobe. Since in (C1) we require each Schützenberger graph relative to $Inv\langle X|R\rangle$ to be effectively constructible, each single application of Construction 1 is effective. Moreover, all the graphs we repeatedly apply Construction 1 to, have only finitely many lobes. Thus the process will require only finitely many applications of Construction 1 and is therefore effective.

Construction 2. $e - f$ completion attaches the Schützenberger graph $S\Gamma$ (X, R, e) to the initial vertex of every t-edge of our graph and the Schützenberger graph $S\Gamma(X, R, f)$ to the terminal vertex of every t-edge. The construction is clearly effective, since each of the Schützenberger graphs $S\Gamma(X, R, e)$ and $S\Gamma(X, R, f)$ is effectively constructible by (C1) and Construction 2 is applied in our procedure only once and only to a graph with finitely many t-edges.

Let us point out again that Construction 2 has to be applied only at the original t-edges of $lin\mathcal{A}(w)$ after Construction 1 has been applied. Each newly added t-edge in subsequent constructions will have a loop labelled by e attached to its initial vertex and a loop labelled by f attached to its terminal vertex.

Construction 3. In order to be able to apply Construction 3 to an automaton $\mathcal{A}$ we need to be able to verify, for every t-edge z, whether the sets $U_A(e(\alpha(z), \Delta))$ and $U_B(e(\omega(z), \Delta'))$ are empty or not. Thus the next condition must be satisfied:

(C4) *There is an algorithm that will decide for every idempotent $e \in E(S)$ whether or not the set $U_A(e) = \{u \in A | u \geq e\}$ is empty and whether or not the set $U_B(e) = \{u \in B | u \geq e\}$ is empty.*

If the sets $U_A(e)$ and $U_B(e)$ are non-empty, then the least elements $f_A(e)$ and $f_B(e)$ exist since we assume that S^* is lower bounded. To effectively carry out Construction 3 we need to be able to compute $f_A(e)$ and $f_B(e)$. Thus the next condition is needed.

(C5) *There is an algorithm to compute $f_A(e)$ for each idempotent $e \in E(S)$ for which $U_A(e) \neq \emptyset$. Similarly, if $U_B(e) \neq \emptyset$, then $f_B(e)$ is computable.*

To decide whether to apply Construction 3 to a t-edge z and whether Construction 3 entails an expansion at $\alpha(z)$ or at $\omega(z)$, we need to check whether or not $[U_A(e(\alpha(z), \Delta))]\varphi \subseteq U_B(e(\omega(z), \Delta'))$ and whether or not $[U_A(e(\alpha(z), \Delta))]\varphi \supseteq U_B(e(\omega(z), \Delta'))$. As follows from Lemmas 3.2.7 and 3.2.8 in [4] and the definitions of $U_A(e), U_B(e), f_A(e)$ and $f_B(e)$, in the case that both $U_A(e(\alpha(z), \Delta))$ and $U_B(e(\omega(z), \Delta'))$ are non-empty, it is equivalent to check whether or not $[f_A(e(\alpha(z), \Delta))]\varphi \geq f_B(e(\omega(z), \Delta'))$, and whether or not $[f_A(e(\alpha(z), \Delta))]\varphi \leq f_B(e(\omega(z), \Delta'))$. We can use the algorithm from (C4) to decide, whether or not the sets $U_A(e(\alpha(z), \Delta))$ and $U_B(e(\omega(z), \Delta'))$ are empty. In the case that they are both non-empty, decidability of these questions is equivalent to decidability whether

$$[f_A(e(\alpha(z), \Delta))]\varphi \cdot f_B(e(\omega(z), \Delta')) = f_A(e(\alpha(z), \Delta))]\varphi,$$

or

$$[f_A(e(\alpha(z), \Delta))]\varphi \cdot f_B(e(\omega(z), \Delta')) = f_B(e(\omega(z), \Delta'))$$

or neither. Thus we can decide these questions using the algorithm for the word problem for S. In the case that both sets are empty, it is clear that they are equal. If one of them is empty, then one of the inclusions is satisfied trivially.

If (C1)–(C5) are satisfied, then each application of Construction 3 can be effectively realized. Construction 3 is used several times throughout our process of building Schützenberger automata, but it is always applied to an automaton with finitely many lobes. Thus, as Lemma 3.2.9 in [4] asserts, repeated applications of Construction 3 to such an automaton results after finitely many steps in an automaton satisfying the equality property.

Let us emphasize that the key argument in the proof of Lemma 3.2.9 is the fact that S^* is a lower bounded HNN-extension. Namely, it is Condition 2 of the definition of a lower bounded HNN-extension that guarantees that repeated application of Construction 3 will terminate after finitely many steps.

Construction 4. One step of the construction removes an occurrence of a pair of non-parallel t-edges whose initial (terminal) vertices are connected by a path labelled by an element from A (from B).

In order to apply Construction 4 we need to be able to recognize such pairs of t-edges. In general, it is possible to have infinitely many paths connecting two vertices of a Schützenberger graph relative to $Inv\langle X \mid R \rangle$. Thus we need the following condition:

(C6) *There is an effective way to decide whether for two given vertices in any Schützenberger graph relative to $Inv\langle X \mid R \rangle$ there is a path from one vertex*

to the other labelled by an element from A (from B).
If such paths exist, the algorithm will produce one of them.

In (C6), the algorithm takes as input a word w and two words s_1 and s_2, and decides first if s_1 and s_2 are $\mathcal{R}$-related to w (i.e. if s_1 and s_2 are vertices of $S\Gamma(w)$), and then decides if there is a path in $S\Gamma(w)$ from s_1 to s_2 labelled by an element of A (of B). Note that (C1) ensures that we can decide if s_1 and s_2 are $\mathcal{R}$-related to w and further ensures that we can decide for any given word whether or not it belongs to $L[(s_1, S\Gamma(w), s_2)]$. However (C6) ensures decidability of whether or not $A \cap L[(s_1, S\Gamma(w), s_2)]$ is empty.

Construction 4 is only applied to automata with finitely many t-edges. Thus there are only finitely many pairs of the initial (terminal) vertices of t-edges to check. Thus using (C6), we can recognize t-edges that have to be separated. As the algorithm produces a path labelled by some $a \in A$ ($b \in B$), and the isomorphisms φ and φ^{-1} are effectively computable, we can effectively find the element $a\varphi$ ($b\varphi^{-1}$) and thus also the vertex v_2 needed to carry out Construction 4. Note that in the proof of Lemma 3.2.12 in [4], we showed that there are two different paths labelled by elements from A each starting at the same initial vertex of some t-edge and each ending at the same vertex v if and only if both images of these paths start at the terminal vertex of the t-edge and end at the same vertex v'. Thus, out of all possible paths between two vertices labelled by elements from A (from B), it is enough to know just one, as it will uniquely determine the vertex v_2.

Since there are only finitely many pairs of t-edges that need to be separated, finitely many repeated applications of Construction 4 are needed.

Construction 5. t-saturation adds all possible t-edges that are equivalent to an existing t-edge of a graph. Two t-edges are equivalent if they are parallel and there is a path connecting their initial vertices labelled by some $a \in A$ and a path connecting their terminal vertices labelled by $a\varphi \in B$.

Since the graphs to which Construction 5 is applied possess the equality property, recognizing whether a new t-edge should be added to a graph at some vertex v is equivalent to recognizing whether there is a path from the initial (terminal) vertex of some already existing t-edge to v labelled by an element of A (of B) and obtaining one such path. Then the terminal (initial) vertex of the new t-edge can be uniquely determined. Thus to realize Construction 5 effectively, we need to be able to decide if any given vertex can be "t-saturated". It follows from the above discussion that (C6) ensures decidability of this.

In [4], we showed that if we start with $lin\mathcal{A}(w)$ for some $w \in (X \cup X^{-1} \cup \{t, t^{-1}\})^+$ and apply of Constructions 1 - 5 as many times as possible in the prescribed order, then the resulting automaton is a T-automaton with finitely many lobes. The lobes in general can be infinite graphs.

Construction 6. It is Construction 6 ("sprouting" new lobes) that causes the Schützenberger automata relative to $Inv\langle X, t \mid R \cup R_{HNN} \rangle$ to possibly have infinitely many lobes. However, all the lobes introduced in Construction 6 are determined by the lobes already existing in the finite-lobe automaton obtained from $linA(w)$ by Constructions 1–5. This fact motivates the concept of a host, that is a subgraph with finitely many lobes that "stores" all important information about the whole Schützenberger graph and that stays unaffected by the consequent steps of the procedure.

In order for our iterative procedure to be effective we need to be able to effectively recognize so called A-buds and B-buds, i.e. for any given vertex v of a Schützenberger graph relative to $Inv\langle X \mid R\rangle$ there must be an effective way to decide whether there is a loop at v labelled by e and whether there is a loop at v labelled by f.

Since we assume that each Schützenberger graph Γ relative to $Inv\langle X \mid R\rangle$ is effectively constructible, there is an algorithm that decides whether or not e, f belongs to $L[(v, \Gamma, v)]$.

Once we can recognize such a vertex v of a graph Γ, we can effectively perform one application of Construction 6:

We check whether or not the set $U_A(e(v, \Delta))$ is empty (effective by (C4)); here Δ denotes a lobe that contains v.

If $U_A(e(v, \Delta)) \neq \emptyset$, we compute $f_A(e(v, \Delta))$ (this can be done by (C5)) and set $f := [f_A(e(v, \Delta))]\varphi$ (effective by (C2)). Otherwise we add a new lobe to Γ. The lobe is isomorphic to $S\Gamma(f)$ and is connected to Γ by a t-edge with the initial vertex at v. $S\Gamma(f)$ is effectively constructible by (C1). After this, Construction 5 is applied to the new t-edge.

The observations made in this section thus far will be used in the proof of the following theorem.

Theorem 5. *Let $S^* = [S; A, B; \varphi]$ be a lower bounded HNN-extension of an inverse semigroup $S = Inv\langle X \mid R\rangle$. Then the word problem for S^* relative to $Inv\langle X, t \mid R \cup R_{HNN}\rangle$ is decidable if the conditions (C1)–(C6) are satisfied.*

Proof. From the above analysis of Constructions 1 - 6, it is clear that if the conditions (C1)–(C6) are satisfied then each Schützenberger automaton relative to $Inv\langle X, t \mid R \cup R_{HNN}\rangle$ is effectively constructible and hence the word problem for S^* is decidable.   □

Let us now describe a bit more directly how our algorithm decides whether or not two words are equal in S^*: Given two words $w_1, w_2 \in (X \cup X^{-1} \cup \{t, t^{-1}\})^+$, we have to decide whether or not $w_1 \in L[A(X \cup \{t\}, R \cup R_{HNN}, w_2)]$ and $w_2 \in L[A(X \cup \{t\}, R \cup R_{HNN}, w_1)]$.

We start with the linear automaton $A_1 = linA(w_1)$. This is an automaton with finitely many lobes where each lobe is the linear automaton of a t-free subword of w_1. In Construction 1 the lobes are replaced by Schützenberger automata of these subwords (with some lobes possibly folded together). Since by (C1) all Schützenberger automata of S are effectively constructible, there is a finite description of each lobe (in fact, each lobe can be represented by the word v from the original alphabet, such that $S\Gamma(X, R, v)$ is isomorphic to the lobe.) This allows a finite description of the whole automaton. Moreover, we are also able to record the adjacency vertices of the automaton. Throughout applications of Constructions 1–4 the lobes are either folded or enlarged by multiplication by a Schützenberger automaton relative to $Inv\langle X \mid R\rangle$. (This clearly corresponds to multiplication of the associated words.) Moreover, only finitely many applications of Constructions 1–4 are needed. Thus we maintain a finite description of each automaton throughout all applications of Constructions 1–4. Denote by $\mathcal{A}_2$ the automaton obtained from $\mathcal{A}_1$ by all possible applications of Constructions 1–4. Note that the subsequent applications of Constructions 5 and 6 will not alter the lobes of $\mathcal{A}_2$. This in particular means that $\mathcal{A}_2$ embeds into automaton $\mathcal{A}(X \cup \{t\}, R \cup R_{HNN}, w_1)$, $\mathcal{A}_2 \hookrightarrow \mathcal{A}(X \cup \{t\}, R \cup R_{HNN}, w_1)$, and moreover, that $\mathcal{A}_2$ is a T-subautomaton of $\mathcal{A}(X \cup \{t\}, R \cup R_{HNN}, w_1)$. We write $\mathcal{A}_2 = (\alpha, \Gamma, \beta)$ and note that w_1 labels a path in $\mathcal{A}_2$ from the initial vertex α to the terminal vertex β.

Consider now the word w_2 and the problem of deciding if $w_2 \in L[\mathcal{A}(X \cup \{t\}, R \cup R_{HNN}, w_1)]$. Suppose, inductively, that we were effectively able to decide that there is a path in $\mathcal{A}_2$ labelled by some prefix u_1 of w_2 starting at the initial root α and ending at some vertex v in some lobe Δ of $\mathcal{A}_2$. Suppose that the next letter read in w_2 is x.

– If $x \in (X \cup X^{-1})$, then since (C1) holds, we can decide whether in the lobe Δ there is an edge labelled by x starting at v. If there is no such edge in Δ, we can conclude that $w_2 \notin L[\mathcal{A}(X \cup \{t\}, R \cup R_{HNN}, w_1)]$ and thus $w_1 \neq w_2$ in S^*. If there is such an edge z, we move to its terminal vertex $\omega(z)$ (i.e. we set $v := \omega(z)$), and repeat the process for the next letter of w_2, if such exists.

– If $x = t$ and v is the initial vertex of some t-edge z of $\mathcal{A}_2$. Then set $v := \omega(z)$ and repeat the process for the next letter of w_2. If v is not the initial vertex of any t-edge of $\mathcal{A}_2$, then we use the algorithm guaranteed by (C6) to check whether or not v is a vertex that would be t-saturated in Construction 5. We have a record of all adjacency vertices of $\mathcal{A}_2$ (there are just finitely many of them), and thus we can decide whether or not there is a path labelled by an element from A between v and one of these adjacency vertices.

• If the output is positive, the algorithm guaranteed by (C6) will produce a word $a \in A$ that labels such a path. Thus we can compute $a\varphi$ (by (C2)) and determine the vertex v', that would be the terminal vertex of the t-edge. We move to vertex v' and repeat the process for the next letter from w_2.

• If the output is negative, we use the algorithm that decides whether or not v is an A-bud. If the answer is positive, then we attach the new t-edge to v and record the new lobe that gets attached to the terminal vertex of this new t-edge (as described in Construction 6). We move to the terminal vertex of this t-edge and repeat the process for the next letter of w_2. If the answer is negative, i.e. v is not an A-bud of Γ then we can conclude that $w_2 \notin L[\mathcal{A}(X \cup \{t\}, R \cup R_{HNN}, w_1)]$ and $w_1 \neq w_2$ in S^*.

– If $x = t^{-1}$ and v is the terminal vertex of some t-edge z of $\mathcal{A}_2$, then set $v := \alpha(z)$ and repeat the process for the next letter of w_2, if such exists. If v is not the terminal vertex of any t-edge of $\mathcal{A}_2$, we use the algorithm guaranteed by (C6) to check whether or not v is the vertex that would be t-saturated in Construction 5.

- If the output is positive, the algorithm will produce a path labelled by some $b \in B$. Then we can compute $b\varphi^{-1}$ (by (C2)) and determine the vertex v' that would be the initial vertex of the t-edge. We move to the vertex v' and repeat the process for the next letter from w_2.
- If the output is negative, we use the algorithm to decide whether or not v is a B-bud. If the answer is positive, then we attach the new lobe to Δ, by a new t-edge with the terminal vertex of t at v. We move to the initial vertex of this t-edge and repeat the process for the next letter of w_2. If the answer is negative, i.e. v is not a B-bud of Γ, then we can conclude that $w_2 \notin L[\mathcal{A}(X \cup \{t\}, R \cup R_{HNN}, w_1)]$ and $w_1 \neq w_2$ in S^*.

If x was the last letter of w_2, then we need to check whether or not the terminal vertex of the edge z labelled by the letter x is equal to the terminal vertex of $\mathcal{A}_2$.

- If $\omega(z) = \beta$, then $w_2 \in L[\mathcal{A}(X \cup \{t\}, R \cup R_{HNN}, w_1)]$.
- If $\omega(z) \neq \beta$, then $w_2 \notin L[\mathcal{A}(X \cup \{t\}, R \cup R_{HNN}, w_1)]$ and $w_1 \neq w_2$ in S^*.

Similarly we can decide whether or not $w_1 \in L[\mathcal{A}(X \cup \{t\}, R \cup R_{HNN}, w_2)]$, by simply interchanging the roles of w_1 and w_2. If both $w_1 \in L[\mathcal{A}(X \cup \{t\}, R \cup R_{HNN}, w_2)]$ and $w_2 \in L[\mathcal{A}(X \cup \{t\}, R \cup R_{HNN}, w_1)]$ are true, then $w_1 = w_2$ in S^*.

3   The Word Problem for HNN-extensions of Free Inverse Semigroups

Throughout this section, we consider an HNN-extension of the form $S^* = [S; A, B; \varphi]$ where A and B are isomorphic finitely generated subsemigroups of $FIS(X)$. By analyzing Conditions (C1) - (C6) from the previous section, we show that our iterative procedure is effective and thus the word problem for HNN-extensions of free inverse semigroup S^* is decidable. This result is well-connected to the general study of decidability and complexity questions on (free) inverse monoids like, solvability of equations, model-checking for some logics, word-problem, membership problem for submonoids, etc.

Recall that each element of $FIS(X)$ can be represented as a birooted Munn tree [9]. Since, for a word w, the birooted Munn tree $(\alpha_w, MT(w), \beta_w)$ is the determinized form of the linear automaton $lin\mathcal{A}(w)$, it is constructible and its underlying graph is a tree. The pleasant consequence for the graphs in our constructions is that all lobes stay finite trees throughout the whole procedure. This is the key point in most of the following arguments.

First notice that $S^* = [S; A, B; \varphi]$ is lower bounded: Every idempotent $e \in S = FIS(X)$ is uniquely associated with its finite rooted tree $(\alpha_e, MT(e), \beta_e)$. The natural partial order of S satisfies $e \leq f$ if and only if $(\alpha_f, MT(f), \beta_f) \hookrightarrow (\alpha_e, MT(e), \beta_e)$. Condition 1 of Definition 4 then follows from the fact that only finitely many birooted Munn tree can be embeded into $(\alpha_e, MT(e), \beta_e)$. For any two idempotents e and f, the birooted Munn tree of their product $(\alpha_e f, MT(ef), \beta_e f)$ has the property that each edge of $MT(ef)$ is covered by an edge from $MT(e)$ or $MT(f)$. Condition 2 of lowerboundedness is somewhat technical but not hard to check, and it follows as in Example 7, Chap. 2 in [4].

To show the decidability of the word problem for S^* it is now enough to check that the conditions of Theorem 5 are satisfied.

(C1) The solution to the word problem for $FIS(X)$ is given by Munn's Theorem [9], that says that two words u and v in $(X \cup X^{-1})^*$ are equal if and only if $(\alpha_u, MT(u), \omega_w) = (\alpha_v, MT(v), \omega_v)$.

(C2) The isomorphisms φ and φ^{-1} are effectively calculable due to the fact that the inverse subsemigroups A and B are finitely generated. In such a case, it is enough to define φ (φ^{-1}) on the generators of A (B).

(C3) That this condition is satisfied follows from the next result of Cherubini, Meakin and Piochi [1]. The proof given there uses the Munn trees associated with elements of $FIS(X)$.

Lemma 6. *If U is a finitely generated inverse subsemigroup of a free inverse semigroup $FIS(X)$ then the membership problem for U in $FIS(X)$ is decidable.*

(C4) and (C5) The algorithms required in these conditions are described again in [1]. Let us briefly give an idea how they work. Let $e \in E(FIS(X))$. Consider the birooted Munn tree of e, $(\alpha_e, MT(e), \beta_e)$. Since e is an idempotent, we have $\alpha_e = \beta_e$. Note that if u is any other element of $FIS(X)$ with the Munn tree of the form $(\alpha_u, MT(u), \beta_u)$, then $u \geq e$ if and only if $MT(u) \subset MT(e)$ and $\alpha_u = \beta_u = \alpha_e$. Thus the set of all elements of $FIS(X)$ that are greater or equal to e corresponds to the set of all subtrees of $MT(e)$ rooted at α_e. Since there are only finitely many such subtrees, we can try all of them and use the algorithm from (C3) to each to check if any of them corresponds to an element of A (of B). Thus we can decide whether or not $U_A(e)$ $(U_B(e))$ is empty. If there is an element from A (from B) such that $u \geq e$, i.e. there is a subtree of $MT(e)$ rooted at α_e corresponding to an element of A (of B), then the maximal such subtree will correspond to the least idempotent $f_A(e)$ $(f_B(e))$.

(C6) Since each Schützenberger graph of $FIS(X)$ is a finite tree, there are only finitely many elements of $FIS(X)$ that can be read in a tree between two vertices. Check this finite set for membership in A (in B).

The next theorem is an immediate corollary of Theorem 5, Lemma 6 and the above arguments.

Theorem 7. *The word problem is decidable for any HNN-extension of the form $S^* = [S; A, B; \varphi]$ where A and B are isomorphic finitely generated inverse subsemigroups of $FIS(X)$.*

Acknowledgments. I wish to thank the referees for very helpful comments and suggestions.

The author acknowledges support from Projects VEGA 1/0577/14 and VEGA 1/0474/15.

References

1. Cherubini, A., Meakin, J., Piochi, B.: Amalgams of free inverse semigroups. Semigroup Forum **54**, 199–220 (1997)
2. Cherubini, A., Rodaro, E.: Decidability versus undecidability of the word problem in amalgams of inverse semigroups. Semigroups Algebras Oper. Theory **142**, 1–25 (2014)
3. Higman, G., Neumann, B., Neumann, H.: Embedding theorems for groups. J. London Math. Soc. **26**, 247–254 (1949)
4. Jajcayova, T.: HNN extensions of inverse semigroups. PhD Thesis at University of Nebraska-Lincoln Department of Mathematics and Statistics (1997)
5. Lawson, M.: Inverse Semigroups. The Theory of Partial Symmetries. World Scientific, River Edge, NJ (1998)
6. Lyndon, R., Schupp, P.: Combinatorial Group Theory. Springer, Heidelberg (1977)
7. Margolis, S., Meakin, J., Sapir, M.: Algorithmic problems in groups, semigroups and inverse semigroups. In: Fountain, J. (eds.) Semigroups, Formal Languages and Groups, pp. 147–214 (1995)
8. Meakin, J.: An invitation to inverse semigroup theory. In: Shum, K.P., et al. (eds.) Ordered Structures and Algebra of Computer Languages, pp. 91–115 (1993)
9. Munn, W.: Free inverse semigroups. Proc. London Math. Soc. **3**(29), 385–404 (1974)
10. Paterson, A.: Grupoids, Inverse Semigroups, and Their Operator Algebras. Birkhauser Boston, Boston (1999)
11. Petrich, M.: Inverse Semigroups. Wiley, New York (1984)
12. Rodaro, E., Silva, P.V.: Amalgams of inverse semigroups and reversible two-counter machines. J. Pure Appl. Algebra **217**, 585–597 (2013)
13. Stephen, J.: Presentation of inverse monoids. J. Pure Appl. Algebra **198**, 81–112 (1990)
14. Yamamura, A.: HNN extensions of inverse semigroups and applications. Int. J. Algebra Comput. **7**(5), 605–624 (1997)

Parsing

Between a Rock and a Hard Place – Uniform Parsing for Hyperedge Replacement DAG Grammars

Henrik Björklund, Frank Drewes, and Petter Ericson[✉]

Department of Computing Science, Umeå University, Umeå, Sweden
{henrikb,drewes,pettter}@cs.umu.se

Abstract. Motivated by applications in natural language processing, we study the uniform membership problem for hyperedge-replacement grammars that generate directed acyclic graphs. Our major result is a low-degree polynomial-time algorithm that solves the uniform membership problem for a restricted type of such grammars. We motivate the necessity of the restrictions by two different NP-completeness results.

Keywords: Graph grammar · Hyperedge replacement · Abstract meaning representation · DAG grammar · Uniform membership problem · Parsing

1 Introduction

Hyperedge-replacement grammars (HRGs [5,7]) are one of the most successful formal models for the generation of graph languages, because their properties resemble those of context-free grammars to a great extent. Unfortunately, polynomial parsing is an exception from this rule: graph languages generated by HRGs may be NP-complete. Thus, not only is the uniform membership problem intractable (unless P $\neq$ NP), but the non-uniform one is as well [1,8].

Recently, Chiang et al. [4] advocated the use of hyperedge-replacement for describing meaning representations in natural language processing (NLP), and in particular the abstract meaning representations (AMRs) proposed by Banarescu et al. [2], and described a general recognition algorithm together with a detailed complexity analysis. Unsurprisingly, the running time of the algorithm is exponential even in the non-uniform case, one of the exponents being the maximum degree of nodes in the input graph. Unfortunately, this is one of the parameters one would ideally not wish to limit, since AMRs may have unbounded node degree. However, AMRs are directed acyclic graphs (DAGs), a fact that is not exploited in [4]. Another recent approach to HRG parsing is [6], where predictive top-down parsing in the style of SLL(1) parsers is proposed. This is a uniform approach yielding parsers of quadratic running time in the size of the input graph, but the generation of the parser from the grammar is not guaranteed to run in polynomial time. (For a list of earlier attempts to HRG parsing, see [6].)

© Springer International Publishing Switzerland 2016
A.-H. Dediu et al. (Eds.): LATA 2016, LNCS 9618, pp. 521–532, 2016.
DOI: 10.1007/978-3-319-30000-9_40

In this paper, we study the complexity of the membership problem for DAG-generating HRGs. Since NLP applications usually involve a machine learning component in which the rules of a grammar are inferred from a corpus, and hence the resulting HRG cannot be assumed to be given beforehand, we are mainly interested in efficient algorithms for the uniform membership problem. We propose restricted DAG-generating HRGs and show, in Sect. 4, that their uniform membership problem is solvable in time $\mathcal{O}(n^2 + nm)$, where m and n are the sizes of the grammar and the input graph, resp. In linguistic applications, where grammars are usually much larger than the structures to be parsed, this is essentially equivalent to $\mathcal{O}(nm)$. To our knowledge, this is the first uniform polynomial-time parsing algorithm for a non-trivial subclass of HRGs. Naturally, the restrictions are rather strong, but we shall briefly argue in Sect. 5 that they are reasonable in the context of AMRs. We furthermore motivate the restrictions with two NP-completeness results for DAG-generating HRGs, in Sect. 6.

To save space, most proofs have been omitted. They are available in [3].

2   Preliminaries

The set of non-negative integers is denoted by $\mathbb{N}$. For $n \in \mathbb{N}$, $[n]$ denotes $\{1, \dots, n\}$. Given a set S, let $S^{\circledast}$ be the set of non-repeating lists of elements of S. If $sw \in S^{\circledast}$ with $s \in S$, we shall also denote sw by (s, w). If $\preceq$ is a (partial) ordering of S, we say that $s_1 \cdots s_k \in S^{\circledast}$ *respects* $\preceq$ if $s_i \preceq s_j$ implies $i \leq j$.

Hypergraphs and DAGs. A *ranked alphabet* is a pair (Σ, rank) consisting of a finite set Σ of symbols and a *ranking function* rank : $\Sigma \to \mathbb{N}$ which assigns a *rank* rank(a) to every symbol $a \in \Sigma$. We usually identify (Σ, rank) with Σ and keep 'rank' implicit.

Let Σ be a ranked alphabet. A (directed hyperedge-labeled) *hypergraph* over Σ is a tuple $G = (V, E, \text{src}, \text{tar}, \text{lab})$ consisting of

- finite sets V and $E \subseteq V \times V^{\circledast}$ of *nodes* and *hyperedges*, respectively
- *source* and *target mappings* src: $E \to V$ and tar: $E \to V^{\circledast}$ assigning to each hyperedge e its source src(e) and its sequence tar(e) of targets, and
- a *labeling* lab: $E \to \Sigma$ such that rank(lab(e)) = |tar(e)| for every $e \in E$.

Below, we call hyperedges edges and hypergraphs graphs, for simplicity. Note that edges have only one source but several targets, similarly to the usual notion of term (hyper)graphs. The DAGs we shall consider below are, however, more general than term graphs in that nodes can have out-degree larger than one.

Continuing the formal definitions, a *path* in G is a (possibly empty) sequence $e_1, e_2, \dots, e_k$ of edges such that for each $i \in [k-1]$ the source of e_{i+1} is a target of e_i. The *length* of a path is the number of edges it contains. A nonempty path is a *cycle* if the source of the first edge is a target of the last edge. If G does not contain any cycle then it is *acyclic* and is called a *DAG*. The *height* of a DAG G is the maximum length of any path in G. A node v is a *descendant* of a node u if $u = v$ or there is a nonempty path $e_1, \dots, e_k$ in G such that $u = \text{src}(e_1)$ and v occurs in tar(e_k). An edge e' is a *descendant edge* of an edge e if there is a path $e_1, \dots, e_k$ in G such that $e_1 = e$ and $e_k = e'$.

The *in-degree* of a node $u \in V$ is the number of edges e such that u is a target of e. The *out-degree* of u is the number of edges e such that u is the source of e. A node with in-degree 0 is a *root* and a node with out-degree 0 is a *leaf*.

For a node u of a DAG $G = (V, E, \text{src}, \text{tar}, \text{lab})$, the *sub-DAG rooted at* u is the DAG $G{\downarrow}_u$ induced by the descendants of u. Thus $G{\downarrow}_u = (U, E', \text{src}', \text{tar}', \text{lab}')$ where U is the set of all descendants of u, $E' = \{e \in E \mid \text{src}(e) \in U\}$, and src', tar', and lab' are the restrictions of src, tar and lab to E'. A leaf v of $G{\downarrow}_u$ is *reentrant* if there exists an edge $e \in E \setminus E'$ such that v occurs in $\text{tar}(e)$.

DAG Grammars. A *marked* graph is a tuple $G = (V, E, \text{src}, \text{tar}, \text{lab}, X)$ where $(V, E, \text{src}, \text{tar}, \text{lab})$ is a graph and $X \in V^{\circledast}$ is nonempty. The sequence X is called the *marking* of G, and the nodes in X are referred to as *external nodes*. If $X = (v, w)$ for some $v \in V$ and $w \in V^{\circledast}$ then we denote them by $\text{root}(G)$ and $\text{ext}(G)$, resp. This is motivated by the form or our rules, which is defined next.

Definition 1 (DAG grammar). *A DAG grammar is a system $H = (\Sigma, N, S, P)$ where Σ and N are disjoint ranked alphabets of terminals and nonterminals, respectively, S is the starting nonterminal with $\text{rank}(S) = 0$, and P is a set of productions. Each production is of the form $A \rightarrow F$ where $A \in N$ and F is a marked DAG over $\Sigma \cup N$ with $|\text{ext}(F)| = \text{rank}(A)$ such that $\text{root}(F)$ is the unique root of F and $\text{ext}(F)$ contains only leaves of F.*

Naturally, a terminal (nonterminal) edge is an edge labeled by a terminal (nonterminal, resp.). We may sometimes just call them terminals and nonterminals if there is no danger of confusion. By convention, we use capital letters to denote nonterminals, and lowercase letters for terminal symbols.

A derivation step of H is described as follows. Let G be a graph with an edge e such that $\text{lab}(e) = A$ and let $A \rightarrow F$ in P be a rule. Applying the rule involves replacing e with an unmarked copy of F in such a way that $\text{src}(e)$ is identified with $\text{root}(F)$ and for each $i \in [|\text{tar}(e)|]$, the ith node in $\text{tar}(e)$ is identified with the ith node in $\text{ext}(F)$. Notice that $|\text{tar}(e)| = |\text{ext}(F)|$ by definition. If the resulting graph is G', we write $G \Rightarrow_H G'$. We write $G \Rightarrow_H^* G'$ if G' can be derived from G in zero or more derivation steps. The *language* $\mathcal{L}(H)$ of H are all graphs G over the terminal alphabet T such that $S^{\bullet} \Rightarrow_H^* G$ where $S^{\bullet}$ is the graph consisting of a single node and a single edge labeled by S.

The graphs produced by DAG grammars are connected, single-rooted, and as the name implies, acyclic. This can be proved in a straightforward manner by induction on the length of the derivation. In the following, we consider only graphs of height at least 1, as the (single) graph of height 0 requires simple but cumbersome special cases.

Ordering the Leaves of a DAG. Let $G = (V, E, \text{src}, \text{tar}, \text{lab})$ be a DAG and let u and u' be leaves of G. We say that an edge e with $\text{tar}(e) = w$ is a *common ancestor edge* of u and u' if there are t and t' in w such that u is a descendant of t and u' is a descendant of t'. If, in addition, there is no edge with its source in w that is a common ancestor edge of u and u', we say that e is a *closest* common ancestor edge of u and u'.

Note that in a DAG, a pair of nodes can have more than one closest common ancestor edge.

Definition 2. *Let $G = (V, E, \mathrm{src}, \mathrm{tar}, \mathrm{lab})$ be a DAG. Then $\preceq_G$ is the partial order on the leaves of G defined by $u \preceq_G u'$ if, for every closest common ancestor edge e of u and u', $\mathrm{tar}(e)$ can be written as wtw' such that t is an ancestor of u and all ancestors of u' in $\mathrm{tar}(e)$ are in w'.*

3   Restricted DAG Grammars

DAG grammars are a special case of hyperedge-replacement grammars. We now define further restrictions that will allow polynomial time uniform parsing. Every rule $A \to F$ of a *restricted DAG grammar* is required to satisfy the following conditions (in addition to the conditions formulated in Definition 1):

1. If a node v of F has in-degree larger than one, then v is a leaf
2. If F consists of exactly two edges e_1 and e_2, both labeled by A, such that $\mathrm{src}(e_1) = \mathrm{src}(e_2)$ and $\mathrm{tar}(e_1) = \mathrm{tar}(e_2)$ we call $A \to F$ a *clone rule*. Clone rules are the only rules in which a node can have out-degree larger than 1 and the only rules in which a nonterminal can have the root as its source.
3. For every nonterminal e in F, all nodes in $\mathrm{tar}(e)$ are leaves.
4. If a leaf of F has in-degree exactly one, then it is an external node or its unique incoming edge is terminal.
5. The leaves of F are totally ordered by $\preceq_F$ and $\mathrm{ext}(F)$ respects $\preceq_F$.

We now demonstrate some properties of restricted DAG grammars.

Lemma 3. *Let $H = (\Sigma, N, S, P)$ be a restricted DAG grammar, G a DAG such that $S^\bullet \Rightarrow_H^* G$, and U the set of nodes of in-degree larger than 1 in G. Then U contains only leaves of G and $\mathrm{tar}(e) \in U^\circledast$ for every nonterminal e of G.*

Note that the lemma implies that leaves with in-degree exactly one are only connected to terminal edges. The lemma is proven by induction on the length of derivations, starting with the observation that $S^\bullet$ has the properties claimed. To simplify the presentation of our algorithm, we introduce a normal form.

Definition 4. *A restricted DAG grammar $H = (\Sigma, N, S, P)$ is on* normal form *if every rule $A \to F$ in P has one of the following three forms.*

(a) The rule is a clone rule.
(b) F has a single edge e, which is terminal.
(c) F has height 2, the unique edge e with $\mathrm{src}(e) = \mathrm{root}(F)$ is terminal, and all other edges are nonterminal.

See Fig. 1 for examples of right-hand sides of the three types. In particular, right-hand sides F of the third type consist of nodes $v, v_1, \ldots, v_m, u_1, \ldots, u_n$, a terminal edge e and nonterminal edges $e_1, \ldots, e_k$ such that

- $v = \mathrm{root}(F) = \mathrm{src}(e)$ and $v_1 \cdots v_m$ is a subsequence of $\mathrm{tar}(e)$,
- $\mathrm{src}(e_i) \in \{v_1, \ldots, v_m\}$ for all $i \in [k]$,
- $\mathrm{ext}(F)$ and $\mathrm{tar}(e_i)$, for $i \in [k]$, are subsequences of $u_1 \cdots u_n$.

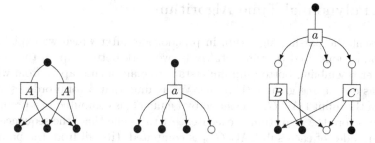

Fig. 1. Examples right-hand sides F of normal form rules of types (a), (b), and (c) for a nonterminal of rank 3. In illustrations such as these, boxes represent hyperedges e, where src(e) is indicated by a line and the nodes in tar(e) by arrows. Filled nodes represent the marking of F. Both tar(e) and ext(F) are drawn from left to right unless otherwise indicated by numbers

The proof of the following lemma follows the standard technique of dividing rules with large right-hand sides into several rules with smaller right-hand sides as in the proof of the Chomsky normal form of context-free grammars. The total size of the grammar does not change (except for a small constant factor).

Lemma 5. *Every restricted DAG grammar H can be transformed in linear time into a restricted DAG grammar H' on normal form such that $\mathcal{L}(H) = \mathcal{L}(H')$.*

One can now show that restrictions 1–5 imply that, in a DAG G generated by a restricted DAG grammar, the orders $\preceq_{G\downarrow_v}$ are consistent for all nodes v, that is, we have the following lemma the proof of which can be found in [3] (like the other proofs left out in this short version).

Lemma 6. *Let H be a restricted DAG grammar and $G = (V, E, \mathrm{src}, \mathrm{tar}, \mathrm{lab})$ a DAG generated by H. Then there is a total order $\trianglelefteq$ on the leaves of G such that $\preceq_G \subseteq \trianglelefteq$ and for every $v \in V$ and every pair u, u' of reentrant nodes of $G\downarrow_v$ we have $u \trianglelefteq u' \Leftrightarrow u \preceq_{G\downarrow_v} u'$.*

If a DAG G has been derived by a restricted DAG grammar in normal form, it is uniquely determined which subgraphs of G have been produced by a nonterminal, and which leaves were connected to it at that point. In particular, given a non-leaf node v in G, consider the subgraph $G\downarrow_v$. Consider the earliest point in the derivation where there was a nonterminal e having v as its source. We say that e generated $G\downarrow_v$. From the structure of G and $G\downarrow_v$, we know that all reentrant nodes of $G\downarrow_v$ are leaves and, by restriction 4, that e must have had exactly these reentrant leaves of $G\downarrow_v$ as targets. By Lemma 6 and restriction 5, the order of these leaves in tar(e) coincides with the total order $\preceq_{G\downarrow_v}$.

In other words, during the generation of G by a restricted DAG grammar, $G\downarrow_v$ must be generated from a nonterminal e such that src(e) = v and tar(e) is uniquely determined by the condition that it consists of exactly the reentrant nodes of $G\downarrow_v$ and respects $\preceq_{G\downarrow_v}$. Therefore, we will from now on view $G\downarrow_v$ as a *marked* DAG, where the marking is $(v, \mathrm{tar}(e))$.

4   A Polynomial Time Algorithm

We present the parsing algorithm in pseudocode, after which we explain various subfunctions used therein. Intuitively, we work bottom-up on the graph in a manner resembling bottom-up finite-state tree automata, apart from where a node has out-degree greater than one. We assume that a total order $\trianglelefteq$ on the leaves of the input DAG G, as ensured by Lemma 6, is computed in a preprocessing step before the algorithm is executed. At the same time, the sequence w_v of external nodes of each sub-DAG $G{\downarrow}_v$ is computed. (Recall from the paragraph above that these are the reentrant leaves of $G{\downarrow}_v$, ordered according to $\preceq_{G{\downarrow}_v}$.) For a DAG G of size n, this can be done in time $O(n^2)$ by a bottom-up process. To explain how, let us denote the set of all leaves of $G{\downarrow}_v$ by U_v for every node v of G. We proceed as follows. For a leaf v, let $\trianglelefteq_v = \{(v,v)\}$ and $w_v = v$. For every edge e with $\mathrm{tar}(e) = u_1 \ldots u_k$ such that u_i has already been processed for all $i \in [k]$, first check if $\trianglelefteq_0 = \bigcup_{i \in [k]} \trianglelefteq_{u_i}$ is a partial order. If so, define $\trianglelefteq_e$ to be the unique extension of $\trianglelefteq_0$ given as follows. Consider two nodes $u, u' \in U_{\mathrm{src}(e)}$ that are not ordered by $\trianglelefteq_0$. If i, j are the smallest indices such that $u \in U_{u_i}$ and $u' \in U_{u_j}$, then $u \trianglelefteq_e u'$ if $i < j$. Note that $\trianglelefteq_e$ is uniquely determined and total. Moreover, let w_e be the unique sequence in $U^{\circledast}_{\mathrm{src}(e)}$ which respects $\trianglelefteq_0$ and contains exactly the nodes in $U_{\mathrm{src}(e)}$ which are targets of edges of which e is not an ancestor edge. Similarly, if v is a node and all edges $e_1, \ldots, e_k$ having v as their source have already been processed, check if $\bigcup_{i \in [k]} \trianglelefteq_{e_i}$ is a partial order. If so, define $\trianglelefteq_e$ to be any total extension of this order. Moreover, check that $w_{e_1} = \cdots = w_{e_k}$, and let w_v be exactly this sequence. The preprocessing may fail for some graphs, but as these may not be part of $L(G)$ for any restricted DAG grammar G, we simply reject.

After this preprocessing, Algorithm 1 can be run. As the sequences w_u of external nodes for each sub-DAG $G{\downarrow}_u$ were computed in the preprocessing step, we consider this information to be readily available in the pseudocode. This, together with the assumption that the DAG grammar H is in normal form allows for much simplification of the algorithm.

Walking through the algorithm step by step, we first extract the root node (line 2) and determine which kind of (sub-)graph we are dealing with (line 4): one with multiple outgoing edges from the root must have been produced by a cloning rule to be valid, meaning we can parse each constituent subgraph (line 5) recursively (line 6) and take the intersection of the resulting nonterminal edges (line 7). Each nonterminal that could have produced all the parsed subgraphs and has a cloning rule is entered into *returns* (line 8). The procedure subgraphs_below is used to partition the sub-DAG $G{\downarrow}_v$ into one sub-DAG per edge having v as its source, by taking each such edge and all its descendant edges (and all their source and target nodes) as the subgraph. Note that the order among these subgraphs is undefined, though they are all guaranteed by the preprocessing to have the same sequence of external nodes w_v.

If, on the other hand, we have a single outgoing edge from the root node (line 9), we iterate through the subgraphs below the (unique) edge below the

Algorithm 1. Parsing of restricted graph grammars

```
 1: function PARSES_TO(restricted DAG grammar H in normal form, DAG G)
 2:     v ← root(G)
 3:     returns ← ∅
 4:     if out_degree(v) > 1 then
 5:         for Gᵢ ← subgraphs_below(v) do
 6:             Xᵢ ← parses_to(Gᵢ)
 7:         X ← ⋂ᵢ Xᵢ
 8:         returns ← {A ∈ X | has_clone_rule(A)}
 9:     else
10:         e ← edge_below(v)
11:         children ← ()
12:         for v′ ← targets(e) do
13:             if leaf(v′) then
14:                 append(children, external_node(v′))
15:             else
16:                 append(children, parses_to(G↓_{v′}))
17:         returns ← {A | (A → F) ∈ P and match(F, e, children)}
18:     return returns
```

root node (line 12). Nodes are marked either with a set of nonterminals (that the subgraph below the nodes can parse to) (line 16), or, if the node is a leaf, with a Boolean indicating whether or not the node is reentrant in the currently processed subgraph G (line 14).

The match function used in line 17 deserves a closer description, as much of the complexity calculations depend on this function taking no more than time linear in the size of the right-hand side. Let $src(e) = v$ and $tar(e) = v_1 \cdots v_k$. Each v_i has an entry in emphchildren. If v_i is a leaf it is a Boolean, otherwise a set of nonterminal labels. From G and $children$, we create a DAG G' as follows. Let T be the union of $\{v, v_1, \ldots, v_k\}$ and the set of leaves ℓ of G such that ℓ is reentrant to G (as indicated by $children$) or there is an $i \in [k]$ with ℓ being external in $G\downarrow_{v_i}$. Let $T = \{v, v_1, \ldots, v_k, t_1, \ldots, t_p\}$. Then G' has the set of nodes $U = \{u, u_1, \ldots, u_k, s_1, \ldots, s_p\}$. Let h be the bijective mapping with $h(v) = u$ and $h(v_i) = u_i$ for every $i \in [k]$ and $h(t_i) = (s_i)$ for every $i \in [p]$. We extend h to sequences in the obvious way. The root of G is u and there is a single edge d connected to it such that $lab(d) = lab(e)$, $src(d) = u$ and $tar(d) = u_1 \cdots u_k$. For every $i \in [k]$ such that v_i is not a leaf, G' has an edge d_i with $src(d_i) = u_i$ and $tar(d_i) = h(w_i)$, where w_i is the subsequence of leaves of $G\downarrow_{v_i}$ that belong to T, ordered by $\trianglelefteq$. The edge is labeled by the set of nonterminals $children[i]$.

Once match has built G' it tests whether there is a way of selecting exactly one label for each nonterminal edge in G' such that the resulting graph is isomorphic to rhs. This can be done in linear time since the leaves of both G' and rhs are totally ordered and, furthermore, the ordering on $v_1 \cdots v_k$ and $u_1 \cdots u_k$ makes the matching unambiguous.

Let us now discuss the running time of Algorithm 1. Entering the if branch of parses_to, we simply recurse into each subgraph and continue parsing.

The actual computation in the if-clause is minor: an intersection of the l sets of nonterminals found. Each time we reach the else clause in parses_to, we consume one terminal edge of the input graph. We recurse once for each terminal edge below this (no backtracking), so the parsing itself enters the else-clause n times, where n is the number of terminal edges in the input graph. For each rule $r = A \to F$, we build and compare at most $|F|$ nodes or edges in the match function. Thus, it takes $\mathcal{O}(nm)$ operations to execute Algorithm 1 in order to parse a graph with n terminal hyperedges according to a restricted DAG grammar H in normal form of size m. If H is not in normal form, Lemma 5 can be used to normalize it in linear time. Since the process does not affect the size of H by more than a (small) linear factor, the time bound is not affected. Finally, a very generous estimation of the running time of the preprocessing stage yields a bound of $\mathcal{O}(n^2)$, because n edges (and at most as many nodes) have to be processed, each one taking no more than n steps. Altogether, we have shown the following theorem, the main result of this paper.

Theorem 7. *The uniform membership problem for restricted DAG grammars is solvable in time $\mathcal{O}(n^2 + mn)$, where n is the size of the input graph and m is the size of the grammar.*

5   Representing and Generating AMRs

An Abstract Meaning Representation (AMR) is an ordinary directed edge-labeled acyclic graph expressing the meaning of a sentence. An example expressing *"Anna's cat is missing her"* is shown in Fig. 2. The root corresponds to the concept "missing", which takes two arguments, the misser and the missed.

In this representation every node has a special "instance edge" that determines the concept represented by its source node (miss, cat, anna). The most important concepts are connected to (specific meanings of) verbs, which have a number of mandatory arguments arg0, arg1 depending on the concept in question. While the representation shown is not directly compatible with the restrictions introduced in Sect. 3 a simple translation helps. Every concept with its k mandatory arguments is turned into a hyperedge of rank $k + 1$, the target nodes of which represent the instance (a leaf) and the roots of the arguments. The resulting hypergraph is shown in Fig. 2 on the right. Note that all shared nodes on the left (corresponding to cross-references) are turned into reentrant leaves. This is important because in a DAG generated by a restricted DAG grammar only leaves can have an in-degree greater than 1.

It might seem that we only need graphs with nodes of out-degree at most 1, and thus no cloning rules for their generation. However, a concept such as miss can typically also have optional so-called modifiers, such as in *"Anna's cat is missing her heavily today"*, not illustrated in the figure. Such modifiers can typically occur in any number. We can add them to the structure by increasing the rank of miss by 1, thus providing the edge with another target v. The out-degree of this node v would be the number of modifiers of miss. Using the notation

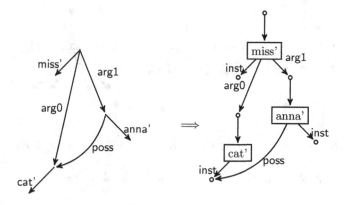

Fig. 2. Example translation of AMR

of Sect. 4, each sub-DAG $G\downarrow_e$ given by one of the outgoing edges e of v would represent one (perhaps complex) modifier. To generate these sub-DAGs $G\downarrow_e$ a restricted DAG grammar would use a nonterminal edge that has v as its source and which can be cloned. The latter makes it possible to generate any number of modifiers all of which can refer to the same shared concepts.

6   NP-Hardness Results

In order to motivate the rather harsh restrictions we impose on our grammars, we present NP-hardness results for two different classes of grammars that are obtained by easing the restrictions in different ways.

Theorem 8. *The uniform membership problem for DAG grammars that conform to restrictions 1–4 is NP-complete.*

Proof. The problem is in NP since the restrictions guarantee that derivations are of linear length in the size of the input graph. It remains to prove NP-hardness.

Let us consider an instance φ of the satisfiability problem SAT, i.e., a set $\{C_1, \ldots, C_m\}$ of clauses C_i, each being a set of literals x_j or $\neg x_j$, where $j \in [n]$ for some $m, n \in \mathbb{N}$. Recall that the question asked is whether there is an assignment of truth values to the variables x_j such that each clause contains a true literal. We have to show how to construct a DAG grammar H conforming to conditions 1–4 and an input graph G such that $G \in L(H)$ if and only if φ is satisfiable.

We first give a construction that violates conditions 4 and 5. It uses nonterminals S, K, K_i, K_{ij} with $i \in [m], j \in [n]$. The terminal labels are c, all $j \in [m]$, and an "invisible" label. The labels K, K_i, K_{ij}, c are of rank $2n$, S is of rank 0 and the remaining ones are of rank 1. Figure 3 depicts the rules of the grammar. Note that the rules are on normal form.

The first row of rules generates $2n$ leaves which, intuitively, represent $x_1, \neg x_1, \ldots, x_n, \neg x_n$ and are targets of a K-labeled nonterminal. The nonterminal is

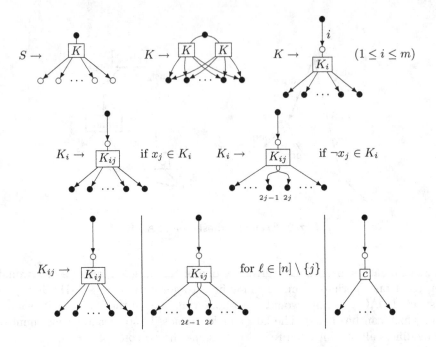

Fig. 3. Reduction of SAT to the uniform membership problem

cloned any number of times (with the intention to clone it m times, once for each clause). Afterwards it "guesses" which clause C_i ($i \in \mathbb{N}$) it should check. The second row of rules lets every K_i "guess" which literal makes C_i true. If the literal is negative, it interchanges the corresponding targets, otherwise it keeps their order. The third row of rules, for all pairs $(x_\ell, \neg x_\ell)$ that are not used to satisfy C_i, interchanges the corresponding targets or keeps their order. Finally, it replaces the nonterminal edge by a terminal one.

Now, consider the input DAG G in Fig. 4 (left). Suppose that G is indeed generated by H. Since the jth outgoing tentacles of all c-labeled edges point to the same node (representing either x_j or $\neg x_j$), a consistent assignment is obtained that satisfies φ. Conversely, a consistent assignment obviously gives rise to a corresponding derivation of G, thus showing that the reduction is correct.

Finally, let us note that changing the initial rule to the one shown in the left part of Fig. 4 (using a new terminal $\diamond$ of rank 2) makes H satisfy condition 4 as well. This change being made, the input graph is changed by including two copies of the original input, both sharing their leaves, and adding a new root with an outgoing $\diamond$-hyperedge targeting the roots of the two copies.   □

If we also disregard restriction 2, the non-uniform membership problem also becomes NP-complete, even if we only consider graphs of height 1.

Theorem 9. *There is a DAG grammar H that conforms to restrictions 1, 3, and 4, such that all graphs in $\mathcal{L}(H)$ have height 1 and $\mathcal{L}(H)$ is NP-complete.*

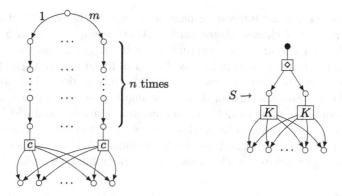

Fig. 4. Input graph in the proof of Theorem 8 (left) and modified starting rule (right)

The proof is by reduction from the membership problem for *context-free grammars with disconnecting* (CFGDs), using a result from [8]. A CFGD is an ordinary context-free grammar in Chomsky normal form, with additional rules $A \to \diamond$, where $\diamond$ is a special symbol that cuts the string into two. Thus, an element in the generated language is a finite multiset of strings rather than a single string. As shown in [8], CFGDs can generate NP-complete languages. We represent a multiset $\{s_1, \ldots, s_k\}$ of strings s_i as a graph consisting of k DAGs of height 1 sharing their roots. If $s_i = a_1 \cdots a_m$ then the DAG representing it consists of the root v, leaves $u_0, \ldots, u_m$, and a_i-hyperedges e_i with $\mathrm{src}(e_i) = r$ and $\mathrm{tar}(e_i) = u_{i-1}u_i$. Moreover, there are two "unlabeled" terminal edges from v to u_0 and u_n, resp. Now, every CFGD can be turned into an equivalent DAG grammar using the schemata in Fig. 5.   □

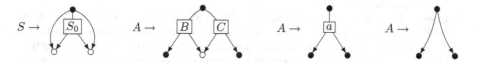

Fig. 5. Rules of a DAG grammar equivalent to a CFGD with initial nonterminal S_0, from left to right: initial rule, $A \to BC$, $A \to a$, $A \to \diamond$.

7   Future Work

A number of interesting questions remain open. Is it the case that lifting any one of our five restrictions, while keeping the others, leads to NP-hardness? It seems that the algorithm we propose leads to a fixed-parameter tractable algorithm, with the size of right-hand sides in the grammar as the parameter, when we lift restriction 5 (enforcing that the marking respects $\preceq_F$). Is this actually the case and are there other interesting parameterizations that give tractability for some less restricted classes of grammars? Another open question is whether the algorithm for checking the structure of the input graph and computing the ordering on the leaves can be optimized to run in linear or $\mathcal{O}(n \log n)$ time.

From a practical point of view, one should study in detail how well suited restricted DAG grammars are for describing linguistic structures such as AMRs. Which phenomena can be modeled in an appropriate manner and which cannot? Are there important aspects in AMRs that can be modeled by general DAG-generating HRGs but not by restricted DAG grammars? If so, can the restrictions be weakened appropriately without sacrificing polynomial parsability?

Acknowledgements. We gratefully acknowledge the support from the Swedish Research Council grant 621-2011-6080 and the EU FP7 MICO project. We are furthermore grateful to the anonymous referees for various helpful comments.

References

1. Aalbersberg, I.J., Ehrenfeucht, A., Rozenberg, G.: On the membership problem for regular DNLC grammars. Discrete Appl. Math. **13**, 79–85 (1986)
2. Banarescu, L., Bonial, C., Cai, S., Georgescu, M., Griffitt, K., Hermjakob, U., Knight, K., Koehn, P., Palmer, M., Schneider, N.: Abstract meaning representation for sembanking. In: Proceedings of 7th Linguistic Annotation Workshop, ACL 2013 (2013)
3. Björklund, H., Drewes, F., Ericson, P.: Between a rock and a hard place - Parsing for hyperedge replacement DAG grammars. Technical report UMINF 15.13, Umeå University (2015). http://www8.cs.umu.se/research/uminf/index.cgi
4. Chiang, D., Andreas, J., Bauer, D., Hermann, K.M., Jones, B., Knight, K.: Parsing graphs with hyperedge replacement grammars. In: Proceedings of 51st Annual Meeting of the Association for Computational Linguistics (ACL 2013), vol. 1: Long Papers, pp. 924–932 (2013)
5. Drewes, F., Habel, A., Kreowski, H.J.: Hyperedge replacement graph grammars. In: Rozenberg, G. (ed.) Handbook of Graph Grammars and Computing by Graph Transformation, vol. 1: Foundations, chap. 2, pp. 95–162. World Scientific (1997)
6. Drewes, F., Hoffmann, B., Minas, M.: Predictive top-down parsing for hyperedge replacement grammars. In: Parisi-Presicce, F., Westfechtel, B. (eds.) ICGT 2015. LNCS, vol. 9151, pp. 19–34. Springer, Heidelberg (2015)
7. Habel, A.: Hyperedge Replacement: Grammars and Languages. LNCS, vol. 643. Springer, Heidelberg (1992)
8. Lange, K.J., Welzl, E.: String grammars with disconnecting or a basic root of the difficulty in graph grammar parsing. Discrete Appl. Math. **16**, 17–30 (1987)

An Error Correcting Parser for Context Free Grammars that Takes Less Than Cubic Time

Sanguthevar Rajasekaran and Marius Nicolae[(✉)]

Department of Computer Science and Engineering, University of Connecticut,
371 Fairfield Way, Unit 4155, Storrs, CT 06269, USA
{sanguthevar.rajasekaran,marius.nicolae}@uconn.edu

Abstract. The problem of parsing has been studied extensively for various formal grammars. Given an input string and a grammar, the parsing problem is to check if the input string belongs to the language generated by the grammar. A closely related problem of great importance is one where the input are a string $\mathcal{I}$ and a grammar G and the task is to produce a string $\mathcal{I}'$ that belongs to the language generated by G and the 'distance' between $\mathcal{I}$ and $\mathcal{I}'$ is the smallest (from among all the strings in the language). Specifically, if $\mathcal{I}$ is in the language generated by G, then the output should be $\mathcal{I}$. Any parser that solves this version of the problem is called an *error correcting parser*. In 1972 Aho and Peterson presented a cubic time error correcting parser for context free grammars. Since then this asymptotic time bound has not been improved under the assumption that the grammar size is a constant. In this paper we present an error correcting parser for context free grammars that runs in $O(T(n))$ time, where n is the length of the input string and $T(n)$ is the time needed to compute the tropical product of two $n \times n$ matrices.

In this paper we also present an $\frac{n}{M}$-approximation algorithm for the *language edit distance problem* that has a run time of $O(Mn^\omega)$, where $O(n^\omega)$ is the time taken to multiply two $n \times n$ matrices. To the best of our knowledge, no approximation algorithms have been proposed for error correcting parsing for general context free grammars.

1 Introduction

Parsing is a well studied problem owing to its numerous applications. For example, parsing finds a place in programming language translations, description of properties of semistructured data [12], protein structures prediction [11], etc. For context free grammars (CFG), two classical algorithms can be found in the literature: CYK [2,6,15] and Earley [3]. Both of these algorithms take $O(n^3)$ time in the worst case. Valiant has shown that context free recognition can be reduced to Boolean matrix multiplication [13].

The problem of parsing with error correction (also known as the *language edit distance problem*) has also been studied well. Aho and Peterson presented an $O(n^3)$ time algorithm for CFG parsing with errors. Three kinds of errors were considered, namely, insertion, deletion, and substitution. This algorithm

© Springer International Publishing Switzerland 2016
A.-H. Dediu et al. (Eds.): LATA 2016, LNCS 9618, pp. 533–546, 2016.
DOI: 10.1007/978-3-319-30000-9_41

depended quadratically on the size of the grammar and was based on Earley parser. Subsequently, Myers [9] presented an algorithm for error correcting parsing for context free grammars that also runs in cubic time but the dependence on the grammar size was linear. This algorithm is based on the CYK parser.

As far as the worst case run time is concerned, to the best of our knowledge, cubic time is the best known for error correcting parsing for general context free grammars. A number of approximation algorithms have been proposed for the DYCK language (which is a very specific context free language). See e.g., [12].

In this paper we present a cubic time algorithm for error correcting parsing that is considerably simpler than the algorithms of [1,9]. This algorithm is based on the CYK parser. Even though the algorithm of [9] is also based on CYK parser, there are some crucial differences between our algorithm and that of [9]. We also show that the language edit distance problem can be reduced to the problem of computing the tropical product (also known as the distance product or the min-plus product) of two $n \times n$ matrices where $n = |\mathcal{I}|$. Using the current best known run time [14] for tropical matrix product, our reduction implies that the language edit distance problem can be solved exactly in $O(n^3/2^{\Omega(\sqrt{\log n})})$ time, improving the cubic run time that has remained the best since 1972.

In many applications, it may suffice to solve the language edit distance problem approximately. To the best of our knowledge, no approximation algorithms are known for general context free grammars. However, a number of such algorithms have been proposed for the DYCK language (which is a specific context free language with a lot of applications). For example, the algorithm of [7] takes subcubic time but its approximation factor is $\Theta(n)$. If $\mathcal{I}'$ is a string in the language generated by the input grammar G that has the minimum language edit distance (say d) with the input string $\mathcal{I}$ and if an algorithm $\mathcal{A}$ outputs a string $\mathcal{I}''$ such that the language edit distance between $\mathcal{I}$ and $\mathcal{I}''$ is no more than $d\beta(n)$, then we say $\mathcal{A}$ is a $\beta(n)$-approximation algorithm. The algorithm of [12] runs in time $O(n^{1+\epsilon})$ for any constant $\epsilon > 0$ and has a polylogarithmic approximation factor. It is noteworthy that DYCK grammar parsing (without error correction) can easily be done in linear time. On the other hand, it is known that parsing of arbitrary context free grammars is as difficult as boolean matrix multiplication [8]. For an extensive discussion on approximation algorithms for the DYCK language, please see [12]. In this paper we present an approximation algorithm for general context free grammars. Specifically, we show that if we are only interested in edit distances of no more than M, then the language edit distance problem can be solved in $O((M + 1)n^\omega)$ time where $O(n^\omega)$ is the time taken to multiply two $n \times n$ matrices. (Currently the best known value for ω is < 2.373 [4]). As a corollary, it follows that there is an $\frac{n}{M}$-approximation algorithm for the language edit distance problem with a run time of $O(Mn^\omega)$.

Some Notations: A context free grammar G is a 4-tuple (N, T, P, S), where T is a set of characters (known as terminals) in the alphabet, N is a set of variables known as nonterminals, S is the start symbol (that is a nonterminal) and P is a set of productions.

We use $L(G)$ to denote the language generated by G. Capital letters such as $A, B, C, \ldots$ will be used to denote nonterminals, small letters such as $a, b, c, \ldots$

will be used to denote terminals, and Greek letters such as $\alpha, \beta, \dots$ will be used to denote any string from $(N \cup T)^*$.

A production of the form $A \rightarrow \epsilon$ is called an ϵ-production. A production of the kind $A \rightarrow B$ is known as a unit production.

Let G be a CFG such that $L(G)$ does not have ϵ. Then we can convert G into Chomsky Normal Form (CNF). A context free grammar is in CNF if the productions in P are of only two kinds: $A \rightarrow a$ and $A \rightarrow BC$.

Let U and V be two $n \times n$ real matrices. Then the tropical (or distance) product Z of X and Y is defined as: $Z_{ij} = \min_{k=1}^{n}(X_{ik} + Y_{kj})$, $1 \leq i, j \leq n$.

The edit distance between two strings $\mathcal{I}$ and $\mathcal{I}'$ from an alphabet T is the minimum number of (insert, delete, and substitution) operations needed to convert $\mathcal{I}$ to $\mathcal{I}'$.

In this paper we assume that the grammar size is $O(1)$ which is a standard assumption made in many works (see e.g., [13]).

A Summary of Aho and Peterson's Algorithm: The algorithm of Aho and Peterson [1] is based on the parsing algorithm of Earley [3]. There are some crucial differences. Let $\mathcal{I} = a_1 a_2 \dots a_n$ be the input string. If $G(N, T, P, S)$ is the input grammar, another grammar $G' = (N', T, P', S')$ is constructed where G' has all the productions of G and some additional productions that can be used to make derivations involving errors. Each such additional production is called an *error production*. Three kinds of errors are considered, namely, insertion, deletion, and substitution. G' also has some additional nonterminals. The algorithm derives the input string beginning with S', minimizing the number of applications of the error productions.

The parser of [1] can be thought of as a modified version of the Earley parser. Like the algorithm of Earley, $n+1$ levels of lists are constructed. Each list consists of items where an item is an object of the form $[A \rightarrow \alpha.\beta, i, k]$. Here $A \rightarrow \alpha\beta$ is a production, . is a special symbol that indicates what part of the production has been processed so far, i is an integer indicating input position at which the derivation of α started, and k is an integer indicating the number of error productions that have been used in the derivation from α. If we use $\mathcal{L}_j$ to denote the list of level $j, 0 \leq i \leq n$, then the item $[A \rightarrow \alpha.\beta, i, k]$ will be in $\mathcal{L}_j$ if and only if for some ν in $(N \cup T)^*$, $S' \stackrel{*}{\Rightarrow} a_1 a_2 \cdots a_i A \nu$ and $\alpha \stackrel{*}{\Rightarrow} a_{i+1} a_{i+2} \cdots a_j$ using k error productions.

The algorithm constructs the lists $\mathcal{L}_0, \mathcal{L}_1, \dots, \mathcal{L}_n$. An item of the form $[S' \rightarrow \alpha., 0, k]$ will be in $\mathcal{L}_n$, for some integer k. In this case, k is the minimum edit distance between $\mathcal{I}$ and any string in $L(G)$.

Note that the Earley parser also works in the same manner except that an item will only have two elements: $[A \rightarrow \alpha.\beta, i]$.

A Synopsis of Valiant's Algorithm: Valiant has presented an efficient algorithm for computing the transitive closure of an upper triangular matrix. The transitive closure is with respect to matrix multiplication defined in a specific way. Each element in a matrix will be a set of items. In the case of context free recognition, each matrix element will be a set of nonterminals. If N_1 and N_2 are two sets of nonterminals, a binary operator $\cdot$ is defined as:

$N_1 \cdot N_2 = \{A | \exists B \in N_1, C \in N_2 \text{ such that } (A \to BC) \in P\}$. If a and b are matrices where each element is a subset of N, the product matrix c of a and b is defined as follows: $c_{ij} = \bigcup_{k=1}^{n} a_{ik} \cdot b_{kj}$. Under the above definition of matrix multiplication, we can define transitive closure for any matrix a as: $a^+ = a^{(1)} \cup a^{(2)} \cup \cdots$ where $a^{(1)} = a$ and $a^{(i)} = \bigcup_{j=1}^{i-1} a^{(j)} \cdot a^{(i-j)}$

Valiant has shown that this transitive closure can be computed in $O(S(n))$ time, where $S(n)$ is the time needed for multiplying two matrices with the above special definition of matrix product. In fact this algorithm works for the computation of transitive closure for generic operators $\odot$ and union as long as these operations satisfy the following properties: The outer operation (i.e., union) is commutative and associative, the inner operation ($\odot$) distributes over union, $\emptyset$ is a multiplicative zero and an additive identity.

2 A Simple Error Correcting Parser

In this section we present a simple error correcting parser for CFGs. This algorithm is based on the algorithm of [10]. We also utilize the concept of error productions introduced in [1]. If $G = (N, T, P, S)$ is the input grammar, we generate another grammar $G' = (N', T, P', S)$ where $N' = N \cup \{H, I\}$. P' has all the productions in P. In addition, P' has some additional error productions. We parse the given input string $\mathcal{I}$ using the productions in P'. For each production, we associate an error count that indicates the minimum number of errors the use of the production will amount to. The goal is to parse $\mathcal{I}$ using as few error productions as possible. Specifically, the sum of error counts of all the error productions used should be minimum. If $A \to \alpha$ is an error production with an error count of k, we denote this rule as $A \xrightarrow{k} \alpha$. If there is no integer above $\to$ in any production, the error count of this production should be assumed to be 0.

2.1 Construction of a Covering Grammar

Let $G = (N, T, P, S)$ be the given grammar and $\mathcal{I} = a_1 a_2 \ldots a_n$ be the given input string. Without loss of generality assume that $L(G)$ does not have ϵ and that G is in CNF. Even if G is not in CNF, we could employ standard techniques to convert G into this form (see e.g., [5]). We construct a new grammar $G' = (N', T, P', S)$ as follows. P' has the following productions in addition to the ones in P: $H \xrightarrow{0} HI$, $H \xrightarrow{0} I$, and $I \xrightarrow{1} a$ for every $a \in T$. Here H and I are new nonterminals. If $A \to a$ is in P, then add the following rules to P': $A \xrightarrow{1} b$ for every $b \in T - \{a\}$, $A \xrightarrow{1} \epsilon$, $A \xrightarrow{0} AH$, and $A \xrightarrow{0} HA$. Each production in P has an error count of 0.

Elimination of ϵ-productions: We first eliminate the ϵ-productions in P' as follows. We say a nonterminal A is nullable if $A \xRightarrow{*} \epsilon$. Let k be the number of errors needed for A to derive ϵ. We denote this as follows: $A \xRightarrow{*,k} \epsilon$. Call k the *nullcount* of A, denoted as *nullcount*(A). We only keep the minimum such *nullcount* for any nonterminal. Let the minimum *nullcount* for any terminal A

be $Mnullcount(A)$. For example, if $A \xrightarrow{0} BC, B \xrightarrow{1} \epsilon$, and $C \xrightarrow{1} \epsilon$ are in P', then $A \overset{*,2}{\Rightarrow} \epsilon$. We identify all the nullable nonterminals in P' using the following procedure. If $B \to CD$ is in P' and if both C and D are nullable, then B is nullable as well. In this case, $nullcount(B) = nullcount(C) + nullcount(D)$.

After identifying all nullable nonterminals and their $Mnullcount$ values, we process each production as follows. Let $A \xrightarrow{k} BC$ be any production in P'. If B is nullable and C is not, and if $Mnullcount(B) = \ell$, then we add the production $A \xrightarrow{k+\ell} C$ to P'. If C is nullable and B is not, and if $Mnullcount(C) = \ell$, then we add the production $A \xrightarrow{k+\ell} B$ to P'. If both or none of B and C are nullable, then we do not add any additional production to P' while processing the production $A \xrightarrow{k} BC$. If there are more than one productions in P' with the same precedent and consequent, we only keep that production for which the error count is the least. Finally, we remove all the ϵ productions.

Elimination of Unit Productions: We eliminate unit productions from P' as follows. Let $A \xrightarrow{k_1} B_1, B_1 \xrightarrow{k_2} B_2, \ldots, B_{q-3} \xrightarrow{k_{q-2}} B_{q-2}, B_{q-2} \xrightarrow{k_{q-1}} B$ be a sequence of unit productions in P' and $B \xrightarrow{k_q} \alpha$ be a non unit production. In this case we add the production $A \xrightarrow{Q} \alpha$ to P', where $Q = \sum_{i=1}^{q} k_i$. After processing all such sequences and adding productions to P' we eliminate duplicates. In particular, if there are more than one rules with the same precedent and consequent, we only keep the production with the least error count. At the end we remove all the unit productions.

Observation: Aho and Peterson [1] indicate that G' is a covering grammar for G and prove several properties of G'. Note that they don't keep any error counts with their productions. Also, the validity of the procedures we have used to eliminate ϵ and unit productions can be found in [5].

An Example. Consider the language $\{a^n b^n : n \geq 1\}$. A CFG for this language has the productions: $S \to aSb | ab$. We can get an equivalent grammar $G = (N, T, P, S)$ in CNF where $N = \{S, A, B, A_1\}, T = \{a, b\}$, and $P = \{S \to AA_1 | AB, A_1 \to SB, A \to a, B \to b\}$.

We can get a grammar $G_1 = (N', T, P_1, S)$ with error productions where $N' = \{S, A, B, A_1, H, I\}$ and $P_1 = \{S \to AA_1 | AB, A_1 \to SB, A \to a, B \to b, H \to HI, H \to I, I \xrightarrow{1} a, I \xrightarrow{1} b, A \xrightarrow{1} b, A \xrightarrow{1} \epsilon, A \to HA, A \to AH, B \xrightarrow{1} a, B \xrightarrow{1} \epsilon, B \to HB, B \to BH\}$. Note that any production with no integer above $\to$ has an error count of zero.

Eliminating ϵ-productions: We identify nullable nonterminals. We realize that the following nonterminals are nullable: A, B, S, and A_1. For example, A_1 is nullable since we have: $A_1 \Rightarrow SB \Rightarrow ABB \overset{*,1}{\Rightarrow} BB \overset{*,1}{\Rightarrow} B \overset{*,1}{\Rightarrow} \epsilon$. We also realize: $A \overset{*,1}{\Rightarrow} \epsilon, B \overset{*,1}{\Rightarrow} \epsilon, S \overset{*,2}{\Rightarrow} \epsilon$, and $A_1 \overset{*,3}{\Rightarrow} \epsilon$.

Now we process every production in P_1 and generate new relevant rules. For instance, consider the rule $S \to AA_1$. Since $A_1 \overset{*,3}{\Rightarrow} \epsilon$, we add the rule $S \xrightarrow{3} A$ to P_1.

When we process the rule $S \to AB$, since $B \overset{*,1}{\Rightarrow} \epsilon$, we realize that $S \overset{1}{\to} A$ has to be added to P_1. However, $S \overset{3}{\to} A$ has already been added to P_1. Thus we replace $S \overset{3}{\to} A$ with $S \overset{1}{\to} A$.

Processing in a similar manner, we add the following productions to P_1 to get P_2: $S \overset{1}{\to} A_1, S \overset{1}{\to} A, S \overset{1}{\to} B, A_1 \overset{1}{\to} S, A_1 \overset{2}{\to} B, A \overset{1}{\to} H$, and $B \overset{1}{\to} H$. We eliminate all the ϵ-productions from P_2.

Eliminating Unit Productions: We consider every sequence of unit productions $A \overset{k_1}{\to} B_1, B_1 \overset{k_2}{\to} B_2, \dots, B_{q-3} \overset{k_{q-2}}{\to} B_{q-2}, B_{q-2} \overset{k_{q-1}}{\to} B$ in P_2 with $B \overset{k_q}{\to} \alpha$ being a non unit production. In this case we add the production $A \overset{Q}{\to} \alpha$ to P_2, where $Q = \sum_{i=1}^{q} k_i$.

Consider the sequence $S \overset{1}{\to} A, A \to a$. This sequence results in a new production: $S \overset{1}{\to} a$. The sequence $S \overset{1}{\to} A_1, A_1 \overset{2}{\to} B, B \overset{1}{\to} a$ suggests the addition of the production $S \overset{4}{\to} a$. But we have already added a better production and hence this production is ignored.

Proceeding in a similar manner we realize that we have to add the following productions to P_2 to get P_3: $S \overset{1}{\to} a, S \overset{1}{\to} b, H \overset{1}{\to} a, H \overset{1}{\to} b, A_1 \overset{2}{\to} a, A_1 \overset{2}{\to} b, S \overset{1}{\to} HB, S \overset{1}{\to} BH, A \overset{1}{\to} HI, B \overset{1}{\to} HI, S \overset{2}{\to} HI, A_1 \overset{3}{\to} HI, A_1 \overset{2}{\to} BH, A_1 \overset{2}{\to} HB, S \overset{1}{\to} AH, S \overset{1}{\to} HA, A_1 \overset{2}{\to} AH$, and $A_1 \overset{2}{\to} HA$. We eliminate all the unit productions from P_3.

The final grammar we get is $G_3 = (N', T, P_3, S)$ where $P_3 = \{S \to AA_1 | AB, A_1 \to SB, A \to a, B \to b, H \to HI, I \overset{1}{\to} a, I \overset{1}{\to} b, A \overset{1}{\to} b, A \to HA, A \to AH, B \overset{1}{\to} a, B \to HB, B \to BH, S \overset{1}{\to} a, S \overset{1}{\to} b, H \overset{1}{\to} a, H \overset{1}{\to} b, A_1 \overset{2}{\to} a, A_1 \overset{2}{\to} b, S \overset{1}{\to} HB, S \overset{1}{\to} BH, A \overset{1}{\to} HI, B \overset{1}{\to} HI, S \overset{2}{\to} HI, A_1 \overset{3}{\to} HI, A_1 \overset{2}{\to} BH, A_1 \overset{2}{\to} HB, S \overset{1}{\to} AH, S \overset{1}{\to} HA, A_1 \overset{2}{\to} AH, A_1 \overset{2}{\to} HA\}$.

2.2 The Algorithm

The algorithm is a modified version of an algorithm given in [10]. This algorithm in turn is a slightly different version of the CYK algorithm. Let $G' = (N', T', P', S')$ be the grammar generated using the procedure given in Sect. 2.1. The basic idea behind the algorithm is the following: The algorithm has n stages. In any given stage we scan through each production in P' and grow larger and larger parse trees. At any given time in the algorithm, each nonterminal has a list of tuples of the form (i, j, ℓ). If A is any nonterminal, $LIST(A)$ will have tuples (i, j, ℓ) such that $A \overset{*,\ell}{\Rightarrow} a_i \dots a_{j-1}$ and there is no $\ell' < \ell$ such that $A \overset{*,\ell'}{\Rightarrow} a_i \dots a_{j-1}$. If $A \overset{\ell_3}{\to} BC$ is a production in P', then in any stage we process this production as follows: We scan through elements in $LIST(B)$ and look for matches in $LIST(C)$. For instance, if (i, k, ℓ_1) is in $LIST(B)$ (for some integer ℓ_1), we check if (k, j, ℓ_2) is in $LIST(C)$, for some j and ℓ_2. If so, we insert $(i, j, \ell_1 + \ell_2 + \ell_3)$ into $LIST(A)$. If $LIST(A)$ for any A has many tuples of the form $(i, j, *)$ we keep only one among these. Specifically, if $(i, j, \ell_1), (i, j, \ell_2), \dots, (i, j, \ell_q)$ are in $LIST(A)$, we keep only (i, j, ℓ) where $\ell = \min_{m=1}^{q} \ell_m$.

We maintain the following data structures: (1) for each nonterminal A, an array (call it X_A) of lists indexed 1 through n, where $X_A[i]$ is the list of all tuples from $LIST(A)$ whose first item is i ($1 \leq i \leq n$); and (2) an $n \times n$ upper triangular matrix $\mathcal{M}$ whose (i,j)th entry will be those nonterminals that derive $a_i a_{i+1} \ldots a_{j-1}$ with the corresponding (minimum) error counts (for $1 \leq i \leq n$ and $2 \leq j \leq (n+1)$). There can be $O(n^2)$ entries in $LIST(B)$, and for each entry (i,k) in this list, we need to search for at most n items in $LIST(C)$.

By induction, we can show that at the end of stage s ($1 \leq s \leq n$), the algorithm would have computed all the nonterminals that span any input segment of length s or less, together with the minimum error counts. (We say a nonterminal spans the input segment $J = a_i a_{i+1} \ldots a_{j-1}$ if it derives J; the nonterminal is said to have a "span-length" of $j - i$.)

A straightforward implementation of the above idea takes $O(n^4)$ time. However, we can reduce the run time of each stage to $O(n^2)$ as follows: In stage s, while processing the production $A \rightarrow BC$, work only with tuples from $LIST(B)$ and $LIST(C)$ whose combination will derive an input segment of length exactly s. That is, for each tuple (i,k,ℓ) in $LIST(B)$, we look for a tuple $(k, i+s, \ell')$ (for any integer ℓ') in $LIST(C)$. Such a tuple can be found in time $O(1)$ by searching for (C,ℓ') in $\mathcal{M}_{k,i+s}$. With this modification, each stage of the above algorithm will only take $O(n^2)$ time and hence the total run time of the algorithm is $O(n^3)$. The pseudocode is given in Algorithm 1.

Algorithm 1. ErrorCorrectingParser$(G, \mathcal{I})$

input : $G = (N, T, P, S)$, a grammar; $\mathcal{I} = a_1 a_2 \ldots a_n$, input string;
output: minimum distance ℓ between $\mathcal{I}$ and any string in $L(G)$;
Generate G' using the procedure in Sect. 2.1;
for $A \in N'$ **and** $i \leftarrow 1$ **to** n **do** $X_A[i] := \{\}$;
for $i \leftarrow 1$ **to** n **and** $j \leftarrow (i+1)$ **to** $(n+1)$ **do** $\mathcal{M}_{i,j} = \{\}$;
for $i \leftarrow 1$ **to** n **and** $(A \xrightarrow{\ell} a_i) \in P'$ **do**
  $\quad$ insert (A, ℓ) into $\mathcal{M}_{i,i+1}$; **and** insert $(i, i+1, \ell)$ into $X_A[i]$;
for $s \leftarrow 2$ **to** n **do**
  $\quad$ **for** $(A \xrightarrow{\ell_3} BC) \in P'$ **and** $(i,k,\ell_1) \in X_B[i]$ **and** $(C, \ell_2) \in \mathcal{M}_{k,i+s}$ **do**
    $\quad\quad \ell := \ell_1 + \ell_2 + \ell_3$;
    $\quad\quad$ insert (A, ℓ) into $\mathcal{M}_{i,i+s}$; **and** insert $(i, i+s, \ell)$ into $X_A[i]$;
return ℓ for which $(1, n+1, \ell) \in X_S[1]$;

Theorem 1. *When the above algorithm completes, for any nonterminal A, $LIST(A)$ has a tuple (i, j, ℓ) if and only if $A \xRightarrow{*,\ell} a_i \ldots a_{j-1}$ and there is no $\ell' < \ell$ such that $A \xRightarrow{*,\ell'} a_i \ldots a_{j-1}$.*

Proof. The proof is by induction on the stage number s.

Base Case: $s = 1$, i.e., $j - i = 1$, for $1 \leq i, j \leq (n+1)$. Note that all the nonterminals, except H and I, are nullable. As a result, P' will have a production of the kind $A \xrightarrow{\ell} b$ for every nonterminal A and every terminal b, for some integer ℓ.

By the way we compute $Mnullcount$ for each nonterminal and eliminate unit productions, it is clear that ℓ is the smallest integer for which $A \overset{*,\ell}{\Rightarrow} b$.

Induction Step: Assume that the hypothesis is true for span lengths up to $s - 1$. We can prove it for a span length of s. Let $A \overset{l_3}{\to} BC$ be any production in P'. Let (i, k, ℓ_1) be a tuple in $LIST(B)$ and (k, j, ℓ_2) be a tuple in $LIST(C)$ with $j - i = s$. Then this means that $A \overset{*,L}{\Rightarrow} a_i \dots a_{j-1}$, where $L = \ell_1 + \ell_2 + \ell_2$. We add the tuple (i, j, L) to $LIST(A)$. Also, the induction hypothesis implies that $B \overset{*,\ell_1}{\Rightarrow} a_i \dots a_{k-1}$ and there is no $\ell < \ell_1$ for which $B \overset{*,\ell}{\Rightarrow} a_i \dots a_{k-1}$. Likewise, ℓ_2 is the smallest integer for which $C \overset{*,\ell_2}{\Rightarrow} a_k \dots a_{j-1}$. We consider all such productions in P that will contribute tuples of the kind $(i, j, *)$ to $LIST(A)$ and from these only keep (i, j, ℓ) where ℓ is the least such integer. □

3   Less Than Cubic Time Parser

In this section we present an error correcting parser that runs in time $O(T(n))$ where n is the length of the input string and $T(n)$ is the time needed to compute the tropical product of two $n \times n$ matrices. There are two main ingredients in this parser, namely, the procedure given in Sect. 2.1 for converting the given grammar into a covering grammar and Valiant's reduction given in [13] (and summarized in Sect. 1).

As pointed out in Sect. 1, Valiant has presented an efficient algorithm for computing the transitive closure of an upper triangular matrix. The transitive closure is with respect to matrix multiplication defined in a special way. Each element in a matrix will be a set of items. In standard matrix multiplication we have two operators, namely, multiplication and addition. In the case of special matrix multiplication, these operations are replaced by $\odot$ and $\cup$ (called *union*). Valiant's algorithm works as long as these operations satisfy the following properties: The outer operation (i.e., union) is commutative and associative, the inner operation $\odot$ distributes over union, $\emptyset$ is a zero with respect to $\odot$ and an identity with respect to union.

Valiant has shown that transitive closure under the above definition of matrix multiplication can be computed in $O(S(n))$ time, where $S(n)$ is the time needed for multiplying two matrices with the above special definition of matrix product.

In the context of error correcting parser we define the two operations as follows. Let $\mathcal{I} = w_1 w_2 \dots w_n$ be the input string. Matrix elements are sets of pairs of the kind (A, ℓ) where A is a nonterminal and ℓ is an integer. We initialize an $(n + 1) \times (n + 1)$ upper triangular matrix a as:

$$a_{i,i+1} = \{(A, \ell) | (A \overset{\ell}{\to} w_i) \in P'\}, \text{ and } a_{i,j} = \emptyset, \text{ for } j \neq i + 1. \tag{1}$$

If N_1 and N_2 are sets of pairs of the kind (A, ℓ), define the union operation as:

$$N_1 \cup N_2 = \{(A, k) : (1) \ (A, k) \in N_1 \text{ and } (A, k') \notin N_2 \text{ for any } k' \text{ or}$$
$$(2) \ (A, k) \in N_2 \text{ and } (A, k') \notin N_1 \text{ for any } k' \text{ or}$$
$$(3) \ (A, k_1) \in N_1, (A, k_2) \in N_2 \text{ and } k = \min(k_1, k_2)\}.$$

It is easy to see that the union operation is commutative and associative since if there are multiple pairs for the same nonterminal with different error counts, the union operation has the effect of keeping only one pair for each nonterminal with the least error count.

If N_1 and N_2 are sets of pairs of the kind (A, ℓ) then $N_1 \odot N_2$ is defined with the procedure in Algorithm 2.

Algorithm 2. $N_1 \odot N_2$

Let $G' = (N', T, P', S)$ be the grammar with error productions;
for $A \in N'$ do $h[A] := +\infty$;
for $(B, k) \in N_1$ and $(C, \ell) \in N_2$ and $(A \overset{m}{\rightarrow} BC) \in P'$ do
$\quad \lfloor \ h[A] := \min(h[A], k + \ell + m);$
return $\{(A, h[A]) | A \in N', h[A] \neq +\infty\};$

We can verify that $\odot$ distributes over union. Let $N_1, N_2,$ and N_3 be any three sets of pairs of the type (A, ℓ). Let $Q = N_1 \odot (N_2 \cup N_3), R = (N_1 \odot N_2) \cup (N_1 \odot N_3), X = (N_1 \odot N_2),$ and $Y = (N_1 \odot N_3)$. If $(B, \ell) \in N_1, (C, k_1) \in N_2, (C, k_2) \in N_3, k = \min(k_1, k_2),$ and $(A \overset{m}{\rightarrow} BC) \in P'$, then $(A, k + \ell + m) \in Q$. Also, $(A, k_1 + \ell + m) \in X$ and $(A, k_2 + \ell + m) \in Y$. Thus, $(A, k + \ell + m) \in R$.

Put together, we get the following algorithm. Given a grammar G and an input string $\mathcal{I}$, generate the grammar G' using the procedure in Sect. 2.1. Construct the matrix a described in Eq. 1. Compute the transitive closure a^+ of a using Valiant's algorithm [13]. (S, ℓ) will occur in $a^+_{1,n+1}$ for some integer ℓ. In this case, the minimum distance between $\mathcal{I}$ and any string in $L(G)$ is ℓ. The pseudocode is given in Algorithm 3.

Algorithm 3. ErrorCorrectingParser2$(G, \mathcal{I})$

---

input : $G = (N, T, P, S)$, a grammar; $\mathcal{I} = w_1 w_2 \ldots w_n$, a string;
output: the minimum distance ℓ between $\mathcal{I}$ and any string in $L(G)$;
Generate $G' = (N', T, P', S)$ using the procedure in Sect. 2.1;
for $i \leftarrow 1$ to n and $j \leftarrow (i+1)$ to $(n+1)$ do $a_{i,j} := \{\}$;
for $i \leftarrow 1$ to n do $a_{i,i+1} := \{(A, \ell) | (A \overset{\ell}{\rightarrow} w_i) \in P'\}$;
$a^+ := \texttt{TransitiveClosure}\ (a);$ // using Valiant's algorithm [13]
return ℓ for which $(S, \ell) \in a^+_{1,n+1};$

Note that, by definition, $a^+ = a^{(1)} \cup a^{(2)} \cup \cdots$ where $a^{(1)} = a$ and $a^{(i)} = \bigcup_{j=1}^{i-1} a^{(j)} \cdot a^{(i-j)}$. It is easy to see that (A, ℓ), where A is a nonterminal and ℓ is an integer in the range $[0, n]$, will be in $a^{(k)}_{i,j}$ if and only if $A \overset{*,\ell}{\Rightarrow} a_i a_{i+1} \cdots a_{j-1}$ such that $(j - i) = k$ and there is no $q < \ell$ such that $A \overset{*,q}{\Rightarrow} a_i a_{i+1} \cdots a_{j-1}$. Also, $a^{(k)}_{i,j} = \emptyset$ if $(j - i) \neq k$.

As a result, (S, ℓ) will be in $a^+_{1,n+1}$ if and only if $S \overset{*,\ell}{\Rightarrow} a_1 a_2 \cdots a_n$ and there is no $q < \ell$ such that $S \overset{*,q}{\Rightarrow} a_1 a_2 \cdots a_n$. We get the following

Theorem 2. *Error correcting parsing can be done in $O(S(n))$ time where $S(n)$ is the time needed to multiply two matrices under the new definition of matrix product.* ☐

It remains to be shown that $S(n) = O(T(n))$ where $T(n)$ is the time needed to compute the tropical product of two matrices.

Let a and b be two matrices where the matrix elements are sets of pairs of the kind (A, ℓ) where A is a nonterminal and ℓ is an integer. Let $c = ab$ be the product of interest under the special definition of matrix product.

For each nonterminal B in G', we define a matrix a_B and for each nonterminal C in G', we define a matrix b_C. $a_B[i,j] = \ell$ if $(B, \ell) \in a[i,j]$, for $1 \le i, j \le n$. Likewise, $b_C[i,j] = \ell$ if $(C, \ell) \in b[i,j]$, for $1 \le i, j \le n$. We call a_B and b_C as distance matrices. The pseudocode for computing a distance matrix is given in Algorithm 6 in the appendix. Compute the tropical product c_{BC} of a_B and b_C for every nonterminal B and every nonterminal C.

For every production $A \overset{k}{\to} BC$ in P' do the following: for every $1 \le i, j \le n$, if $c_{BC}[i,j] = \ell$ then add $(A, k + \ell)$ to $c[i,j]$, keeping only the smallest distance if A is already present in $c[i,j]$. The pseudocode is given in Algorithm 4.

Algorithm 4. MatrixMultiplication(a, b, G')

input : a, b: $m \times q$ and $q \times n$ matrices respectively; each entry of a and b is a
    set of pairs (A, ℓ) such that A is a nonterminal and ℓ is an integer;
    $G' = (N', T, P', S)$, grammar with error productions;
output: $c = ab$ where multiplication is done with respect to $(\odot, \cup)$;
for $i \leftarrow 1$ **to** m **and** $j \leftarrow 1$ **to** n **and** $A \in N'$ **do** $\text{minDist}_{i,j}[A] := +\infty$;

for $(A \overset{k}{\to} BC) \in P'$ **do**
 $a_B := \text{DistMatrix}\,(a, B)$; **and** $b_C := \text{DistMatrix}\,(b, C)$;
 $c_{BC} := \text{TropicalProduct}\,(a_B, b_C)$;
 for $i \leftarrow 1$ **to** m **and** $j \leftarrow 1$ **to** n **do**
  $\text{minDist}_{i,j}[A] := \min(k + c_{BC}[i,j], \text{minDist}_{i,j}[A])$;

for $i \leftarrow 1$ **to** m **and** $j \leftarrow 1$ **to** n **do**
 $c[i,j] := \{(A, \ell) | A \in N', \text{minDist}_{i,j}[A] = \ell, \ell \ne \infty\}$;

return c;

Clearly, the time spent in computing ab (under the new special definition of matrix product) is $O(T(n) + n^2)$ assuming that the size of the grammar is $O(1)$ (in fact, it suffices that the number of productions of the form $A \overset{k}{\to} BC$ is $O(1)$). Put together we get the following theorem.

Theorem 3. *Error correcting parsing can be done in $O(T(n))$ time where $T(n)$ is the time needed to compute the tropical product of two matrices.* ☐

Using the currently best known algorithm for tropical products [14], we get the following theorem.

Theorem 4. *Error correcting parsing can be done in $O\left(\frac{n^3}{2^{\Omega(\log n)^{1/2}}}\right)$ time.* ☐

Furthermore, consider the case of error correcting parsing where we know a priori that there exists a string $\mathcal{I}'$ in $L(G)$ such that the distance between $\mathcal{I}$ and $\mathcal{I}'$ is upper bounded by m. We can solve this version of the language edit distance problem using the tropical matrix product algorithm of Zwick [16]. This algorithm multiplies two $n \times n$ integer matrices in $O(Mn^\omega)$ time if the matrix elements are in the range $[-M, M]$ [16]. Here $O(n^\omega)$ is the time taken to multiply two $n \times n$ real matrices. Recall that when we reduce the matrix multiplication under $\odot$ and $\cup$ to tropical matrix multiplication, we have to compute the tropical product c_{BC} of a_B and b_C for every nonterminal B and every nonterminal C. Elements of a_B and b_C are integers in the range $[0, n]$. Note that even if all the elements of c_{BC} are $\leq m$, some of the elements of a_B and b_C could be larger than m. Before using the algorithm of [16] we have to ensure that all the elements of a_B and b_C are less than M (where M is some function of m). This can be done as follows. Before invoking the algorithm of [16], we replace every element of a_B and b_C by $m + 1$ if the element is $> m$. $m + 1$ is 'infinity' as far as this multiplication is concerned. By doing this replacement, we are not affecting the final result of the algorithm and at the same time, we are making sure that the elements of a_B and b_C are $\leq M = (m + 1)$. As a result, we get the following theorem.

Theorem 5. *Error correcting parsing can be done in $O(mn^\omega)$ time where m is an upper bound on the edit distance between the input string $\mathcal{I}$ and some string $\mathcal{I}'$ in $L(G)$, G being the input CFG. $O(n^\omega)$ is the time it takes to multiply two $n \times n$ matrices.* $\qquad\square$

As a corollary to the above theorem we can also get the following theorem.

Theorem 6. *There exists an $\frac{n}{m}$-approximation algorithm for the language edit distance problem that has a run time of $O(mn^\omega)$, where $O(n^\omega)$ is the time taken to multiply two $n \times n$ matrices.*

Proof. Here again, we replace every element of a_B and b_C by $m+1$ if the element is $> m$. In this case the elements of c_{BC} will be $\leq (2m + 2)$. We replace any element in c_{BC} that is larger than m with $(m + 1)$. In general whenever we generate or operate on a matrix, we will ensure that the elements are $\leq (m+1)$. If $S \overset{\ell}{\Rightarrow} \mathcal{I}$ for some $\ell \leq m$, then the final answer output will be exact. If $\ell > m$, then the algorithm will always output n. Thus the theorem follows. $\qquad\square$

Retrieving $\mathcal{I}'$. In all the algorithms presented above, we have focused on computing the minimum edit distance between the input string $\mathcal{I}$ and any string $\mathcal{I}'$ in $L(G)$. We have also shown that $\mathcal{I}'$ can be found in $O(n^2)$ time, where $n = |\mathcal{I}|$. Please see the Appendix for details.

4   Conclusions

In this paper we have presented an error correcting parser for general context free languages. This algorithm takes less than cubic time, improving the 1972 algorithm of Aho and Peterson that has remained the best until now. We have also shown that if M is an upper bound on the edit distance between the input string $\mathcal{I}$ and some string of $L(G)$, then we can solve the parsing problem in $O(Mn^\omega)$ time, where $O(n^\omega)$ is the time it takes to multiply two $n \times n$ matrices. As a corollary, we have presented an $\frac{n}{M}$-approximation algorithm for the general context free language edit distance problem that runs in time $O(Mn^\omega)$.

Acknowledgments. This work has been supported in part by the following grant: NIH R01LM010101. The first author thanks Barna Saha for the introduction of this problem and Alex Russell for providing pointers to tropical matrix multiplication.

Appendix: Retrieving $\mathcal{I}'$

In all the algorithms presented above, we have focused on computing the minimum edit distance between the input string $\mathcal{I}$ and any string $\mathcal{I}'$ in $L(G)$. In this section we address the problem of finding $\mathcal{I}'$. We show that $\mathcal{I}'$ can be found in $O(n^2)$ time, where $n = |\mathcal{I}|$. Let $S \overset{*,\ell}{\Rightarrow} \mathcal{I}$ such that there is no $k < \ell$ such that $S \overset{*,k}{\Rightarrow} \mathcal{I}$. Let $\mathcal{I} = a_1 a_2 \cdots a_n$.

Realize that in the algorithms given in Sects. 2.2 and 3 we compute, for every i and j (with $j > i$), all the nonterminals A such that A spans $a_i a_{i+1} \ldots a_{j-1}$ and we also determine the least k such that $A \overset{*,k}{\Rightarrow} a_i a_{i+1} \ldots a_{j-1}$. In this case, there will be an entry for A in the matrix $\mathcal{M}$. Specifically, (A, k) will be in $\mathcal{M}(i, j)$. We can utilize this information to identify an $\mathcal{I}'$ such that the edit distance between $\mathcal{I}$ and $\mathcal{I}'$ is equal to ℓ. Note that we can deduce $\mathcal{I}'$ if we know the sequence of productions used to derive $\mathcal{I}'$. The pseudocode is given in Algorithm 5. We will invoke the algorithm as $ParseTree(\mathcal{M}, S, 1, n + 1, \ell)$.

Algorithm 5 finds the first production in $O(n)$ time. Having found the first production, we can proceed in a similar manner to find the other productions needed to derive $\mathcal{I}'$. In the second stage we have to find a production that can be used to derive $a_1 a_2 \ldots a_j$ from A and another production that can be used to derive $a_{j+1} a_{j+2} \ldots a_n$ from B. Note that the span length of A plus the span length of B is n and hence both the productions can be found in a total of $O(n)$ time.

We can think of a tree $\mathcal{T}$ where S is the root and S has two children A and B. If $A \to CD$ is the first production that can be used to derive $a_1 a_2 \ldots a_j$ from A and $B \to EF$ is the first production that can be used to derive $a_{j+1} a_{j+2} \ldots a_n$ from B, then A will have two children C and D and B will have two children E and F.

The rest of the tree is constructed in the same way. Clearly, the total span length of all the nonterminals in any level of the tree is $\leq n$ and hence the time spent at each level is $O(n)$. Also, there can be at most n levels. As a result, we get the following theorem.

Theorem 7. *We can identify $\mathcal{I}'$ in $O(n^2)$ time.*   □

Algorithm 5. ParseTree($\mathcal{M}, D, i, j, \ell$)

---

input : $\mathcal{M}$, transitive closure matrix as discussed in the text;
D, a non terminal;
$i, j - 1$, start and end position in the input string $\mathcal{I}$, where
$\mathcal{I} = a_1 a_2 \ldots a_n$;
ℓ, an edit distance;
output: a parse tree which can derive $\mathcal{I}[i..(j - 1)]$ from D with ℓ errors;
begin

   if $i = j - 1$ **then**
      find ℓ for which $(D, \ell) \in \mathcal{M}_{i,j}$;
      return new node$(i, j, D \xrightarrow{\ell} a_i)$;

   for $k \leftarrow (i + 1)$ **to** $(j - 1)$ **do**
      if $\exists (A, q_1) \in \mathcal{M}(i, k)$ **and** $\exists (B, q_2) \in \mathcal{M}(k, j)$ **then**
         if $\exists (D \xrightarrow{q_3} AB) \in P'$ **and** $q_1 + q_2 + q_3 = \ell$ **then**
            break;

   $T_1 :=$ **ParseTree**$(\mathcal{M}, A, i, k, q_1)$;
   $T_2 :=$ **ParseTree**$(\mathcal{M}, B, k, j, q_2)$;
   $r :=$ **new node**$(i, j, D \xrightarrow{q_3} AB)$;
   left$(r) := T_1$;
   right$(r) := T_2$;
   return r;

Appendix: Computing the distance matrices

Algorithm 6. DistMatrix(a, B)

---

input : a, $m \times n$ matrix where each entry is a set of pairs (A, ℓ) such
that A
is a nonterminal and ℓ is an integer;
B, a nonterminal;
output: a_B, $m \times n$ matrix where each entry is an integer ℓ such that
either $(B, \ell) \in a[i, j]$ or $l = +\infty$;
begin

   for $i \leftarrow 1$ **to** m **and** $j \leftarrow 1$ **to** n **do**
      if $\exists \ell$ for which $(B, \ell) \in a[i, j]$ **then**
         $a_B[i, j] := \ell$;
      else
         $a_B[i, j] := +\infty$;

   return a_B;

References

1. Aho, A., Peterson, T.: A minimum distance error-correcting parser for context-free languages. SIAM J. Comput. **1**(4), 305–312 (1972)
2. Cocke, J., Schwartz, J.: Programming languages and their compilers: preliminary notes. Courant Institute of Mathematical Sciences, New York University (1970)
3. Earley, J.: An efficient context-free parsing algorithm. Commun. ACM **13**(2), 94–102 (1970)
4. Gall, F.L.: Powers of tensors and fast matrix multiplication (2014). arXiv:1401.7714
5. Hopcroft, J.E., Motwani, R., Ullman, J.D.: Introduction to Automata Theory, Languages, and Computation. Addison Wesley, Reading (2006)
6. Kasami, T.: An efficient recognition and syntax analysis algorithm for context-free languages. Technical report, DTIC Document (1965)
7. Korn, F., Saha, B., Srivastava, D., Ying, S.: On repairing structural problems in semi-structured data. Proc. VLDB Endowment **6**(9), 601–612 (2013)
8. Lee, L.: Fast context-free grammar parsing requires fast boolean matrix multiplication. J. ACM **49**(1), 1–15 (2002)
9. Myers, G.: Approximately matching context-free languages. Inf. Process. Lett. **54**(2), 85–92 (1995)
10. Rajasekaran, S.: Tree-adjoining language parsing in $o(n^6)$ time. SIAM J. Comput. **25**(4), 862–873 (1996)
11. Rajasekaran, S., Al Seesi, S., Ammar, R.A.: Improved algorithms for parsing esltags: a grammatical model suitable for rna pseudoknots. IEEE/ACM Trans. Comput. Biol. Bioinf. (TCBB) **7**(4), 619–627 (2010)
12. Saha, B.: Efficiently computing edit distance to dyck language (2013). arXiv:1311.2557
13. Valiant, L.G.: General context-free recognition in less than cubic time. J. Comput. Syst. Sci. **10**(2), 308–315 (1975)
14. Williams, R.: Faster all-pairs shortest paths via circuit complexity. In: Proceedings of the 46th Annual ACM Symposium on Theory of Computing, STOC 2014, pp. 664–673. ACM, New York, NY, USA (2014)
15. Younger, D.H.: Recognition and parsing of context-free languages in time n^3. Inf. Control **10**(2), 189–208 (1967)
16. Zwick, U.: All pairs shortest paths in weighted directed graphs-exact and almost exact algorithms. In: Proceedings of 39th Annual Symposium on Foundations of Computer Science, pp. 310–319. IEEE (1998)

System Analysis and Program Verification

Accurate Approximate Diagnosability
of Stochastic Systems

Nathalie Bertrand[1], Serge Haddad[2], and Engel Lefaucheux[1,2]($\boxtimes$)

[1] Inria Rennes, Rennes, France
nathalie.bertrand@inria.fr, engel.lefaucheux@irisa.fr
[2] LSV, ENS Cachan and CNRS and Inria, Université Paris-Saclay, Cachan, France
serge.haddad@lsv.fr

Abstract. Diagnosis of partially observable stochastic systems prone to faults was introduced in the late nineties. Diagnosability, *i.e.* the existence of a diagnoser, may be specified in different ways: (1) exact diagnosability (called A-diagnosability) requires that almost surely a fault is detected and that no fault is erroneously claimed while (2) approximate diagnosability (called ε-diagnosability) allows a small probability of error when claiming a fault and (3) accurate approximate diagnosability (called AA-diagnosability) requires that this error threshold may be chosen arbitrarily small. Here we mainly focus on approximate diagnoses. We first refine the almost sure requirement about finite delay introducing a uniform version and showing that while it does not discriminate between the two versions of exact diagnosability this is no more the case in approximate diagnosis. Then we establish a complete picture for the decidability status of the diagnosability problems: (uniform) ε-diagnosability and uniform AA-diagnosability are undecidable while AA-diagnosability is decidable in PTIME, answering a longstanding open question.

Keyword: Automata for system analysis and programme verification

1 Introduction

Diagnosis and diagnosability. The increasing use of software systems for critical operations motivates the design of fast automatic detection of malfunctions. In general, diagnosis raises two important issues: deciding whether the system is *diagnosable* and, in the positive case, synthesizing a *diagnoser* possibly satisfying additional requirements about memory size, implementability, etc. One of the proposed approaches consists in modelling these systems by partially observable labelled transition systems (LTS) [11]. In such a framework, diagnosability requires that the occurrence of unobservable faults can be deduced from the previous and subsequent observable events. Formally, an LTS is diagnosable if there

---

S. Haddad—This author was partly supported by ERC project EQualIS (FP7-308087).

A.-H. Dediu et al. (Eds.): LATA 2016, LNCS 9618, pp. 549–561, 2016.
DOI: 10.1007/978-3-319-30000-9_42

exists a diagnoser that satisfies *reactivity* and *correctness* contraints. Reactivity requires that if a fault occurred, the diagnoser eventually detects it. Correctness asks that the diagnoser only claims the existence of a fault when there actually was one. Diagnosability for LTS was shown to be decidable in PTIME [7] while the diagnoser itself could be of size exponential w.r.t. the size of the LTS. Diagnosis has been extended to numerous models (Petri nets [3], pushdown systems [8], etc.) and settings (centralized, decentralized, distributed), and have had an impact on important application areas, *e.g.* for telecommunication network failure diagnosis. Also, several contributions, gathered under the generic name of active diagnosis, focus on enforcing the diagnosability of a system [4,5,10,13].

Diagnosis of stochastic systems. Diagnosis was also considered in a quantitative setting, and namely for probabilistic labelled transition systems (pLTS) [1,12], that can be seen as Markov chains in which the transitions are labelled with events. Therefore, one can define a probability measure over infinite runs. In that context, the specification of reactivity and correctness can be relaxed. Here, reactivity only asks to detect faults almost surely (*i.e.* with probability 1). This weaker reactivity constraint takes advantage of probabilities to rule out negligible behaviours. For what concerns correctness, three natural variants can be considered. *A-diagnosability* sticks to strong correctness and therefore asks the diagnoser to only claim fault occurrences when a fault is certain. ε-*diagnosability* tolerates small errors, allowing to claim a fault if the conditional probability that no fault occurred does not exceed ε. *AA-diagnosability* requires the pLTS to be ε-diagnosable for all positive ε, allowing the designer to select a threshold according to the criticality of the system. A-diagnosability and AA-diagnosability were introduced in [12]. Recently, we focused on semantical and algorithmic issues related to A-diagnosability, and in particular we established that A-diagnosability is PSPACE-complete [1]. When it comes to approximate diagnosability (*i.e.* ε and AA-diagnosability), up to our knowledge, a (PTIME-checkable) sufficient condition for AA-diagnosability [12] has been given, but no decidability result is known.

Contributions. Our contributions are twofold. From a semantical point of view, we investigate the specification of reactivity, introducing *uniform reactivity* which requires that once a fault occurs, the probability of detection when time elapses converges to 1 uniformly w.r.t. faulty runs. Uniformity provides the user with a stronger guarantee about the delay before detection. We show that uniform A-diagnosability and A-diagnosability coincide while this is no longer the case for approximate diagnosability. From an algorithmic point of view, we first show that ε-diagnosability and its uniform version are undecidable. Then we characterize AA-diagnosability as a separation property between labelled Markov chains (LMC), precisely a *distance* 1 between appropriate pairs of LMCs built from the pLTS. Thanks to [6], this yields a polynomial time algorithm for AA-diagnosability. AA-diagnosability can thus be checked more efficiently than A-diagnosability (PTIME vs PSPACE), yet, surprisingly, contrary to A-diagnosers, AA-diagnosers may require infinite memory. Finally, we show that uniform AA-diagnosability is undecidable.

Organization. In Sect. 2, we introduce the different variants of diagnosability and establish the full hierarchy between these specifications. In Sect. 3, we address the decidability and complexity issues related to approximate diagnosis. Full proofs can be found in the companion research report [2].

2   Specification of Diagnosability

2.1   Probabilistic Labelled Transition Systems

To represent stochastic discrete event systems, we use transition systems labelled with events and in which the transition function is probabilistic.

Definition 1. *A* probabilistic labelled transition system *(pLTS) is a tuple* $\mathcal{A} = \langle Q, q_0, \Sigma, T, \mathbf{P} \rangle$ *where:*

- *Q is a finite set of states with $q_0 \in Q$ the initial state;*
- *Σ is a finite set of events;*
- *$T \subseteq Q \times \Sigma \times Q$ is a set of transitions;*
- *$\mathbf{P} : T \rightarrow \mathbb{Q}_{>0}$ is the probability function fulfilling for every $q \in Q$:*
  $\sum_{(q,a,q') \in T} \mathbf{P}[q, a, q'] = 1.$

Observe that a pLTS is a labelled transition system (LTS) equipped with transition probabilities. The transition relation of the underlying LTS is defined by: $q \xrightarrow{a} q'$ for $(q, a, q') \in T$; this transition is then said to be *enabled* in q.

Let us now introduce some important notions and notations that will be used throughout the paper. A *run* ρ of a pLTS $\mathcal{A}$ is a (finite or infinite) sequence $\rho = q_0 a_0 q_1 \ldots$ such that for all i, $q_i \in Q$, $a_i \in \Sigma$ and when q_{i+1} is defined, $q_i \xrightarrow{a_i} q_{i+1}$. The notion of run can be generalized, starting from an arbitrary state q. We write Ω for the set of all infinite runs of $\mathcal{A}$ starting from q_0, assuming the pLTS is clear from context. When it is finite, ρ ends in a state q and its *length*, denoted $|\rho|$, is the number of actions occurring in it. Given a finite run $\rho = q_0 a_0 q_1 \ldots q_n$ and a (finite or infinite) run $\rho' = q_n a_n q_{n+1} \ldots$, we call concatenation of ρ and ρ' and we write $\rho\rho'$ for the run $q_0 a_0 q_1 \ldots q_n a_n q_{n+1} \ldots$; the run ρ is then a *prefix* of $\rho\rho'$, which we denote $\rho \preceq \rho\rho'$. The *cylinder* generated by a finite run ρ consists of all infinite runs that extend ρ: $\mathsf{Cyl}(\rho) = \{\rho' \in \Omega \mid \rho \preceq \rho'\}$. The sequence associated with $\rho = q a_0 q_1 \ldots$ is the word $\sigma_\rho = a_0 a_1 \ldots$, and we write indifferently $q \xRightarrow{\rho}$ or $q \xRightarrow{\sigma_\rho}$ (resp. $q \xRightarrow{\rho} q'$ or $q \xRightarrow{\sigma_\rho} q'$) for an infinite (resp. finite) run ρ. A state q is *reachable* (from q_0) if there exists a run such that $q_0 \xRightarrow{\rho} q$, which we alternatively write $q_0 \Rightarrow q$. The language of pLTS $\mathcal{A}$ consists of all infinite words that label runs of $\mathcal{A}$ and is formally defined as $\mathcal{L}^\omega(\mathcal{A}) = \{\sigma \in \Sigma^\omega \mid q_0 \xRightarrow{\sigma}\}$.

Forgetting the labels and merging (and summing the probabilities of) the transitions with same source and target, a pLTS yields a discrete time Markov chain (DTMC). As usual for DTMC, the set of infinite runs of $\mathcal{A}$ is the support of a probability measure defined by Caratheodory's extension theorem from the probabilities of the cylinders:

$$\mathbb{P}_{\mathcal{A}}(\mathsf{Cyl}(q_0 a_0 q_1 \ldots q_n)) = \mathbf{P}[q_0, a_1, q_1] \cdots \mathbf{P}[q_{n-1}, a_{n-1}, q_n] \ .$$

When $\mathcal{A}$ is fixed, we may omit the subscript. To simplify, for ρ a finite run, we will sometimes abuse notation and write $\mathbb{P}(\rho)$ for $\mathbb{P}(\mathsf{Cyl}(\rho))$. If R is a (denumerable) set of finite runs (such that no run is a prefix of another one), we write $\mathbb{P}(R)$ for $\sum_{\rho \in R} \mathbb{P}(\rho)$.

2.2 Partial Observation and Ambiguity

Beyond the pLTS model for stochastic discrete event systems, in order to formalize problems related to fault diagnosis, we partition Σ into two disjoint sets Σ_o and Σ_u, the sets of *observable* and of *unobservable events*, respectively. Moreover, we distinguish a special *fault* event $\mathbf{f} \in \Sigma_u$. Let σ be a finite word over Σ; its length is denoted $|\sigma|$. The projection of words onto Σ_o is defined inductively by: $\pi(\varepsilon) = \varepsilon$; for $a \in \Sigma_o$, $\pi(\sigma a) = \pi(\sigma)a$; and $\pi(\sigma a) = \pi(\sigma)$ for $a \notin \Sigma_o$. We write $|\sigma|_o$ for $|\pi(\sigma)|$. When σ is an infinite word, its projection is the limit of the projections of its finite prefixes. As usual the projection mapping is extended to languages: for $L \subseteq \Sigma^*$, $\pi(L) = \{\pi(\sigma) \mid \sigma \in L\}$. With respect to the partition of $\Sigma = \Sigma_o \uplus \Sigma_u$, a pLTS $\mathcal{A}$ is *convergent* if, from any reachable state, there is no infinite sequence of unobservable events: $\mathcal{L}^\omega(\mathcal{A}) \cap \Sigma^* \Sigma_u^\omega = \emptyset$. When $\mathcal{A}$ is convergent, for every $\sigma \in \mathcal{L}^\omega(\mathcal{A})$, $\pi(\sigma) \in \Sigma_o^\omega$. In the rest of the paper we assume that pLTS are convergent. We will use the terminology *sequence* for a word $\sigma \in \Sigma^* \cup \Sigma^\omega$, and an *observed sequence* for a word $\sigma \in \Sigma_o^* \cup \Sigma_o^\omega$. The projection of a sequence to Σ_o is thus an observed sequence.

The *observable length* of a run ρ denoted $|\rho|_o \in \mathbb{N} \cup \{\infty\}$, is the number of observable events that occur in it: $|\rho_o| = |\sigma_\rho|_o$. A *signalling run* is a finite run ending with and observable event. Signalling runs are precisely the relevant runs w.r.t. partial observation issues since each observable event provides an external observer additional information about the execution. In the sequel, SR denotes the set of signalling runs, and SR_n the set of signalling runs of observable length n. Since we assume pLTS to be convergent, for every $n > 0$, SR_n is equipped with a probability distribution defined by assigning measure $\mathbb{P}(\rho)$ to each $\rho \in \mathsf{SR}_n$. Given ρ a finite or infinite run, and $n \leq |\rho|_o$, $\rho_{\downarrow n}$ denotes the signalling subrun of ρ of observable length n. For convenience, we consider the empty run q_0 to be the single signalling run, of null length. For an observed sequence $\sigma \in \Sigma_o^*$, we define its cylinder $\mathsf{Cyl}(\sigma) = \sigma \Sigma_o^*$ and the associated probability $\mathbb{P}(\mathsf{Cyl}(\sigma)) = \mathbb{P}(\{\rho \in \mathsf{SR}_{|\sigma|} \mid \pi(\rho) = \sigma\})$, often shortened as $\mathbb{P}(\sigma)$.

Let us now partition runs depending on whether they contain a fault or not. A run ρ is *faulty* if σ_ρ contains $\mathbf{f}$, otherwise it is *correct*. For $n \in \mathbb{N}$, we write F_n (resp. C_n) for the set of faulty (resp. correct) signalling runs of length n, and further define the set of all faulty and correct signalling runs $\mathsf{F} = \cup_{n \in \mathbb{N}} \mathsf{F}_n$ and $\mathsf{C} = \cup_{n \in \mathbb{N}} \mathsf{C}_n$. W.l.o.g., by considering two copies of each state, we assume that the state space Q is partitioned into correct states and faulty states: $Q = Q_f \uplus Q_c$ such that faulty (resp. correct) states, *i.e.* states in Q_f (resp. Q_c) are only reachable by faulty (resp. correct) runs. An infinite (resp. finite) observed sequence $\sigma \in \Sigma_o^\omega$ (resp. Σ_o^*) is *ambiguous* if there exists a correct infinite (resp. signalling) run ρ and a faulty infinite (resp. signalling) run ρ' such that $\pi(\rho) = \pi(\rho') = \sigma$.

2.3   Fault Diagnosis

Whatever the considered notion of diagnosis in probabilistic systems, *reactivity* requires that when a fault occurs, a diagnoser almost surely will detect it after a finite delay. We refine this requirement by also considering *uniform reactivity* ensuring that given any positive probability threshold α there exists a delay n_α such that the probability to exceed this delay is less or equal than α. Here uniformity means "independently of the faulty run".

Similarly, *correctness* of the diagnosis may be specified in different ways. Since we focus on approximate diagnosis, a fault can be claimed after an ambiguous observed sequence. This implies that ambiguity should be quantified in order to assess the quality of the diagnosis. To formalise this idea, with every observed sequence $\sigma \in \Sigma_o^*$ we associate a *correctness proportion*

$$\mathsf{CorP}(\sigma) = \frac{\mathbb{P}(\{\rho \in \mathsf{C}_{|\sigma|} \mid \pi(\rho) = \sigma\})}{\mathbb{P}(\{\rho \in \mathsf{C}_{|\sigma|} \cup \mathsf{F}_{|\sigma|} \mid \pi(\rho) = \sigma\})} \ ,$$

which is the conditional probability that a signalling run is correct given that its observed sequence is σ. Thus approximate diagnosability also denoted ε-*diagnosability* allows the diagnoser to claim a fault when the correctness proportion does not exceed ε while accurate approximate diagnosability denoted AA-*diagnosability* ensures that ε can be chosen as small as desired but still positive.

Definition 2 (Diagnosability notions). *Let $\mathcal{A}$ be a pLTS and $\varepsilon \geq 0$.*

- *$\mathcal{A}$ is ε-diagnosable if for all faulty run $\rho \in \mathsf{F}$ and all $\alpha > 0$ there exists $n_{\rho,\alpha}$ such that for all $n \geq n_{\rho,\alpha}$:*

$$\mathbb{P}(\{\rho' \in \mathsf{SR}_{n+|\rho|_o} \mid \rho \preceq \rho' \wedge \mathsf{CorP}(\pi(\rho')) > \varepsilon\}) \leq \alpha \mathbb{P}(\rho).$$

$\mathcal{A}$ is uniformly ε-diagnosable if $n_{\rho,\alpha}$ does not depend on ρ.
- *$\mathcal{A}$ is (uniformly) AA-diagnosable if it is (uniformly) ε-diagnosable for all $\varepsilon > 0$.*

Two variants of diagnosability for stochastic systems were introduced in [12]: AA-diagnosability and A-diagnosability. *A-diagnosability*, which corresponds to exact diagnosis, is nothing else but 0-diagnosability in Definition 2 wording. By definition, A-diagnosability implies AA-diagnosability which implies ε-diagnosability for all $\varepsilon > 0$. Observe also that since the faulty run ρ (and so $\mathbb{P}(\rho)$) is fixed, ε-diagnosability can be rewritten:

$$\lim_{n \to \infty} \mathbb{P}(\{\rho' \in \mathsf{SR}_{n+|\rho|_o} \mid \rho \preceq \rho' \wedge \mathsf{CorP}(\pi(\rho')) > \varepsilon\}) = 0.$$

We now provide examples that illustrate these notions. Consider $\mathcal{A}_1$, the pLTS represented on Fig. 1. We claim that $\mathcal{A}_1$ is AA-diagnosable but neither A-diagnosable, nor uniformly AA-diagnosable. We only give here intuitions on these claims, and refer the reader to the proof of Proposition 3 in [2]. First an

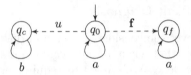

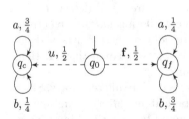

Fig. 1. An AA-diagnosable pLTS $\mathcal{A}_1$, that is neither A-diagnosable, nor uniformly AA-diagnosable

Fig. 2. An uniformly AA-diagnosable pLTS $\mathcal{A}_2$, that is not A-diagnosable

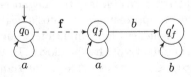

Fig. 3. An A-diagnosable pLTS $\mathcal{A}_3$

ε-diagnoser will look at the proportion of b occurrences and if the sequence is "long" enough and the proportion is "close" to $\frac{3}{4}$, it will claim a fault. However, the delay $n_{\alpha,\rho}$ before claiming a fault cannot be selected independently of the faulty run. Indeed, given the faulty run $\rho_n = q_0 \mathbf{f} q_f (a q_f)^n$, we let $p_{n,m}$ for the probability of extensions of ρ_n by m observable events and with correctness proportion below ε. In order for $p_{n,m}$ to exceed $1 - \alpha$, m must depend on n. So $\mathcal{A}_1$ is not uniformly AA-diagnosable. $\mathcal{A}_1$ is neither A-diagnosable since all observed sequences of faulty runs are ambiguous.

Consider now the pLTS $\mathcal{A}_2$ depicted in Fig. 2, for which we consider a uniform distribution on the outgoing edges from q_0. First note that every faulty run $(q_0 a)^i q_0 \mathbf{f} (q_f a)^j q_f$ has a correct run, namely $q_0 (a q_0)^{i+j}$ with the same observed sequence. So $\mathcal{A}_2$ is not A-diagnosable. Yet, we argue that it is uniformly AA-diagnosable. The correctness proportion of a faulty run (exponentially) decreases with respect to its length. So the worst run to be considered for the diagnoser is $q_0 \mathbf{f} q_f a q_f$ implying uniformity.

Consider the pLTS $\mathcal{A}_3$ from Fig. 3, with uniform distributions in q_0 and q_f. Viewed as an LTS, it is not diagnosable, since the observed sequence a^ω is ambiguous and forbids the diagnosis of faulty runs without any occurrence of b. On the contrary, let $\rho = q_0 (a q_0)^x \mathbf{f} q_f (a q_f)^y (b q_f')^z$ be an arbitrary faulty run. If $z > 0$ then $\mathsf{CorP}(\pi(\rho)) = 0$. Otherwise $\mathbb{P}(\{\rho' \in \mathsf{SR}_{n+|\rho|_o} \mid \rho \preceq \rho' \wedge \mathsf{CorP}(\pi(\rho')) > 0\}) = \frac{1}{2^n} \mathbb{P}(\rho)$ and so $\mathcal{A}_3$ is A-diagnosable.

Proposition 3 establishes the exact relations between the different specifications. Observe that uniform AA-diagnosability is strictly stronger than AA-diagnosability while A-diagnosability and uniform A-diagnosability are equivalent.

Proposition 3. – *A pLTS is A-diagnosable if and only if it is uniformly A-diagnosable.*
  – *There exists an AA-diagnosable pLTS, not uniformly $\frac{1}{2}$-diagnosable and so not uniformly AA-diagnosable.*

– *There exists a uniformly AA-diagnosable pLTS, not A-diagnosable.*

Although we have not explicitly defined diagnosers for diagnosable pLTS, given a fixed threshold $\varepsilon > 0$, a simple diagnoser would monitor the sequence of observed events σ, compute the current correctness proportion, and output "faulty" if $\mathsf{CorP}(\sigma)$ is below ε. However such an ε-diagnoser may need an infinite memory. This contrasts with the case of A-diagnosability for which finite-memory diagnosers suffice [12].

Proposition 4. *There exists an AA-diagnosable pLTS, thus ε-diagnosable for every $\varepsilon > 0$, that admits no finite-memory diagnoser when $0 < \varepsilon \leq \frac{1}{2}$.*

Proof. Consider $\mathcal{A}_1$ the AA-diagnosable pLTS of Fig. 1 and assume there exists a diagnoser with m states for some threshold $0 < \varepsilon \leq \frac{1}{2}$. After any sequence a^n, it cannot claim a fault. So there exist $1 \leq i < j \leq m+1$ such that the diagnoser is in the same state after observing a^i and a^j.

Consider the faulty run $\rho = q_0 \mathsf{f} q_f (a q_f)^i$. Due to the reactivity requirement, there must be a run $\rho\rho'$ for which the diagnoser claims a fault. This implies that for all n, the diagnoser claims a fault after $\rho_n = \rho(a q_f)^{n(j-i)}\rho'$ but $\lim_{n\to\infty} \mathsf{CorP}(\pi(\rho_n)) = 1$, which contradicts the correctness requirement.   □

3   Analysis of Approximate Diagnosability

A-diagnosability was proved to be a PSPACE-complete problem [1]. We now focus on the other notions of approximate diagnosability introduced in Definition 2, and study their decidability and complexity.

Reducing the emptiness problem for probabilistic automata [9] (PA), we obtain the following first result:

Theorem 5. *For any rational $0 < \varepsilon < 1$, the ε-diagnosability and uniform ε-diagnosability problems are undecidable for pLTS.*

We now turn to the decidability status of AA-diagnosability and uniform AA-diagnosability. We prove that AA-diagnosability can be solved in polynomial time by establishing a characterization in terms of distance on labelled Markov chains; this constitutes the most technical contribution of this section.

A *labelled Markov chain* (LMC) is a pLTS where every event is observable: $\Sigma = \Sigma_o$. In order to exploit results of [6] on LMC in our context of pLTS, we introduce the mapping $\mathcal{M}$ that performs *in polynomial time* the probabilistic closure of a pLTS w.r.t. the unobservable events and produces an LMC. For sake of simplicity, we denote by $\mathcal{A}_q$, the pLTS $\mathcal{A}$ where the initial state has been substituted by q.

Definition 6. *Given a pLTS $\mathcal{A} = \langle Q, q_0, \Sigma, T, \mathbf{P} \rangle$ with $\Sigma = \Sigma_o \uplus \Sigma_u$, the labelled Markov chain $\mathcal{M}(\mathcal{A}) = \langle Q, q_0, \Sigma_o, T', \mathbf{P}' \rangle$ is defined by:*

– $T' = \{(q, a, q') \mid \exists \rho \in \mathsf{SR}_1(\mathcal{A}_q)\ \rho = q \cdots a q'\}$ *(and so a is observable).*
– *For all $(q, a, q') \in T'$, $\mathbf{P}'(q, a, q') = \mathbb{P}\{\rho \in \mathsf{SR}_1(\mathcal{A}_q) \mid \rho = q \cdots a q'\}$.*

Let E be an *event* of Σ^ω (*i.e.* a measurable subset of Σ^ω for the standard measure), we denote by $\mathbb{P}^{\mathcal{M}}(E)$ the probability that event E occurs in the LMC $\mathcal{M}$. Given two LMC $\mathcal{M}_1$ and $\mathcal{M}_2$, the (probabilistic) distance between $\mathcal{M}_1$ and $\mathcal{M}_2$ generalizes the concept of distance for distributions. Given an event E, $|\mathbb{P}^{\mathcal{M}_1}(E) - \mathbb{P}^{\mathcal{M}_2}(E)|$ expresses the absolute difference between the probabilities that E occurs in $\mathcal{M}_1$ and in $\mathcal{M}_1$. The distance is obtained by getting the supremum over the events.

Definition 7. *Let $\mathcal{M}_1$ and $\mathcal{M}_2$ be two LMC over the same alphabet Σ. Then $d(\mathcal{M}_1, \mathcal{M}_2)$ the distance between $\mathcal{M}_1$ and $\mathcal{M}_2$ is defined by:*

$$d(\mathcal{M}_1, \mathcal{M}_2) = \sup(\mathbb{P}^{\mathcal{M}_1}(E) - \mathbb{P}^{\mathcal{M}_2}(E) \mid E \text{ event of } \Sigma^\omega).$$

The *distance 1 problem* asks, given labelled Markov chains $\mathcal{M}_1$ and $\mathcal{M}_2$, whether $d(\mathcal{M}_1, \mathcal{M}_2) = 1$. We summarize in the next proposition, the results by Chen and Kiefer on LMC that we use later.

Proposition 8 ([6]).

– *Given two LMC $\mathcal{M}_1, \mathcal{M}_2$, there exists an event E such that:*

$$d(\mathcal{M}_1, \mathcal{M}_2) = \mathbb{P}^{\mathcal{M}_1}(E) - \mathbb{P}^{\mathcal{M}_2}(E).$$

– *The distance 1 problem for LMC is decidable in polynomial time.*

Towards the decidability of AA-diagnosability, let us first explain how to solve the problem on a subclass of pLTS called *initial-fault pLTS*. Informally, an initial-fault pLTS $\mathcal{A}$ consists of two disjoint pLTS $\mathcal{A}^f$ and $\mathcal{A}^c$ and an initial state q_0 with an outgoing unobservable correct transition leading to $\mathcal{A}^c$ and a transition labelled by $\mathbf{f}$ leading to $\mathcal{A}^f$ (see the figure below). Moreover no faulty transitions occur in $\mathcal{A}^c$. We denote such a pLTS by $\mathcal{A} = \langle q_0, \mathcal{A}^f, \mathcal{A}^c \rangle$.

The next lemma establishes a strong connection between distance of LMC and diagnosability of initial-fault pLTS.

Lemma 9. *Let $\mathcal{A} = \langle q_0, \mathcal{A}^f, \mathcal{A}^c \rangle$ be an initial-fault pLTS. Then $\mathcal{A}$ is AA-diagnosable if and only if $d(\mathcal{M}(\mathcal{A}^f), \mathcal{M}(\mathcal{A}^c)) = 1$.*

Proof. We write $\mathbb{P}$, $\mathbb{P}_f$ and $\mathbb{P}_c$ for the probability distributions of pLTS $\mathcal{A}$, $\mathcal{A}^f$ and $\mathcal{A}^c$. By construction of $\mathcal{M}(\mathcal{A}^f)$ and $\mathcal{M}(\mathcal{A}^c)$, for every observed sequence σ, $\mathbb{P}^{\mathcal{M}(\mathcal{A}^f)}(\sigma) = \mathbb{P}_f(\sigma)$ and similarly $\mathbb{P}^{\mathcal{M}(\mathcal{A}^c)}(\sigma) = \mathbb{P}_c(\sigma)$. In words, the mapping $\mathcal{M}$ leaves unchanged the probability of occurrence of an observed sequence.

• If $\mathcal{A}$ is AA-diagnosable, for every $\varepsilon > 0$ and every faulty run ρ:

$$\lim_{n \to \infty} \mathbb{P}(\{\rho' \in SR_{n+|\rho|_o} \mid \rho \preceq \rho' \wedge \mathsf{CorP}(\pi(\rho')) > \varepsilon\}) = 0. \tag{1}$$

Pick some $0 < \varepsilon < 1$. By applying Eq. (1) on the faulty run $\rho_f = q_0 f q_f$ with $|\pi(\rho_f)| = 0$, there exists some $n \in \mathbb{N}$ such that:

$$\mathbb{P}(\{\rho \in \mathsf{SR}_n \mid \rho_f \preceq \rho \wedge \mathsf{CorP}(\pi(\rho)) > \varepsilon\}) \leq \varepsilon.$$

Let $\mathfrak{S}$ be the set of observed sequences of faulty runs with length n and correctness proportion not exceeding theshold ε:

$$\mathfrak{S} = \{\sigma \in \Sigma_o^n \mid \exists \rho \in \mathsf{SR}_n, \pi(\rho) = \sigma \wedge \rho_f \preceq \rho \wedge \mathsf{CorP}(\sigma) \leq \varepsilon\}.$$

$E = Cyl(\mathfrak{S})$ is the event consisting of the infinite suffixes of those sequences. Let us show that $\mathbb{P}_c(E) \leq \frac{\varepsilon}{1-\varepsilon}$ and $\mathbb{P}_f(E) \geq 1 - 2\varepsilon$.

$$\mathbb{P}_f(E) = 1 - 2\mathbb{P}(\{\rho \in \mathsf{SR}_n \mid \rho_f \preceq \rho \wedge \mathsf{CorP}(\pi(\rho)) > \varepsilon\}) \geq 1 - 2\varepsilon.$$

The factor 2 comes from the probability $\frac{1}{2}$ in $\mathcal{A}$ to enter $\mathcal{A}^f$ that $\mathbb{P}_f$ does not take into account contrary to $\mathbb{P}$.

Moreover, for every observed sequence $\sigma \in \mathfrak{S}$, there exists a faulty run ρ such that $\pi(\rho) = \sigma$. Thus, $\mathsf{CorP}(\sigma) \leq \varepsilon$. Using the definition of CorP:

$$\mathsf{CorP}(\sigma) = \frac{\mathbb{P}(\{\rho \in C_n \mid \pi(\rho) = \sigma\})}{\mathbb{P}(\{\rho \in \mathsf{SR}_n \mid \pi(\rho) = \sigma\})} = \frac{\mathbb{P}_c(\sigma)}{\mathbb{P}_c(\sigma) + \mathbb{P}_f(\sigma)} \leq \varepsilon.$$

Thus, $\mathbb{P}_c(\sigma) \leq \frac{\varepsilon}{1-\varepsilon}\mathbb{P}_f(\sigma)$. Hence:

$$\mathbb{P}_c(E) = \sum_{\sigma \in \mathfrak{S}} \mathbb{P}_c(\sigma) \leq \sum_{\sigma \in \mathfrak{S}} \frac{\varepsilon}{1-\varepsilon}\mathbb{P}_f(\sigma) = \frac{\varepsilon}{1-\varepsilon}\mathbb{P}_f(E) \leq \frac{\varepsilon}{1-\varepsilon}.$$

Therefore $d(\mathcal{M}(\mathcal{A}^c), \mathcal{M}(\mathcal{A}^f)) \geq \mathbb{P}_f(E) - \mathbb{P}_c(E) \geq 1 - \varepsilon(2 + \frac{1}{1-\varepsilon})$. Letting ε go to 0, we obtain $d(\mathcal{M}(\mathcal{A}^c), \mathcal{M}(\mathcal{A}^f)) = 1$.

• Conversely assume that $d(\mathcal{M}(\mathcal{A}^f), \mathcal{M}(\mathcal{A}^c)) = 1$. Due to Proposition 8, there exists an event $E \subseteq \Sigma_o^\omega$ such that $\mathbb{P}_f(E) = 1$ and $\mathbb{P}_c(E) = 0$.
For all $n \in \mathbb{N}$, let $\mathfrak{S}_n$ be the set of prefixes of length n of the observed sequences of E: $\mathfrak{S}_n = \{\sigma \in \Sigma_o^n \mid \exists \sigma' \in E, \sigma \preceq \sigma'\}$.
For all $\varepsilon > 0$, let $\mathfrak{S}_n^\varepsilon$ be the subset of sequences of $\mathfrak{S}_n$ whose correctness proportion exceeds threshold ε: $\mathfrak{S}_n^\varepsilon = \{\sigma \in \mathfrak{S}_n \mid \mathsf{CorP}(\sigma) > \varepsilon\}$.
As $\bigcap_{n \in \mathbb{N}} Cyl(\mathfrak{S}_n) = E$, $\lim_{n \to \infty} \mathbb{P}_c(\mathfrak{S}_n) = \mathbb{P}_c(E) = 0$.
So $\lim_{n \to \infty} \mathbb{P}_c(\mathfrak{S}_n^\varepsilon) = 0$.
On the other hand for all $n \in \mathbb{N}$,

$$\mathbb{P}_c(\mathfrak{S}_n^\varepsilon) = \sum_{\sigma \in \mathfrak{S}_n^\varepsilon} \mathbb{P}_c(\sigma) > \sum_{\sigma \in \mathfrak{S}_n^\varepsilon} \frac{\varepsilon}{1-\varepsilon}\mathbb{P}_f(\sigma) = \frac{\varepsilon}{1-\varepsilon}\mathbb{P}_f(\mathfrak{S}_n^\varepsilon).$$

Therefore we have $\lim_{n \to \infty} \mathbb{P}_f(\mathfrak{S}_n^\varepsilon) = 0$.
Let ρ be a faulty run and $\alpha > 0$. There exists $n_\alpha \geq |\rho|_o$ such that for all $n \geq n_\alpha$, $\mathbb{P}_f(\mathfrak{S}_n^\varepsilon) \leq \alpha$. Let $n \geq n_\alpha$, and $\tilde{\mathfrak{S}}_n$ be the set of observed sequences of length n triggered by a run with prefix ρ and whose correctness proportion exceeds ε:

$$\tilde{\mathfrak{S}}_n = \{\sigma \in \Sigma_o^n \mid \exists \rho' \in \mathsf{SR}_n, \rho \preceq \rho' \wedge \pi(\rho') = \sigma \wedge \mathsf{CorP}(\sigma) > \varepsilon\}.$$

Let us prove that $\mathbb{P}(\widetilde{\mathfrak{S}}_n) \leq \alpha$. On the one hand, since $\mathbb{P}_f(\mathfrak{S}_n) \geq \mathbb{P}_f(E) = 1$, $\mathbb{P}_f(\widetilde{\mathfrak{S}}_n \cap (\Sigma_o^n \setminus \mathfrak{S}_n)) = 0$. On the other hand, since $\mathbb{P}_f(\mathfrak{S}_n^\varepsilon) < \alpha$, $\mathbb{P}_f(\widetilde{\mathfrak{S}}_n \cap \mathfrak{S}_n) \leq \mathbb{P}_f(\mathfrak{S}_n^\varepsilon) \leq \alpha$. Thus $\mathbb{P}_f(\widetilde{\mathfrak{S}}_n) = \mathbb{P}_f(\widetilde{\mathfrak{S}}_n \cap \mathfrak{S}_n) + \mathbb{P}_f(\widetilde{\mathfrak{S}}_n \cap (\Sigma_o^n \setminus \mathfrak{S}_n)) \leq \alpha$. Because α was taken arbitrary, we obtain that $\lim_{n \to \infty} \mathbb{P}_f(\widetilde{\mathfrak{S}}_n) = 0$.

Observe now that $\mathbb{P}(\{\rho' \in \mathsf{SR}_n \mid \rho \preceq \rho' \wedge \mathsf{CorP}(\pi(\rho')) > \varepsilon\}) = \frac{1}{2}\mathbb{P}_f(\widetilde{\mathfrak{S}}_n)$. Therefore, $\lim_{n \to \infty} \mathbb{P}(\{\rho' \in \mathsf{SR}_n \mid \rho \preceq \rho' \wedge \mathsf{CorP}(\pi(\rho')) > \varepsilon\}) = 0$. So $\mathcal{A}$ is AA-diagnosable.   □

In order to understand why characterizing AA-diagnosability for general pLTS is more involved, let us study the pLTS $\mathcal{A}_2$ presented in Fig. 2 where outgoing transitions of any state are equidistributed. Recall that $\mathcal{A}_2$ is AA-diagnosable (and even uniformly AA-diagnosable).

Let us look at the distance between pairs of a correct and a faulty states of $\mathcal{A}$ that can be reached by runs with the same observed sequence. On the one hand, $d(\mathcal{M}(\mathcal{A}_{q_0}), \mathcal{M}(\mathcal{A}_{q_f})) \leq \frac{1}{2}$ since for any event E either (1) $a^\omega \in E$ implying $\mathbb{P}^{\mathcal{M}(\mathcal{A}_{q_f})}(E) = 1$ and $\mathbb{P}^{\mathcal{M}(\mathcal{A}_{q_0})}(E) \geq \frac{1}{2}$ or (2) $a^\omega \notin E$ implying $\mathbb{P}^{\mathcal{M}(\mathcal{A}_{q_f})}(E) = 0$ and $\mathbb{P}^{\mathcal{M}(\mathcal{A}_{q_0})}(E) \leq \frac{1}{2}$. On the other hand, $d(\mathcal{M}(\mathcal{A}_{q_c}), \mathcal{M}(\mathcal{A}_{q_f})) = 1$ since $\mathbb{P}^{\mathcal{M}(\mathcal{A}_{q_f})}(a^\omega) = 1$ and $\mathbb{P}^{\mathcal{M}(\mathcal{A}_{q_c})}(a^\omega) = 0$.

We claim that the pair (q_0, q_f) is irrelevant, since the correct state q_0 does not belong to a bottom strongly connected component (BSCC) of the pLTS, while (q_c, q_f) is relevant since q_c belongs to a BSCC triggering a "recurrent" ambiguity.

The next theorem characterizes AA-diagnosability, establishing the soundness of this intuition. Moreover, it states the complexity of deciding AA-diagnosability.

Theorem 10. *Let $\mathcal{A}$ be a pLTS. Then, $\mathcal{A}$ is AA-diagnosable if and only if for every correct state q_c belonging to a BSCC and every faulty state q_f reachable by runs with same observed sequence, $d(\mathcal{M}(\mathcal{A}_{q_c}), \mathcal{M}(\mathcal{A}_{q_f})) = 1$.*
The AA-diagnosability problem is decidable in polynomial time for pLTS.

The full proof of Theorem 10 is given [2]. Let us sketch the key ideas to establish the characterization of AA-diagnosability in terms of the distance 1 problem. The left-to-right implication is the easiest one, and is proved by contraposition. Assume there exist two states in $\mathcal{A}$, $q_c \in Q_c$ belonging to a BSCC and $q_f \in Q_f$ reachable resp. by ρ_c and ρ_f with $\pi(\rho_c) = \pi(\rho_f)$, and with $d(\mathcal{M}(\mathcal{A}_{q_c}), \mathcal{M}(\mathcal{A}_{q_f})) < 1$. Applying Lemma 9 to the initial-fault pLTS $\mathcal{A}' = \langle q_0', \mathcal{A}_{q_f}, \mathcal{A}_{q_c} \rangle$, one deduces that $\mathcal{A}'$ is not AA-diagnosable. First we relate the probabilities of runs in $\mathcal{A}$ and $\mathcal{A}'$. Then we show that considering the additional faulty runs with same observed sequence as ρ_f does not make $\mathcal{A}$ AA-diagnosable.

The right-to-left implication is harder to establish. For ρ_0 a faulty run, $\alpha > 0, \varepsilon > 0$, $\sigma_0 = \pi(\rho_0)$ and $n_0 = |\sigma_0|$, we start by extending the runs with observed sequences σ_0 by n_b observable events where n_b is chosen in order to get a high probability that the runs end in a BSCC. For such an observed sequence

$\sigma \in \Sigma_o^{nb}$, we partition the possible runs with observed sequence $\sigma_0 \sigma$ into three sets: $\mathfrak{R}_\sigma^F$ is the subset of faulty runs; $\mathfrak{R}_\sigma^C$ (resp. $\mathfrak{R}_\sigma^T$) is the set of correct runs ending (resp. not ending) in a BSCC. At first, we do not take into account the "transient" runs in $\mathfrak{R}_\sigma^T$. We apply Lemma 9 to obtain an integer n_σ such that from $\mathfrak{R}_\sigma^F$ and $\mathfrak{R}_\sigma^C$ we can diagnose with (appropriate) high probability and low correctness proportion after n_σ observations. Among the runs that trigger diagnosable observed sequences, some exceed the correctness proportion ε, when taking into account the runs from $\mathfrak{R}_\sigma^T$. Yet, we show that the probability of such runs is small, when cumulated over all extensions σ, leading to the required upper bound α.

Using the characterization, one can easily establish the complexity of AA-diagnosability. Indeed, reachability of a pair of states with the same observed sequence is decidable in polynomial time by an appropriate "self-synchronized product" of the pLTS. Since there are at most a quadratic number of pairs to check, and given that the distance 1 problem can be decided in polynomial time, the PTIME upper-bound follows.

In constrast, uniform AA-diagnosability is shown to be undecidable by a reduction from the emptiness problem for probabilistic automata, that is more involved than the one for Theorem 5.

Theorem 11. *The uniform AA-diagnosability problem is undecidable for pLTS.*

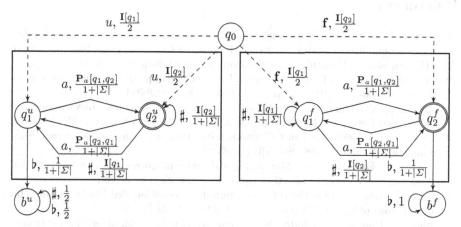

The reduction is illustrated above, and we sketch here the undecidability proof. Assuming there exists a word $w \in \Sigma^*$ accepted with probability greater than $\frac{1}{2}$ in the probabilistic automaton. We pick arbitrary $\alpha < 1$ and n_α. Then, one can exhibit a faulty signalling run ρ_n with $\pi(\rho_n) = (w\sharp)^n$ for some appropriate n, such that for every extension $\rho_n \preceq \rho$ with $|\rho| = |\rho_n| + n_\alpha$, one has $\mathsf{CorP}(\rho) > \frac{1}{2}$. This shows that the constructed pLTS is not uniformly $\frac{1}{2}$-diagnosable.

Assuming now that all words are accepted with probability less than $\frac{1}{2}$. Then for any observed sequence $\sigma \in (\Sigma \cup \{\sharp\})^*$, $\mathsf{CorP}(\sigma) \leq \frac{1}{2}$. After reaching a BSCC, the correctness proportion decreases uniformly, due to the $\sharp$-loop on b^u. Given positive α and ε, one can thus find integers n_0 and n_1 such that a BSCC is reached after n_0 observable events with probability at least $1 - \alpha$, and after n_1 more the correctness proportion, which was at most $\frac{1}{2}$, decreases below ε. This shows the uniform AA-diagnosability.

4   Conclusion

This paper completes our previous work [1] on diagnosability of stochastic systems, by giving here a full picture on approximate diagnosis. On the one hand, we performed a semantical study: we have refined the reactivity specification by introducing a uniform requirement about detection delay w.r.t. faults and studied its impact on both the exact and approximate case. On the other hand, we established decidability and complexity of all notions of approximate diagnosis: we have shown that (uniform) ε-diagnosability and uniform AA-diagnosability are undecidable while AA-diagnosability can be solved in polynomial time.

There are still interesting issues to be tackled, to continue our work on monitoring of stochastic systems. For example, prediction and prediagnosis, which are closely related to diagnosis and were analyzed in the exact case in [1], should be studied in the approximate framework.

References

1. Bertrand, N., Haddad, S., Lefaucheux, E.: Foundation of diagnosis and predictability in probabilistic systems. In: Proceedings of FSTTCS 2014. LIPIcs, vol. 29, pp. 417–429. Schloss Dagstuhl - Leibniz-Zentrum fuer Informatik (2014)
2. Bertrand, N., Haddad, S., Lefaucheux, E.: Accurate approximate diagnosability of stochastic systems (2015). https://hal.inria.fr/hal-01220954
3. Cabasino, M., Giua, A., Lafortune, S., Seatzu, C.: Diagnosability analysis of unbounded Petri nets. In: Proceedings of CDC 2009, pp. 1267–1272. IEEE (2009)
4. Cassez, F., Tripakis, S.: Fault diagnosis with static and dynamic observers. Fundamenta Informaticae **88**, 497–540 (2008)
5. Chanthery, E., Pencolé, Y.: Monitoring and active diagnosis for discrete-event systems. In: Proceedings of SP 2009, pp. 1545–1550. Elsevier (2009)
6. Chen, T., Kiefer, S.: On the total variation distance of labelled Markov chains. In: Proceedings of CSL-LICS 2014, pp. 33:1–33:10. ACM (2014)
7. Jiang, S., Huang, Z., Chandra, V., Kumar, R.: A polynomial algorithm for testing diagnosability of discrete-event systems. IEEE Trans. Autom. Control **46**(8), 1318–1321 (2001)
8. Morvan, C., Pinchinat, S.: Diagnosability of pushdown systems. In: Namjoshi, K., Zeller, A., Ziv, A. (eds.) HVC 2009. LNCS, vol. 6405, pp. 21–33. Springer, Heidelberg (2011)
9. Paz, A.: Introduction to Probabilistic Automata. Academic Press, Orlando (1971)
10. Sampath, M., Lafortune, S., Teneketzis, D.: Active diagnosis of discrete-event systems. IEEE Trans. Autom. Control **43**(7), 908–929 (1998)
11. Sampath, M., Sengupta, R., Lafortune, S., Sinnamohideen, K., Teneketzis, D.: Diagnosability of discrete-event systems. IEEE Trans. Autom. Control **40**(9), 1555–1575 (1995)

12. Thorsley, D., Teneketzis, D.: Diagnosability of stochastic discrete-event systems. IEEE Trans. Autom. Control **50**(4), 476–492 (2005)
13. Thorsley, D., Teneketzis, D.: Active acquisition of information for diagnosis and supervisory control of discrete-event systems. J. Discrete Event Dyn. Syst. **17**, 531–583 (2007)

Proof–Based Synthesis of Sorting Algorithms for Trees

Isabela Drămnesc[1], Tudor Jebelean[2], and Sorin Stratulat[3(✉)]

[1] Department of Computer Science, West University, Timisoara, Romania
idramnesc@info.uvt.ro
[2] Research Institute for Symbolic Computation, Johannes Kepler University,
Linz, Austria
Tudor.Jebelean@jku.at
[3] LITA, Department of Computer Science, Université de Lorraine, Metz, France
sorin.stratulat@univ-lorraine.fr

Abstract. We develop various proof techniques for the synthesis of sorting algorithms on binary trees, by extending our previous work on the synthesis of algorithms on lists. Appropriate induction principles are designed and various specific prove-solve methods are experimented, mixing rewriting with assumption-based forward reasoning and goal-based backward reasoning *à la* Prolog. The proof techniques are implemented in the *Theorema* system and are used for the automatic synthesis of several algorithms for sorting and for the auxiliary functions, from which we present few here. Moreover we formalize and check some of the algorithms and some of the properties in the *Coq* system.

Keywords: Algorithm synthesis · Sorting · Theorem proving

1 Introduction

Program synthesis is currently a very active area of programming language and verification communities. Generally speaking, the program synthesis problem consists in finding an algorithm which satisfies a given specification. We focus on the proof-based synthesis of functional algorithms, starting from their formal specification expressed astwo predicates: the input condition $I[X]$ and the output condition $O[X, T]$, where X and T are vectors of universal and existential variables, respectively.The desired function F must satisfy the correctness condition $(\forall X)(I[X] \implies O[X, F[X]])$.[1]

We are interested to develop proof-based methods for finding F and to build formal tools for mechanizing and (partially) automatizing the proof process, by following constructive theorem proving and program extraction techniques to deductively synthesize F as a functional program [5]. The way the constructive proof is built is essential since the definition of F can be extracted as a side effect

---

[1] The square brackets have been used for function and predicate applications instead of round brackets.

© Springer International Publishing Switzerland 2016
A.-H. Dediu et al. (Eds.): LATA 2016, LNCS 9618, pp. 562–575, 2016.
DOI: 10.1007/978-3-319-30000-9_43

of the proof. For example, case splits may generate conditional branches and induction steps may produce recursive definitions. Hence, the use of different case reasoning techniques and induction principles may output different definitions of F. The extraction procedure guarantees that F satisfies the specification.

Non-trivial algorithms, as for sorting [14], are generated when X is a recursively-defined unbounded data structure, as lists and trees. In this paper, we apply the deductive approach to synthesize binary tree algorithms, extending similar results for lists [8]. In order to do this, we introduce new induction principles, proof strategies and inference rules based on properties of binary trees. Numerous new algorithms have been synthesized. For lack of space, we fully present the synthesis process for one of these algorithms; the proofs for the other algorithms are only summarized but can be found in the technical report [9]. The correctness of the discovered algorithms is ensured by the soundness of the induction principles, the specific inference rules and proof strategies introduced in this paper.

The implementations of the new prover and extractor, as well as of the case studies presented in this paper are carried out in the frame of the *Theorema* system[2] and e.g., [4] which is itself implemented in Mathematica [20]. *Theorema* offers significant support for automatizing the algorithm synthesis; in particular, the new proof strategies and inference rules have been quickly prototyped, tested and integrated in the system thanks to its extension features. Also, the proofs are easier to understand since they are presented in a human-oriented style. Moreover the synthesized algorithms can be directly executed in the system. The implementation files can be accessed in the technical report [9].

Additionally we have formalized part of the theory presented here and mechanically checked that some extracted algorithms satisfy the correctness condition in the frame of the *Coq* system [3].

1.1 Related Work

For an overview of the most common approaches used to tackle the synthesis problem, the reader may consult [12]. Synthesis methods and techniques similar to our proof-based approach are extensively presented in [8]. It can be noticed that most of the proof methods are based on expressive and undecidable logics that integrate induction principles.

The proof environments underlying deductive synthesis frameworks are usually supporting both automated and interactive proof methods. Those based on abstract datatype and computation refinements [2,19] integrate techniques that are mainly executed manually and implemented by higher-order proof assistants like Isabelle/HOL [15] or more synthesis-oriented tools as Specware [16]. On the other hand, automated proof steps can be performed with decision procedures, e.g., for linear arithmetics, or SAT and SMT solvers as those integrated in Leon [13]. The generated algorithms can be checked for conformity with the input specification by validating the proof trails for each refinement process,

---

[2] https://www.risc.jku.at/research/theorema/software/.

for example using the Coq library Fiat [7] to ensure the soundness of the validation step by certification with the Coq kernel. [6] presents a different Coq library using datatype refinement to verify parameterized algorithms for which the soundness proof of some version can be deduced from that of a previous (less efficiently implemented) version. Implementing inference rules directly in Coq may be of interest if one can prove that every generated synthesized algorithm is sound. In general, this is a rather difficult task, therefore this approach does not fit for rapid prototyping and testing new ideas.

2   The Proof-Based Synthesis Method

This section introduces the algorithm synthesis problem and presents the proof-based synthesis techniques that we use, by adapting and improving inference rules and induction principles from [8].

2.1   Our Approach

Basic notions and notations. According to the *Theorema* style, we use square brackets for function and for predicate application (e.g., f[x] instead of f(x) and P[a] instead of P(a)). Moreover the quantified variables appear under the quantifier: $\underset{X}{\forall}$ ("for all X") and $\underset{T}{\exists}$ ("exists T"). We consider binary trees over a totally ordered domain. In our formulae there are two kinds of objects: domain objects which are tree members (usually denoted by lower-case letters – e.g. a, b, n), and binary trees (usually represented by upper-case letters – e.g. X, T, Y, Z). However the formulae do not indicate explicitly the types of the objects, but our specific predicate and function symbols are not overloaded[3]. Furthermore the meta–variables are starred (e.g., T^*, T_1^*, Z^*) and the Skolem constants have integer indices (e.g., X_0, X_1, a_0).

The ordering between tree elements is denoted by the usual $\leq$, and the ordering between a tree and an element is denoted by: $\preceq$ (e.g., $T \preceq z$ states that all the elements from the tree T are smaller or equal than the element z, $z \preceq T$ states that z is smaller or equal than all the elements from the tree T). We use two constructors for binary trees, namely: ε for the empty tree, and the triplet $\langle L, a, R \rangle$ for non-empty trees, where L and R are trees and a is the root element.

A tree is a *sorted* (or *search*, or *ordered*) tree if it is either ε or of the form $\langle L, a, R \rangle$ such that i) $L \preceq a \preceq R$, and ii) L and R are sorted trees.

Functions: RgM, LfM, Concat, Insert, Merge have the following interpretations, respectively: $RgM[\langle L, n, R \rangle]$ (resp. $LfM[\langle L, n, R \rangle]$) returns the last (resp. first) visited element by traversing the tree $\langle L, n, R \rangle$ using the in-order (symmetric) traversal, i.e., the rightmost (resp. leftmost) element; $Concat[X, Y]$ concatenates X with Y (namely, when X is of the form $\langle L, n, R \rangle$ adds Y as a right subtree of the element $RgM[\langle L, n, R \rangle]$); $Insert[n, X]$ inserts an element n in a tree X (if X

---

[3] Each predicate and function symbol applies to a certain combination of types of argument.

is sorted, then the result is also sorted); $Merge[X, Y]$ combines trees X and Y into a new tree (if X, Y are sorted then the result is also sorted).

Predicates: $\approx$ and *IsSorted* have the following interpretations, respectively: $X \approx Y$ states that X and Y have the same elements with the same number of occurrences (but may have different structures), i.e., X is a *permutation* of Y; *IsSorted*$[X]$ states that X is a sorted tree.

The formal definitions of these functions and predicates are:

Definition 1. $\quad \underset{n,m,L,R,S}{\forall} \left(\begin{array}{c} RgM[\langle L, n, \varepsilon \rangle] = n \\ RgM[\langle L, n, \langle R, m, S \rangle \rangle] = RgM[\langle R, m, S \rangle] \end{array} \right)$

Definition 2. $\quad \underset{n,m,L,R,S}{\forall} \left(\begin{array}{c} LfM[\langle \varepsilon, n, R \rangle] = n \\ LfM[\langle \langle L, n, R \rangle, m, S \rangle] = LfM[\langle L, n, R \rangle] \end{array} \right)$

Definition 3. $\quad \underset{n,L,R,S}{\forall} \left(\begin{array}{c} Concat[\varepsilon, R] = R \\ Concat[\langle L, n, R \rangle, S] = \langle L, n, Concat[R, S] \rangle \end{array} \right)$

Definition 4.

$$\underset{L,m,R}{\forall} \left(\begin{array}{c} IsSorted[\varepsilon] \\ (IsSorted[L] \wedge IsSorted[R] \wedge RgM[L] \leq m \leq LfM[R]) \Longleftrightarrow IsSorted[\langle L, m, R \rangle] \end{array} \right)$$

A formal definition of $\approx$ is not given, however we use the properties of $\approx$ as equivalence implicitly in our inference rules and strategies. In particular, we use in our prover the fact that equivalent trees have the same multiset of elements, which translates into equivalent tree–expressions having the same multiset of constants and variables.

The functions LfM and RgM do not have a definition for the empty tree, however we assume that: $\underset{m}{\forall} \left(RgM[\varepsilon] \leq m \leq LfM[\varepsilon] \right)$.

An example of simple property which can be proven inductively from Definition 3 is the following:

Property 5. $\quad \underset{z,T}{\forall} \left(IsSorted[T] \Longrightarrow (T \preceq z \Longleftrightarrow RgM[T] \leq z) \right)$

All the statements used at object level in our experiments are formally just predicate logic formulae, however for this presentation we will call them differently depending on their role: a *definition* or an *axiom* is given as an initial piece of the theory, considered to hold; a *property* is a logical consequence of the definitions and axioms; a *proposition* is a formula which we sometimes assume, and sometimes prove, depending of the current experiment scenario; and a *conjecture* is something we want to prove.

The synthesis problem. As stated in the introduction, the *specification* of the target function F consists of two predicates: the input condition $I[X]$ and the output condition $O[X, T]$, and the correctness property for F is $\underset{X}{\forall}(I[X] \Rightarrow O[X, F[X]])$. The synthesis problem is expressed by the conjecture: $\underset{XT}{\forall \exists}(I[X] \Rightarrow O[X, T])$. Proof-based synthesis consists in proving this conjecture in a constructive way and then extracting the algorithm for the computation of F from this proof.

In the case of sorting the input condition specifies the type of the input, therefore it is missing since the type is implicit using the notations presented above (e.g., X is a tree). The output condition $O[X,T]$ is $X \approx T \wedge IsSorted[T]$ thus the synthesis conjecture becomes:

Conjecture 6. $\underset{X\,T}{\forall\exists}(X \approx T \wedge IsSorted[T])$

This conjecture can be proved in several ways. Each constructive proof is different depending on the applied induction principle and the content of the knowledge base. Hence, different algorithms are extracted from different proofs.

Synthesis scenarios. The simple scenario is when the proof succeeds, because the properties of the auxiliary functions which are necessary for the implementation of the algorithm are already present in the knowledge base. An example of knowledge base is given in [10]. The auxiliary algorithms used for tree sorting are $Insert[a, A]$ (insert element a into sorted tree A, such that the result is sorted) and $Merge[A, B]$ (merge two sorted trees into a sorted tree). Some of their necessary properties are:

Proposition 7. $\underset{T}{\forall}\Big(IsSorted[T] \implies IsSorted[Insert[n, T]]\Big)$

Proposition 8. $\underset{L,R}{\forall}\Big((IsSorted[L] \wedge IsSorted[R]) \implies IsSorted[Merge[L, R]]\Big)$

More complex is the scenario where the auxiliary functions are not present in the knowledge base. In this case the prover fails and on the failing proof situation we apply *cascading*: we create a conjecture which would make the proof succeed, and it also expresses the synthesis problem for the missing auxiliary function. In this scenario, the functions $Insert$ and $Merge$ are synthesized in separate proofs, and the main proof is replayed with a larger knowledge base which contains their properties.

2.2 Induction Principles

The illustration of the induction principles and algorithm extraction in this subsection is similar to the one from [8], but the induction principles are adapted for trees and the extracted algorithms are more complex.

The following induction principles are direct *term-based* instances of the Noetherian induction principle [17] and can be represented using *induction schemas*. Consider the domain of binary trees with a well-founded ordering $<_t$ and denote by $\ll_t$ the multiset extension [1] of $<_t$ as a well-founded ordering over vectors of binary trees. An induction schema to be applied to a predicate $\underset{\overline{x}}{\forall}P[\overline{x}]$ defined over a vector of tree variables $\overline{x}$ is a conjunction of instances of $P[\overline{x}]$ called *induction conclusions* that 'cover' $\underset{\overline{x}}{\forall}P[\overline{x}]$, i.e., for any value $\overline{v}$ from the domain of $\overline{x}$, there is an instance of an induction conclusion $P[\overline{t}]$ that equals

$P[\overline{v}]$, where $\overline{t}$ is a vector of trees. An induction schema may attach to an induction conclusion $P[\overline{t}]$, as *induction hypotheses*, any instance $P[\overline{t'}]$ of $\forall_{\overline{x}} P[\overline{x}]$ as long as $\overline{t'} \ll_t \overline{t}$. The induction conclusions without (resp., with) attached induction hypotheses are *base* (resp., *step*) cases of the induction schema.

In the current presentation we will use the *number of elements* as the measure of binary trees. Checking strict ordering $E <_t E'$ between two expressions E, E' representing trees reduces to check strict inclusion between the multisets of symbols (constants and variables except ε) occurring in the expressions. This is because the expressions representing trees contain only functions which preserve the number of elements in the tree (*Concat, Insert, Merge*).

In our experiments we used different induction principles for proving P as unary predicate over binary trees. A first example is:

Induction-1:

$$\left(P[\varepsilon] \bigwedge_{n,L,R} \forall ((P[L] \wedge P[R]) \Longrightarrow P[\langle L, n, R \rangle]) \right) \Longrightarrow \forall_X P[X]$$

The 'covering' property of the two induction conclusions $P[\varepsilon]$ and $P[\langle L, n, R \rangle]$ is satisfied since any binary tree is either ε or of the form $\langle L, n, R \rangle$. $P[L]$ and $P[R]$ are induction hypotheses attached to $P[\langle L, n, R \rangle]$, and it is very easy to see that their terms are smaller than the one of the induction conclusion.

Induction-2:

$$\left(P[\varepsilon] \bigwedge_{n,L} \forall (P[L] \Longrightarrow P[\langle L, n, \varepsilon \rangle]) \bigwedge_{n,L,R} \forall ((P[\langle L, n, \varepsilon \rangle] \wedge P[R]) \Longrightarrow P[\langle L, n, R \rangle]) \right) \Longrightarrow$$
$$\forall_X P[X]$$

Induction-3:

$$\left(P[\varepsilon] \bigwedge_n \forall (P[\langle \varepsilon, n, \varepsilon \rangle]) \bigwedge_{n,L} \forall (P[L] \Longrightarrow P[\langle L, n, \varepsilon \rangle]) \bigwedge_{n,R} \forall (P[R] \Longrightarrow P[\langle \varepsilon, n, R \rangle]) \bigwedge \right.$$
$$\left. \bigwedge_{n,L,R} \forall ((P[L] \wedge P[R]) \Longrightarrow P[\langle L, n, R \rangle]) \right) \Longrightarrow \forall_X P[X]$$

In the formula above, L and R are assumed to be nonempty. In order to encode this conveniently during the proof, they are replaced by $\langle A, a, B \rangle$ and $\langle C, b, D \rangle$, respectively.

Induction schema discovery. In some examples (e.g., synthesis of $Merge[X, Y]$ [11]), the induction principles are generated in a lazy way, especially when it is not possible to find a witness term using only the constants and functions present in the proof situation. In such cases the prover allows the use of terms containing the function to be synthesized, by assuming that it fulfils the desired specification. However, the call of this function must apply to arguments which are strictly smaller (w.r.t. $\ll_t$) then the arguments of the main call of the function which is currently synthesized.

2.3   Special Inference Rules and Proof Strategies

We summarize here the main inference rules and proof strategies. More details are given in [9].

Inference rules:

IR-1: *Generate Microatoms.* Certain atoms can be transformed into a conjunction of (micro)atoms, depending on the specific properties of our functions and predicates. E.g., *IsSorted*[$\langle T_1, n, T_2 \rangle$] is transformed into (*IsSorted*[T_1] $\wedge$ *IsSorted*[T_2] $\wedge$ *RgM*[T_1] $\leq n \wedge n \leq$ *LfM*[T_2]). Similarly, we get $x \preceq A \wedge x \leq b \wedge x \preceq C$ from $x \preceq \langle A, b, C \rangle$.

IR-2: *Eliminate-Ground-Formulae-from-Goal.* The ground formulas from any goal are deleted if they are assumption instances.

IR-3: *Replace-Equivalent-Term-in-Goal.* Let $t_1 \approx t_2$ be an assumption and assume that t_1 occurs in a goal as argument of a predicate which is preserved by equivalence ($\approx$, $\preceq$). The rule replaces t_1 by t_2.

IR-4: *Generate permutations and expressions.* This rule applies combinatorial techniques that are widely explained in [11]. Given a goal of the form *Expression* $\approx T^* \wedge$ *IsSorted*[T^*], it generates all *permutations* of the list of non-empty symbols from *Expression*. Then, for each permutation it generates all possible witnesses as a tree *expressions* containing these symbols. E.g., if *Expression* is $\langle L, x, \varepsilon \rangle$, then the generated trees are: $\langle L, x, \varepsilon \rangle$, $\langle \varepsilon, x, L \rangle$, and *Insert*[$x, L$].

Strategies:

S-1: *Quantifier reduction.* The strategy organizes the inference rules for quantifiers (see **IR-1**), in situations where it is clear that several such rules are to be performed in sequence (e.g., when applying an induction principle).

S-2: *Priority-of-Local-Assumptions.* The strategy consists in using with priority the local assumptions, usually ground formulae generated during the current proof and considered as "true" in the context of the proof, w.r.t. the global assumptions consisting of definitions and propositions from the database, considered as being always "true".

3   Experiments

3.1   Synthesis of Sort-1

In this subsection we present the automatically generated proof of Conjecture 6 in the *Theorema* system. Note that the statement which has to be proven by induction is:

$$P[X] : \underset{T}{\exists}(X \approx T \wedge \textit{IsSorted}[T]).$$

Proof. Start to prove Conjecture 6 using the current knowledge base and by applying **Induction-3**, then **S-1** to eliminate the existential quantifier.

Base case 1: Prove: $\varepsilon \approx T^* \wedge IsSorted[T^*]$.
One obtains the substitution $\{T^* \to \varepsilon\}$ and the new goal is $IsSorted[\varepsilon]$, which is true by Definition 4.

Base case 2: Prove: $\langle \varepsilon, n, \varepsilon \rangle \approx T^* \wedge IsSorted[T^*]$.
One obtains the substitution $\{T^* \to \langle \varepsilon, n, \varepsilon \rangle\}$. The new goal is $IsSorted[\langle \varepsilon, n, \varepsilon \rangle]$ which is true by Definition 4.

Induction case 1: Assume:

$$\exists_{T}(L_0 \approx T \wedge IsSorted[T]) \tag{1}$$

and prove:

$$\exists_{T}(\langle L_0, n, \varepsilon \rangle \approx T \wedge IsSorted[T]) \tag{2}$$

Apply **S-1** on (1) and (2) to eliminate the existential quantifiers. The induction hypotheses are:

$$L_0 \approx T_1, \quad IsSorted[T_1] \tag{3}$$

and the goal is:

$$\langle L_0, n, \varepsilon \rangle \approx T^* \wedge IsSorted[T^*] \tag{4}$$

Apply **IR-3** and rewrite our goal (4) by using the first conjunct of the assumption (3). The goal becomes:

$$\langle T_1, n, \varepsilon \rangle \approx T^* \wedge IsSorted[T^*] \tag{5}$$

Apply **IR-4** (to generate permutations of $\langle T_1, n, \varepsilon \rangle$) and prove alternatives:

Alternative-1: One obtains the substitution $\{T^* \to \langle T_1, n, \varepsilon \rangle\}$ to get:

$$IsSorted[\langle T_1, n, \varepsilon \rangle] \tag{6}$$

Apply **IR-1** on (6) and prove:

$$IsSorted[T_1] \wedge RgM[T_1] \leq n \tag{7}$$

Apply **IR-2** using (3) and the new goal is:

$$RgM[T_1] \leq n \tag{8}$$

Apply **IR-5** and the goal (8) becomes the conditional assumption on this branch.

Alternative-2: One obtains the substitution $\{T^* \to \langle \varepsilon, n, T_1 \rangle\}$. The proof is similar and one has to prove:

$$n \leq LfM[T_1] \tag{9}$$

which becomes the conditional assumption on this branch.

Alternative-3: Since the disjunction of the conditions (8) and (9) is not provable, the prover generates a further alternative. This depends on the synthesis scenario (see the end of Sect. 2.1). If the properties of the function *Insert* are present in the knowledge base, then the prover generates the substitution $\{T^* \rightarrow Insert[n, T_1]\}$ based on these properties.

If the properties of *Insert* are not present, then the prover generates a failing branch. A new conjecture is further generated which is used for the synthesis of *Insert*. Then we replay the current proof with knowledge about this auxiliary function and the proof will proceed further.

Induction case 2: Similar to *Induction case 1* one obtains:

Alternative-1: $\{T^* \rightarrow \langle \varepsilon, n, T_2 \rangle\}$ and the conditional assumption is: $n \leq LfM[T_2]$.

Alternative-2: $\{T^* \rightarrow \langle T_2, n, \varepsilon \rangle\}$ and the conditional assumption is: $RgM[T_2] \leq n$.

Alternative-3: Since the auxiliary function *Insert* is already known, the proof will succeed with the substitution: $\{T^* \rightarrow Insert[n, T_2]\}$.

Induction case 3: Assume:

$$L_1 \approx T_3, \quad IsSorted[T_3], \quad R_1 \approx T_4, \quad IsSorted[T_4] \tag{10}$$

and prove:

$$\langle L_1, n, R_1 \rangle \approx T^* \wedge IsSorted[T^*] \tag{11}$$

Apply **IR-3** and rewrite our goal (11) by using the first and the third conjunct of the assumption (10) and the new goal is:

$$\langle T_3, n, T_4 \rangle \approx T^* \wedge IsSorted[T^*] \tag{12}$$

Apply **IR-4** and obtain the permutations of the list $\langle T_3, n, T_4 \rangle$, for each permutation a number of possible tree expressions as witness for T^*, and for each witness an alternative possibly generating a condition as goal.

If the function *Merge* is not present, then the branch corresponding to *Concat* will be followed by a failing branch which has the same witness. For the purpose of this presentation we use only the alternative branch generated by the list $\langle n, T_3, T_4 \rangle$ with expression $Insert[n, Concat[T_3, T_4]]$. This generates the same conjecture for the synthesis of *Merge* and also the last branch in the following sorting algorithm, knowing that if the proof succeeds to find a witness $T^* = \Im[n, L_0, R_0, T_1, T_2]$ (term depending on n, L_0, R_0, T_1 and T_2), then a new branch $F[\langle L, n, R \rangle] = \Im[n, L, R, F[L], F[R]]$ of the synthesized algorithm is generated (T_1 and T_2 are replaced by $F[L]$ and $F[R]$, respectively) by using as conditions the conditional assumptions required by the witness:

$$\underset{n,L,R}{\forall} \left(\begin{array}{l} F_1[\varepsilon] = \varepsilon \\ F_1[\langle \varepsilon, n, \varepsilon \rangle] = \langle \varepsilon, n, \varepsilon \rangle \\ F_1[\langle L, n, \varepsilon \rangle] = \left\{ \begin{array}{l} \langle F_1[L], n, \varepsilon \rangle, \text{ if } RgM[F_1[L]] \leq n \\ \langle \varepsilon, n, F_1[L] \rangle, \text{ if } n \leq LfM[F_1[L]] \\ Insert[n, F_1[L]], \text{ otherwise} \end{array} \right. \\ F_1[\langle \varepsilon, n, R \rangle] = \left\{ \begin{array}{l} \langle \varepsilon, n, F_1[R] \rangle, \text{ if } n \leq LfM[F_1[R]] \\ \langle F_1[R], n, \varepsilon \rangle, \text{ if } RgM[F_1[R]] \leq n \\ Insert[n, F_1[R]], \text{ otherwise} \end{array} \right. \\ F_1[\langle L, n, R \rangle] = Insert[n, Merge[F_1[L], F_1[R]]] \end{array} \right)$$

3.2 Additional Certification of the Synthesized Algorithm F_1

The theoretical basis and the correctness of this proof-based synthesis scheme is well known – see for instance [5]. However, the implementation of the presented rules in *Theorema* is error-prone. To check the soundness of the implementation, we have mechanically verified that the algorithm F_1 satisfies the correctness condition, by using the Coq proof assistant (https://coq.inria.fr). The Coq formalization of the *LfM* and *RfM* functions has slightly changed from the partial definitions given here, as Coq requires that the functions be total. The conversion into total functions is possible if the components of the triplet given as argument are represented as the new arguments, as below.

Definition 9. $\underset{n,m,L,R,S}{\forall} \left(\begin{array}{l} RgM[L, n, \varepsilon] = n \\ RgM[L, n, \langle R, m, S \rangle] = RgM[R, m, S] \end{array} \right)$

Definition 10. $\underset{n,m,L,R,S}{\forall} \left(\begin{array}{l} LfM[\varepsilon, n, R] = n \\ LfM[\langle L, n, R \rangle, m, S] = LfM[L, n, R] \end{array} \right)$

The proof effort was non-trivial, involving significant user interaction. The certification proofs used rules and proof strategies completely different from those generating the synthesized algorithms, requiring additionally 2 induction schemas and 15 lemmas.[4]

3.3 Synthesis of Other Sorting Algorithms

Sort-2. The prover generated automatically the proof of Conjecture 6 by applying **Induction-2** and by using the current knowledge base (Definition 4, Propositions 7, and 8), including the following property:

Proposition 11.
$$\underset{n,L,R,A,B}{\forall} (((\langle L, n, \varepsilon \rangle \approx A \wedge R \approx B) \Longrightarrow \langle L, n, R \rangle \approx Merge[A, B])$$

The proof is similar with the ones presented above and from this proof the following algorithm is extracted automatically:

---

[4] The full Coq script is available at: http://web.info.uvt.ro/~idramnesc/LATA2016/coq.v.

$$\forall_{n,L,R} \left(\begin{array}{c} F_2[\varepsilon] = \varepsilon \\ F_2[\langle L, n, \varepsilon \rangle] = \left\{ \begin{array}{l} \langle F_2[L], n, \varepsilon \rangle, \text{ if } RgM[F_2[L]] \leq n \\ \langle \varepsilon, n, F_2[L] \rangle, \text{ if } n \leq LfM[F_2[L]] \\ Insert[n, F_2[L]], \text{ otherwise} \end{array} \right. \\ F_2[\langle L, n, R \rangle] = Merge[F_2[\langle L, n, \varepsilon \rangle], F_2[R]] \end{array} \right)$$

Sort-3. The proof of Conjecture 6 is generated automatically by applying **Induction-3** and by using properties from the knowledge base (including properties of *Concat*).

The corresponding algorithm which is extracted automatically from the proof is similar to F_1 excepting the last branch, which is:

$$F_3[\langle L, n, R \rangle] = Insert[n, F_3[Concat[L, R]]]$$

Sort-4. The prover automatically generates the proof of Conjecture 6 by applying **Induction-3** and by using properties from the knowledge base (including properties of *Insert*, *Merge*) and applies the inference rule **IR-4** which generates permutations.

The automatically extracted algorithm is similar to F_1 excepting the last branch, where F_4 has three branches:

$$F_4[\langle L, n, R \rangle] = \left\{ \begin{array}{l} \langle F_4[L], n, F_4[R] \rangle, \text{ if } (RgM[F_4[L]] \leq n \wedge n \leq LfM[F_4[R]]) \\ \langle F_4[R], n, F_4[L] \rangle, \text{ if } (RgM[F_4[R]] \leq n \wedge n \leq LfM[F_4[L]]) \\ Insert[n, Merge[F_4[L], F_4[R]]], \text{ otherwise} \end{array} \right.$$

Sort-5. The prover generates automatically the proof of Conjecture 6 by applying **Induction-3** and by using properties from the knowledge base (including properties of *Insert*, *Concat*) and applies the inference rule **IR-4** which generates permutations.

The algorithm which is extracted automatically from the proof is similar to F_3 excepting the last branch, where F_5 has three branches:

$$F_5[\langle L, n, R \rangle] = \left\{ \begin{array}{l} \langle F_5[L], n, F_5[R] \rangle, \text{ if } RgM[F_5[L]] \leq n \wedge n \leq LfM[F_5[R]] \\ \langle F_5[R], n, F_5[L] \rangle, \text{ if } RgM[F_5[R]] \leq n \wedge n \leq LfM[F_5[L]] \\ Insert[n, F_5[Concat[L, R]]], \text{ otherwise} \end{array} \right.$$

The automatically generated proofs corresponding to these algorithms, their extraction process and the computations with the extracted algorithms in *Theorema* are fully presented in the technical report [9].

The following table presents the synthesized sorting algorithms. For each of them Conjecture 6 has been proved using the induction principles from the first column. The second column specifies the auxiliary function used and the third column shows whether the rule **IR-4** (which generates the permutations and witnesses) is used or not.

Induction principle	Auxiliary used functions	Uses **IR-4**	Extracted algorithm
Induction-2	*LfM, RgM, Insert, Merge*	No	F_2
Induction-3	*LfM, RgM, Insert, Merge*	No	F_1
	LfM, RgM, Insert, Merge	Yes	F_4
	LfM, RgM, Insert, Concat	No	F_3
	LfM, RgM, Insert, Concat	Yes	F_5

4   Conclusions and Further Work

Our results are: a new theory of binary trees, an arsenal of special strategies and specific inference rules based on properties of binary trees, a new prover in the *Theorema* system which generates all the presented synthesis proofs, an extractor in the *Theorema* system which is able to extract from a proof the corresponding algorithms (including if-then-else algorithms), the synthesis of numerous sorting algorithms and auxiliary algorithms. We have also certified by *Coq* the soundness property of F_1 with the current implementation of the auxiliary functions. The certification proof is more complex and its generation less automatic than for the *Theorema* proof that helped for extracting F_1, by using different inference rules and additional properties.

The problem of sorting binary trees does not appear to have an important practical significance, and in fact the algorithms we synthesize are not very efficient. (For instance it appears to be more efficient to extract the elements of the tree in a list, to sort it by a fast algorithm, and then to construct the sorted tree.) However, the problem itself poses interesting algorithmic problems, and also the proof techniques are more involved than the ones from lists. This is very relevant for our research, because our primary goal is not to generate the most efficient algorithms, but to study interesting examples of proving and synthesis, from which we can discover *new proof methods for algorithm synthesis*.

Our experiments done in the *Theorema* system and presented in detail in the technical report [9] show that by applying different induction principles and by choosing different alternatives in the proofs one can discover numerous algorithms for the same functions, differing in efficiency and complexity. This case study illustrates that the automation of the synthesis problem is not a trivial one.

As further work, for a fully automatization of the synthesis process, we want to use other systems in order to automatically generate the induction principles, which in the *Theorema* system are given as inference rules in the prover. For example, we can apply induction schemas that are issued from recursive data structures and functions defined in Coq [18]. We also want to use the method presented in this paper on more complex recursive data structures (e.g. red-black trees). In the near future, we intend to certify the correctness property for the other synthesized sorting algorithms, using a similar approach as for F_1. One of our long-term goals is to define procedures for translating the *Theorema* proofs directly into Coq scripts, by following similar translation procedures as those used for implicit induction proofs [18].

Acknowledgments. Isabela Drămnesc: This work was partially supported by the strategic grant POSDRU/159/1.5/S/137750, Project Doctoral and Postdoctoral programs support for increased competitiveness in Exact Sciences research cofinanced by the European Social Fund within the Sectoral Operational Programme Human Resources Development 2007–2013.

References

1. Baader, F., Nipkow, T.: Term Rewriting and All That. Cambridge University Press, Cambridge (1998)
2. Back, R.J., von Wright, J.: Refinement Calculus. Springer Verlag, New York (1998)
3. Bertot, Y., Casteran, P.: Interactive theorem proving and program development Coq'Art: the calculus of inductive constructions. Texts in Theoretical Computer Science An EATCS, vol. XXV. Springer, Heidelberg (2004)
4. Buchberger, B., Craciun, A., Jebelean, T., Kovacs, L., Kutsia, T., Nakagawa, K., Piroi, F., Popov, N., Robu, J., Rosenkranz, M., Windsteiger, W.: Theorema: towards computer-aided mathematical theory exploration. J. Appl. Logic **4**(4), 470–504 (2006)
5. Bundy, A., Dixon, L., Gow, J., Fleuriot, J.: Constructing induction rules for deductive synthesis proofs. Electron. Notes Theor. Comput. Sci. **153**, 3–21 (2006)
6. Cohen, C., Dénès, M., Mörtberg, A.: Refinements for free!. In: Gonthier, G., Norrish, M. (eds.) CPP 2013. LNCS, vol. 8307, pp. 147–162. Springer, Heidelberg (2013)
7. Delaware, B., Claudel, C.P., Gross, J., Chlipala, A.: Fiat: deductive synthesis of abstract data types in a proof assistant. In: Proceedings of the 42nd Annual ACM SIGPLAN-SIGACT Symposium on Principles of Programming Languages, POPL 2015, pp. 689–700. ACM, New York (2015)
8. Dramnesc, I., Jebelean, T.: Synthesis of list algorithms by mechanical proving. J. Symbolic Comput. **68**, 61–92 (2015)
9. Dramnesc, I., Jebelean, T., Stratulat, S.: Synthesis of some algorithms for trees: experiments in Theorema. Technical report 15–04, RISC Report Series, Johannes Kepler University, Linz, Austria (2015)
10. Dramnesc, I., Jebelean, T., Stratulat, S.: Theory exploration of binary trees. In: 13th IEEE International Symposium on Intelligent Systems and Informatics (SISY 2015), pp. 139–144. IEEE Publishing (2015)
11. Dramnesc, I., Jebelean, T., Stratulat, S.: Combinatorial techniques for proof-based synthesis of sorting algorithms. In: Proceedings of the 17th International Symposium on Symbolic and Numeric Algorithms for Scientific Computing, SYNASC 2015 (to appear)
12. Gulwani, S.: Dimensions in program synthesis. In: Proceedings of the 12th International ACM SIGPLAN Symposium on Principles and Practice of Declarative Programming, PPDP 2010, pp. 13–24. ACM, New York (2010)
13. Kneuss, E., Kuraj, I., Kuncak, V., Suter, P.: Synthesis modulo recursive functions. In: Proceedings of the 2013 ACM SIGPLAN International Conference on Object Oriented Programming Systems Languages and Applications, OOPSLA 2013, pp. 407–426. ACM, New York (2013)
14. Knuth, D.E.: The Art of Computer Programming, Volume 3: Sorting and Searching, 2nd edn. Addison Wesley Longman Publishing, Redwood City (1998)
15. Nipkow, T., Paulson, L.C., Wenzel, M.: Isabelle/HOL - A Proof Assistant for Higher-Order Logic. LNCS, vol. 2283. Springer, Heidelberg (2002)

16. Smith, D.R.: Generating programs plus proofs by refinement. In: Meyer, B., Woodcock, J. (eds.) VSTTE 2005. LNCS, vol. 4171, pp. 182–188. Springer, Heidelberg (2008)

17. Stratulat, S.: A unified view of induction reasoning for first-order logic. In: Voronkov, A. (ed.) Turing-100 (The Alan Turing Centenary Conference). EPiC Series, vol. 10, pp. 326–352. EasyChair (2012)

18. Stratulat, S.: Mechanically certifying formula-based noetherian induction reasoning. J. Symbolic Comput. (accepted). http://lita.univ-lorraine.fr/~stratula/jsc2016.pdf

19. Wirth, N.: Program development by stepwise refinement. Commun. ACM **14**(4), 221–227 (1971)

20. Wolfram, S.: The Mathematica Book. Wolfram Media Inc., Champaign (2003)

Unconventional Models
of Computation

Reversible Shrinking Two-Pushdown Automata

Holger Bock Axelsen[1][(✉)], Markus Holzer[2], Martin Kutrib[2],
and Andreas Malcher[2]

[1] Department of Computer Science, University of Copenhagen,
Universitetsparken 5, 2100 Copenhagen E, Denmark
`funkstar@di.ku.dk`
[2] Institut für Informatik, Universität Giessen, Arndtstr. 2,
35392 Giessen, Germany
`{holzer,kutrib,malcher}@informatik.uni-giessen.de`

Abstract. The deterministic shrinking two-pushdown automata characterize the deterministic growing context-sensitive languages, known to be the Church-Rosser languages. Here, we initiate the investigation of *reversible* two-pushdown automata, RTPDAs, in particular the shrinking variant. We show that as with the deterministic version, shrinking and length-reducing RTPDAs are equivalent. We then give a separation of the deterministic and reversible shrinking two-pushdown automata, and prove that these are incomparable with the (deterministic) context-free languages. We further show that the properties of emptiness, (in)finiteness, universality, inclusion, equivalence, regularity, and context-freeness are not even semi-decidable for shrinking RTPDAs.

Keywords: Unconventional models of computation · Reversible computing · Shrinking two-pushdown automata · Church-Rosser languages

1 Introduction

Reversible variants of universal computation models, e.g., Turing machines, are usually equal in power to irreversible (deterministic) ones.[1] For subuniversal models, however, equality is very model-dependent. For example, one-way reversible (multihead) finite automata, reversible pushdown automata, and reversible Turing machines with run-times between real-time and linear time, are *not* equal to their deterministic variants in expressive power, but two-way reversible multihead finite automata and reversible linear-bounded automa *are*. This motivates further study of various limited reversible automata models in order to better understand how reversibility affects computational capacity.

In this paper we initiate the study of reversible (shrinking) two-pushdown automata, RTPDAs for short. Two-pushdown automata have the input placed in one pushdown, and perform computations by inspecting and rewriting words at

---

[1] At least when considered as language acceptors—for functions the picture is somewhat more complex [1].

© Springer International Publishing Switzerland 2016
A.-H. Dediu et al. (Eds.): LATA 2016, LNCS 9618, pp. 579–591, 2016.
DOI: 10.1007/978-3-319-30000-9_44

the top of the pushdowns. These automata are of particular interest as the deterministic shrinking two-pushdown automata (s-DTPDAs) are known to characterize the class of Church-Rosser languages [6]. The unrestricted RTPDAs are easily shown to be Turing-complete. We then turn to the shrinking variant, s-RTPDAs, and show that the s-RTPDAs are strictly weaker than the deterministic version. The separation is achieved using the mirror language with a binary regular language as infix, i.e., words of form $w\$v\$w^R$, using the idea that a shrinking reversible computation cannot "pass over" the infix word v enough times to verify the mirror condition without also compressing it, so a Kolmogorov complexity argument shows that this language is not accepted reversibly. However, in a deterministic computation we can erase the infix, so this language separates the reversible and deterministic models. Further, even though the s-RTPDAs accept some non-context-free languages, we can exploit the separation to show that the class of accepted languages is actually incomparable with the deterministic context-free languages, and the context-free languages. Following this, we use a reduction from the emptiness problem for linear bounded automata to show that emptiness is not semi-decidable for s-RTPDAs, by encoding valid computations of LBAs into words with sufficient character repetitions to enable a shrinking checking of validity. This result is then further used to show that finiteness, infiniteness, universality, inclusion, equivalence, regularity, and context-freeness are not semi-decidable problems for s-RTPDAs.

2   Preliminaries

Throughout the paper we will use λ to denote the empty word. Moreover, $\mathscr{L}(\mathsf{A})$ will denote the class of languages accepted by the automata of type A.

Definition 1. *A* two-pushdown automaton (TPDA) *with pushdown windows of size k is a nondeterministic automaton with two pushdown stores. Formally, it is defined by a 7-tuple $M = (Q, \Sigma, \Gamma, \bot, q_0, F, \delta)$, where*

- *Q is a finite set of internal states,*
- *Σ is a finite input alphabet,*
- *Γ is a finite tape alphabet containing Σ such that $\Gamma \cap Q = \emptyset$,*
- *$\bot \notin \Gamma$ is a special symbol used to mark the bottom of the pushdown stores,*
- *$q_0 \in Q$ is the initial state,*
- *$F \subseteq Q$ is the set of final states, and*
- *$\delta : Q \times {}_{\bot}\Gamma^{\leq k} \times \Gamma_{\bot}^{\leq k} \to \mathcal{P}_{fin}(Q \times {}_{\bot}\Gamma^* \times \Gamma_{\bot}^*)$ is the partial transition relation, where ${}_{\bot}\Gamma^{\leq k} = \Gamma^k \cup \{{}_{\bot}u \mid |u| \leq k-1\}$, $\Gamma_{\bot}^{\leq k} = \Gamma^k \cup \{v\bot \mid |v| \leq k-1\}$, ${}_{\bot}\Gamma^* = \Gamma^* \cup {}_{\bot}\Gamma^*$, $\Gamma_{\bot}^* = \Gamma^* \cup \Gamma^*\bot$, and $\mathcal{P}_{fin}(Q \times {}_{\bot}\Gamma^* \times \Gamma_{\bot}^*)$ denotes the set of finite subsets of $Q \times {}_{\bot}\Gamma^* \times \Gamma_{\bot}^*$.*

The automaton M is a deterministic two-pushdown automaton (DTPDA), *if δ is a (partial) function from $Q \times {}_{\bot}\Gamma^{\leq k} \times \Gamma_{\bot}^{\leq k}$ to $Q \times {}_{\bot}\Gamma^* \times \Gamma_{\bot}^*$.*

When constructing a two-pushdown automaton of a certain window size k, we sometimes do not specify the whole content of the windows, in a slight abuse of notation. For instance, if $k = 3$ and we write $\delta(q_0, \#a, \#) \ni (q_r, \#, \#\#)$, we take this to be shorthand notation for the transitions $\delta(q_0, x\#a, \#y) \ni (q_r, x\#, \#\#y)$, for $x \in {}_\perp\Gamma^{\leq 1}$ and $y \in \Gamma^{\leq 2}_\perp$.

A configuration of a (D)TPDA is described as (u, q, v), where $q \in Q$ is the current state, $u \in {}_\perp\Gamma^*$ is the content of the first pushdown store with the first letter of u at the bottom and the last letter of u at the top, and $v \in \Gamma^*_\perp$ is the content of the second store with the last letter of v at the bottom and the first letter of v at the top. For an input string $w \in \Sigma^*$, the corresponding *initial configuration* is $(\perp, q_0, w\perp)$, that is, the input is given as the initial content of the second pushdown store, while the other pushdown store just contains the bottom marker. The (D)TPDA M induces a computation relation $\vdash^*_M$ on the set of configurations, which is the reflexive transitive closure of the single-step computation relation $\vdash_M$ defined by

$$(uu_1, q_1, v_1v) \vdash_M (uu_2, q_2, v_2v) \qquad \text{if} \qquad \delta(q_1, u_1, v_1) \ni (q_2, u_2, v_2).$$

Note that the restrictions on δ for a DTPDA make $\vdash_M$ a deterministic function on configurations. A (D)TPDA *halts* if the transition function is undefined for the current configuration. An input word w is *accepted* if the machine halts at some time in an accepting state, otherwise it is *rejected*. The *language accepted* by M is $L(M) = \{\, w \in \Sigma^* \mid w \text{ is accepted by } M \,\}$.

Definition 2. *1. A (D)TPDA M is called* shrinking *if there exists a weight function $\varphi : Q \cup \Gamma \cup \{\perp\} \to \mathbb{N}_+$ such that, for all $q \in Q$, $u \in {}_\perp\Gamma^{\leq k}$, and $v \in \Gamma^{\leq k}_\perp$, the transition $(p, u', v') \in \delta(q, u, v)$ implies $\varphi(u'pv') < \varphi(uqv)$.[2] By s-TPDA and s-DTPDA we denote the corresponding classes of shrinking automata.*

2. A (D)TPDA M is called length-reducing *if, for all $q \in Q$, $u \in {}_\perp\Gamma^{\leq k}$, and $v \in \Gamma^{\leq k}_\perp$, the transition $(p, u', v') \in \delta(q, u, v)$ implies $|u'v'| < |uv|$. We denote the corresponding classes of length-reducing automata by* lr-TPDA *and* lr-DTPDA.

The definitions used here are those of Niemann and Otto [6], except that in their definition of TPDAs the pushdown stores contain preassigned contents. In the original definition by Buntrock and Otto [2], the automaton only sees the topmost symbol on each pushdown, but both define the same language classes.

It is well known that TPDAs and DTPDAs characterize the class RE of all recursively enumerable languages. The weight and length reducing variants of (D)TPDAs coincide [6], that is $\mathscr{L}(\text{s-TPDA}) = \mathscr{L}(\text{lr-TPDA})$ and $\mathscr{L}(\text{s-DTPDA}) = \mathscr{L}(\text{lr-DTPDA})$. Moreover, it is known that the s-TPDA characterize the class GCSL of growing context-sensitive languages, and that the s-DTPDA characterizes the class CRL of Church-Rosser languages [6].

Now we turn to *reversible* two-pushdown automata. Basically, reversibility is meant with respect to the possibility of deterministically stepping the computation back and forth. So, the DTPDA has to also be backward deterministic.

---

[2] The extension to $(Q \cup \Gamma \cup \{\perp\})^*$ is defined by $\varphi(xy) = \varphi(x) + \varphi(y)$ and $\varphi(\lambda) = 0$.

That is, any configuration must have at most one predecessor which, in addition, is computable by a DTPDA. So, for reversible DTPDAs there must exist a *reverse transition function* δ^- that maps a configuration to its *predecessor configuration*. One step from a configuration to its *predecessor configuration* is denoted by $\vdash^-$. We denote the corresponding class of reversible deterministic automata by RTPDA. Note that a DTPDA is reversible when δ^{-1} extended to a transition function fulfills the determinism criteria of Definition 1. Moreover, we denote the corresponding classes of shrinking and length-reducing reversible automata by s-RTPDA and lr-RTPDA, respectively. Observe that when a shrinking (length-reducing) reversible automaton is run backward, its backward simulation is growing (length-increasing).

Example 3. We present an s-RTPDA $M = (Q, \Sigma, \Gamma, \bot, q_0, F, \delta)$ for the non-semilinear language $L = \{ a^{2^n} \mid n \geq 0 \}$. The idea for the construction is to iteratively move the content of one pushdown store to the other, while dividing the number of a's by two, until only one single a is left. In order to make this process reversible in addition a special symbol is placed into one of the pushdowns which indicates that one copying sweep from one pushdown store to the other was performed. This is realized as follows.

We define M by taking $Q = \{q_0, q_1, q_\ell, q_r\}$, $\Sigma = \{a\}$, $\Gamma = \Sigma \cup \{\#\}$, $F = \{q_\ell, q_r\}$, and δ is defined as follows: in order to accept the word a the automaton uses the transition

$$(q_0, \bot, a\bot) \rightarrow (q_\ell, \bot\#, \bot).$$

Whenever there is more than one a in the second pushdown, the automaton starts a loop which divides the number of a's by two and copies the context from the second pushdown into the first, accordingly. To become reversible, the start of this loop is remembered by a special symbol # in the first pushdown. Thus, the following three transitions take care of this behavior:

$$(q_0, \bot, aa) \rightarrow (q_0, \bot\#a, \lambda)$$
$$(q_0, \#, aa) \rightarrow (q_0, \#\#a, \lambda)$$

and

$$(q_0, a, aa) \rightarrow (q_0, aa, \lambda).$$

The copying ends if only one single a is left on the first pushdown, which is controlled by

$$(q_0, \#a, \bot) \rightarrow (q_r, \#, \#\bot)$$
$$(q_0, \#a, \#) \rightarrow (q_r, \#, \#\#)$$

and leads to acceptance. If this is not the case the copying restarts, now from left to right and by placing a # in the second pushdown at the very beginning of this cycle. To this end the transitions

$$(q_0, aa, \bot) \rightarrow (q_1, \lambda, a\#\bot)$$
$$(q_0, aa, \#) \rightarrow (q_1, \lambda, a\#\#)$$

are used. The corresponding rules for state q_1 for copying from left to right, controlling acceptance, and restarting the copying now from right to left are

$$(q_1, aa, a) \rightarrow (q_1, \lambda, aa)$$
$$(q_1, \#, a\#) \rightarrow (q_\ell, \#\#, \#)$$

and

$$(q_1, \#, aa) \rightarrow (q_0, \#\#a, \lambda),$$

respectively. It is easy to see that the function $\varphi : Q \cup \Gamma \cup \{\bot\} \rightarrow \mathbb{N}_+$ defined by $\varphi(q) = 1$, for $q \in Q$, and $\varphi(a) = 2$, $\varphi(\#) = 1$, and $\varphi(\bot) = 1$, makes the automaton M shrinking (weight reducing).

Consider the example computation on input $w = a^8$, which reads as follows:

$$(\bot, q_0, a^8 \bot) \vdash_M (\bot \# a, q_0, a^6 \bot)$$
$$\vdash_M (\bot \# a^2, q_0, a^4 \bot)$$
$$\vdash_M (\bot \# a^3, q_0, a^2 \bot)$$
$$\vdash_M (\bot \# a^4, q_0, \bot)$$
$$\vdash_M (\bot \# aa, q_1, a\# \bot)$$
$$\vdash_M (\bot \#, q_1, a^2 \# \bot)$$
$$\vdash_M (\bot \# \# a, q_0, \# \bot) \vdash_M (\bot \# \#, q_r, \# \# \bot).$$

From this example computation it is easy to see that the constructed DTPDA is reversible, since with the help of the #-symbols one can reverse the computation until the input word is produced in the second pushdown store, by undoing the described forward computation. The exact details are left to the reader.

Moreover, by a slight adaption of the given construction, one can always enforce that the #-symbols are collected in one pushdown store only. To keep the presentation simple we did not use this more complicated construction here. □

3   On the Accepting Power of RTPDAs

In this section we first investigate the accepting power of RTPDAs, showing that these machines accept all recursively enumerable languages, and that the shrinking and length-reducing variants are equivalent. Then, lr-RTPDAs are shown to be strictly less powerful than their deterministic variants. Finally, it is shown that the class of languages accepted by lr-RTPDAs is incomparable with the deterministic context-free languages and the context-free languages.

3.1   Basic Results on RTPDAs

By directly simulating a reversible Turing machine we get the following result.

Theorem 4. $\mathscr{L}(\text{RTPDA}) = \text{RE}$. □

Next, we consider shrinking and length-reducing variants of reversible two-pushdown automata. For ordinary DTPDAs we have the equality $\mathscr{L}(\text{s-DTPDA}) = \mathscr{L}(\text{lr-DTPDA})$ by [6], and this is equal to the class CRL of Church-Rosser languages. In fact, the proof given in [6] generalizes to the reversible two-pushdown automata.

Theorem 5. $\mathscr{L}(\text{s-RTPDA}) = \mathscr{L}(\text{lr-RTPDA})$. $\qquad\qquad\qquad\qquad$ $\square$

This equivalence motivates the following definition: let RevCRL refer to the class of languages accepted by length-reducing or equivalently shrinking reversible deterministic two-pushdown automata.

3.2   Separation of DTPDAs and RTPDAs

We now consider whether the languages accepted by shrinking and length-reducing two-pushdown automata are the same under determinism and reversibility. Since the reversible automata as defined here are also deterministic, the inclusion $\mathscr{L}(\text{lr-RTPDA}) \subseteq \mathscr{L}(\text{lr-DTPDA})$ is immediate. We show that this inclusion is, in fact, proper. So, the languages classes are different. The basic idea of the proof is to use the mirror language with a binary regular language in the center. Irreversible two-pushdown automata can check the infix from the regular language whereby it is successively deleted. However, if the infix is an incompressible string, it cannot be deleted by a length-reducing *reversible* two-pushdown automaton. So, we will use Kolmogorov complexity and incompressibility arguments. General information on this technique can be found, e.g., in the textbook [4, Chap. 7]. Let $w \in \{a, b\}^*$ be an arbitrary binary string. The Kolmogorov complexity $C(w)$ of w is defined to be the minimal size of a binary program (Turing machine) describing w. The following key component for using the incompressibility method is well known: there are binary strings w of *any* length such that $|w| \le C(w)$.

Theorem 6. RevCRL $= \mathscr{L}(\text{lr-RTPDA}) \subset \mathscr{L}(\text{lr-DTPDA}) = $ CRL.

Proof. We use the witness language $L = \{\, w\$v\$w^R \mid v, w \in \{a, b\}^* \,\}$. Since L is deterministic context-free, it is a Church-Rosser language [5] and, thus, accepted by some lr-DTPDA.

Now we turn to show that L cannot be accepted by any lr-RTPDA. Assume for the purpose of contradiction that L is accepted by some lr-RTPDA $M = (Q, \Sigma, \Gamma, \bot, q_0, F, \delta)$ with pushdown window size k. We consider accepting computations on inputs of the form $w\$v\$w^R$, where $|w| = n$, $|v| = n^2$, and $C(v) \ge |v|$, that is, v is incompressible.

In the following we will imagine that there may be invisible tokens between the symbols in the pushdown stores. These tokens are for the sake of easier writing. The automaton cannot see the tokens and operates as usual. Now assume that there are tokens $\{t_0, t_1, \ldots, t_m, t_{m+1}\}$ with $m = \lfloor n^2 / \lfloor \log(n) \rfloor \rfloor$ which are initially placed between the symbols of v. More precisely, the tokens are placed as $v = t_0 v_1 t_1 v_2 t_2 \cdots t_{m-1} v_m t_m v_{m+1} t_{m+1}$, where $|v_1| = |v_2| = \cdots = |v_m| = \lfloor \log(n) \rfloor$

and $|v_{m+1}| = n^2 - m \cdot \lfloor \log(n) \rfloor$. We say that M operates on a token whenever the token appears between symbols accessed by M in a transition. What happens with the token if M operates on it? If the transition pops symbols from the pushdown the token moves up by the number of symbols deleted, where it moves to the opposite pushdown when these positions are not available. Tokens never pass each other. If there are not enough positions on both pushdown stores the token remains at the top. For example, if $\delta(p, x_3 x_2 x_1, y_1 y_2 y_3 y_4) = (q, x_2' x_1', y_1' y_2')$ and there is a token between y_3 and y_4, it is placed between y_1' and y_2'. If the token appears between y_1 and y_2, it moves to the opposite pushdown between x_2' and x_1'. On the other hand, if the transition pushes symbols to the pushdown store where the token is, the token moves down by the number of symbols additionally pushed. For example, if $\delta(p, x_3 x_2 x_1, y_1 y_2 y_3 y_4) = (q, x_1', y_1' y_2' y_3' y_4' y_5')$ and there is a token between y_3 and y_4, it is placed between y_4' and y_5'. Finally, if the number of symbols in the pushdown store is unchanged, the token remains at its position. For example, if $\delta(p, x_3 x_2 x_1, y_1 y_2 y_3) = (q, x_2' x_1', y_1' y_2' y_3')$ and there is a token between y_2 and y_3, it remains in position, which is now between y_2' and y_3'.

Next assume that, for n large enough, there is an accepting computation so that M halts in a configuration where there are more than $2k$ symbols in between two neighboring tokens, say, in between t_i and t_{i+1}, where $0 \leq i \leq m - 1$. Since M is length-reducing, this implies in particular that M performs at most $O(\log(n))$ transitions in which only pushdown symbols in between t_i and t_{i+1} are accessed. In order to derive an upper bound on the number of possibilities for these transitions we have to consider the accessible contents of the pushdown stores, the state, and the position in between t_i and t_{i+1} where the operation is performed. There are no more than $(|\Gamma| + 1)^{2k}$ accessible contents of the pushdown stores, at most $|Q|$ states, and no more than $\log(n)$ positions. Altogether, we derive at most

$$\left(\log(n) \cdot (|\Gamma| + 1)^{2k} \cdot |Q|\right)^{O(\log(n))} = \log(n)^{O(\log(n))} \cdot c_1^{O(\log(n))}$$
$$= 2^{\log \log(n) \cdot O(\log(n))} \cdot 2^{O(\log(n))} = 2^{O(\log \log(n) \cdot \log(n))} = o(2^n)$$

possibilities for these transitions, where c_1 is a constant.

Since there are 2^n different words w of length n, for n large enough, there are two words $w \neq w'$ with $|w| = |w'| = n$ so that the transitions in which only pushdown symbols in between t_i and t_{i+1} are accessed are the same. But this implies that $w\$v\$w'^R$ is accepted as well, a contradiction to the assumption that M halts with more than $2k$ symbols in between t_i and t_{i+1}.

We conclude that M halts in a configuration with at most

$$O\left(2n + 2k \cdot \frac{n^2}{\log(n)}\right) = O\left(\frac{n^2}{\log(n)}\right)$$

symbols in its pushdown stores.

Recall that v is incompressible. We now derive a contradiction by showing that v can be compressed. The basic idea is to use the halting and accepting configuration of M on input $w\$v\$w^R$. Since M is reversible, it can be run backward

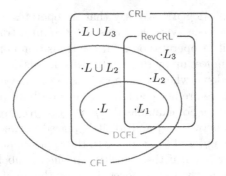

Fig. 1. Graphical representation of the language classifications in Lemma 7

until the initial configuration and, thus, the word v is reconstructed. To this end, we have to overcome the problem that M might run through several initial configurations. However, this can be handled by using a binary counter that gives the number of initial configurations passed through by M on input $w\$v\$w^R$. When the counter is decreased to zero during the reconstruction, the correct configuration is reached. Since M is length-reducing, it halts after at most $O(n^2)$ time steps. So, the counter can be represented with size $O(\log(n^2)) = O(\log(n))$. Altogether, we have

$$C(v) \leq C(M) + O\left(\frac{n^2}{\log(n)}\right) + O(\log(n)) + |p| = O\left(\frac{n^2}{\log(n)}\right) = o(n^2) = o(|v|)$$

where $|p|$ is the constant size of the program above reconstructing v. We conclude $C(v) < |v|$, for v long enough, contradicting that $C(v) \geq |v|$. Therefore, M cannot accept L.   □

3.3 RevCRL is Incomparable with DCFL and CFL

From Theorem 6 we now know that the shrinking reversible two-pushdown automata are weaker than shrinking deterministic two-pushdown automata. On the other hand, we also know from Example 3 that RevCRL contains non-semilinear and thus non-context-free languages, so their acceptance power is not trivial. We here consider further how the class RevCRL compares to the (deterministic) context-free languages (D)CFL, and the Church-Rosser languages CRL.

Lemma 7. *The following language memberships hold.*

1. $L_1 = \{\, a^n b^n \mid n \geq 0 \,\} \in \mathsf{RevCRL} \cap \mathsf{DCFL}$
2. $L_2 = \{\, a^n b^n c \mid n \geq 0 \,\} \cup \{a^n b^{2n} d \mid n \geq 0\} \in (\mathsf{RevCRL} \cap \mathsf{CFL}) - \mathsf{DCFL}$
3. $L_3 = \{\, a^n b^n c^n \mid n \geq 0 \,\} \in \mathsf{RevCRL} - \mathsf{CFL}$
4. $L = \{\, w\$v\$w^R \mid v, w \in \{a, b\}^* \,\} \in \mathsf{DCFL} - \mathsf{RevCRL}$
5. $L \cup L_2 \in (\mathsf{CRL} \cap \mathsf{CFL}) - (\mathsf{DCFL} \cup \mathsf{RevCRL})$
6. $L \cup L_3 \in \mathsf{CRL} - (\mathsf{CFL} \cup \mathsf{RevCRL})$ □

Corollary 8. RevCRL *is incomparable with* DCFL *and* CFL.   □

We note that Lemma 7 shows that all the intersections of interest are in fact non-empty. Figure 1 provides a graphical representation of the relationships.

4   Decidability Problems

In this section, we study decidability questions for reversible two-pushdown automata. The undecidability results are obtained by reductions of the emptiness problem for deterministic, linearly space bounded, one-tape, one-head Turing machines, so-called linear bounded automata (LBAs). It is well known that emptiness for LBAs is not semidecidable. See, for example, [3] where also the notion of *valid computations* is used. These are, basically, histories of LBA computations which are encoded into single words. We may safely make the following assumptions on LBAs. We suppose that they get their inputs in between two endmarkers, can halt only after an odd number of moves, accept by halting in a unique accepting state q_f, and make at least three moves.

Let M be an LBA where Q is the state set, q_0 is the initial state, T is the tape alphabet satisfying $T \cap Q = \emptyset$, and $\Sigma \subset T$ is the input alphabet. Then a configuration of M can be written as a string of the form $\triangleright T^* Q T^* \triangleleft$ such that the word $t_0 t_1 \cdots t_i q t_{i+1} \cdots t_{n+1}$ is used to express that M is in the state q, scanning tape symbol t_{i+1}, and the string $t_0 t_1 \cdots t_{n+1} \in \triangleright T^* \triangleleft$ is the tape inscription with t_0 being the left endmarker and t_{n+1} being the right endmarker.

Next, configurations are written more concisely by packing the current state q and the currently scanned symbol into a new symbol from the set $Q' = Q \times (T \cup \{\triangleright, \triangleleft\})$. Thus, a configuration is a string $t_0 t_1 \cdots t_i [q, t_{i+1}] t_{i+2} \cdots t_{n+1}$ of the form $\triangleright T^* Q' T^* \triangleleft$, if M's head scans a symbol from T. If the head scans $\triangleright$ or $\triangleleft$, the configurations are of the form $[q, \triangleright] t_1 \cdots t_{n+1}$ or $t_0 t_1 \cdots t_n [q, \triangleleft]$.

Now we consider words of the form

$$w_1 \$_1 w_3 \$_1 \cdots \$_1 w_{2m-1} \$_2 w_{2m-1}^R \$_1 w_{2m-3}^R \$_1 \cdots \$_1 w_1^R \$_3$$
$$w_2 \$_1 w_4 \$_1 \cdots \$_1 w_{2m} \$_2 w_{2m}^R \$_1 w_{2m-2}^R \$_1 \cdots \$_1 w_2^R,$$

where $\$_1, \$_2, \$_3 \notin T \cup Q'$, $w_i \in T^* Q' T^*$ are configurations of M, w_1 is an initial configuration of the form $\triangleright [q_0, \triangleleft]$ if the input is empty, or of the form $\triangleright [q_0, \sigma_1] \sigma_2 \cdots \sigma_\ell \triangleleft$, if the input $\sigma_1 \sigma_2 \cdots \sigma_\ell \in \Sigma^+$ is non-empty, $w_{2m} \in \{q_f\} T^*$ is a halting (accepting) configuration, and w_{i+1} is the successor configuration of w_i.

These configurations are now encoded as follows. Let A be an alphabet and $\overline{A}$ be a disjoint copy of A. Furthermore, let $\hat{A}$ be a disjoint copy of $A \cup \overline{A}$. We consider maps $f, f_R : A^+ \to (A \cup \overline{A})^+$ and $g, g_R : (A \cup \overline{A})^+ \to (A \cup \overline{A} \cup \hat{A})^+$ which are defined on words of even length such that $f(a_1 a_2 \cdots a_\ell) = a_1^2 \overline{a}_2^4 a_3^8 \overline{a}_4^{16} \cdots \overline{a}_\ell^{2^{\ell+1}}$ and $f_R(a_1 a_2 \cdots a_\ell) = \overline{a}_1^{2^{\ell+1}} a_2^{2^\ell} \cdots \overline{a}_{\ell-1}^4 a_\ell^2$. Similarly, we define the mappings $g(a_1 a_2 \cdots a_\ell) = a_1^2 \hat{a}_2^4 a_3^8 \hat{a}_4^{16} \cdots \hat{a}_\ell^{2^{\ell+1}}$ and $g_R(a_1 a_2 \cdots a_\ell) = \hat{a}_1^{2^{\ell+1}} a_2^{2^\ell} \cdots \hat{a}_{\ell-1}^4 a_\ell^2$.

Finally, consider $h : (A \cup \overline{A})^+ \to (A \cup \overline{A})^+$ such that $h(a_1 a_2 \cdots a_\ell) = a_1^4 a_2^4 \cdots a_\ell^4$. Now, let $A = T \cup Q' \cup \{\$_1, \$_2, \$_3\}$. Then, the set of encodings

$$h(g(f(w_1 \$_1 w_3 \$_1 \cdots \$_1 w_{2m-1}) \$_2 f_R(w_{2m-1}^R \$_1 w_{2m-3}^R \$_1 \cdots \$_1 w_1^R)) \$_3$$
$$g_R(f(w_2 \$_1 w_4 \$_1 \cdots \$_1 w_{2m}) \$_2 f_R(w_{2m}^R \$_1 w_{2m-2}^R \$_1 \cdots \$_1 w_2^R)))$$

is defined to be the set of *valid computations* of M. We denote it by VALC(M). It should be noted that the restriction to words of even length is done to keep the following constructions somewhat simpler. It is straightforward to extend the definitions and constructions also to words of not necessarily even length.

Our goal is to show that VALC(M) is accepted by some lr-RTPDA. Disregarding the maps f, f_R, g, g_R, and h, the set VALC(M) contains subwords of the form $x_1 x_2 \cdots x_m \$ x_m x_{m-1} \cdots x_1$. Thus, in a first step we show how the language $\{ f(x) \$ f_R(x^R) \mid x \in A^+, |x| \text{ is even} \}$ can be accepted by some lr-RTPDA. This construction will be used several times in the later construction of an lr-RTPDA accepting VALC(M).

Lemma 9. *Let A be an arbitrary alphabet such that $\$ \notin A$. An lr-RTPDA accepting $\{ f(x) \$ f_R(x^R) \mid x \in A^+, |x| \text{ is even} \}$ can effectively be constructed.*   □

The idea of the construction is that the repetition of symbols by f and f_R "pays" for the (linear) number of times we have to move the stacks back and forth to check the palindrome structure; ending with $\perp x$, $\$ x^R \perp$ in the pushdowns.

Remark 10. The intended construction stores the matched words in both pushdown stores. However, it is easy to modify the construction such that matched symbols are stored only in the second pushdown store. This is still length-reducing and reversible, and we use both variants below.

A slight variation of Lemma 9 also proves useful. We again have to check a palindrome structure, but only for a clearly marked prefix of the input.

Lemma 11. *Let A and B be alphabets such that $\$ \notin A \cup B$ and $A \cap B = \emptyset$. Then, an lr-RTPDA accepting the language*

$$\{ f(xy) \$ f_R(x^R z) \mid x \in A^+, y, z \in B^+, |x|, |y|, |z| \text{ are even} \}$$

can effectively be constructed.   □

Returning to valid computations, our next step is to compute to each (encoded) configuration its successor configuration and to store this in a second component by using a suitable alphabet including pairs of symbols.

Lemma 12. *An lr-RTPDA can effectively be constructed that constructs the following string in its second pushdown store*

$$g \left(f \begin{pmatrix} w_1 \\ w_2 \end{pmatrix} \$_1 \begin{pmatrix} w_3 \\ w_4 \end{pmatrix} \$_1 \cdots \$_1 \begin{pmatrix} w_{2m-1} \\ w_{2m} \end{pmatrix} \right) \$_2 f_R \left(\begin{pmatrix} w_{2m-1}^R \\ w_{2m}^R \end{pmatrix} \$_1 \begin{pmatrix} w_{2m-3}^R \\ w_{2m-2}^R \end{pmatrix} \$_1 \cdots \$_1 \begin{pmatrix} w_1^R \\ w_2^R \end{pmatrix} \right) \right) \$_3$$
$$g_R \left(f \begin{pmatrix} w_2 \\ w_3 \end{pmatrix} \$_1 \begin{pmatrix} w_4 \\ w_5 \end{pmatrix} \$_1 \cdots \$_1 \begin{pmatrix} w_{2m} \\ n^{|w_{2m}|} \end{pmatrix} \right) \$_2 f_R \left(\begin{pmatrix} w_{2m}^R \\ n^{|w_{2m}|} \end{pmatrix} \$_1 \begin{pmatrix} w_{2m-2}^R \\ w_{2m-1}^R \end{pmatrix} \$_1 \cdots \$_1 \begin{pmatrix} w_2^R \\ w_3^R \end{pmatrix} \right) \right),$$

provided with an initial input (with $|w_i| = |w_j|$ for all w_i, w_j) of the form

$$h(g(f(w_1\$_1 w_3\$_1 \cdots \$_1 w_{2m-1})\$_2 f_R(w_{2m-1}^R\$_1 w_{2m-3}^R\$_1 \cdots \$_1 w_1^R))\$_3$$
$$g_R(f(w_2\$_1 w_4\$_1 \cdots \$_1 w_{2m})\$_2 f_R(w_{2m}^R\$_1 w_{2m-2}^R\$_1 \cdots \$_1 w_2^R))). \qquad \square$$

Lemma 13. *Let M be an* LBA. *Then an* lr-RTPDA *accepting* $VALC(M)$ *can effectively be constructed.*

Proof. The proof is a combination of Lemmata 9, 11, and 12. We assume that the input is correctly formatted, since this property can easily be tested by simulating a reversible deterministic finite automaton in the state set and blocking the computation in case of any error. Then, with the help of encoding map h and Lemma 12, we introduce two components in the pushdown alphabet, preserve the original input in the upper component, and compute to each configuration its successor configuration in the lower component. Then, we utilize Lemma 9 and the encoding mappings g and g_R to compare the lower component of $f\left(\begin{smallmatrix} w_1 \\ w_2 \end{smallmatrix}\$_1 \begin{smallmatrix} w_3 \\ w_4 \end{smallmatrix}\$_1 \cdots \$_1 \begin{smallmatrix} w_{2m-1} \\ w_{2m} \end{smallmatrix}\right)$ with the upper component of $f_R\left(\begin{smallmatrix} w_{2m}^R \\ n^{|w_{2m}|} \end{smallmatrix}\$_1 \begin{smallmatrix} w_{2m-2}^R \\ w_{2m-1}^R \end{smallmatrix}\$_1 \cdots \$ \begin{smallmatrix} w_2^R \\ w_3^R \end{smallmatrix}\right)$. Subsequently, we utilize Lemma 11 to compare the upper component of $f_R\left(\begin{smallmatrix} w_{2m-1}^R \\ w_{2m}^R \end{smallmatrix}\$_1 \begin{smallmatrix} w_{2m-3}^R \\ w_{2m-2}^R \end{smallmatrix}\$_1 \cdots \$_1 \begin{smallmatrix} w_3^R \\ w_4^R \end{smallmatrix}\$_1 \begin{smallmatrix} w_1^R \\ w_2^R \end{smallmatrix}\right)$ with the lower component of $f\left(\begin{smallmatrix} w_2 \\ w_3 \end{smallmatrix}\$_1 \begin{smallmatrix} w_4 \\ w_5 \end{smallmatrix}\$_1 \cdots \$_1 \begin{smallmatrix} w_{2m-2} \\ w_{2m-1} \end{smallmatrix}\$_1 \begin{smallmatrix} w_{2m} \\ n^{|w_{2m}|} \end{smallmatrix}\right)$ where each last configuration is ignored, that is, the comparison starts by checking w_{2m-1}^R in the upper component against w_{2m-1} in the lower component and ends by checking w_3^R in the upper component against w_3 in the lower component. Additionally, note that w_{2m} is an accepting and halting configuration. Thus, there is no successor configuration, and this is marked by using the special symbol n. Furthermore, it should be noted that the "ignored" configurations are suitably marked in the phase where the successor configurations are computed so that Lemma 11 can be applied. The latter two phases of computations ensure that each successor configuration from an odd to an even step and from an even to an odd step has been computed correctly. It remains to be checked that both substrings divided by $\$_3$ have a palindrome structure. To this end, we want to apply Lemma 9 again to the strings in both upper components and observe that the first resp. second pushdown store contains the following strings:

$$\bot f\left(\begin{smallmatrix} w_1 \\ w_2 \end{smallmatrix}\$_1 \begin{smallmatrix} w_3 \\ w_4 \end{smallmatrix}\$_1 \cdots \$_1 \begin{smallmatrix} w_{2m-1} \\ w_{2m} \end{smallmatrix}\right)\$_2 f_R\left(\begin{smallmatrix} w_{2m-1}^R \\ w_{2m}^R \end{smallmatrix}\$_1 \begin{smallmatrix} w_{2m-3}^R \\ w_{2m-2}^R \end{smallmatrix}\$_1 \cdots \$_1 \begin{smallmatrix} w_2^R \\ w_2^R \end{smallmatrix}\right), \text{ resp.}$$

$$\$_3 f_R\left(\begin{smallmatrix} w_{2m}^R \\ n^{|w_{2m}|} \end{smallmatrix}\$_1 \begin{smallmatrix} w_{2m-2}^R \\ w_{2m-1}^R \end{smallmatrix}\$_1 \cdots \$ \begin{smallmatrix} w_2^R \\ w_3^R \end{smallmatrix}\right)\$_2 f\left(\begin{smallmatrix} w_2 \\ w_3 \end{smallmatrix}\$_1 \begin{smallmatrix} w_4 \\ w_5 \end{smallmatrix}\$_1 \cdots \$_1 \begin{smallmatrix} w_{2m-2} \\ w_{2m-1} \end{smallmatrix}\$_1 \begin{smallmatrix} w_{2m} \\ n^{|w_{2m}|} \end{smallmatrix}\right)\bot.$$

Following Remark 10 we apply the suitably adapted variant of Lemma 9 to the string in the upper component of the second pushdown store and check its palindrome structure. After the check, the given string in the first pushdown store is still untouched and we apply another suitably adapted variant of Lemma 9 to the string in the upper component of the first pushdown store to check its palindrome structure. Finally, the following strings are stored in the first resp. second pushdown store:

$$\bot \; \genfrac{}{}{0pt}{}{w_1}{w_2}\$_1 \genfrac{}{}{0pt}{}{w_3}{w_4}\$_1 \cdots \$_1 \genfrac{}{}{0pt}{}{w_{2m-1}}{w_{2m}}\$_2, \text{ resp. } \genfrac{}{}{0pt}{}{w_{2m}^R}{\mathsf{n}^{|w_{2m}|}}\$_1 \genfrac{}{}{0pt}{}{w_{2m-2}^R}{w_{2m-1}^R}\$_1 \cdots \$ \genfrac{}{}{0pt}{}{w_2^R}{w_3^R}\bot.$$

If an error occurs while the computation in still in some phase, the automaton halts in a non-accepting state. If all phases have successfully been ended, an accepting and halting state is entered. Since all phases are reversible and length-reducing, it is clear that the overall computation is reversible and length-reducing. Moreover, each phase can effectively be realized by an lr-RTPDA. Thus, an lr-RTPDA can effectively be constructed that accepts the set VALC(M) for a given LBA M.   □

Theorem 14. *Emptiness, finiteness, infiniteness, universality, inclusion, equivalence, regularity, and context-freeness are not semidecidable for* lr-RTPDA.

Proof. Let M be an LBA. According to Lemma 13, we can effectively construct an lr-RTPDA M' accepting VALC(M). Clearly, $L(M') = $ VALC(M) is empty if and only if $L(M)$ is empty, and emptiness is not semidecidable for LBAs. Further, an lr-RTPDA that accepts nothing can effectively be constructed, so non-semidecidability of equivalence and inclusion follows immediately from this. Also, $L(M') = $ VALC(M) is finite if and only if $L(M)$ is finite and finiteness is not semidecidable for LBAs either.

Like lr-DTPDAs, also lr-RTPDAs always halt, either an accepting or a non-accepting state. By interchanging accepting and non-accepting states of the lr-RTPDA from Lemma 13, we constructively obtain an lr-RTPDA that accepts exactly the complement of the set VALC(M), called INVALC(M), the set of invalid computations. Using INVALC(M), we obtain that infiniteness and universality are not semidecidable, since finiteness and emptiness are not semidecidable.

For regularity and context-freeness we consider again the set VALC(M). By a simple application of the pumping lemma for context-free languages [3] it can be shown that VALC(M) is regular or context free if and only if $L(M)$ is finite. Thus, also non-semidecidability of regularity and context-freeness follows from the non-semidecidability of finiteness.   □

Acknowledgments. The authors acknowledge partial support from COST Action IC1405 *Reversible Computation*. H. B. Axelsen was supported by the Danish Council for Independent Research | Natural Sciences under the *Foundations of Reversible Computing* project, and by an IC1405 STSM (short-term scientific mission) grant.

References

1. Axelsen, H.B., Glück, R.: What do reversible programs compute? In: Hofmann, M. (ed.) FOSSACS 2011. LNCS, vol. 6604, pp. 42–56. Springer, Heidelberg (2011)
2. Buntrock, G., Otto, F.: Growing context-sensitive languages and Church-Rosser languages. Inform. Comput. **141**(1), 1–36 (1998)
3. Hopcroft, J.E., Ullman, J.D.: Introduction to Automata Theory, Languages, and Computation. Addison-Wesley, Reading (1979)
4. Li, M., Vitányi, P.M.B.: An Introduction to Kolmogorov Complexity and Its Applications. Springer, New York (1993)
5. McNaughton, R., Narendran, P., Otto, F.: Church-Rosser Thue systems and formal languages. J. ACM **35**(2), 324–344 (1988)
6. Niemann, G., Otto, F.: The Church-Rosser languages are the deterministic variants of the growing context-sensitive languages. Inform. Comput. **197**(1–2), 1–21 (2005)

Reachability in Resource-Bounded Reaction Systems

Alberto Dennunzio[1], Enrico Formenti[2], Luca Manzoni[1],
and Antonio E. Porreca[1(✉)]

[1] Dipartimento di Informatica, Sistemistica e Comunicazione,
Università Degli Studi di Milano-Bicocca,
Viale Sarca 336/14, 20126 Milano, Italy
{dennunzio,luca.manzoni,porreca}@disco.unimib.it
[2] University of Nice Sophia Antipolis, CNRS, I3S, UMR 7271,
06900 Sophia Antipolis, France
enrico.formenti@unice.fr

Abstract. Reaction systems, a formalism describing biochemical reactions in terms of sets of reactants, inhibitors, and products, are known to have a **PSPACE**-complete configuration reachability problem. We show that the complexity of the problem remains unchanged even for some classes of resource-bounded reaction systems, where we disallow either inhibitors or reactants. We also prove that the complexity decreases to **NP** in the specific case of inhibitorless reaction systems using only *one* reactant per reaction.

Keywords: Unconventional models of computation · Natural computing · Reaction systems · Discrete dynamical systems · Reachability

1 Introduction

During the last decades, many new computing models have been introduced. Each one was meant to more clearly illustrate some features or provide new settings for developing new computing technologies. In most cases, nature has been the main source of inspiration. In 2004, Ehrenfeucht and Rozenberg introduced *reaction systems* (RS in the following) as an abstract model of chemical reactions in living cells [5,6]. Indeed, in living cells, a biochemical reaction takes place only whenever *reactants* are present and *inhibitors* are missing. Hence, a reaction can be represented by a triple (R, I, P) where R is the set of reactants, I the inhibitor and P is the set of products which are left once the reaction is finished. Of course, one has to require that $R \cap I = \varnothing$. Informally, a RS is a (finite) collection of reactions.

The simple definition of the model contrasts with its computing capabilities. In fact, RS are capable of simulating any space-bounded Turing machine computation (many constructions have been provided, one is also given in Sect. 3). Moreover, they provide new examples of natural problems in higher levels of the

© Springer International Publishing Switzerland 2016
A.-H. Dediu et al. (Eds.): LATA 2016, LNCS 9618, pp. 592–602, 2016.
DOI: 10.1007/978-3-319-30000-9_45

polynomial hierarchy [7,8]. As a third argument in favour of studying RS, one may advocate that they are a reference model for other finite systems. Indeed, in [7], it is proved that RS provide lower complexity bounds for Boolean automata networks (BAN). We remark that in the context of BAN, the complexity of relatively few problems is known (see for instance [14]).

This paper pursues the study of complexity problems for RS in the same vein as [8] and subsequent papers [3,7,9]. The new focus is on resource-bounded computation. The idea is to take a classical and important reference problem, namely the *reachability problem*, and try to see how its complexity varies according to constraints that are put on reactions. The constraints we impose consist in limiting the maximum number of reactants and inhibitors involved in each reaction; this changes how much of the current state can be "observed" by a single reaction. In principle, the resulting dynamical behaviours of reaction systems are less rich than for unrestricted systems, although this does not necessarily reduce the complexity of the reachability problem.

From [11], it is known that result functions (state transition functions) of RS can be completely classified into five classes of functions over lattices, which correspond to specific limitations on the number of reactants and inhibitors allowed in each reaction of the corresponding RS. Theorems 7 and 9 prove that the reachability problem is **PSPACE**-complete for three out of the five classes. The class of result functions computed by RS with no reactants and no inhibitors corresponds to constant functions, making the reachability problem very simple. Concerning the fifth class, we only succeeded in proving that reachability is in **NP** (Theorem 12) but we suspect that it is also **NP**-hard. Indeed, a slight variant of the reachability problem is **NP**-complete for this class. The proof of this last result is also of some interest in its own. Indeed, it uses the Prime Number Theorem to precisely evaluate the complexity of the reduction.

2   Basic Notions

This section briefly recalls the basic notions about RS as introduced in [6]. We remark that in this paper the set of reactants and inhibitors of a reaction are allowed to be empty, unlike what is often required in literature. The reason for this generalised definition is that, as it will be shown later (Corollary 10), the reachability problem for "minimal" RS [13], having exactly one reactant and one inhibitor per reaction, is already **PSPACE**-complete.

Definition 1. *Consider a finite set S, whose elements are called* entities. *A reaction a over S is a triple (R_a, I_a, P_a) of subsets of S. The set R_a is the set of* reactants, *I_a the set of* inhibitors, *and P_a is the nonempty set of* products. *The set of all reactions over S is denoted by* rac(S).

Definition 2. *A reaction system (RS) is a pair $\mathcal{A} = (S, A)$ where S is a finite set, called the* background set, *and $A \subseteq$ rac(S).*

Given a *state* $T \subseteq S$, a reaction a is said to be *enabled* in T when $R_a \subseteq T$ and $I_a \cap T = \varnothing$. The *result function* res$_a : 2^S \to 2^S$ of a, where 2^S denotes the power set of S, is defined as $\mathrm{res}_a(T) = \mathrm{P}_a$ if a is enabled in T, and $\mathrm{res}_a(T) = \varnothing$ otherwise. The definition of res_a naturally extends to sets of reactions: given $T \subseteq S$ and $A \subseteq \mathrm{rac}(S)$, define $\mathrm{res}_A(T) = \bigcup_{a \in A} \mathrm{res}_a(T)$. The result function $\mathrm{res}_{a,\mathcal{A}}$ of a RS $\mathcal{A} = (S, A)$ is res_A, i.e., the result function on the whole set of reactions. In this way, any RS $\mathcal{A} = (S, A)$ induces a discrete dynamical system where the state set is 2^S and the next state function is res_A. The set of reactions of $\mathcal{A}$ enabled in a state T is denoted by $\mathrm{en}_A(T)$.

The *orbit* or *state sequence* of a given state T of a RS $\mathcal{A}$ is defined as the sequence of states obtained by iterations of res_A starting from T, namely the sequence $(T, \mathrm{res}_A(T), \mathrm{res}_A^2(T), \ldots)$. Being finite systems, RS only admit ultimately periodic orbits, i.e., orbits ending up in a cycle.

We now recall the classification of RS in terms of number of resources employed per reaction [11].

Definition 3. *Let* $r, i \in \mathbb{N}$. *The class* $\mathcal{RS}(r, i)$ *consists of all RS having at most* r *reactants and* i *inhibitors for reaction. We also define the unbounded classes* $\mathcal{RS}(\infty, i) = \bigcup_{r=0}^{\infty} \mathcal{RS}(r, i)$, $\mathcal{RS}(r, \infty) = \bigcup_{i=0}^{\infty} \mathcal{RS}(r, i)$, *and* $\mathcal{RS}(\infty, \infty) = \bigcup_{r=0}^{\infty} \bigcup_{i=0}^{\infty} \mathcal{RS}(r, i)$.

We remark that this classification does not include the number of products as a parameter, since RS can always be assumed to be in *singleton product normal form* [2]: any reaction $(R, I, \{p_1, \ldots, p_m\})$ can be replaced by the set of reactions $(R, I, \{p_1\}), \ldots, (R, I, \{p_m\})$, since they produce the same result.

Several of the above defined classed have a characterisation in terms of functions over the Boolean lattice 2^S [11]. Recall that a function $f \colon 2^S \to 2^S$ is *antitone* if $X \subseteq Y$ implies $f(X) \supseteq f(Y)$, *monotone* if $X \subseteq Y$ implies $f(X) \subseteq f(Y)$, *additive* (or an *upper-semilattice endomorphism*) if $f(X \cup Y) = f(X) \cup f(Y)$. We say that the RS $\mathcal{A} = (S, A)$ computes the function $f \colon 2^S \to 2^S$ if $\mathrm{res}_A = f$. Furthermore, we say that the RS $\mathcal{A} = (S', A)$ computes a function $f \colon 2^S \to 2^S$ via k-simulation if $S \subseteq S'$ and $\mathrm{res}_A^k(T) \cap S = f(T)$ for all $T \subseteq S$. The (distinct) classes of functions computed by restricted classes of RS are illustrated in Fig. 1. These results show that, in a sense, the classes $\mathcal{RS}(1, 1)$, $\mathcal{RS}(0, 1)$, and $\mathcal{RS}(2, 0)$ capture the expressiveness of the whole classes $\mathcal{RS}(\infty, \infty)$, $\mathcal{RS}(0, \infty)$, and $\mathcal{RS}(\infty, 0)$, respectively (i.e., they simulate the more generic RS with a polynomial slowdown).

We conclude this section by recalling the formulation of the problem addressed in this paper.

Class of RS	Subclass of $2^S \to 2^S$	(via k-simulation)
$\mathcal{RS}(\infty, \infty)$	all	$\mathcal{RS}(1, 1)$
$\mathcal{RS}(0, \infty)$	antitone	$\mathcal{RS}(0, 1)$
$\mathcal{RS}(\infty, 0)$	monotone	$\mathcal{RS}(2, 0)$
$\mathcal{RS}(1, 0)$	additive	$\mathcal{RS}(1, 0)$
$\mathcal{RS}(0, 0)$	constant	$\mathcal{RS}(0, 0)$

Fig. 1. Functions computed by restricted classed of RS.

Definition 4. *The* reachability problem *for the class* $\mathcal{RS}(i,r)$, *with* i *and* r *possibly infinite, consists of deciding, given* $\mathcal{A} \in \mathcal{RS}(i,r)$ *and two of its states* T, U, *whether* U *is reachable from* T, *i.e., whether* $\mathrm{res}_{\mathcal{A}}^t(T) = U$ *for some* $t \geq 0$.

For the notions of complexity theory, such as the definitions of the classes of problems **NP** and **PSPACE**, we refer the reader to any relevant textbook, such as [12].

3   Inhibitorless Classes $\mathcal{RS}(\infty, 0)$ and $\mathcal{RS}(2, 0)$

We begin by describing how inhibitorless RS are able to efficiently simulate Turing machines with bounded tape. A similar simulation was previously published [7], but it required both reactants and inhibitors.

In the following, let M be any single-tape deterministic Turing machine using m tape cells during its computation; let Σ be the tape alphabet, Q the set of states, and $\delta\colon Q \times \Sigma \to Q \times \Sigma \times \{-1, 0, +1\}$ the transition function of M. We are going to define a RS $\mathcal{M} = (S, A) \in \mathcal{RS}(2, 0)$ simulating M.

Entities. The set of entities of $\mathcal{M}$ is

$$S = \{a_j : a \in \Sigma, 1 \leq j \leq m\} \cup \{q_j : q \in Q, 1 \leq j \leq m\} \cup \{\spadesuit_j : 1 \leq j \leq m\}$$

that is, it consists of all symbols of the alphabet, states, and the extra item $\spadesuit$, each of them indexed by every possible tape position.

In this way, the generic configuration where M is in state $q \in Q$, its tape head is located on cell i, and its tape contains the string $x = x_1 \cdots x_m$, is encoded as the following $2m$-entity state:

$$T = \{x_{j,j} : 1 \leq j \leq m\} \cup \{q_i\} \cup \{\spadesuit_j : 1 \leq j \leq m, j \neq i\} \subseteq S$$

In other terms, T contains each symbol x_j of the string x indexed by its position on the tape as element $x_{j,j}$, an entity q_i storing both the current state and the head position, and $m - 1$ entities $\spadesuit_j$, one for each position $j \neq i$ on which the tape head is *not* located.

Example 5. Consider a Turing machine M working in space $m = 4$ and the configuration where M is in state q, its tape head is located on cell 3, and its tape contains the string $abba$. The state of the RS $\mathcal{M}$ encoding such a configuration of M is then $T = \{a_1, b_2, b_3, a_4, \spadesuit_1, \spadesuit_2, q_3, \spadesuit_4\}$.

Reactions. Each transition $\delta(q, a) = (r, b, d)$ of M, with $q, r \in Q$, $a, b \in \Sigma$, and $d \in \{-1, 0, +1\}$, gives rise to the following two sets of reactions:

$$(\{q_i, a_i\}, \varnothing, \{r_{j+d}, b_i\}) \qquad\qquad \text{for } 1 \leq i \leq m \qquad (1)$$

$$(\{q_i, a_i\}, \varnothing, \{\spadesuit_j : 1 \leq j \leq m, j \neq i + d\}) \qquad \text{for } 1 \leq i \leq m. \qquad (2)$$

If the tape head of M is located on cell i, then the i-th reaction from (1) produces the entity encoding both the new state and the new tape head position of M, as well as the symbol written in the position i over the head. The production of one $\spadesuit_j$ for all the tape positions $j \neq i + d$, i.e., those on which the tape head is *not* located after the transition of M, is assured by the i-th reaction from (2).

Finally, the following reactions preserve the encoding of the tape cells j where the head is *not* located, i.e., those indicated by the presence of $\spadesuit_j$:

$$(\{\spadesuit_j, a_j\}, \varnothing, \{a_j\}) \qquad\qquad \text{for } 1 \leq j \leq m. \qquad (3)$$

The set A of the reactions of $\mathcal{M}$ is defined as the union of the three sets of reactions from (1), (2), and (3).

It is easy to see that if $T \subseteq S$ is an encoding of a configuration of M using space m (i.e., it contains, for all $1 \leq j \leq m$, a single entity a_j for some $a \in \Sigma$, a single entity q_i for some $q \in Q$ and some $1 \leq i \leq m$, and entities $\spadesuit_j$ for all $1 \leq j \leq m$ with $j \neq i$), then the next state $\mathrm{res}_{\mathcal{M}}(T)$ encodes the next configuration of M.

We remark that all reactions of $\mathcal{M}$ have exactly two reactants and no inhibitors, that is $\mathcal{M} \in \mathcal{RS}(2,0)$. We are now able to prove the following:

Lemma 6. *Reachability for* $\mathcal{RS}(2,0)$ *is* **PSPACE**-*hard*.

Proof. We reduce reachability of configurations of polynomial-space Turing machines (one of the canonical **PSPACE**-complete problems [12]) to this problem. Given a Turing machine M working in space m and two configurations $\mathcal{C}_1$, $\mathcal{C}_2$ of M, it is possible to build the RS $\mathcal{M} \in \mathcal{RS}(2,0)$ simulating M as described above; the construction can be done in polynomial time, since the reactions can be built by iterating over all entries of the transition table of M and the range of the m possible tape positions. The question then becomes whether in the RS $\mathcal{M}$ the encoding of the configuration $\mathcal{C}_2$ is reachable from the encoding of the configuration $\mathcal{C}_1$; the construction of $\mathcal{M}$ assures that this happens if and only if $\mathcal{C}_2$ is reachable in M from $\mathcal{C}_1$. Therefore, the reduction holds and reachability for $\mathcal{RS}(2,0)$ is then **PSPACE**-hard. $\qquad\square$

We conclude this section with the complexity result for the inhibitorless classes.

Theorem 7. *Reachability for* $\mathcal{RS}(\infty,0)$ *and for* $\mathcal{RS}(2,0)$ *is* **PSPACE**-*complete*.

Proof. Recall that reachability for $\mathcal{RS}(\infty,\infty)$ can be decided in polynomial space, by storing the current configuration of the involved RS and applying the reactions one by one at each time step. The thesis follows as a consequence of this fact and Lemma 6. $\qquad\square$

4   Reactantless Classes $\mathcal{RS}(0,1)$ and $\mathcal{RS}(0,\infty)$

It is known [11] that each RS from $\mathcal{RS}(\infty,0)$ can be simulated with a linear slowdown by a RS from $\mathcal{RS}(0,1)$. Since the two classes of RS are equivalent from this point of view, it is reasonable to assume that their reachability problems have the same complexity, with reachability for $\mathcal{RS}(0,1)$ being **PSPACE**-complete as well. However, the original simulation does not directly imply this result, since each state of the simulating RS contains a number of auxiliary entities, and it is not obvious which auxiliary entities must appear in the target state.

Therefore, in the next Lemma, we provide a construction, with no auxiliary entities appearing, of an RS from $\mathcal{RS}(0,1)$ simulating a given RS from $\mathcal{RS}(\infty,0)$. In this simulation, the states at even time steps of the former coincide exactly with the states of the simulated RS.

Lemma 8. *Let* $\mathcal{A} = (S,A) \in \mathcal{RS}(\infty,0)$ *be a RS such that* $\bigcup\{P_a : a \in A\} = S$ *and* $\mathrm{res}_{\mathcal{A}}(S) = S$. *Then, there exists a RS* $\mathcal{B} = (S',A') \in \mathcal{RS}(0,1)$ *such that, for any* $T \subseteq S$, *the following condition holds:*

$$\forall t \in \mathbb{N} \qquad \mathrm{res}_{\mathcal{B}}^{2t}(T) = \mathrm{res}_{\mathcal{A}}^{t}(T) \wedge S \subseteq \mathrm{res}_{\mathcal{B}}^{2t+1}(T).$$

Proof. Set $S' = S \cup \{\bar{a} : a \in A\}$, that is, S' is obtained by adding to S one barred entity for each reaction of $\mathcal{A}$. For each reaction $a = (R_a, \varnothing, P_a) \in A$, the set A' contains the reactions

$$(\varnothing, \{s\}, \{\bar{a}\}) \qquad\qquad \text{for } s \in R_a \qquad\qquad (4)$$

which produce the entity $\bar{a}$ if at least one of the reactants of a is missing in the current state (i.e., if a is not enabled in it). Furthermore, for each $a = (R_a, \varnothing, P_a) \in A$ the set A' also contains the reaction

$$(\varnothing, \{\bar{a}\}, P_a) \qquad\qquad (5)$$

which gives the same products as a when $\bar{a}$ is missing in the current state, or, equivalently, when a is enabled.

Thus, for any state $T \subseteq S$ and any state $T' \subseteq \{\bar{a} : a \in A\}$ it holds that $\mathrm{res}_{\mathcal{B}}(T) = \{\bar{a} : a \notin \mathrm{en}_{\mathcal{A}}(T) \cup S\}$ and $\mathrm{res}_{\mathcal{B}}(T') = \bigcup\{P_a : a \in A, \bar{a} \notin T'\}$.

Choose now an arbitrary state $T \subseteq S$. We are going to prove the thesis condition by induction on t. Clearly, $\mathrm{res}_{\mathcal{B}}^{2\cdot 0}(T) = T = \mathrm{res}_{\mathcal{A}}^{0}(T)$. Furthermore, since T contains no entity $\bar{a}$, all reactions of type (5) are enabled, and so $\mathrm{res}_{\mathcal{B}}^{2\cdot 0+1}(T) \supseteq \bigcup\{P_a : a \in A\} = S$.

Assume now that the thesis condition holds for t. Then,

$$\mathrm{res}_{\mathcal{B}}^{2(t+1)}(T) = \mathrm{res}_{\mathcal{B}}^{2}(\mathrm{res}_{\mathcal{B}}^{2t}(T))$$
$$= \mathrm{res}_{\mathcal{B}}^{2}(\mathrm{res}_{\mathcal{A}}^{t}(T))$$
$$= \mathrm{res}_{\mathcal{B}}(\{\bar{a} : a \notin \mathrm{en}_{\mathcal{A}}(\mathrm{res}_{\mathcal{A}}^{t}(T))\} \cup S)$$

Since S disables all reactions of type (4), it follows that

$$\mathrm{res}_{\mathcal{B}}^{2(t+1)}(T) = \mathrm{res}_{\mathcal{B}}(\{\bar{a} : a \notin \mathrm{en}_{\mathcal{A}}(\mathrm{res}_{\mathcal{A}}^{t}(T))\})$$
$$= \bigcup\{P_a : a \in \mathrm{en}_{\mathcal{A}}(\mathrm{res}_{\mathcal{A}}^{t}(T))\}$$
$$= \mathrm{res}_{\mathcal{A}}^{t+1}(T)$$

In particular, due to the reactions of type (5), one obtains that $S \subseteq \mathrm{res}_{\mathcal{B}}^{2t+1}(\mathrm{T})$ for all $t \in \mathbb{N}$.   □

By exploiting Lemma 8 we can finally show the complexity of reachability for reactantless RS.

Theorem 9. *Reachability for* $\mathcal{RS}(0,1)$, *and thus for* $\mathcal{RS}(0,\infty)$, *is* **PSPACE-**
complete.

Proof. It is enough to prove the **PSPACE**-hardness of the problem for $\mathcal{RS}(0,1)$, which will be accomplished by reduction from reachability for $\mathcal{RS}(\infty,0)$ as follows. Given a RS $\mathcal{A} = (S, A) \in \mathcal{RS}(\infty,0)$, let $\mathcal{A}' = (S', A')$ be the RS with $S' = S \cup \{\spadesuit\}$ for some $\spadesuit \notin A$, and $A' = A \cup \{(S', \varnothing, S')\}$. Clearly, $\mathcal{A}' \in \mathcal{RS}(\infty,0)$ and it has the same behaviour of $\mathcal{A}$ as long as its initial state does not contain $\spadesuit$, i.e., $\mathrm{res}_{\mathcal{A}'}(\mathrm{T}) = \mathrm{res}_{\mathcal{A}}(\mathrm{T})$ whenever $\spadesuit \notin T$.

Moreover, $\mathcal{A}'$ satisfies the hypotheses of Lemma 8 as a consequence of the changes made to build it from $\mathcal{A}$. Let $\mathcal{B} \in \mathcal{RS}(0,1)$ be then the RS obtained from $\mathcal{A}'$ using that lemma. Notice that that the mapping $\mathcal{A}' \mapsto \mathcal{B}$ can be computed in polynomial time.

For any two states $U, V \subseteq S$, it holds that $\mathrm{res}_{\mathcal{A}}^{t}(U) = V$ for some $t \in \mathbb{N}$ if and only if $\mathrm{res}_{\mathcal{B}}^{2t}(U) = V$. Furthermore, we have $\mathrm{res}_{\mathcal{B}}^{2s+1}(U) \neq V$ for all $s \in \mathbb{N}$, because $\spadesuit \in S' \subseteq \mathrm{res}_{\mathcal{B}}^{2s+1}(U)$, while $\spadesuit \notin S$ and, in particular, $\spadesuit \notin V$. Hence, the state V is reachable from U in the RS $\mathcal{B}$ if and only if the same occurs in $\mathcal{A}$. Therefore, reachability for $\mathcal{RS}(\infty,0)$ is reducible to reachability for $\mathcal{RS}(0,1)$ in polynomial time and the thesis then follows from Theorem 7.   □

As a consequence of Theorem 9, we also obtain the **PSPACE**-completeness of the reachability problem for $\mathcal{RS}(1,1)$ and for the class $\mathcal{RS}(\infty,\infty)$ (the latter having already been proved in a different way [7]).

Corollary 10. *Reachability   for $\mathcal{RS}(1,1)$,   and   thus   for $\mathcal{RS}(\infty,\infty)$,*
is **PSPACE**-*complete.*   □

5   Single-Reactant Inhibitorless Class $\mathcal{RS}(1,0)$

We proved that disallowing either reactants or inhibitors does not decrease the complexity of reachability problems. However, reducing the number of reactants to 1 in inhibitorless RS makes the evolution of each single entity "context-free", i.e., not influenced by the presence or absence of other entities. Indeed, the result functions of the RS from $\mathcal{RS}(1,0)$ are always upper-semilattice endormorphisms, that is, $\mathrm{res}_{\mathcal{A}}(U \cup V) = \mathrm{res}_{\mathcal{A}}(U) \cup \mathrm{res}_{\mathcal{A}}(V)$ for all $\mathcal{A} \in \mathcal{RS}(1,0)$ and arbitrary states U, V [11]. It is thus reasonable to conjecture that reachability for $\mathcal{RS}(1,0)$ might be easier than for other variants, as the entities of the target configuration can be traced back to a set of originating entities independently one from another, and the only difficulty is to find a common number of backwards steps. In this section we show that this is actually the case, under the assumption that **NP** $\neq$ **PSPACE**.

We begin by recalling the notion of *influence graph* [1], which describes static causality relations in a RS. Given a RS $\mathcal{A} = (S, A)$, the associated influence graph is the directed graph $G = (S, E)$, where $(x, y) \in E$ if and only if there exists a reaction $(R_a, I_a, P_a) \in A$ such that $x \in R_a \cup I_a$ and $y \in P_a$. In other words, there is an edge (x, y) whenever the presence or absence of x contributes to the appearance of y.

In particular, if $\mathcal{A} \in \mathcal{RS}(1, 0)$ we have $(x, y) \in E$ if and only if $y \in \text{res}_{\mathcal{A}}(\{\text{x}\})$. An entity y appearing at time t can thus be recursively traced back to a single entity occurring in the initial state of the RS (or to multiple independent entities, only one of which needs to occur in order that y appears at time t).

Recall that the powers G^t of the Boolean adjacency matrix of any graph G can be computed in polynomial time even for exponential values of t, by repeated squaring; the entry $G^t_{i,j}$ is 1 if and only if a (not necessarily simple) path of length t exists between v_i and v_j.

Let $\mathcal{A} = (S, A) \in \mathcal{RS}(1, 0)$ be a RS with $S = \{s_1, \ldots, s_n\}$ and let G be its influence graph. Any state $U \subseteq S$ can be viewed as a column vector in $\{0, 1\}^n$, where $U_i = 1$ if and only if $s_i \in U$. Then, a state $V \subseteq S$ is reachable from a state $U \subseteq S$ in $\mathcal{A}$ if and only if $G^t U = V$, where $G^t U$ is the product of the matrix G^t and the vector U. This observation allows us to prove the following result:

Theorem 11. *Reachability for* $\mathcal{RS}(1, 0)$ *is in* **NP**.

Proof. Consider a RS $\mathcal{A} \in \mathcal{RS}(1, 0)$ and two states U, V. Let G be the influence graph of $\mathcal{A}$. Since G^t can be computed in polynomial time, the validity of the equation $G^t U = V$ can be checked in polynomial time for any fixed t, even when the latter is exponential with respect to the number n of entities. It is enough to use the guessing power of a nondeterministic Turing machine to choose an n-bit integer $0 \leq t < 2^n$, since state V is either reached within $2^n - 1$ steps, or it is never reached. $\square$

It is unknown whether this problem is also **NP**-hard. The variant where the target state V consists of a single entity is in **NL**, being the reachability problem (with several possible source vertices) for the influence graph in **NL** too. It is actually **NL**-complete, since every graph is the influence graph of a RS (the one just having the vertex set as background set and reactions $(\{u\}, \varnothing, \{v\})$, one for each edge (u, v) of the graph). On the other hand, the variant where we check if a *superset* of V is reachable is **NP**-complete.

Theorem 12. *It is* **NP***-complete to decide, given a RS* $\mathcal{A} = (S, A) \in \mathcal{RS}(1, 0)$ *and two states* $U, V \subseteq S$, *whether* $V \subseteq \text{res}^t_{\mathcal{A}}(\text{U})$ *for some* $t \in \mathbb{N}$.

Proof. Membership in **NP** is proved in a similar way to what has been done in the proof of Theorem 11, i.e., for a RS $\mathcal{A} = (S, A) \in \mathcal{RS}(1, 0)$ and two states $U, V \subseteq S$, by guessing $0 \leq t < 2^n$ and checking whether $V \leq G^t U$, where G is the influence graph of $\mathcal{A}$ and the comparison is made element-wise.

The **NP**-hardness of the problem is proved by reduction from Boolean satisfiability of CNF formulae [12]. Let $\varphi = \varphi_1 \wedge \cdots \wedge \varphi_m$ be a CNF formula with m clauses $C = \{\varphi_1, \ldots, \varphi_m\}$ over the n variables $X = \{x_1, \ldots, x_n\}$.

Denote by p_i the i-th prime number and set $X_i = \{x_{i,j} : 0 \leq j < p_i\}$. Define the RS $\mathcal{A} = (S, A) \in \mathcal{RS}(1, 0)$, where $S = X_1 \cup \cdots \cup X_n \cup C$ and A consists of the reactions described in the following.

For each variable x_i, we build a cycle of prime length p_i iterating across all elements of X_i by means of the reactions

$$(\{x_{i,j}\}, \varnothing, \{x_{i,(j+1) \mod p_i}\}) \qquad \text{for } 0 \leq j < p_i. \qquad (6)$$

We define a "well-formed" state T of $\mathcal{A}$ as a state containing exactly one entity $x_{i,j}$ for each $1 \leq i \leq n$. Such a state T is interpreted as the truth assignment $v : X \to \{0, 1\}$ to φ defined as

$$v(x_i) = \begin{cases} 1 & \text{if } x_{i,0} \in T \\ 0 & \text{otherwise} \end{cases}$$

that is, all elements $x_{i,j}$ with $j > 0$ denote a false value of x_i.

Since the lengths of the cycles associated to the variables of φ are pairwise coprime, all 2^n assignments of φ will be eventually reached in $\mathcal{A}$, possibly with several repetitions (since distinct states encode the same truth assignment). Therefore, if the initial state of $\mathcal{A}$ is $U = \{x_{1,0}, \ldots, x_{n,0}\}$, then a state encoding the assignment $v : X \to \{0, 1\}$ will be reached at time step $\prod_{i=1}^{n} p_i^{v(x_i)}$.

We are going to introduce the remaining reactions in A. They have the role of evaluating formula φ under the assignment encoded by the $x_{i,j}$'s. We map each entity $x_{i,j}$ to the set of clauses satisfied by $v(x_i) = 1$ (if $j = 0$) or by $v(x_i) = 0$ (if $j > 0$) by the following reactions:

$$(\{x_{i,0}\}, \varnothing, \{\varphi_k\}) \qquad \text{if } x_i \text{ implies } \varphi_k \qquad (7)$$
$$(\{x_{i,j}\}, \varnothing, \{\varphi_k\}) \qquad \text{for } j > 0 \text{ if } \neg x_i \text{ implies } \varphi_k \qquad (8)$$

The influence graph of the resulting RS for a sample Boolean formula is shown in Fig. 2.

As a consequence of the above construction, given a well-formed assignment $Y \subseteq X_1 \cup \cdots \cup X_n$, it follows that $\mathrm{res}_{\mathcal{A}}(Y) = D \cup Y'$ where $D \subseteq C$ is exactly the set of clauses satisfied by Y, and Y' is the next truth assignment in the order given by the reactions of type (6). Since they do not appear as reactants in any reaction, the entities representing the clauses of φ appear only if the previous state encodes an assignment satisfying them.

Consider now the states $U = \{x_{1,0}, \ldots, x_{n,0}\}$ and $V = C$. According to the above reasoning, a superset of V is reachable from U, or, equivalently, $V \subseteq \mathrm{res}_{\mathcal{A}}^t(U)$ for some $t \in \mathbb{N}$, if and only if there exists a truth assignment for X satisfying all clauses, i.e., the entire formula φ.

It remains to be proved that the mapping $\varphi \mapsto (\mathcal{A}, U, V)$ can be computed in polynomial time. In particular, we need to show that we can find in polynomial time n primes of polynomially bounded value.

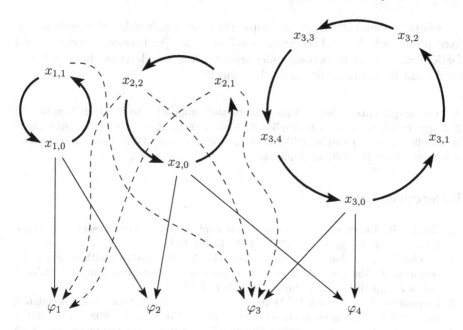

Fig. 2. The influence graph for the RS encoding the formula $\varphi \equiv \varphi_1 \wedge \varphi_2 \wedge \varphi_3 \wedge \varphi_4$, where $\varphi_1 \equiv x_1 \vee \neg x_2$, $\varphi_2 \equiv x_1 \vee x_2$, $\varphi_3 \equiv \neg x_1 \vee \neg x_2 \vee x_3$, and $\varphi_4 \equiv x_2 \vee x_3$. The thick edges represent the reactions of type (6), the continuous thin ones those of type (7), and the dashed thin ones those of type (8).

First of all, the Prime Number Theorem [4] implies that p_n is asymptotically $n \ln n$; thus, we only need to check the first $O(n \ln n)$ integers. These can be checked for primality in polynomial time via a brute force algorithm, since they are polynomial in value with respect to n. The reactions of type (6) are then simple to compute, while those of types (7) and (8) only require to check whether variable x_i appears as a positive or negated literal in φ_k.

Therefore, Boolean satisfiability of CNF formulae is reducible to the considered problem in polynomial time and then the thesis follows.

6   Conclusions

We proved that the reachability problem for RS remains **PSPACE**-complete, as in the general case, even when inhibitors or reactants are disallowed in each reaction. The problem only becomes easier (assuming **NP** $\neq$ **PSPACE**) for inhibitorless RS using only one reactant per reaction: this variant has been proved to be in **NP**. It is left as an open problem to establish whether this problem is also **NP**-hard, as it is in the case where we check the reachability a *superset* of a given state.

It would also be interesting to examine further problems related to the dynamics of RS, such as the detection of fixed points, global and local attractors, and "Gardens of Eden", in order to check whether these become easier for resource-bounded RS. As a simple example, establishing the existence of fixed points, which is **NP**-complete for $\mathcal{RS}(\infty, \infty)$ [8], becomes entirely trivial for $\mathcal{RS}(\infty, 0)$, since monotonic functions over complete lattices always admit a fixed point by the Knaster-Tarski theorem [10].

Acknowledgments. This work has been partially supported by Fondo di Ateneo (FA) 2013 grants of Università degli Studi di Milano-Bicocca: "Complessità computazionale in modelli di calcolo bioispirati: Sistemi a membrane e sistemi a reazioni" and "Sistemi complessi e incerti: Teoria ed applicazioni".

References

1. Brijder, R., Ehrenfeucht, A., Main, M., Rozenberg, G.: A tour of reaction systems. Int. J. Found. Comput. Sci. **22**(7), 1499–1517 (2011)
2. Brijder, R., Ehrenfeucht, A., Rozenberg, G.: Reaction systems with duration. In: Kelemen, J., Kelemenová, A. (eds.) Computation, Cooperation, and Life. LNCS, vol. 6610, pp. 191–202. Springer, Heidelberg (2011)
3. Dennunzio, A., Formenti, E., Manzoni, L., Porreca, A.E.: Ancestors, descendants, and gardens of Eden in reaction systems. Theor. Comput. Sci. **608**, 16–26 (2015)
4. Dudley, U.: Elementary Number Theory. W.H. Freeman and Company, New York (1969)
5. Ehrenfeucht, A., Rozenberg, G.: Basic notions of reaction systems. In: Calude, C.S., Calude, E., Dinneen, M.J. (eds.) DLT 2004. LNCS, vol. 3340, pp. 27–29. Springer, Heidelberg (2004)
6. Ehrenfeucht, A., Rozenberg, G.: Reaction systems. Fundamenta Informaticae **75**, 263–280 (2007)
7. Formenti, E., Manzoni, L., Porreca, A.E.: Cycles and global attractors of reaction systems. In: Jürgensen, H., Karhumäki, J., Okhotin, A. (eds.) DCFS 2014. LNCS, vol. 8614, pp. 114–125. Springer, Heidelberg (2014)
8. Formenti, E., Manzoni, L., Porreca, A.E.: Fixed points and attractors of reaction systems. In: Beckmann, A., Csuhaj-Varjú, E., Meer, K. (eds.) CiE 2014. LNCS, vol. 8493, pp. 194–203. Springer, Heidelberg (2014)
9. Formenti, E., Manzoni, L., Porreca, A.E.: On the complexity of occurrence and convergence problems in reaction systems. Nat. Comput. **14**(1), 185–191 (2015)
10. Granas, A., Dugundji, J.: Fixed point theory. Springer Monographs on Mathematics. Springer, New York (2003)
11. Manzoni, L., Poças, D., Porreca, A.E.: Simple reaction systems and their classification. Int. J. Found. Comput. Sci. **25**(4), 441–457 (2014)
12. Papadimitriou, C.H.: Computational Complexity. Addison-Wesley, Boston (1993)
13. Salomaa, A.: Minimal and almost minimal reaction systems. Nat. Comput. **12**(3), 369–376 (2013)
14. Tošić, P.T.: On the complexity of enumerating possible dynamics of sparsely connected Boolean network automata with simple update rules. In: Fatès, N., Kari, J., Worsch, T. (eds.) Discrete Mathematics and Theoretical Computer Science Proceedings, DMTCS, Automata 2010–16th International Workshop on CA and DCS, 14–16 June 2010, Nancy, France, pp. 125–144 (2010)

Canonical Multi-target Toffoli Circuits

Hans-Jörg Kreowski, Sabine Kuske, and Aaron Lye$^{(\boxtimes)}$

Department of Computer Science, University of Bremen,
33 04 40, 28334 Bremen, Germany
{kreo,kuske,lye}@informatik.uni-bremen.de

Abstract. In this paper, we study reversible circuits as cascades of multi-target Toffoli gates. This new type of gates allows to shift parts of a gate to the preceding gate within a circuit provided that a certain independence condition holds. It turns out that shifts decrease the so-called waiting degree such that shifting as long as possible always terminates and yields shift-reduced circuits. As the main result, we show that shift-reduced circuits are unique canonical representatives of their shift equivalence classes. Canonical circuits are optimal with respect to maximal and as-early-as-possible parallelism of targets within gates.

Keywords: Canonical form · Multi-target Toffoli circuits · Reversible computation · Shift equivalence

1 Introduction

Reversible computation is an alternative to conventional computing motivated by the fact that the integration density of circuits reaches physical limits in scale and power dissipation. Due to the fact that energy dissipation is significantly reduced or even eliminated in reversible circuits [1], reversible computing is a very promising research area.

Reversible circuits are cascades of reversible gates that compute invertible functions on Boolean vectors. To specify reversible circuits, the gate model introduced by Toffoli [9] is frequently used. In the past this model has been generalized in different ways. In this paper we want to generalize this model further by introducing multi-target Toffoli gates.

A (single-target multi-controlled) Toffoli gate consists of a target line and a set of control lines each of which is different from the target line. The lines represent Boolean variables. The target line gets negated if and only if all control lines are carrying the value 1. All other values are kept invariant by the evaluation of the gate. In particular, a Toffoli gate is reversed by itself. Consider now a set of Toffoli gates such that the target lines are pairwise different and all control lines are disjoint from all target lines. Such gates may be called independent because their evaluation in every sequential order yields the same Boolean function. Moreover, their evaluation can be done in parallel because the various negations cannot interfere with each other. This motivates us to introduce such sets of independent Toffoli gates as a multi-target Toffoli gate.

© Springer International Publishing Switzerland 2016
A.-H. Dediu et al. (Eds.): LATA 2016, LNCS 9618, pp. 603–616, 2016.
DOI: 10.1007/978-3-319-30000-9_46

As the parallel as well as each sequential evaluation of independent gates yield the same result, a multi-target Toffoli gate can be sequentialized with respect to every partition of the set of target lines. Conversely, two gates can be parallelized into one gate if their sets of target lines are disjoint and no target line is control line of the other gate. There is a weaker form of combining two multi-target Toffoli gates that can be applied much more frequently than the full parallelization: One of the gates is sequentialized first and then only one of the parts is parallelized with the other gate if possible. In this case, a part of a gate is shifted to another gate. The shifts (together with the parallelization) define a relation on multi-target Toffoli circuits with quite significant properties. First of all, the shift relation has the local Church-Rosser property meaning that the circuits resulting from two shifts on a given circuit can be further shifted into a common result. Secondly, shifts decrease the so-called waiting degree. For each target line of some gate, there is a number of preceding gates. If evaluation is done gate by gate, this is the number of steps a negation must wait before it is executed. The waiting degree sums up all these numbers. As the waiting degree decreases with each shift, the lengths of shift sequences are bounded by the maximum waiting degree (which is $\frac{m(m-1)}{2}$ for the number m of target lines of a circuit). In particular, the iteration of shifts as long as possible terminates always with a circuit reduced with respect to shifting. Combining both results, the shift-reduced circuits turn out to be unique normal forms within the classes of shift-equivalent circuits. Therefore, it is justified to call shift-reduced multi-target Toffoli gates canonical. Canonical circuits are optimal with respect to maximal and as-early-as-possible parallelism of targets within gates.

Shifts, shift equivalence and shift-reduced normal forms as unique canonical representatives of their shift equivalence classes were studied by the first author quite some time ago for parallel derivations in graph grammars (see [5–7]). Although multi-target Toffoli circuits as considered in this paper provide a setting quite different from parallel graph grammar derivations, the same ideas work.

The paper is organized as follows. Section 2 introduces the characteristics of reversible functions and circuits. In Sect. 3 multi-target Toffoli circuits are defined, followed by considering sequentialization, parallelization and shift in Sect. 4. Section 5 introduces the waiting degree. Section 6 covers our theorem on canonical circuits. Finally, Sect. 7 contains a conclusion.

2   Reversible Circuits

In this section we introduce the background on reversible functions and their relation to reversible circuits.

2.1   Reversible Functions

Reversible logic can be used for realizing reversible functions. Reversible functions are special multi-output functions and defined as follows.

Definition 1. Let $\mathbb{B} = \{0, 1\}$ be the set of truth values with the negations $\overline{0} = 1$ and $\overline{1} = 0$ and *ID* be a set of identifiers serving as a reservoir of Boolean variables. Let $\mathbb{B}^X$ be the set of all mappings $a\colon X \to \mathbb{B}$ for some $X \subseteq ID$ where the elements of $\mathbb{B}^X$ are called *assignments*. If the set of variables is ordered, each assignment corresponds to a Boolean vector. Then a bijective Boolean (multi-output) function $f\colon \mathbb{B}^X \to \mathbb{B}^X$ is called *reversible*.

2.2   Reversible Circuits

Reversible circuits are used for representing reversible functions because a reversible function can be realized by a cascade of reversible gates. Reversible circuits differ from conventional circuits: while conventional circuits rely on the basic binary operations and also fanouts are applied in order to use signal values on several gate inputs, in reversible logic fanouts and feedback are not directly allowed because they would destroy the reversibility of the computation. Also the logic operators AND and OR cannot be used since they are irreversible. Instead a reversible gate library is applied. Since the Boolean operator NOT is inverse, the NOT-gate is part of this reversible library. To increase the expressiveness the universal *Toffoli gate* has been introduced, which is a multi-controlled NOT-gate. Since the Toffoli gate is universal, all reversible functions can be realized by cascades of this gate type alone (cf. [9]).

A *(multiple-control) Toffoli gate* consists of a *target line* $t \in ID$ and a set $C \subseteq ID - \{t\}$ of *control lines* and is denoted by $TG(t, C)$. The gate defines the function $f_{t,C}\colon \mathbb{B}^X \to \mathbb{B}^X$ for each $X \subseteq ID$ with $\{t\} \cup C \subseteq X$ which maps an assignment $a\colon X \to \mathbb{B}$ to $f_{t,C}(a)\colon X \to \mathbb{B}$ given by $f_{t,C}(a)(t) = \overline{a(t)}$ if $a(c) = 1$ for all $c \in C$. In all other cases, $f_{t,C}(a)$ is equal to a. Hence, $f_{t,C}(a)$ inverts the value of the target line if and only if all control lines are set to 1. Otherwise the value of the target line is passed through unchanged. The values of all other lines always pass through a gate unchanged. Consequently, $f_{t,C}$ is a mapping on $\mathbb{B}^X$ which is inverse to itself and, therefore, reversible in particular. A multiple-control Toffoli gate can be realized by a sequence of Toffoli gates with two control lines.

Example 2. The four simplest multi-controlled Toffoli gates are *NOT*, *CNOT*, *CCNOT*, and C^3NOT.

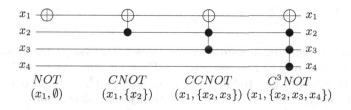

Fig. 1. The four simplest multi-controlled Toffoli gates

In the graphical representation, the target line is indicated by $\oplus$ and the control lines by $\bullet$ vertically connected with their target line (Fig. 1).

In addition to positive control lines, in the recent past also negative- and mixed-control Toffoli gates have been considered [8]. This gains smaller circuits in general. Nevertheless, the expressiveness remains the same, since each negative control can be replaced by a positive one with a negation before and after the control. For this reason, in this work we focus on positive control Toffoli gates.

An *extended-target Toffoli gate*, as proposed in [2], with multiple control lines and multiple target lines, denoted by the sets C and T, respectively, holding $C \cup T \subseteq X$, $T \neq \emptyset$ and $C \cap T = \emptyset$, realizes the function $f(a)(x) = \overline{a(x)}$ if $x \in T$ and $a(y) = 1$ for all $y \in C$, and $a(x)$ otherwise. This means that the values of all target lines are negated if the value of each control line is 1. We discuss a further generalization in the following section.

3　Multi-target Toffoli Circuits

In this section, we introduce the notion of multi-target Toffoli circuits as cascades of multi-target Toffoli gates. Such a gate has a set of target lines where each target line is controlled by a set of control lines which is disjoint from the set of target lines.

Definition 3. 1. A *multi-target Toffoli gate* over a set X of lines is a pair $mtg = (T, c\colon T \to 2^X)$ with $T \subseteq X$, $T \neq \emptyset$ and $T \cap c(T) = \emptyset$ where $c(T) = \bigcup_{t \in T} c(t)$. T is the *set of target lines*, and $c(t)$ is the *set of control lines* of t for $t \in T$.

2. A multi-target Toffoli gate $mtg = (T, c)$ models the following semantic function f_{mtg} on $\mathbb{B}^X$:

$$f_{mtg}(a)(x) = \begin{cases} \overline{a(x)} & \text{if } x \in T \text{ and } a(y) = 1 \text{ for all } y \in c(x), \\ a(x) & \text{otherwise.} \end{cases}$$

3. A *multi-target Toffoli circuit* $mtc = mtg_1 \ldots mtg_n$ is a sequence of multi-target Toffoli gates. Its length n is denoted by $|mtc|$.

4. Let mtc be a multi-target Toffoli circuit. It models the semantic function f_{mtc} defined as the sequential composition of the semantic functions of the gates, i.e.

$$f_{mtc} = f_{mtg_n} \circ \cdots \circ f_{mtg_1}.$$

If a multi-target Toffoli gate mtg has the set T of target lines and T' is a subset of T, then mtg can be restricted to T' and its complement $T'' = T - T'$ yielding the multi-target Toffoli gates mtg' and mtg''. It turns out that the sequential composition $mtg'mtg''$ is semantically equivalent to mtg.

Proposition 4. Let $mtg = (T, c)$ be a multi-target Toffoli gate, let $T' \subseteq T$ with $\emptyset \neq T' \neq T$.

1. Then $mtg' = (T', c')$ with $c'(t') = c(t')$ for $t' \in T'$ is a multi-target Toffoli gate. This gate may be denoted by $mtg|_{T'}$, called the *restriction of mtg to T'*.
2. Accordingly, $mtg'' = (T'', c'')$ with $T'' = T - T'$ and $c''(t'') = c(t'')$ for $t'' \in T''$ is also a multi-target Toffoli gate.
3. The sequential composition $mtg'mtg''$ is semantically equivalent to mtg, i.e. $f_{mtg} = f_{mtg'mtg''}$.

Proof. 1. $T' \cap c'(T') = T' \cap \bigcup\limits_{t' \in T'} c'(t') = T' \cap \bigcup\limits_{t' \in T'} c(t') \subseteq T \cap \bigcup\limits_{t \in T} c(t) = T \cap c(T) = \emptyset$.

2. $T' \subseteq T$ and $\emptyset \neq T' \neq T$ imply $T - T' \subseteq T$ and $\emptyset \neq T - T' \neq T$ such that Point 1 applies to $T'' = T - T'$.

3. By definition, we get the following equations for all $a \in \mathbb{B}^X$ and $x \in X$:

$$f_{mtg'mtg''}(a)(x) = (f_{mtg''} \circ f_{mtg'})(a)(x) = f_{mtg''}(f_{mtg'}(a))(x)$$
$$= \begin{cases} \overline{f_{mtg'}(a)(x)} & \text{if } x \in T'' \text{ and } f_{mtg''}(a)(y) = 1 \text{ for all } y \in c''(x), \\ f_{mtg'}(a)(x) & \text{otherwise,} \end{cases}$$

as well as

$$f_{mtg'}(a)(x) = \begin{cases} \overline{a(x)} & \text{if } x \in T' \text{ and } a(y) = 1 \text{ for all } y \in c'(x), \\ a(x) & \text{otherwise.} \end{cases}$$

Combining these results and using $T' \cap T'' = \emptyset$, $T' \cup T'' = T$ and the disjointness of control and target lines, we get:

$$f_{mtg'mtg''}(a)(x) = \begin{cases} \overline{a(x)} & \text{if } x \in T'' \text{ and } a(y) = 1 \text{ for all } y \in c''(x), \\ \overline{a(x)} & \text{if } x \in T' \text{ and } (y) = 1 \text{ for all } y \in c'(x), \\ a(x) & \text{otherwise,} \end{cases}$$
$$= \begin{cases} \overline{a(x)} & \text{if } x \in T \text{ and } a(y) = 1 \text{ for all } y \in c(x), \\ a(x) & \text{otherwise,} \end{cases}$$
$$= f_{mtg}(a)(x).$$

This proves the statement.

Given the situation of Proposition 4, the circuit $mtg'mtg''$ may be seen as a sequentialization of mtg and mtg as a parallelization of $mtg'mtg''$. In the next section, both operations are considered within arbitrary circuits.

4 Sequentialization, Parallelization and Shift

Sequentialization and parallelization can be done within large circuits inducing an equivalence relation on multi-target Toffoli circuits. As parallelization, a particular composition of a sequentialization and a parallelization shifts some target lines of a gate to the preceding gate.

Shifts are defined formally as a generalization of parallelization. The shift operation is quite nondeterministic as there may be many gates within a circuit that allow shifting. But it turns out that shifting has the local Church-Rosser property meaning that two circuits obtained by two shifts on a circuit can always be shifted into a common circuit.

Definition 5. Let $mtg = (T, c)$ be a multi-target Toffoli gate and $T' \subseteq T$ with $\emptyset \neq T' \neq T$. Let mtg' be the restriction of mtg to T' and mtg'' the restriction of mtg to $T'' = T - T'$. Then

1. $mtg'mtg''$ is called *sequentialization* of mtg wrt T' and mtg *parallelization* of $mtg'mtg''$. The parallelization is also denoted by $mtg' + mtg''$.
2. Let $mtc = mtc'mtgmtc''$ be a multi-target Toffoli circuit and $mtg'mtg''$ be the sequentialization of mtg wrt T'. Then mtc and $\overline{mtc} = mtc'mtg'mtg''mtc''$ are in *seq*-relation wrt T' in gate $i = |mtc'| + 1$, denoted by

$$mtc \xrightarrow[seq(i,T')]{} \overline{mtc}$$

as well as in *par*-relation after gate $i - 1 = |mtc'|$, denoted by

$$\overline{mtc} \xrightarrow[par(i-1)]{} mtc.$$

Let $\sim_{seq}$ be the equivalence relation induced by *seq*, i.e. the reflexive, symmetric, and transitive closure of *seq* and $\sim_{par}$ the corresponding equivalence relation induced by *par*. Then, obviously, $\sim_{seq}$ and $\sim_{par}$ are equal because *seq* and *par* are inverse to each other.

Definition 6. Let mtc and $\widetilde{mtc}$ be two multi-target Toffoli circuits. Then $\widetilde{mtc}$ is a *shift* of mtc if $mtc \xrightarrow[par(i-1)]{} \widetilde{mtc}$ or $mtc \xrightarrow[seq(i+1,T')]{} \overline{mtc} \xrightarrow[par(i-1)]{} \widetilde{mtc}$ for some $i \geq 1$ and $T' \subseteq X$, denoted by

$$mtc \xrightarrow[sh(i,T')]{} \widetilde{mtc}$$

where T' is the set of target lines of the gate $i + 1$ in case that the shift is just a parallelization. If i and T' are clear from the context, then we may write $mtc \xrightarrow[sh]{} \widetilde{mtc}$.

Example 7. Consider the mtc over four lines x_1 to x_4 including the gates $mtg_1 = (\{x_3\}, c_1)$ with $c_1(x_3) = \{x_1\}$, $mtg_2 = (\{x_2\}, c_2)$ with $c_2(x_2) = \{x_1\}$, $mtg_3 = (\{x_4\}, c_3)$ with $c_3(x_4) = \{x_3\}$, $mtg_4 = (\{x_4\}, c_4)$ with $c_4(x_4) = \{x_1, x_3\}$ and $mtg_5 = (\{x_2\}, c_5)$ with $c_5(x_2) = \{x_3\}$ as depicted in Fig. 2(a).

Obviously, mtg_1 and mtg_2 can be parallelized because the target line of the one gate is no target or control line of the other gate. The same holds for gates mtg_4 and mtg_5. Hence, we get $mtc \xrightarrow[par(0)]{} mtc' \xrightarrow[par(2)]{} mtc''$ with $mtc' = mtg_1'mtg_3mtg_4mtg_5$ and $mtc'' = mtg_1'mtg_3mtg_4'$ where $mtg_1' = (\{x_2, x_3\}, c_1')$

with $c_1'(x_2) = c_1'(x_3) = \{x_1\}$ and $mtg_4' = (\{x_2, x_4\}, c_4')$ with $c_4'(x_2) = \{x_3\}$ and $c_4'(x_4) = \{x_1, x_3\}$. Afterwards we can apply the shift $sh(2, \{x_2\})$ to mtc'' by sequentializing mtg_4' wrt $\{x_2\}$ and parallelizing the resulting circuit after mtg_1'. This yields the circuit depicted in Fig. 2(b) where $mtg_2' = (\{x_2, x_4\}, c_2')$ with $c_2'(x_2) = c_2'(x_4) = \{x_3\}$.

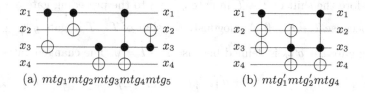

(a) $mtg_1 mtg_2 mtg_3 mtg_4 mtg_5$ (b) $mtg_1' mtg_2' mtg_4$

Fig. 2. Shifting a multi-target Toffoli circuit

Proposition 8. The shift relation has the local Church-Rosser property meaning that two shifts on a circuit mtc

$$mtc \overset{sh}{\underset{sh}{\lessgtr}} \genfrac{}{}{0pt}{}{mtc_1}{mtc_2} \quad \text{imply} \quad \genfrac{}{}{0pt}{}{mtc_1}{mtc_2} \overset{*}{\underset{sh}{\gtrless}} \overline{mtc}$$

for some circuit $\overline{mtc}$ where $\xrightarrow[sh]{*}$ is the reflexive and transitive closure of the shift relation sh.

Proof. A shift changes two successive gates of a circuit and keeps the rest invariant. Hence two shifts that change four different gates cannot interfere with each other so that they can be applied in any order yielding the same result. The situation becomes more complicated if the two shifts change two or three successive gates. Then various cases can occur. They are listed in Fig. 3.

Let us start with shifts on the same two gates. Then both shifts may be proper shifts of different parts of the second gate or one of the shifts is the parallelization of the two gates. If both shifts are proper, the parts shifted may be incomparable or one may be a subpart of the other. Hence there are three cases to be considered. As an abbreviation, we write g for mtg.

Case 1: Let $g = (T, c)$ and $g' = (T', c')$. Then the given shifts of $\hat{T}$ and $\hat{\hat{T}}$ in gate $|mtc'| + 2$ with $\hat{T} - \hat{\hat{T}} \neq \emptyset \neq \hat{\hat{T}} - \hat{T}$ are defined because $\hat{T} \cap T = \emptyset = \hat{\hat{T}} \cap T$ and $T \cap c'(\hat{T}) = \emptyset = T \cap c'(\hat{\hat{T}})$. The changed gates after the shifts are:

$$g + g'|_{\hat{T}} = (T \cup \hat{T}, \hat{c}), \qquad g'|_{T-\hat{T}} = (T - \hat{T}, c'|_{T-\hat{T}}),$$

$$g + g'|_{\hat{\hat{T}}} = (T \cup \hat{\hat{T}}, \hat{\hat{c}}), \qquad g'|_{T-\hat{\hat{T}}} = (T - \hat{\hat{T}}, c'|_{T-\hat{\hat{T}}})$$

with $\hat{c}(x) = \tilde{\hat{c}}(x) = c(x)$ for $x \in T$, $\hat{c}(x) = c'(x)$ for $x \in \hat{T}$, $\tilde{\hat{c}}(x) = c'(x)$ for $x \in \tilde{\hat{T}}$. Moreover, the following holds:

$$(\tilde{\hat{T}} - \hat{T}) \cap (T \cup \hat{T}) = ((\tilde{\hat{T}} - \hat{T}) \cap T) \cup ((\tilde{\hat{T}} - \hat{T}) \cap \hat{T}) \subseteq \tilde{\hat{T}} \cap T = \emptyset,$$

$$(T \cup \hat{T}) \cap c'(\tilde{\hat{T}} - \hat{T}) = (T \cap c'(\tilde{\hat{T}} - \hat{T})) \cup (\hat{T}) \cap c'(\tilde{\hat{T}} - \hat{T}) \subseteq T \cap c'(\tilde{\hat{T}}) = \emptyset.$$

Therefore the shift of $\tilde{\hat{T}} - \hat{T}$ in gate $g'|_{T-\hat{T}}$ to the preceding gate $g + g'|_{\hat{T}}$ is defined because $\tilde{\hat{T}} - \hat{T} \neq \emptyset$. Analogously the shift of $\hat{T} - \tilde{\hat{T}}$ in gate $g'|_{T-\tilde{\hat{T}}}$ to the preceding gate $g + g'|_{\tilde{\hat{T}}}$ is defined because $\hat{T} - \tilde{\hat{T}} \neq \emptyset$. The changed gates are:

$$(g + g'|_{\hat{T}}) + (g'|_{T-\hat{T}})|_{\tilde{\hat{T}}} = g + g'|_{\hat{T} \cup \tilde{\hat{T}}}, \qquad (g'|_{T-\hat{T}})|_{(T-\hat{T})-\tilde{\hat{T}}} = g'|_{T-(\hat{T} \cup \tilde{\hat{T}})},$$

$$(g + g'|_{\tilde{\hat{T}}}) + (g'|_{T-\tilde{\hat{T}}})|_{\hat{T}} = g + g'|_{\hat{T} \cup \tilde{\hat{T}}}, \qquad (g'|_{T-\tilde{\hat{T}}})|_{(T-\tilde{\hat{T}})-\hat{T}} = g'|_{T-(\hat{T} \cup \tilde{\hat{T}})}.$$

This proves that the two further shifts yield the same circuit.

Case 2: The situation is similar to Case 1 with the exception that $\hat{T} \subseteq \tilde{\hat{T}}$ implies $\hat{T} - \tilde{\hat{T}} = \emptyset$. But then the shift of $\tilde{\hat{T}} - \hat{T}$ after the shift of $\hat{T}$ yields the same result as the shift of $\tilde{\hat{T}}$ in the first place using arguments similar to Case 1.

Case 3: Given a shift and a parallelization as in Fig. 3(c), the parallelization after the shift is defined and yields the same result as the parallelization directly using arguments similar to Case 1.

Cases 4–7: Now we consider two shifts changing three successive gates. Then both shifts may be parallelization or one is a parallelization and the other one a proper shift or both are proper shifts.

The argumentation that the given shifts can be continued by further shifts into the same result is in all four cases similar to the argumentation in Case 1. Nevertheless, we go into the details of Case 5 because, in this very case, two further shifts are applied after the given proper shift to keep up with the given parallelization. The circuit after the shift has the form $mtc'g(g'+g''|_{\hat{T}})g''|_{T''-\hat{T}}mtc''$ where the gate $|mtc'| + 2$ can be sequentialized wrt T' yielding the circuit $mtc'gg'g''|_{\hat{T}}g''|_{T''-\hat{T}}mtc''$. By assumption g and g' can be parallelized. Both together establish the shift of T' in gate $|mtc'| + 2$ yielding the circuit $mtc'(g+g')g''|_{\hat{T}}g''|_{T''-\hat{T}}mtc''$ where $g''|_{\hat{T}}g''|_{T''-\hat{T}}$ is a sequentialization of g'' so that the parallelization is defined yielding $mtc'(g+g')g''mtc''$ as stated. As there are no cases left, the local Church-Rosser property of shifts is proved.

5 Waiting Degree

Besides the local Church-Rosser property, the shift operation has a second significant property: It does not allow infinite shift sequences. In other words, the lengths of shift sequences starting in some circuit are bounded. Consequently, shifting as long as possible always terminates in a circuit that is reduced with respect to shifting. To prove this, we introduce the waiting degree and show that

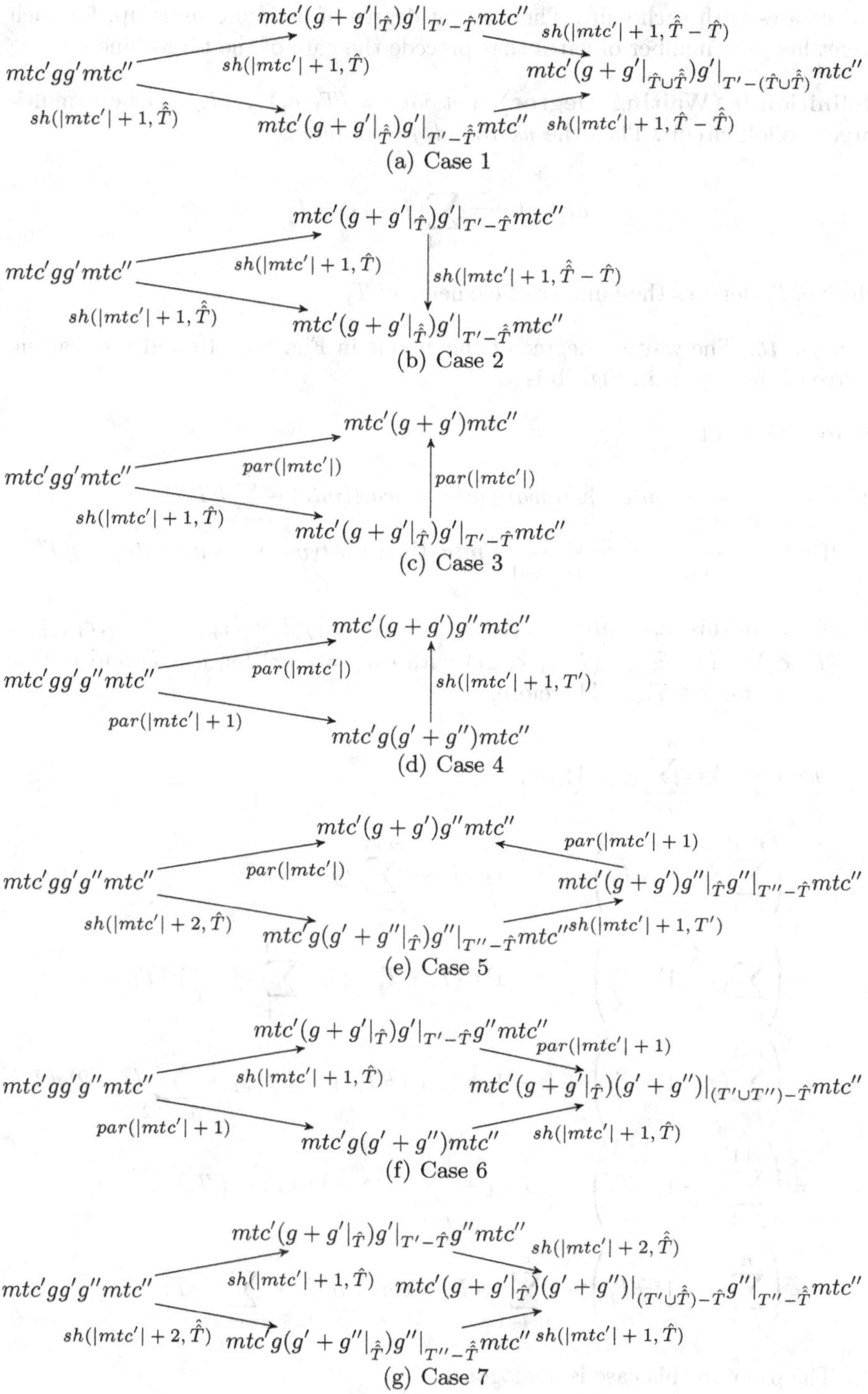

$$mtc'(g + g'|_{\hat{T}})g'|_{T'-\hat{T}}mtc''$$

$$mtc'gg'mtc'' \xrightarrow{sh(|mtc'|+1,\hat{T})} \xrightarrow{sh(|mtc'|+1,\hat{\hat{T}}-\hat{T})} mtc'(g + g'|_{\hat{T}\cup\hat{\hat{T}}})g'|_{T'-(\hat{T}\cup\hat{\hat{T}})}mtc''$$

$$sh(|mtc'|+1,\hat{\hat{T}}) \qquad mtc'(g + g'|_{\hat{\hat{T}}})g'|_{T'-\hat{\hat{T}}}mtc'' \quad sh(|mtc'|+1,\hat{T}-\hat{\hat{T}})$$

$$\text{(a) Case 1}$$

$$mtc'(g + g'|_{\hat{T}})g'|_{T'-\hat{T}}mtc''$$

$$mtc'gg'mtc'' \xrightarrow{sh(|mtc'|+1,\hat{T})} \Big\downarrow sh(|mtc'|+1,\hat{\hat{T}}-\hat{T})$$

$$sh(|mtc'|+1,\hat{\hat{T}}) \qquad mtc'(g + g'|_{\hat{\hat{T}}})g'|_{T'-\hat{\hat{T}}}mtc''$$

$$\text{(b) Case 2}$$

$$mtc'(g + g')mtc''$$

$$mtc'gg'mtc'' \xrightarrow{par(|mtc'|)} \Big\uparrow par(|mtc'|)$$

$$sh(|mtc'|+1,\hat{T}) \qquad mtc'(g + g'|_{\hat{T}})g'|_{T'-\hat{T}}mtc''$$

$$\text{(c) Case 3}$$

$$mtc'(g + g')g''mtc''$$

$$mtc'gg'g''mtc'' \xrightarrow{par(|mtc'|)} \Big\uparrow sh(|mtc'|+1,T')$$

$$par(|mtc'|+1) \qquad mtc'g(g' + g'')mtc''$$

$$\text{(d) Case 4}$$

$$mtc'(g + g')g''mtc'' \xleftarrow{par(|mtc'|+1)}$$

$$mtc'gg'g''mtc'' \xrightarrow{par(|mtc'|)} \qquad mtc'(g + g')g''|_{\hat{T}}g''|_{T''-\hat{T}}mtc''$$

$$sh(|mtc'|+2,\hat{T}) \quad mtc'g(g' + g''|_{\hat{T}})g''|_{T''-\hat{T}}mtc'' \xrightarrow{sh(|mtc'|+1,T')}$$

$$\text{(e) Case 5}$$

$$mtc'(g + g'|_{\hat{T}})g'|_{T'-\hat{T}}g''mtc''$$

$$mtc'gg'g''mtc'' \xrightarrow{sh(|mtc'|+1,\hat{T})} \qquad mtc'(g + g'|_{\hat{T}})(g' + g'')|_{(T'\cup T'')-\hat{T}}mtc''$$

$$par(|mtc'|+1) \qquad mtc'g(g' + g'')mtc'' \xrightarrow{sh(|mtc'|+1,\hat{T})}$$

$$\text{(f) Case 6}$$

$$mtc'(g + g'|_{\hat{T}})g'|_{T'-\hat{T}}g''mtc'' \xrightarrow{sh(|mtc'|+2,\hat{T})}$$

$$mtc'gg'g''mtc'' \xrightarrow{sh(|mtc'|+1,\hat{T})} \quad mtc'(g + g'|_{\hat{T}})(g' + g'')|_{(T'\cup\hat{T})-\hat{T}}g''|_{T''-\hat{T}}mtc''$$

$$sh(|mtc'|+2,\hat{T}) \quad mtc'g(g' + g''|_{\hat{T}})g''|_{T''-\hat{T}}mtc'' \xrightarrow{sh(|mtc'|+1,\hat{T})}$$

$$\text{(g) Case 7}$$

Fig. 3. The 7 cases of the shift relation

it decreases with each shift. The waiting degree of a circuit sums up, for each target line, the number of gates that precede the gate of the target line.

Definition 9 (Waiting Degree). Let $mtc = (T_1, c_1) \ldots (T_n, c_n)$ be a multi-target Toffoli circuit. Then the *waiting degree* of mtc is

$$wait(mtc) = \sum_{j=1}^{n} (j-1) \cdot \#T_j$$

where $\#T_j$ denotes the number of elements of T_j.

Example 10. The waiting degree of the circuit in Fig. 2a is 10 and the waiting degree of the circuit in Fig. 2b is 4.

Proposition 11

1. If $mtc \xrightarrow{par(i-1)} \widetilde{mtc}$, then $wait(\widetilde{mtc}) = wait(mtc) - \sum_{j=i+1}^{n} \#T_j$.

2. If $mtc \xrightarrow{seq(i+1,T')} \overline{mtc} \xrightarrow{par(i-1)} \widetilde{mtc}$, then $wait(\widetilde{mtc}) = wait(mtc) - \#T'$.

Proof. 1. In this case, $\widetilde{mtc} = (T_1, c_1) \ldots (T_{i-1}, c_{i-1})(T_i + T_{i+1}, c)(T_{i+2}, c_{i+2}) \ldots (T_n, c_n) = (\tilde{T}_1, \tilde{c}_1) \ldots (\tilde{T}_{n-1}, \tilde{c}_{n-1})$ with $c(x) = c_i(x)$ for $x \in T_i$ and $c(x) = c_{i+1}(x)$ for $x \in T_{i+1}$. Therefore,

$$wait(\widetilde{mtc}) = \sum_{j=1}^{n-1} (j-1)\#\tilde{T}_j$$

$$= \left(\sum_{j=1}^{i-1} (j-1)\#\tilde{T}_j \right) + (i-1)\#\tilde{T}_i + \sum_{j=i+1}^{n-1} (j-1)\#\tilde{T}_j$$

$$= \left(\sum_{j=1}^{i-1} (j-1)\#T_j \right) + (i-1)\#(T_i + T_{i+1}) + \sum_{j=i+1}^{n-1} (j-1)\#T_{j+1}$$

$$= \left(\sum_{j=1}^{i-1} (j-1)\#T_j \right) + (i-1)\#T_i + i\#T_{i+1} - \#T_{i+1} + \sum_{j=i+2}^{n} (j-2)\#T_j$$

$$= \left(\sum_{j=1}^{i+1} (j-1)\#T_j \right) - \#T_{i+1} + \sum_{j=i+2}^{n} ((j-1)\#T_j - \#T_j)$$

$$= \left(\sum_{j=1}^{n} (j-1)\#T_j \right) - \sum_{j=i+1}^{n} \#T_j = wait(mtc) - \sum_{j=i+1}^{n} \#T_j$$

2. The proof in this case is analogously.

Proposition 12. Let $mtc = (T_1, c_1) \ldots (T_n, c_n)$ be a multi-target Toffoli circuit. Then $wait(mtc) \leq \frac{m(m-1)}{2}$ for $m = \sum\limits_{j=1}^{n} \#T_j$.

Proof. Sequentialize mtc as long as possible. Then the result has length m and waiting degree $\frac{m(m-1)}{2}$. But $wait(mtc)$ is not greater because sequentialization increases the waiting degree.

The two properties imply the following corollary.

Corollary 13.   1. Let $mtc \xrightarrow[sh]{n} \overline{mtc}$ be a shift sequence of n shifts. Then $n \leq wait(mtc)$.

2. Let $mtc = (T_1, c_1) \ldots (T_n, c_n)$ be a multi-target Toffoli circuit. Let $m = \sum\limits_{i=1}^{n} \#T_i$. Then shifting as long as possible terminates with a circuit that is reduced wrt shifts after at most $\frac{m(m-1)}{2}$ shifts.

Let $\sim$ be the equivalence relation generated by the shift relation, called shift equivalence. Then $\sim$ is equal to $\sim_{seq} = \sim_{par}$ as $par \subseteq shift$ and $shift \subseteq par \cup par \circ seq \subseteq par \cup (par \circ par^{-1}) \subseteq (par \cup par^{-1})^* = \sim_{par}$.

6   Canonical Circuits

Circuits that are reduced wrt shifts are called canonical. They are local optima wrt the waiting degree. But this result can be tremendously improved by combining the termination with the local Church-Rosser property. The shifting defines an equivalence relation on circuits. Each equivalence class contains only circuits that are semantically equivalent. Moreover, it turns out that each canonical circuit is a unique representative of its shift equivalence class so that it is a global optimum within its class. To show this, we prove first that shift equivalence is confluent meaning that each two equivalent circuits can be shifted into a common circuit.

Theorem 14. *Shift-equivalent canonical circuits are equal.*

Proof. Let mtc and $\overline{mtc}$ be two shift-equivalent canonical circuits. Due to the following Lemma, there is a circuit $\widetilde{mtc}$ and there are shift sequences from mtc and $\overline{mtc}$ into $\widetilde{mtc}$. Because mtc and $\overline{mtc}$ are canonical and hence shift-reduced, both shift sequences have length 0 yielding $mtc = \widetilde{mtc} = \overline{mtc}$ as stated.

Lemma 15. $mtc \sim \overline{mtc}$ implies $\begin{matrix} mtc \\ \overline{mtc} \end{matrix} \begin{matrix} \scriptstyle * \\ \scriptstyle sh \ * \\ \scriptstyle sh \end{matrix} \widetilde{mtc}$ for some multi-target Toffoli

circuit $\widetilde{mtc}$.

Proof. $mtc \sim \overline{mtc}$ iff there is a sequence $zz = mtc_0 \ldots mtc_n$ such that $mtc_0 = mtc, mtc_n = \overline{mtc}, mtc_i \xrightarrow{sh} mtc_{i+1}$ or $mtc_{i+1} \xrightarrow{sh} mtc_i$ for all $i = 0, \ldots, n-1$, i.e. a zigzag of shifts.

Let $MTC(zz) = \{mtc_i \mid i = 0, \ldots, n\}$ and let X be a finite set of multi-target Toffoli circuits. Let $reach(X) = \{\overline{mtc} \mid mtc \xrightarrow{*}{sh} \overline{mtc}, mtc \in X\}$. Note that $reach(X)$ is finite. Then, $mtc_i \begin{smallmatrix} \xrightarrow{\;\;\;} mtc_{i-1} \\ sh \\ \xrightarrow{sh} mtc_{i+1} \end{smallmatrix}$ is a critical pair of zz if $mtc_i \notin reach(MTC(zz) - \{mtc_i\})$, i.e. mtc_i is a critical element of zz.

Induction on $\#reach(CE(zz))$, where $CE(zz)$ denotes the set of critical elements of zz.

Base: $\#reach(CE(zz)) = 0$. Then there is no critical element because each critical element is reachable by itself by 0 shifts and belongs to $reach(CE(zz))$. Therefore, zz must contain a multi-target Toffoli circuit mtc_{i_0} with $mtc_i \xrightarrow{sh} mtc_{i+1}$ for all $i < i_0$ and $mtc_{i+1} \xrightarrow{sh} mtc_i$ for all $i \geq i_0$.

Step: Let $\#reach(CE(zz)) = k$ with $k > 0$. Let mtc_i be a critical element of zz i.e. $mtc_i \in CE(zz)$. Then one can replace $mtc_{i-1} \xleftarrow{sh} mtc_i \xrightarrow{sh} mtc_{i+1}$ in zz by the shifts that make the shift relation locally Church-Rosser due to Proposition 8 defining a new zz'. The new elements of zz' are not critical as none of them has branching shifts. Hence, $CE(zz') \subseteq MTC(zz) \subseteq reach(CE(zz))$. This implies $reach(CE(zz')) \subseteq reach(reach(CE(zz)) = reach(CE(zz))$. The inclusion is proper as $mtc_i \notin reach(CE(zz'))$ because of the following reason. Assuming $mtc_i \in reach(CE(zz'))$ then $mtc_j \xrightarrow{*}{sh} mtc_i$ for some $mtc_j \in CE(zz')$. As $mtc_j \in MTC(zz) - \{mtc_i\}$ we get $mtc_i \in reach(MTC(zz) - \{mtc_i\})$ in contradiction to the choice of mtc_i.

Therefore, $\#reach(CE(zz')) < k$ so that by induction hypothesis, the lemma holds for zz' and therefore for zz too.

7 Conclusion

In this paper, we have studied a generalized class of Toffoli circuits that are sequentially composed of multi-target Toffoli gates. Under certain independence conditions parts of a gate can be shifted to the preceding gate within a circuit. It has turned out that shift-reduced circuits are unique canonical representatives of their shift equivalence classes. To shed more light on the significance of these considerations, further research on the following topics may be helpful.

1. In the case considered in this paper, the negation on a target line takes place if and only if all control lines are 1. More generally, there may be two types of control lines where the lines of one type must be 1 as before, but the other lines must be 0 to trigger the negation (see e.g., [8]). We are confident that all the results in this paper still hold if one considers this more general kind of control with positive and negative control lines.

2. Canonical circuits have minimal waiting degree within their shift equivalence classes. But they may be the starting point for further optimizations. For example, it is clear that the sequential composition of a Toffoli gate with itself yields the identity. Therefore, two identical parts in successive multi-target Toffoli gates can be removed without changing the semantics. Afterwards another round of shift optimization can be started. And there are other operations with such a perspective.

3. Drechsler et al. [4] study exclusive sums of products (ESOPs) which are a special kind of Toffoli circuits where the target lines and the control lines stem from disjoint sets. Therefore, the independence check for ESOPs concerns only the disjointness of target lines and shifting may become more efficient.

4. Chen et al. [2] and Wille et al. [10] study a special case of our multi-target Toffoli gates where all target lines have the same set of control lines. In both cases, the authors relate the special case with quantum circuits. Hence is may be interesting whether our more general case may yield further improvements in this line of research.

5. As mentioned in the introduction, shifts on parallel graph grammar derivations behave like the shifts on multi-target Toffoli circuits (see, e.g., [3,5]). Therefore, we wonder whether there is a way to represent Toffoli circuits as parallel derivations.

References

1. Bennett, C.: Logical reversibility of computation. IBM J. Res. Dev. **17**(6), 525–532 (1973)
2. Chen, J.-L., Zhang, X.-Y., Wang, L.-L., Wei, X.-Y., Zhao, W.-Q.: Extended Toffoli gate implementation with photons. In: Proceeding of the 9th International Conference on Solid-State and Integrated-Circuit Technology, pp. 575–578. ICSICT (2008)
3. Corradini, A., Montanari, U., Rossi, F., Ehrig, H., Heckel, R., Löwe, M.: Algebraic approaches to graph transformation part I: basic concepts and double pushout approach. In: Rozenberg, G. (ed.) Handbook of Graph Grammars and Computing by Graph Transformation, vol. 1: Foundations, pp. 163–245. World Scientific, Singapore (1997)
4. Drechsler, R., Finder, A., Wille, R.: Improving ESOP-based synthesis of reversible logic using evolutionary algorithms. In: Di Chio, C., et al. (eds.) EvoApplications 2011, Part II. LNCS, vol. 6625, pp. 151–161. Springer, Heidelberg (2011)
5. Kreowski, H.-J.: Transformation of derivation sequences in graph grammars. Proceeding of Conference Fundamentals of Computation Theory (Poznan-Kornik, Sept. 1977). LNCS, vol. 56, pp. 275–286. Springer, Heidelberg (1977)
6. Kreowski, H.-J.: Manipulationen von Graphmanipulationen. Ph.D. thesis, Technische Universität Berlin , Fachbereich Informatik (1978)
7. Kreowski, H.-J.: An axiomatic approach to canonical derivations. In: Proceedings of IFIP World Computer Congress, IFIP-Transactions, vol. A-51, pp. 348–353. North-Holland (1994)
8. Soeken, M., Thomsen, M.K.: White dots do matter: rewriting reversible logic circuits. In: RC 2013 Proceedings of the 5th International Conference on Reversible Computation, pp. 196–208, RC 2013, Springer, Heidelberg (2013)

9. Toffoli, T.: Reversible computing. In: de Bakker, W., van Leeuwen, J. (eds.) Automata, Languages and Programming. Lecture Notes in Computer Science, vol. 85, pp. 632–644. Springer, Heidelberg (1980)

10. Wille, R., Soeken, M., Otterstedt, C., Drechsler, R.: Improving the mapping of reversible circuits to quantum circuits using multiple target lines. In: Asia and South Pacific Design Automation Conference (ASP-DAC), pp. 145–150. IEEE (2013)

Author Index

Printed in the United States
By Bookmasters